Handbook of Geosynthetic Engineering

Second edition

Institution of Civil Engineers
publishing

Handbook of Geosynthetic Engineering

Geosynthetics and their applications

Second edition

Edited by
Sanjay Kumar Shukla
School of Engineering, Edith Cowan University, Perth, Australia

Published by ICE Publishing, 40 Marsh Wall, London E14 9TP.

Full details of ICE Publishing sales representatives and distributors can be found at:
www.icevirtuallibrary.com/info/printbooksales

Also available from ICE Publishing

Ground and Soil Improvement.
C. Raison (ed.). ISBN 978-0-7277-3170-8
A Short Course in Soil and Rock Slope Engineering.
N. Simons, B. Menzies and M. Matthews. ISBN 978-0-7277-2871-5
ICE Manual of Geotechnical Engineering.
J. Burland, T. Chapman, H. Skinner and M. Brown (eds).
ISBN 978-0-7277-3652-9 (2 volumes)
UK Specification for Ground Investigation, Second edition
(Site Investigation in Construction Series).
Site Investigation Steering Group. ISBN 978-0-7277-3506-5
Monitoring Underground Construction.
British Tunnelling Society. ISBN 978-0-7277-4118-9

www.icevirtuallibrary.com

A catalogue record for this book is available from the British Library

ISBN 978-0-7277-4175-2

ICE Publishing is a division of Thomas Telford Ltd, a wholly-owned subsidiary of the Institution of Civil Engineers (ICE).

FSC
www.fsc.org
MIX
Paper from responsible sources
FSC® C013604

Associate Commissioning Editor: Jennifer Barratt
Production Editor: Imran Mirza
Market Development Executive: Catherine de Gatacre

Typeset by Academic + Technical, Bristol
Index created by Indexing Specialists (UK) Ltd, Hove, East Sussex
Printed and bound by CPI Group (UK) Ltd, Croydon, CR0 4YY

Contents

Preface ix
Acknowledgements x
List of contributors xi
About the handbook xiii

01 **Fundamentals of geosynthetics** **1**
S. K. Shukla
1.1. Introduction 1
1.2. Definitions and classification 1
1.3. Historical development 5
1.4. Basic functions and selection 7
1.5. Raw materials and manufacturing processes 10
1.6. Properties and test methods 14
1.7. Application areas 33
1.8. Standards 35
1.9. Summary of main points 35
1.10. Some common questions and their answers 39
References 41

02 **Soil–geosynthetic interaction** **45**
M. L. Lopes
2.1. Introduction 45
2.2. Granular soil behaviour 45
2.3. Soil–geosynthetic interaction mechanisms 46
2.4. Soil–geosynthetic interface resistance 47
2.5. Factors influencing soil–geosynthetic interaction 49
2.6. Laboratory tests for quantification of soil–geosynthetic interface resistance 55
2.7. Summary of main points 64
2.8. Some common questions and their answers 64
References 65

03 **Filters and drains** **67**
A. H. Aydilek
3.1. Filters 67
3.2. Geotextile filters 67
3.3. Geotextile filtration mechanism 67
3.4. Factors affecting geotextile filter behaviour 70
3.5. Laboratory filtration performance tests 74
3.6. Filtration design 76
3.7. Filtration of high water content geomaterials with geotextiles 78
3.8. Drains 80
3.9. Summary of main points 83
3.10. Common questions and answers 83
References 84

04 **Retaining walls** **89**
B. M. Das
4.1. Introduction 89
4.2. Design considerations 89
4.3. Design procedure for retaining walls with geotextile reinforcement 91
4.4. Design procedure for retaining walls with geogrid reinforcement 96
4.5. Self-evaluation questions 97
4.6. Summary of main points 98
4.7. Some common questions and their answers 98
References 98

05 **Embankments** **101**
E. M. Palmeira
5.1. Introduction 101

5.2. Geosynthetics as a basal reinforcement in embankments 101
5.3. Geosynthetics for drainage in embankments 113
5.4. Combined use of geosynthetic reinforcement and vertical drainage 116
5.5. Piled embankments 117
5.6. Summary of main points 124
5.7. Some common questions and their answers 124
References 125

06 **Shallow foundations** **129**
S. K. Shukla
6.1. Introduction 129
6.2. Reinforcing mechanisms and functions 129
6.3. Reinforcing patterns 131
6.4. Modes of failure 131
6.5. Model tests 132
6.6. Load-bearing capacity analysis 139
6.7. Settlement analysis 145
6.8. Prestressed geosynthetic-reinforced foundations 147
6.9. Geosynthetic deflection profiles and strain analysis 151
6.10. Field applications 153
6.11. Summary of main points 157
6.12. Some common questions and their answers 157
References 158

07 **Unpaved roads** **163**
P. L. Bourdeau and A. K. Ashmawy
7.1. Introduction 163
7.2. Unpaved road reinforcement 164
7.3. Summary of main points 175
7.4. Some common questions and their answers 175
References 175

08 **Paved roads** **179**
S. W. Perkins, R. R. Berg and B. R. Christopher
8.1. Introduction 179
8.2. Distress features and their relationship to geosynthetics 179
8.3. Geosynthetic functions 181
8.4. History and experimental evidence for base reinforcement 185
8.5. Summary of critical design variables for base reinforcement 186
8.6. Design solutions and approaches for base reinforcement 187
8.7. Summary of main points 189
8.8. Some common questions and their answers 189
References 190

09 **Railway tracks** **193**
S. A. Tan and S. K. Shukla
9.1. Introduction 193
9.2. Track components and substructure 193
9.3. Functions of geosynthetics 196
9.4. Properties of geosynthetics 197
9.5. Design procedure 199
9.6. Installation of geosynthetics 200
9.7. Case histories in railway track stabilisation 200
9.8. Geosynthetic drains for track drainage applications 202
9.9. Summary of main points 205
9.10. Some common questions and their answers 206
References 207

10 **Slopes – erosion control** **209**
T. S. Ingold and M.-H. Li
10.1. Introduction 209
10.2. Interaction of rain and river erosion 209

10.3. Mechanics of surface erosion 210
10.4. Classification of erosion control systems 210
10.5. Design approach 211
10.6. Study of short-term yield factors 213
10.7. Erosion control performance testing of geosynthetics 215
10.8. Lifetime cost and the performance of erosion control products 218
10.9. Summary of main points 219
10.10. Some common questions and their answers 219
References 221

11 .... **Slopes – stabilisation** **223**
S. K. Shukla, N. Sivakugan and B. M. Das
11.1. Introduction 223
11.2. Types and orientations of geosynthetics 224
11.3. Function of reinforcement against slope failure 224
11.4. Stability analysis of reinforced slopes 225
11.5. Design aspects of reinforced slopes 227
11.6. Model tests 229
11.7. Stabilisation methods in practice 231
11.8. Summary of main points 240
11.9. Some common questions and their answers 240
References 241

12 .... **Landfills** **245**
H. Zanzinger, E. Gartung and W. P. Hornsey
12.1. Introduction 245
12.2. Multi-barrier concept 246
12.3. Landfill categories 246
12.4. Basal lining systems 247
12.5. Components of the composite liner 250
12.6. Construction of liners 253
12.7. Leachate collection and removal 255
12.8. Cover system 256
12.9. Summary of main points 259
12.10. Some common questions and their answers 260
References 260

13 .... **Earth dams** **263**
D. N. Singh, S. K. Shukla and D. N. Arnepalli
13.1. Introduction 263
13.2. Use of conventional materials 264
13.3. Use of geosynthetics 266
13.4. River bed and bank protection 274
13.5. Design considerations 274
13.6. Summary of main points 275
13.7. Some common questions and their answers 275
References 276

14 .... **Containment ponds, reservoirs and canals** **279**
D. Poulain, N. Touze-Foltz, L. Peyras and C. Duquennoi
14.1. Introduction 279
14.2. Historical background 279
14.3. Design of geosynthetic systems 281
14.4. Evaluation of the durability of geomembranes 284
14.5. Case studies 285
14.6. Summary of main points 295
14.7. Some common questions and their answers 295
References 300

15 .... **Hydraulic tunnels** **303**
D. Cazzuffi, A. Scuero and G. Vaschetti
15.1. Introduction 303

15.2. Geomembrane systems 304
15.3. The exposed liner system 307
15.4. The covered liner system 310
15.5. Summary of main points 313
15.6. Some common questions and their answers 313
References 315

16 .... **Geosynthetic-reinforced soil walls and slopes – seismic aspects** **317**
R. J. Bathurst, K. Hatami and M. C. Alfaro

16.1. Introduction 317
16.2. Material properties under dynamic loading 318
16.3. Seismic analysis and the design of walls and slopes 326
16.4. Physical testing of model walls and slopes 348
16.5. Seismic buffers 353
16.6. Observed performance of reinforced soil walls and slopes during earthquakes 354
16.7. Summary of main points 355
16.8. Some common questions and their answers 356
References 357

17 .... **Geosynthetic applications – general aspects and selected case studies** **365**
S. K. Shukla

17.1. Introduction 365
17.2. General guidelines 365
17.3. Quality control and *in situ* monitoring 369
17.4. Cost analysis 370
17.5. General problems 373
17.6. Selected case studies 374
17.7. Summary of main points 383
17.8. Some common questions and their answers 383
References 384

18 .... **Geosynthetic applications – sustainability aspects** **387**
A. Bouazza and G. Heerten

18.1. Introduction 387
18.2. Economical and environmental advantages of using geosynthetics 388
18.3. Life-cycle assessment 389
18.4. Comparison of different construction methods 390
18.5. Summary of main points 395
18.6. Some common questions and their answers 395
References 396

Index **397**

Preface

In present-day civil engineering practice, geosynthetics are being used extensively in several areas, including geotechnical, transportation, environmental and hydraulic engineering, to provide technically efficient, cost-effective, environment-friendly and/or energy-efficient solutions to a myriad of civil engineering problems throughout the world. Rational design methods, based on sound concepts and standardised test techniques to determine properties of technical interest to geosynthetics, are now available. This places the use of geosynthetics on a firm base for the sustainable development of infrastructural projects. Geosynthetics have great potential as a construction material, and can function for hundreds of years, if they are properly handled and protected. The continued growth of geosynthetic applications at a rapid pace attests to the fact that geosynthetics have arrived as a viable and widely used construction material, and they have now been added to the list of traditional materials such as soil, rock, brick, lime, cement, bitumen, glass, timber, steel and aluminium. The subject of geosynthetics and their applications is today known as *geosynthetic engineering*, which is defined here as follows:

> Geosynthetic engineering deals with the application of scientific principles and methods to the acquisition, interpretation and use of knowledge of geosynthetic products for the solution to geotechnical, transportation, environmental, hydraulic and other civil-engineering-related problems.

Although geosynthetics is a continually growing field in civil engineering, it is generally not taught, especially at the undergraduate level, in engineering and technical colleges as a full unit/subject. Therefore, students, researchers, and practising engineers need the basic concepts of geosynthetics and their applications presented in a simple, easily accessible form. Updated details of geosynthetic applications with a description of case studies and practical aspects are required for quick reference in connection with the study and solution to specific field and research problems. Keeping these pressing needs in view as key features, this handbook of geosynthetic engineering is written as a single volume that contains all the basic features of the book titled *Geosynthetics and Their Applications*, published in 2002. An abstract, keywords, a summary of main points, and some common questions and their answers with updated details for each chapter are some additional special features that make this book more effective than other available good books on geosynthetics. I sincerely hope that this handbook will be a useful tool for civil engineering students, practising engineers and others interested in the field of civil engineering.

Sanjay Kumar Shukla

Acknowledgements

I would like to extend special thanks and recognition to all the contributors for their valuable input. The contributors have worked sincerely to present their chapters with updated details of the topics to meet the need of readers and users of geosynthetics and their applications. I am truly grateful to them.

I am grateful to the following for their help and support at various stages of preparation of the manuscript of this handbook: Professor Richard J. Bathurst, Royal Military College of Canada, Canada; Professor Braja M. Das, California State University, USA; and Associate Professor Nagaratnam Sivakugan, James Cook University, Australia.

I am grateful to Professor Daryoush Habibi, Head, School of Engineering, Edith Cowan University, Australia, for his encouragement to produce useful publications such as this book.

I wish to acknowledge the following for providing useful information and materials required for the preparation of the manuscript of the book titled *Geosynthetics and Their Applications* (2002), which was the main source for this handbook: Terram Limited, Gwent, UK; Netlon Limited, Blackburn, UK; Naue Fasertechnik GmbH & Co., Lubbecke, Germany; Huesker Synthetic GmbH & Co., Gescher, Germany; and Netlon India, Vadodara, India.

I extend special thanks to the staff of ICE Publishing for their support and co-operation at all stages of the preparation and production of this handbook, especially the following: Mr Chris Gibson, Managing Director; Mr Daniel Keirs, Commissioning Editor; Miss Jennifer Barratt, Associate Commissioning Editor; and Mr Imran Mirza, Production Editor.

I wish to extend appreciation to my wife, Sharmila, for her encouragement and support throughout the preparation of the manuscript. Thanks also go to my daughter, Sakshi, and my son, Sarthak, for their patience during my work on this book at home.

I welcome suggestions from readers and users of this book for improving its content in future editions.

Sanjay Kumar Shukla

List of contributors

A.H. Aydilek PhD
Associate Professor, Department of Civil and Environmental Engineering, University of Maryland, College Park, MD 20742, USA

M.C. Alfaro PhD
Associate Professor and Associate Head, Department of Civil Engineering, University of Manitoba, Winnipeg, Canada MB R3T 5V6

A.K. Ashmawy PhD
School of Engineering, American University in Dubai, UAE

D.N. Arnepalli PhD
Assistant Professor, Department of Civil Engineering, Indian Institute of Technology Madras, Chennai 600 036, India

R.J. Bathurst PhD, PEng, FEIC, FCAE
Professor, GeoEngineering Centre at Queen's-RMCC, Department of Civil Engineering, Royal Military College of Canada, Kingston, Ontario, Canada K7K 7B4

R.R. Berg PE
Ryan R. Berg and Associates, Inc., Woodbury, Minnesota, USA

P.L. Bourdeau PhD
Professor, School of Engineering, Purdue University, 550 Stadium Mall Drive, West Lafayette, IN 47907-2051, USA

A. Bouazza PhD, FIEAust
Professor, Department of Civil Engineering, Monash University, Melbourne, VIC 3800, Australia

D. Cazzuffi Dr. Ing.
IGS Past President, CESI SpA, Via Rubattino 54, I-20134 Milano, Italy

B.R. Christopher PhD, PE
Christopher Consultants, Roswell, Georgia, USA

B.M. Das PhD, FASCE
Dean Emeritus, College of Engineering and Computer Science, California State University, Sacramento, CA 95819-6023, USA

C. Duquennoi PhD
Hydrosystems and Bioprocesses Research Unit, Cemagref, Parc de Tourvoie, BP 44, 92163 ANTONY Cedex, France

N. Touze-Foltz PhD
Hydrosystems and Bioprocesses Research Unit, Cemagref, Parc de Tourvoie, BP 44, 92163 ANTONY Cedex, France

E. Gartung Dr.-Ing.
LGA, Geotechnical Institute, Tillystrasse 2, 90431 Nuremberg, Germany

K. Hatami PhD, PEng, MASCE
Assistant Professor, School of Civil Engineering and Environmental Science, University of Oklahoma, 202 West Boyd Street, Norman, OK 73019-1024, USA

G. Heerten
NAUE Gmbh & Co KG, Espelkamp, Germany

W.P. Hornsey
National Technical Manager, Geofabrics Australasia Pty Ltd, 11 Production Avenue, Gold Coast, QLD 4214, Australia

T.S. Ingold BSc, MSc, PhD, DIC Eur. Ing, CEng, FICE, FIHT, FGS, FASCE, MSoc.IS(France), MAE
Professor and Founding Editor of *Geotextiles and Geomembranes*, UK, and *Geosynthetics International*, UK, Mulberry Lodge, St Peters Close, St Albans, Hertfordshire AL1 3ES, UK

M.-H. Li PhD, PE, RLA
Associate Professor, Department of Landscape Architecture and Urban Planning, College of Architecture, Texas A&M University, College Station, TX 77843-3137, USA

M.L. Lopes PhD
Professor, Department of Civil Engineering, Faculty of Engineering, Porto University, Porto, Portugal

E.M. Palmeira PhD
Professor, Department of Civil and Environmental Engineering – FT, University of Brasília, 70910-900 Brasília, DF, Brazil

S.W. Perkins, PhD, PE
Professor, Department of Civil Engineering, Montana State University, Bozeman, MT 59717, USA

L. Peyras PhD
Hydrosystems and Bioprocesses Research Unit, Cemagref, Parc de Tourvoie, BP 44, 92163 ANTONY Cedex, France

D. Poulain PhD
Hydrosystems and Bioprocesses Research Unit, Cemagref, Parc de Tourvoie, BP 44, 92163 ANTONY Cedex, France

A. Scuero Dr. Ing.
Carpi Tech BV, Via Passeggiata 1, CH-6828 Balerna, Switzerland

S.K. Shukla PhD, FIEAust, FIE (India), FIGS, MIRC, MISRMTT, MISTE, MCAII, MIGS
Associate Professor and Program Leader, Discipline of Civil Engineering, School of Engineering, Edith Cowan University, Joondalup, Perth, WA 6027, Australia;
Associate Professor, Department of Civil Engineering, Institute of Technology, Banaras Hindu University, Varanasi 221 005, India

D.N. Singh PhD, FINA, FASCE
Professor, Department of Civil Engineering, Indian Institute of Technology Bombay, Powai, Mumbai 400 076, India

N. Sivakugan PhD, RPEQ, FIEAust, CPEng., MASCE
Associate Professor and Head, Civil and Environmental Engineering, School of Engineering and Physical Sciences, James Cook University, Townsville, QLD 4811, Australia

S.A. (Harry) Tan PhD
Associate Professor, Department of Civil Engineering, National University of Singapore, 1 Engineering Drive 2, E1A 07-03, Singapore 117576

G. Vaschetti Dr. Ing.
Carpi Tech BV, Via Passeggiata 1, CH-6828 Balerna, Switzerland

H. Zanzinger Dipl.-Ing.
SKZ – TeConA GmbH, Friedrich-Bergius-Ring 22, 97076 Wuerzburg, Germany

About the handbook

The *Handbook of Geosynthetic Engineering* is an updated and extended version of *Geosynthetics and Their Applications* (2002). Filters and drains, hydraulic tunnels and geosynthetic applications – sustainability aspects are the new chapters. This handbook is designed for use by civil engineering students and practising engineers who are looking for novel approaches to solutions to civil engineering problems using geosynthetics. The chapters presented cover a wide range of application areas of geosynthetics in civil engineering, and, therefore, readers and users of the handbook can find information related to solutions to their specific problems easily, which is one of the most important key features of this handbook. An abstract, keywords, a summary of main points, some common questions and their answers, and a description of several case studies and practical aspects provided in each chapter are some of the additional key features of this new handbook. The inclusion of recent developments with their references will be very useful, especially to researchers. All the chapters are written by authors who are well recognised in their areas of speciality.

Chapter 1 provides basic information on geosynthetics, including definitions and classification, historical development, functions and selections, raw materials and manufacturing processes, properties and testing, areas of applications, and available standards.

Chapter 2 discusses soil–geosynthetic interaction, considering only the reinforcement function of geosynthetics, complemented by the description of the methods, namely direct shear and pull-out tests, for the evaluation of interaction properties along with a detailed discussion on parameters affecting these properties.

Chapter 3 presents an extensive literature pertinent to the filtration of natural soils and high-water-content geomaterials using geotextiles. Factors affecting the behaviour of filtration processes are described. A step-by-step methodology for designing filters and drains is also explained.

Chapter 4 provides general guidelines for designing retaining walls using geotextiles and geogrids as reinforcing materials along with a few example problems.

Chapter 5 deals with the design of embankments on soft soils using geosynthetics as basal reinforcing materials as well as drains. The combination of geosynthetic layers and piles is also addressed for cases where minimum levels of settlement are required.

Chapter 6 covers various aspects of shallow footings resting on geosynthetic-reinforced foundation soils, including reinforcing mechanisms and functions, reinforcing patterns, modes of failure, model test results, methods of analyses for load-bearing capacity and settlement, the concepts of prestressing, geosynthetic profiles and strain analysis, and selected case histories.

Chapter 7 addresses the application of geosynthetics in unpaved roads; that is, roadway structures that are not capped by concrete slabs or an asphaltic concrete wearing course, with a detailed discussion on the interaction between soil masses and geosynthetics, and design approaches.

Chapter 8 presents material related to the use of geosynthetics in paved roads. The functions of reinforcement, separation, drainage and filtration are discussed, with emphasis placed on the application of reinforcement. Guidelines for the selection and use of geosynthetics for paved road reinforcement are presented. New methods incorporating mechanistic–empirical design principles are also discussed.

Chapter 9 introduces the components of conventional track structures and their functions, and describes the properties, design and installation of geosynthetics for the stabilisation and drainage of railway tracks, along with a few case histories.

Chapter 10 begins with an overview of soil erosion, erosion mechanics and types of erosion control systems, followed by updated design approaches and performance testing methods. The ageing and cost-effectiveness of geosynthetic erosion control products also are considered.

Chapter 11 deals with several aspects of slopes stabilised with geosynthetics as a major component, such as types and orientations of geosynthetic layers, reinforcing mechanism, methods of slope stability analysis and design, model tests, and stabilization methods in practice.

Chapter 12 presents the details of basal liners and covers as essential barriers of solid waste landfills, with an emphasis on the German and South African practice in the design and construction of landfills.

Chapter 13 discusses various features related to the use of geosynthetics in earth dams along with a brief review of conventional earth dam construction practices. Several case studies are included to highlight the construction process and efficiency of installation of geosynthetics.

Chapter 14 illustrates the use of geosynthetics, mainly geomembranes and geosynthetic clay liners, in hydraulic applications such as containment ponds, reservoirs and canals. The presentation of case studies illustrates the recommendations.

Chapter 15 provides a complete description of the engineering aspects of geosynthetic-based lining systems for new and old hydraulic tunnels.

Chapter 16 focuses on seismic aspects of geosynthetic-reinforced walls and slopes. It describes the properties and behaviour of cohesionless soils, geosynthetic reinforcement and facing components under dynamic and cyclic loading, and summarises the important features of current analytical and numerical methods for the seismic analysis and design of walls and slopes along with their field performance observations.

Chapter 17 provides information on application-related general aspects, namely general guidelines, quality control and *in situ* monitoring, cost analysis and general problems, and descriptions of selected case histories.

Chapter 18 discusses the sustainability aspects of geosynthetic applications based on a comparison between classical construction techniques and geosynthetic construction alternatives. The cases are presented to indicate that geosynthetics can provide environment-friendly and energy-efficient solutions to the construction industry.

Handbook of Geosynthetic Engineering
ISBN 978-0-7277-4175-2

http://dx.doi.org/10.1680/hge.41752.001

Chapter 1
Fundamentals of geosynthetics

S. K. Shukla
Discipline of Civil Engineering, School of Engineering, Edith Cowan University, Perth, Australia

In the present-day civil engineering practice, geosynthetics are used as one of the major construction materials. Their use offers technically efficient, cost-effective, environment-friendly and/or energy-efficient alternative solutions to many civil engineering problems. This chapter describes the fundamental aspects of geosynthetics, including definitions and classification, historical development, functions and selections, raw materials and manufacturing processes, properties and testing, areas of applications, and available standards.

1.1. Introduction
In the past four decades, geosynthetics have been used successfully worldwide in several areas of civil engineering, and are now a well-accepted construction material. Currently, the geosynthetic industry deals with a large number of different geosynthetic products, and it greatly controls the economic aspects of many construction projects. The main reason is that the utilisation of geosynthetics offers excellent technical, economic, environment-friendly and/or energy-efficient alternatives to the conventional solutions for many civil engineering problems, and thus allows sustainable development of infrastructural projects. Therefore, students as well as practising engineers require an exposure to the fundamentals of geosynthetics as a construction material. This chapter fulfils this requirement by providing the basic information on geosynthetics.

1.2. Definitions and classification
'Geosynthetics' is a generic term for all synthetic materials used with soil, rock and/or any other civil-engineering-related material as an integral part of a man-made project, structure or system. It includes a broad range of synthetic products; the most common ones are

- geotextiles
- geogrids
- geonets
- geomembranes
- geofoams
- geocomposites.

These products are almost exclusively polymeric, and those based on natural fibres (jute, cotton, wool, silk, etc.) are generally not included. They are available nowadays in numerous varieties in the market, under different trade names/designations for their use mainly in geotechnical, environmental, hydraulic and transportation engineering applications.

Geotextiles are permeable, polymeric textile products in the form of flexible sheets (Figure 1.1). Currently available geotextiles are classified into the following categories based on the manufacturing process.

- *Woven geotextiles* – these are made from yarns (made of one or several fibres) by a conventional weaving process with a regular textile structure.
- *Non-woven geotextiles* – these are made from directionally or randomly oriented fibres in a loose web by bonding with partial melting, needle punching or chemical binding agents (glue, rubber, latex, cellulose derivative, etc.).
- *Knitted geotextiles* – these are produced by interlooping one or more yarns together.
- *Stitch-bonded geotextiles* – these are formed by the stitching together of fibres or yarns.

Geogrid is a polymeric, mesh-like planar product formed by intersecting elements, called ribs, joined at the junctions (Figure 1.2). The ribs can be linked by extrusion, bonding or interlacing, and the resulting geogrids are called *extruded geogrid*, *bonded geogrid* and *woven geogrid*, respectively. Extruded geogrids are classified into the following two categories based on the direction of stretching during their manufacture.

- *Uniaxial geogrids* – these are made by the longitudinal stretching of regularly punched polymer sheets and, therefore, possess a much higher tensile strength in the longitudinal direction than in the transverse direction.
- *Biaxial geogrids* – these are made by both the longitudinal and the transverse stretching of regularly punched polymer sheets, and, therefore, possess equal tensile strength in both the longitudinal and the transverse directions.

The key feature of geogrids is that the openings between the longitudinal and transverse ribs, called apertures, are large

Figure 1.1 Typical geotextiles: (a) woven, (b) non-woven and (c) knitted

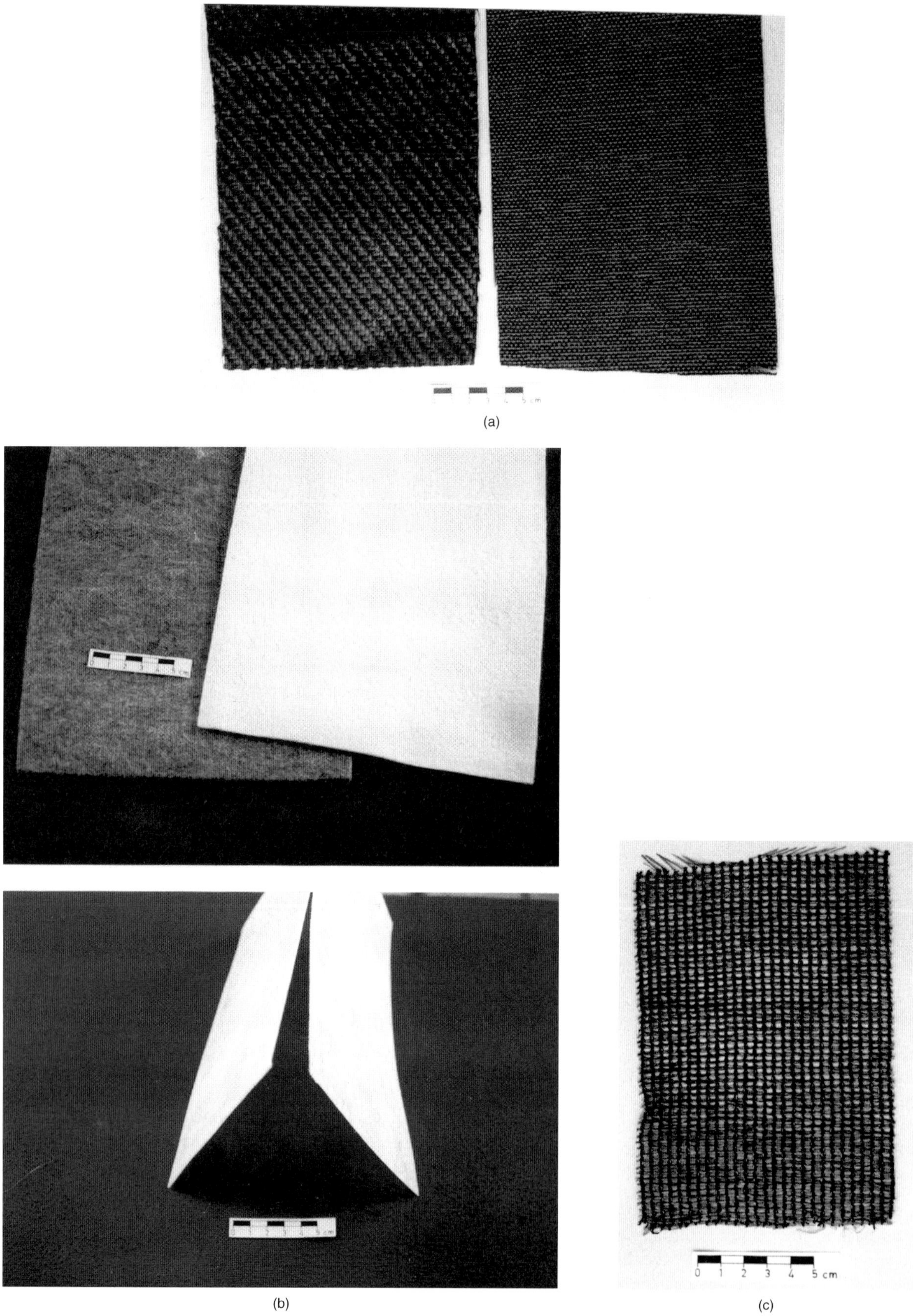

Figure 1.2 Typical geogrids: extruded ((a) uniaxial and (b) biaxial), (c) bonded and (d) woven

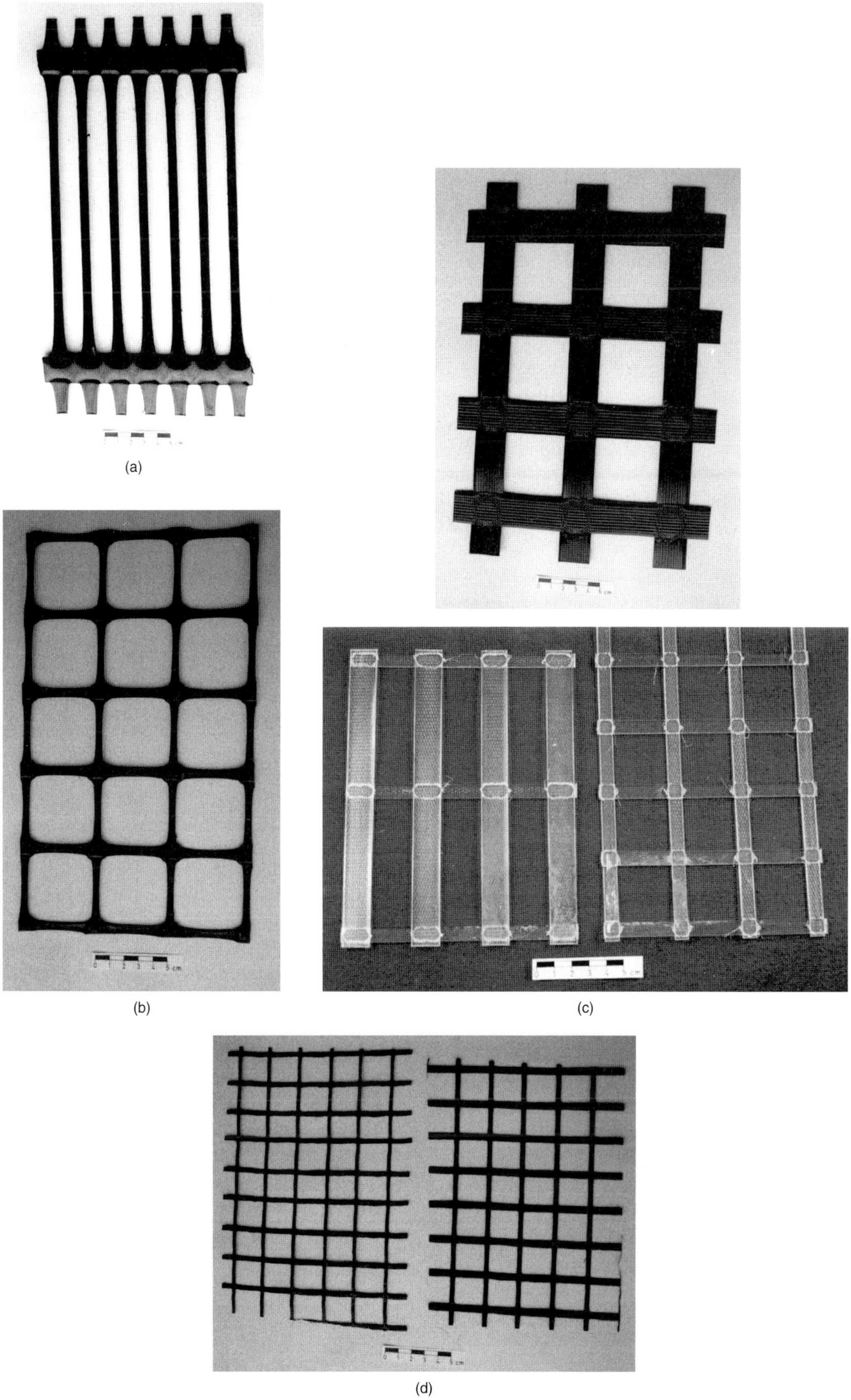

Figure 1.3 The interlocking mechanism in geogrid-reinforced soil

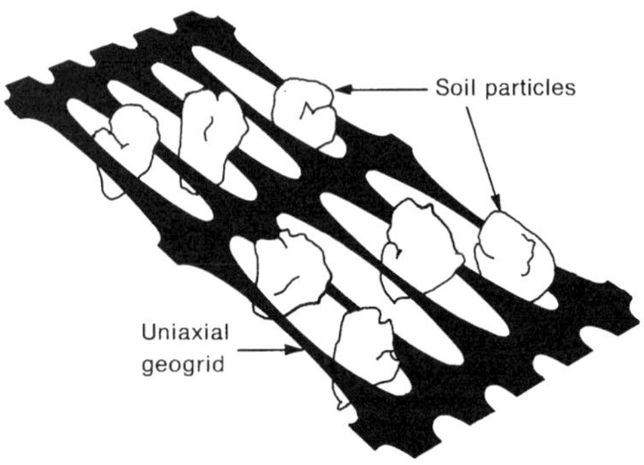

enough to create interlocking with the surrounding soil particles (Figure 1.3). The shapes of the apertures are either elongated ellipses, near-squares with rounded corners, squares or rectangles. The dimensions of the apertures vary from about 2.5 to 15 cm. The ribs of geogrids are often quite stiff compared with the fibres of geotextiles. Also, the junction strength is important in the case of geogrids because, through these junctions, loads are transmitted from one rib to the other rib when geogrid layers are installed within the soil in field applications.

Geonets are planar, polymeric products consisting of a regular dense network of integrally connected parallel sets of ribs overlying similar sets at various angles (Figure 1.4). At first glance, geonets appear similar to geogrids; however, they are different from each other, not mainly in the material or their configuration but in their functions (see Section 1.4). Geonets generally have diamond-shaped apertures that are typically 12 mm long and 8 mm wide. The resulting angles between sets of ribs are of the order of 70–110°.

Geomembrane is a continuous membrane-type barrier/liner composed of materials of low permeability to control fluid migration (Figure 1.5). The materials may be asphaltic or polymeric, or a combination thereof. The term *barrier* applies when the geomembrane is used inside an earth mass. The term *liner* is usually reserved for the cases where the geomembrane is used as an interface or a surface revetment.

Geofoam is a lightweight product in slab or block form with a high void content, and has applications primarily as lightweight fills, thermal insulators and drainage channels. It is manufactured by the application of the polymer in semi-liquid form through the use of a foaming agent.

The term *geocomposites* is applied to products that are manufactured in laminated or composite form from two or more geosynthetic materials (geotextiles, geogrids, geonets, geomembranes, etc.) that, in combination, perform specific functions more effectively than when used separately. There can be several combinations, such as geotextile–geonet, geotextile–geogrid, geotextile–geomembrane, geonet–geomembrane, geomembrane–clay and geomembrane–geonet–geomembrane, which are used in different civil engineering applications. Figure 1.6 shows some typical geocomposites.

There are many other terms for products used in the field of geosynthetic manufacture and applications. Some of them are mentioned below.

- *Geofabric* – a planar flat sheet of geotextile or geotextile-related products.
- *Geomat* – a three-dimensional, permeable, polymeric structure made of coarse and rigid filaments bonded at their junctions, used to reinforce roots of vegetation such as grass and small plants, and extend the erosion control limits of vegetation for permanent installation (Figure 1.6(d)).
- *Geomesh* – a geosynthetic or geonatural generally with a planar woven structure having large pore sizes, which

Figure 1.4 Typical geonets

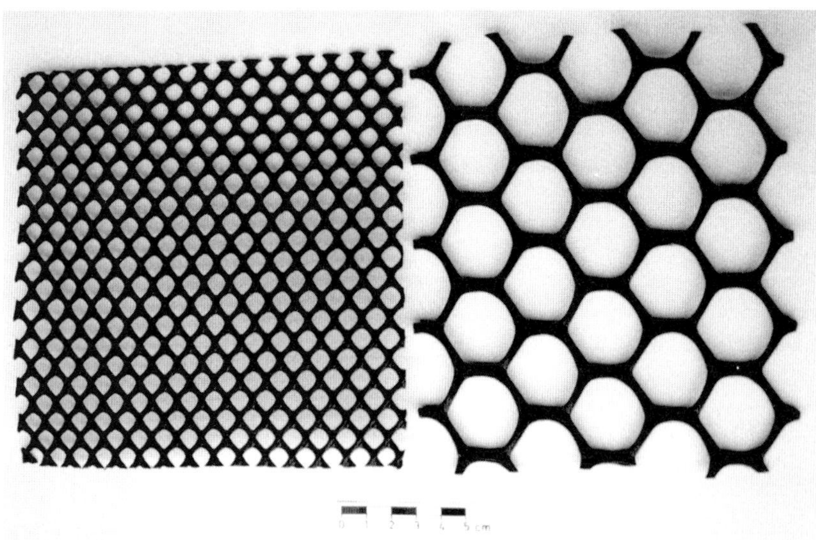

Figure 1.5 Typical geomembranes

vary from several millimetres to several centimetres, for use in mainly erosion control works.

- *Geocell* – a three-dimensional, permeable, polymeric honeycomb or web structure assembled from geogrids and special bodkins couplings in triangular or square cells on the construction site, or produced in the factory using strips of needle-punched polyester or solid high-density polyethylene (HDPE).
- *Geopipe* – a plastic pipe (smooth or corrugated with or without perforations) placed beneath the ground surface and subsequently backfilled.
- *Geotube* – a factory-assembled geosynthetic product, made from high-strength woven geotextile/fabric in tube form and hydraulically filled with sand or fines at the shoreline protection/dewatering projects.
- *Geoproducts* – a term meaning geosynthetics, or geotextile-related products, made from natural fibres and metals.
- *Geonatural* – a product manufactured from natural fibres (jute, coir, cotton, wool, etc.) having a short lifespan when used with soil, rock and/or other civil-engineering-related materials.
- *Geospacer* – a three-dimensional polymeric moulded structure consisting of cuspidated or corrugated plates with large void spaces.
- *Geostrip* – a polymeric material in the form of a strip.
- *Electrokinetic geosynthetic (EKG)* – a mesh made from a metal wire stringer coated in a conductive polymer; it resembles a reinforcing geomesh, and is available in the form of sheets, strips or tubes.

1.3. Historical development

The oldest historical examples of the use of fabrics as an aid to road construction over soft ground include the use of woven reed mats by the ancient Romans. In a style remarkably similar to the present-day techniques, they used to lay the mats over marshy ground before overlaying with stone (Rankilor, 1981). The modern concept of soil reinforcement using a membrane was proposed by Casagrande, who idealised the problem in the form of a weak soil reinforced by high-strength membranes laid horizontally in layers (Westergaard, 1938).

Woven cotton fabrics were used as an early form of geotextile/geomembrane in a series of road construction field tests started in 1926 by the South Carolina Highways Department (John, 1987). In the late 1950s, Terzaghi made use of filter fabrics (today termed geotextiles) as flexible forms. They were filled with a cement grout, thereby making closure between steel sheet piling and rock abutments at the Mission Dam (now the Terzaghi Dam) in British Columbia, Canada. During this same project, Terzaghi used pond liners (today known as geomembranes) to keep an upstream clay seepage control liner for desiccating (Terzaghi and Lacroix, 1964). It is believed that the first applications of polymer-based geotextiles were woven industrial fabrics used beneath concrete block revetments in the late 1950s (Barratt, 1966). A polyvinyl chloride (PVC) monofilament woven geotextile was first used in 1958 at the base of the riprap under the sea dykes of Florida. A woven polyvinyl fabric was first used, instead of straw bags, in the early 1950s in Japan (Fukuoka, 1990). Agerschou (1961) described the use of woven materials to protect coastal structures from soil migration and eventual collapse, and this is most probably the earliest published work.

During the early 1970s, the Japanese-developed filter fabrics (based on their available weaving resources) were being used in, and exerting influences on, Southeast Asia designs for coastal works. By the mid-1970s, the UK had started to produce geotextiles, and, at this time, firms such as ICI in the UK, Rhone Poulence in France, Chemie Linz in Austria and DuPont in the USA started to promote the use of

Figure 1.6 Typical geocomposites: (a) a reinforced drainage separator, (b) a geosynthetic clay liner, (c) drainage composites and (d) a surface erosion control mat

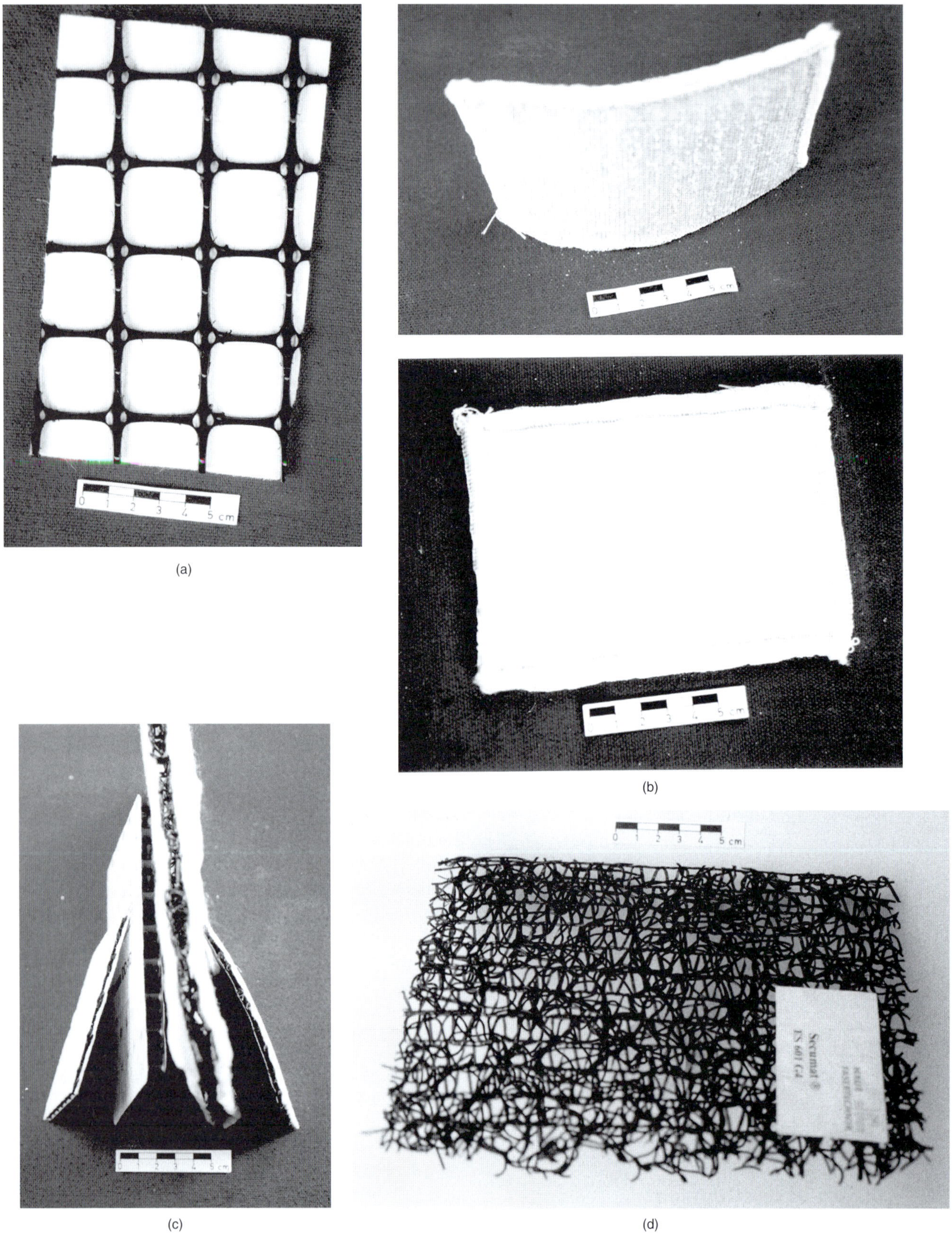

non-woven geotextiles. In 1970, for the first time, a non-woven geotextile was used in an earth dam (Valcros Dam in France) (Giroud *et al.*, 1977). Around 1971, three other areas of geotextile application first appeared, namely the first fin drains (Healy and Long, 1971), the first woven geotextile basal reinforcement beneath embankments (Holtz, 1975) and the first geotextile-reinforced soil wall (Puig *et al.*, 1977). The first composite geotextiles to appear were those used in fin drain systems during the period from 1969 to 1974. Composite geotextiles were also developed during the 1970s as types of band drain used to accelerate the consolidation of clay deposits by providing vertical drainage. Geonets were invented by F. B. Mercer in the UK in 1958. During the late 1970s, Netlon Ltd in the UK developed a more efficient means of utilising the basic polymer raw material, to yield a product with greater strength and elastic modulus in the form of polymer grids to be used in many soil reinforcement applications. The first samples of Tensar grid were made in the Blackburn laboratories of Netlon Ltd in July 1978. Geofoams, which are basically foamed geosynthetic products, were used as a lightweight fill in Norway in 1972.

Soil confinement systems based on cellular geotextile nets were first developed and evaluated in France during 1980 (Simon *et al.*, 1982) (Figure 1.7) and subsequently marketed under the trade name Armater, to be used in the control of surface erosion as well as in temporary road bases. In fully stretched form, the cellular geotextile net forms a honeycomb structure about 200 mm deep with either hexagonal or diamond-shaped apertures. Netlon developed a similar concept, but on a larger scale, with the introduction of the Tensar Geocell Mattress in 1982 (Mercer, 1982). The geocell mattresses are assembled from the Tensar geogrids and special bodkins couplings on the construction site in order to form triangular or square cells, 1 or 0.5 m deep, and they are used basically as a foundation layer beneath embankments, roads and buildings constructed over soft soils.

Geosynthetics were introduced to Indian engineers by the Central Board of Irrigation and Power (CBIP), New Delhi, in 1985 by organising the first National Workshop on Geomembranes and Geotextiles. The first state-of-the-art volume, *Use of Geosynthetics in India: Experiences and Potential*, was brought out by the CBIP in 1989 (Venkatappa Rao and Saxena, 1989). This was a compilation of the field trials in the country, which helped the Indian engineers to gain confidence in the use of geosynthetics.

Figure 1.7 A cellular geotextile net
Reproduced from Simon *et al.* (1982); Industrial Fabrics Association International

Many geotechnical/civil engineering books published or revised as late as 1969 did not provide reference to the use or design of geosynthetics in soil structures. However, despite the lack of general recognition of geosynthetic technology at this time, a few papers were published somewhat sporadically, the earliest known being that by Agerschou (1961). The number of publications suddenly increased from 1971. The first conference on geosynthetics was held in Paris in 1977. The first book on geosynthetics was written in 1980 (Koerner and Welsh, 1980). The international technology exchange has become active after the establishment of the International Geosynthetics Society in 1983. At present, there are two specialist international journals (*Geotextiles and Geomembranes* and *Geosynthetics International*), and several magazines and newsletters. In line with many other conferences, there are regularly scheduled national, regional and international conferences on geosynthetics. The field of geosynthetics has thus established itself in civil and environmental engineering. Various geosynthetic manufacturers have consistently been involved in pushing forward the frontiers of the geosynthetics technology. Many academicians, researchers and practising engineers have made efforts to present the subject in the form of research papers and books written with different objectives for the spread of the concepts of the geosynthetics and their applications (John, 1987; Raymond and Giroud, 1993; Holtz *et al.*, 1997; Shukla, 2002; Koerner, 2005; Shukla and Yin, 2006).

1.4. Basic functions and selection

Geosynthetics have numerous application areas in civil engineering. They always perform at least one of the following major functions when used in conjunction with soil, rock and/or any other civil-engineering-related material

- separation
- reinforcement
- filtration
- drainage (or fluid transmission)
- fluid barrier.

If a geosynthetic prevents intermixing of adjacent soil layers with different properties during construction and the projected service period of the geosynthetic-reinforced soil structure, it is said to have a *separation* function. Figure 1.8 shows that the geosynthetic layer prevents the intermixing of soft soil with granular fill, thereby maintaining the structural integrity of the granular fill.

A geosynthetic shows its *reinforcement* function by increasing the strength of a soil mass as a result of its inclusion, and thus

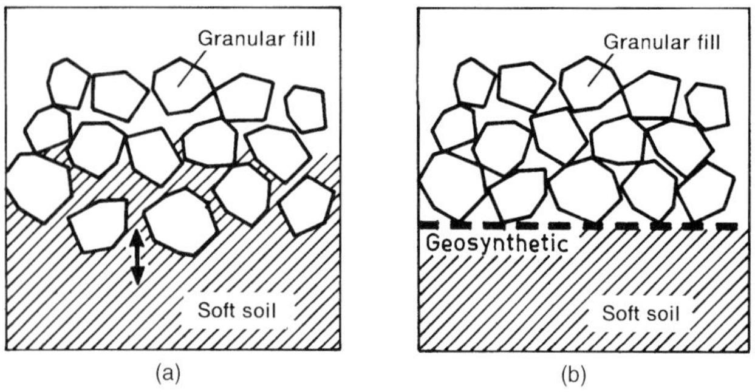

Figure 1.8 The separation function: (a) granular fill–soft soil system without a geosynthetic layer and (b) granular fill–soft soil system with a geosynthetic layer

maintains the stability of the soil mass. In this process, the geosynthetic layer carries tensile loads (Figure 1.9). More details on reinforcing mechanisms (Shukla, 2004; Shukla *et al.*, 2009a) are described in Section 6.2 of Chapter 6.

In the past two decades, research works have been carried out extensively to investigate the behaviour of soils reinforced with synthetic and/or natural fibres oriented randomly. Preparation of randomly distributed fibre-reinforced soils mimics soil stabilisation by admixtures. Randomly distributed fibres within the soil mass offer strength isotropy and limit the potential planes of weakness that can develop parallel to the oriented reinforcement, as included in geosynthetic-reinforced soils (Gray and Maher, 1989; Yetimoglu and Salbas, 2003; Shukla *et al.*, 2009a, Lovisa *et al.*, 2010; Shukla and Sivakugan, 2010; Shukla *et al.*, 2010). Although fibres are polymeric materials and concepts of their reinforcing applications can be included in geosynthetic engineering, their details – including their types and applications – are not covered within this book.

Selection of the major reinforcement or separation function is governed by the ratio of the applied stress on the soft soil to its shear strength. This aspect is discussed further in Section 6.2 of Chapter 6.

A geosynthetic may function as a *filter* that allows for adequate flow of fluids across its plane while preventing the migration of soil particles, along with fluid flow, during

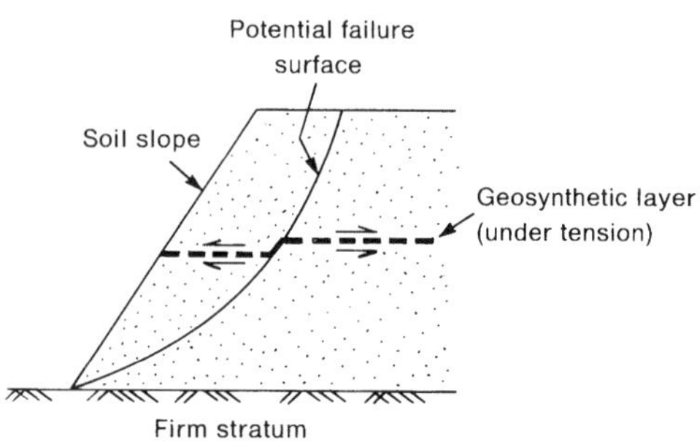

Figure 1.9 The reinforcement function

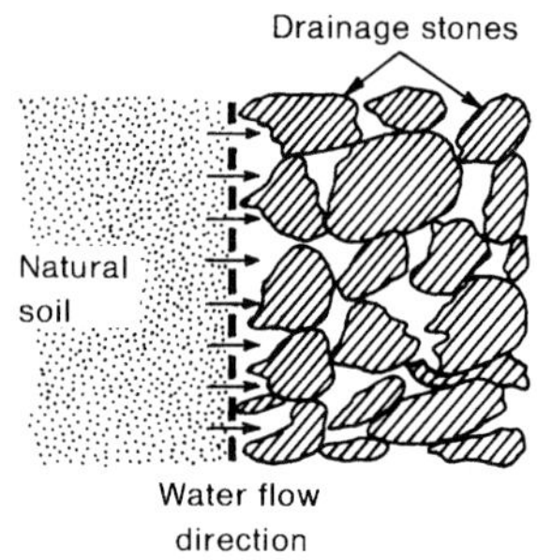

Figure 1.10 The filtration function

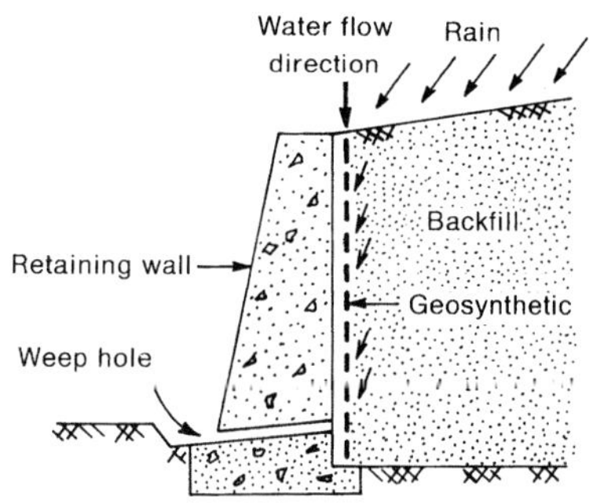

Figure 1.11 The drainage function

the projected service period of application under consideration (Figure 1.10).

If a geosynthetic allows for adequate flow of fluids within its plane from the surrounding soil mass to various outlets during the projected service period of the application under consideration, it is said to have a *drainage* or *fluid transmission* function. Figure 1.11 shows that a geosynthetic layer adjacent to the retaining wall collects water from the backfill and conveys it to the weep hole made in the retaining wall.

A geosynthetic may also act like an almost impermeable membrane as far as the flow of fluids is concerned: this is the *fluid barrier* function. Figure 1.12 shows that the geosynthetic layer, kept at the base of a pond, prevents the infiltration of liquid waste into the natural soil.

In addition to the basic functions described above, in some specific field applications, a geosynthetic may also perform one or more than one of the following functions, which are basically dependent on the basic functions.

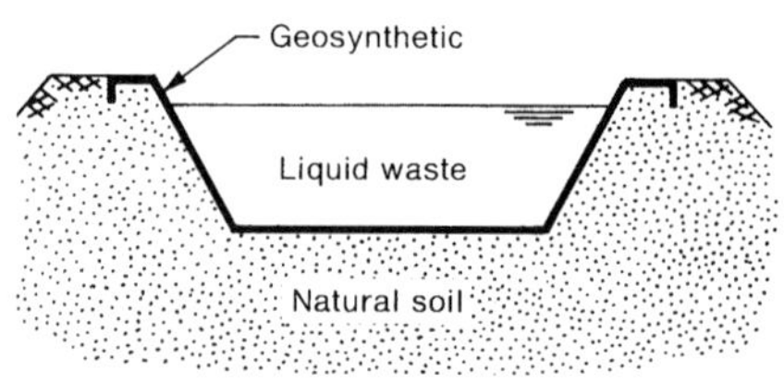

Figure 1.12 The fluid barrier function

- *Protection* – where a geosynthetic is used as a localised stress reduction layer to prevent damage to a given surface or layer (e.g. geomembrane layer), it is said to perform a protection function.
- *Cushioning* – where a geosynthetic is used to control and eventually to damp dynamic mechanical actions, it is said to perform a cushioning function. This function has to be emphasised particularly for applications in canal revetments, in shore protections and in geosynthetic strip layers as seismic base isolation of earth structures.
- *Absorption* – the process of fluid being assimilated or incorporated into a geotextile. This function may be considered for two specific environmental aspects: water absorption in erosion control applications; and the recovery of floating oil from surface waters following ecological disasters.
- *Interlayer* – a function performed by a geosynthetic to improve the shear resistance between two layers of geosynthetic products and/or earth materials.
- *Containment* – with this function, a geosynthetic encapsulates or contains a civil-engineering-related material such as soil, rock or fresh concrete to a specific geometry, and prevents its loss.
- *Insulation* – a geosynthetic provides insulation when it is used to reduce the passage of electricity, heat or sound.
- *Screening* – a geosynthetic provides screening when it is placed across the path of a flowing fluid carrying fine particles in suspension to retain some or all particles, while allowing the fluid to pass through.

When installed, a geosynthetic may perform more than one of the above-listed functions simultaneously, but generally one of them will result in a lower factor of safety, and thus it becomes the primary function. The use of a geosynthetic in a specific application needs classification of its functions as primary or secondary. Table 1.1 shows such a classification, which is useful when selecting the appropriate type of geosynthetic to solve the problem in hand. The function concept is generally used in the design with the formulation of a factor of safety, FS, in the traditional manner as

$$FS = \frac{\text{value of allowable (or test) property}}{\text{value of required (or design) property}} \qquad (1.1)$$

Factors of safety must be greater than 1; the actual magnitude depends upon the implication of failure, which is always site-specific. The value of the allowable property is obtained from a stimulated *performance test* (or an *index test* modified by site-specific reduction factors), whereas the required property is obtained from an appropriate design model. Such models are generally modifications of existing geotechnical or hydraulic models. The entire process, generally called 'design by function', is widespread in its use. However, as might be anticipated with a young technology, universally accepted values of minimum factors of safety have not yet been established, and conservativeness in this regard is still warranted (Koerner, 2000).

It is to be noted that only geotextiles and geocomposites perform most of the functions and, hence, they are used in many applications. Geotextiles are porous to water flow, both normal to the manufactured plane and within the plane. The degree of porosity, which may vary widely, is used to determine the selection of specific geotextiles. Geotextiles can also be used as a fluid barrier on impregnation with materials such as bitumen. The geotextiles vary with the type of polymer used, the type of fibre and the fabric style.

Geogrids are used mainly for reinforcement (separation may occasionally be a function, especially when soils with very large particle sizes are involved). The performance of the geogrid for reinforcement relies on its rigidity or high tensile modulus, and on its open geometry, which accounts for its high capacity for interlocking with soil particles (see Figure 1.3).

It has been observed that for geotextiles to function properly as reinforcement, friction must develop between the soil and the reinforcement to prevent sliding, whereas for geogrids

Table 1.1 Selection of geosynthetics based on their functions

Function(s) to be served by the geosynthetic		Geosynthetics that can be used
Separation	Primary	Geotextiles, geocomposites, geofoams
	Secondary	Geotextiles, geogrids, geonets, geomembranes, geocomposites, geofoams
Reinforcement	Primary	Geotextiles, geogrids, geocomposites
	Secondary	Geotextiles, geocomposites
Filtration	Primary	Geotextiles, geocomposites
	Secondary	Geotextiles, geocomposites
Drainage	Primary	Geotextiles, geonets, geocomposites, geopipes
	Secondary	Geotextiles, geocomposites, geofoams
Fluid barrier	Primary	Geomembranes, geocomposites
	Secondary	Geocomposites

it is the interlocking of the soil through the apertures of the geogrid that achieves an efficient anchoring effect. In this respect, geotextiles are frictional resistance-dependent reinforcement, whereas geogrids are passive resistance-dependent reinforcement. Laboratory studies have shown that geogrids are a superior form of reinforcement owing to the interlocking of the soil with the grid membrane.

Geotextiles may be used to serve such functions as protection, cushioning, interlayering or absorption. Protection and cushion functions may also be performed by geonets.

Geonets, unlike geotextiles, are relatively stiff, net-like material with large open spaces (0.9–5.0 cm) between structural ribs. It should be noted that geogrids are not geonets, which are used exclusively for their in-plane drainage capability. For a fair drainage function, geonets should not be laid in contact with soils or waste material, but should be used as drainage cores with geotextile, geomembrane or other materials on their upper and lower surfaces, thus avoiding the soil particles from obstructing the drainage net channels. Geonets as drainage materials, in their flow capability, fall between thick needle-punched non-woven geotextiles and drainage composites.

Geomembranes are used as a fluid barrier/liner only. The permeability of a typical thermoplastic or thermoset geomembrane is 10^{-13} to 10^{-15} m/s. In this regard, we speak of it as being relatively impermeable.

Since geofoams are used as lightweight fills, thermal insulators and drainage channels, their primary function is dictated by the specific application in the field projects. However, in general, a geofoam is considered to function as a separator in most of its applications.

Geocomposites can be manufactured to perform a combination of the functions described above. For example, a geomembrane–geonet–geomembrane composite can be made where the interior net acts as a drain to the leak detection system. Similarly, a geotextile–geonet composite improves separation, filtration and drainage. Geocomposites are generally, but certainly not always, completely polymeric. Other options include using fibreglass or steel for tensile reinforcement, sand in compression or as a filler, dried clay for subsequent expansion as a liner, or bitumen as a waterproofing agent. Geomembrane–clay composites are used as the liners, where the geomembrane decreases the leakage rate while the clay layer increases the breakthrough time. In addition, the clay layer reduces the leakage rate from any holes that might develop in the geomembrane, while the geomembrane will prevent cracks in the clay layer due to changes in moisture content. Geocomposites have been used in trapping and conveying leachate in landfills, and in collecting gases from beneath geomembrane liners of various types.

Electrokinetic geosynthetics (EKGs) are created as geocomposites by incorporating electrokinetic phenomena with the existing traditional functions of geosynthetics. This type of geocomposite basically consists of a metallic wire mesh coated in a conductive polymer, and is available in the form of sheets, strips or tubes. When used as a cathode, the tubular electrode is enveloped in a geotextile sleeve such as a non-woven, heat-bonded product. Its great advantage over conventional metallic electrodes used in soil improvement applications is that it overcomes the difficulties of corrosion, insufficient electrical contact and physical removal of water from systems that require dewatering applications (Hamir *et al.*, 2001; Fourie and Jones, 2010).

The selection of a geosynthetic for a particular application is governed by several other factors, such as specification, durability, availability and cost.

1.5. Raw materials and manufacturing processes

The polymers generally used as raw materials for geosynthetics are polyester (PET), polypropylene (PP), polyethylene (PE) (very low-density polyethylene (VLDPE), medium-density polyethylene (MDPE) and high-density polyethylene (HDPE)), chlorinated polyethylene (CPE), chlorosulphonated polyethylene (CSPE), polyamid (PA), polyvinyl chloride (PVC), expanded polystyrene (EPS), extruded polystyrene (XPS), etc. Table 1.2 provides a list of raw materials used for manufacturing different geosynthetics.

There are a wide number of variables that affect the material properties of these polymers, including polymer density, melt flow rate, draw ratio and polymer additives. The properties of geosynthetics are governed by these variables, and their effects have been the subject of much investigation.

Most geotextiles are manufactured from polypropylene or polyester. The primary reason for polypropylene usage in geotextile manufacturing is its low cost. For non-critical structures, it provides an excellent, cost-effective raw material. It exhibits a second advantage in that it has excellent chemical and pH resistance. Additives and stabilisers (such as carbon black) must be added to give polypropylene ultraviolet light resistance. As the critical nature of the structure increases, or the long-term anticipated loads go up, polypropylene tends to lose its effectiveness. This is

Table 1.2 Polymers used as raw materials for manufacturing geosynthetics

Geosynthetics	Raw materials
Geotextiles	PP, PET, PA, PE
Geogrids	PET, PP, HDPE
Geonets	MDPE, HDPE
Geomembranes	HDPE, LLDPE, VLDPE, PVC, CPE, CSPE, PP
Geofoams	EPS, XPS
Geopipes	HDPE, PVC, PP

Table 1.3 A comparison of properties of polymers used in manufacturing the geosynthetics
Data taken from John (1987)

Properties		Polymers			
		PP	PET	PA	PE
Strength		Low	High	Medium	Low
Modulus		Low	High	Medium	Low
Strain at failure		High	Medium	Medium	High
Creep		High	Low	Medium	High
Unit weight		Low	High	Medium	Low
Cost		Low	High	Medium	Low
Resistance to ultraviolet light	Stabilised	High	High	Medium	High
	Unstabilised	Medium	High	Medium	Low
Resistance to alkalis		High	Low	High	High
Resistance to fungus, vermin, insects		Medium	Medium	Medium	High
Resistance to fuel		Low	Medium	Medium	Low
Resistance to detergents		High	High	High	High

because of relatively poor creep deformation characteristics under long-term sustained load.

Polyester is increasingly being used to manufacture reinforcing geosynthetics, such as geogrids, because of its high strength and resistance to creep. The chemical resistance of polyester is generally excellent, with the exception of very high pH environments. It is inherently stable to ultraviolet light. The properties of some of the polymers mentioned above are compared in Table 1.3.

Although most of the geosynthetics are made from synthetic polymers, a few specialist geosynthetics, especially geotextiles, may also incorporate either steel wire or natural biodegradable fibres such as jute, coir, paper, cotton, wool or silk. Biodegradable geotextiles are usually limited to erosion control applications, where natural vegetation will replace the role of the geotextile as it degrades. Jute nets are marketed under various trade names, including *Geojute*, *Soil Saver* and *Anti-wash*. They are usually in the form of a woven net with a mesh open size of about 11 × 18 mm, a typical thickness of about 5 mm, and an open area of about 65%. Vegetation can easily grow through the openings and use the fabric matrix as support. The jute, which is about 80% natural cellulose, should completely degrade in about two years. An additional advantage of these biodegradable products is that the decomposed jute improves the quality of the soil for vegetation growth.

The manufacturing process of a geotextile includes two steps (Giroud and Carroll, 1983). The first step consists of making linear elements such as fibres and yarns. The second step consists of combining these linear elements to make a planar structure, usually called a fabric.

The basic elements of a geotextile are its fibres. There are mainly four types of synthetic fibres: *filaments* (produced by extruding melted polymer through dies or spinnerets, and subsequently drawing it longitudinally); *staple fibres* (obtained by cutting filaments to a short length, typically 2–10 cm); *slit films* (flat tape-like fibres, typically 1–3 mm wide, produced by slitting an extruded plastic film with blades and subsequently drawing it); and *strands* (a bundle of tape-like fibres that can be partially attached to each other). During the drawing process, the molecules become oriented in the same direction, resulting in an increase of the modulus of the fibres. A *yarn* is made of one or more fibres. Several types of yarn are used to construct woven geotextiles: *monofilament yarn* (made from a single filament), *multifilament yarn* (made from fine filaments aligned together), *spun yarn* (made from staple fibres interlaced or twisted together), *slit film yarn* (made from a single slit-film fibre) and *fibrillated yarn* (made from strands). It should be noted that synthetic fibres are very efficient load-carrying elements, with tensile strengths equivalent to prestressing steel in some cases (e.g. in the case of polyaramid fibres).

As the name implies, *woven geotextiles* are obtained by conventional weaving processes, using a mechanical loom (Figure 1.13). This weaving process gives these geotextiles their characteristic appearance of two sets of parallel yarns interlaced at right angles to each other, as shown in Figure 1.14. The terms 'warp' and 'weft' are used to distinguish between the two different directions of yarn. The yarn running along the length of the loom and hence along the length of the geotextile roll is known as the *warp*. The yarn running along the transverse direction, across the width of both the loom and the geotextile roll, is known as the *weft*. The type of weave described is *plain weave*, of which there are many variations, such as *twill*, *satin* and *serge*; however, plain weave is the one most commonly used in geotextiles. Resulting structures are typically 1–2 mm thick with a comparatively regular distribution of pore or mesh openings, which vary in dimension over a

Figure 1.13 Main components of a weaving loom
Reproduced from Rankilor (1981), © Wiley-Blackwell

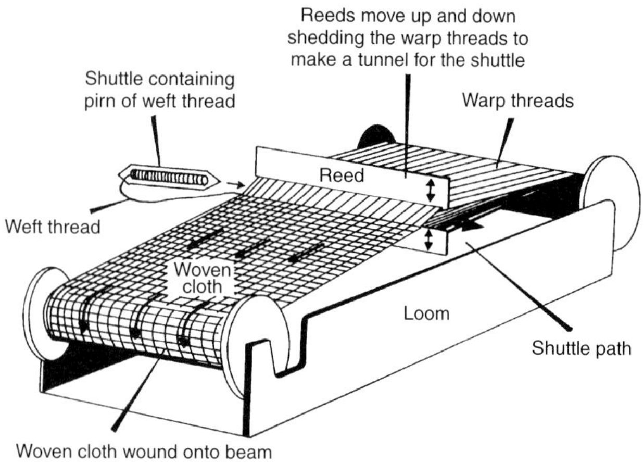

reasonably small size band. Kaswell (1963) gives an excellent review of weaving technology with clear illustrations of various fabric weaves.

There are no rigid criteria relating polymer type to structure; however, tapes are most commonly polypropylene and monofilaments are most commonly polyethylene, whereas the finer multifilaments or multifilament yarns are commonly polyester (Ingold and Miller, 1988). The numerous variations of weaving structure have a major influence on the physical, mechanical and hydraulic properties of the resulting geotextile. The highly anisotropic properties shown by woven geotextiles are also the influence of the weaving structure.

Non-woven geotextiles are obtained by processes other than weaving. Continuous monofilaments are usually employed; these may, however, be cut into short staple fibres before processing. The processing involves continuous laying of the fibres or filaments on to a moving conveyor belt to form a loose web slightly wider than the finished product. This passes along the conveyor to be bonded by *mechanical bonding* (obtained by punching thousands of small barbed needles through the loose web), *thermal bonding* (obtained

Figure 1.14 A typical woven geotextile having a plain weave

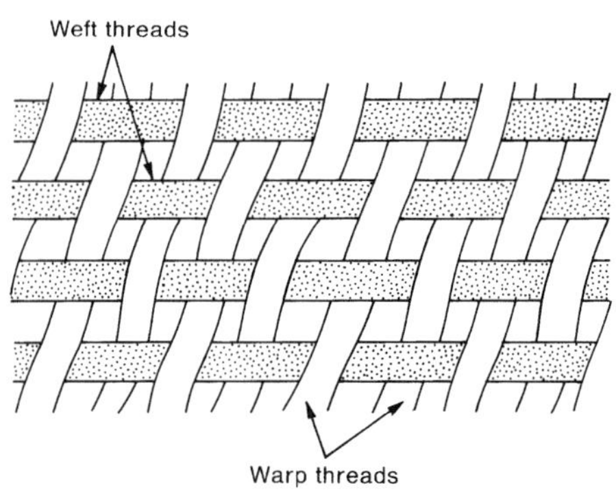

Figure 1.15 A typical knitted geotextile

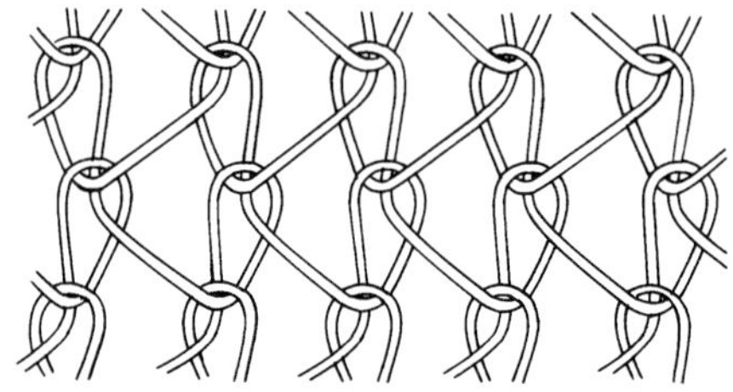

by partial melting of the fibres) or *chemical bonding* (obtained by fixing the fibres with a cementing medium, such as glue, latex, cellulose derivative or synthetic resin), resulting in three different types: *mechanically bonded non-woven geotextile* (or *needle-punched geotextiles*), *thermally bonded non-woven geotextile* and *chemically bonded non-woven geotextile*, respectively. These geotextiles are usually relatively thick, with a typical thickness in the region of 0.5–5 mm.

Knitted geotextiles are manufactured using a knitting process that involves interlocking a series of loops of one or more yarns together to form a planar structure. There is a wide range of different types of knit used, one of which is illustrated in Figure 1.15. These geotextiles are used in very limited quantities.

Extruded geogrids are manufactured by the method of processing sheet polymer in two or three stages. The first stage involves feeding a sheet of polymer, several millimetres thick, into a punching machine, which punches out holes on a regular grid pattern. Following this, the punched sheet is heated and stretched, or drawn, in the machine direction. This distends the holes, to form an elongated grid opening. In addition to changing the initial geometry of the holes, the drawing process orients the randomly oriented long-chain polymer molecules in the direction of drawing. The degree of orientation will vary along the length of the grid; however, the overall effect is an enhancement of tensile strength and tensile stiffness. The process may be halted at this stage, in which case the end product is a uniaxially oriented geogrid. Alternatively, the uniaxially oriented grid may proceed to a third stage of processing to be warm drawn in the transverse direction, in which case a biaxially oriented geogrid is obtained. Although the temperatures used in the drawing process are above ambient, this is effectively a cold drawing process, as the temperatures are significantly below the melting point of the polymer. Netlon Ltd manufactures geogrids using its patented Tensar manufacturing process (Figure 1.16). Other types of geogrids are manufactured by weaving and knitting, as well as by bonding the mutually perpendicular high-strength strips together at their cross-over points ultrasonically or thermally.

The most common manufacturing technique for *geonets* is to extrude the molten polymer through slits in counter-rotating

Figure 1.16 The Tensar manufacturing process
Courtesy of Netlon® all rights reserved

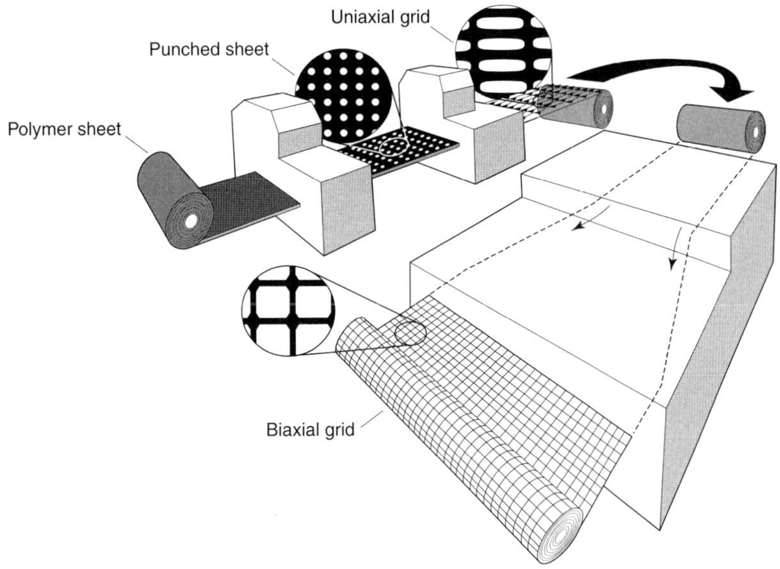

dies, which forms a tight net of closely spaced ribs. This net is then opened up by forcing it over a tapered mandrel until it reaches its final configurations, when it is cooled and rolled. The resulting geonet has intersecting sets of ribs at 60–75° apart, with the cross-over points being integrally bonded to one another. A slight variation of the above technique is to add a foaming agent to the polymer mix and then process it as just described. The foaming agent is released and forms micrometer-sized gas-filled spheres within the rib cross-sections. Geonets formed in this manner can have very high ribs (resulting in increased flow capability) in comparison with the solid-formed ribs. The width of the geonet depends on the production facilities and the processes. Currently, geonets up to 4.5 m wide are available on the market, and their thickness typically ranges between 5.0 and 9.0 mm (Koerner, 2005).

Most *geomembranes* are made in a plant using one of the following manufacturing processes: extrusion, spread-coating or calendering (Giroud and Frobel, 1983). The *extrusion* process is a method whereby a molten polymer is extruded into an unreinforced sheet. Immediately after extrusion, when the sheet is still warm, it can be laminated with a geotextile; the geomembrane thus produced is reinforced. The *spread-coating* process usually consists of coating a geotextile (woven, non-woven, knitted) by spreading a polymer or asphalt compound on it. The geomembranes thus produced are therefore reinforced. Non-reinforced geomembranes can be made by spreading a polymer on a sheet of paper that is removed and discarded at the end of the manufacturing process. *Calendering* is the most frequently used manufacturing process in which a heated polymeric compound is passed through a series of heated rollers (calender). Typical thicknesses of geomembranes range from 0.25 to 7.5 mm (10 to 300 mils, 1 mil = 0.001 in.), and they are produced in rolls approximately 1.5–10 m long.

The raw material for the manufacture of geofoams is polystyrene (PS), which is known as packaging material and insulating material by the common people. Its genesis is from ethylene and is available in two forms: expanded polystyrene (EPS) and extruded polystyrene (XPS). The manufacturing process of geofoam starts with introducing styrene with various additives, water and a blowing agent (pentane). Use of steam helps to soften the resin, and the blowing agent causes an enormous increase in its volume, resulting in a unit weight value ranging between 0.1 and 0.5 kN/m^3, which is extremely low when compared with the unit weight of water and soils. In the final stage of the manufacturing process, geofoams are obtained as slabs or blocks. It should be noted that EPS-based geofoams are relatively thick as well as cheap in comparison with XPS-based geofoams. Both the thickness and price aspects of the EPS-based geofoams make them suitable for lightweight fill applications in civil engineering, whereas XPS-based geofoams are generally used as insulators. A typical size of a geofoam block may be $1.0 \times 0.5 \times 2.0$ m, having its weight between 0.1 and 0.5 kN (or 10 and 50 kgf).

Geocomposites can be manufactured from two or more of the geosynthetic types described above. A geocomposite can therefore combine the properties of the constituent members in order to meet the needs of a specific application. Some examples of geocomposites are sheet drains, strip (wick) drains, fin drains and geosynthetic clay liners. A *strip drain* usually consists of plastic fluted or nubbed cores that are surrounded by a geotextile filter. A *fin drain* comprises a vertical water-conducting core; that is, a drainage net, sandwiched between outer layers of the geotextile. Geotextiles are commonly used in conjunction with geomembranes for puncture protection, drainage and improved tensile strength.

A *geosynthetic clay liner* is used in lieu of compacted soil for the low-permeability soil component of the composite liner. It consists of a thin layer of sodium bentonite (mass per unit area $\approx 5\,kg/m^2$), which is either sandwiched between two geotextiles or mixed with an adhesive and attached to a geomembrane. Geosynthetic clay liners are manufactured in panels that measure 4–5 m in width and 30–60 m in length, and are placed on rolls for shipment to the jobsite. When a geosynthetic clay liner comes into contact with water, the bentonite swells in the pores, thereby forming a watertight sheet or offering protection to geomembranes (Venkatappa Rao, 1996). A geosynthetic clay liner has many advantages over a compacted soil liner, including the following (Snow *et al.*, 1994)

- simple installation and lower installed cost
- low water consumption, dust generation and vehicular traffic during construction
- low susceptibility to desiccation cracking
- self-healing capabilities if punctured
- material quality maintained in a controlled environment
- lower construction quality assurance costs
- tensile strength developed by the geotextiles or geomembranes
- reduced loss of valuable waste disposal facility.

The combination of the geotextile (filtering action), geomembranes (waterproofing properties) and geonets (drainage and load distribution) offers a complete system of filter-drainage protection, which is very compact and easy to install.

1.6. Properties and test methods

This section deals with the properties of geosynthetics and highlights the basic concepts of their measurement. Geosynthetics, being polymer-based products, are viscoelastic, which means that, under working conditions, their performance is dependent on the ambient temperature, the level of stress, the duration of the applied stress, the rate at which the stress is applied, etc. The properties of geosynthetics should therefore be used to keep these factors in view.

1.6.1 Physical properties

The physical properties of geosynthetics that are of prime interest are *specific gravity*, *mass per unit area*, *thickness* and *stiffness*. There are some more physical properties, which are important in the case of geogrids and geonets only, and these are *type of structure*, *junction type*, *aperture size* and *shape*, *rib dimensions*, *planar angles made by intersecting ribs* and *vertical angles made at the junction point*. The physical properties are more dependent on temperature and humidity than those of soils and rocks. In order to achieve consistent results in the laboratory, good environmental control during the testing is therefore important.

Typical values of the specific gravity of commonly used polymeric materials are given in Table 1.4. It is to be noted that the specific gravity of some of the polymers is less than 1.0, which is a drawback when working with geosynthetics underwater; that is, some of them may float.

Table 1.4 Specific gravity of polymeric materials

Polymers	Specific gravity
Polypropylene	0.90–0.91
Polyester	1.22–1.38
Polyethylene	0.91–0.96
Polyvinyl chloride	1.3–1.5
Polyamide	1.05–1.15

The *mass per unit area* of a geosynthetic is usually given in units of gram per square metre. Sometimes it is referred to as the 'basis weight'. It can be a good indicator of cost and several other properties such as tensile strength, tear strength and puncture strength, which are defined in Section 1.6.2. It is also necessary for quality control, and thus it is the most useful basic property of geosynthetics. For commonly used geosynthetics, it varies in the order of magnitude from typically 100 to $1000\,g/m^2$. For Tensar SR2 and SS2 grids, the mass per unit area was estimated to be 930 and $345\,g/m^2$, respectively.

The *thickness* of geosynthetics, particularly of geotextiles, is measured as the distance between the upper and the lower surfaces of the material at a specified normal pressure (generally 2.0 kPa). The thickness of commonly used geosynthetics ranges from 10 to 300 mils. Most geomembranes used today are 20 mils (0.50 mm).

The *stiffness* or *flexibility* of a geosynthetic is related to its bending under its own weight, and indicates the feasibility of providing a suitable working surface for installation. It can be measured by its capacity to form a cantilever beam without exceeding a certain amount of downward bending under its own weight. It should be noted that the workability of a geosynthetic (the ability of the geosynthetic to support personnel in an uncovered state and construction equipment during the initial stages of cover fill placement) also depends on other factors, such as water absorption and buoyancy. When placing a geotextile or geogrid on extremely soft soils, a high stiffness is desirable.

Properties such as aperture size and shape, and rib dimensions, can be measured directly, and are relatively easily determined.

1.6.2 Mechanical properties

Mechanical properties are important in those applications where a geosynthetic is required to perform a structural role, or where it is required to survive installation damage and localised stresses. There are several mechanical properties; however, some of them are only important in the case of particular geosynthetics.

The *compressibility* of a geosynthetic is measured by the decrease in its thickness at varying applied normal pressures.

Figure 1.17 Variation in the thickness of geotextiles with applied pressure
Reproduced from Shamsher (1992)

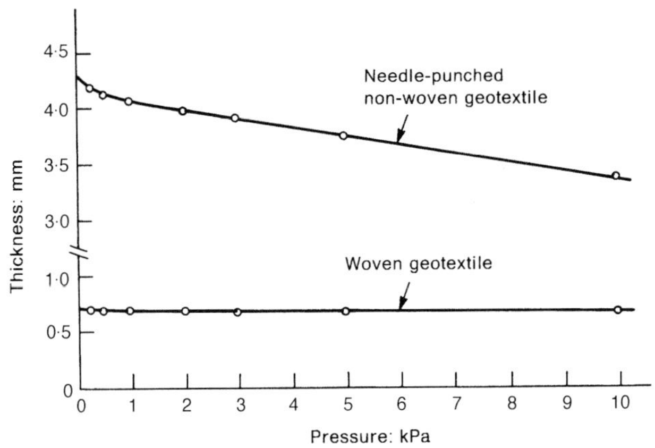

This mechanical property is very important for non-woven geotextiles because they are often used to convey liquid within the plane of their structure. Figure 1.17 shows changes in thickness under pressure for typical woven and needle-punched nonwoven geotextiles. For most geotextiles, except needle-punched non-woven geotextiles, the compressibility is relatively very low.

Due to the specific geometry and irregular cross-sectional area, the *tensile strength* of geosynthetics cannot be expressed conveniently in terms of stress. It is, therefore, defined as the peak load that can be applied per unit width. Tensile strength is generally determined by a wide-width strip tensile test on a 200 mm wide strip, because, by approximating plane strain conditions, this test more closely simulates the deformation experienced by a geosynthetic embedded in soil (Figure 1.18). The test provides parameters such as peak strength, elongation and tensile modulus. The measured strength and the rupture strain are a function of many test variables, including the sample geometry, the gripping method, the strain rate, the temperature, the initial preload, conditioning, and the amount of any normal confinement applied to the geosynthetic. Figure 1.19 shows the influence of the geotextile specimen width on the tensile strength. To minimise the effects of these factors, the test sample should have a width-to-gauge length ratio (aspect ratio) of at least

Figure 1.18 The wide-width strip tensile test (note: $B = 200$ mm, $L = 100$ mm)

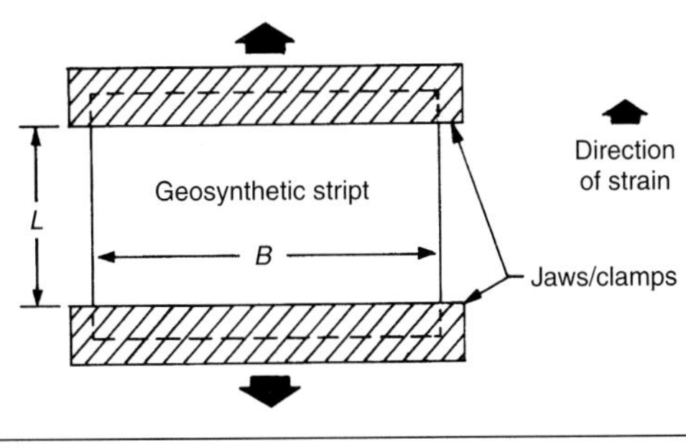

Figure 1.19 The influence of the geotextile sample width on its tensile strength
Reproduced from Myles and Carswell (1986); Industrial Fabrics Association International

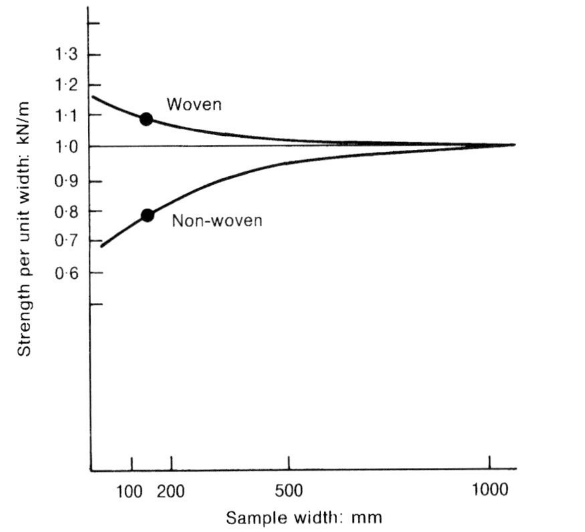

2, and the test should be carried out at a standard temperature. The tensile strength is closely related to the mass per unit area (Figure 1.20). Other forms of tensile tests, such as the grab test and the biaxial test, are shown schematically in Figure 1.21.

For the determination of the tensile strength of geomembranes, the multi-axial tension test is a well-established experimental procedure. The test is performed by clamping a

Figure 1.20 Variation in the tensile strength with the mass per unit area for polypropylene geotextiles
Reproduced from Ingold and Miller (1988)

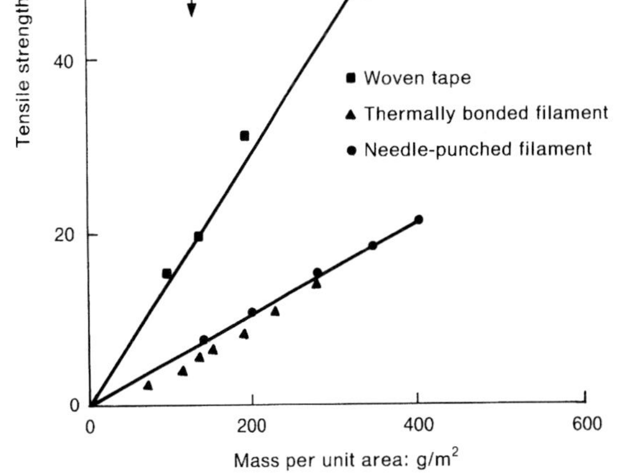

Figure 1.21 (a) The grab tensile test, (b) the biaxial tensile test and (c) the plain strain tensile test

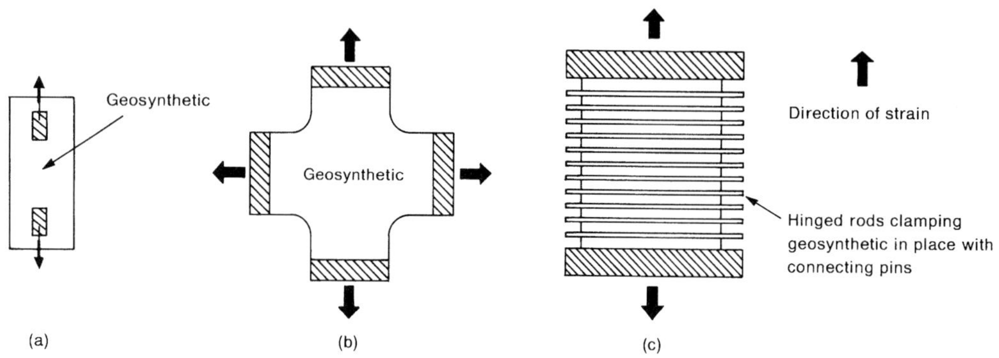

circular specimen around its perimeter and then introducing pressurised air or water to one side of the specimen, to deform it in the out-of-plane direction. The measured pressure and centreline deflection of the specimen are then used to compute the multi-axial stress–strain response of the material (ASTM, 2010). Andrejack and Wartman (2010) presented a new large-diameter experimental device capable of applying multi-axial, out-of-plane loading to pervious geosynthetics such as geotextiles. They also presented equations to interpret the experimental data. A key advantage of the multi-axial tensile test is that, in contrast to the conventional wide-width tensile test, it better simulates in-service boundary conditions of geosynthetics used in a range of typical applications.

Previously, it was pointed out that the strength of woven geotextiles is governed by the weaving structure. It has been found that the strength of a woven geotextile is higher at 45° to the warp and weft directions, but is lower parallel to the warp/weft, whereas non-woven geotextiles tend to have a lower but more uniform strength in all directions. One should obtain the minimum strength of the geosynthetic product and ensure that this stress is never exceeded in practical applications.

The *tensile modulus* is the slope of the geosynthetic stress–strain or load–strain curve, as determined from wide-width tensile test procedures. This is equivalent to the Young's modulus for other construction materials, such as concrete, steel, timber and structural plastic. It depicts the deformation required to develop a given stress (load) in the material. Figure 1.22 shows typical load–strain curves for geotextiles and interpretation methods for the tensile modulus (Myles and Carswell, 1986). At the commencement of the test, the load will be zero unless a preload is used. As the test is begun, the geotextile strains without loading until it reaches

Figure 1.22 Load-strain curves for geotextiles exhibiting: (a) linear behaviour and (b) non-linear behaviour
Reproduced from Myles and Carswell (1986), Industrial Fabrics Association International

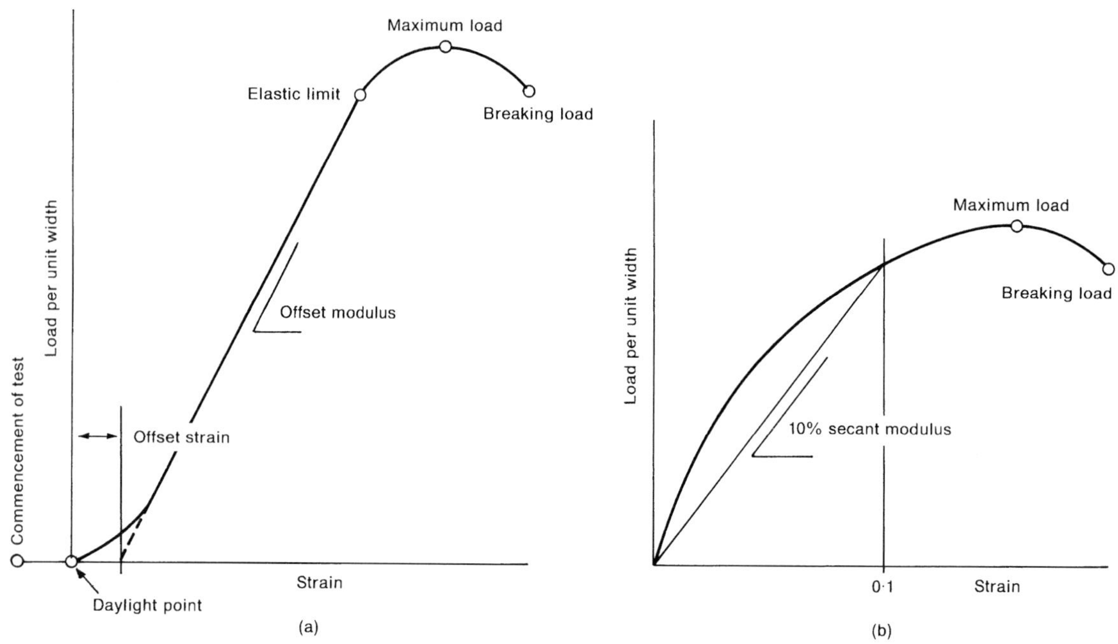

Figure 1.23 Typical strength properties of some geosynthetics (note: overlapping zones have not been shown for clarity – some non-typical geosynthetics may lie outside the zones indicated, and some geosynthetics are more sensitive to the test method)
Reproduced from John (1987), Blackie and Son Ltd

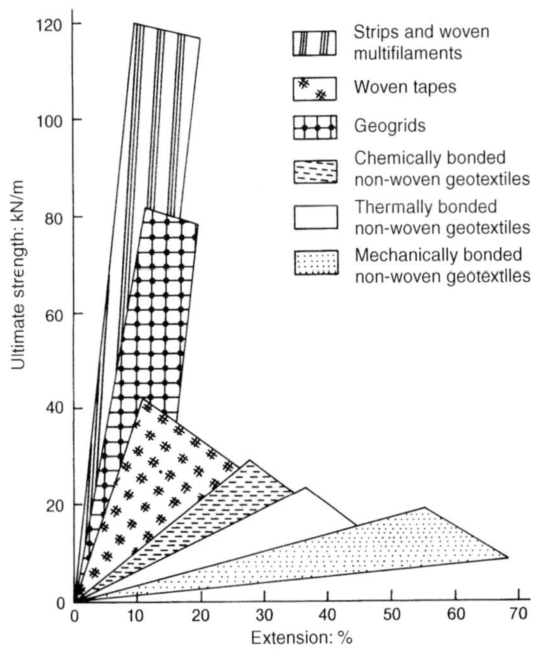

the daylight point (a point where the load extension curve parts from the strain). The *offset modulus* (*working modulus*) is obtained from the slope of the linear portion of the load–extension data. An offset strain is then defined by extending the linear portion of the data back to the zero load line. It is important to understand that the (unknown) strain from the indicated start of the test to the daylight point is eliminated by preloading and that the amount of offset strain is influenced by the amount of preloading. For geotextiles that do not have a linear range, the modulus is typically defined as the *secant modulus* at 5 or 10% strain. The designer and specifier must have a clear understanding of the interpretation of these moduli. Figure 1.23 shows typical strength properties of some geosynthetics. It is noted that woven geotextiles display generally the lowest extensibility and highest strengths of all the geotextiles. Geogrids have relatively high dimensional stability, high tensile strength and high tensile modulus at low strain levels. They develop reinforcing strength even at a strain equal to 2% (Carroll, 1988). The high tensile modulus results from prestressing during manufacture, which also creates integrally formed structures without weak points either in ribs or junctions. In the case of geonets, there is a preferential direction in strength between the machine and cross-machine directions. Geonets have the greatest strength in the machine direction.

The viscoelastic behaviour of geosynthetics can produce misleading results for both short-term and rapid-rate tensile tests. Tests conducted to provide design data should also consider long-term conditions and account for the effect of the surrounding soil. The geosynthetic confinement within the soil in the field and the resultant interlocking of soil particles with the geosynthetic structure are found to have a significant effect on the stress–strain properties. It is generally found that the modulus of a geosynthetic confined in soil is likely to be higher than when tested in isolation. The mechanism of this enhancement is simply the frictional force development. The deformation of a geosynthetic structure is, therefore, likely to be overestimated if the in-isolation modulus is used in the calculations (Hausmann, 1990). This fact tends to support the use of a working modulus as an appropriate interpretation method. The confined tensile test methods have been presented by McGown *et al.* (1982) and El-Fermaoui and Nowatzki (1982). Due to the high costs involved, confined tensile testing is not carried out on a routine basis. Keeping these facts in view, it should be noted that the wide-width tensile test is essentially an index test.

Geosynthetics in field applications are subjected to loadings that may cause longitudinal and lateral normal strains simultaneously. The ratio of the lateral normal strain to the longitudinal normal strain is known as Poisson's ratio, which varies in the range 0 to 0.5 for most linear elastic engineering materials, but the theoretical range is −1.0 to 0.5. Although Poisson's ratio is often assumed as a material constant, it is strain dependent. It is a critical parameter in some engineering applications where geosynthetics are being used as reinforcements (e.g. the capping of contaminated high-water-content geomaterials or hydraulic filling of geosynthetic containers). The uniaxial tensile strength test data reported by Wesseloo *et al.* (2004) indicate that Poisson's ratio for a 0.2 mm thick high-density polyethylene geomembrane decreases from 0.50 to 0.24 when the axial strain increases from 0 to 200% at strain rates, ranging from 50 to 0.05%/min at 22 ± 1°C. Based on the wide-width tensile strength tests, Kutay *et al.* (2006) reported that the lateral strain was greater than the axial strain at failure for one needle-punched polypropylene non-woven geotextile (mass per unit area = 278 g/m^2) and two polyester-woven geotextiles (mass per unit area = 290/1907 g/m^2, ultimate tensile strength = 70/632 kN/m, tensile strength at 5% strain = 16/316 kN/m), where Poisson's ratio at failure ranged from 1.0 to 2.1, whereas Poisson's ratio was in the range 0.3–0.81 for four polyester geogrids (mass per unit area = 305/387/617/1284 g/m^2, ultimate tensile strength = 63/102/137/371 kN/m, tensile strength at 5% strain = 25/37/65/146 kN/m), depending on the clamping technique used. Giroud (2004) presented analytical expressions for Poisson's ratio (ν) as a function of axial strain (ε) for both incompressible and compressible materials. The expression for compressible material is based on an assumption where the small-strain theory is extended to large strains. Shukla *et al.* (2009b) presented an improved understanding of the theoretical expressions for the Poisson's ratio of geosynthetics, and proposed a simple method by presenting a graphical chart for estimating the Poisson's ratio of geosynthetics at zero strain (ν_0), which is required to calculate the realistic values of the Poisson's ratio of geosynthetics at other non-zero strains using the analytical expression.

At this stage, it is worthwhile mentioning index and performance tests. *Index tests* are carried out under standardised conditions used to compare the basic properties of geosynthetic products (wide-width tensile strength, creep under load, friction properties, etc.). They are generally used in quality control and quality assurance. They are also used to monitor changes that may occur after a geosynthetic has had some sort of exposure. Index tests generally do not reflect design features or applications. *Performance tests*, on the other hand, are carried out by placing the geosynthetic in contact with a soil/fill under standardised conditions in the laboratory, to provide better simulation of site conditions than index testing. Performance testing, if possible, should also be carried out at full scale at the site. It is to be noted that geosynthetics vary randomly in thickness and weight in any given sample roll due to normal manufacturing techniques. Tests must be conducted on representative samples collected as per the guidelines of available standards, which ensure that all areas of the sample roll and a full variation of the product are represented within each sample group.

When two pieces of similar or dissimilar geosynthetics (or related material) are attached to each other, this is known as a 'joint', and when a geosynthetic is physically linked to, or cast into, another material (e.g. the facing panel of a retaining wall – see Chapter 4), this is known as a 'connection'. When no physical attachment is involved between two geosynthetics or a geosynthetic and another material, this is known as an 'overlap'.

Where geosynthetic widths or lengths greater than those supplied on one roll are required, jointing becomes necessary, and the same may be effected by one of the jointing methods, such as *overlapping*, *sewing*, *stapling* or *gluing*. Different joints currently in use may be classified into prefabricated joints and joints made during field applications. In the vast majority of cases, the geosynthetic width or length is extended simply by overlapping, which is usually found to be the easiest field method (Figure 1.24). Geotextiles may be jointed mechanically, by sewing or stapling, or chemically using an adhesive bond. Figure 1.25(a) shows the most suitable seam configuration, known as the prayer seam. Another type of seam, known as the lapped ('J') seam, is shown in Figure 1.25(b). Depending on the critical nature of the construction, either a single or double stitch is used. For jointing the geotextiles by the stapling method, corrosion-resistant staples should be used. Figure 1.26 shows the stapled seam configuration. For geosynthetics such as geonets, and geogrids, on the other hand, a bodkin joint may be employed, whereby two overlapping sections are coupled together using a bar passed through the apertures (Figure 1.27). Geogrids can also be sewn using a robust cord threaded through the grid apertures. Hog rings, staples, threaded loops, wires, etc. are also used for jointing geosynthetics. The list of field-seaming techniques for plastomeric geomembranes (made from high-density polyethylene, low-density polyethylene, polyethylene, polypropylene or polyvinyl chloride) includes fillet extrusion, flat extrusion, hot air, hot wedge, ultrasonic and electric welding methods. Figure 1.28 shows some typical seams in geomembranes. Elastomeric geomembranes (made from rubbers of various types as the barrier component) require seaming by means of solution or adhesives. Geosynthetic clay liners are jointed by the application of bentonite at the panel joints.

An important criterion for assessing joint performance is the load transmission between the two pieces of the geosynthetics. In some applications, it may be essential that the load transfer capability is equal to that of the parent material. For other situations, a more important criterion may be the magnitude of the deformation of the joint under load. The *seam strength* is the load transfer capability from one geosynthetic roll to another when the ends of both the rolls are joined together by any method. The efficiency (E) of a seam joint, between geosynthetic sheets, is generally defined as the percentage of the ultimate tensile strength of the geosynthetic that the joint can bear before rupture. It is therefore expressed as

$$E = \left(\frac{T_{\text{seam}}}{T_{\text{geosynthetic}}} \times 100\right)\% \qquad (1.2)$$

where T_{seam} is the wide-width seam strength and $T_{\text{geosynthetic}}$ is the wide-width (unseamed) geosynthetic strength.

Ideally, the joint would be stronger than the geosynthetic being jointed, and would thus never fail in tension. In practice, in the field, high efficiencies are rarely obtained. Publications generally mention that laboratory-obtained efficiencies are usually higher than field efficiencies. This is of little help to the field engineer trying to meet a consultant's specification. However, efforts should be made to make seam efficiency near to 100%. As the fabric strength becomes higher, seams become less efficient. Above a geosynthetic strength of 44 kN/m, even the best seams have an efficiency of less than 100%, and beyond 440 kN/m the best one can have approximately 50% efficiency (Koerner, 1990).

Figure 1.24 A simple overlap
Courtesy of Terram®, all rights reserved

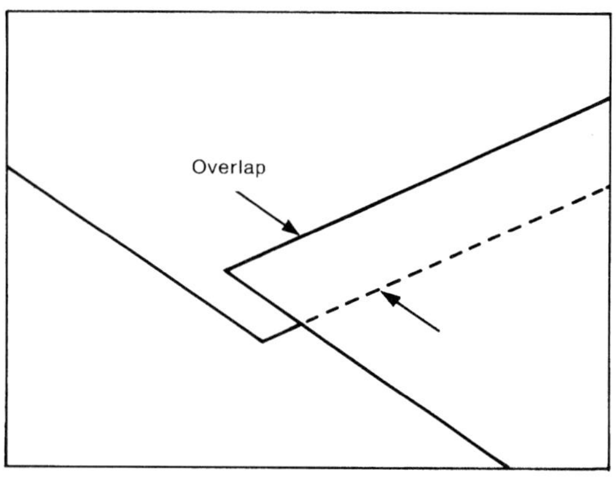

Figure 1.25 (a) Face-to-face ('prayer') seams: (i) single-stitch line and (ii) double-stitch line. (b) a lapped ('J') seam

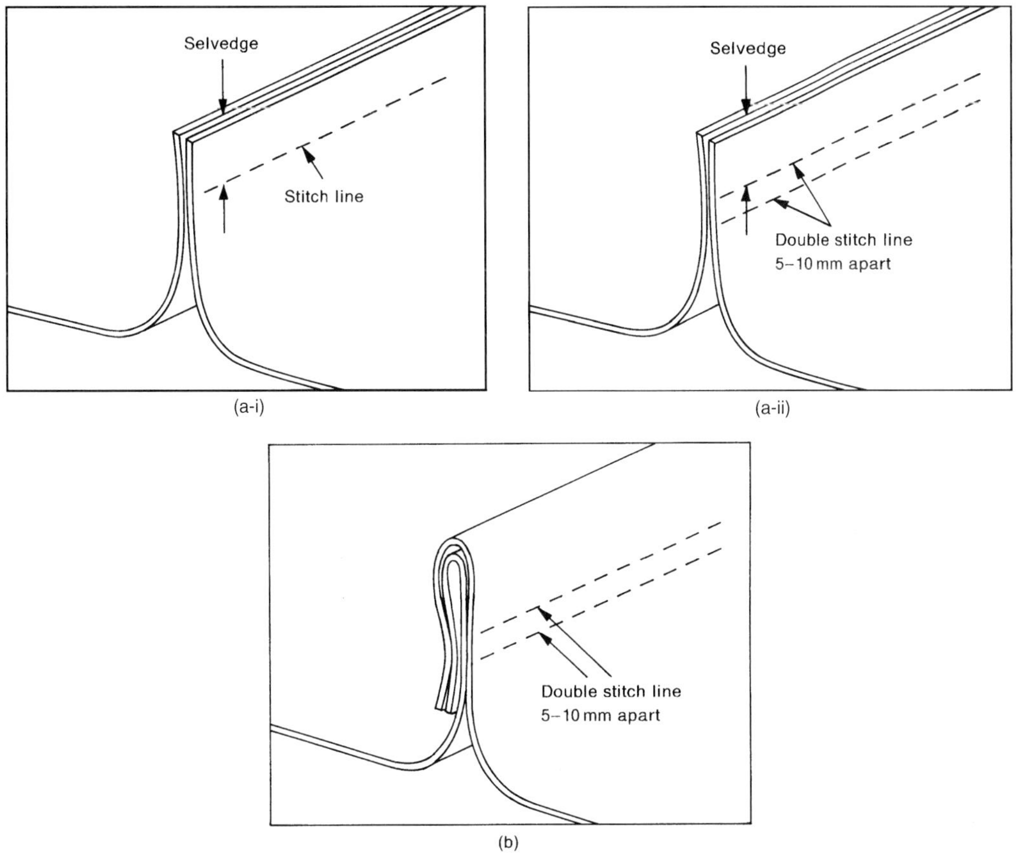

Murray *et al.* (1986) undertook research work into the seam strengths obtained from both sewn and adhesive-bonded seams. Their work was comprehensive and stated that 100% efficiency could be obtained using adhesives. With sewn joints, they described efficiencies up to 90%, but they drew attention to the large deformations that are experienced. The technique of jointing geogrids by means of a bodkin joint proved to be an effective procedure, whereby load-carrying efficiencies of about 90% were obtained. Rankilor and Heiremans (1996) reported that the use of adhesives can reduce seam extension dramatically.

Figure 1.26 A stapled seam

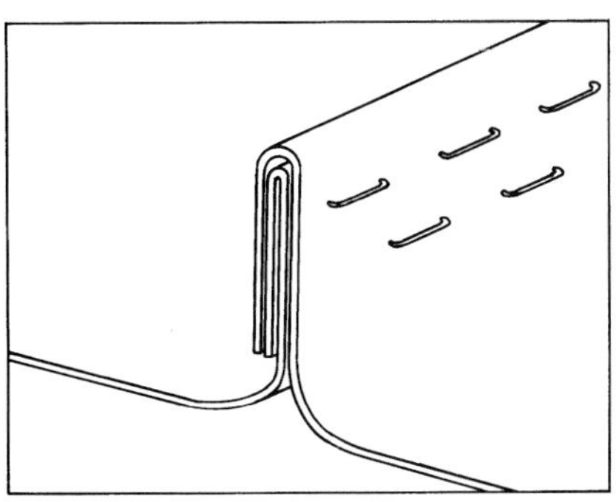

Figure 1.27 A bodkin joint

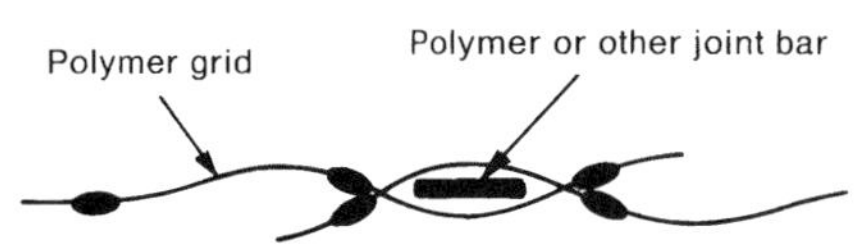

Figure 1.28 Some typical seams in geomembranes
Reproduced from Giroud (1994), Industrial Fabrics Association International

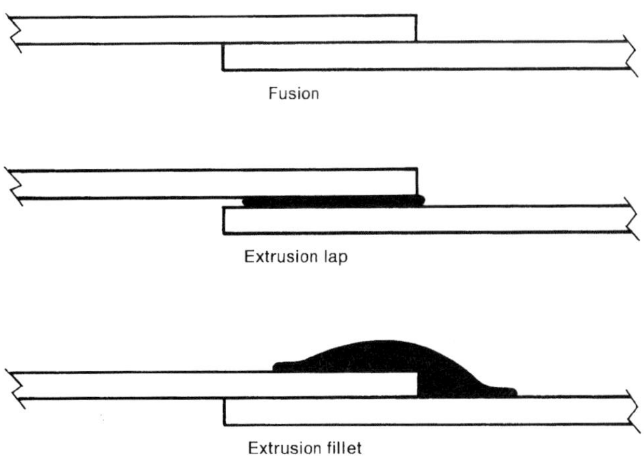

Figure 1.29 Laboratory tests related to the geosynthetic survivability and the separation function
Reproduced from Hausmann (1990), The McGraw-Hill Companies

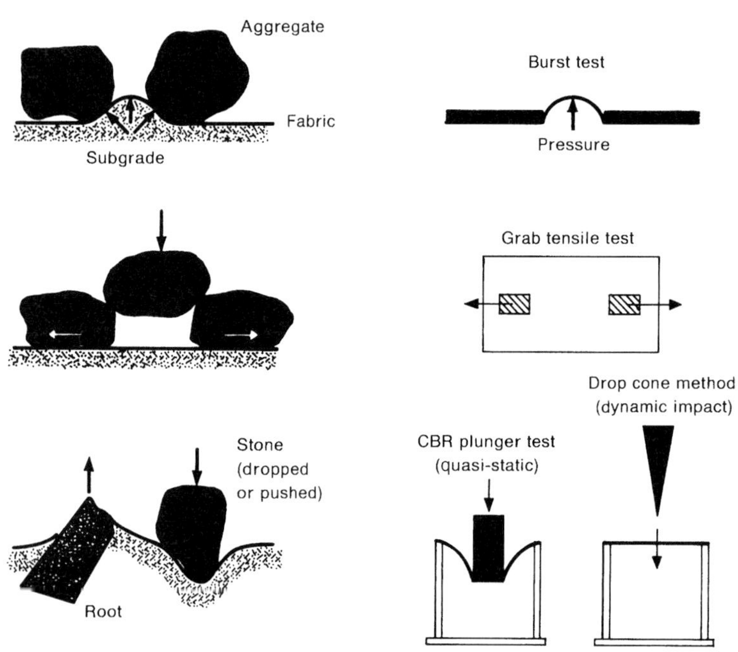

In the past, it has been observed that defects of geomembrane seams are the cause of 65% of leaks through uncovered liners of landfills, tanks, basins and ponds (Rollin *et al.*, 2002). In the case of covered geomembranes, only 6% of leaks have been observed due to seaming failures (Colucci and Lavagnolo, 1995; Nosoko and Touze-Foltz, 2000). It is therefore very important to assess geomembrane seam quality. This is accomplished by both destructive tests (shear and peel tests) and non-destructive tests (visual observation, the mechanical probe test, the electrical method, etc.). Geomembrane seams can also be tested using the gas permeation method (Barroso *et al.*, 2006).

There are some mechanical properties of geosynthetics that are related to geosynthetic survivability and the separation function. Such tests are known as *integrity tests*, and are as follows.

- *Fatigue strength* – the ability of geosynthetics to withstand repetitive loading before undergoing failure.
- *Burst strength* – the ability of geosynthetics to withstand loading when no further deformation is possible.
- *Tear strength* – the ability of geosynthetics to withstand tearing stresses often generated during their installation.
- *Impact strength* – the ability of geosynthetics to withstand stresses generated by falling objects, such as rock pieces, tools and other construction items.
- *Puncture strength* – the ability of geosynthetics to withstand stresses generated by penetrating objects, such as pieces of rock or wood, under quasi-static condition.

Figure 1.29 explains the fundamental concepts of some laboratory integrity tests.

When a geosynthetic is used to reinforce a soil mass, it is important that the bond developed between the soil and the geosynthetic is sufficient to stop the soil from sliding over the geosynthetic or the geosynthetic from pulling out of the soil when the tensile load is mobilised in the geosynthetic. The bond between a geosynthetic and a soil depends on the interaction of their contact surfaces. The *soil–geosynthetic interaction* (*interface friction* and/or *bearing resistance* that develops against transversal elements in geosynthetics such as geogrids) is thus the key element in the performance of most geosynthetic-reinforced soil structures. It is used to determine the bond length of the geosynthetic needed beyond the critical zone.

Two common test procedures currently used to evaluate soil–geosynthetic interaction are the direct shear test and the pull-out (anchorage) test. In the direct shear test, the upper half of the box is fixed, while the lower half is subjected to a horizontal force. In this case, the geosynthetic sample is anchored along the edge of the box where the tensile (horizontal) force is applied. In a pull-out test, the two halves of the box are fixed, and one end of the geosynthetic specimen is subjected to a horizontal force. Figure 1.30 depicts both the tests and the corresponding reinforcing applications with the type of failure mode. A designer of

Figure 1.30 Comparison of pull-out and direct shear tests with corresponding reinforcing application
Reproduced from Paulson (1987), with permission from Elsevier

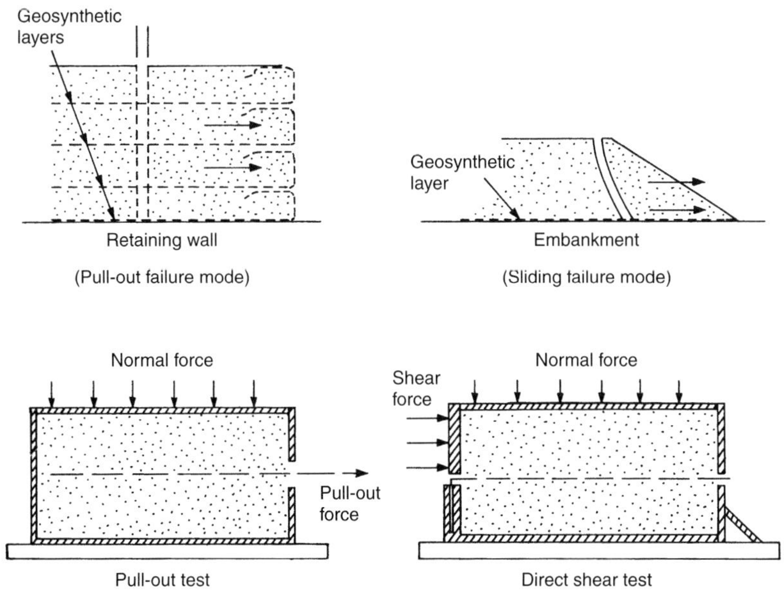

geosynthetic-reinforced soil structures must consider the potential failure mode, then the appropriate test procedure should be used to evaluate the soil–geosynthetic interaction properties. In the case of an unpaved road, the recommended test should be a combination of direct shear and pull-out tests conducted simultaneously (Giroud, 1980). Ingold and Miller (1988) reported that, for most geotextiles, the ratio of the angle of the shearing resistance of geotextile-reinforced granular soil to the angle of shearing resistance of unreinforced granular soil, as determined from direct shear tests, rarely drops below 0.75 and is often close to unity; however, there may be exceptions.

In several geosynthetic applications (landfills, canals, ponds, reservoirs, river banks, sea coasts, etc.), geosynthetics are installed on sloping ground where the normal stress levels are generally low. To assess the soil–geosynthetic interaction for such applications, direct shear tests are not easy to carry out. In these cases, the inclined plane test (IPT) is commonly performed to measure the interface shear strength. The IPT apparatus consists of two steel plates hinged at one end: one plate is kept stationary and horizontal, while the other is tilted by using a threaded rod (Figure 1.31). The details of this test in slightly different forms have been described by several researchers (Palmeira *et al.*, 2002; Narejo, 2003; Pitanga *et al.*, 2009), and they have also been presented in the test standard ISO 12957-2 (ISO, 2005). The test results reported by Narejo (2003) indicate that a simple IPT apparatus (tilt table device) can be used to measure the index friction angle of geosynthetic materials. The sensitivity and precision of the test suggest that the test can be used as a quality control or conformance testing tool, to ensure the required level of performance or help with interpretation of results. It should be noted that the index friction angle value obtained through the IPT apparatus relates only to the peak friction resistance of geosynthetics. The residual shear resistance of certain geosynthetics may be significantly different from the static or peak behaviour indicated by this apparatus.

Figure 1.31 Sketch of the inclined plane test apparatus
Reproduced from Narejo (2003), with permission from Elsevier

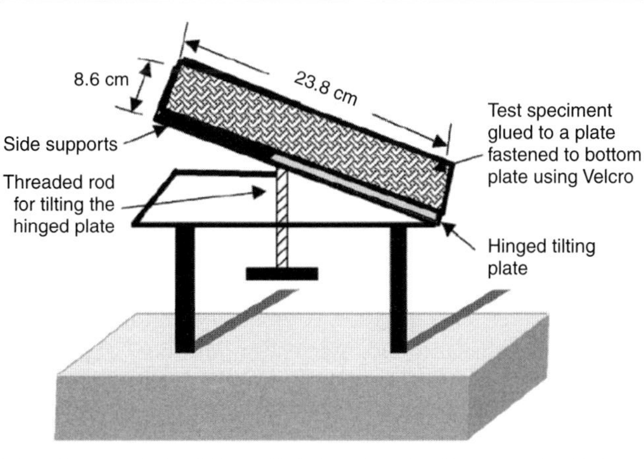

1.6.3 Hydraulic properties

The hydraulic testing of geosynthetics is completely based on new and original concepts, methods, devices, interpretation and databases, unlike physical and mechanical testing, discussed in previous sections of this chapter. The reason behind this is that the traditional textile tests rarely have hydraulic applications. *Porosity*, *permittivity* and *transmissivity* are the most important hydraulic properties of geosynthetics, especially of geotextiles, geonets and some drainage composites, which are explained below.

Table 1.5 Comparison of methods for determining the pore size distribution of geotextiles

Test method	Test mechanism	Test material	Sample size: cm^2	Time for one test
Dry sieving	Sieving dry	Glass beads fraction	434	2 hours
Hydrodynamic sieving	Alternating water flow	Glass beads mixture	257	24 hours
Wet sieving	Sieving wet	Glass beads mixture	434	2 hours
Bubble point	Comparison of air flow, dry versus saturated	Porewick	22.9	20 minutes
Mercury intrusion	Intrusion of a liquid in a pore	Mercury	1.77	35 minutes
Image analysis	Direct measurement of pore spaces in a cross-section of the geotextile	None	1.5	2–3 days

Geosynthetic porosity is related to the ability to allow liquid to flow through it, and is defined as the ratio of the void volume to the total volume. It may be indirectly calculated for geotextiles using the relationship (Koerner, 1990)

$$\eta = 1 - \frac{m}{\rho t} \tag{1.3}$$

where η is the porosity, m is the mass per unit area, ρ is the overall geotextile density and t is the thickness of the geotextile.

The *per cent open area* (POA) of a geosynthetic is the ratio of the area of its openings to its total area, and is expressed as a percentage. The pores in a given geosynthetic, especially in a geotextile, are not of one size but are of a range of sizes. The pore size distribution can be represented in much the same way as the particle size distribution for a soil. Various methods are available for evaluating the pore size distribution of geotextiles. Bhatia *et al.* (1994) made a comparison of six methods, as presented in Table 1.5.

In the *dry sieving test method*, glass beads of a known size are sieved in dry condition through a screen made of the geotextile (Figure 1.32). Sieving is done by using beads of successively coarser size until 5% or less, by weight, pass through the geotextile. The *hydrodynamic test method* is based on hydrodynamic filtration, where a glass bead mixture is sieved with a basket with a geotextile bottom by alternating the water flow that occurs as a result of the immersion and emersion of the basket several times in water. In the *wet sieving method*, a glass bead mixture is sieved through a screen made of a geotextile while a continuous water spray is applied. The *bubble point test (BPT) method* is based on

- a dry porous material will pass air through all of its pores when any amount of air pressure is applied to one side of the material
- a saturated porous material will only allow a fluid to pass when the pressure applied exceeds the capillary attraction of the fluid in the largest pore.

The development and operation of a simple and inexpensive BPT apparatus was described by Elton *et al.* (2007). This apparatus evaluates the pore sizes of geotextiles according to ASTM D6767 (ASTM, 2008a) and gives consistent results.

The *mercury intrusion method* is based on the relationship between the pressure required to force a non-wetting fluid (mercury) into the pores of a geotextile and the radius of the pores intruded. *Image analysis* is a technique used for the direct measurement of pore spaces within a cross-sectional plane of a geotextile with the help of a microscope. The pore openings, which are obtained experimentally, are dependent on the technique used for their determination. It is believed that, despite some limitations, both wet and hydrodynamic sieving methods are better techniques than dry sieving.

In the case of most geogrids, the open areas of the grids are greater than 50% of the total area. In this respect, a geogrid may be looked at as a highly permeable polymeric structure.

Figure 1.33 shows pore size distribution curves for typical woven and non-woven geotextiles. The pore size at which 95% of the pores in the geotextile are finer was originally termed the *equivalent opening size* (EOS), designated as O_{95}. If a geotextile has an O_{95} value of 300 mm, then 95% of the geotextile pores are 300 mm or smaller. In other words, 95% of particles with a diameter of 300 mm are retained on the geotextile during sieving for a constant period of time. This notation is similar to that used for soil particle size distributions where, for instance, D_{10} is the sieve size through which 10% by weight of the soil passes. The *apparent opening size* (AOS) is equivalent to the EOS, but is also quoted for other percentages retained, such as

Figure 1.32 Details of the dry sieving method
Courtesy of Terram®, all rights reserved

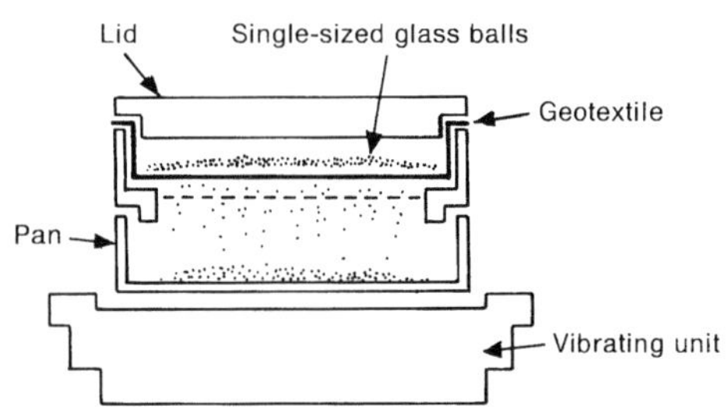

Figure 1.33 Pore size distributions of typical geotextiles
Reproduced from Ingold and Miller (1988)

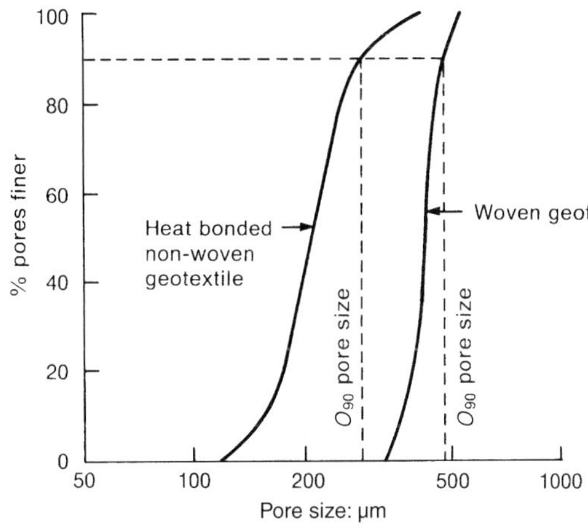

O_{50} or O_{90}. The EOS is used in many filter criteria established to prevent piping and erosion. It should be noted that the meaning of EOS and AOS values and their determination in the laboratory are still not uniform throughout the engineering profession, and, hence, filter criteria developed in different countries may not be directly comparable.

In Figure 1.33, it is noted that the pores in a woven geotextile tend to be fairly uniform in size and regularly distributed. In general, non-woven geotextiles exhibit smaller O_{90} pore sizes than woven geotextiles; however, there is a degree of overlap in the commonly employed O_{90} sizes, which vary from approximately 50 to 350 µm for non-woven geotextiles and from 150 to 600 µm for woven geotextiles (Ingold and Miller, 1988). For filtration application, a geotextile with a high POA should be selected, with a controlled opening size to suit the soil being filtered. Most non-woven geotextiles, and some woven geotextiles, will suit this application.

The permeability of a geosynthetic to water flow may be expressed by *Darcy's coefficient*, by *permittivity* (as defined below) or by a *volume flow rate*. The advantage in expressing geosynthetic permeability in terms of Darcy's coefficient is that it is easy to relate geosynthetic permeability directly with soil permeability. A major disadvantage is that Darcy's law assumes laminar flow, whereas geosynthetics, especially geotextiles, are often characterised as exhibiting semi-turbulent or turbulent flows. The simplest method of describing the permeability characteristics of geosynthetics is in terms of the volume flow rate at a specific constant water head (generally 10 cm) (Figure 1.34). The advantage of this method is that it is the simplest test to carry out, it does not rely on Darcy's law for its authenticity, and it can easily be used to compare different geosynthetics used for drainage and filtration applications.

The measurement of in-plane water permeability is important if the geosynthetic, such as a geotextile or a fin drain, is being used to drain water within itself; that is, its water-transporting capability is of prime importance. The

Figure 1.34 Details of the apparatus for measuring the permeability of a geosynthetic for water flow normal to its plane
Courtesy of Terram®, all rights reserved

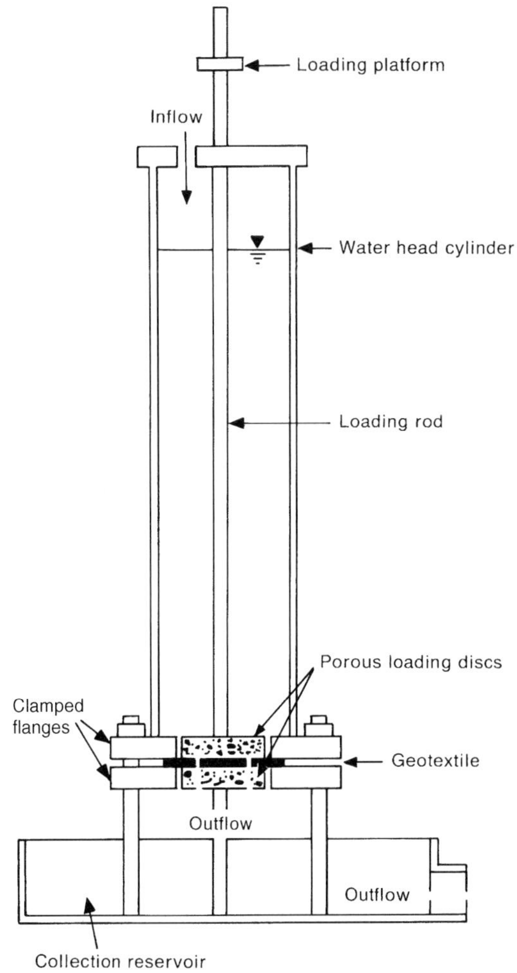

in-plane water permeability is normally described in terms of *transmissivity*, as defined below. The types of test used to measure the in-plane drainage characteristics of geosynthetics are essentially the same as those used to measure the water permeability normal to the plane of the geosynthetic (Figure 1.34), except that the hydraulic gradient is applied along the length of the geosynthetic (Figure 1.35) rather than across the thickness of the geosynthetic.

The *permittivity* of a geosynthetic (generally geotextile) is simply the coefficient of the permeability for water flow normal to its plane (Figure 1.36(a)) divided by its thickness. This property is the preferred measure of water flow capacity across the geosynthetic plane, and quite useful in filter applications. Darcy's law, in terms of permittivity, can be expressed as

$$Q_n = k_n \frac{\Delta h}{\Delta x}(LB) = \psi \Delta h A_n \tag{1.4}$$

where Q_n is the cross-plane volumetric rate of flow (m^3/s) (i.e. the volumetric rate of flow for flow across the plane of the geosynthetic), k_n is the coefficient of cross-plane permeability

Figure 1.35 Details of the apparatus for measuring the permeability of a geosynthetic for water flow within its plane
Courtesy of Terram®, all rights reserved

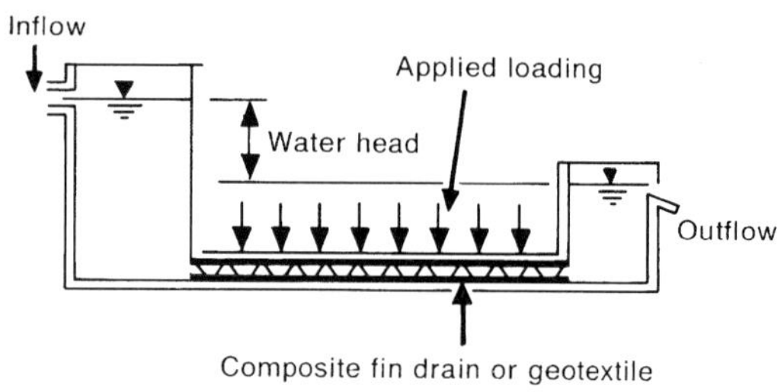

(m/s), Δh is the head causing flow (m), Δx is the thickness of the strip of geosynthetic measured along the flow direction under a specified normal stress (m), L is the length of the strip of geosynthetic (m), B is the width of the strip of geosynthetic (m), $\psi = k_n/\Delta x$, which is the permittivity of the geosynthetic (s^{-1}), and $A_n = LB$, which is the area of cross-section of geosynthetic for cross-plane flow (m^2). Permittivity may thus be defined as the volumetric rate of flow of water per unit cross-sectional area, per unit head, under laminar flow conditions in a direction normal to the plane of the geosynthetic.

The *transmissivity* of a geosynthetic (thick non-woven geotextile, geonet or geocomposite) is simply the product of the permeability for in-plane water flow (Figure 1.36(b)) and its thickness. This property is the preferred measure of the in-plane water flow capacity of a geosynthetic, and is widely used in drainage applications. Darcy's law in terms of transmissivity can be expressed as

$$Q_p = k_p \frac{\Delta h}{L} A_p = k_p \frac{\Delta h}{L} (B\,\Delta x) = \theta\, iB \qquad (1.5)$$

Figure 1.36 The flow of water through a geotextile strip: (a) normal flow and (b) in-plane flow

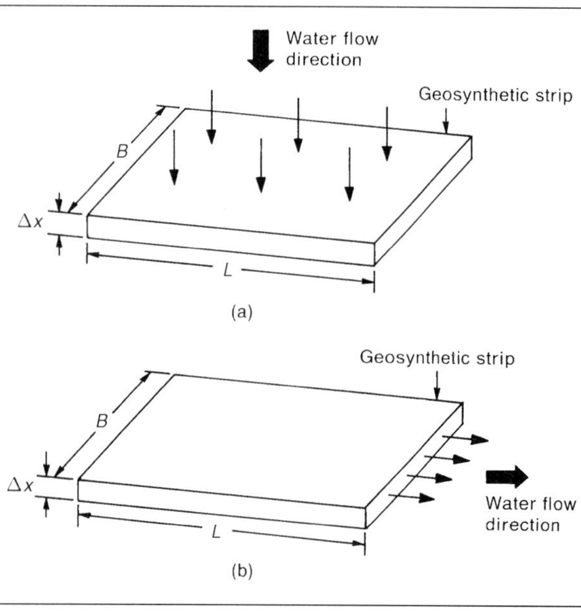

where Q_p is the in-plane volumetric rate of flow (m^3/s) (i.e. the volumetric rate of flow for flow within the plane of the geosynthetic), k_p is the coefficient of in-plane permeability, $\theta = k_p\,\Delta x$, which is the transmissivity of the geosynthetic (m^2/s), $i = \Delta h/L$, which is the hydraulic gradient, and $A_p = B\,\Delta x$, which is the area of the cross-section of the geosynthetic for in-plane flow (m^2). The transmissivity may thus be defined as the volumetric rate of flow of water per unit width of the geosynthetic, per unit hydraulic gradient, under laminar flow conditions within the plane of the geosynthetic. To exhibit a large transmissivity, a geotextile must be thick and/or have a large permeability in its plane.

Equations 1.4 and 1.5 indicate that once the permittivity (ψ) and the transmissivity (θ) are successfully determined, the flow rates Q_n and Q_p do not depend on the thickness of the strip of geosynthetic, Δx, which is highly dependent on the applied pressures, and, therefore, it is difficult to measure the thickness accurately.

Typical values of the permeability are 10^{-5} to 1 m/s for geotextiles, and 10^{-13} m/s or less for geomembranes. The permeability of geotextiles is of the same order of magnitude as the permeability of highly permeable soils, such as sands and gravels. Woven geotextiles and thermally bonded non-woven geotextiles have almost no transmissivity, and cannot be used as drains. The permeability of geomembranes is much smaller than the permeability of clay, which is the least permeable soil. Needle-punched geotextiles have permeability values of the order of 10^{-4}–10^{-3} m/s, and geonets have permeability values of the order of 10^{-2}–10^{-1} m/s. A maximum saturated hydraulic conductivity ranging from 5×10^{-11} to 1×10^{-12} m/s is typical of geosynthetic clay liners over the range of confining pressures typically encountered in practice.

It should be noted that Darcy's law is valid only for laminar flow. This means that permeability, permittivity and transmissivity are constants; that is, independent of the gradient only if the water flow is not turbulent. These properties are governed by several other factors, such as the fibre type, size and orientation, the porosity or void ratio, the confining pressure, repeated loading, contamination and ageing. When dry, some fabrics exhibit resistance to wetting. In such cases, the initial permeability is low but it rises until the fabric reaches saturation. Permeability may also be reduced through air bubbles trapped in the geosynthetic. This is the reason why testing standards usually require careful saturation of geosynthetic specimens before they are subjected to water flow. In addition, permeability measurements will be more consistent with the use of de-aired water rather than tap water (Hausmann, 1990). Woven geotextiles are much less affected by the stress level, but their permeability is dramatically controlled by the structure of the fabric. The common, and generally less expensive, tape-on-tape fabrics have a low open area ratio, and, in consequence, exhibit water permeabilities typically in the range 10–30 l/s/m^2 for a 10 cm head. In contrast, the woven monofilament-on-monofilament geotextiles have much larger open area

Figure 1.37 The influence of compressive stress on: (a) the thickness, (b) the permeability, (c) the permittivity and (d) the transmissivity of a needle-punched non-woven geotextile
Reproduced from Giroud (1980)

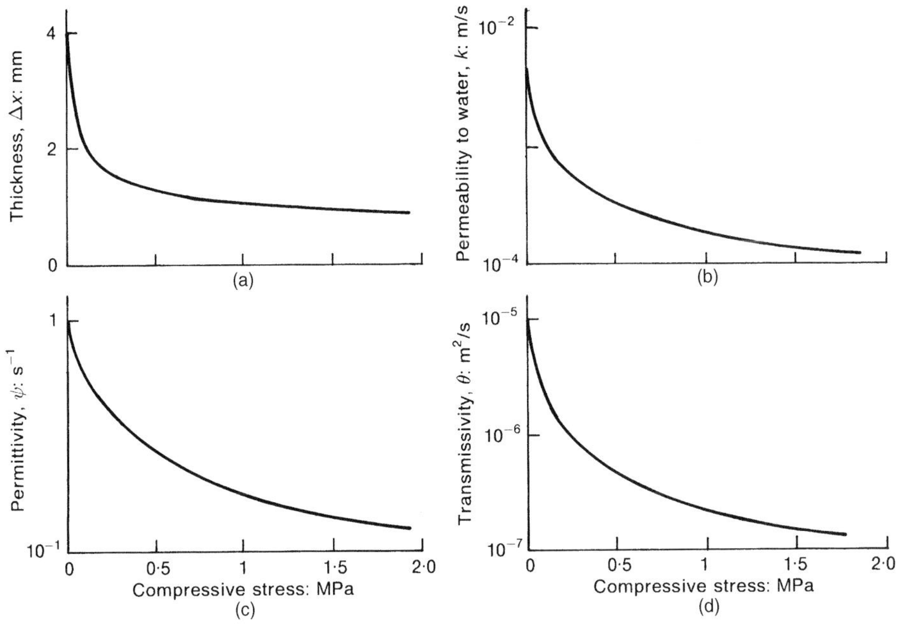

ratios, giving water permeabilities in the range 100–1000 l/s/m^2 for a 10 cm head (Ingold and Miller, 1988). Tests performed at the University of Grenoble (France) have shown that the thickness and the permeability of needle-punched non-woven geotextiles are significantly affected by the confining pressure, as shown in Figure 1.37 (Giroud, 1980). In this figure, the values of k_n and k_p were close for the considered geotextile; therefore, only an average value, k, is presented. It may be noted that the flow in the plane of the geotextile is more affected by the confining pressure than a normal flow.

During the early development of drainage applications of geosynthetics, the researchers attempted to assess the drainage performance of geotextiles under the influence of various parameters that may affect their drainage performance (Masounave *et al.*, 1980; Ling *et al.*, 1993). In most early research works, the seating time for the normal compressive stress was not given much priority, although standards such as ASTM D4716-08 (ASTM, 2008b) on the test method for transmissivity suggests that the geotextile specimen should be seated under the minimum normal compressive stress for a minimum period of 15 minutes. Shukla and Sharma (2006) made an attempt to study the effect of the seating time on the in-plane permeability (transmissivity) and the cross-plane permeability (permittivity) of three different geotextiles using constant-head permeability test methods. The results obtained from the experimental study show that the transmissivity of geotextiles decreases as the seating time increases. Permittivity tests showed a similar trend as the variation of transmissivity. It was observed that, on application of a certain normal compressive stress on geotextile specimens, the permittivity and transmissivity of geotextiles do not reduce to a final value immediately, but take about 25–35 minutes to attain a stabilised value under the influence of that load. It was also observed that a decrease in the transmissivity and permittivity was higher for the geotextile having larger opening size (O_{90}), for the same load increment. The study recommends that the observations of the permittivity and the transmissivity of non-woven geotextiles in laboratory tests should be recorded after at least 30 minutes following a particular normal stress applied to the geotextile specimen.

When a geotextile is placed adjacent to a base soil (the soil to be filtered), a discontinuity arises between the original soil structure and the structure of the geotextile. This discontinuity allows some soil particles to migrate through the geotextile under the influence of seepage flows. This condition is shown in an idealised manner in Figure 1.38(a). For a geotextile to act as a filter, it is essential that a condition of equilibrium is established at the soil–geotextile interface as soon as possible after installation, to prevent soil particles from being piped indefinitely through the geotextile; if this were to happen, the drain would eventually become blocked. The time taken to establish equilibrium conditions with geotextile filters varies, but it is normally between one and four months. The structure, or stratification, of the soil immediately adjacent to the geotextile at the onset of equilibrium conditions dictates the filtering efficiency of the system. The stratification is dependent on the type of soil being filtered, the size and frequency of the pores of geotextile, and the magnitude of the seepage forces present. Figure 1.38(b) shows typical stratification occurring with three different soil types – single-sized soil, well-graded soil and cohesive soil. When the soil is well-graded, considerable

Figure 1.38 (a) Idealised soil–geotextile interface conditions immediately following geotextile installation. (b) Idealised interface conditions at equilibrium between three different soil types and a geotextile filter: (i) single-sized soil and a geotextile filter, (ii) well-graded soil and geotextile filter and (iii) cohesive soil and a geotextile filter
Courtesy of Terram®, all rights reserved

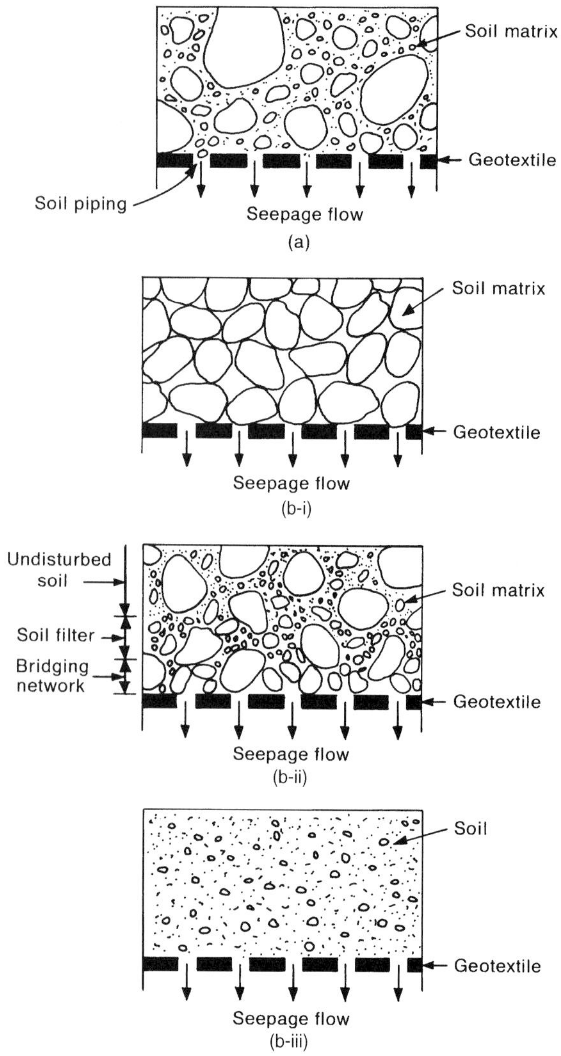

rearrangement of the soil takes place. At equilibrium, three zones may be identified: the undisturbed soil; a 'soil filter' layer, which consists of progressively smaller particles as the distance from the geotextile increases; and a bridging layer, which is a porous, open structure. Once the stratification process is complete, it is actually the soil filter layer that actively filters the soil. If the geotextile is chosen correctly, it is possible for the soil filter layer to be more permeable than the undisturbed soil. The function of the geotextile is to ensure that the soil remains in an undisturbed state without any soil piping (Figure 1.39).

To achieve satisfactory filter performance by geosynthetics, especially geotextiles, the following functions are required during the design life of the application under consideration.

Figure 1.39 Overall requirements for optimal filter performance
Reproduced from Lawson (1986), Industrial Fabrics Association International

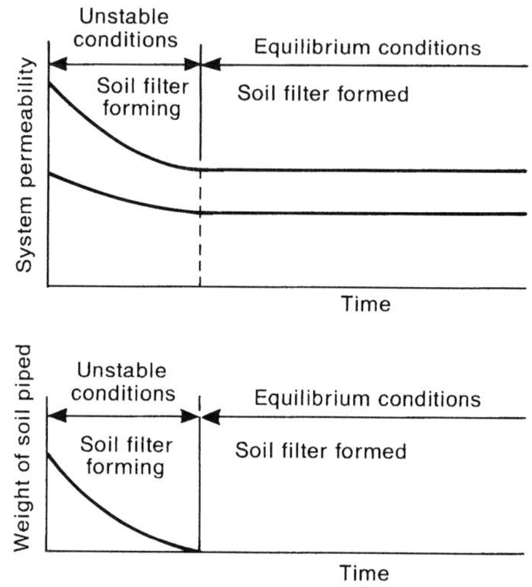

- Maintain adequate permeability to allow flow of water from the soil layer without significant flow impedance so as not to build up excess hydrostatic porewater pressure behind the geosynthetic (*permeability criterion*).
- Prevent significant wash out of soil particles (i.e. soil piping) (*retention* or *soil-tightness criterion*).
- Avoid accumulation of soil particles within the geosynthetic structure, called fabric clogging, resulting in complete shut-off of the water flow (*clogging criterion*).

It may be noted that the permeability criterion places a lower limit on the pore size of a geotextile, whereas the retention criterion places an upper limit on the pore size of a geotextile. These two criteria are, to some extent, contradictory, because the permeability of a geosynthetic filter increases with its increasing pore size. However, in the majority of cases, it is possible to find a filter that meets both the permeability criterion and the retention criterion. Several different geosynthetic filter criteria have been developed (Giroud, 1982; Hoare, 1982; Lawson, 1982, 1986; Wang, 1994), largely based on the conventional granular filter criteria that were first formulated by Terzaghi and Peck (1948). All of these criteria use soil permeability and compare it with the geotextile permeability for establishing the permeability criterion, whereas they compare soil particle size distribution with the geotextile pore size distribution for establishing the retention criterion. All these criteria are applicable for specific filter applications.

It is a general misconception that the pore sizes of a filter should be smaller than the smallest particle size of the soil to be protected, because this would result in the use of quasi-impermeable filters (which, of course, would not meet the permeability criterion). In some cases, the filter openings can be larger than the largest soil particles, and the filter will still retain the soil (Giroud, 1994). In fact, soil retention does not require that the migration of all soil particles be prevented. Soil retention simply requires that the soil behind the filter remains stable; in other words, some small particles may migrate into and/or through the filter, provided that this migration does not affect the soil structure; that is, it does not cause any movement of the soil mass. At the same time, the filter and the drainage medium located downstream of the filter should be such that they can accommodate the migrating particles without clogging.

The criteria for a geotextile filter, as commonly used, are given in the following forms:

$$k_n \geq \lambda k_s \quad \text{(permeability criterion)} \tag{1.6}$$

$$O_i \leq \beta D_j \quad \text{(retention criterion)} \tag{1.7}$$

where k_n is the coefficient of the cross-plane permeability of the geotextile, k_s is the coefficient of the permeability of the protected soil, λ is the constant of permeability varying over a wide range, say 0.1–100, β is the dimensionless parameter of soil conservation, varying over a certain range, and i and j are integers.

The permeability criterion $k_n \geq k_s$ has long been advocated by many researchers on the assumption that the geotextile needs to be no more permeable than the protected soil. Carroll (1983) and Christopher and Holtz (1985) recommend the criterion $k_n \geq 10k_s$ for critical soil and hydraulic conditions in which clogging has been shown to cause, roughly, an order of magnitude decrease in the geotextile permeability. The criterion $k_n \geq 0.1k_s$ was proposed by Giroud (1982) on the premise that a geotextile with only 10% of the permeability of the soil would still have a much greater flow capacity than the soil because the length of the flow path is directly related to the flow rate through a porous media.

All the existing retention criteria for geotextile filters are functions of various opening sizes of the geotextile, such as O_{95}, O_{90}, O_{50} and O_{15}, and the diameter of soil particles, such as D_{90}, D_{85}, D_{50} and D_{15}, depending mostly on the uniformity coefficient of the soil, C_U (D_{60}/D_{10}). Most retention criteria are given in the form of the O_i/D_j ratio not exceeding a certain value or range. Typical ranges in the variation of O_{95}/D_{50}, O_{95}/D_{85} and O_{90}/D_{90} are, respectively, 1–6, 1–3 and 1–2.

The filter criteria, for a particular filter application in the field, should be developed on the basis of data obtained from detailed soil–geotextile performance investigations. However, in the absence of such data, the following criteria can be considered for soils with predominantly granular fractions.

Permeability criterion:

$$O_{90} \geq D_{15} \tag{1.8a}$$

$$O_{90} \geq 0.05\,\text{mm} \tag{1.8b}$$

Retention criterion:

$$O_{90} \leq D_{85} \tag{1.9}$$

In the absence of suitable data, for soils with significant cohesive fractions, it is suggested that the following criteria can be used.

Permeability criterion:

$$O_{90} \geq 0.05\,\text{mm} \tag{1.10a}$$

$$\text{minimum volume water permeability} = 30\,\text{l/m}^2\text{/s}-10\,\text{cm head} \tag{1.10b}$$

Retention criterion:

$$O_{90} \leq 0.12\,\text{mm} \tag{1.11}$$

The clogging criterion is described in the following section.

Geosynthetics such as geotextile tubes (Figure 1.40) provide containment while still allowing water to pass through them. They are excellent design strategies for both shoreline protection and dewatering of fine materials. A difficulty with regard to designing geotextile tubes is the matching of the appropriate fabric with the site-specific infilled material. This task is generally carried out at the project site by conducting the hanging bag test (HBT). This test is basically a performance test used to determine the flow rate of water passing through the geofabric, as well as any sediment passing through over a specific time period. Figure 1.41 shows an HBT apparatus presented by Koerner and Koerner (2006). The test bag is fabricated by sewing a fabric along its sides and bottom to form an enclosure that will support and contain a measured amount of saturated infill material. The bag opening is fitted with evenly spaced metal grommets around its upper perimeter. It is hung from the inside of a wooden frame. A pan is placed under the container to collect sediment that passes through the bag. Ten five-gallon

Figure 1.40 A geotextile tube
Reproduced from Koerner and Koerner (2006), with permission from Elsevier

Figure 1.41 Hanging bag test apparatus
Reproduced from Koerner and Koerner (2006), with permission from Elsevier

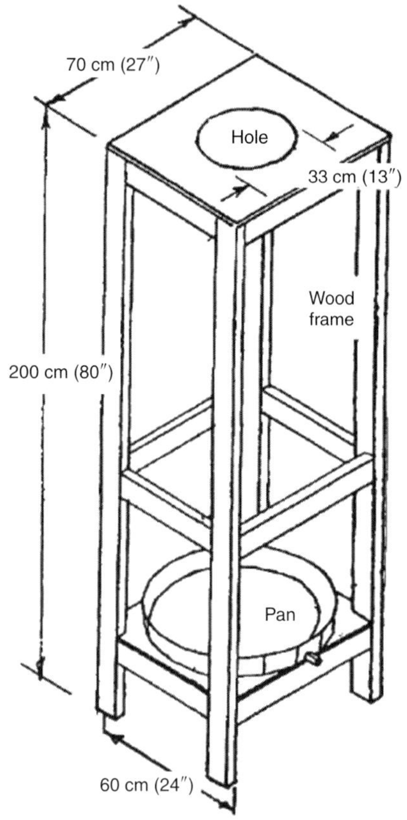

Figure 1.42 Results of creep tests on various yarns of different polymers. PA, polyamid, PE, polyethylene; PET, polyester; PP, polypropylene: (a) creep at 20% load and (b) creep at 60% load
Reproduced from den Hoedt (1986), with permission from Elsevier

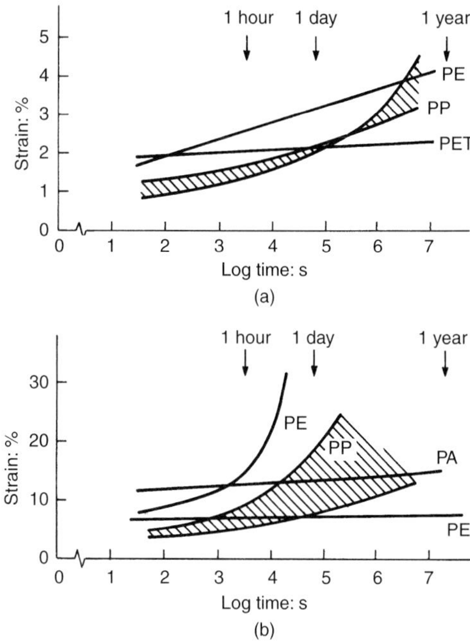

buckets, a stop watch, a balance and a tape measure are needed for the test. Moisture contents of the soil retained within the bag and the particle size of the soil passing through the bag are determined in the laboratory. Koerner and Koerner (2006) looked at four different geotextiles and how they interacted with three different infill materials. Their findings showed that the flow rates were directly proportional to the permeability of the soil contained within the tube. In this regard, the HBT results adequately predicted the geotube field performance, which was not dependent on any conventional parameters, including the fabric AOS values. In this regard, the HBT results were superior to the fabric properties in assessing the performance of the various fabrics to site-specific infilled materials.

1.6.4 Endurance and degradation properties

The endurance and degradation properties (creep behaviour, abrasion resistance, long-term flow capability, durability – construction survivability and longevity, etc.) of geosynthetics are related to their behaviour during service conditions, including time.

Creep is the continued strain or elongation of a geosynthetic when subjected to a sustained load. It is an important factor in the design and performance of geosynthetic-reinforced structures having longer design life, say for about 50 years. Depending on the type of polymer and ambient temperature, creep may be significant at stress levels as low as 20% of the ultimate strength. Figure 1.42 compares the strain versus time behaviour of various yarns of different polymers. As shown, both the total strain and the rate of strain differ markedly. The understanding of geosynthetic creep helps the design engineer in the selection of allowable loads for use in designs. Two approaches to evaluate the allowable loads are the following.

- Allowable load based on limiting creep strains – this requires the analysis of creep strains versus time plots for various stress levels. Details of this procedure were described by Jewell (1986) and Bonaparte and Berg (1987).
- Allowable loads using factors of safety – it is necessary to reduce the geosynthetic strength by a factor of safety corresponding to the specific polymer type to obtain the allowable load. Values for the factor of safety are given in Table 1.6. Although not as technically accurate as the previous method, this approach is sometimes the only one available to designers.

Since polymers are viscoelastic materials, the strain rate and the temperature are important parameters while testing geosynthetics (Andrawes *et al.*, 1986). When a low strain is applied in a wide-width tensile test, the geosynthetic sample takes longer to come to failure, and, therefore, the creep strain is greater. High rates of strain (which can be as much as 100% per minute) tend to produce lower failure strains and sometimes yield higher strengths than do low rates of strain. The creep rate of geosynthetics depends on temperature. Higher creep rates are associated with higher

Table 1.6 Factors of safety
Reproduced from den Hoedt (1986), with permission from Elsevier

Polymer	Factor of safety
Polypropylene	4.0
Polyester	2.0
Polyamide (nylon)	2.5
Polyethylene	4.0

temperatures, resulting in larger strains to rupture geosynthetics. The rate of creep is also related to the level of the load to which the polymer is subjected (Greenwood and Myles, 1986; Mikki *et al.*, 1990). Chang *et al.* (1996) reported that under the same confining pressure on geotextiles, the amount of creep increases as the creep load rises; and where the creep load is the same, the increases in the confining pressure decrease the amount of creep, which may even be reduced to nil. The creep is minimised by a pre-stretching operation of the Tensar process.

It has been found that the tensile and creep properties of some non-woven geotextiles can be improved by confinement in soil (McGown *et al.*, 1982). This has a greater effect on the tensile properties of mechanically bonded geotextiles and those of heat-bonded geotextiles. Creep reduction under confinement may be due to a significant increase in frictional resistance between the soil and the geosynthetic. However, there is little effect from confining pressure on the performance of woven geotextiles.

In some applications, increases in soil strength result in the reduction of geosynthetic stresses with time. An example of this type is the foundation support for a permanent embankment over soft deposits (Figure 1.43(a)). When a basal geosynthetic is used beneath an embankment, the strain becomes fairly constant once most of the settlement has taken place. In such a situation, there may be a loss of tensile stress with time experienced by the geosynthetic. This phenomenon is called *stress relaxation*, and is closely related to creep. Fortunately, during this period, the underlying soil is consolidating and increasing in strength. The subsoil is therefore able to offer greater resistance to failure as time passes. The factor of safety should not be compromised unless the geosynthetic sheds load faster than the gain in soil strength.

In a permanent geosynthetic-reinforced retaining wall or steep embankment, the geosynthetic load remains constant throughout the life of the structure. In this case, the creep strain may be very high and, therefore, the factor of safety should not be compromised (Figure 1.43(b)).

Dynamic creep or the repeated loading behaviour of geosynthetics is of paramount importance in a number of applications, including reinforced paved and unpaved roads, railway tracks, and reinforced retaining structures and slopes under large repetitive live loads, such as traffic and wave action. The behaviour of geogrids under repeated loading is generally different from those of geotextiles (Figure 1.44). Zanzinger *et al.* (2010) have presented the

Figure 1.43 Examples of: (a) time-dependent and (b) time-independent reinforcement applications
Reproduced from Paulson (1987), with permission from Elsevier

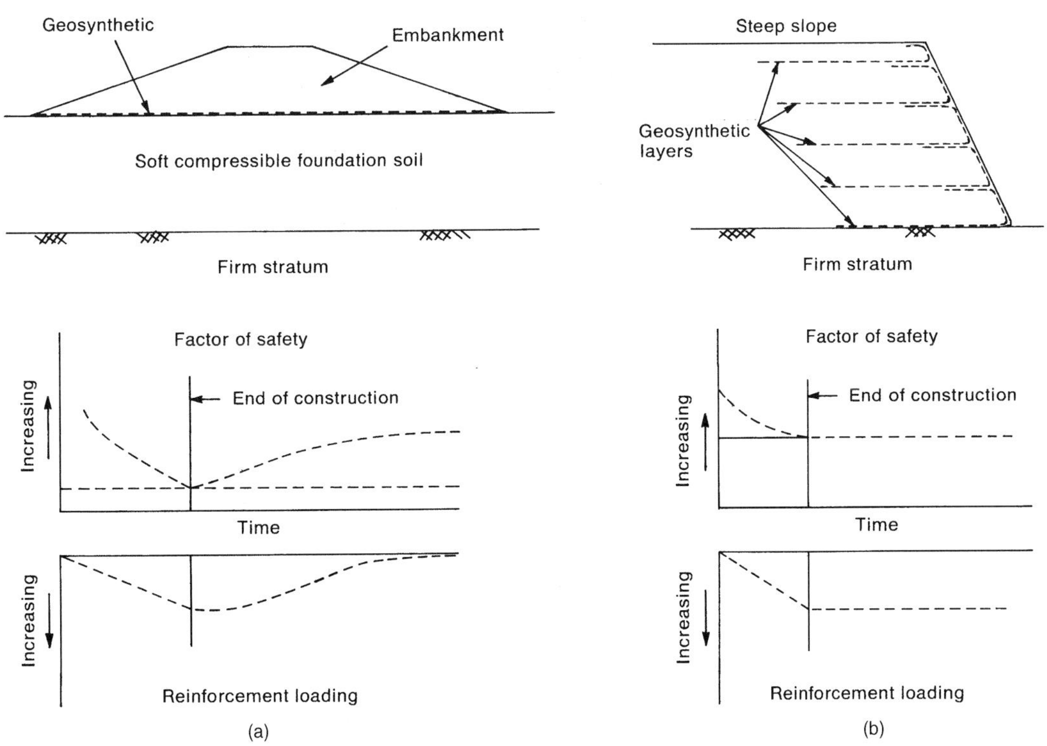

Figure 1.44 Total strain versus log *N* plots for: (a) a woven geotextile and (b) a geogrid
Reproduced from Kabir and Ahmed (1994), Industrial Fabrics Association International

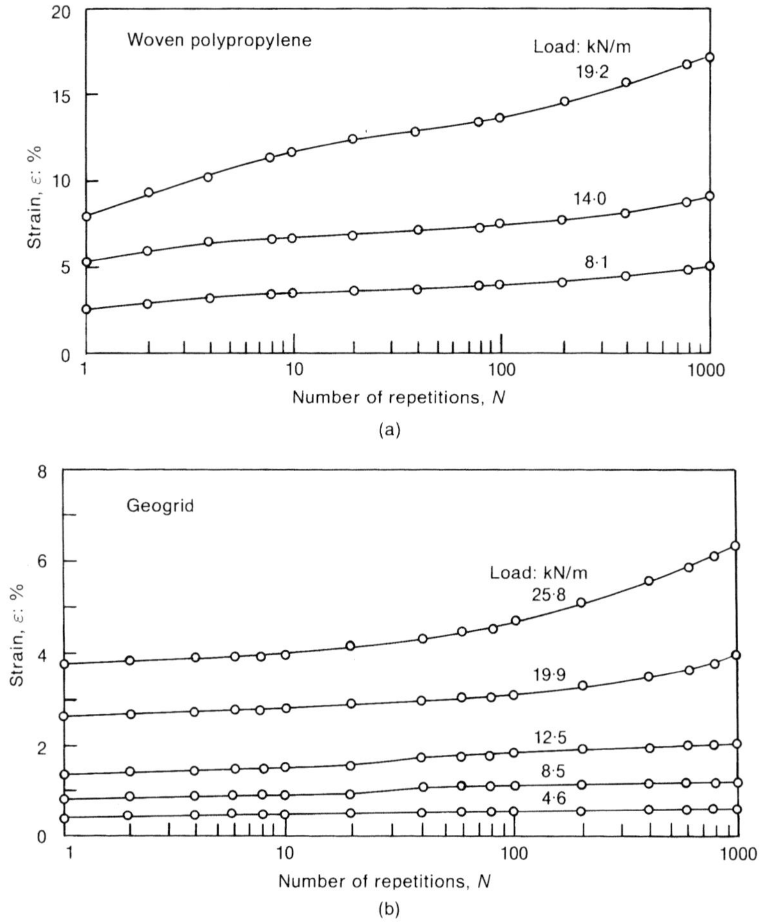

behaviour of the polyester geogrid under cyclic loading in detail. Due care must be paid to such applications, which are described in other chapters of this book.

In the recent past, constitutive models have been developed to describe direction- and time-dependent, non-linear, inelastic stress–strain behaviour. More details on this type of model can be found in the work of Perkins (2000).

Abrasion of a geosynthetic is defined as the wearing away of any part of it by rubbing against any surface. The abrasion damage occurs during both the installation and service phases of geosynthetic applications such as the use of geosynthetics for coast and river bank protection. The abrasion damage is reported as the percentage weight loss or strength/elongation retained under specified tests (sliding block/sandpaper tests) in particular conditions. In testing the abrasion resistance of geosynthetics, it is important to simulate the actual type of abrasion that a geosynthetic would meet in the field. Van Dine *et al.* (1982), Gray (1982) and Huang (2008) suggested the test procedures for the evaluation of resistance to abrasion caused by different processes such as wear and impact.

The long-term flow capability of geosynthetics (generally geotextiles) with respect to the hydraulic load from the upstream soil is of significant practical interest. The compatibility between the pore size openings of a geotextile and retained soil particles in filtration and/or drainage applications can be assessed by the *gradient ratio test*. This test is basically used to evaluate the clogging resistance of geotextiles. Figure 1.45 shows the constant-head-type permeameter developed by the US Army Corps of Engineers. This permeameter allows the measurement of the head loss along a soil–geotextile system. After the test is run for some hours (or days), the piezometer readings stabilise, and the so-called *gradient ratio* (GR) is determined. This is defined as the hydraulic gradient through the lower 25 mm of the soil, plus the geotextile divided by the hydraulic gradient through the adjacent 50 mm of soil. The long-term flow rate behaviour of geotextile filters can also be assessed by the accelerated filtration test method (Mikki *et al.*, 1994). In such filtration tests, the following condition should be satisfied in order to ensure satisfactory performance in the field:

$$\text{GR} \leq 3.0 \quad \text{(clogging criterion)} \tag{1.12}$$

Figure 1.45 The gradient ratio permeameter developed by the US Army Corps of Engineers
Reproduced from Haliburton and Wood (1982), Industrial Fabrics Association International

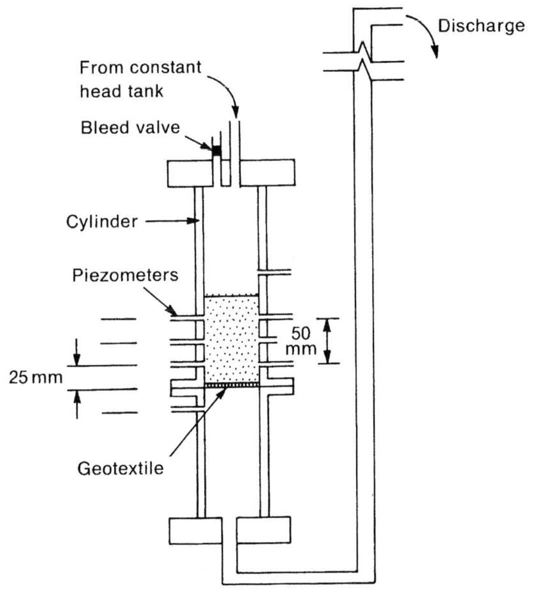

Fannin *et al.* (1994) presented a critical evaluation of the gradient ratio test by addressing issues of equipment design, specimen preparation and test procedure. They reported that the problem of biological clogging in the test may be eliminated by treatment of the water with an algicide.

Geotextiles are used as filters in several applications where the hydraulic conditions impose either a unidirectional or a reversing flow. The flow conditions exert a very strong influence on the stability of the soil adjacent to the filter. The reversing/cyclic flow may occur in pavement edge drains and marine protection works. Therefore, it becomes essential to evaluate the long-term flow capability of geotextiles under cyclic flow conditions. Hameiri and Fannin (2002) modified the gradient ratio test to enable the imposition of a dynamic cyclic flow, typical of that in marine protection works, using head control. The application of head control was proved to be simple and effective for achieving cyclic flow conditions.

It is imperative that geotextile filters should be placed in close contact with the soil, so that there will be no space between the soil and the geotextile where particles could move or accumulate to form a thin layer of soil (often called *soil cake*) at the surface of a geotextile filter. The mechanism for the formation of a soil cake is known as *blinding* of the filter, and this is far more detrimental than clogging. It has been reported by Giroud (1994) that geotextiles with a tortuous surface in contact with the soil, such as needle-punched non-woven geotextiles, do not favour the development of a continuous cake of fine soil particles, whereas geotextile filters with a smooth surface may favour the development of such a cake. Furthermore, geotextiles with a tortuous surface do not favour the mechanism of *blocking*, because they do not have individual openings.

Situations, such as those involving poorly graded soils, gap-graded soils, cohesionless sands and silts, high hydraulic gradients and very high-alkalinity ground waters, have been identified as creating severe clogging problems. Under such situations, one should avoid the use of a geotextile filter, and instead use a granular filter or open up the geotextile to the point where some soil loss will occur, if the upstream conditions permit such soil loss. Situations involving biological elements may also create clogging problems in filter applications.

The *durability* of a geosynthetic may be regarded as its ability to maintain the requisite properties against environmental or other influences over the selected design life. It is dependent, to a great extent, on the composition of the polymers from which it is made. To quantify the properties of polymers, knowledge of their structures at the chemical, molecular and super-molecular level is necessary, which was provided by Cassidy *et al.* (1992). The durability of geosynthetics has traditionally been assessed on the basis of mechanical property test results; it has not been assessed on the microstructural changes that cause the changes in the mechanical properties. From an engineering point of view, it may be expressed in terms of a loss of strength, and studied as construction survivability and longevity.

Construction survivability addresses how geosynthetics survive during installation. Geosynthetics may suffer damage during installation due to placement and compaction of the overlying fill. In some cases, the installation stresses might be more severe than the actual design stresses for which the geosynthetic is intended. Before installation of geosynthetics in specific field applications, field trials should be performed to investigate the type and quantity of the damage during installation. If field trials are not possible, laboratory simulations of installation damage should be performed (Hsieh and Wu, 2001; Hufenus *et al.*, 2005; Huang, 2006). In general, installation damage is taken into account during design by reducing the characteristic strength by a partial safety factor, varying from 1.05 to 1.70 for thin to thick types, respectively.

Longevity addresses how the geosynthetic properties change over the life of the structure.

All geosynthetics are likely to be exposed to sunlight and weather during construction on site, and must therefore have a limited resistance to weathering. In service life, most of them will be covered by soil; those that remain exposed during their entire life may need a far greater degree of resistance.

Generally, as the ambient temperature is increased, the strength, creep and durability characteristics of geosynthetics deteriorate. Geosynthetics are likely to encounter high temperatures only in paving applications, where they come into contact with hot bituminous materials. This application favours the use of polypropylene grids in preference to polyethylene grids, because of their greater temperature resistance. Polypropylene melts at 165°C and polyester at 250°C. These high-temperature extremes should be avoided.

Geosynthetics may degrade when exposed to the ultraviolet (UV) component (wavelength shorter than 400 nm) of sunlight. As geosynthetics are generally kept buried throughout their life, in most applications, UV degradation is not a major cause of concern, provided sensible placement procedures are followed. Generally, those geosynthetics that are white or grey in colour are likely to be the most vulnerable to UV degradation. Carbon black and other stabilisers are used to provide the polymer with UV protection. Polypropylene products perform worst, with the majority having a 50% strength loss in about 4–24 weeks. The study of the long-term performance of geosynthetics in sunlight can be carried out by exposing geosynthetics to natural UV radiation (the outdoor weathering test) or artificial radiation such as carbon (or xenon) arc lighting (the accelerated weathering test). The outdoor weathering test (Leflaive, 1988; Suits and Hsuan, 2003; ASTM, 2009) is more reliable but time consuming. But the accelerated weathering test (Suits and Hsuan, 2003; Li and Hsuan, 2004; Yang and Ding, 2006) has the advantage that not only can testing be accelerated by increasing the mean irradiance level and the temperature, and eliminating the cycles of night and day, winter and summer, but, equally important, that the exposure parameters can be controlled. The difficulty with the accelerated weathering test lies in interpreting the results with respect to the actual performances of geosynthetics in outdoor applications. In fact, the accelerated weathering/UV degradation test results must be analysed for practical applications, keeping in mind the geographic location, radiation angle, temperature, humidity, rainfall, wind, air pollution, etc. associated with a particular construction site.

Geosynthetics may come into contact with chemicals/leachates that are not normally part of the soil environment. Site-specific tests must be performed to ascertain the chemical degradation of geosynthetics, resulting in deterioration in their engineering behaviour. Index tests are generally used in chemical-resistance studies. Chemical-resistance testing of geosynthetics should employ worst-case scenario conditions. This is necessary to ensure that, when in actual use, geosynthetics will not be subjected to conditions worse than those experienced in the testing laboratory. Chemical resistance can be evaluated, at least initially, by comparing chemicals anticipated in the application with manufacturers' published resistivity charts. Tests on geotextiles in contact with uncured concrete indicate that polypropylene products are largely unaffected, whereas polyester products can lose about 50% of their strength in two months of prolonged exposure (Wewerka, 1982). In ocean environments, seawater immersion brings the least weathering effect on the engineering properties of geotextiles when the resulting degradation is compared with UV degradation (Hsieh *et al.*, 2006).

All polymeric materials have a tendency to gain moisture over time. This moisture causes some swelling, but probably not enough to cause measurable changes in mechanical or hydraulic properties.

All geosynthetic resins are very high in molecular weight with relatively few chain endings for the biological degradation to initiate. *Biological degradation* differs in that, so far, there has been no evidence of any such degradation in geosynthetics. Only those based on natural fibres degrade, as is the intention. Biological degradation cannot be accelerated beyond the selection of optimum soil conditions and temperature: if one tries to accelerate it further, the microorganisms will be destroyed. It can be studied by conducting soil burial tests in which geosynthetic specimens are buried in a microbially active soil. There is no need to inoculate the soil with specific bacteria or fungi: all relevant species are assumed to be already present, and those that benefit from the nutrients in the geosynthetic, if any, will multiply and accelerate the attack. The soil must be allowed to stabilise before the specimens are placed in it. A sample of untreated cotton is used to test the soil – if the tensile strength of the cotton strips is less than 25% of the original tensile strength after a seven-day exposure, the soil is regarded as biologically active. A good-quality horticultural compost should be sufficient for soil burial tests (Greenwood *et al.*, 1996).

Ozone attack, termite attack, etc. may also degrade geosynthetics. The research work needs more attention in this direction, as well as with regard to the combination of other degradation mechanisms.

Ageing and burial test procedures and results are becoming more critical as the long-term demands on geosynthetics increase. This is an area that requires much research work. The knowledge and understanding of the long-term, in-service behaviour of geosynthetics are vital to the continued growth of this industry, and to the science of geosynthetic design.

It must be noted that there can be many ways to perform a given test in the laboratory, depending on the case to be designed. The recommended method is that which best simulates the actual performance of the geosynthetic at the site. Usually, a laboratory test simulates the field situation at only one point of the geosynthetic. When the whole field situation can be simulated in a laboratory test, the test results can be applied to the field situation either directly or using a minor mathematical treatment to deal with the difference in scale between the laboratory and the field. In the later case, the test is a model test, and an analogical method of design is used. This method of design is the simplest one, but it can rarely be used, so other methods, such as the analytical method (based on mathematical theories that use the basic parameters of geosynthetics) and empirical methods (based on experience and, sometimes, systematic testing, including full-scale tests), are needed (Giroud, 1980).

There are presently a large number of geosynthetic products available commercially, each having different properties, and their inclusion in this chapter is beyond the scope of the

Table 1.7 A general range of some specific properties of commercially available geosynthetics
Based on data compiled by Lawson and Kempton (1995)

Type of geosynthetic		Tensile strength: kN/m	Extension at max. load: %	Apparent opening size: mm	Water flow rate (volume permeability): l/m²/s	Mass per unit area: g/m²
Geotextiles	Non-wovens					
	Heat bonded	3–25	20–60	0.02–0.35	10–200	60–350
	Needle punched	7–90	30–80	0.03–0.20	30–300	100–3000
	Resin bonded	5–30	25–50	0.01–0.25	20–100	130–800
	Wovens					
	Monofilament	20–80	20–35	0.07–4.0	80–2000	150–300
	Multifilament	40–1200	10–30	0.05–0.90	20–80	250–1500
	Flat tape	8–90	15–25	0.10–0.30	5–25	90–250
	Knitteds					
	Weft	2–5	300–600	0.20–2.0	60–2000	150–300
	Warp	20–800	12–30	0.40–1.5	80–300	250–1000
	Stitch bonded	30–1000	10–30	0.07–0.50	50–100	250–1000
Geogrids	Extruded	10–200	20–30	15–150	NA	200–1100
	Textile based					
	Knitted	20–400	3–20	20–50	NA	150–1300
	Woven	20–250	3–20	20–50	NA	150–1100
	Bonded cross-laid strips	30–200	3–15	50–150	NA	400–800
Geomembranes	Natural					
	Reinforced (made from bitumen and non-woven geotextile)	20–60	30–60	0	0	1000–3000
	Plastomeric (made from plastomers, such as HDPE, LDPE, PP or PVC)					
	Unreinforced	10–50	50–200	0	0	400–3500
	Reinforced	30–60	15–30	0	0	600–1200
	Elastomeric (made from elastomers, i.e. rubbers of various types)					
	Reinforced	30–60	15–30	0	0	500–1500
Geocomposites	Geosynthetic clay liners	10–20	10–30	0	0	5000–8000
	Linked structures (geostrip based)[a]	100–1500	3–15	NA	NA	400–4500

HDPE, high-density polyethylene; LDPE, low-density polyethylene; PP, polypropylene; PVC, polyvinyl chloride.
[a] Geostrips are geocomposites having a tensile strength in the range 20–200 kN and extension at maximum load in the range 3–15%. Geobars are geocomposites having a tensile strength in the range 20–1000 kN if reinforced internally, and in the range 20–300 kN if reinforced externally, and extension at maximum load in the range 3–15% for both cases.

book. Users can obtain specific values for various properties of the geosynthetic products from the manufacturers or suppliers. Some representative properties of commercially available geosynthetics are listed in Table 1.7.

If the test method for determining the geosynthetic properties is not completely field-simulated, the test values must be adjusted. For example, the laboratory-generated tensile strength is usually an ultimate value, which must be reduced before being used in design. This can be carried out using the following equation (Koerner, 1990):

$$T_{\text{allow}} = T_{\text{ult}}\left(\frac{1}{\text{FS}_{\text{ID}}\,\text{FS}_{\text{CR}}\,\text{FS}_{\text{CD}}\,\text{FS}_{\text{BD}}}\right) \tag{1.13}$$

where T_{allow} is the allowable tensile strength to be used in Equation 1.1 for final design purposes, T_{ult} is the ultimate tensile strength from the test, FS_{ID} is the factor of safety for installation damage (1.1–3.0 for geotextiles, 1.1–1.6 for geogrids), FS_{CR} is the factor of safety for creep (1.0–4.0 for geotextiles, 1.5–3.5 for geogrids), FS_{CD} is the factor of safety for chemical degradation (1.0–2.0 for geotextiles) and FS_{BD} is the factor of safety for biological degradation (1.0–1.3).

1.7. Application areas

Geosynthetics are versatile in use, adaptable to many field situations, and can be combined with several traditional and new building materials. They are utilised in a range of

Figure 1.46 Growth of geosynthetics in North America based on quantity and sales
Reproduced from Koerner (2000), with permission from the ASCE

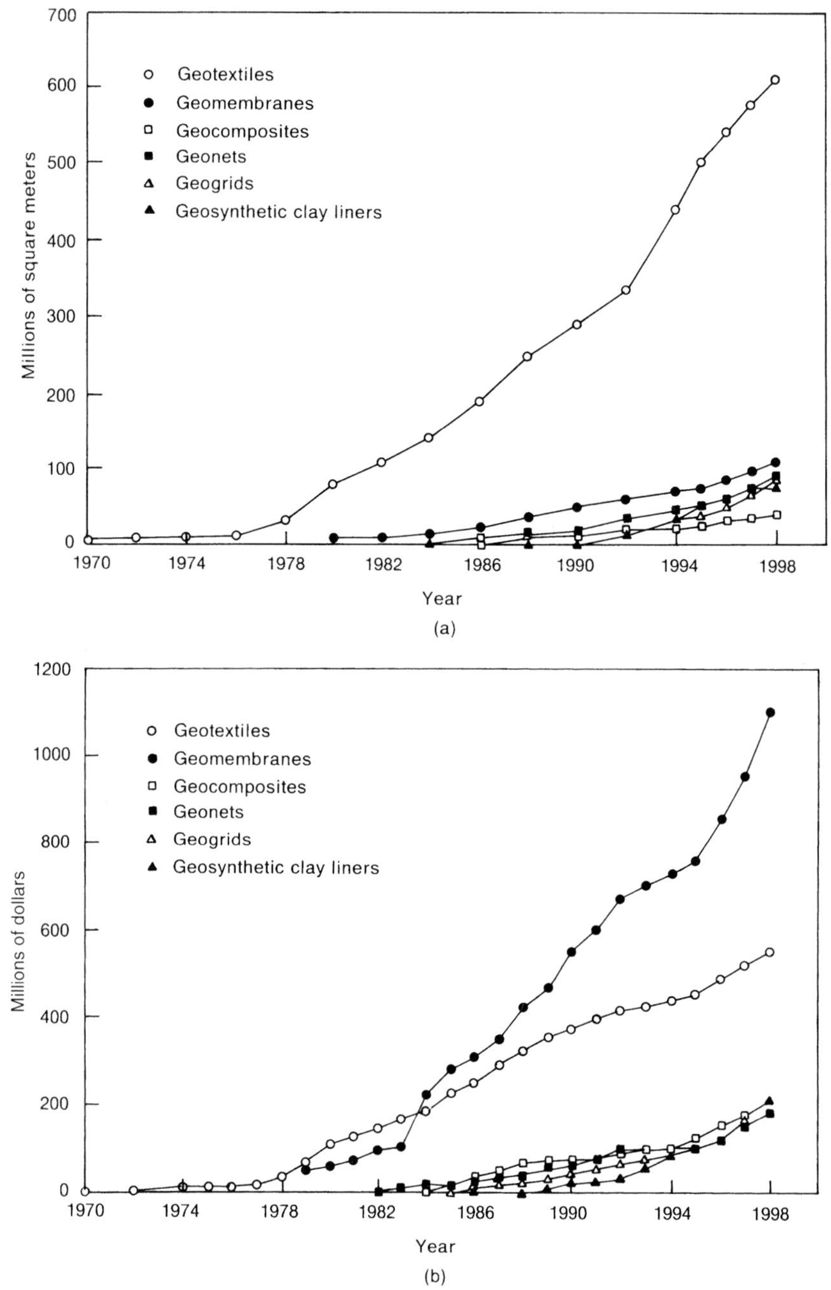

applications in many areas of civil engineering, especially geotechnical, transportation, hydraulic and environmental engineering, in which geosynthetics are widely used for achieving technical benefits and/or economic advantages. The early growth of geosynthetic applications in North America and Western Europe is shown in Figure 1.46 and Figure 1.47, respectively (Lawson and Kempton, 1995; Koerner, 2000). The growth of geosynthetic applications is still continuing at a rapid pace, which is due mainly to the following favourable basic characteristics of geosynthetics

- non-corrosiveness
- highly inert to biological and chemical degradation
- long-term durability under soil cover
- high flexibility
- minimum volume
- lightness
- robustness (geosynthetics can withstand the stresses that may be induced during installation and throughout the life of the structure)
- factory produced to have specific quality-controlled standards, and do not exhibit the inherent variability of naturally occurring materials
- ease of storing and transportation
- simplicity of installation, even by unskilled personnel
- ease in control of execution

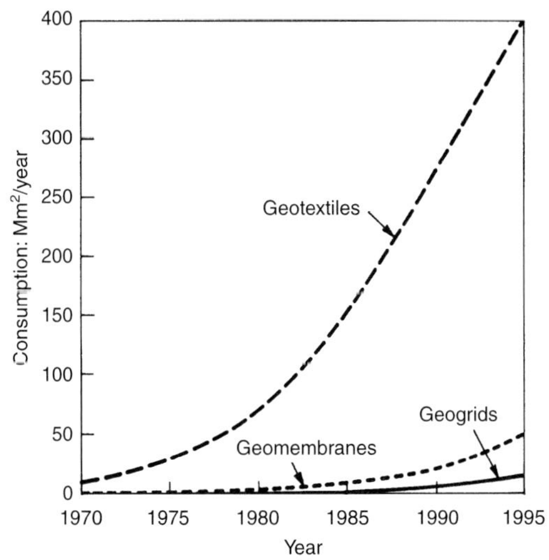

Figure 1.47 Estimated consumption of geosynthetics in Western Europe
Reproduced from Lawson and Kempton (1995); Terram®, all rights reserved

- rapid installation, even in adverse environmental conditions and thus speeding up the construction process
- useable, even with unsuitable soils
- replace soil/mineral construction materials – conserving scarce resources
- cause less wear and tear on equipment
- available in a wide range of products, in numerous configurations and weights, to perform a wide range of functions when placed in soils
- capacity to solve problems that cannot be addressed by traditional techniques
- technically effective and economical solutions (the cost should be estimated to include the initial construction cost, continuing maintenance cost, cost related to production losses, in the case of roads as a result of their closure, etc.)
- provide environment-friendly and energy-efficient solutions, thus allowing for sustainable development
- improve the performance of structures
- improve the aesthetics of structures.

The technical acceptance of geosynthetics has been achieved in a large number of application areas. Table 1.8 lists the major application areas for the geosynthetics that are described, in detail, in later chapters of this book. Most of these applications are illustrated schematically in Figure 1.48. It rarely happens, in any field application, that a geosynthetic performs only one function. However, for simplicity, the most important function is identified as per the site conditions, and the same is considered in the design.

1.8. Standards

The application of geosynthetics is a subject area where improvement in the basic concepts and the developments related to raw materials, manufacturing processes, and methods of analysis, design and construction, has been continuing for the past three to four decades. Preparation of new standards on different aspects of geosynthetics and the revision of existing standards have been continuing throughout the world, so that geosynthetics can be used on a scientific basis for economical solutions to civil engineering problems.

Table 1.9 provides a list of some of the standards on the terminology and testing of geosynthetics from the USA, UK, Switzerland, India and Australia. When dealing with the solution of a civil engineering problem using geosynthetics, the standards listed in Table 1.9, or other applicable standards, may be consulted.

1.9. Summary of main points

- Geosynthetics are almost exclusively polymeric, and their major types are geotextiles, geogrids, geonets, geomembranes, geofoams and geocomposites.
- In civil engineering applications, geosynthetics perform their functions primarily as separation, reinforcement, filtration, drainage and fluid barrier. They also show several other performance characteristics, which are basically dependent on these major functions.
- Polyester (PET), polypropylene (PP), polyethylene (PE), polyamide (PA) and polystyrene (PS) are the major polymers used as raw materials to manufacture most geosynthetics. Polyester is notable for its high strength and high resistance to creep.
- Geosynthetics are characterised by their physical, mechanical, hydraulic, endurance and degradation properties. Realistic values for these properties are estimated by performance tests, which simulate field conditions closely.
- Most design specifications use a variety of short-term tests to predict long-term performance. Therefore, it is important to assess the long-term performance of geosynthetics by exhuming and testing installed geosynthetics after substantial field exposure.
- The laboratory-measured allowable property values must be adjusted for design, to reflect the anticipated *in situ* behaviour, such as installation damage, long-term creep, ultraviolet degradation, chemical degradation and environmental degradation.
- Many manufacturers and marketing groups designate geosynthetics with trade names, and they all have different properties that are not included in this chapter. However, users of geosynthetics can obtain a wealth of information directly from the manufacturers.
- Geosynthetics have numerous applications in geotechnical, transportation, hydraulic and environmental engineering projects. They are used mainly to achieve technical and/or economic benefits, and can also provide environment-friendly and energy-efficient solutions to construction problems in order to provide sustainable development.
- Many countries have developed test standards and codes of practice, keeping local conditions and needs in view.

Table 1.8 Major application areas for geosynthetics

Sl no.	Application areas	Main purpose of using geosynthetics	Major functions	Major geosynthetic products	Most important properties	Special consideration
1	Retaining walls and steep-sided embankments	Reinforce and protect backfill/soil	Reinforcement	Geotextiles Geogrids	Strength Soil–geosynthetic friction	Creep
2	Embankments on soft ground	Improve stability, provide drainage	Reinforcement Separation Drainage	Geotextiles Geogrids Geocells Geocomposites Geofoams	Strength Soil–geosynthetic friction Pore size Permeability	Creep/stress relaxation
3	Shallow foundations	Increase load-bearing capacity, reduce settlement	Reinforcement Separation	Geotextiles Geogrids Geocells Geocomposites	Strength Soil–geosynthetic friction Pore size	Elongation
4	Unpaved roads	Increase bearing capacity, reduce degree of rutting	Reinforcement Separation	Geotextiles Geogrids Geocomposites	Strength Soil–geosynthetic friction Pore size	Repeated loading Elongation
5	Paved roads	Inhibit crack propagation, improve cyclic fatigue behaviour	Separation Drainage	Geotextiles Geogrids Geocomposites	Pore size Permeability Abrasion resistance	Repeated loading Elongation
6	Railway tracks	Prevent ballast contamination, distribute load on subgrade	Separation Filtration Drainage	Geotextiles Geogrids Geocomposites	Pore size Permeability	Repeated loading Elongation Resistance to impact and wear abrasions
7	Slopes	Protect soil slope against erosion reinforce soil, provide drainage	Filtration Drainage Reinforcement	Geotextiles Geogrids Geocomposites	Pore size Permeability Strength Soil–geosynthetic friction Abrasion resistance	Rapid changes in water level Clogging Construction stresses
8	Landfills	Extract leachate out of waste and retain the same	Fluid barrier Drainage Filtration Reinforcement	Geomembranes Geotextiles Geogrids Geocomposites	Pore size Permeability Strength Abrasion resistance	Leachate characteristics Construction stresses Elongation
9	Dams	Reduce seepage through the dam embankment, prevent internal erosion/piping, provide drainage, protect slope against erosion	Fluid barrier Filtration Drainage	Geomembranes Geotextiles Geonets Geocomposites Geogrids	Pore size Permeability Abrasion resistance	Clogging Construction stresses

10	Containment ponds, reservoirs and canals	Reduce seepage of water/liquid into the ground	Fluid barrier	Geomembranes Geocomposites	Permeability Abrasion resistance	Construction stresses
11	Filters	Prevents migration of soil particles without impeding the water flow	Filtration	Geotextiles	Permittivity	Clogging
12	Pipeline and drainage facilities	Protect the drainage medium, provide drainage	Drainage Filtration	Geonets Geotextiles Geocomposites	Pore size Transmissivity Permittivity	Clogging
13	Tunnels and underground structures	Prevent seepage, provide drainage of seepage water	Fluid barrier Protection Drainage	Geomembranes Geotextiles Geocomposites	Permeability Transmissivity Permittivity	Construction stresses Clogging

Figure 1.48 Typical geosynthetic applications: (a) retaining wall, (b) steep-sided embankment, (c) embankment on soft foundation soil, (d) bridge pier, (e) unpaved road, (f) paved road, (g) railway track, (h) slope (erosion control), (i) slope (stabilisation), (j) landfill, (k) dam, (l) liquid reservoir, (m) water channel and (n) trench drain

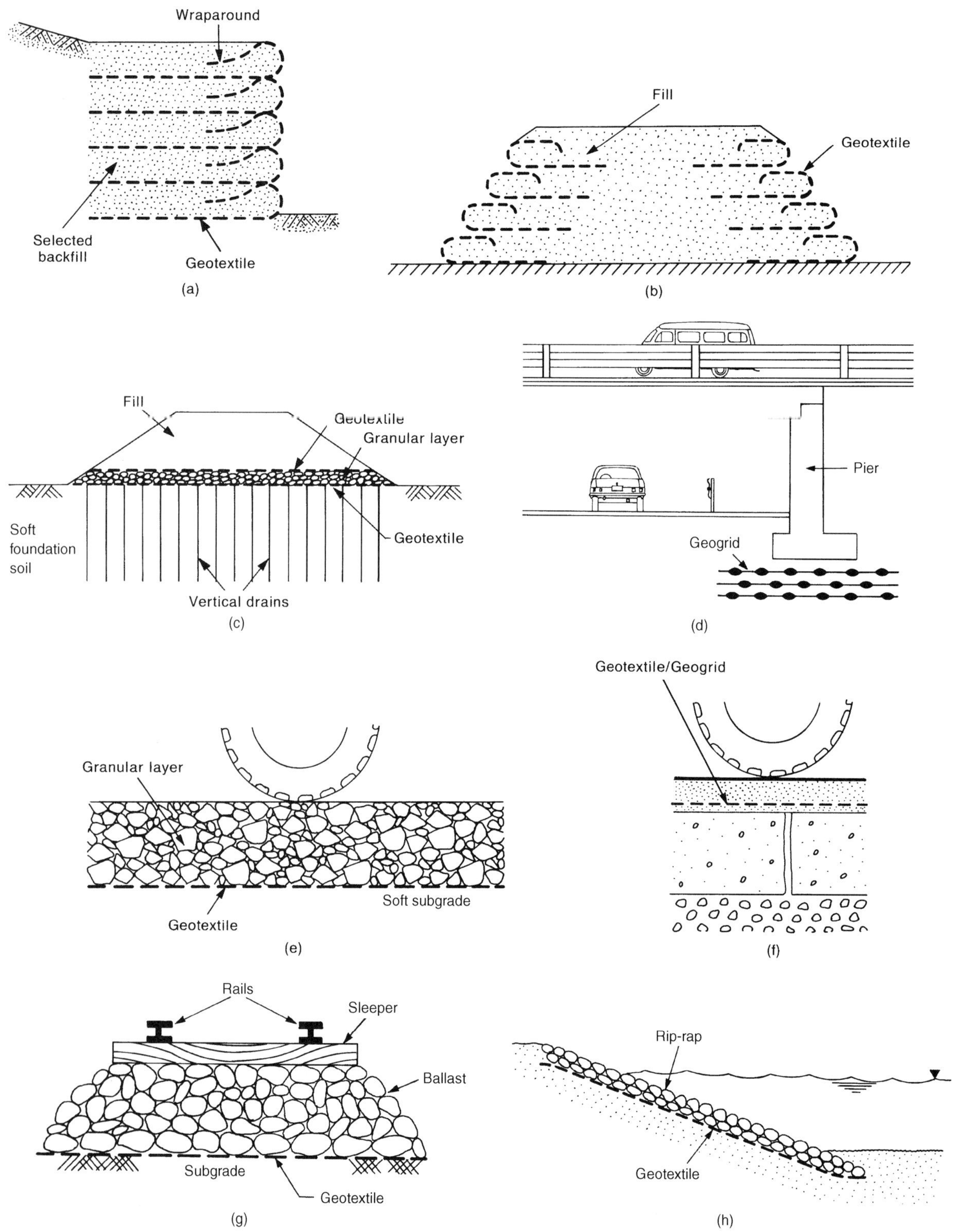

Figure 1.48 Continued

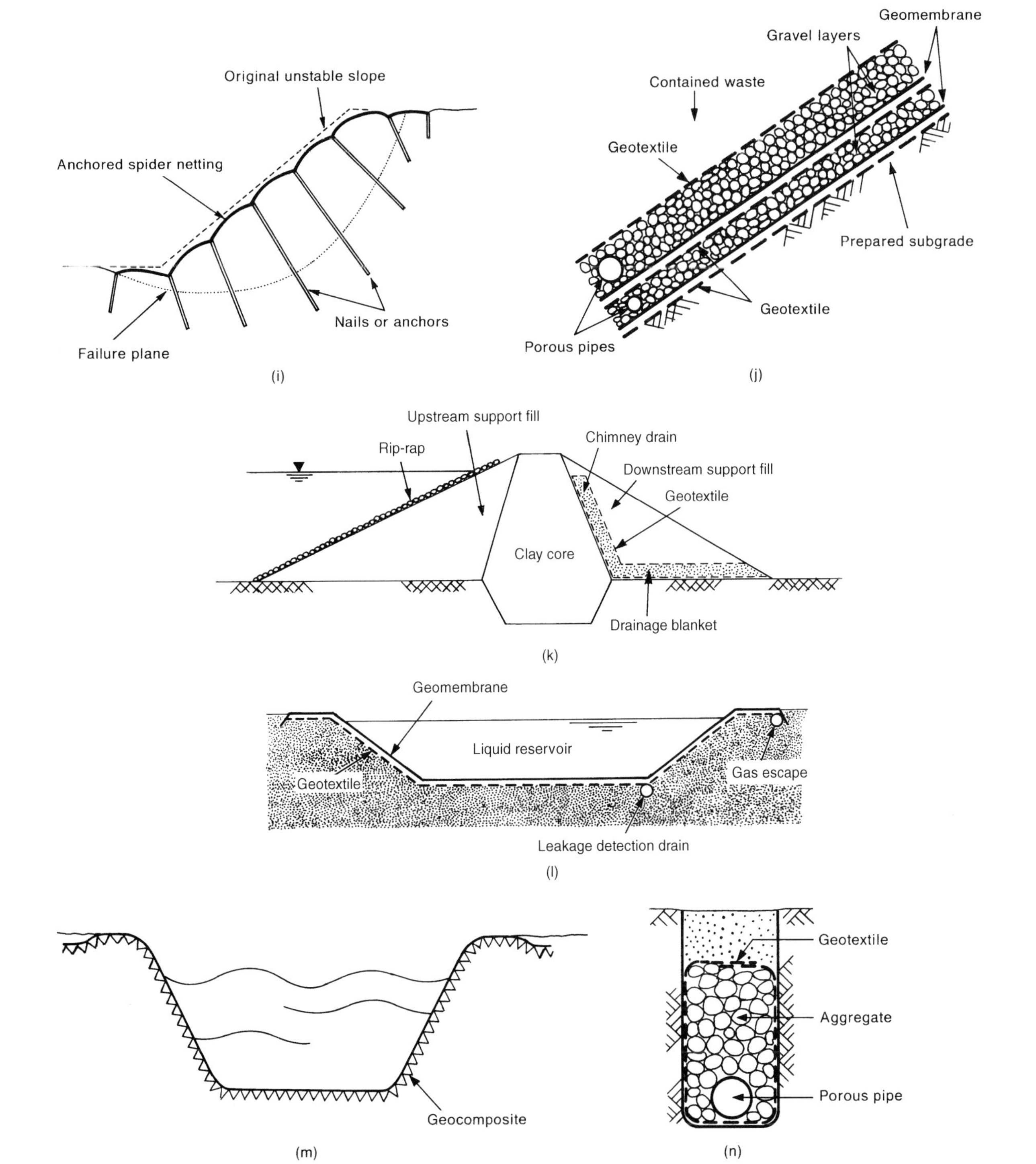

- The basic information provided in this chapter will help in understanding the application-oriented concepts of geosynthetics that are discussed in detail in later chapters of this book.

1.10. Some common questions and their answers

Questions

1. Are geosynthetics manufactured from polymeric materials?
2. Do geosynthetics have a corrosion problem?
3. What is the lifespan of geosynthetics when used in geotechnical structures?
4. Which polymer is generally used to manufacture geosynthetics for reinforcement applications?
5. What are the ideal and practical ways of assessing the service life of geosynthetics in field applications?
6. Why are geofoams lightweight materials?
7. What is the primary parameter for the selection of a geotextile for filtration?

Table 1.9 Some standards on the terminology and testing of geosynthetics (in 2011) published by the American Society of Testing Materials (ASTM), the British Standards (BS) Institution, the International Organization for Standardization (ISO), the Bureau of Indian Standards (IS) and Standards Australia (AS)

Topics	Designations of relevant standards			
	ASTM	BS/ISO	IS	AS
Standard terminology for geosynthetics	D4439-04	BS EN ISO 10318: 2005	13321-1: 1992	3704: 2005
Sampling and preparation of test specimens	D4354-99	BS EN ISO 9862: 2005	14706: 1999	3706.1: 2003
Determination of mass per unit area	D5261-10 D5993-99 (2009)	EN ISO 9864: 2005 BS EN ISO 14196: 2003	14716: 1999	
Determination of thickness	D5199-01 (2006) D6525-00 (2006)	BS EN ISO 9863-1: 2005	13162-3: 1992	
Determination of compression behaviour	D6364-06 D6244-06	BS EN ISO 25619-2: 2008		
Determination of tensile properties – wide-width strip tensile test	D4595-09 D4885-01 (2006)	ISO 10319: 2008	13162-5: 1992 13325: 1992	3706.2: 2000
Determination of joint/seam strength	D4884-09	BS EN ISO 10321: 2008	15060: 2001	3706.6: 2000
Determination of trapezoid tearing strength	D4533-04 (2009) D5884-04a (2010)		14293: 1995	3706.3: 2000
Determination of puncture resistance – CBR test	D6241-04 (2009)	BS EN ISO 12236: 2006		3706.4: 2001
Determination of puncture resistance – falling cone method		BS EN ISO 13433: 2006	13162-4: 1992	3706.5: 2000
Determination of interface friction – direct shear method	D5321-08	BS EN 13738: 2004	13326-1: 1992	
Determination of apparent opening size	D4751-04	BS EN ISO 12956: 2010	14294: 1995	3706.7: 2003
Determination of water permeability – permittivity without load	D4491-99a (2009)	ISO 11058: 2010	14324: 1995	3706.9: 2001
Determination of permittivity under load	D5493-06			
Determination of the (in-plane) flow rate per unit width and hydraulic transmissivity	D4716-08	BS EN ISO 12958: 2010		3706.10: 2001
Determination of hydraulic conductivity ratio (HCR) of soil/geotextile systems	D5567-94 (2006)			
Determination of durability/resistance to weathering	D5819-05	BS EN 12224: 2000 BS EN 12226: 2000		
Determination of the creep properties	D5262-07	EN ISO 25619-1: 2008 BS EN ISO 13431: 1999	14739: 1999	
Determination of abrasion resistance (sand paper/sliding block method)	D4886-10	BS EN ISO 13427: 1998	14714: 1999	
Evaluation of soil-geotextile system clogging potential by the gradient ratio	D5101-01 (2006)			
Evaluation of biological clogging of geotextile or soil/geotextile filters	D1987-07			
Determination of resistance to the exposure of UV light and water (xenon-arc type apparatus)	D4355-07		13162-2: 1991	3706.11: 2004
Evaluation of the chemical resistance	D6389-99 (2005) D6213-97 (2009) D6388-99 (2005) D5747-08			3706.12: 2001
Determination of microbiological resistance – soil burial test		BS EN 12225: 2000		3706.1: 2003

8. Can the gradient ratio test be used to evaluate the long-term performance of a geotextile filter in marine protection works?
9. What is the effect of temperature variation on geosynthetics?
10. Can a geosynthetic have a unit weight less than the unit weight of water?

Answers

1. Most geosynthetics are manufactured almost exclusively using polymeric materials, but geocomposites such as geosynthetic clay liners and some erosion control products are manufactured using natural materials (clays, geojutes, coirs, etc.) in combination with polymeric materials.

2. No.

3. On the basis of accelerated performance tests in the laboratory and experience gained during the past four decades, it is expected that geosynthetics can perform their functions under soil covers for about 120 years.

4. Polyester (PET).

5. The service life of geosynthetics is ideally assessed by examining actual field samples over the service life. However, this is not feasible because it would take too long to obtain results under field conditions. Therefore, the service life is generally assessed using laboratory-accelerated ageing/wreathing tests.

6. A large amount of air is encapsulated within the cellular structure of geofoams, which represents about 95% of the total volume of the material.

7. The pore size of the geotextile.

8. No, the modified form of the gradient ratio test, called the cyclic gradient ratio test, should be used.

9. When geosynthetics are subjected to high temperatures, they lose mass. What remains above 500°C is probably carbon black and ash.

10. Yes. Examples are geofoams, and geosynthetics manufactured from polypropylene (PP) and polyethylene (PE).

REFERENCES

Agerschou HA (1961) Synthetic material filters in coastal protection. *Journal of Waterways and Harbours Division*, ASCE **87**: 111–124.

Andrawes KZ, McGown A and Murray RT (1986) The load–strain–time–temperature behaviour of geotextiles and geogrids. *Proceedings of the 3rd International Conference on Geotextiles*, Vienna, pp. 707–712.

Andrejack TL and Wartman J (2010) Development and interpretation of a multi-axial tension test for geotextiles. *Geotextiles and Geomembranes* **28(6)**: 559–569.

ASTM (American Society of Testing and Materials) (2008a) ASTM D6767-08. Standard test method for pore size characteristics of geotextiles by capillary flow test. ASTM, West Conshohocken, PA.

ASTM (2008b) ASTM D4716-08. Standard test method for determining the (in-plane) flow rate per unit width and hydraulic transmissivity of a geosynthetic using a constant head. ASTM, West Conshohoken, PA.

ASTM (2009) ASTM D5970-09. Standard test method for deterioration of geotextiles from outdoor exposure. ASTM, West Conshohoken, PA.

ASTM (2010) ASTM D5617-04(2010). Standard test method for multi-axial tension test for geosynthetics. ASTM, West Conshohoken, PA.

Barratt RJ (1966) Use of plastic filters in coastal structures. *Proceedings of the 10th International Conference on Coastal Engineering*, Tokyo, pp. 1048–1067.

Barroso M, Pierson P and Lopes MG (2006) A non-destructive method for testing non-flexible dual geomembranes seams using gas permeation. *Geosynthetics International* **13(1)**: 15–22.

Bhatia SK, Smith JL and Christopher BR (1994) Interrelationship between pore openings of geotextiles and methods of evaluation. *Proceedings of the 5th International Conference on Geotextiles, Geomembranes and Related Products*, Singapore, pp. 705–710.

Bonaparte R and Berg R (1987) Long term allowable tension for geosynthetic reinforement. *Proceedings of the Geosynthetics '87*, New Orleans, LA.

Carroll Jr RG (1983) Geotextile filter criteria. *Transportation Research Record* **916**: 46–53.

Carroll Jr RG (1988) Specifying geogrids. *Geotechnical Fabrics Report*. IFAI, St Paul, MN.

Cassidy PE, Mores M, Kerwick DJ, Koeck DJ, Verschoor KL and White DF (1992) Chemical resistance of geosynthetic materials. *Geotextiles and Geomembranes* **11**: 61–98.

Chang DTT, Chen CA and Fu YC (1996) The creep behaviour of geotextiles under confined and unconfined conditions. *Proceedings of the International Symposium on Earth Reinforcement*, Fukuoka, pp. 19–24.

Christopher BR and Holtz RD (1985) *Geotextile Engineering Manual*. US Federal Highway Administration, Washington, DC. Report No. FHWA-TS-86/203.

Colucci P and Lavagnolo MC (1995) Three years' field experience in electrical control of synthetic landfill liners. *Proceedings of 5th International Landfill Symposium*, Cagliari, pp. 437–452.

den Hoedt G (1986) Creep and relaxation of geotextiles fabrics. *Geotextiles and Geomembranes* **4(2)**: 83–92.

El-Fermaoui A and Nowatzki E (1982) Effect of confining pressure on performance of geotextiles in soils. *Proceedings of the 2nd International Conference on Geotextiles*, Las Vegas, NV, pp. 799–804.

Elton DJ, Hayes DW and Adanur S (2007) Bubblepoint testing of geotextiles: apparatus and operation. *Geotechnical Testing Journal* **30(1)**: 1–8.

Fannin RJ, Vaid YP and Shi Y (1994) A crtical evaluation of the gradient ratio test. *Geotechnical Testing Journal* **17(1)**: 35–42.

Fourie AB and Jones CJFP (2010) Improved estimates of power consumption during dewatering of mine tailings using electrokinetic geosynthetics (EKGs). *Geotextiles and Geomembranes* **28(2)**: 181–190.

Fukuoka M (1990) Earth reinforcement for foundations – east and west. *Geotextiles and Geomembranes* **9**: 3–9.
Giroud JP (1980) Introduction to geotextiles and their applications. *Proceedings of the 1st Canadian Symposium on Geotextiles*, Calgary, pp. 3–31.
Giroud JP (1982) Filter criteria for geotextiles. *Proceedings of the 2nd International Conference on Geotextiles*, Las Vegas, NV, pp. 103–108.
Giroud JP (1994) Quantification of geosynthetic behaviour. *Proceedings of the 5th International Conference on Geotextiles, Geomembranes and Related Products*, Singapore, pp. 1249–1273.
Giroud JP (2004) Poisson's ratio of unreinforced geomembranes and nonwoven geotextiles subjected to large strains. *Geotextiles and Geomembranes* **22(4)**: 297–305.
Giroud JP and Carroll RG (1983) Geotextile products. *Geotechnical Fabrics Report*. IFAI, St Paul, MN, pp. 12–15.
Giroud JP and Frobel RK (1983) Geomembrane products. *Geotechnical Fabrics Report*, IFAI, St Paul, MN, pp. 38–42.
Giroud JP, Gourc JP, Bally P and Delmas P (1977) Behaviour of a non-woven geotextile in an earth dam. *Proceedings of the International Conference on the Use of Fabrics in Geotechnics*, Paris, pp. 213–218. (In French.)
Gray CG (1982) Abrasion resistance of geotextiles fabrics. *Proceedings of the 2nd International Conference on Geotextiles*, Las Vegas, NV, pp. 817–821.
Gray DH and Maher MH (1989) Admixture stabilization of sands with discrete, randomly distributed fibers. *Proceedings of the XIIth International Conference on Soil Mechanics and Foundation Engineering*, Rio de Janeiro, pp. 1363–1366.
Greenwood JH and Myles B (1986) Creep and stress relaxation of geotextiles. *Proceedings of the 3rd International Conference on Geotextiles*, Vienna, pp. 821–826.
Greenwood JH, Trubiroha P, Schroder HF, Frank P and Hufenus R (1996) Durability standards for geosynthetics: the tests for weathering and biological resistance. *Proceedings of the 1st European Geosynthetics Conference. – Eurogeo 1*, Maastricht, pp. 637–641.
Haliburton TA and Wood PD (1982) Evaluation of the US Army Corps of Engineers' gradient ratio test for geotextile performance. *Proceedings of the 2nd International Conference on Geotextiles*, Las Vegas, NV, pp. 97–101.
Hameiri A and Fannin RJ (2002) A cyclic gradient ratio test device. *Geotechnical Testing Journal* **25(3)**: 1–11.
Hamir RB, Jones CJFP and Clarke BG (2001) Electrically conductive geosynthetics for consolidation and reinforced soil. *Geotextiles and Geomembranes* **19(8)**: 455–482.
Hausmann MR (1990) *Engineering Principles of Ground Modification*. McGraw-Hill, Singapore.
Healy KA and Long RP (1971) Prefabricated subsurface drains. *Highway Research Record* **360**: 57–64.
Hoare DJ (1982) Synthetic fabrics as soil filters. *Journal of Geotechnical Engineering Division*, ASCE **108**(GT10): 1230–1245.
Holtz RD (1975) Recent developments in reinforced earth. *Proceedings of the 7th Scandinavian Geotechnical Meeting*, Polyteknish Forlag, pp. 281–291.
Holtz RD, Christopher BR and Berg RR (1997) *Geosynthetic Engineering*. BiTech, Richmond.
Hsieh CW and Wu JH (2001) Installation survivability of flexible geogrids in various pavement subgrade materials. *Transportation Research Record* **1772**: 190–196.
Hsieh C, Wang JB and Chiu YF (2006) Weathering properties of geotextiles in ocean environments. *Geosynthetics International* **13(5)**: 210–217.
Huang CC (2006) Laboratory simulation of installation damage of a geogrid. *Geosynthetics International* **13(3)**: 120–132.
Huang CC (2008) Comparative study on performance and index tests simulating abrasion damage of a woven geotextile. *Geosynthetics International* **15(5)**: 358–368.
Hufenus R, Ruegger R, Flum D and Sterba IJ (2005) Strength reduction factors due to installation damage of reinforcing geosynthetics. *Geotextiles and Geomembranes* **23(5)**: 401–424.
Ingold TS and Miller KS (1988) *Geotextiles Handbook*. Thomas Telford, London.
ISO (International Organisation for Standardization) (2005) ISO 12957-2-2005. Geosynthetics – Determination of friction characteristics – Part 2: Inclined plane test. ISO.
Jewell RA (1986) Material properties for design of geotextile reinforced slopes. *Geotextiles and Geomembranes*, **2(2)**: 83–109.
John NWM (1987) *Geotextiles*. Blackie, London.
Kabir MH and Ahmed K (1994) Dynamic creep behaviour of geosynthetics. *Proceedings of the 5th International Conference on Geotextiles, Geomembranes and Related Products*, Singapore, pp. 1139–1144.
Kaswell ER (1963) *Handbook of Industrial Textiles*. Industrial Fabrics Division, West Point Pepperell, New York.
Koerner GR and Koerner RM (2006) Geotextile tube assessment using a hanging bag test. *Geotextiles and Geomembranes* **24(2)**: 129–137.
Koerner RM (1990) *Designing with Geosynthetics*, 2nd edn. Prentice Hall, Upper Saddle River, NJ.
Koerner RM (2000) Emerging and future developments of selected geosynthetic applications. *Journal of Geotechnical and Geoenvironmental Engineering*, ASCE **126(4)**: 291–306.
Koerner RM (2005) *Designing with Geosynthetics*, 5th edn. Prentice Hall, Upper Saddle River, NJ.
Koerner RM and Welsh JP (1980) *Construction and Geotechnical Engineering Using Synthetic Fabrics*. Wiley, New York.
Kutay ME, Guler M and Aydilek AH (2006) Analysis of factors affecting strain distribution in geosynthetics. *Journal of Geotechnical and Geoenvironmental Engineering* **132(1)**: 1–11.
Lawson CR (1982) Filter criteria for geotextiles: relevance and use. *Journal of Geotechnical Engineering Division*, ASCE **108(GT10)**: 1300–1317.
Lawson CR (1986) Geotextile filter criteria for tropical residual soils. *Proceedings of the 3rd International Conference on Geotextiles*, Vienna, pp. 557–562.

Lawson CR and Kempton GT (1995) *Geosynthetics and their Use in Reinforced Soil*. Terram, Maldon.

Leflaive E (1988) Durability of geotextiles: the French experience. *Geotextiles and Geomembranes* **12(1–2)**: 23–31.

Li MJ and Hsuan YG (2004) Temperature and pressure effects on the degradation of polypropylene tape yarns–depletion of antioxidants. *Geotextiles and Geomembranes* **22(6)**: 511–530.

Ling HI, Tatsuoka F, Wu JTH and Nishimura J (1993) Hydraulic conductivity of geotextiles under typical operational conditions. *Geotextiles and Geomembranes* **12(6)**: 509–542.

Lovisa J, Shukla SK and Sivakugan N (2010) Shear strength of randomly distributed moist fiber-reinforced sand. *Geosynthetics International* **17(2)**: 100–106.

Masounave J, Denis R and Rollin AL (1980) Prediction of hydraulic properties of synthetic nonwoven fabrics used in geotechnical work. *Canadian Geotechnical Journal* **17**: 517–525.

McGown A, Andrawes KZ and Kabir MH (1982) Load–extension testing of geotextiles confined in soil. *Proceedings of the 2nd International Conference on Geotextiles*, Las Vegas, NV, pp. 793–796.

Mercer FB (1982) Retaining fill in a geotechnical structure. British Patent 2,078,833A.

Mikki H, Hayashi Y and Sato M (1994) Accelerated filtration test of geotextile filters. *Proceedings of the 5th International Conference on Geotextiles, Geomembranes and Related Products*, Singapore, pp. 643–646.

Mikki H, Hayashi Y, Yamada K, Takasago and Shido H (1990) Plane strain tensile strength and creep of spunbonded nonwovens. *Proceedings of the 4th International Conference on Geotextiles, Geomembranes and Related Products*, The Hague, pp. 667–672.

Murray RT, McGown A, Andrawes KZ and Swan D (1986) Testing joints in geotextiles and geogrids. *Proceedings of the 3rd International Conference on Geotextiles*, Vienna, pp. 731–736.

Myles B and Carswell IG (1986) Tensile testing of geotextiles. *Proceedings of the 3rd International Conference on Geotextiles*, Vienna, pp. 713–718.

Narejo DB (2003) A simple tilt table device to measure index friction angle geosynthetics. *Geotextiles and Geomembranes* **21**: 49–57.

Nosko V and Touze-Foltz N (2000) Geomembrane liner failure: modelling of its influence on containment transfer. *Proceedings of EuroGeo 2*, Bologna, pp. 557–560.

Palmeira EM, Lima Jr NR and Mello LGR (2002) Interaction between soils and geosynthetic layers in large-scale ramp tests. *Geosynthetics International* **9(2)**: 149–187.

Paulson JN (1987) Geosynthetic material and physical properties relevant to soil reinforcement applications. *Geotextiles and Geomembranes* **6**: 211–223.

Perkins SW (2000) Constitutive modeling of geosynthetics. *Geotextiles and Geomembranes* **18**: 273–292.

Pitanga HN, Gourc JP and Vilar OM (2009) Interface shear strength of geosynthetics: evaluation and analysis of inclined plane tests. *Geotextiles and Geomembranes* **27**: 435–446.

Puig J, Bilvet JC and Pasquet P (1977) Earth fill reinforced with synthetic fabric. *Proceedings of the International Conference on the Use of Fabrics in Geotechnics*, Paris, pp. 85–90. (In French.)

Rankilor PR (1981) *Membranes in Ground Engineering*. Wiley, Chichester.

Rankilor PR and Heiremans F (1996) Properties of sewn and adhesive bonded joints between geosynthetic sheets. *Proceedings of the 1st European Geosynthetics Conference: Eurogeo 1*, Maastricht, pp. 261–269.

Raymond GP and Giroud JP (1993) *Geosynthetics Case Histories*. ISSMFE Technical Committee TC9, Geotextiles and Geosynthetics, London.

Rollin AL, Marcotte M and Chaput L (2002) Lessons learned from geo-electrical leaks surveys. *Proceedings of 7th International Conference on Geosynthetics*, Nice, pp. 527–530.

Shamsher FH (1992) Ground improvement with oriented geotextiles and randomly distributed geogrid micromesh. PhD thesis, Indian Institute of Technology Delhi, New Delhi.

Shukla SK (2002) *Geosynthetics and their Applications*. Thomas Telford, London.

Shukla SK (2004) Discussion of 'Applications of geosynthetics for soil reinforcement' by M.I.M. Pinto. *Ground Improvement* **8(4)**: 179–181.

Shukla SK and Sharma U (2006) Effect of seating time on transmissivity and permittivity of geotextiles. *Proceedings of the 8th International Conference on Geosynthetics*, Yokohama, pp. 403–406.

Shukla SK and Sivakugan N (2010) Discussion of 'Fiber-reinforced fly ash subbases in rural roads' by P. Kumar and S.P. Singh. *Journal of Transportation Engineering* **136(4)**: 400–401.

Shukla SK and Yin JH (2006) *Fundamentals of Geosynthetic Engineering*. Taylor and Francis, London.

Shukla SK, Sivakugan N and Das BM (2009a) Fundamental concepts of soil reinforcement – an overview. *International Journal of Geotechnical Engineering* **3(3)**: 329–342.

Shukla SK, Sivakugan N and Mahto S (2009b) A simple method for estimating Poisson's ratio of geosynthetics at zero strain. *Geotechnical Testing Journal*, ASTM **32(2)**: 181–185.

Shukla SK, Sivakugan N and Singh AK (2010) Analytical model for fiber-reinforced granular soils under high confining stresses. *Journal of Materials in Civil Engineering*, ASCE **22(9)**: 935–942.

Simon A, Payany M and Puig J (1982) The use of honeycombed geotextile lap to combat erosion. *Proceedings of the 2nd International Conference on Geotextiles*, Las Vegas, pp. 247–252. (In French.)

Snow MS, Kavazanjian Jr E and Sanglerat TR (1994) Geosynthetic composite liner system for steep canyon landfill side slopes. *Proceedings of the 5th International Conference on Geotextiles, Geomembranes and Related Products*, Singapore.

Suits LD and Hsuan YG (2003) Assessing the photodegradation of geosynthetics by outdoor exposure and laboratory weatherometer. *Geotextiles and Geomembranes* **21(2)**: 111–122.

Terzaghi K and Lacroix Y (1964) Misson dam: an earth and rockfill dam on a highly compressible foundation. *Géotechnique* **14(1)**: 13–50.

Terzaghi K and Peck RB (1948) *Soil Mechanics in Engineering Practice*. Wiley, New York.

Van Dine D, Raymond G and Williams SE (1982) An evaluation of abrasion tests for geotextiles. *Proceedings of the 2nd International Conference on Geotextiles*, Las Vegas, NV, pp. 811–816.

Venkatappa Rao G (1996) Geosynthetics in the Indian environment. *Indian Geotechnical Journal* **26(1)**: 3–94.

Venkatappa Rao G and Saxena KR (eds) (1989) *Use of Geosynthetics in India: Experiences and Potential*. CBIP, New Delhi.

Wang DW (1994) Filter criteria of woven geotextiles for protective works. *Proceedings of the 5th International Conference on Geotextiles, Geomembranes and Related Products*, Singapore, pp. 763–766.

Wesseloo J, Visser AT and Rust E (2004) A mathematical model for the strain rate dependent stress-strain response of HDPE geomembranes. *Geotextiles and Geomembranes* **22(4)**: 273–295.

Westergaard HM (1938) *A Problem of Elasticity Suggested by a Problem in Soil Mechanics, Soft Material Reinforced by Numerous Strong Horizontal Sheets*. Harvard University, Cambridge, MA.

Wewerka M (1982) Practical experience in the use of geotextiles. *Proceedings of the Symposium on Recent Developments in Ground Engineering Techniques*, Bangkok, pp. 167–175.

Yang X and Ding X (2006) Prediction of outdoor weathering performance of polypropylene filaments by accelerated weathering tests. *Geotextiles and Geomembranes* **24(2)**: 103–109.

Yetimoglu T and Salbas O (2003) A study on shear strength of sands reinforced with randomly distributed discrete fibers. *Geotextiles and Geomembranes* **21(2)**: 103–110.

Zanzinger H, Hangen H and Alexiew D (2010) Fatigue behaviour of a PET-geogrid under cyclic loading. *Geotextiles and Geomembranes* **28(3)**: 251–261.

Handbook of Geosynthetic Engineering
ISBN 978-0-7277-4175-2

http://dx.doi.org/10.1680/hge.41752.045

Chapter 2
Soil–geosynthetic interaction

M. L. Lopes
Department of Civil Engineering, University of Porto, Portugal

Soil–geosynthetic interaction mechanisms and soil–geosynthetic interface resistance are referred to in this chapter. Factors influencing soil–geosynthetic interaction, such as soil particle size, confinement stress, soil density and geosynthetic structure, are analysed. The main relevance of soil particle size when the reinforcement is a geogrid is put forward. Direct shear and pull-out laboratory tests for the quantification of soil–geosynthetic interface resistance are described in detail. Emphasis is given to an explanation of the limitations of these tests, test procedures, and the test results and their analysis.

2.1. Introduction

Soil–geosynthetic interaction is of the utmost importance in several applications of geosynthetics, especially when they act as soil reinforcement.

Soil reinforcement consists of the placement of elements duly oriented in the soil, which, by their character, improve the mechanical properties of the new material (reinforced soil) when compared with that of the unreinforced soil.

The main target of reinforcement is to inhibit the development of tensile strains in the soil and, consequently, to support the tensile stresses that the soil cannot withstand. The tensile stress supported by reinforcement improves the soil mechanical properties by reducing the shear stress that has to be carried by the soil and by increasing its available shearing resistance, as the normal stress acting on potential shear surfaces increases. The effectiveness of reinforcement depends on its alignment, it being most effective when aligned in a direction of tensile strain in the soil, so that tensile reinforcement stress develops (McGown *et al.*, 1978; Jewell and Wroth, 1987; Jewell, 1996).

The behaviour of the reinforced soil depends on several factors, such as

- the soil and reinforcement mechanical characteristics
- the soil–reinforcement interaction mechanism and properties
- the geometry of the reinforced system
- the shape, number, location and alignment of reinforcements
- the process of construction.

As examples, the shear strength of the reinforced soil relies on the mobilised shear resistance in the soil and the mobilised tensile stress in the reinforcement, and the relative values of these mobilised resistances are dependent on the deformation properties of the soil and reinforcement. On the other hand, the rate of mobilisation of reinforcement tensile stress is determined by its stress–strain properties, and, to prevent failure, the maximum mobilised tensile stress cannot exceed the reinforcement bond stress.

Although all the mentioned determinant factors on soil-reinforced behaviour are important, particular attention must be given to soil–reinforcement interaction mechanisms, and to the factors that can influence them, because the effectiveness of the transference of tensile stresses from soil to reinforcements rely on it and, so too, the behaviour of the reinforced soil system. This subject will be analysed in the following sections by considering geosynthetic materials as a reinforcement only, complemented by the study of the methods for the evaluation of soil–reinforcement interaction properties.

2.2. Granular soil behaviour

Granular soils are widely considered in studies of soil–geosynthetic interaction because fill materials are, usually, of that type, and their characteristics are determinant on the effectiveness of soil–geosynthetic interaction.

The strength and stiffness of granular soils are extremely dependent on density. Dense soils show greater stiffness and resistance than loose soils because of greater grain interlocking. During soil shearing, when inter-grain sliding starts, the mobilised forces (due to the rearrangement of grains) can be high if the soil is dense; on the other hand, inter-granular friction forces are almost independent of soil density. When the shear process starts, the void ratio of dense soils is lower than the critical value, and shear stresses induce a volume increase. For small deformations, the stress–strain curve of dense soils shows a peak (maximum strength) that depends on volume increase and initial density. For large deformations, when grain interlocking is cancelled, the soil void ratio is equal to the critical value, and the soil strength is constant (soil strength at a constant volume).

At the beginning of shearing of loose soils, the void ratio is greater than the critical value, and the shear stresses induce volume reduction. The soil stress–strain curve does not show a peak, the maximum soil strength is equal to the soil strength at a constant volume of dense soils, and is mobilised at large deformations, when the void ratio of the soil equals its critical value.

Besides the density, other factors can influence the behaviour of granular soils, such as the confinement stress, grain shape and size distribution. An increase in the confinement stress leads to a reduction in the soil critical void ratio, implying a decrease in the soil dilatant behaviour and an approach of its peak and constant volume strengths. The grain shape and size distribution can affect the soil density, as denser or looser arrangements of particles are determined by them.

Although the grain size is not very important in relation to the behaviour of granular soils, it is of the utmost relevance in the soil–geosynthetic interaction mechanism, especially when the geosynthetic is a geogrid.

It must be emphasised that the characteristics that influence granular soil behaviour on reinforced soil do not change; however, its strength is improved by the presence of reinforcement (Figure 2.1), especially when it is aligned in a direction of tensile strain in the soil, so that tensile reinforcement stress develops (McGown *et al.*, 1978).

2.3. Soil–geosynthetic interaction mechanisms

Although factors such as the geometry of a reinforced soil system and the process of its construction can influence the soil–reinforcement interaction properties, they are strongly determined by the interaction mechanism, the physical and mechanical properties of soil (density, grain shape and size, grain size distribution, water content, etc.), and the mechanical properties, shape and geometry of reinforcements.

Three mechanisms of interaction can be identified in reinforced systems

- skin friction along the reinforcement
- soil–soil friction
- passive thrust on the bearing members of the reinforcement.

Skin friction is the only mechanism with geotextiles and strips. In the case of geogrids, the passive thrust on the bearing members of the grids must also be considered as soil–soil friction, if relative movement occurs in the soil along the grids' apertures.

Shear strength mobilisation between granular soils and geotextiles is a two-dimensional phenomenon, where soil dilatance is allowed, strongly affected by the extensibility of geotextiles. In the case of strips, the phenomenon is three-dimensional and greatly dependent on the characteristics of soil dilatance and on the roughness of the reinforcement surfaces. In fact, the volume of soil shearing around the reinforcement is influenced by its geometry and roughness. With regard to geogrids, the phenomenon can also be considered three-dimensional, mobilising skin friction for

Figure 2.1 Stress–strain curves of a reinforced sand: (a) loose and (b) dense
Adapted from McGown *et al.* (1978)

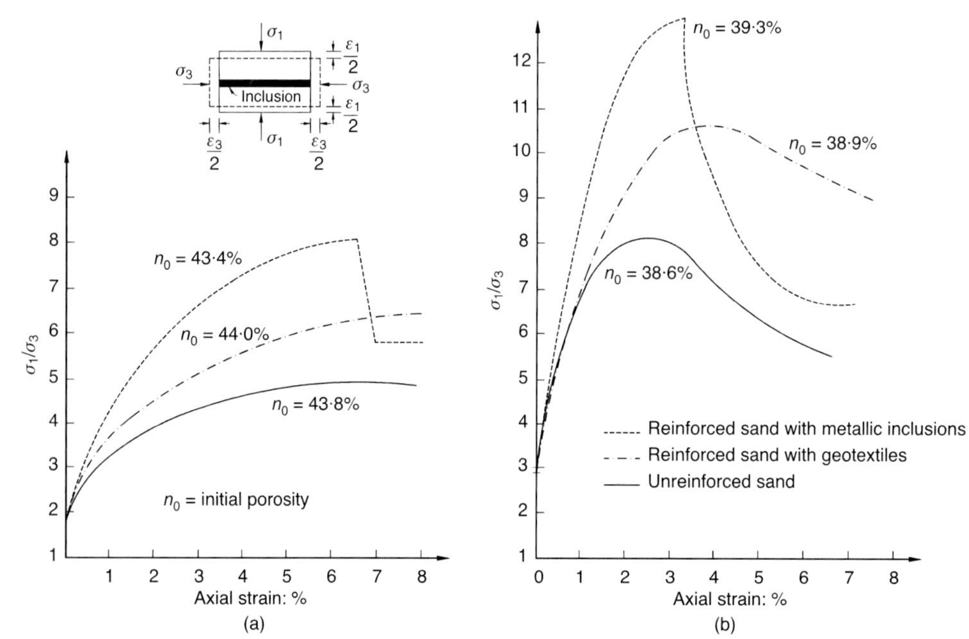

Figure 2.2 Stress conditions in reinforced soil: (a) free dilatance and (b) restricted dilatance
Adapted from Hayashi *et al.* (1994), Industrial Fabrics Association International

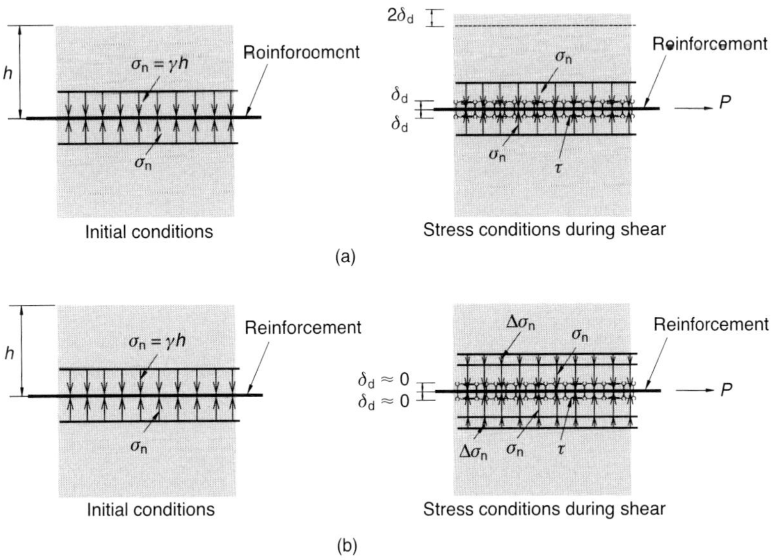

small displacements and progressively mobilising the passive thrust on the bearing members of the grid as displacement increases.

Figure 2.2 shows the stress distribution in the cases of free soil dilatance (two-dimensional phenomenon) and restricted soil dilatance (three-dimensional phenomenon).

Since geogrids are less extensible than geotextiles, the improvement in soil strength and the mobilisation of shear resistance along the interface with the soil increase when the reinforcement used is a geogrid.

2.4. Soil–geosynthetic interface resistance

The stability of reinforced soil is strongly related to the effectiveness of stress transference from the soil to the reinforcement, which is dependent on the available reinforcement length to shear. In fact, as shown in Figure 2.3 for reinforced slopes, the reinforcement length beyond the failure line must be enough to mobilise the required shear stresses to balance the maximum tensile force of reinforcement. The ratio of stress mobilisation is affected by the resistance of the soil–reinforcement interface.

Figure 2.3 Force distribution along the reinforcement
Adapted from Jewell *et al.* (1984)

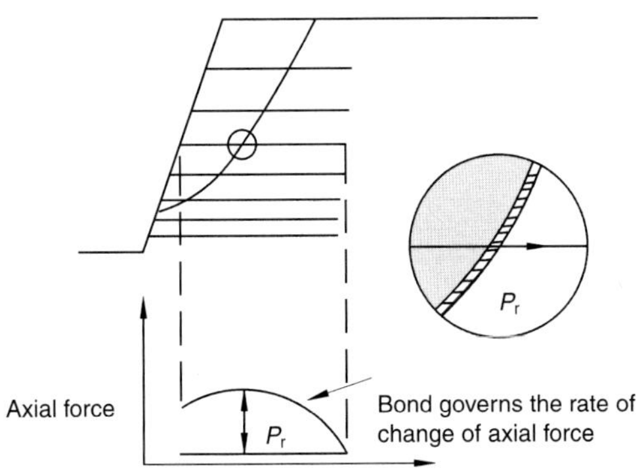

As stated above, with geotextiles and strips only the skin friction mechanism contributes to soil–geosynthetic interface resistance. However, with geogrids, two other interaction mechanisms must be added: the passive thrust mobilisation on the bearing members of the grid; and soil–soil friction, in the case of relative displacement in the soil along the grid apertures. Figure 2.4 shows the soil–geogrid interaction mechanisms.

Two relative movements can be responsible for the mobilisation of strength in soil–reinforcement interfaces

- a block of soil slides across one side of the reinforcement that is 'linked' on the other side to the other block of soil (direct sliding)

Figure 2.4 Soil–geogrid interaction mechanisms: (a) shear between soil and plane surfaces and (b) soil bearing on reinforcement surfaces
Adapted from Jewell *et al.* (1984)

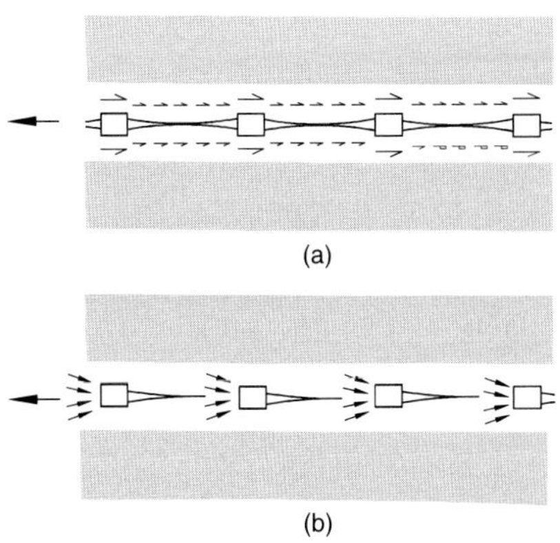

Figure 2.5 Reinforcement dimensions
Adapted from CIRIA SP123, Jewell (1996), http://www.ciria.org

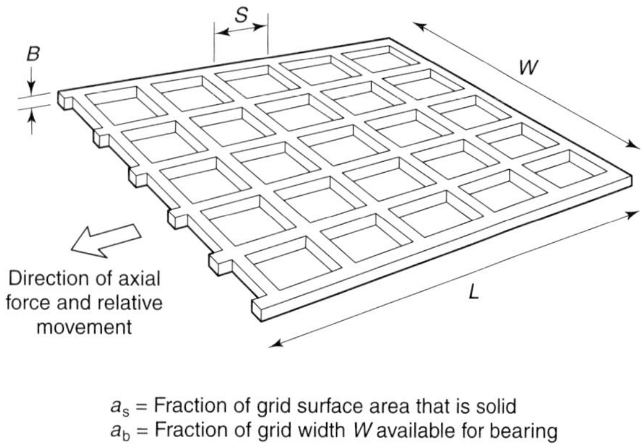

- the reinforcement moves in relation to the surrounding soil (pull-out).

In the first case, when the shear strength of the soil–reinforcement interface is exceeded, the failure occurs by direct shear, and, in the second case, by pull-out. In each case, the soil–reinforcement interface coefficient, f, has a different definition, as will be seen below.

The soil–reinforcement interface shear strength in direct shear and pull-out movements can be defined by Equations 2.1 and 2.2, respectively:

$$T = WL\sigma'_n f \tan\phi' \quad (2.1)$$

$$T = 2WL\sigma'_n f \tan\phi' \quad (2.2)$$

where $0 < f < 1$, f is the soil–reinforcement interface coefficient, ϕ' is the soil friction angle in terms of effective stresses (peak or at constant volume, depending on the soil density), σ'_n is the effective normal stress in the interface, and W and L are the width and the length of the reinforcement, respectively (Figure 2.5).

In Equations 2.1 and 2.2, the main problem lies with the definition of f. In fact, f depends on the interaction mechanism mobilised on the soil–reinforcement interface. So, if the mechanism mobilised is only skin friction, as with geotextiles, f is very similar, if not identical, for both direct sliding and pull-out movements (Jewell, 1996), and is

$$f = f_{ds} = f_b = \frac{\tan\delta}{\tan\phi'} \quad (2.3)$$

where δ is the friction angle at the soil–reinforcement interface, and f_{ds} and f_b are the interface coefficients of direct sliding and of bond, corresponding, respectively, to the direct sliding and to the pull-out movements in the soil–reinforcement interface (Jewell, 1996).

In the case of geogrids, the shear strength of the soil–reinforcement interface for direct sliding movement is the sum of two items, corresponding to

- the skin friction mechanism (T_s)
- the soil–soil friction mechanism (T_{s-s}).

The contribution of the passive thrust mobilisation on the bearing members of the geogrid mechanism is almost negligible in the case of direct sliding.

$$T = T_s + T_{s-s} \quad (2.4)$$

with

$$T_s = a_s WL\sigma'_n \tan\delta \quad (2.5)$$

and

$$T_{s-s} = (1 - a_s)WL\sigma'_n \tan\phi'_n \quad (2.6)$$

where a_s is the fraction of the geogrid surface area that is solid (Figure 2.5).

Using Equations 2.1, 2.4, 2.5 and 2.6, the interface coefficient of direct sliding is obtained as

$$f = f_{ds} = a_s \frac{\tan\delta}{\tan\phi} + (1 - a_s) \quad (2.7)$$

In the case of pull-out movement, the contribution of a soil–soil friction mechanism on soil–reinforcement interface resistance is almost nil – that resistance results from the contribution of the skin friction (T_s) at both sides of the reinforcement and from the passive thrust mobilisation on the bearing members of the geogrid mechanism (T_p):

$$T = 2T_s + T_p \quad (2.8)$$

where T_s is expressed by Equation 2.5, and

$$T_p = \frac{L}{S} a_b WB\sigma'_p \quad (2.9)$$

where S, B and a_b are, respectively, the distance between bearing members, the thickness of the geogrid bearing members and the fraction of the width of the geogrid available for bearing (Figure 2.5), and σ'_p is the effective passive stress mobilised.

Using Equations 2.2, 2.5, 2.8 and 2.9, the interface coefficient (coefficient of bond) is obtained as

$$f = f_b = a_s \frac{\tan\delta}{\tan\phi'} + \frac{\sigma'_p}{\sigma'_n}\frac{a_b B}{S}\frac{1}{2\tan\phi'} \quad (2.10)$$

If $a_s = 1$ and $a_b = 0$, Equations 2.3 and 2.10 become equal, representing the coefficient of bond of a reinforcement where the only interaction mechanism mobilised is skin friction, as in geotextiles.

If, in the soil–reinforcement interface, direct sliding and pull-out movements occur, the interface coefficient that needs to be considered will be the minimum between the coefficients of direct sliding and of bond. However, the movement that starts first must be considered, especially when the soil is dense. When the soil is loose, the equations deduced above may be applicable, considering the soil strength at constant volume.

When the soil is dense, and the first movement that occurs is direct sliding, the soil peak strength can be considered to define the coefficient of direct sliding, but on the deduction of the coefficient of bond, the strength at constant volume must be considered on the side where direct sliding occurred, as well as the peak strength on the other side. When pull-out is the first movement that occurs, the soil peak strength can be considered in the definition of the coefficient of bond, but only the strength at constant volume in the definition of the coefficient of direct sliding.

Excluding the relation σ'_p/σ'_n, the other parameters in Equation 2.10 are easy to obtain. According to Jewell *et al.* (1984) and Jewell (1990, 1996), the passive resistance mobilised on the bearing members of the grids is limited by the theoretical values shown in Figure 2.6 (Equations 2.14 and 2.16).

The theoretical values are defined by the general theory of bearing capacity, with the bearing members of the grids considered similar to strip footings turned 90°. The passive resistance mobilised is

$$\sigma'_p = c'N_c + \sigma'_n N_q \tag{2.11}$$

where σ'_p is the effective passive resistance, σ'_n is the effective normal stress on the interface, c' is soil cohesion, and N_c and N_q are the bearing capacity factors similar to those used for footings considering the failure mechanism by bearing capacity (Peterson and Anderson, 1980). N_c and N_q are given by

$$N_q = \tan^2\left(\frac{\pi}{4}+\frac{\phi'}{2}\right)e^{\pi \tan\phi'} \tag{2.12}$$

$$N_c = \frac{N_q - 1}{\tan\phi'} \tag{2.13}$$

where ϕ' is the friction angle of the soil in terms of effective stresses.

In soils without cohesion, the passive resistance mobilised on the bearing members of the grid has as an upper bound

$$\frac{\sigma'_p}{\sigma'_n} = \tan^2\left(\frac{\pi}{4}+\frac{\phi'}{2}\right)e^{\pi \tan\phi'} \tag{2.14}$$

For the lower bound of the passive resistance mobilised on the bearing members of the grids, Jewell *et al.* (1984) suggested the adoption of the shear failure mechanism by puncture on deep foundations. The bearing capacity factor, N_c, is given by Equation 2.13, and the bearing capacity factor by

$$N_q = \tan\left(\frac{\pi}{4}+\frac{\phi'}{2}\right)e^{[(\pi/2)+\phi']\tan\phi'} \tag{2.15}$$

In soils without cohesion, the passive resistance mobilised on the bearing members of the grid has as a lower bound

$$\frac{\sigma'_p}{\sigma'_n} = \tan\left(\frac{\pi}{4}+\frac{\phi'}{2}\right)e^{[(\pi/2)+\phi']\tan\phi'} \tag{2.16}$$

which is recommended for design.

Figure 2.6 Bearing stresses on geogrids
Adapted from CIRIA SP123, Jewell (1996), http://www.ciria.org; data from Jewell *et al.* (1984) and Palmeira and Milligan (1989)

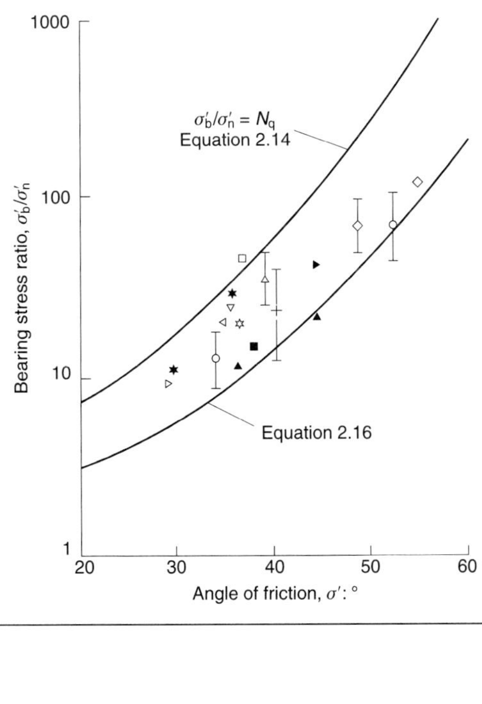

Ospina (1988) observed that for dry sand under small confinement stresses, failure followed the mechanism underlying the lower bound, and for high confinement stresses it followed the mechanism underlying the upper bound. Palmeira and Milligan (1989) have shown, based on data from pull-out tests performed with metallic grids and Leighton Buzzard sands, that when the ratio B/D_{50} (where B is the thickness of the bearing members of the grid and D_{50} is the mean soil particle size) is lower than 10, the failure mechanism changes from that underlying the lower bound to that underlying the upper bound.

The reinforcement geometry and the soil particle size affect the passive resistance on the bearing members of grids, which can, partially, justify the scatter in the published data. In the following sections the influence of these factors will be analysed.

2.5. Factors influencing soil–geosynthetic interaction

2.5.1 Soil particle size

The soil particle size has an important role in soil–geosynthetic interaction, especially when the geosynthetic is a geogrid.

By studying the influence of the soil particle size on soil–geogrid interaction by direct sliding movement, Jewell *et al.* (1984) concluded that the coefficient of direct sliding increases with the soil particle size, and has its maximum value when the grain size is similar to that of the geogrid apertures. In fact, when the soil has particles of the size of silt and fine sand, the failure surface adapts to the lateral surface of the grid; when the grain size increases, but stays lower than the dimensions of the grid apertures, the failure surface is at a tangent to the bearing members of the grid; and when the grain size is similar to that of the grid apertures, the soil particles placed against the bearing members overtopping them drive away the failure surface to the soil mass. The soil–geogrid interface coefficient is at a maximum in this case. The lower value of this coefficient is achieved when the soil particle size is large enough to inhibit penetration of the particles into the grid apertures, and the interface resistance is mobilised only at the points of contact between the soil and the grid.

Jewell *et al.* (1984) recommended the ratio

$$\frac{\text{minimum aperture dimension}}{\text{averages soil particle size}} \geq 3 \tag{2.17}$$

for geogrids used as a soil reinforcement.

Based on the results of pull-out tests carried out with metallic grids and Leighton Buzzard sands, Palmeira and Milligan (1989) showed the influence of the mean soil particle size on the mobilised bearing stress on the grid bars (Figure 2.7).

It can be seen that when the ratio, B/D_{50}, between the bearing member grid size and the mean soil particle size, is less than 10, the mobilised bearing stress can be improved more than twofold, depending on the bearing member shape.

Figure 2.7 Influence of particle size B/D_{50} on mobilised bearing stress
Adapted from CIRIA SP123, Jewell (1996), http://www.ciria.org

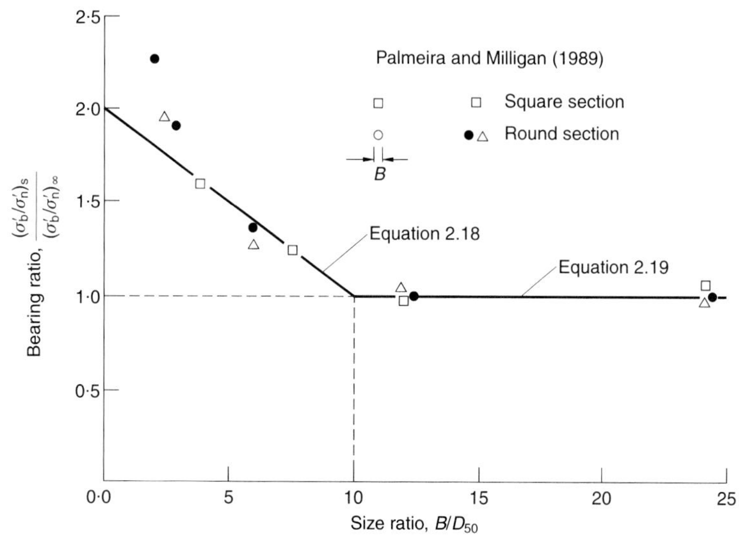

Based on the results of Palmeira and Milligan (1989), Jewell (1990) suggests consideration of the influence of the soil grain size on the mobilised bearing stress, in terms of B/D_{50}, as

$$\frac{\sigma'_p}{\sigma'_n} = \left(\frac{\sigma'_p}{\sigma'_n}\right)_\infty \left(2 - \frac{B}{10D_{50}}\right) \quad \text{when } \frac{B}{D_{50}} < 10 \tag{2.18}$$

and

$$\frac{\sigma'_p}{\sigma'_n} = \left(\frac{\sigma'_p}{\sigma'_n}\right)_\infty \quad \text{when } \frac{B}{D_{50}} > 10 \tag{2.19}$$

In Equations 2.18 and 2.19, and in Figure 2.7, $(\sigma'_p/\sigma'_n)_\infty$ is the bearing stress that is mobilised where the soil particle size is unimportant, assuming a continuum given by Equation 2.16 for grid bearing members of circular cross-section. The improvement in the bearing stress when the grid bars are rectangular is about 20% greater than when they are circular.

Jewell (1996) suggests that Equation 2.10 can be rewritten as

$$f = f_b = \alpha_s \frac{\tan\delta}{\tan\phi'} + F_1 F_2 \left(\frac{\sigma'_p}{\sigma'_s}\right)_\infty \frac{\alpha_b B}{S} \frac{1}{2\tan\phi'} \tag{2.20}$$

where F_1 is the scale effect due to the soil mean particle size, D_{50}, and F_2 is the shape factor.

When $B/D_{50} < 10$:

$$F_1 = 2 - \frac{B}{10D_{50}} \tag{2.21}$$

and when $B/D_{50} > 10$:

$$F_1 = 1.00 \tag{2.22}$$

Figure 2.8 Grain size distribution for: (a) sand 1 and (b) sand 2

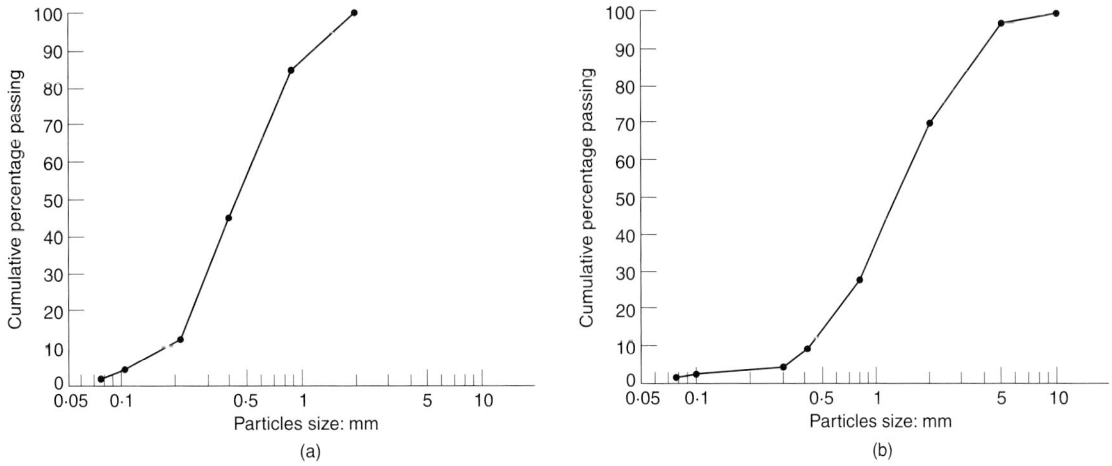

For circular bars, $F_2 = 1.0$, and for rectangular bars, $F_2 = 1.2$.

Lopes (1998) studied the influence of the soil grain size on soil–geogrid interaction by carrying out pull-out tests with geogrids in sand. Test conditions and procedures were similar throughout the test programme; however, different soils and geogrids were used.

Figure 2.8 shows the particle size distribution of the soils (sand 1 and sand 2) used in the study, and Table 2.1 presents their properties.

In Table 2.1:

$$C_u = \frac{D_{60}}{D_{10}} \tag{2.23}$$

$$C_c = \frac{D_{30}^2}{D_{10} D_{60}} \tag{2.24}$$

$$I_D = \left(\frac{\gamma_{max}}{\gamma} \frac{\gamma - \gamma_{min}}{\gamma_{max} - \gamma_{min}} \right) \times 100\% \tag{2.25}$$

where D_{10}, D_{30} and D_{60} are the grain size corresponding to 10, 30 and 60% of the soil passed during sieving (see Figure 2.8), γ_{min} and γ_{max} are the minimum and maximum soil unit weight, and γ is the soil unit weight for the relative density I_D.

The characteristics of the two high-density polyethylene (HDPE) uniaxial geogrids (GG1 and GG2) tested are presented in Table 2.2.

Figure 2.9 shows the variation in the pull-out force with front displacement and the displacements by strain along the geogrid GG1 tested in sand 1 and sand 2. Similar behaviour was observed with geogrid GG2.

With both geogrids, Lopes observed an increase in the soil–geogrid interface strength when they were tested with sand 2. That is, when the soil grain size increases and the B/D_{50} ratio decreases. Table 2.3 shows, for geogrids GG1 and GG2 in both sands, the ratio B/D_{50} and the scale factor F_1 defined by Equations 2.21 and 2.22.

Lopes (1998) found an increase in the global strength of the soil–geogrid interface of approximately 24 and 27% for GG1 and GG2, respectively, which is about one-half of that suggested by the values of the scale factor (due to the mean particle size effect) proposed by Jewell (1990, 1996), based on the results of Palmeira and Milligan (1989) for the increase in the bearing strength mobilised at the interface (48 and 56% for GG1 and GG2, respectively).

Among other reasons, such as different test procedures and conditions, some of the following can be given to explain the difference in the results

Table 2.1 Soil characteristics

Soil	D_{min}	D_{10}	D_{30}	D_{50}	D_{60}	D_{max}	C_u	C_c	γ_{min}	γ_{max}	$\gamma_{(ID=50\%)}$	ϕ° [a]
	(mm)						–		(kN/m^3)			
Sand 1	0.074	0.18	0.30	0.43	0.53	2.00	2.94	0.94	15.00	17.90	16.32	35.7
Sand 2	0.074	0.44	0.84	1.30	1.60	9.54	3.64	1.00	15.60	18.70	17.01	44.2

[a] Soil friction angle (at 38 kPa vertical pressure) in direct shear.

Table 2.2 Geogrids tested (dimensions of the pull-out apparatus: 1.53 m length, 1.00 m width and 0.80 m height)

Geogrid	Material	L (mm)	W (mm)	B (mm)	S (mm)	Tensile strength: kN/m
GG1	HDPE	960	330	3.55	16	80
GG2	HDPE	960	330	5.70	16	120

- the geogrids tested by Lopes (1998) are constructed of HDPE, while those tested by Palmeira and Milligan (1989) are made of mild and galvanised steel
- inextensible materials, such as steel grids, move in relation to the surrounding soil during pull-out, and the resistance is mobilised simultaneously along the reinforcement and at all the bearing members of the grids
- extensible materials, such as HDPE geogrids, deform at the same time as they move in relation to the surrounding soil during pull-out, owing to different degrees of strength mobilisation along the reinforcement and at the bearing members of the geogrids

Figure 2.9 Influence of soil grain size: (a) pull-out force versus front displacement and (b) displacements by strain along the geogrid
Adapted from Lopes (1998)

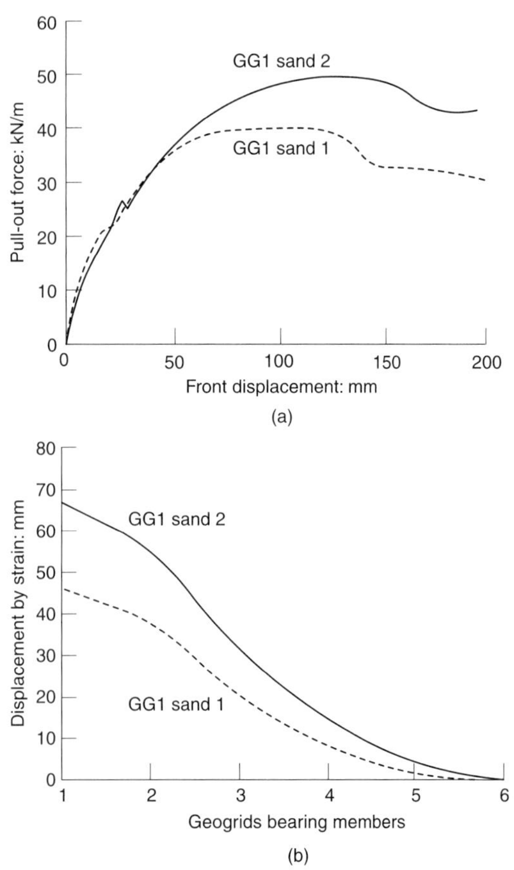

Table 2.3 B/D_{50} and F_1

Geogrid	Sand 1		Sand 2	
	B/D_{50}	F_1	B/D_{50}	F_1
GG1	8.26	1.17	2.73	1.73
GG2	13.26	1.00	4.38	1.56

- with extensible materials the increase in the passive thrust, mobilised on the bearing members of the grid due to the soil grain size, can be responsible for an increase in material deformation during pull-out
- for the reasons given above, extensible materials can mobilise a lower interface strength during pull-out than inextensible materials can.

Considering the results obtained, Lopes (1998) suggests an adoption of a scale factor F_1 lower than that proposed by Jewell (1996) when grids are in extensible materials.

2.5.2 Confinement stress

The confinement stress has an important role in soil–geosynthetic interface resistance because it affects the soil friction angle, and both are directly related (see Equations 2.1 and 2.2). The influence of the confinement stress is even more notable when the strength mobilisation in the interface is a three-dimensional phenomenon, as can be considered in strips and, partially, in geogrids. In this case, an increase in the confinement stress can inhibit, more efficiently, the dilatance that tends to occur, in dense soils, at the interface, leading to a greater improvement in the soil–geosynthetic interface strength.

Several authors have studied the influence of the confinement stress on soil–geosynthetic interaction. As an example, results from the study carried out by Lopes (1998), concerning the influence of the confinement stress on soil–geogrid interaction during pull-out, are presented. Lopes performed pull-out tests with geogrid GG1 (see Table 2.2) in sand 1 (see Table 2.1), with two values for the confinement stress: 24.5 and 38.0 kPa. The friction angle of the sand defined in direct shear tests at 24.5 kPa was 38.4°.

Figure 2.10 presents the variation in the pull-out force with front displacement and the displacements by strain along the geogrid. As can be seen, an increase of about 55% in the confinement stress leads to an increase in the shear strength mobilised at the interface (Figure 2.10(b)) and to an improvement in the soil–geogrid interface resistance of about 11%. It must be emphasised that the soil relative density (I_D) was 50%, which cannot be considered a dense state; if the sand were denser, a greater increase in the interface resistance would be expected. However, it must be kept in mind that the pulling out of geogrids leads to a different degree of interface strength mobilisation along the reinforcement, with the soil, in some areas, at its strength at constant volume and in other areas at peak strength.

Figure 2.10 Influence of the confinement stress: (a) pull-out force versus front displacement and (b) displacements by strain along the geogrid
Adapted from Lopes (1998)

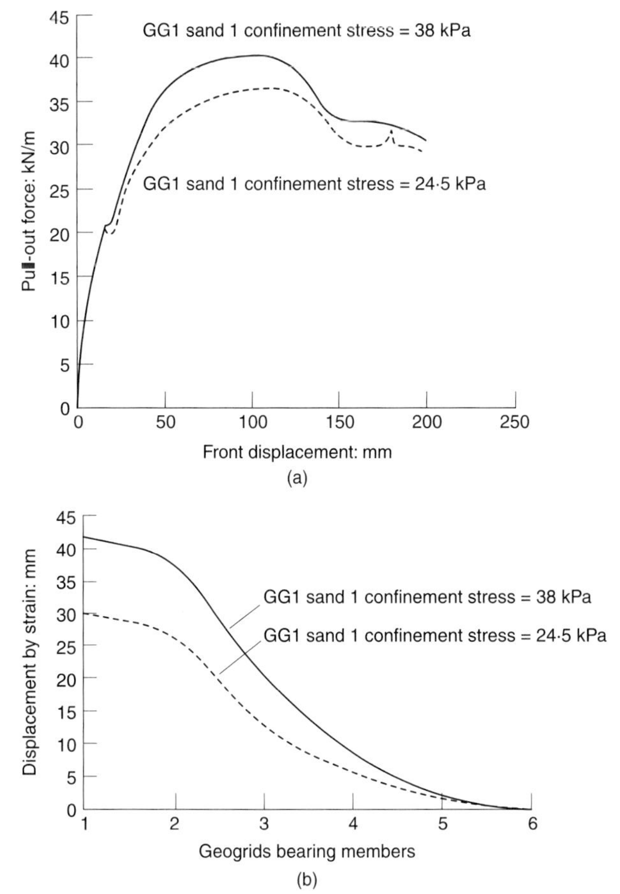

Figure 2.11 Grain size distribution for sand 3

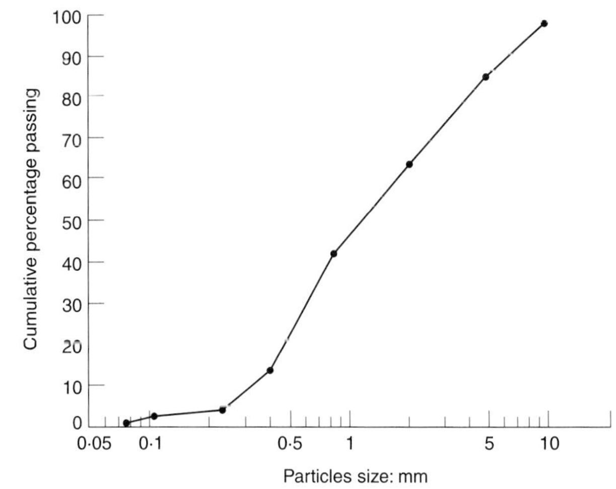

Table 2.4 Geogrid (dimensions of the pull-out apparatus: 1.53 m length, 1.00 m width and 0.80 m height)

Geogrid	Material	L (mm)	W	B	S	Tensile strength: kN/m	Peak strain: %
GG3	HDPE	960	330	2.6	16	55	11.5

size distribution of the sand. The main characteristics of the geogrid and of the sand used during the study are presented in Tables 2.4 and 2.5, respectively.

Figure 2.12(a) shows the variation in the pull-out force with the frontal displacement recorded in the tests. It can be observed that, for the looser state of the sand, the geogrid fails by lack of adherence at a pull-out force of about 32.2 kN/m, failing by lack of tensile strength for the denser state. The maximum pull-out force achieved with the denser sand was about 45.3 kN/m, which would certainly have been higher if the reinforcement had not failed through lack of tensile strength.

An increase higher than 40% of the strength of the soil–geogrid interface when the relative density of the sand changes from 50 to 86% is due to the greater soil and soil–geogrid strength in the denser sand. The displacement of the geogrid decreases, increasing the interface stiffness modulus and the

2.5.3 Soil density

The soil density affects the soil–geosynthetic interface strength in the same way as the confinement stress. Dense granular soils are more resistant and stiffer than loose ones, presenting dilatant behaviour and inducing higher confinement stresses.

Lopes and Ladeira (1996a) studied the influence of the soil density on the soil–geogrid interface resistance by performing pull-out tests with a uniaxial geogrid in the sand. The tests were carried out in similar conditions, except with regard to the sand relative density (I_D), which was 50% in one test and 86% in the other. Figure 2.11 shows the grain

Table 2.5 Soil characteristics

Soil	D_{min}	D_{10}	D_{30}	D_{60}	D_{max}	C_u	C_c	γ_{min}	γ_{max}	$\gamma_{(ID=50\%)}$	$\gamma_{(ID=86\%)}$	$\phi_{(ID=50\%)}$ ° [a]	$\phi_{(ID=86\%)}$ ° [a]
	mm					–		kN/m^3					
Sand 3	0.074	0.34	0.63	1.67	11.67	4.91	0.70	16.10	18.90	17.39	18.45	35.2	35.7

[a] Soil friction angle (at 46.7 kPa vertical pressure) in direct shear.

Figure 2.12 Influence of soil density: (a) pull-out force versus front displacement and (b) displacements by strain along the geogrid
Adapted from Lopes and Ladeira (1996b), with permission from Elsevier

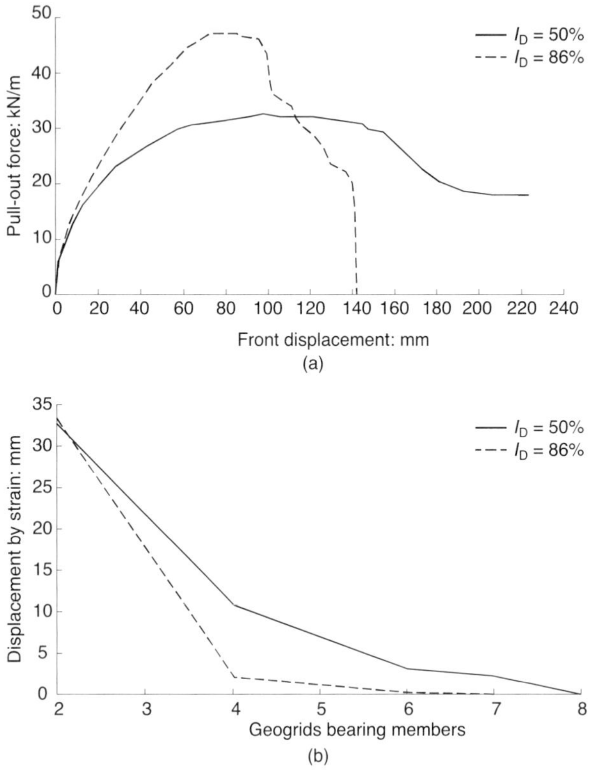

pull-out force. As the soil density increases, the length of adherence decreases. In fact, for the higher soil density tested, only one-third of the inclusion length contributes to the resistance (Figure 2.12(b)).

2.5.4 Geosynthetic structure

The distance (S) between the bearing members of grids is an important parameter with regard to soil–grid interaction. In particular, if that distance is lower than an optimum value, there is interference between members, each one being less effective.

Assuming the limiting case where the strength of the soil–grid interface derives only from the passive thrust mobilised on the bearing members of the reinforcement, and that there is an upper limit for the interface strength equivalent to that of a completely rough sheet ($\delta = \phi'$), Jewell *et al.* (1984) and Jewell (1990) consider that the maximum strength of the interface is achieved for an optimum geometry of the grid $(S/(a_b B))_\infty$. Taking into account the authors' assumptions, from the general Equation 2.10, f_b is

$$f_b = \frac{\sigma'_p}{\sigma'_n} \frac{a_b B}{S} \frac{1}{2 \tan \phi'} \tag{2.26}$$

and, taking $(f_b)_{max} = 1.00$,

$$\left(\frac{S}{a_b B}\right)_{\phi'} = \frac{\sigma'_p}{\sigma'_n} \frac{1}{2 \tan \phi'} \tag{2.27}$$

From Equations 2.26 and 2.27, the coefficient of bond can be expressed as

$$f_b = \frac{(S/a_b B)_{\phi'}}{S/a_b B} \tag{2.28}$$

The authors concluded that when the grid geometry is lower than the optimum, the bond strength mobilised in each single member is proportionately lower as the coefficient of bond cannot increase above $(f_b)_{max} = 1.00$, as shown in Figure 2.13.

Palmeira and Milligan (1989) studied the influence of the distance between bearing members of grids on the resistance of the soil–grid interface by carrying out pull-out tests with metallic grids with different distances between the bearing members in sand. The authors concluded that, as this distance decreases, the interface resistance also decreases, thereby correlating the increase in the interference of bearing members with the reduction in the distance between them. The authors suggest that the concept of interference between the grid members, as the ratio of the mobilised bearing stress to the maximum possible value, should be introduced explicitly into the analysis of bond resistance in terms of

Figure 2.13 Influence of reinforcement geometry on the coefficient of bond
Adapted from Jewell *et al.* (1984)

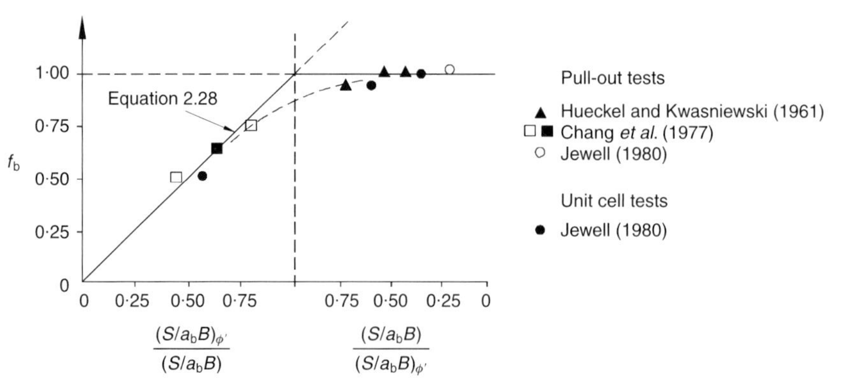

the distance of interference:

$$DI = 1 = \frac{S/a_b B}{(S/a_b B)_{\phi'}} \tag{2.29}$$

where DI ≥ 0. In this case,

$$f_b = \frac{\sigma'_p}{\sigma'_n} \frac{a_b B}{S} \frac{1}{2 \tan \phi'} \tag{2.30}$$

and

$$\left(\frac{S}{a_b B}\right)_{\phi'} = \frac{\sigma'_p}{\sigma'_n} \frac{1}{2 \tan \phi'} \tag{2.31}$$

which is another expression for Equation 2.28.

Palmeira and Milligan (1989) remark that the soil properties (including particle shape and surface characteristics) and the diameter, spacing and number of bearing members are the main factors that control the interference between bearing members. Although the results obtained obey the general pattern of variation, the authors stated that deviations from that pattern are likely to occur when

- there is tangency between members
- the reinforcements are very long
- the reinforcements are extensible
- the friction along the longitudinal members in a grid is a significant part of the soil–reinforcement interface resistance.

Lopes (1998) studied the influence of the structure of geosynthetics on the soil–reinforcement interface strength by carrying out pull-out tests with sand 1 (see Table 2.1). The test procedures and conditions were similar, but the reinforcements used were a uniaxial geogrid (Table 2.4) and a spunbounded non-woven geotextile. The main characteristics of the geotextile tested are presented in Table 2.6.

Figure 2.14 presents the variation in the pull-out force with front displacement for the two materials. It can be seen that, although both geosynthetics had similar tensile strength (50 kN/m for the non-woven geotextile and 55 kN/m for the uniaxial geogrid), the strength mobilised on the soil–geogrid interface is about 2.6 times greater than that mobilised on the soil–geotextile interface, the first being achieved for front displacements of about one-quarter of the second.

Figure 2.14 Influence of geosynthetic structure on soil–geosynthetic interface strength
Adapted from Lopes (1998)

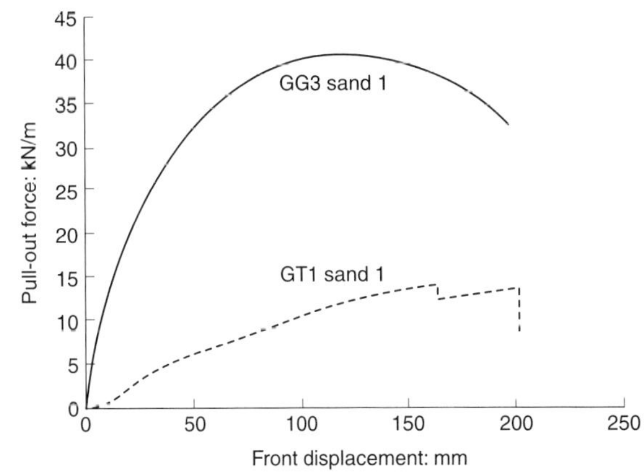

The structure of the geosynthetics and their extensibility had greater influence on the observed behaviour. The structure of the spunbounded non-woven geotextile is much more extensible than that of the geogrid (see Tables 2.4 and 2.6). The degree of deformation of the geotextile during pull-out is much higher than that of the geogrid, and the mobilisation of the interface resistance is less effective.

On the other hand, it should be noted that the interaction mechanisms mobilised in both geosynthetics are different, for the geotextile skin friction and for the geogrid skin friction plus the passive thrust on the bearing members, which can also contribute to the effectiveness of the interface strength mobilisation when geogrids are used as reinforcement.

2.6. Laboratory tests for quantification of soil–geosynthetic interface resistance

The laboratory tests available for quantification of the soil–geosynthetic interface strength are the direct shear and pull-out tests. The adequacy of each test for the definition of interface characteristics relies on the relative movement that tends to occur in the soil–geosynthetic interface – for direct sliding movement, the direct shear test is more suitable, whereas for pull-out movement, the pull-out test should be conducted. Quite often, the interface characteristics defined by the direct shear and pull-out tests are different, and are even sometimes inconsistent as a result of different test procedures, stress paths, failure mechanisms and boundary conditions.

Table 2.6 Geotextile (dimensions of the pull-out apparatus: 1.53 m length, 1.00 m width and 0.80 m height)

Geotextile	Type	Material	L	W	Thickness	Unit weight: g/m²	Tensile strength: kN/m	Peak strain: %
			mm					
GT1	Non-woven	Polypropylene	960	330	6	800	50	65

Figure 2.15 (a) For geotextiles: the direct shear test with a solid block in the lower half of the apparatus. (b) For geogrids: the direct shear test with soil in the lower half of the apparatus. (c) The pull-out test

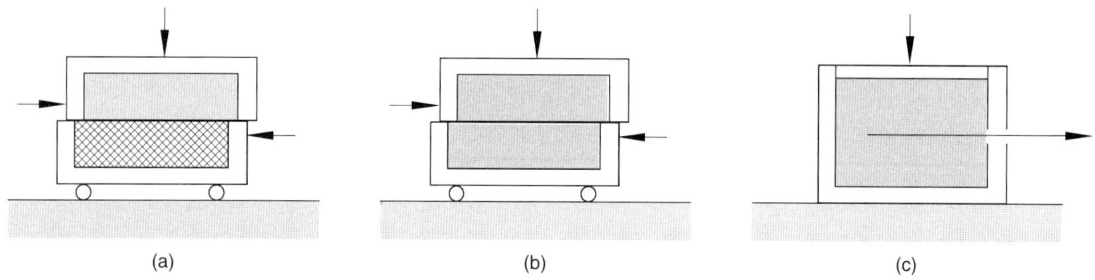

2.6.1 Direct shear test

The coefficient of direct sliding is quite well defined for geotextiles and geogrids, using the modified direct shear test, adapted from the test used in soil mechanics. Gourc *et al.* (1996) recommend the use of a direct shear apparatus with a sample area of 300 mm × 300 mm or greater.

In the most common direct shear tests performed with geotextiles, these materials are fixed to a rigid plane support placed on the lower half of the apparatus (Figure 2.15(a)). This procedure models, with sufficient accuracy, the interaction mechanism that occurs in the soil–geotextile interface during direct sliding; that is, skin friction between the full plan area of contact. Alternatively, the geotextiles can be supported on soil (Figure 2.15(b)).

Direct shear tests with geogrids can also be performed with the geosynthetic fixed to a rigid plane support, placed on the lower half of the apparatus. However, for these materials, especially for those with large apertures and a high percentage of openings, it is suggested that the tests be carried out with the geogrids supported on soil, placed on the lower half of the apparatus (Figure 2.15(b)). In fact, the direct sliding strength for geogrids is generated by two mechanisms: soil–soil friction along the apertures of the grid; and skin friction over the geogrid itself. Soil sliding over soil through the apertures of the grid is not modelled when using the first-mentioned test procedure, and it can reach a significant percentage of the interface resistance, namely for grids with large apertures and a high percentage of openings.

The direct shear tests are relatively easy to interpret, although the results are dependent on certain factors, such as the relative position of the soil and reinforcement; the methods used for normal stress control; the thickness of the soil layer; and the roughness of the rigid plane (Gourc *et al.*, 1996). Since the appearance of the first standards on the soil–geosynthetic direct shear test (BS 6906:1991 (British Standards Institution, 1991) and ASTM D5321-92 (American Society for Testing and Materials, 1992)), many authors have been studying the influence on the test results of the geometry and size of the direct shear apparatus, as well as of the test procedures (Takasumi *et al.*, 1991; Gourc and Lalarokotson, 1997; Bemben and Schulze, 1998; Jones and Dixon, 1998; Nakamura *et al.*, 1999; Blümel and Stoewahse, 1998; Blümel *et al.*, 2000; Jones and Dixon, 2000; Wasti and Özdüzgün, 2001; Stoewahse *et al.*, 2002; Hsieh and Hsieh, 2003; Dixon *et al.*, 2006). A study of the influence of some of the direct shear test procedures and conditions on the results was performed by Silvano (2005), and Lopes and Silvano (2010). For this study, a prototype of a soil–geosynthetic direct shear apparatus was developed, based on the recommendations of the EN ISO 12957-1 (International Organization for Standardization, 2004) and ASTM D5321-92 standards (Figure 2.16). A description of this apparatus can be found in Silvano (2005), and Silvano and Lopes (2005, 2006).

EN ISO 12957-1 allows both constant-area and reduced-area tests to be carried out. The equipment was developed in order to perform constant-area soil–geosynthetic direct shear tests. The upper half of the shear box is 0.30 m wide, 0.60 m long and 0.15 m high. The lower half of the shear box is 0.34 m wide, 0.80 m long and 0.10 m high. The

Figure 2.16 General view of the direct shear apparatus

geosynthetic specimen is clamped on the lower-half shear box. A rigid base can be placed inside the lower half of the box to perform direct shear tests with geosynthetics (e.g. geotextiles) with a full-plane lateral contact area with the soil. The horizontal movement is applied through a hydraulic system and recorded by an internal displacement transducer located inside the hydraulic system. A load cell measures the horizontal load. The vertical stress is applied through two actuators on a rigid metallic plate (0.60 m × 0.30 m). The load is measured indirectly through pressure transducers that register the oil internal pressure in the actuators. Control software was developed, enabling the application of constant linear actions both in the vertical and horizontal axis. The equipment is suitable for testing the soil–geosynthetic interface in a wide range of test conditions.

The European standard (EN ISO 12957-1) states that four specimens for each direction to be tested must be used in order to characterise the soil–geosynthetic interface in direct shear movement. If the geosynthetic has different faces, the soil–geosynthetic interface characterisation must be done at both faces of the geosynthetic. For the prototype developed, the size of the specimen is 0.33 m wide and 0.85 m long. The specimen is placed and clamped over the soil that fills the lower half of the box (or over the rigid base) (Figure 2.17). Then, the upper half of the box is positioned over the geosynthetic with a 0.5 mm gap between its base and the specimen. A 50 mm-thick compacted soil layer is placed inside the upper half of the box. Then, one of the standardised vertical stresses (50, 100 or 150 kPa) is applied (the test at 100 kPa is performed twice). The horizontal movement starts, and the test finishes when 0.10 m (≈16.5% × 0.60 m) horizontal displacement is reached. At the end of the test, the movement of the load plate must be noted. During disassembly of the test, the soil must be removed with care, and the geosynthetic specimen examined and any unexpected phenomena noted.

For each test, the vertical stress, σ (kPa), and the shear stress, τ (kPa), are given by

$$\sigma = \frac{N}{A} \tag{2.32a}$$

$$\tau = \frac{S}{A} \tag{2.32b}$$

where N is the vertical load (kN), S is the horizontal load (kN) and A is the geosynthetic contact area with the soil (m^2).

For each vertical stress, a plot of the horizontal displacement versus the shear stress is drawn, and the maximum soil/geosynthetic shear stress ($\tau^{\max(\sigma_i)}$) should be defined. The direct sliding interface coefficient for each vertical stress, $f_{\mathrm{ds}}(\sigma_i)$, can be defined by

$$f_{\mathrm{ds}}(\sigma_i) = \frac{\tau^{\max(\sigma_i)}}{\tau_3^{\max(\sigma_i)}} \tag{2.33}$$

where $\tau_{\mathrm{s}}^{\max(\sigma_i)}$ is the maximum soil–soil shear stress at a vertical stress σ_i.

2.6.1.1 Influence of the type of test

It is not clear in the EN ISO 12957-1 standard if a constant-area-type direct shear test should only be performed with the geosynthetic placed on a rigid base. Anomalous behaviour was found in several tests carried out with the soil in the lower half of the box. In fact, during the test the soil in the lower half of the box blisters. The phenomenon is similar to that of soil failure under a footing load. During the test there is soil settlement under the load plate and vertical blistering at the free end of the lower half of the box (Figure 2.18).

However, for the definition of the soil–geosynthetic direct sliding coefficient of some geosynthetics (e.g. geogrids), it is fundamental to carry out tests with the soil in the lower half of the box. On the other hand, as this coefficient is the ratio between the soil–geosynthetic and the soil–soil shear stresses, it was decided to define the friction properties of the soil in the prototype developed. To define the friction properties of the soil, reduced-area shear-type tests must be performed: to achieve that objective, a metallic box was placed inside the lower half of the box (Figure 2.19). The shear area at the beginning of the test was 0.60 m × 0.30 m.

Figure 2.17 Specimen clamping
Adapted from Lopes and Silvano (2010), with permission from Springer Science+Business Media

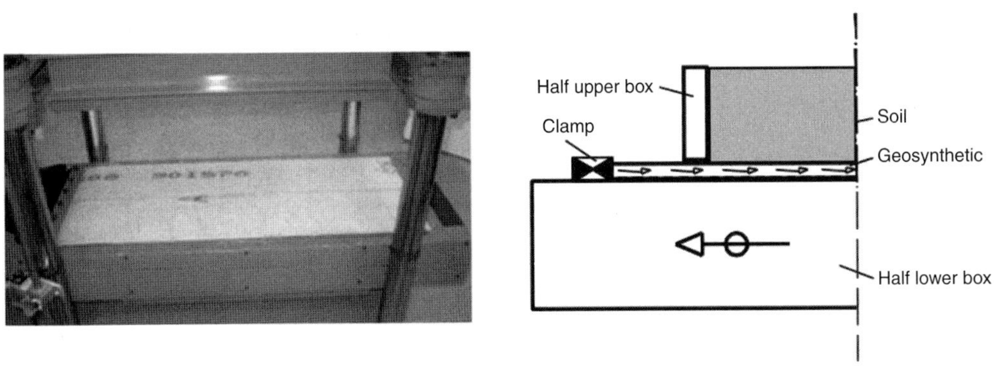

Figure 2.18 Problem identification: (a) schematic representation of the phenomenon and (b) confirmation of the phenomenon
Adapted from Lopes and Silvano (2010), with permission from Springer Science+Business Media

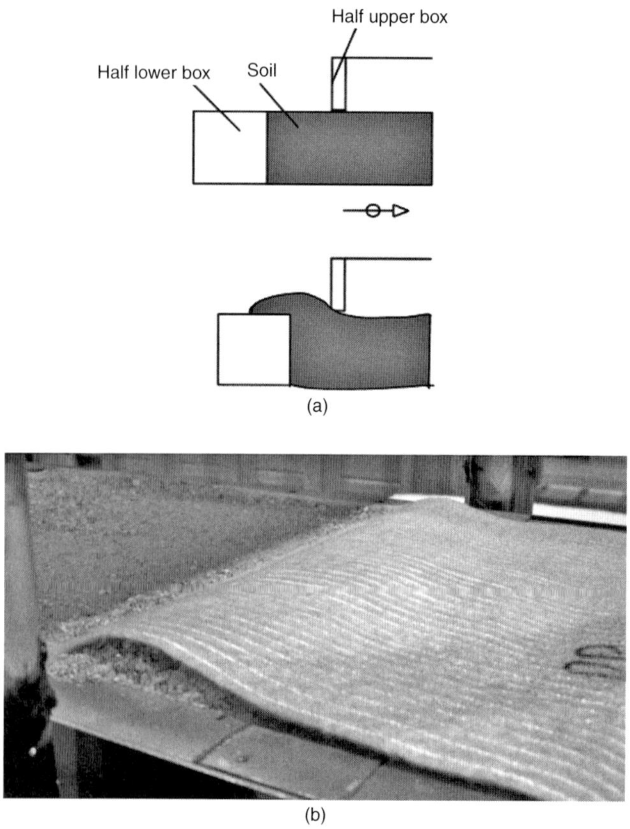

As the imposed horizontal displacement in the direct shear test is quite large (0.10 m), there is a fraction of the soil–metallic surface friction that could influence the test data. In fact, in the case of soil–soil direct shear, the shear resistance (F_{total}) is a sum of two parts: soil/soil friction ($F_{soil-soil}$) and soil/metallic surface friction ($F_{soil-iron\ sheet,RA}$).

Two tests were carried out (with 150 kPa of applied vertical stress) in order to quantify the influence of the soil–metallic surface friction: a soil–soil shear test (both halves of the box filled with soil) and a soil–iron sheet shear test (an iron sheet with identical roughness to that of the surface of the metallic box was placed underneath the upper half of the box, which was filled with soil). The physical and mechanical characteristics of the soil used are presented in Tables 2.7 and 2.8, respectively. Figure 2.20 shows the horizontal displacement/shear stress relationship for the two tests. It can be observed that the mobilised friction at the soil–iron sheet interface ($F_{soil-iron\ sheet,CA}$) is much lower than that mobilised in the other test ($F_{soil\ total}$). Similar values for the peak and residual shear stress were obtained for the soil–iron sheet interface, and no soil dilatancy was identified.

Table 2.7 Soil physical characteristics
Data taken from Lopes and Silvano (2010)

D_{30}	D_{50}	D_{60}	D_{max}	γ_{min}	γ_{max}	$\gamma_{ID=50\%}$
mm				kN/m^3		
0.18	0.38	0.53	9.51	13.59	17.20	15.18

Table 2.8 Soil mechanical characteristics
Data taken from Lopes and Silvano (2010)

Relative density = 50%	Consolidation vertical stress = 50 kPa			ϕ': °	c': kPa
	σ'_ν: kPa	τ: kPa	τ/σ'_ν		
Peak	53.11	45.35	0.85	41.1	0.0
Constant volume	56.73	39.71	0.70	36.6	0.0

Based on the results obtained, the soil–soil friction parcel is

$$F_{soil\ total} = F_{soil-soil} + F_{soil-iron\ sheet,RA} \tag{2.34}$$

where

$$F_{soil-iron\ sheet,RA} = \frac{A_{soil-iron\ sheet,RA}}{A_0} F_{soil-iron\ sheet,\ CA} \tag{2.35}$$

Figure 2.19 Reduced area direct shear test adaptation
Adapted from Lopes and Silvano (2010), with permission from Springer Science+Business Media

Figure 2.20 Reduced-area direct shear tests: shear stress versus horizontal displacement for the soil–soil shear test and for the soil–iron sheet shear test
Adapted from Lopes and Silvano (2010), with permission from Springer Science+Business Media

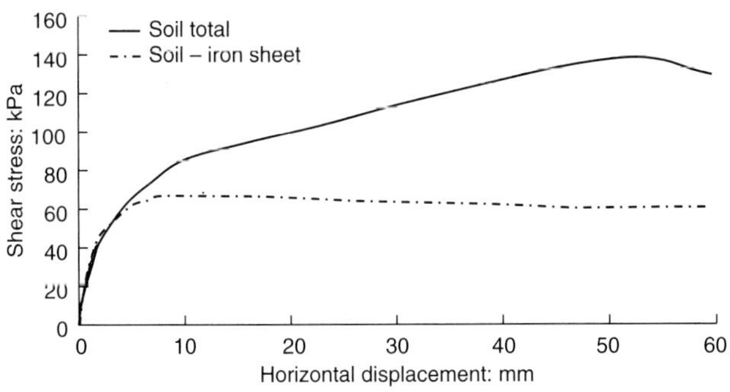

$F_{\text{soil–iron sheet,RA}}$ is the soil–iron sheet shear force in a reduced-area direct shear test (kN), $F_{\text{soil–iron sheet,CA}}$ is the soil–iron sheet shear force in a constant-area direct shear test (kN), A_0 is the shear area in a constant-area direct shear test (0.18 m^2) and $A_{\text{soil–iron sheet}}$ is the contact area between the soil and the metallic surface in a reduced-area direct shear test (0–0.3 m^2).

Figure 2.21 shows horizontal displacement versus shear force plots of the parts identified in a reduced-area direct shear test. The shape of the soil–soil curve can be related to the progressive mobilisation of the shear.

It can be concluded that the influence of the soil–iron sheet friction on the soil–soil direct shear results is small, due, probably, to its reduced shear resistance (Figure 2.21) and to the small shear area (compared with the total area of shear).

2.6.1.2 Influence of specimen fixation

When, in direct shear tests, a rigid base is used, inadequate fixation of the geosynthetic specimen along the shear area can lead to mistaken data, as not only soil–geosynthetic shear occurs, but also rigid base–geosynthetic shear and geosynthetic tensile stresses. In order to achieve an effective fixation of the geosynthetic inside the shear area, an aluminium oxide abrasive sheet (P80) is glued over the rigid base, as suggested in EN ISO 12957-1.

2.6.1.3 Influence of the vertical stress on the registered horizontal force

The ASTM D5321-2 standard is clear about a horizontal force correction, resulting from the internal resistance to shear inherent to the shear device. However, in the EN ISO 12957-1 standard, nothing is mentioned about this effect. According to ASTM D5321-92, a shear test without the geosynthetic and with no vertical stress applied should be carried out at a displacement rate of 10 mm/min for at least 50 mm. The maximum value of the horizontal force measured during the reference test will be the correction applied to the horizontal force.

Based on the ASTM standard recommendation, it was decided to perform shear tests without the geosynthetic, but in such a way that the test programme conditions were reproduced. In order to fulfil this requirement, after confirming the repeatability of results with the equipment (Silvano and Lopes, 2005), shear tests were carried out at a 2 mm/min horizontal displacement rate for each of the three vertical stresses to be analysed (50, 100 and 200 kPa), with the rigid base placed in the lower half of the box. In order to apply vertical stress without any additional friction between the loading plate and the moving lower half of the box, three rolling bars were placed between these two elements.

It can be seen in Figure 2.22, where the horizontal displacement/shear stress relationship is shown, that there is a small resistance to the movement dependent on the applied vertical stress. The maximum value of the shear stress recorded in each of these tests (τ_c) was considered to be the correction to apply to tests performed at the corresponding vertical stress (Table 2.9). The real value for shear stress should be obtained as follows:

$$\tau_{\text{real}} = \tau_{\text{read}} - \tau_c \tag{2.36}$$

Figure 2.21 Reduced-area direct shear test: shear force versus horizontal displacement, showing the influence of soil–soil and soil–iron sheet friction on the overall behaviour
Adapted from Lopes and Silvano (2010), with permission from Springer Science+Business Media

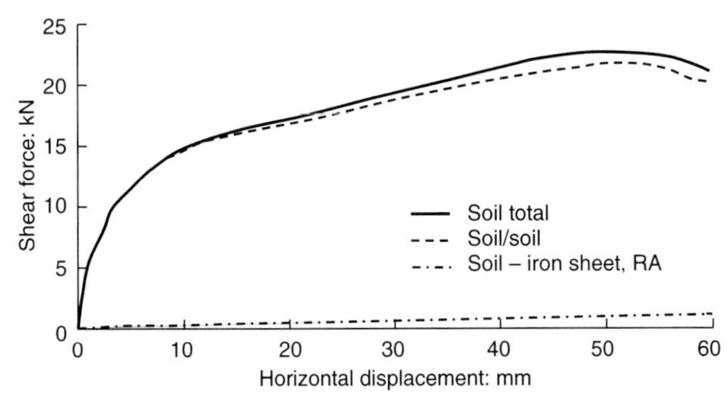

Figure 2.22 Shear stress correction curves: horizontal displacement/shear stress
Adapted from Lopes and Silvano (2010), with permission from Springer Science+Business Media

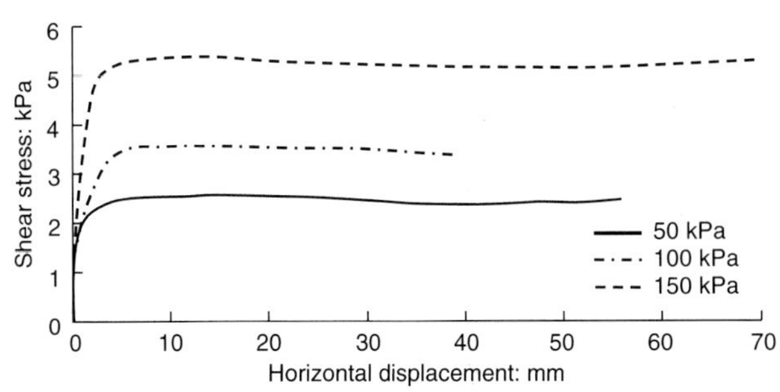

Table 2.9 Maximum shear stress value

Reproduced from Lopes and Silvano (2010), with permission from Springer Science + Business Media

σ_n: kPa	$\delta(\tau_{max})$: mm	$\tau_{max} = \tau_c$: kPa
50	14.18	2.83
100	30.87	3.91
150	14.37	5.90

It should be pointed out that the correction values for the shear stress determined as described above should only be considered valid for tests performed at similar displacement rates and with a rigid base in the lower half of the box. For other test procedures the correction values will need to be re-evaluated.

2.6.1.4 Measurement of vertical displacement

Although the EN ISO 12957-1 standard recommends load plate lift and rotation measurement only at the end of the test, it was decided to register load plate movements from the start of the horizontal displacement until the end of the test. In order to accomplish vertical displacement measurement, three dial gages were placed on the load plate, one in front of the top of the upper half of the box and another at the back of the top of the same half of the box. As mentioned by Silvano and Lopes (2006), significant differences were observed in the relation between the horizontal displacement and the interface shear stress compared with tests performed in similar conditions, where the main difference was the measuring of vertical displacement during tests.

The authors justified these differences by the increased period of time for which the soil is subjected to vertical stress before the horizontal displacement starts, and by an inadequate definition of the gap between the upper half of the box and the geosynthetic in the EN ISO 12957-1 standard. In fact, as the vertical displacement measurement system can only be placed after applying vertical stress, the period of time for which the soil was subjected to vertical stress, before the horizontal displacement started, increased. Although the EN ISO 12957-1 standard defines the value of the gap between the upper half of the box and the geosynthetic without any vertical stress applied, the resultant gap at the beginning of shear depends on the value of the vertical stress at the interface.

To overcome the problems resulting from the identified causes and consequently to improve the consistency of test results, two modifications were introduced to the test procedures

- a fixed period of time (20 minutes) for the vertical stress application before starting the shear
- gap definition, with the vertical stress applied over the geosynthetic (during the definition of the gap, a thick geotextile is placed between the loading plate and the geosynthetic to be tested, in order to avoid damage).

It should be clarified that the fixed period of time for the vertical stress before the start of the horizontal displacement (20 minutes) was chosen in such a way that correct placement of the vertical displacement measurement system was permitted. As the influence of this aspect of the test procedure on the test results is considered to be related to the time-dependent settlement of soil under vertical stress, it is considered that, in future tests, there is a need to establish criteria associated with control of the vertical displacement before starting the shear test by, for example, observing the vertical displacement rate.

2.6.2 Pull-out test

The coefficient of bond can be defined by performing pull-out tests (Figure 2.15(c)). However, in contrast to what happens in direct shear tests, the pull-out tests are difficult to interpret, as the results are greatly influenced by the boundary conditions, and test procedures and conditions (Lopes and Ladeira, 1996a,b).

Jewell (1996) suggests that, since the bond mechanism on both sides of geotextiles is identical to that in direct sliding (skin friction), there is no need to perform pull-out tests with these materials to define the coefficient of bond, as it is very similar to that of direct sliding obtained from the much simpler modified direct shear test.

However, for geogrids, the interaction mechanisms mobilised during pull-out are significantly different from those in direct shear, developing resistance by mobilising passive thrusts on the transverse members of the grid, by soil–soil friction through the grid apertures and by skin friction over the planar geogrid surface areas. Thus, the coefficient of bond for geogrids can only be measured by pull-out tests.

Considerable care is required in both the execution and interpretation of pull-out tests, as the results can be greatly affected by the use of different test apparatus associated with distinct boundary effects, different test procedures, different schemes of placement and compaction of the soil, etc. (Juran *et al.*, 1988).

A comprehensive study on the influence of the displacement rate, specimen size, soil height and sleeve length on the pull-out behaviour of geogrids was developed by Lopes and Ladeira (1996a,b). In their studies, the pull-out apparatus used was, internally, 1.53 m long, 1.00 m wide and 0.80 m high (Figure 2.23). The use of an apparatus with large dimensions aimed to minimise the influence of the lateral, base and top boundaries. To reduce the influence of the top boundary and to achieve a uniform distribution of the applied vertical stresses on the sample top, a 0.025 m-thick smooth neoprene slab was used between the top soil and the wood plate where the confining stress was applied. To reduce the influence of the front wall, a steel sleeve 0.20 m long was used inside the box.

The soil was poured into the box from a constant height of 0.50 m, and placed in 0.15 m-thick layers. Each layer was

Figure 2.23 Pull-out apparatus: (a) front view and hydraulic system and (b) lateral view

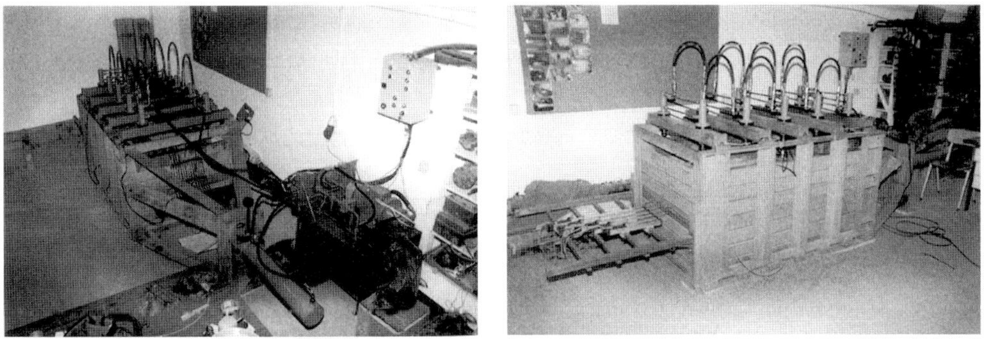

(a) (b)

levelled and compacted to the required relative density (50%) using an electric vibratory hammer. The soil density was controlled with a gamma densimeter. When the steel sleeve level was reached (located at mid-height of the box), the reinforcement was placed over the compacted soil, introduced through the sleeve and fixed to the clamps outside the box. The inextensible wires used to measure the displacements along the reinforcement were then put in place, and connected to the linear potentiometers at the back and at the front of the box. Six potentiometers were used in the tests. Finally, two 0.15 m-thick soil layers were placed, levelled and compacted, resulting in a soil height of 0.60 m, with the reinforcement at the middle.

The pull-out force, applied by a hydraulic system, was measured by a load cell placed in the clamping system that transmits the force to the reinforcement. The confining stress (46.7 kPa at the sample level), applied by ten small cylindrical masses, was kept constant through the test, and was measured by a load cell located between one of the cylindrical masses and the wood plate below. The tests were carried out at a constant displacement rate and volume, and the results were recorded using an automatic data acquisition system.

The main characteristics of the sand and of the uniaxial geogrid used in the tests are presented in Table 2.5 and Figure 2.11, and in Table 2.4, respectively. The test programme carried out to study the influence of the displacement rate, specimen size, soil height and sleeve length on the results of the pull-out tests with the geogrids in sand is presented in Table 2.10.

It is important to note the relevance of the measurement of displacements along extensible reinforcements such as geosynthetics. In fact, as mentioned before, in this type of material, the displacement during pull-out has two components, one corresponding to the relative movement between the reinforcement and the surrounding soil (displacement by shear strain on the soil–reinforcement interface) and another due to the reinforcement elongation, this component being more significant at the front part of the specimen. The

Table 2.10 Testing programme

Parameter under study	Sample length/sample width: m	Displacement rate: mm/min	Soil height above the sample/soil height below the sample: m	Sleeve length: m
Displacement rate	0.96/0.33	5.4 1.8 11.8 22.0	0.30/0.30	0.20
Sample length	0.80/0.33 0.96/0.33 1.12/0.33	5.4	0.30/0.30	0.20
Sample width	0.96/0.33 0.96/0.47 0.96/0.60	5.4	0.30/0.30	0.20
Soil height	0.96/0.33	5.4	0.20/0.20 0.30/0.30	0.20
Sleeve length	0.96/0.33	5.4	0.30/0.30	Without sleeve 0.20

Figure 2.24 Influence of the displacement rate: (a) pull-out force versus front displacement and (b) displacements by strain along the geogrid
Adapted from Lopes and Ladeira (1996a), with permission from Elsevier

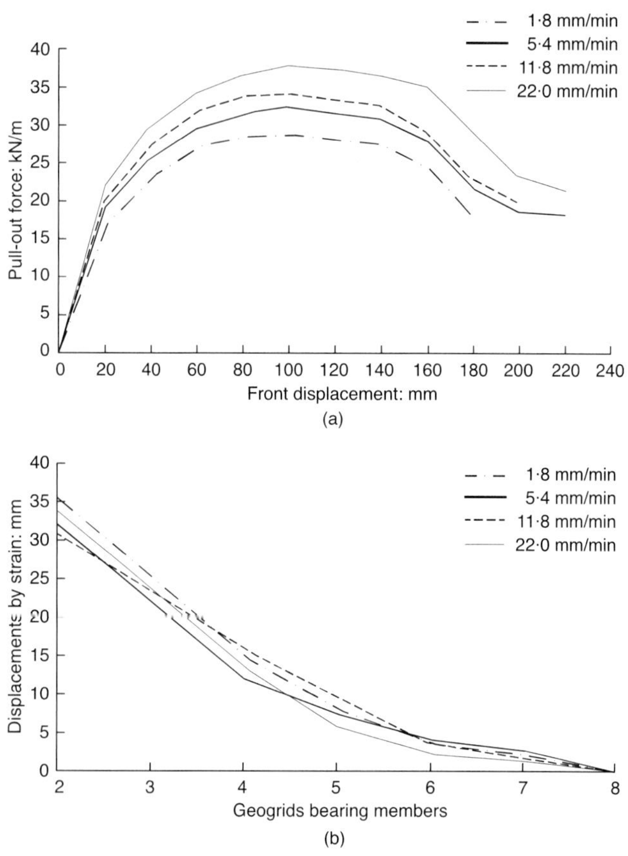

Figure 2.25 Influence of the specimen length on the geogrid pull-out behaviour
Data taken from Lopes and Ladeira (1996b)

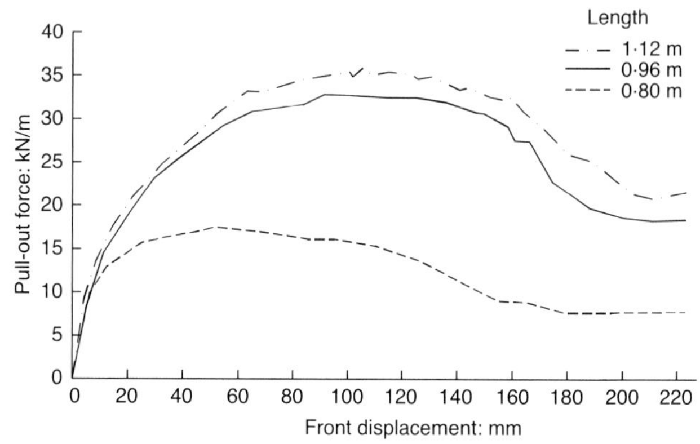

possibility of knowing each one of the displacement components of the movement separately allows for a better understanding of the soil–geosynthetic interaction phenomenon.

2.6.2.1 Influence of displacement rate

The influence of the displacement rate on the pull-out behaviour of the geogrid under study is shown in Figure 2.24. It can be seen that the maximum pull-out force of the reinforcement increases with the displacement rate, decreasing the front displacement necessary to mobilise the same pull-out force (Figure 2.24(a)). On the other hand, the displacement along the reinforcement induced by strain shows a tendency to reduce as the displacement rate increases (Figure 2.24(b)).

The observed behaviour leads to the conclusion that the increase in the pull-out force with the displacement rate results, at least, partially, from the increase in the reinforcement stiffness with the velocity, and not from the increase in the mobilised tangential stresses in the soil–reinforcement interface. Another factor responsible for the increase in the pull-out force with the displacement rate is the lower soil capacity for rearrangement with the increment in the velocity.

2.6.2.2 Influence of specimen size

The study of the influence of the specimen size on the pull-out test results is important for the definition of the relation between the dimensions of the apparatus and of the specimen in order to minimise the influence of the lateral boundaries on the test results. This relation is difficult to define because it depends not only on the dimensions of the equipment, but also on factors such as the characteristics of the reinforcement, the density and type of soil, the confinement pressure, the displacement rate and the height of the soil.

The influence of the specimen length on the pull-out test results is presented in Figure 2.25 in terms of the variation in the pull-out force, with front displacement for the three lengths tested (0.80, 0.96 and 1.12 m). It can be seen that the results are similar for the three tests in qualitative terms, but not in quantitative terms.

The maximum pull-out force increases from 15.3 to 32.2 kN/m (about 105%) when the specimen length changes from 0.80 to 0.96 m. However, when this parameter changes to 1.12 m, the increase in the maximum pull-out force is much smaller, from 32.2 to 35.6 kN/m (about 10.6%). The results show that there is no significant increase in the interface resistance when the length of the inclusion is greater than a certain value.

In order to define the maximum specimen width for its use in the apparatus, allowing for consideration of the negligible influence of its lateral boundaries on the test results, three tests were carried out with three different specimen widths (0.33, 0.47 and 0.60 m). The variation in the pull-out force with the front displacement is presented in Figure 2.26 for the three widths tested.

As the specimen width decreases, the maximum pull-out force increases slightly (28.8, 30.3 and 32.2 kN/m, respectively, for samples 0.60, 0.47 and 0.33 m wide). Taking into account the difference between the results of the present study (about 12% for the maximum pull-out force), it can be said that the influence of the specimen width on the test results is small. However, it can be seen that the reduction in the distance between the lateral boundaries of the

Figure 2.26 Influence of the specimen width on the geogrid pull-out behaviour
Data taken from Lopes and Ladeira (1996b)

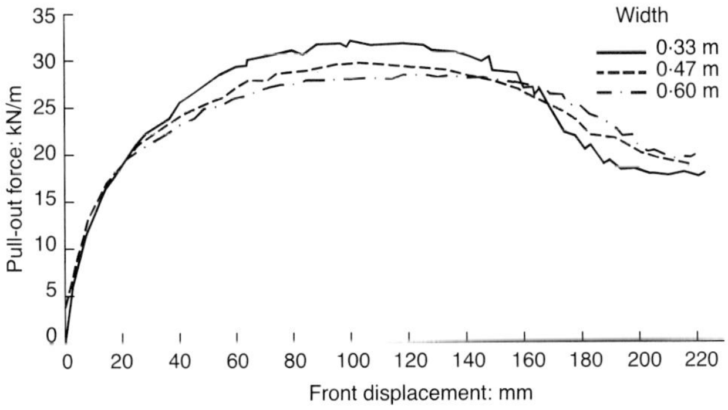

equipment and the specimen tends to reduce the soil–reinforcement interface resistance.

2.6.2.3 Influence of soil height

The distances between the upper and bottom boundaries of the apparatus and the reinforcement level can affect the resistance of the soil–reinforcement interface. This influence leads to an increase of the confining stress in the reinforcement, especially when the soil height is small and soil dilatancy is forbidden.

In the tests performed, the friction angle (ϕ) was defined as the arch of the tangent of the relation between the measured shear (τ) and normal (σ_n) stresses on the reinforcement interface ($=\tau/\sigma_n$) (the normal stress was considered to be the sum of the applied surcharge and the pressure due to the height of the soil above the reinforcement). In fact, the defined friction angle can be higher than the real one, as a result of the increase in the confining pressure that can occur in the pull-out tests due to the restriction of soil dilatancy. Another phenomenon that can occur in these tests is the development of shear forces between the soil and the stiff horizontal boundaries, especially the bottom one (Farrag *et al.*, 1993).

In order to study the influence of the soil height placed above and below the specimen on the pull-out response of the geogrid under consideration, two tests were carried out. In one test, the reinforcement was placed in the middle of soil 0.40 m high and, in the other test, the reinforcement was placed in the middle of soil 0.60 m high.

Figure 2.27 shows the geogrid pull-out response for both situations tested. For the test with soil 0.60 m high, the maximum pull-out force was 32.2 kN/m, and for the test with soil 0.40 m high, it was 33.9 kN/m (about 5.3% higher) (Figure 2.27(a)). The displacements by strain along the reinforcement for the maximum pull-out force measured were higher for the lower soil height tested (Figure 2.27(b)).

Based on the present test results, it can be said that for the soil heights considered there is no evidence of a significant influence of this parameter on the pull-out response of the geogrid. However, different soil heights must be tested to confirm the present results. Unfortunately, owing to geometrical limitations of the equipment used, it was not possible to test soil heights over and below the reinforcement higher than 0.30 m. Nevertheless, published pull-out test results, carried out with higher heights of soil surrounding specimens with similar dimensions to those tested in the present study, show an insignificant influence of soil heights greater than 0.30 m above and 0.30 m below the reinforcement (Farrag *et al.*, 1993).

Figure 2.27 Influence of soil height: (a) pull-out force versus front displacement and (b) displacements by strain along the geogrid
Data taken from Lopes and Ladeira (1996b)

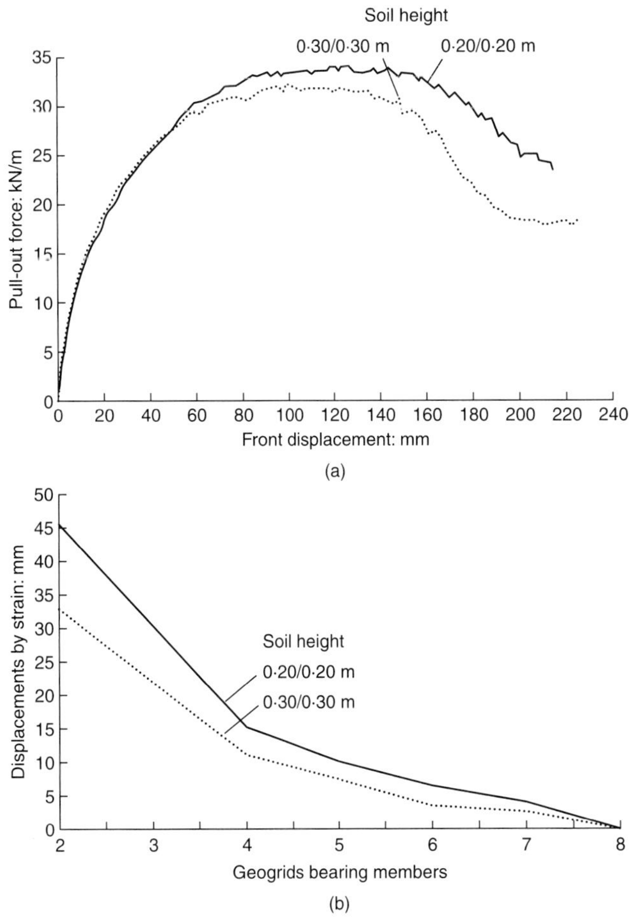

It must be emphasised that the length of the reinforcement plays an important role with regard to the influence of the top and bottom boundaries of the apparatus. In fact, as the length of the reinforcement increases in relation to a fixed height of the box, the influence of the horizontal boundaries also increases (Palmeira and Milligan, 1989).

From the test results obtained, it can be said that when the soil height above and below the reinforcement increases, there is a tendency for a reduction to occur in the resistance of the soil–reinforcement interface.

2.6.2.4 Influence of a sleeve

The interaction between the soil–reinforcement system and the stiff frontal wall of the pull-out box can affect the pull-out resistance of the reinforcement, especially when friction along that wall is high (Palmeira and Milligan, 1989). When the reinforcement is pulled out inside the box, there is an increase in the lateral pressure of the soil against the frontal wall that leads to an apparent increase in the pull-out resistance of the reinforcement.

There are some procedures, such as the lubrication of the internal face of the wall, or the placement of a sleeve inside the equipment at the level of the existing aperture in the front box wall, that can be followed in order to minimise the influence of the frontal wall of the equipment. The placement of the sleeve allows the transfer of the point of application of the pull-out force away from the frontal wall, in the interior of the soil mass (Farrag *et al.*, 1993).

In order to study the influence of the sleeve on the pull-out response of the geogrid, two tests were carried out, one without a sleeve and another with a 0.20 m-long steel sleeve. The geogrid pull-out responses for the two tests are presented in Figure 2.28.

Figure 2.28 Influence of sleeve: (a) pull-out force versus front displacement and (b) displacements by deformation along the geogrid
Data taken from Lopes and Ladeira (1996b)

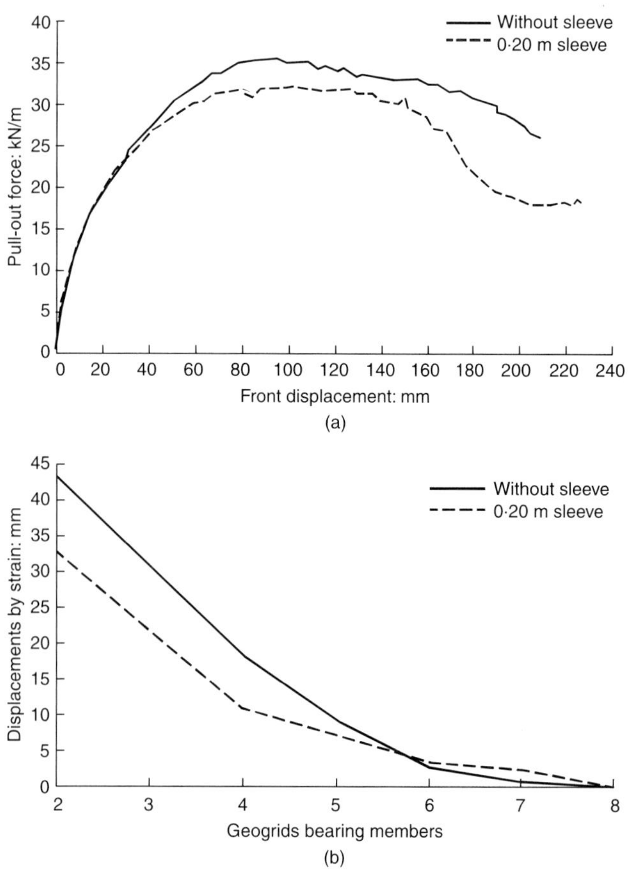

The absence of a sleeve leads to an increase in the maximum pull-out force by about 10%. This behaviour is due to the influence of the roughness of the frontal wall in the tests without a sleeve. In fact, the frictional forces developed on the front face of the box lead to an increase in the average vertical stresses on the reinforcement responsible for the increase in the pull-out forces.

The lateral pressures developed on the front wall are higher in the test without a sleeve. In this case, the restriction of geogrid movement is more effective, resulting in a higher mobilisation of shear stresses in the soil–geogrid interface near to the point of application of the pull-out force (Figure 2.28(b)) and to a lower mobilisation of the stresses in the back part of the reinforcement. The results showed (Lopes and Ladeira, 1996b) a lower pull-out movement in the test without a sleeve, which is in agreement with the higher mobilised resistance due to the increase in the state of stress, mainly, in the front part of the reinforcement, resulting from the influence of the stiff frontal wall of the equipment.

From the results obtained, it can be concluded that, for the equipment used, it is advisable to use a sleeve at least 0.20 m long in order to minimise the influence of the front wall on the pull-out test results.

2.7. Summary of main points

- Soil reinforcement with geosynthetics relies on the efficacy of soil–geosynthetic interaction, which is governed largely by the properties of the soil and the geosynthetic.
- Soil particle size assumes special importance when the reinforcement is a geogrid.
- Direct shear and pull-out are the most common laboratory tests available for the quantification of soil–geosynthetic interaction.
- The direct shear test standard, EN ISO 12957-1, needs to be modified in order to consider the dependence of the measured horizontal force from the vertical stress applied.
- The EN ISO 12957-1 standard should clarify that constant-area direct shear tests are not able to be performed with the lower half of the test box filled with soil.
- The EN ISO 12957-1 standard should recommend measurement of the vertical displacement of the load plate along the test and not only at the end of it.
- Pull-out tests performed in order to study the influence on the results of some test conditions and procedures highlighted the important roles of the displacement rate, the specimen length and the placement of a sleeve in the front wall of the test box.

2.8. Some common questions and their answers

Questions

1. Why is soil reinforcement with metallic inclusions more efficient than with geosynthetics?

2. Can only granular soils be used in soil reinforcement with geosynthetics?
3. Why does only the skin friction mechanism contribute to the soil–geosynthetic interface resistance when geotextiles and strips are used?
4. According to Equations 2.1 and 2.2, the soil–reinforcement interface shear strength in pull-out movement is double that in direct shear movement. Why?
5. To improve soil–geogrid interaction, the soil grain size should be as large as possible. Is it essential?
6. Can direct shear tests be used to characterise soil–geosynthetic interaction in pull-out movement?

Answers

1. Soil reinforcement with metallic inclusions is more efficient than that with geosynthetics because metallic inclusions are stiffer than geosynthetics, which leads to different soil states of stress in the reinforced zone: at the rest state of stress for metallic reinforcements and at the active state of stress for geosynthetic reinforcements.

2. No, other types of soils can be reinforced with geosynthetics (e.g. residual soils and cohesive soils). However, as soil reinforcement theories were first developed for granular soils and metallic reinforcements (by Terre Armée), the use of geosynthetics as soil reinforcements followed the same trend at the beginning. This is why mainly granular soils were reinforced with geosynthetics.

3. For geotextiles and strips, only the skin friction mechanism contributes to the soil–geosynthetic interface resistance because the lateral area of these materials can be considered completely solid, in contrast to what happens with geogrids, where the area of the apertures is much larger than the solid lateral area.

4. Equation 2.2 is twice the magnitude of Equation 2.1 because in pull-out movement both lateral areas of the reinforcement contribute to the interface shear resistance, while in direct shear movement, only one lateral area of the reinforcement contributes to that resistance.

5. No, there is a limit to the soil grain size, over which the grains cannot penetrate the geogrid apertures and mobilise the passive thrust against the grid-bearing members. In this situation, soil–geogrid interaction reduces.

6. No, because in direct shear tests the deformability of the geosynthetic cannot be considered, and this parameter is of the utmost relevance for soil–geosynthetic interaction in pull-out movement.

REFERENCES

American Society for Testing and Materials (1992) ASTM D5321-92. Standard test method for determining the coefficient of soil and geosynthetic or geosynthetic and geosynthetic friction by the direct shear method. ASTM, West Conshohoken, PA.

Bemben SM and Schulze DA (1998) The influence of equipment style and setup dimensions on sand/geomembrane direct shear test measurements. *Proceedings of the Sixth International Conference on Geosynthetics*, vol. 1, Atlanta, pp. 453–458.

Blümel W and Stoewahse C (1998) Geosynthetic interface friction testing in Germany – effect of test setups. *Proceedings of the Sixth International Conference on Geosynthetics*, vol. 1, Atlanta, pp. 447–452.

Blümel W, Stoewahse C, Dixon N, Kamugisha P and Jones DRV (2000) British–German cooperative research on geosynthetic friction testing methods. *Proceedings of the 2nd European Geosynthetics Conference*, vol. 2, Bologna, pp. 923–927.

British Standards Institution (1991) BS 6906. Methods of test for geotextiles, part 8: Determination of sand geotextile frictional behaviour by direct shear. BSI, London.

Chang JC, Hannon JB and Forsyth RA (1977) Pull resistance and interaction of earth work reinforcement soil. *Transportation Research Record* **640**: 1–7.

Dixon N, Jones DRV and Fowmes GJ (2006) Interface shear strength and its use in reliability-based landfill stability analysis. *Geosynthetics International* **13(1)**: 1–14.

Farrag K, Acar YB and Juran I (1993) Pull-out resistance of geogrid reinforcements. *Geotextiles and Geomembranes* **12(2)**: 33–159.

Gourc JP and Lalarokotson S (1997) *Research and Intercomparison Tests for the Harmonization of Standards on Geotextiles.* EC Measurement and Testing Programme Project 0169, Task 3.2 – Friction, Report No. 3.

Gourc JP, Lalarakotoson S, Muëller-Rochholtz H and Bronstein Z (1996) Friction measurement by direct shear or tilting process – development of a European standard. *Proceedings of the 1st European Conference on Geosynthetics – Eurogeo*, Maastricht, pp. 1039–1046.

Hayashi S, Makiuchi K and Ochiai H (1994) Testing methods for soil–geosynthetic frictional behaviour – Japanese standard. *Proceedings of the 5th International Conference on Geotextiles, Geomembranes and Related Products*, Singapore, pp. 411–414.

Hsieh C and Hsieh MW (2003) Load plate rigidity and scale effects on the frictional behavior of sand/geomembrane interfaces. *Geotextiles and Geomembranes* **21(1)**: pp. 25–47.

Hueckel SM and Kwasniewski JK (1961) Essais sur modele reduit de la capacite d'Ancrage d'elements rigides horizontaux, enfouis dans le sable. *Proceedings of the 5th International Conference on Soil Mechanics and Foundation Engineering*, Paris, pp. 431–434.

International Organization for Standardization (2004) EN ISO 12957-1. Geotextiles and geotextile-related products – Determination of friction characteristics – part 1: Direct shear test. European Committee for Standardization in collaboration with ISO, Brussels.

Jewell RA (1980) Some effects of reinforcement on the mechanical behaviour of soils. DPhil thesis, Cambridge University, Cambridge.

Jewell RA (1990) Reinforced bond capacity. *Géotechnique* **40(3)**: 513–518.

Jewell RA (1996) *Soil Reinforcement with Geotextiles.* Thomas Telford, London.

Jewell RA and Wroth CP (1987) Direct shear tests on reinforced sand. *Géotechnique* **37(1)**: 53–68.

Jewell RA, Milligan GWE, Sarsby RW and Dubois D (1984) Interaction between soil and geogrids. *Proceedings of the Conference on Polymer Grid Reinforcement*. Thomas Telford, London, pp. 18–30.

Jones DRV and Dixon N (1998) Shear strength properties of geomembrane/geotextile interfaces. *Geotextiles and Geomembranes* **16(1)**: 45–71.

Jones DRV and Dixon N (2000) A comparison of geomembrane/geotextile interface shear strength by direct shear and ring shear. *Proceedings of the 2nd European Geosynthetics Conference*, vol. 2, Bologna, pp. 929–932.

Juran I, Knochenmus G, Acar YB and Arman A (1988) Pull-out response of geotextiles and geogrids. *Proceedings of the Symposium on Geotextiles for Soil Improvement*. ASCE Geotechnical Special Publication, No. 18. American Society of Civil Engineers, Reston, VA, pp. 92–111.

Lopes MJFP (1998) Study of the influence of soil grain size and reinforcement structure on soil–geosynthetic interaction mechanisms. MSc thesis, University of Porto, Porto. (In Portuguese.)

Lopes ML and Ladeira M (1996a) Influence of the confinement, soil density and displacement rate on soil–geogrids interaction. *Geotextiles and Geomembranes* **14(10)**: 543–554.

Lopes ML and Ladeira M (1996b) Role of the specimen geometry, soil height, and sleeve length on the pull-out behaviour of geogrids. *Geosynthetics International* **3(6)**: 701–719.

Lopes ML and Silvano R (2010) Soil/geotextile interface behaviour in direct shear and pullout movements. *Geotechnical and Geological Engineering* **28(6)**: 791–804.

McGown A, Andrawes KZ and Al-Hasani MM (1978) Effect of inclusion properties on the behaviour of sand. *Géotechnique* **28(3)**: 327–346.

Nakamura T, Mitachi T and Ikeura I (1999) Direct shear testing method as a means for estimating geogrid–sand interface shear-displacement behaviour. *Soils and Foundations* **39(4)**: 1–8.

Ospina RI (1988) An investigation on the fundamental interaction mechanism of non-extensible reinforcement embedded in sand. MSc thesis, Georgia Institute of Technology, Atlanta, GA.

Palmeira EM and Milligan GWE (1989) Scale and other factors affecting the results of pull-out tests of grids buried in sand. *Géotechnique* **39(3)**: 511–524.

Peterson LM and Anderson LR (1980) *Pullout Resistance of Welded Wire Mesh Embedded in Soil*. Department of Civil Engineering, Utah State University, Logan UT.

Silvano RPSX (2005) Soil reinforced with geosynthetics – study of the relevant parameters of the behaviour. MSc thesis, University of Porto, Porto. (In Portuguese.)

Silvano R and Lopes ML (2005) Soil/geosynthetic interface characterization through direct shear tests. *Proceedings of the Sixth International Conference on Ground Improvement Techniques*, Coimbra, pp. 555–560.

Silvano R and Lopes ML (2006) Soil/geosynthetic interface characterization – the influence of some aspects of the test procedure. *Proceedings of the 8th International Conference on Geosynthetics*, vol. 4, Yokohama, pp. 1459–1462.

Stoewahse C, Dixon N, Jones DRV, Blümel W and Kamugisha P (2002) Geosynthetic interface shear behaviour: part 1. Test methods. *Ground Engineering* **35(2)**: 35–41.

Takasumi DL, Green KR and Holtz RD (1991) Soil–geosynthetics interface strength characteristics: a review of state-of-the-art testing procedures. *Proceedings of Geosynthetics '91*, vol. 1, Atlanta, pp. 87–101.

Wasti Y and Özdüzgün ZB (2001) Geomembrane–geotextile interface shear properties as determined by inclined board and shear box tests. *Geotextiles and Geomembranes* **19(1)**: 45–57.

Handbook of Geosynthetic Engineering
ISBN 978-0-7277-4175-2

http://dx.doi.org/10.1680/hge.41752.067

ice
Institution of Civil Engineers
publishing

Chapter 3
Filters and drains

A. H. Aydilek
Department of Civil and Environmental Engineering, University of Maryland, College Park, Maryland, USA

Geosynthetics are composed of plastic polymers, and have been commonly employed in civil and environmental engineering projects for the last 30 years to prevent erosion of soils without impeding the flow of water seeping through the soils. Geotextiles are increasingly being used to replace graded granular filters in many applications, because of their economy, consistent properties, ease of placement and comparable performance. Various different geosynthetics can also provide lateral drainage of percolating water in a construction application, and prevent the build-up of excess pore-water pressure in the soil. Several empirical criteria incorporating varying factors of safety are available for selection of geotextile filters; however, they are not directly applicable to high water content geomaterials. An extensive literature review has been made to gather information pertinent to filtration of natural soils and high water content geomaterials using geotextiles. Factors affecting the behaviour of filtration processes are presented. This chapter also provides a step-by-step methodology for designing filters and drains.

3.1. Filters

The purpose of a filter used in geotechnical engineering applications is to allow the passage of the water while maintaining the stability of the soil structure by preventing excessive migration of soil particles. Two types of filter are used in geotechnical engineering: granular filters and geotextile filters. Granular filters consist of quasi-randomly organised mineral particles of different sizes. Geotextiles consist of fibres organised in a quasi-regular pattern, in woven geotextiles, or a quasi-random pattern, in non-woven geotextiles.

There are many similarities between granular filters and geotextile filters. Both are porous and permeable. Geotextiles contain pores, which play the same role as the voids in soil, and filaments/fibres, which act in the same way as soil particles. Both types of filter can have different pore sizes, the size of the pores decreasing with increasing thickness of the filter. On the other hand, there are many differences between the two types of filter. The geometrical relationship between the filaments/fibres and pores in geotextile filters is more complex than the relationship between the particles and spaces in granular filters (soils), due to the shape and compressibility of the filaments/fibres (Christopher and Fischer, 1992). The estimation of the pore sizes in granular materials is very complex, and therefore they are given as size distributions. Another difference lies in the porosity of the filters; void spaces surround the moving particles in non-woven geotextiles, whereas these voids are more likely to act as channels in granular filters (Giroud, 1996).

3.2. Geotextile filters

In recent years the use of geotextile filters has increased due to their comparable performance, economy, consistent properties and ease of placement, and these materials are being used extensively to replace soil filters in transportation, geotechnical and environmental applications (Figure 3.1).

Geotextiles were first used as filters in 1958, when graded-soil filters and gravel blankets were replaced with geotextiles by the Carthage Mills Company (Rankilor, 1981). Today, use of geotextiles is required by regulations in several applications. Holtz *et al.* (1998) summarised the advantages of geotextile filters over granular filters as

- use of less drainage aggregate
- use of smaller drains
- possible elimination of collector pipes
- expediency of construction
- reduced risk of filter contamination
- reduced excavation
- less material waste.

3.3. Geotextile filtration mechanism

The classical criteria used in soil mechanics for granular filters such as sand filters were established as a result of the combination of laboratory findings and theoretical considerations. Designing geotextile filters is essentially the same as designing graded granular filters, as a geotextile is similar to a soil in that it has voids (pores) and particles (filaments in woven geotextiles, fibres in non-woven geotextiles).

Figure 3.1 Various areas of use for geotextiles in filtration applications
Adapted from Holtz *et al.* (1998)

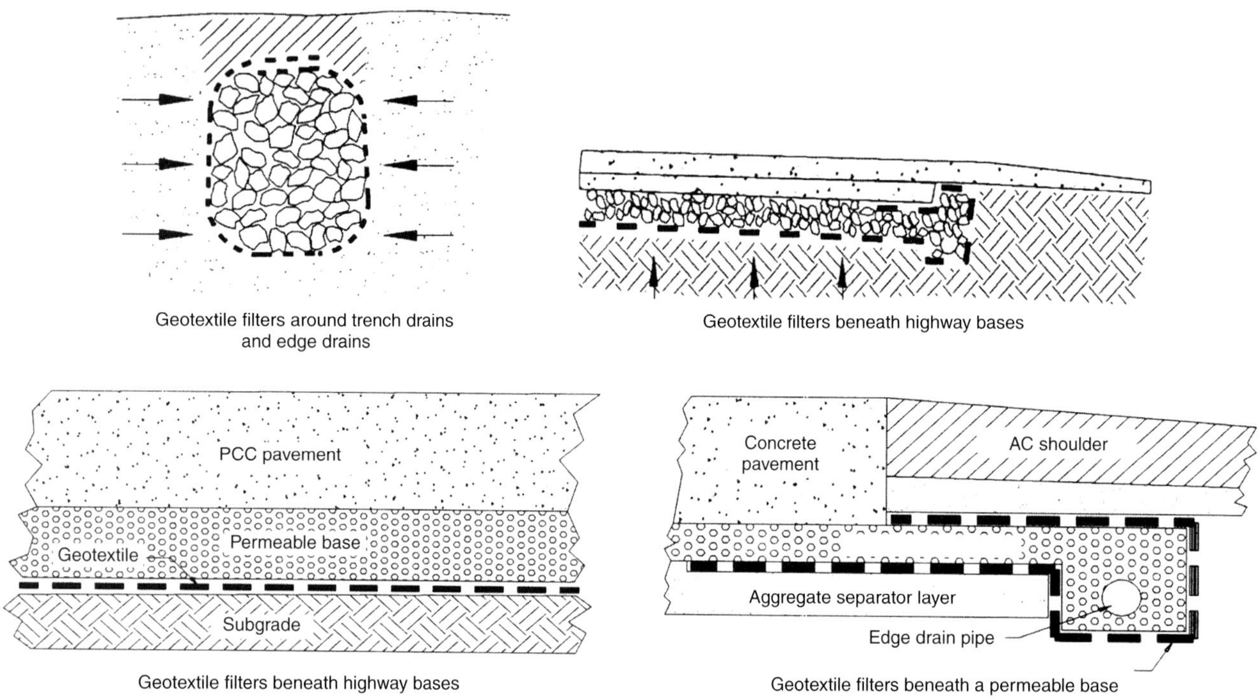

The estimation of pore sizes in granular materials is complex, and therefore pore sizes are determined from the particle size (grain size) distribution curve; whereas in geotextiles, pore sizes can be measured directly or indirectly. In general, non-woven geotextiles have a porosity of 90–95% compared with the much lower porosity of soil filters of 10–50% (Aydilek, 2006). As the pore size of a non-woven geotextile can be directly or indirectly measured, a relatively simple relationship can be developed between the pore size of the geotextile O_x and the particle size of the soil retained D_y; that is the filter retention criterion O_x/D_y. In the past, considerable effort has been devoted to experimental studies of geotextile filters. Although the mechanism for retention by geotextiles is similar to that in granular materials, the criteria for hydraulic conductivity and clogging are different.

There are three basic filter criteria used for the proper selection of a geotextile

- a retention requirement, to prevent the migration of soil particles through the geotextile
- a clogging requirement, to ensure free flow of liquids through the geotextile
- a hydraulic conductivity requirement, to ensure that the geotextile will adequately meet the clogging and retention criteria throughout the lifetime of the structure.

However, survivability and durability criteria, which ensure that the geotextile will survive during the construction and perform its function throughout the project, should also be considered (Christopher and Fischer, 1992).

When a geotextile filter is used, the hydraulic conductivity, survivability and durability criteria are considered in addition to the retention criteria for a granular aggregate filter. Carroll (1983) reported that geotextiles would fail due to a lack of sufficient hydraulic conductivity, even if retention and clogging criteria were met. The survivability criterion relates to the protection of the geotextile and materials adjacent to it from damage during installation. In addition, contact stresses, which will be exerted on the geotextile by the adjacent materials due to compacting processes or confining pressures, should be considered. For example, angular drainage media (e.g. angular riprap) typically exert high contact stresses on the geotextile filter, whereas rounded drainage media exert moderate contact stresses.

Durability considerations address the resistance of a geotextile to ultraviolet light and adverse chemical environments. Geotextile filters that are located near the ground surface (i.e. in shoreline protection), or exposed to sunlight during construction, must be protected using antioxidants such as carbon black or titanium dioxide. Geotextile filters that are exposed to chemicals (i.e. in landfill leachate collection systems) should be composed of chemically resistant fibres.

3.3.1 Retention criterion

A filter must retain the soil and allow the water to pass through. Therefore, a good filter has openings large enough to allow water to flow almost freely, but small

Figure 3.2 A graded filter zone (bridging network)

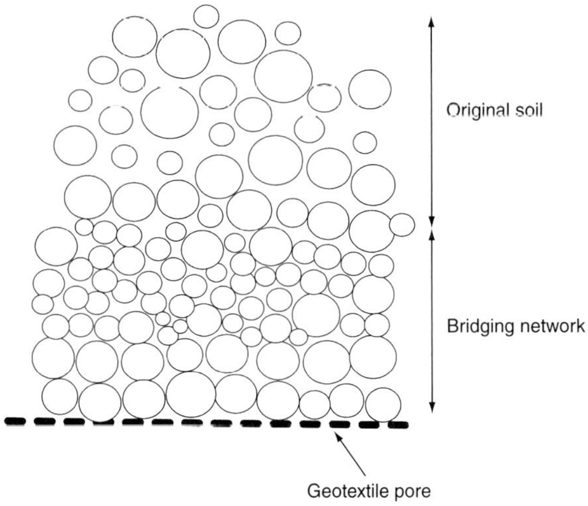

enough that particle skeleton, which gives the soil structure its stability, is not disturbed as a result of the loss of fine particles. Experiments have shown that, if a filter can prevent the movement of the larger particles of the soil, these particles will collect over the openings of the filter and create smaller openings to trap the smaller particles of the soil (Kutay and Aydilek, 2005). As a practical limit, the filter needs to restrain only 15% of the soil, or the D_{85} size (Cedergren, 1989).

Bhatia (1995) reported that the retention performance of a geotextile depends on the direction of flow. When flow is unidirectional (i.e. in retaining walls and earth dams), particles that are smaller than the geotextile pores can migrate into and through the geotextile under the influence of seepage forces. If the soil has a range of particle sizes, the larger particles adjacent to the surface of the geotextile may, over time, form a bridging network, and this will tend to impede the migration of the smaller particles. This structure formed on the geotextile is called 'bridging network' or 'graded filter zone' (Figure 3.2). In this case, especially in non-woven geotextiles, the large particles bridging across the geotextile pores retain slightly smaller soil particles, which, in turn, retain even smaller soil particles. Piping of soil through the geotextile occurs if the pore openings of the geotextile (i.e. O_{90}, O_{95}) are larger than the largest particles of soil. This process is usually called 'internal erosion', and it changes the internal stability of the soil. Lafleur *et al.* (1989) suggested that a piping rate of less than 2500 g/m^2 through a granular filter does not affect the soil stability. This rate is also widely accepted for geotextile filters (Fischer, 1994; Aydilek, 2006).

In the case of reverse flow conditions (i.e. in edge drains or coastal erosion control structures) and/or challenging soil conditions (e.g. high water content geomaterials, including sludges and ash slurries), the large opening sizes are not the only critical pore sizes, and the relatively smaller pore sizes may also be critical. In these conditions, soil bridges cannot form. Soil slurry may develop near the geotextile and native soil is not filtered by this slurry. The fine pores of the geotextile are the most critical in preventing the soil slurry from washing through the geotextile.

Numerous retention criteria have been proposed for geotextiles, these being mainly adapted from the existing granular soil filter criteria. The retention criteria have been developed either from laboratory methods (empirical) or by converting the granular soil grain size to the geotextile equivalent pore size. Table 3.1 summarises the existing geotextile retention criteria.

3.3.2 Clogging criterion

The second duty expected of a geotextile filter is the anti-clogging performance. If the largest pore openings in the geotextile are much smaller than the smallest particles of the soil, then the fines in the soil close to the geotextile surface will be unable to pass through the geotextile. This will prevent the formation of an effective filter zone, and may lead to blinding, blocking or clogging of the geotextile. These mechanisms are shown in Figure 3.3. Blocking is a phenomenon encountered in woven geotextiles. Coarse particles migrate to the upstream face of the geotextile and stop at the entrance of the pores (Figure 3.3(a)). Blocking is expected when a geotextile is selected to retain particles contained in very low concentrated suspensions, or when there is a lack of contact between the soil and the geotextile (Rollin and Lombard, 1988). Clogging occurs in non-woven geotextiles mostly due to particles, sediments, organic matter migrating between its fibres, or bacterial growth (Figure 3.3(b)). Clogging relates to the ability of soil particles to clog a majority of the geotextile pores. Here the word 'majority' is important, as, even with a low porosity, a clogged geotextile will still have a higher hydraulic conductivity than that of soil (be up to 80% clogging according to Bhatia (1995)).

Blinding also occurs in non-woven geotextiles due to a build up of soil fines on the upstream side of the geotextile. It is mostly encountered when the largest pore openings in the geotextile are much smaller than the smallest particles of the soil (Figure 3.3(c)). In this case, an impervious layer of soil particles is established between two layers of coarse particles, preventing the formation of a filter zone (Bhatia, 1995). Unlike clogging, which causes a change in the geotextile pore sizes, blinding is a process that occurs at the soil–geotextile interface.

Even though the potential for clogging is closely related to the hydraulic conductivity criterion, research has shown that satisfying the retention and hydraulic conductivity criteria does not necessarily preclude clogging (Haliburton and Wood, 1982; Carroll, 1983). The long-term performance of geotextiles is based on their clogging potential, and the existing clogging criteria are based on either soil–geotextile filtration tests, the porosity or percent open area of the geotextile, or correlations between the apparent opening size of

Table 3.1 Existing geotextile retention criteria
Data taken from Fischer *et al.* (1990), Christopher *et al.* (1993) and Holtz *et al.* (1998)

Source	Criterion	Remarks
Calhoun (1972)	$O_{95}/D_{85} \leq 1$	Woven, soils with 50% passing #200
	$O_{95} \leq 0.2$ mm	Woven, cohesive soils
Zitscher (1975) (from Rankilor, 1981)	$O_{50}/D_{50} \leq 1.7$–2.7	Woven, soils with $C_u < 2$, $D_{50} = 0.1$–2 mm
	$O_{50}/D_{50} \leq 25$–37	Non-woven, cohesive soils
Ogink (1975)	$O_{95}/D_{90} \leq 1$	Woven
	$O_{95}/D_{90} \leq 1.8$	Non-woven
Sweetland (1977)	$O_{15}/D_{85} \leq 1$	Non-woven, soils with $C_u = 1.5$
	$O_{15}/D_{15} \leq 1$	Non-woven, soils with $C_u = 4.0$
Schober and Teindl (1979)	$O_{90}/D_{50} \leq 2.5$–4.5	Woven and thin non-woven, C_u dependent
	$O_{90}/D_{50} \leq 4.5$–7.5	Thick non-woven, C_u dependent
Millar *et al.* (1980)	$O_{50}/D_{85} \leq 1$	Woven and non-woven
Rankilor (1981)	$O_{50}/D_{85} \leq 1$	Non-woven, and soils with $0.02 < D_{85} < 0.25$
	$O_{15}/D_{15} \leq 1$	Non-woven, and soils with $D_{85} < 0.25$
Giroud (1982, 1988)	$O_{95}/D_{50} \leq (9$–$18)/C_u$	Dependent on C_u and density, assumes fines migrate for large C_u
Carroll (1983)	$O_{95}/D_{85} \leq 2$–3	Woven and non-woven
Faure *et al.* (1986)	$O_f/D_{85} \leq 1$–2	Dependent on critical nature of project
French Committee of Geotextiles and Geomembranes (1986)	$O_f/D_{85} \leq 0.38$–1.25	Dependent on soil type, compaction, hydraulic and application conditions
Fischer *et al.* (1990)	$O_{50}/D_{85} < 0.8$	Based on geotextile pore size and distribution, dependent on C_u of soil
	$O_{50}/D_{15} < 1.8$–7.0	
	$O_{50}/D_{50} < 0.8$–2.0	
Holtz *et al.* (1998)	$O_{95}/D_{85} \leq 1.0$	$1 < C_u < 2$; $D_{50} > 74$ mm
	$O_{95}/D_{85} \leq 1.8$	$1 < C_u < 2$; $D_{50} < 74$ mm
	$O_{95}/D_{85} \leq 1$–2	$2 < C_u < 8$; $D_{50} > 74$ mm
	$O_{95}/D_{85} \leq 1.8$	$2 < C_u < 8$; $D_{50} < 74$ mm
	$O_{95}/D_{15} \leq 1$ or	Dynamic pulsating and cyclic flow, if soil can move beneath geotextile
	$O_{50}/D_{85} \leq 0.5$	

C_u, coefficient of uniformity $= D_{60}/D_{10}$; O_{15}, O_{50}, O_{90} and O_{95} are 15%, 50%, 90% and 95% of the opening size of the geotextile, respectively; D_{85}, D_{50} and D_{15} are diameters corresponding to fines of 85%, 50% and 15% respectively; O_f, filtration opening size of the geotextile, generally referred to as O_{95}.

the geotextile and the surrounding soil. The latter two were developed as alternatives to filtration tests, and correlations are used from granular filter criteria. Table 3.2 summarises the existing geotextile clogging criteria.

3.3.3 Hydraulic conductivity criterion

The hydraulic conductivity criterion requires that the hydraulic conductivity of the composite layer, comprising the geotextile and bridging network, prevents excess pore-water pressure from developing and allows for an adequate flow capacity. Flow is not impeded at the soil–geotextile interface if the permeabilities are at least equal. Such a condition allows the hydraulic gradient in the geotextile to be equal to that throughout the soil, as long as there is a continuous flow, and thereby no volume changes occur in the soil (i.e. the flux remains constant with time). In general, the hydraulic conductivity of a soil–geotextile system is controlled by the soil, because of its lower hydraulic conductivity compared with that of the geotextile. It is known that geotextiles having the lowest hydraulic conductivity are equivalent in hydraulic conductivity to medium sands (Bhatia, 1995). The existing hydraulic conductivity criteria (Table 3.3) are ratios between the permeability of the soil and that of the geotextile.

Giroud (1982) suggested that hydraulic conductivity of a geotextile should be approximately ten times larger than the hydraulic conductivity of a soil. Later, this statement has been argued by others (Christopher *et al.*, 1993) and geotextile is required to be at least more permeable than the soil it is filtering.

3.4. Factors affecting geotextile filter behaviour

There are many factors affecting the filtration performance of geotextiles. These factors can be summarised as (Aydilek

Figure 3.3 (a) Blocking, (b) clogging and (c) blinding mechanisms

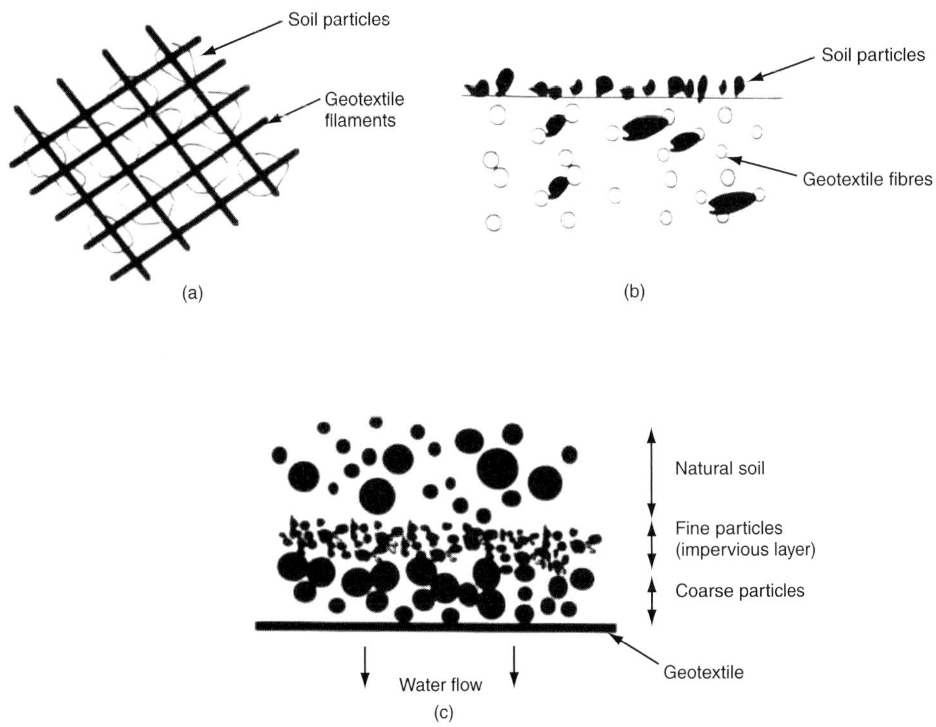

and Edil, 2008) geometrical factors (e.g. soil grain size and shape, and geotextile pore structure parameters), physical factors (e.g. soil friction, cohesion, overburden and shear stresses, and geotextile thickness and wetting characteristics) and external factors (e.g. viscosity of water, hydraulic gradient, flow capacity and biological effects). Among these factors, the pore-structure parameters of a geotextile (i.e. percent open area and pore-size distribution) have a profound effect on filtration. They have been included in all the existing filter selection criteria and are discussed briefly below.

3.4.1 Pore-size distribution of a geotextile

Because the flow of water through a soil and a geotextile is really a pore phenomenon, geotextile filter design should fundamentally be based on the pore-size distribution of the filter. Thus, if the filtration performance of a geotechnical filter is to be assessed properly, accurate information is needed on the distribution of pore openings or pore channels in the filter (Aydilek *et al.*, 2007). Another complicating factor is the constriction pore size, especially in non-woven geotextiles. As defined by Fischer (1994), the constriction pore size C_x is the minimum pore size of a flow channel, and it is the critical parameter in the filtration response of a geotextile. Due to their tortuous flow paths, the constriction pore sizes of non-woven geotextiles may be located at any depth inside the geotextile. Vermeersch and Mlynarek (1996), Giroud *et al.* (1998) and Aydilek *et al.* (2005) have confirmed that the constriction pore size of a geotextile impacts its filtration performance, and if accurately measured, it should be used in designing a geotextile filter.

The pore-size distribution is usually established by hydrodynamic sieving, in which case it is called the 'filtration opening size' (FOS, by the Canadian General Standards Board), or from dry sieving, in which case it is called the 'apparent opening size' (AOS) (in ASTM D4751-04 (ASTM, 2004)). Sometimes dry sieving is also used to determine pore sizes smaller than the AOS or O_{95}. However, this method is not very accurate due to problems in the testing procedure (Van der Sluys and Dierickx, 1991; Giroud, 1996). The values obtained with the two methods are not the same, and a comparison (Faure *et al.*, 1986) has shown that the dry sieving method gives values 60% larger than those measured using the hydrodynamic method for non-woven geotextiles. Other methods used to determine the pore size of geotextiles are wet sieving, the bubble point method and the mercury intrusion test (Prapaharan *et al.*, 1989). Aydilek *et al.* (2007) showed that the bubble point test provides highly accurate values, and the sizes measured using this method can be called 'constriction sizes'. An alternative method for determining geotextile pore and constriction sizes is image analysis. This method is generally used for non-woven geotextiles and captures an image of the microstructure of the geotextile. Various measurements, including the number of fibres, void spaces and pore sizes, can then be inferred from the image (Aydilek *et al.*, 2005).

3.4.2 Porosity (per cent open area) of a geotextile

The minimum porosity (for non-woven geotextiles), or percent open area (POA) (for woven geotextiles), of a geotextile is also important for determining the clogging potential in filtration applications. Porosity is the ratio of void volume

Table 3.2 Existing geotextile clogging criteria
Data taken from Christopher *et al.* (1993)

Source	Criterion	Remarks
Calhoun (1972), Haliburton and Wood (1982), Giroud (1982), Haliburton *et al.* (1982), Carroll (1983), Holtz *et al.* (1998), Koerner (2005)	Perform soil–geotextile tests	Critical or severe applications
Millar *et al.* (1980)	$O_{50}/D_{15} \geq 1$	Less critical or non-severe conditions
Haliburton *et al.* (1982)	Perform soil–geotextile tests	Less critical or non-severe conditions
Holtz *et al.* (1998)	$O_{95} \geq 3$, D_{15} for $C_u \geq 3$ Specify geotextile with maximum opening size from retention criteria for $C_u \leq 3$	Less critical or non-severe conditions
French Committee of Geotextiles and Geomembranes (1986)	$O_f \geq 4D_{15}$	Less critical or non-severe conditions
Fischer *et al.* (1990), Fischer (1994)	Specific minimum pore size for soils containing fines: $O_{15}/D_{15} \geq 0.8$–1.2 $O_{50}/D_{50} \geq 0.2$–1.0	Less critical or non-severe conditions
Luettich *et al.* (1992)	$C_u > 3$ Use a geotextile with a maximum AOS from the retention criterion	Less critical or non-severe conditions
Heerten (1993)	$O_{90\ \mathrm{actual}} \geq O_{90\ \mathrm{design}}$	Less critical or non-severe conditions
Calhoun (1972), Koerner (2005)	Specific minimum open area Woven, POA $\geq$ 4–6%	Less critical or non-severe conditions
Holtz *et al.* (1998), Koerner (2005)	Specific minimum open area Non-woven, porosity $\geq$ 30%	Less critical or non-severe conditions

C_u, coefficient of uniformity $= D_{60}/D_{10}$; O_{15}, O_{50}, O_{90} and O_{95} are 15%, 50%, 90% and 95% of the opening size of the geotextile, respectively; D_{85}, D_{50} and D_{15} are diameters corresponding to fines of 85%, 50% and 15% respectively; O_f, filtration opening size of the geotextile, generally referred to as O_{95}; AOS, apparent opening size; POA, percent open area.

to total volume, and is related to the ability of the liquid to flow through the geotextile. However, it is rarely measured directly, and is therefore calculated from the geotextile properties using the equation

$$n = 1 - m/\rho t \tag{3.1}$$

where n is the porosity, m is the mass per unit area, ρ is the density of the geotextile polymer, and t is the thickness of the geotextile. Research has supported the use of Equation 3.1 and has indicated that the porosity of a geotextile decreases with increasing mass per unit area (Koerner, 2005).

The POA for woven geotextiles is a comparison of the total open area to the sample area. For woven (primarily monofilament) geotextiles, the total porosity may not represent the pore volume influencing the flow through the geotextile, due to the surficial porosity of these geotextiles (Fischer, 1994). Therefore, the POA, a comparison of the total open area (the void spaces between adjacent threads) to the total specimen area, is a better indicator of pore space (Calhoun, 1972). A contemporary way of determining POA is to use image analysis (Aydilek and Edil, 2004). POA of woven geotextiles varies from 0 to 36%, with a common range of 6–12%.

In order to obtain an acceptable filtration performance, a porosity of greater than 30% is required for non-woven geotextiles, whereas woven geotextiles should have a minimum POA of 4–6% (Fischer *et al.*, 1990; Holtz *et al.*, 1998; Koerner, 2005). It is reported by DeBerardino (1993) that the porosity of needle-punched geotextiles is sensitive to applied normal loads. Both the porosity and the POA can be determined by means of standard tests. However, these parameters should not be used alone as filtration criteria for geotextiles, and without addressing the soil retention and clogging criteria, the porosity and POA are meaningless.

3.4.3 Long-term performance of geotextile filters

The long-term performance of a geotextile filter can be defined as the prevention of its failure (blocking, blinding, clogging or piping) as a filter. Geotextiles, particularly the non-woven type, have been successfully used in different geotechnical and hydraulic applications (e.g. coastal structures, water liners, railroads, dams, edge drains, landfills), and

Table 3.3 Existing geotextile hydraulic conductivity criteria
Data taken from Christopher and Fischer (1992)

Source	Criterion	Remarks
Calhoun (1972), Schober and Tendl (1979), Haliburton *et al.* (1982), Carroll (1983), Holtz *et al.* (1998)	$k_g \geq k_s$	Steady-state flow, non-critical non-severe conditions
Carroll (1983), Holtz *et al.* (1998)	$k_g \geq 10k_s$	Critical applications and severe soil or hydraulic conditions
Giroud (1982)	$k_g \geq 0.1k_s$	No factor of safety
French Committee of Geotextiles and Geomembranes (1986)	$\psi \geq 10^3 k_s$ to $10^5 k_s$	Critical 10^5, less critical 10^4, clean sand 10^3
Giroud (1988)	$k_g \geq ik_s$ where $i \geq 1.0$ $k_g \geq k_s$ where $i \leq 1.0$	
Koerner (2005)	$\psi_{geotextile} \geq \psi_{required}/FS$	The factor of safety (FS) is based on the application and the soil conditions, and ranges from 6 to 25
Lafleur *et al.* (1993)	$k_g \geq 20k_s$	
De Berardino (1993)	$\psi_{geotextile} \geq i\psi_{required}$	

k_g, hydraulic conductivity of the filter (geotextile); k_s, hydraulic conductivity of the soil; ψ, permittivity.

studies conducted to investigate their long-term performance after years indicated very little to no mechanical or hydraulic degradation (Black and Holtz, 1999; Aydilek and Edil, 2003).

There are short- and long-term factors that can lead to a reduction in the hydraulic conductivity or to clogging of a geotextile. Short-term factors include the magnitude of stress acting on the geotextile and the type of geotextile. Different types of geotextile have different responses to loading conditions and have different compressibilities; for example, non-woven needle-punched geotextiles are more compressible than non-woven heat-bonded geotextiles. Long-term factors include chemical and biological clogging of geotextiles or the soil–geotextile system. Chemical clogging occurs when various compounds are present in the permeant precipitate. The deposition of salts such as carbonates and sulphates can be a cause of chemical clogging during filtration of water/suspended soils through the geotextile. Saline water in coastal structures and highly alkaline groundwater can also lead to the deposition of a chemical layer (sodium, calcium or magnesium) on a geotextile (Sansone and Koerner, 1992).

Biological clogging occurs as a result of bacterial growth, both on and in a geotextile. It commonly occurs in landfills; there are sufficient nutrients and temperature for bacterial growth. Geotextiles in solid-waste landfill structures are exposed to an environment that affects their filtration performance. Little is known about the long-term functioning of geotextile filters at the base of landfills where biological and chemical processes cause the blocking or clogging of the geotextile. In leachate filtration in particular, the access of bacteria to nutrients can impede the flow due to the formation of ochre, bacterial adhesion and biofilms, which may ultimately clog the filters.

The ASTM D1987-07 (ASTM, 2007) test method is generally used to determine the compatibility of soil–geotextile systems; however, the test requires parallel chemical and bacteriological studies (Almeida *et al.*, 1995). With leachate as the permeating liquid instead of water, the filtration becomes more complex due to the ingredients of the leachate. Suspended solids in solid waste and sludges can be as high as 50 000 mg/l, and microorganisms in domestic waste leachate can result in biochemical oxygen demand (BOD_5) values as high as 40 000 mg/l (typical values in landfills are 3000–20 000 mg/l). This problematic behaviour can be solved temporarily by backflushing. However, reclogging cannot be prevented and, in general, water backflushing is more effective than leachate, nitrogen gas or vacuum backflushing (Koerner and Koerner, 1992). Woven geotextiles with a POA less than 4% are the worst and non-woven needle-punched geotextiles are usually the best in terms of clogging resistance, and sand–geotextile systems usually exhibit more biological clogging than the geotextiles alone. The flow reduction due to clogging is usually of the order of 20% in anaerobic conditions and 12–100% in aerobic conditions. The various processes of biologically-produced sediment formation in filtration/drainage systems are: slime and sheath formation, biomass formation, ochering, iron sulphide formation and carbonate deposition (Moore, 1980; Shuckrow *et al.*, 1981; Young *et al.*, 1982; Bass *et al.*, 1984; Rohde and Gribb, 1990; Mlynarek and Rollin, 1995; Palmeira *et al.*, 2008).

These findings are supported by the research of Corcoran and Bhatia (1996), who looked at the five-year performance of geotextiles in Fresh Kills Landfill, and the laboratory study by Kossendey *et al.* (1996), in which 14 different landfill leachates were used to determine the biological clogging of geotextiles. These studies found decreases in the permittivity of one and three orders of magnitude, respectively. Both studies indicate that the growth of microbes reduces the permittivity of geotextiles, but does not lead to complete clogging and degradation of the geotextile. The BOD_5 value (457 mg/l) of the Fresh Kills Landfill leachate was many times lower than typical landfill values (3000–20 000 mg/l), and no biofilm was observed on micrographs of exhumed geotextile, which is comparable to the findings of Koerner and Koerner (1992).

In summary, studies of the biological clogging of geotextiles indicated that

- microorganisms will easily grow on geotextiles, as commercially available geotextiles are not coated with biocides (chemicals used to prevent the formation or growth of bacteria)
- microorganisms that cause the biological clogging of geotextile filters can be present in nature, and cannot be eliminated from leachate collection systems
- the amount of biological clogging can be reduced by minimising the supply of nutrients, and this can be done by pretreating the waste in order to decrease the carbon content of the leachate (Kossendey *et al.*, 1996).

3.5. Laboratory filtration performance tests

Compared to soil filters, geotextile filters are more delicate, due to their relatively low thickness. The geotextile filter acts to force the upstream soil layer to become its own filter; that is, the upstream layer forms a 'self-filter' or 'filter cake'. If such a filter is not formed, the fine fraction of the soil may get trapped (clogging) or pass through the geotextile, leading to soil losses (piping). A number of laboratory tests have been developed to examine these unacceptable situations: the gradient ratio test, the long-term flow test, the filter press test, the biological clogging test, the hydraulic conductivity ratio test, the permittivity test, the fine-fraction filtration test and the dynamic filtration test. The advantages and disadvantages of these tests are summarised by Sansone and Koerner (1992) and given in Table 3.4. The biological clogging test (ASTM D1987-07 (ASTM, 2007)) is a version of the long-term flow test. The tests generally involve a sample of the soil and geotextile of some dimension and condition, and monitoring of the performance of the system after application of a hydraulic condition. The performance tests usually reflect the physical (compaction characteristics) and hydraulic conditions (type and direction of flow) in the field. Microscopic analyses (image analyses and scanning electron microscope analyses, SEM) are also widely used in assessment of the clogging and/or the blocking and blinding processes (Aydilek 2006).

3.5.1 Gradient ratio test

The gradient ratio (GR) test (ASTM, 2006) is a short-term test that can be performed in 24 hours. In the test, the geotextile and soil are placed in a specially designed apparatus, and flow is maintained under a constant hydraulic gradient. A GR (clogging ratio) is defined as the ratio of the hydraulic gradient of the soil–geotextile system to the hydraulic gradient of the soil itself ($GR = i_{soil\text{-}GT}/i_{soil}$). The gradient ratio varies with the type of geotextile and soil gradation. A ratio greater than 1.0 signifies clogging according to the American Society for Testing and Materials (ASTM) standards (ASTM, 2006); however, the Civil Works Construction Guide (1977) established a maximum acceptable ratio of 3.0. Many researchers agreed with this ratio, and a value of 3.0 was confirmed in several testing programmes (Haliburton and Wood, 1982; Koerner and Ko, 1982; Carroll, 1983; Aydilek and Edil, 2003; Kutay and Aydilek, 2005).

Studies have shown that the soil retention and hydraulic conductivity criteria are important in the proper selection of a geotextile for a filtration design, and these can be evaluated by means of gradient ratio tests. Fannin *et al.* (1994) performed gradient ratio tests on non-woven geotextiles and soils having a range of gradations. They modified the ASTM test apparatus by providing additional ports to define the water-head distribution in a better way, and to identify any blinding that occurs at the top of the sample. Results obtained in this study showed excellent to reasonable agreement with existing soil retention criteria for uniformly graded soils, but poor agreement with those for broadly graded soils. In the design criteria, for both gradations of soil, D_{85} was used as the representative soil particle size and D_{50} was considered suitable as an indicator of the piping potential. It was also noted that it would be reasonable to accept a hydraulic conductivity in the soil–geotextile composite that was less than that in the protected soil ($GR > 1$) if it could be demonstrated that no overall impeded seepage of flow would result, as the soil–geotextile composite layer was relatively thin.

Haliburton and Wood (1982), Aydilek and Edil (2003, 2004) and Kutay and Aydilek (2005) conducted gradient ratio tests, and reported that the POAs of woven geotextiles were almost directly related to the clogging potential, whereas the apparent opening sizes of non-woven geotextiles have no relationship to the clogging potential. Monofilament woven geotextiles were the best in terms of clogging resistance, and the highest level of clogging was observed for non-woven heat-bonded geotextiles.

The gradient ratio test is useful for understanding the filtration performance of soil–geotextile systems. However, the test has its problems and complications, including: the poor long-term stability of the gradient ratio value; piping along the test cylinder wall; air bubbles in the soil and manometer ports, geotextile and tubing system; and the need to use a high-capacity water-de-airing system.

3.5.2 Long-term flow test

In the long-term flow test, a permeameter is used to measure the soil retention and water flow through a soil–geotextile system (Koerner and Ko, 1982). The test is a simplified version of the gradient ratio test; the piezometer ports are removed and only a single hydraulic gradient is used. The soil sample is placed on a geotextile installed at the bottom

Table 3.4 Advantages and disadvantages of soil–geotextile filtration tests
Data taken from Sansone and Koerner (1992)

Method	Principle	Advantages	Disadvantages
Gradient ratio test	Flow downward through the soil and geotextile at a constant head to measure the hydraulic gradients	■ ASTM D5101-01 (ASTM, 2006) ■ Rapid results	■ Air pockets in the tubes ■ Instability for fine-grained soils
Long-term flow test	Flow downward through the soil and geotextile at a constant head to measure the flow rate over time	■ Straightforward set up ■ Simple measurements ■ Direct conclusions	■ Long time to reach equilibrium ■ Biological growth in the system ■ Uncontrolled test conditions
Fine fraction filtration test	Slurry of the fine fraction of soil is directed at a saturated geotextile system	■ Fines are examined directly ■ Soil piping is readily assessed ■ Rapid results	■ Unrepresentative of the entire soil ■ High hydraulic conditions ■ Challenging test conditions
Filter press test	Slurry of the soil is filtered through a saturated geotextile system, under the desired air pressure	■ Fines are examined directly ■ Filter-cake formation can be assessed ■ Rapid results	■ Unrepresentative of the entire soil
Permittivity test	The hydraulic conductivity of water of the geotextile is measured by determining the permittivity, using the constant or falling head method	■ Permittivity is readily assessed ■ Easy interpretation of results	■ Requirements such as preconditioning of the geotextile, temperature and the use of de-aired water ■ No soil specimen
Hydraulic conductivity ratio test	A triaxial permeameter is used without, and then with, the geotextile to calculate the hydraulic conductivity ratio	■ Controlled stress conditions ■ Backpressure saturation ■ Integrated into the design method	■ Requires a triaxial soil specimen ■ Requires time to establish flow equilibrium
Dynamic filtration test	Slurry of the soil is added to a saturated geotextile system under dynamic conditions to determine the flow-rate behaviour	■ Simulates pulse loads ■ Soil can cover the geotextile ■ Rapid results	■ Test set-up is quite complex ■ Challenging test conditions

of a cylinder, and the test is performed at a constant head for a sufficiently long period of time to assess the equilibrium flow rate or non-equilibrium condition of excessive clogging or piping. During the test, the hydraulic conductivity of the soil–geotextile composite and the piping rate are measured (Rollin and Lombard, 1988; Siva and Bhatia, 1993; Bhatia and Huang, 1995). Typical hydraulic conductivity curves observed in the long-term flow test are shown in Figure 3.4 (Rollin and Lombard, 1988).

The Type 1 curve in Figure 3.4 indicates normal behaviour, where the hydraulic conductivity decreases with time, until it reaches a constant due to the formation of a filter cake on the geotextile. The time needed for the cake to form varies between 24 and 100 hours (Koerner and Ko, 1982). The Type 2 curve results when there is a loss of particles. As more soil particles are lost, the hydraulic conductivity increases, until it reaches a point where it is equal to that of the clogged geotextile. The Type 3 curve implies the clogging of the geotextile due to a continuous decrease in its hydraulic conductivity. According to long-term filtration tests conducted by Faure *et al.* (1986), and Lombard and Rollin (1987), an O_{95}/D_{85} ratio of 1.5–3 is necessary to prevent piping (see Table 3.1), and the observed filtration behaviour is a function of the filtration opening size, porosity and density of the geotextile, and the gradation of the soil.

Figure 3.4 Results of the long-term flow test

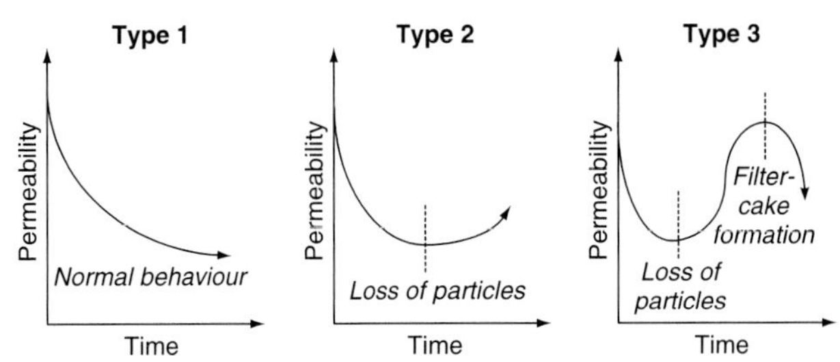

3.5.3 Filter press test

The filter press test is generally used to determine the hydraulic conductivity k_c and the thickness of the filter cake t_c in the filtration of bentonite slurries. It can also be used (or modified) to determine the filtration characteristics of liquid wastes (Kutay and Aydilek, 2004), as these materials usually contain a high percentage of fine particles. Therefore, similarly to bentonite slurries, the filtration of liquid wastes can be expected to be provided by the filter cake, which may form on the geotextile. The idea behind the test is very similar to that for the fine-fraction filtration test, but not as sophisticated. A slurry or sludge specimen is placed onto the geotextile and an air pressure (mostly similar to the pressure expected in the field) is applied. The amount of outflow is determined and the formation of a filter cake noted. The formation of a filter cake is similar to the formation of a bridging network in other filtration tests (i.e. gradient ratio tests). The formation of a filter cake is desirable, as it contributes to the retention of the solids and the filtration of the water. For bentonite slurries the amount of outflow is usually monitored for 30 minutes; however, hours (or days) can be needed to evaluate the long-term performance of high water content waste materials (Kutay and Aydilek, 2004).

Important parameters with regard to the filtration and dewatering of sludges are the rate of water discharge (specific resistance) and the moisture content of the dewatered solids (cake solids). Another parameter, which is often overlooked in the characterisation of sludges for filtration and dewatering, is the filtrate clarity. The measurement of filtrate solids is usually considered important in centrifugal dewatering (Novak and Langford, 1977; Novak *et al.*, 1979), but is frequently ignored when considering other processes and can be measured using the filter press test.

3.5.4 Other test methods

The permittivity of a geotextile is the volumetric flow rate of water per unit cross-sectional area per unit head, under laminar flow conditions in the normal direction through the geotextile. The use of permittivity has advantages over the 'hydraulic conductivity', as it eliminates the thickness dependency. The ASTM D4491-99a (ASTM, 2009) test method was developed to measure the water hydraulic conductivity (and thus permittivity) of geotextiles by the constant or falling head method. The test method has some requirements, such as preconditioning of the geotextile, temperature and the use of de-aired water. The ASTM permittivity test is not a soil–geotextile hydraulic conductivity test, and is used either to determine the permittivity of a virgin geotextile itself or to compare the permittivities of the virgin and the clogged geotextile (e.g. after the gradient ratio test or long-term flow test) (Corcoran and Bhatia, 1996; Aydilek and Edil, 2002).

The hydraulic conductivity ratio (HCR) test is a soil–geotextile interaction test based on laboratory simulation of controlled field conditions. The test is conducted in a triaxial test cell (Williams and Abouzakhm, 1989), and measures the ratio of the hydraulic conductivity of the soil K_{soil} to the hydraulic conductivity of the soil–geotextile system $K_{soil\text{-}GT}$ ($HCR = K_{soil\text{-}GT}/K_{soil}$). The soil specimen is backpressured and consolidated before application of a small hydraulic gradient in order to determine the hydraulic conductivity of the soil alone. Then the flow is reversed and $K_{soil\text{-}GT}$ calculated. Large hydraulic gradients (up to 200) can be applied for fine-grained soils, without any danger of piping.

The fine-fraction filtration test is used to determine the hydraulic conductivity characteristics of slurry–geotextile systems. In the test, the soil is mixed with water to form a slurry, which is introduced incrementally into a column containing the geotextile under a constant hydraulic head. The hydraulic head is increased to maintain constant flow as the flow diminishes during the test. The soil fraction selected to make the slurry depends on the soil and design specifics, but is usually the portion of the soil finer than D_{15}. If the soil used is mostly in the fine-fraction range (e.g. silts or clays), the entire soil should be used (Sansone and Koerner, 1992). The resulting decrease in the flow rate is dependent on the type of soil and geotextile used, and is largest for fine soils, possibly leading to excessive clogging.

The dynamic filtration test is a flow rate (or permittivity) test that is conducted after incremental periods of dynamic pulsing, usually at a rate of 2–3 pulse/s (Narejo and Koerner, 1992). Between each cycle of adding soil and pulsing, flow rates are determined to elucidate the permittivity change of the geotextile. The test may be used to determine the flow rate change under the following conditions: dynamic loads on geotextile filters under railroads; impact loads on geotextiles under aircraft pavements; erosion control filters in coastal waterways with turbulent flow; and various types of groundwater surges on geotextile filters. However, the test results obtained under conditions of either excessive soil loss through the geotextile (piping), or excessive clogging of the geotextile, represent the worst-case hydraulic conditions (Narejo and Koerner, 1992), which are usually harsher than the conditions encountered in the field.

3.6. Filtration design

Several filtration criteria have been proposed for the selection of geotextile filters. These criteria are typically expressed in terms of a ratio of a characteristic geotextile pore size to a characteristic soil particle (grain) size, and the criteria give limiting values for this ratio at which the filter is not clogged but the filtered materials are retained. However, there are very few rational filtration design guidelines, and the most comprehensive one was developed by Luettich *et al.* (1992). The method recommends the consideration of survivability and durability of the geotextile during construction and throughout the life of the structure, in addition to the clogging, retention and hydraulic conductivity criteria, and defined a nine-step filter design procedure as follows.

(*a*) Define the filter requirements for the application (i.e. define retention versus hydraulic conductivity trade-off).

Figure 3.5 Soil-retention criteria for: (a) steady-state and (b) dynamic-flow conditions (Luettich *et al.*, 1992)
Parts of (a) modified from Giroud (1988)

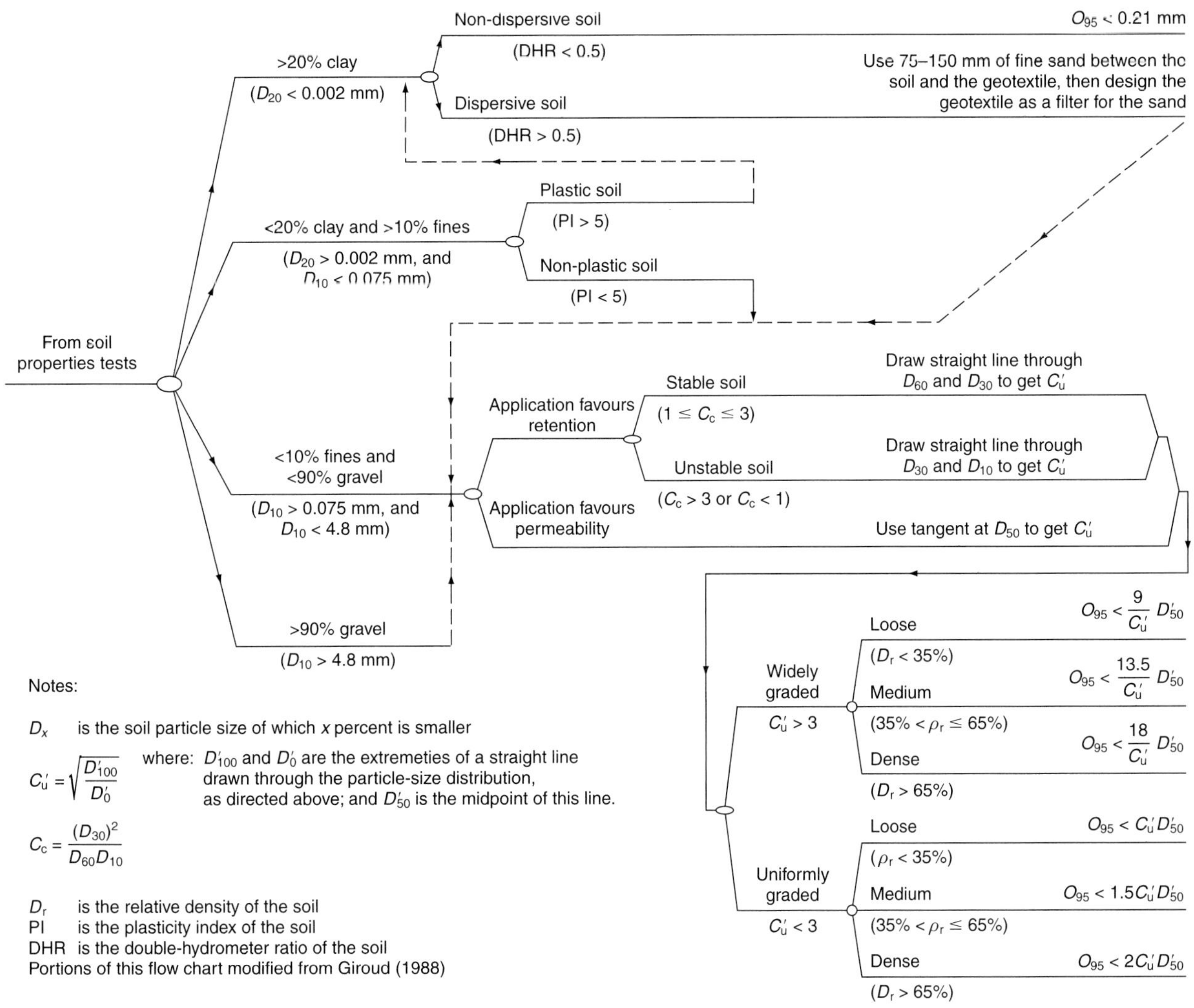

(a)

(*b*) Define the boundary conditions (i.e. evaluate the confining stress and flow type).

(*c*) Determine the soil-retention requirements (i.e. define the soil index properties and the geotextile AOS). Decision trees for the retention criteria for two different flow conditions are given in Figure 3.5. The double-hydrometer ratio required for some steps can be determined using the procedures outlined in ASTM D4221-11 (ASTM, 2011).

(*d*) Determine the geotextile hydraulic conductivity requirements (i.e. define the soil hydraulic conductivity and the hydraulic gradient for the application). These parameters can then be used to determine the minimum required geotextile permittivity ψ_{req} (ASTM, 2009):

$$\psi_{required} = \frac{k}{t} = \frac{q}{\Delta h/A} \tag{3.2}$$

where q is the flow rate, Δh is the hydraulic head loss and A is the cross-sectional area of the geotextile. The available allowable permittivity is

$$\psi_{allowable} = \frac{k_g}{t}\left(\frac{1}{FS_{SCB} \times FS_{IN} \times FS_{CR} \times FS_{CC} \times FS_{BC}}\right) \tag{3.3}$$

where FS_{SCB}, FS_{IN}, FS_{CR}, FS_{CC} and FS_{BC} are the partial factors of safety (factory values) for soil clogging–blinding, intrusion of adjacent materials into the geotextile, creep reduction, chemical clogging and biological clogging, respectively. All factors of safety should theoretically be greater than 1.0; however, much greater values are typically recommended (Koerner, 2005). Thus, the corresponding factor of safety FS is

$$FS = \frac{\psi_{allowable}}{\psi_{required}} \tag{3.4}$$

(*e*) Determine the anti-clogging requirements (i.e. use an AOS that satisfies the retention criteria). The criterion

Figure 3.5 Continued

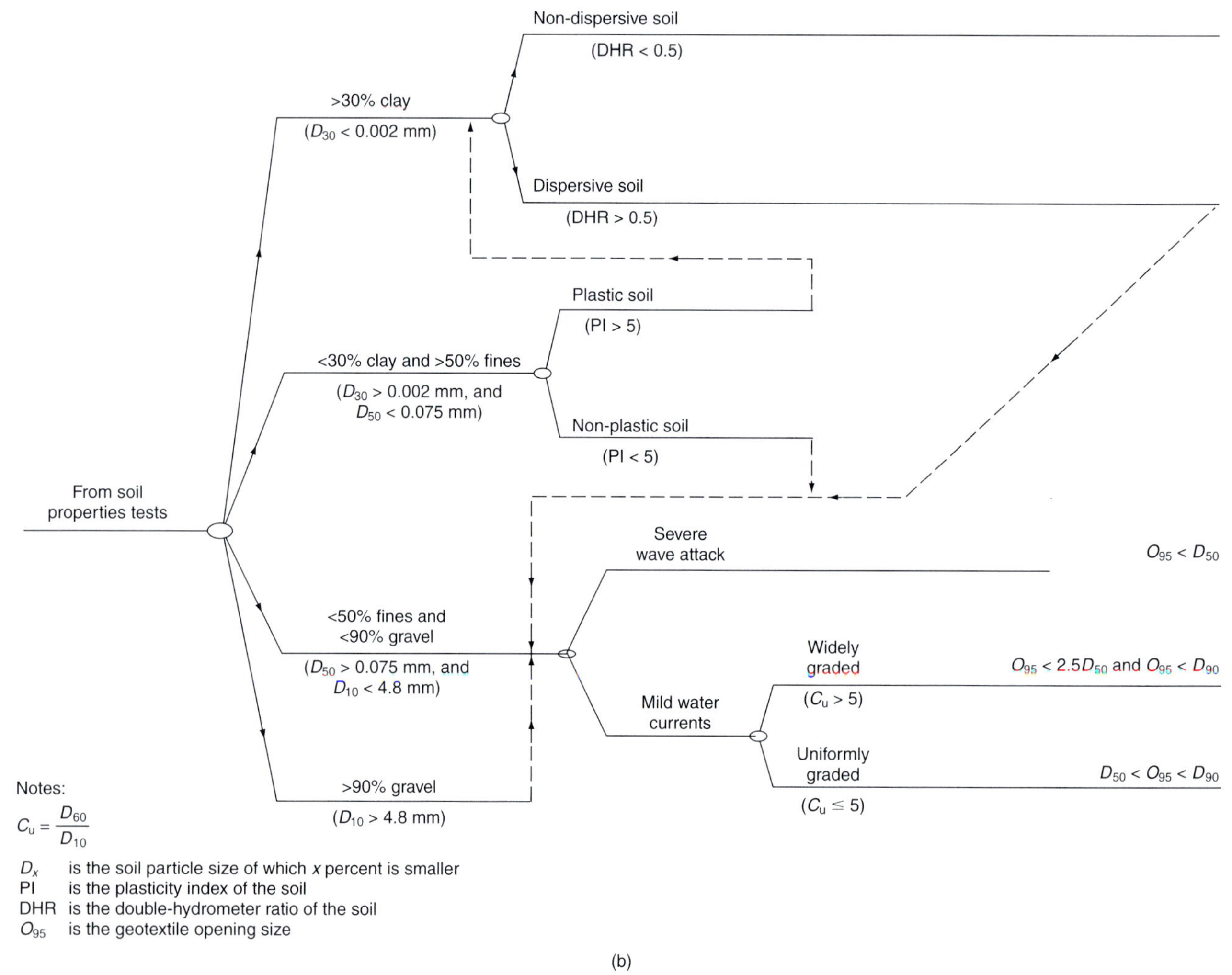

developed by Carroll (1983) is a commonly used one: $O_{95} < 2.5D_{85}$.

(*f*) Moreover, for non-woven geotextiles, use the porosity $n > 30\%$, and for woven geotextiles use POA > 4%.

(*g*) Determine the survivability requirements (i.e. specify the minimum index strength properties for potential damage to the geotextile due to adjacent materials and the construction technique to be used).

(*h*) Determine the durability requirements (i.e. use chemical additives to minimise degradation of the geotextile due to exposure to sunlight or chemicals, or check the chemical compatibility of the filter if it is to be placed next to contaminants).

(*i*) Other design considerations (i.e. abrasion of the geotextile due to dynamic action, and biological and biochemical clogging factors).

3.7. Filtration of high water content geomaterials with geotextiles

The retirement of large industrial-waste storage facilities, in accordance with environmental regulations, has become a critical cost issue for industry and a challenge to the geotechnical community. Many facilities were constructed prior to the emergence of modern environmental regulations and contain a variety of contaminated high water content materials. Capping of the surface impoundments, such as lagoons, ponds and old quarries, where these materials are often stored, is the least costly remediation alternative. Of the various types of capping options, composite caps are usually preferred, in which a geotextile component serves as the filtration layer.

Another alternative, which is gaining in popularity, is the beneficial reuse of the waste materials, but this can only be accomplished after a dewatering procedure has been completed. Dewatering is an important step in the remediation of these geomaterials, as their natural water content may be up to 800%, and they may have low shear strengths (Pilarczyk, 2000; Moo-Young *et al.*, 2002). An efficient method of dewatering high water content materials is to use geotextile containers (Mori *et al.*, 2002). These are constructed by sewing together high-strength woven geotextile sheets, and can hold relatively large amounts of high water content geomaterials. The geomaterials are hydraulically pumped in slurry form into these containers, and allowed to settle and temporally dissipate their fluid through the pore openings of the container material. The use of geotextile containers provides a fast, cost-effective and efficient approach, as it reduces the time required for dewatering and the surface area of the materials. The containers are simple to construct, and labour and equipment costs are eliminated (Pilarczyk, 2000).

Whether dewatering high water content materials via geotextile containers or filtering them in capping applications, the hydraulic compatibility of the geotextile with the contact soil is important, and this should be considered during the design. The compatibility between the geotextile and the soil, which ensures long-term performance, is usually determined through laboratory soil filtration tests. The first requirement for ensuring hydraulic compatibility is that the geotextile should not become clogged during the dewatering process. The second requirement is that the amount of soil piped through the geotextile is minimal (i.e. successful retention performance), so the internal stability and modulus of the soil are not adversely affected.

Recent advances in geotextile filter design have suggested that the filtration characteristics of some geomaterials (e.g. fly ash, dredged sediments) are highly different from those of natural soils (Akram and Gabr, 1997; Aydilek and Edil, 2002; Moo-Young and Tucker, 2002; Liao and Bhatia, 2005). Retention is particularly critical when the fill materials are contaminated, as some contaminants tend to adsorb onto soil solids due to their low solubility (e.g. polychlorinated biphenyls). Existing criteria developed for sands and clays do not directly address the retention capacity of geotextiles used to confine non-traditional geomaterials (Aydilek and Edil, 2008).

McKelvey (1995) conducted gradient ratio tests on sludge-drying beds in order to determine the sludge–geotextile clogging behaviour in the remediation process. The sludge contained 95% fines and its plasticity varied (i.e. non-plastic to a plasticity index PI of >5). The hydraulic conductivity of the system decreased by two orders of magnitude, and the gradient ratio increased with increasing hydraulic gradient. A hydraulic gradient of 9.0 seemed to be the maximum value before clogging occurred (GR > 3). A light non-woven geotextile performed well in the tests, but it did not work in field applications.

Ziani *et al.* (1995) investigated the suspension filtration capacity of non-woven geotextiles, which may be used in the testing of liquid-waste–geotextile systems. The geotextiles were tested with suspension (silty soil) concentrations of up to 5 g/l, and no evidence of clogging was observed. A steady-state flow occurred after nearly two hours. It was found that the filtration efficiency was highly dependent on the D_{50}/O_{95} ratio.

Application of the gradient ratio test was extended to the testing of fly ash slurry with non-woven geotextiles (Gabr and Akram, 1996a). Unlike in the ASTM procedure, the permeameter had piezometer locations and a funnel section at the bottom to trap piped particles. The test duration was 5000 hours, in order to evaluate the long-term performance of the geotextiles. The results showed that the variation in the hydraulic conductivity and the gradient ratio are significant. Clogging of up to 88% was observed, but this did not influence the filtration performance, as the permittivity of the clogged geotextile was still four orders of magnitude higher than that of fly ash itself (Akram, 1995). The measured piping rate of 100–400 g/m^2 was below the limit of 2500 g/m^2 suggested by Lafleur *et al.* (1989).

Fine-fraction filtration and dynamic filtration tests have also been used to test fly ash–geotextile and slurry–geotextile systems (Hoover, 1982; Legge, 1990; Narejo and Koerner, 1992; Sansone and Koerner, 1992). However, these tests had some of the disadvantages known to be associated with such tests, and the results obtained were not as conclusive as those determined using the gradient ratio and long-term flow tests.

Geotextiles have been successfully used to filter fine-grained wastewater-treatment sludges contaminated with polychlorinated biphenyls at a United States Environmental Protection Agency (US EPA) Superfund site (Aydilek and Edil, 2002, 2003). Long-term laboratory filtration tests and five-year field observations showed no apparent physical or chemical clogging of the filters under harsh environmental conditions (i.e. temperatures as low as −46°C and pH < 6.0). Kutay and Aydilek (2004, 2005) studied the hydraulic compatibility between the geotextile and fly ash slurries and bottom-sea dredged sediments. The results indicated that both fly ash and dredged sediments could be successfully retained by a variety of non-woven/woven geotextile combinations. In addition, it was found that the use of a double-layer geotextile system, rather than a single-layer woven geotextile, significantly increased the retention capacity, and that the hydraulic properties of the geotextile, such as permittivity and AOS, directly influenced the retention performance. A relatively low initial water content of the slurry and the formation of a filter cake at the soil–geotextile interface increased the retention performance. In a similar study, Huang and Luo (2007) conducted a series of falling head column tests on reservoir sediments and four types of geotextile, to investigate the factors that control the hydraulic conductivity and piping of dewatering systems and the efficiency of dewatering ponds. It was found that the hydraulic conductivity of the geotextile and the thickness of the filter cake, which progressively deposited on the upstream side of the geotextile, affected the dewatering rate. The height of the potential filter cake was predicted based on the total applied suspended solids loading and the measured void ratios.

The geometrical, physical and external factors that affect the performance of geotextiles also influence the filtration properties of high water content geomaterials. A summary of these factors is given in Table 3.5. Studies have shown that pore sizes smaller than the AOS (i.e. O_{95}) may have a significant impact on the filtration performance of a geotextile, and a complete pore-size distribution curve (such as one

Table 3.5 Studies on the geotextile filtration of high water content geomaterials and the factors reported to affect the filtration process

Study	Geotextile and soil type	Factors affecting the filtration process
Montero and Overmann (1990)	Non-woven geotextiles; Clayey slurry	Soil GSD; geotextile AOS
Sansone and Koerner (1992)	Woven and non-woven geotextiles; fly ash and sand slurries	Soil grain shape and GSD; geotextile permittivity
McKelvey (1995)	Woven and non-woven geotextiles; wastewater-treatment sludge	Plasticity index; applied hydraulic gradient
Gabr and Akram (1996a,b)	Non-woven geotextiles; fly ash slurry	Soil GSD; geotextile permittivity, thickness and AOS
Moo-Young *et al.* (1999)	Non-woven geotextiles; contaminated harbour sediments	Flow rate; geotextile AOS
Moo-Young *et al.* (2002)	Woven geotextiles; pulp and paper-mill sludges; lake-bottom silty sediments	Biological factors
Baker *et al.* (2002)	Woven geotextiles; dairy-waste sludge	Flow rate and capacity
Moo-Young and Tucker (2002)	Woven geotextiles; sandy and clayey slurries; wastewater-treatment sludge	Plasticity index; geotextile AOS, mass/unit area
Aydilek and Edil (2002, 2003)	Woven and non-woven geotextiles; PCB-contaminated wastewater-treatment sludge	Fines content; geotextile permittivity, POA/porosity, geotextile PSD; biological factors
Kutay and Aydilek (2004)	Woven and non-woven geotextiles; fly ash slurry; dredged harbour sediments	Plasticity, fines content; geotextile permittivity, AOS and PSD
Liao and Bhatia (2005)	Woven geotextiles; sandy sediments	Fines content; geotextile AOS
Kutay and Aydilek (2005)	Woven and non-woven geotextiles; fly ash slurry; dredged harbour sediments	Plasticity, fines content; geotextile permittivity, applied hydraulic gradient
Muthukumaran and Ilmaparuthi (2006)	Woven and non-woven geotextiles; fly ash slurry; harbour sediments	Soil GSD; geotextile PSD and AOS
Huang and Luo (2007)	Woven geotextiles; silty sediments	Geotextile permittivity and AOS

AOS, apparent opening size; GSD, grain size distribution; PCB, polychlorinated biphenyl; POA, percent open area; PSD, pore size distribution.

obtained from bubble point tests) should be used in development of the filtration criteria (Aydilek *et al.*, 2007; Aydilek and Edil, 2008). Retention is usually related to larger pore opening sizes (e.g. sizes greater than O_{50}), whereas small pore opening sizes are usually responsible for clogging (e.g. sizes smaller than O_{30}). It should be noted that biological factors may also affect the process, as most high water content materials are contaminated, although the effects of these factors have not been widely studied. Microorganisms can easily grow on geotextiles, as commercially available geotextiles are not coated with biocides (chemicals used to prevent the formation or growth of bacteria), and may cause biological clogging of the geotextile filters (Kossendey *et al.*, 1996).

Considering the critical nature of the filter design in the case of contaminated materials, the selection of the geotextile is crucial. There is a broad consensus that a parametric study is required to evaluate O_x/D_x ratios and find the value that clearly discriminates between the retention performance of such geomaterials. As a result, several criteria have been developed to predict the hydraulic performance of geotextiles when used to filter high water content materials, and these are summarised in Table 3.6.

3.8. Drains

Drainage refers to an in-plane flow of water. In this regard, drainage is different from filtration, which refers to the flow of water in the cross-plane direction. Several geosynthetics serve as drainage media, the most commonly used ones being geotextiles, geonets and geocomposites. Typical applications are shown in Figure 3.6.

3.8.1 Drainage with geotextile

Of the various geotextiles available on the market, non-woven geotextiles are the only ones that have a meaningful drainage capacity. Two general applications of geotextiles for drainage are their use in gravity drainage (e.g. chimney drains, and drainage galleries in earth dams, behind retaining walls and in landfill gas collection systems) and in pressure drainage (vertical drains to accelerate consolidation beneath surcharge fills and earth embankment dams).

Table 3.6 Geotextile filter selection criteria for high water content geomaterials

Criteria	Soil and geotextile type	Reference
Retention: $O_{85}/D_{75} < 1$ for all POA Clogging: $O_{40}/D_{15} > 1$ for $0.5 < \text{POA} < 8$; $O_{40}/D_{30} > 1$ for $\text{POA} > 8$	PCB-contaminated wastewater treatment sludge; woven geotextile	Aydilek (2000)
Retention: $O_{95}/D_{85} < 1$ Clogging: POA = 1–5% Retention: $O_{95} < 0.3$ mm or $O_{90}/D_{90} = 2–5$	Sandy or silty slurry; woven geotextile Clayey slurry; woven geotextile	Moo-Young and Tucker (2002)
Retention: $O_{85}/D_{50} < 1$ for all POA Clogging: $O_{40}/D_{15} > 1$ for $\text{POA} < 8$; $O_{40}/D_{30} > 1$ for $\text{POA} > 8$	Silty sand slurry; woven geotextile	Aydilek (2006)
AOS < 0.425 mm $40 < \text{AOS}/D_{15} < 125$ $0.3 < \text{AOS} \times D_{85}/D_{15} < 1.7$	Silty slurries; woven geotextile	Muthukumaran and Ilmaparuthi (2006)

AOS, apparent opening size; POA, percent open area.

Figure 3.6 Various areas of use for geosynthetics in drainage applications
Adapted from Holtz *et al.* (1998)

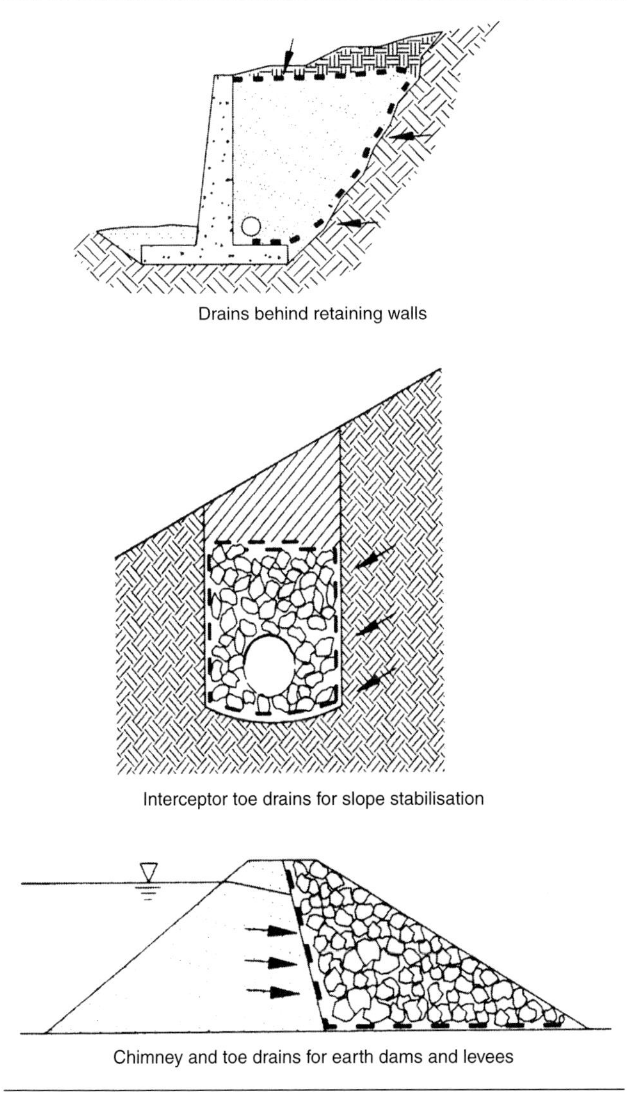

In gravity drainage applications, the most important hydraulic property of geotextiles is transmissivity. Transmissivity is the amount of water flow within the plane of a geotextile under a certain hydraulic gradient:

$$\theta_{\text{required}} = k_{\text{p}}t = \frac{q}{iW} \tag{3.5}$$

where θ_{required} is the transmissivity, q is the flow rate (amount of flow per unit area), k_{p} is the in-plane hydraulic conductivity, and W and t are the width and thickness of the geotextile, respectively. Available transmissivity in the system and the corresponding flow factor of safety can be calculated as

$$\theta_{\text{allowable}} = k_{\text{p}}t\left(\frac{1}{\text{FS}_{\text{SCB}} \times \text{FS}_{\text{IN}} \times \text{FS}_{\text{CR}} \times \text{FS}_{\text{CC}} \times \text{FS}_{\text{BC}}}\right) \tag{3.6}$$

$$\text{FS} = \frac{\theta_{\text{allowable}}}{\theta_{\text{required}}} \tag{3.7}$$

where FS_{SCB}, FS_{IN}, FS_{CR}, FS_{CC} and FS_{BC} are the partial safety factors for soil clogging–blinding, intrusion of adjacent materials into the geotextile, creep reduction, chemical clogging and biological clogging, respectively. All factors of safety should theoretically be greater than 1.0; however, typically much greater values are recommended (Koerner, 2005). Transmissivity is equally important for the design of both gravity and pressure drainage, as water will flow from a high- to a low-pressure zone, regardless of the geotextile orientation. Giroud (1984) identified a situation in which a geotextile is placed above a soft foundation underlying a surcharge fill, and developed the following formula for designing the drain:

$$\theta_{\text{required}} = k_{\text{p}}t = \frac{B^2k}{\sqrt{c_{\text{v}}T}} \tag{3.8}$$

where k and c_{v} are the hydraulic conductivity and coefficient of consolidation of the foundation soil, respectively, B is the

Figure 3.7 Geonet use in municipal solid-waste landfills

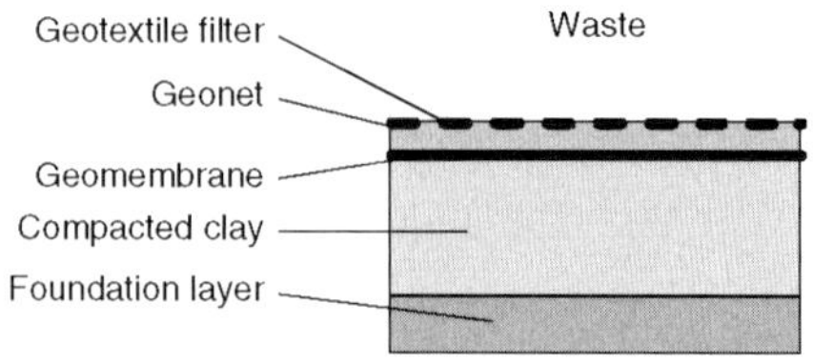

(a)

(b)

Figure 3.8 Geocomposite sheet drains in landfills

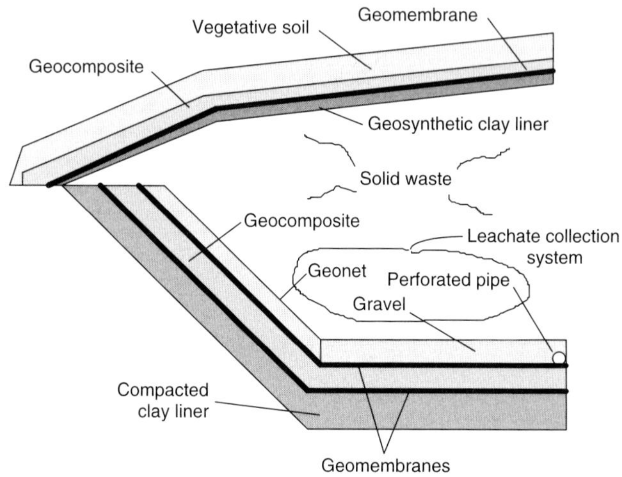

width of the surcharge layer, and T is the time for the surcharge fill to be placed. Koerner (2005) provides design examples for gravity and pressure drainage.

3.8.2 Drainage with geonets and geocomposites

Geonets are grid-like materials and are primarily used in drainage applications. The most important hydraulic property of geotextiles is transmissivity, and the same methods used for geotextiles are followed for designing geonets for drainage. Geonets are mostly used with other geosynthetics, typically with a geotextile or geomembrane, to prevent soil intrusion into their apertures (Figure 3.7). The intrusion of soil particles into geonets can decrease their drainage capacity tremendously. Flow rate, applied normal stress and hydraulic gradient play a significant role in the degree of intrusion, and therefore laboratory drainage tests should be performed to evaluate the geonet performance.

Geocomposites are prefabricated geosynthetics that consist of geonets sandwiched between two geotextiles. They are used extensively in drainage applications, and typically in prefabricated vertical (wick) drains (PVDs), highway edge drains and sheet drains. PVDs are used to reduce the post-construction settlement of soft soils to a minimum. Koerner (2005) listed the advantages of PVDs over traditional sand drains as their: straightforward analysis; nominal resistance to the flow of water entering the drain; requirement for low-weight construction equipment; and simple installation and low cost. PVDs are exposed to kinking (necessary shortening of the drain during the consolidation process), and typical decreases in the flow rate due to kinking are given by Koerner (2005). This reduction is used to calculate the required flow rates, and the factor of safety can be calculated using Equation 3.7 and the relevant partial safety factors.

Sheet drains are used in a variety of applications involving planar sheet flow, such as retaining structures and landfills (Figure 3.8). They are typically constructed of a geotextile and a semi-rigid drainage system (Figure 3.9). The main duty expected of a sheet drain is to collect and transport liquid off site. In order to accomplish this, the drain must: allow water to seep perpendicularly to the plane of the geotextile into the core; allow water to flow in the plane of

Figure 3.9 Two types of sheet drain material

Figure 3.10 Configuration of a prefabricated highway edge drain (Koerner, 2005)

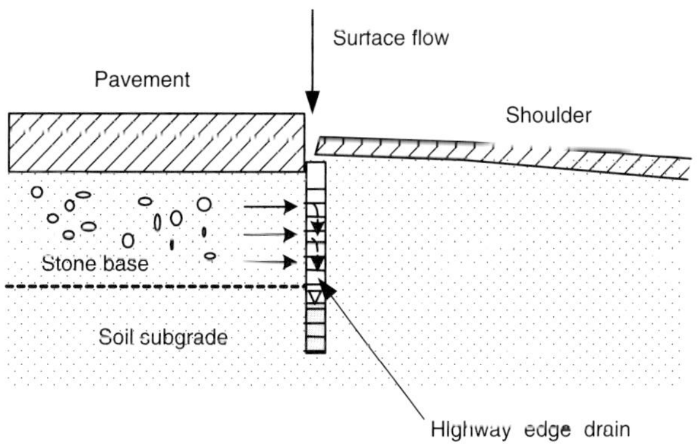

the core; and, if required, collect the water flowing from the core and conduct it to some discharge points (Bhatia, 1995). Koerner (2005) has reported that geocomposite sheet drains have significantly higher in-plane flow capacity than do geotextiles, and triplanar sheet drains tend to offer better flow rate capabilities than biplanar ones.

Prefabricated highway edge drains are used regularly by engineers for highway systems. A prefabricated subsurface geocomposite drainage product, consisting of a geotextile filter material supported by a core, net, mesh or spacer, collects liquids and drains them off. In addition to their drainage capacity, edge drains also provide filtration to the surrounding soil (Figure 3.10). The main advantages of edge drains are that less soil needs to be excavated and removed, there is less weight on the subsurface, and the potential use of reclaimed materials (Koerner and Hwu, 1991).

The design of a prefabricated edge drain focuses on the drainage core and geotextile filter (Figure 3.11). The core segment should have the sufficient compressive strength to sustain the field stresses and meet the transmissivity requirements, while the geotextile should satisfy the filtration criteria (i.e. retention, clogging and hydraulic conductivity). A series of studies has been conducted to evaluate the field performance of highway edge drains. Dempsey (1988) observed that surface of seepage of water and frost heave problems did not occur after the installation of edge drains. Dempsey (1988) recommended that the geotextile in a geocomposite edge drain should not be loosely attached, as it may sag into the core and decrease the overall drainage capacity. Koerner *et al.* (1996) obtained field core samples from 41 geocomposite edge drains and found that the cores generally performed quite well. However, the geotextile filter separated from the edge drain, most probably due to poor contact between the upstream soil and the geotextile during construction.

Figure 3.11 A prefabricated highway edge drain material

3.9. Summary of main points

- A major aspect of the design of geosynthetic materials for filtration and drainage applications is to ensure that the hydraulic requirements are met. These requirements are typically the retention, anti-clogging and hydraulic conductivity performance for filters, and the transmissivity for drains, in addition to other requirements of the project.
- Proper analysis of the retention or clogging performance is especially important when the geosynthetic material is to be used in connection with high water content geomaterials, such as sludges and fly ash slurries, as these materials are often contaminated.
- The factors that affect the filtration performance of high water content geomaterials are similar to those that affect the filtration performance of natural soils.
- Soil plasticity, fines content and grain-size distribution, geotextile permittivity and characteristic pore opening sizes, as well as the applied hydraulic gradient and flow conditions, are the main factors that have been reported to directly affect the retention and clogging of geotextiles.
- Biological factors can have a profound effect on the filtration performance of geosynthetic filters and drains, and more research is needed to study these effects.

3.10. Common questions and answers

Questions

1. What is the difference between drainage and filtration?

2. What are typical geosynthetic products used in filtration and drainage?

3. What are the advantages of geotextile filters over granular filters?

4. What is the basis for the existing geotextile filter selection criteria? Is there any theoretical basis?

5. What are the important considerations when selecting a filter?

6. Should one use the existing criteria or perform laboratory tests to select the proper geotextile filter in a design?

7. Are the existing filter-selection criteria applicable to high water content geomaterials, such as sludges and slurries?

Answers

1. Drainage refers to the in-plane flow of water, while filtration refers to the flow of water in the cross-plane direction.

2. Geotextiles are the only geosynthetics used in filtration applications. Geotextiles, geonets and geocomposites are used in drainage applications.

3. The use of geotextile filters rather than granular filters can be justified mainly due to their cost advantages arising from

- the use of less costly drainage aggregate
- the possible use of smaller sized drains
- the possible elimination of collector pipes
- fast construction
- lower risk of contamination and segregation of drainage aggregate during construction
- reduced excavation (Holtz *et al.*, 1998).

4. Most of the existing geotextile filter criteria have been developed either from laboratory methods (empirical) or by converting the granular soil grain size to the geotextile equivalent pore size. There is no theoretical basis, even though some attempts have been made to come up with theory-based methods (Faure *et al.*, 1986; Giroud, 1996). On the other hand, most criteria provide conservative estimates of the filtration behaviour, and thus are commonly used.

5. The considerations include the grain-size distribution of the soil, the permittivity and pore structure characteristics of the geotextile filter, the water content and consistency of the soil (e.g. highly plastic clay, sludge or uniform sand), the potential for biological clogging, the criticality of the project (e.g. contaminated soil, high-pressure applications that may require high strength), the requirements of the project (e.g. capping applications often require fast and uniform settlement of the foundation soil), the viscosity of the liquid to be drained, and the flow characteristics (dynamic or steady-state flow).

6. Existing geotextile filtration criteria typically provide conservative estimates and should be adequate for natural soils. Critical applications (e.g. high flow rates, high hydraulic gradient) require laboratory testing. Laboratory tests should be selected to best mimic the field conditions in a laboratory setting. A detailed description of the common soil–geotextile filtration tests is given in Section 3.5.

7. Most of the time they are not, because the index and engineering properties, as well as the filtration characteristics, of these geomaterials are highly different from those of natural soils. Most of these geomaterials are contaminated and, considering the critical nature of the design, the selection of the correct geotextile is crucial. A list of existing geotextile filter selection criteria for these materials is provided in Section 3.7.

REFERENCES

Akram MH (1995) Long-term filtration behavior of nonwoven geotextiles with fly ash. *Proceedings of Geosynthetics '95*, Nashville, Industrial Fabrics Association International, St Paul, MN, pp. 1195–1206.

Akram MH and Gabr MA (1997) Filtration of fly ash using nonwoven geotextiles: effect of sample preparation technique and testing method. *Geotechnical Testing Journal, ASTM* **20(3)**: 263–271.

ASTM (2004) ASTM D4751-04. Standard test method for determining apparent opening size of a geotextile. ASTM, West Conshohocken, PA.

ASTM (2006) ASTM D5101-01. Standard test method for measuring the soil–geotextile system clogging potential by the gradient ratio. ASTM, West Conshohocken, PA.

ASTM (2007) ASTM D1987-07. Standard test method for biological clogging of geotextile or soil/geotextile filters. ASTM, West Conshohocken, PA.

ASTM (2009) ASTM D4491-99a. Test methods for water permeability of geotextiles by permittivity. ASTM, West Conshohocken, PA.

ASTM (2011) ASTM D4221-11. Standard test method for dispersive characteristics of clay soil by double hydrometer. ASTM, West Conshohocken, PA.

Aydilek AH (2000) Filtration performance of geotextile–wastewater sludge systems. PhD dissertation, University of Wisconsin–Madison, Madison, WI.

Aydilek AH (2006) A semi-analytical model for development of woven geotextile filter selection criteria. *Geosynthetics International* **13(2)**: 59–72.

Aydilek AH and Edil TB (2002) Filtration performance of woven geotextiles with wastewater treatment sludge. *Geosynthetics International* **9(1)**: 41–69.

Aydilek AH and Edil TB (2003) Long-term filtration performance of nonwoven geotextile–sludge systems. *Geosynthetics International* **10(4)**: 110–123.

Aydilek AH and Edil TB (2004) Evaluation of woven geotextile pore structure parameters using image analysis. *Geotechnical Testing Journal, ASTM* **27(1)**: 99–110.

Aydilek AH and Edil TB (2008) *Solidification/Stabilization of PCB-Contaminated Wastewater Treatment Sludges* (Khire MV, Alshawabkeh AN and Reddy KR (eds)). ASCE, New Orleans, LA, pp. 724–731. Geotechnical Special Publication 177.

Aydilek AH, Oguz SH and Edil TB (2005) Constriction size of geotextile filters. *Journal of Geotechnical and Geoenvironmental Engineering, ASCE* **131(1)**: 28–38.

Aydilek AH, D'Hondt D and Holtz RD (2007) Comparative evaluation of geotextile pore sizes using bubble point test and image analysis. *Geotechnical Testing Journal, ASTM* **30(3)**: 173–181.

Baker KB, Chastain JP and Dodd RB (2002) *Treatment of Lagoon Sludge and Liquid Animal Manure Utilizing Geotextile Filtration.* American Society of Agricultural and Biological Engineers, St Joseph, MI. ASABE Paper No. 024128.

Bass JM, Ehrenfeld JR and Valentine JM (1984) *Potential Clogging of Landfill Drainage Systems.* US EPA, Washington, DC. Report EPA-600/S2-83-109.

Bhatia SK (1995) IFAI Professor Training Course for Geosynthetics Lecture Notes.

Bhatia SK and Huang Q (1995) Geotextile filters for internally stable/unstable soils. *Geosynthetics International* **2(3)**: 537–565.

Black PJ and Holtz RD (1999) Performance of geotextile separators five years after installation. *Journal of Geotechnical and Geoenvironmental Engineering* **125(5)**: 404–412.

Calhoun CC (1972) *Development of Design Criteria and Acceptance Specifications for Plastic Filter Cloth*. US Army Waterways Experiment Station, Vicksburg, MS, p. 55. Technical Report S-72-7.

Carroll RG Jr (1983) Geotextile filter criteria. In *Engineering Fabrics in Transportation Construction*. Transportation Research Board, Washington, DC, pp. 46–53. Transportation Research Record 916.

Cedergren HR (1989) Seepage, Drainage and Flow Nets, 3rd edn. Wiley, New York.

Christopher BR and Fischer GR (1992) Geotextile filtration principles, practices and problems. *Geotextiles and Geomembranes* **11(4–6)**: 337–353.

Christopher BR, Holtz RD and Fischer GR (1993) Research needs in geotextile filter design. In *Filters in Geotechnical and Hydraulic Engineering* (Brauns J, Heibaum M and Schuler U (eds)). Balkema, Rotterdam, pp. 19–26.

Civil Works Construction Guide (1977) *Specification for Plastic Filter Fabric*. US Army Corps of Engineers, Office of the Chief of Engineers, Washington, DC. CW 02214.

Corcoran BW and Bhatia SK (1996) Evaluation of geotextile filter in a collection system at Fresh Kills Landfill. In *Recent Developments in Geotextile Filters and Prefabricated Drainage Geocomposites* (Bhatia SK and Suits LD (eds)). ASTM, Philadelphia, PA, pp. 182–195. ASTM STP 1281.

DeBerardino S (1993) Filtration design examples: implementing the state of the practice. *Geotechnical Fabrics Report* **10(9)**: 19–26.

Dempsey BJ (1988) Core flow capacity requirements of geocomposite fin-drain materials utilized in pavement subdrainage. *Proceedings of the 67th Annual Transportation Research Board Meeting*, Washington, DC.

Fannin RJ, Vaid YP and Shi YC (1994) A critical evaluation of the gradient ratio test. *Geotechnical Testing Journal, ASTM* **17(1)**: 35–42.

Faure YH, Gourc JP, Millot F and Sunjoto S (1986) Theoretical and experimental determination of the filtration opening size of geotextiles. *Proceedings of the Third International Conference on Geotextiles*, vol. 4, Vienna. Balkema, Rotterdam, pp. 1275–1280.

Fischer GR (1994) The influence of geotextile pore structure on the behavior of geotextile filters. PhD dissertation, University of Washington, Seattle, WA.

Fischer GR, Holtz RD and Christopher BR (1990) Filter criteria based on pore size distribution. *Proceedings of the Fourth International Conference on Geotextiles*, vol. 1, The Hague. Balkema, Rotterdam, pp. 289–294.

French Committee of Geotextiles and Geomembranes (1986) Recommendations for the use of geotextiles in drainage and filtration systems. In *Geotextile Manual*. Institut Textile de France, Boulogne-Billancourt, pp. 43–54 (as cited by Fischer, 1994).

Gabr M and Akram MH (1996a) Sample preparation techniques for filtration testing of fly ash with nonwoven geotextiles. In *Sampling Environmental Media* (Morgan JH (ed.)). ASTM, West Conshohocken, PA. *ASTM STP* **1282**: 376–388.

Gabr MA and Akram MH (1996b) Clogging and piping criteria for geotextile filters for fly ash. *3rd International Symposium on Environmental Technology*, San Diego, pp. 836–847.

Giroud JP (1982) Filter criteria for geotextiles. *Proceedings of the Second International Conference on Geotextiles*, vol. 1, Las Vegas. Balkema, Rotterdam, pp. 103–108.

Giroud JP (1984) Designing with geotextiles. *Geotextiles and Geomembranes* **1(1)**: 5–40.

Giroud JP (1988) Review of geotextile filter criteria. *Proceedings of the 1st Indian Geotextiles Conference on Reinforced Soil and Geotextiles*, Bombay. Oxford & IBH, Delhi, pp. 1–6.

Giroud JP (1996) Granular filters and geotextile filters. *Proceedings of Geofilters '96*, Montreal, pp. 565–680.

Giroud JP, Delmas P and Artières O (1998) Theoretical basis for the development of a two-layer geotextile filter. *Proceedings of the Sixth International Conference on Geosynthetics*, Atlanta, pp. 1037–1044.

Haliburton TA and Wood PD (1982) Evaluation of the US Army Corps of Engineers gradient ratio test for geotextile performance. *Proceedings of the Second International Conference on Geotextiles*, vol. 1, Las Vegas. Balkema, Rotterdam, pp. 97–101.

Haliburton TA, Lawmaster JD and McGuffy VE (1982) *Use of engineering fabric in transportation related applications*. US Federal Highway Administration Training Manual-Contract No. DTFH-80-C-0094.

Heerten G (1993) A contribution to the improvement of dimensioning analogies for grain filters and geotextile filters. In *Filters in Geotechnical and Hydraulic Engineering* (Brauns J, Heibaum M and Schuler U (eds)). Balkema, Rotterdam, pp. 121–127.

Holtz RD, Christopher BR and Berg RR (1998) *Geosynthetic design and construction guidelines – participant notebook*. US Federal Highway Administration, FHWA HI-95-038, McLean, VA.

Hoover TP (1982) Laboratory testing of geotextile filters. *Proceedings of the Second International Conference on Geotextiles*, vol. 3, Las Vegas. Balkema, Rotterdam, pp. 839–843.

Huang CC and Luo SY (2007) Dewatering of reservoir sediment slurry using woven geotextiles. *Geosynthetics International* **14(5)**: 253–263.

Koerner GR and Koerner RM (1992) Leachate flow behavior through geotextile and soil filters and possible remediation methods. *Geotextiles and Geomembranes* **11**: 401–430.

Koerner RM (2005) *Designing with Geosynthetics*, 5th edn. Prentice Hall, Upper Saddle River, NJ.

Koerner RM and Ko FK (1982) Laboratory studies on long term drainage capability of geotextiles. *Proceedings of the Second International Conference on Geotextiles*, vol. 1, Las Vegas. Balkema, Rotterdam, pp. 91–96.

Koerner RM and Hwu B-L (1991) *Prefabricated Highway Edge Drains*. Transportation Research Board, Washington, DC, pp. 14–20. Transportation Research Record No. 1329.

Koerner GR, Koerner RM and Wilson-Fahmy RF (1996) Field performance of geosynthetic highway drainage systems. In *Recent Developments in Geotextile Filters and Prefabricated Drainage Geocomposites* (Bhatia SK and Suits LD (eds)). ASTM, Philadelphia, PA, pp. 165–181. ASTM STP 1281.

Kossendey TH, Gartung E and Schmidt St (1996) Microbiological influences on the long-term performance of geotextile filters. *Geofilters '96*, Montreal, pp. 115–124.

Kutay ME and Aydilek AH (2004) Retention performance of geotextile containers confining geomaterials. *Geosynthetics International* **11(2)**: 100–113.

Kutay ME and Aydilek AH (2005) Filtration performance of two-layer geotextile systems. *Geotechnical Testing Journal, ASTM* **28(1)**: 79–91.

Lafleur J, Mlynarek J and Rollin AL (1989) Filtration of broadly graded cohesionless soils. *Journal of Geotechnical Engineering, ASCE* **115(12)**: 1747–1768.

Lafleur J, Mlynarek J and Rollin AL (1993) Filter criteria for well graded cohesionless soil. In *Filters in Geotechnical and Hydraulic Engineering* (Brauns J, Heibaum M and Schuler U (eds)). Balkema, Rotterdam, pp. 97–106.

Legge KR (1990) A new approach to geotextile selection. *Proceedings of the Fourth International Conference on Geotextiles*, vol. 1, The Hague. Balkema, Rotterdam, pp. 269–272.

Liao K and Bhatia SK (2005) Geotextile tube: filtration performance of woven geotextiles under pressure. *Proceedings of the Geosynthetics '05*, Las Vegas, NV.

Lombard G and Rollin A (1987) Filtration behavior analysis of thin heat-bonded geotextiles. *Proceedings of Geosynthetics '87*, New Orleans. Industrial Fabrics Association International, St Paul, MN, pp. 482–492.

Luettich SM, Giroud JP and Bachus RC (1992) Geotextile filter design guide. *Geotextiles and Geomembranes* **11(4–6)**: 355–370.

McKelvey JA (1995) Filtration system design for sludge drying beds by gradient ratio performance. *Proceedings of Geosynthetics '95*, vol. 1, Nashville. Industrial Fabrics Association International, St Paul, MN, pp. 203–215.

Millar PJ, Ho KW and Turnbull HR (1980) *A Study of Filter Geotextiles for Geotechnical Applications in New Zealand.* Ministry of Works and Development, Upper Hutt, Wellington. Central Laboratories Report No. 2-80/5.

Mlynarek J and Rollin AL (1995) Bacterial clogging of geotextiles – Overcoming engineering concerns. *Proceedings of Geosynthetics '95*, vol. 1, Nashville. Industrial Fabrics Association International, St Paul, MN, pp. 177–188.

Montero CM and Overmann LK (1990) Geotextile filtration performance test. In *Geosynthetic Testing for Waste Containment Applications* (Koerner RK (ed.)), pp. 273–284. ASTM, Philadelphia, PA. ASTM STP 1081.

Moore CA (1980) *Landfill and Surface Impoundment Performance Evaluation.* US EPA, Washington, DC. Report SW-869.

Moo-Young H, Myers TE, Townsend D and Ochola C (1999) The migration of contaminants utilized in dredgings operations. *Engineering Geology* **53**: 167–176.

Moo-Young HK and Tucker WR (2002) Evaluation of vacuum filtration testing for geotextile tubes. *Geotextiles and Geomembranes* **20(3)**: 191–212.

Moo-Young HK, Gaffney DA and Mo X (2002) Testing procedures to assess the viability of dewatering with geotextile tubes. *Geotextiles and Geomembranes* **20(5)**: 289–303.

Mori H, Miki H and Tsuneoka N (2002) The geotube method for dioxin-contaminated soil. *Geotextiles and Geomembranes* **20(5)**: 279–342.

Muthukumaran AE and Ilmaparuthi K (2006) Laboratory studies on geotextile filters as used in geotextile tube dewatering. *Geotextiles and Geomembranes* **24(4)**: 210–219.

Narejo DB and Koerner RM (1992) A dynamic filtration test. *Geotextiles and Geomembranes* **11(4–6)**: 395–400.

Novak JT and Langford M (1977) The use of polymers for improving chemical sludge dewatering on sand beds. *Journal of American Water Works Association* **69(2)**: 106–115.

Novak JT, Haugan BE and Knocke WR (1979) Dewatering of sludges that penetrate through filtering media. *Proceedings of the 34th Purdue International Waste Conference.* Ann Arbor, Lafayette, IN, pp. 615–627.

Ogink HJM (1975) *Investigations on the Hydraulic Characteristics of Synthetic Geotextiles.* Delft Hydraulics Laboratory. Publication No. 146. (As cited by Fischer (1994).)

Palmeira EM, Remigio AFN, Ramos MLG and Bernardes RS (2008) Study on biological clogging of nonwoven geotextiles under leachate flow. *Geotextiles and Geomembranes* **26(3)**: 205–219.

Pilarczyk KW (2000) *Geosynthetics and Geosystems in Hydraulic and Coastal Engineering.* Balkema, Rotterdam.

Prapaharan S, Holtz RD and Luna JD (1989) Pore size distribution of nonwoven geotextiles. *Geotechnical Testing Journal, ASTM* **12(4)**: 261–268.

Rankilor PR (1981) *Membranes in Ground Engineering.* Wiley, New York.

Rohde J and Gribb MM (1990) Biological and particular clogging of geotextile/soil filter systems. In *Geosynthetics Testing for Waste Containment Application* (Koerner RM (ed.)). ASTM, Philadelphia, PA, pp. 299–312. ASTM STP 1081.

Rollin AL and Lombard G (1988) Mechanisms affecting long-term filtration behavior of geotextiles. *Geotextiles and Geomembranes* **7**: 119–145.

Sansone LJ and Koerner RM (1992) Fine fraction filtration test to assess geotextile filter performance. *Geotextiles and Geomembranes* **11(4–6)**: 371–393.

Schober W and Teindl H (1979) Filter criteria for geotextiles. *Proceedings of 7th European Conference on Soil Mechanics and Foundation Engineering*, Brighton, vol. 2, pp. 121–129.

Shuckrow JL, Pajak AP and Osheka JW (1981) *Concentration Technologies for Hazardous Aqueous Waste Treatment.* US EPA, Washington, DC. Report EPA-600/2-81-019. (As cited by Koerner and Koerner (1992).)

Sweetland DB (1977) The performance of nonwoven fabrics as drainage screens in subdrains. Master of Science Thesis, University of Strathclyde, Glasgow.

Van der Sluys L and Dierickx W (1991) Comparative studies of different porometry determination methods for geotextiles. *Geotextiles and Geomembranes* **9**. 183–198.

Vermeersch OG and Mlynarek J (1996) Determination of the pore size distribution of nonwoven geotextiles by a modified flow porometry technique. In *Recent Developments in Geotextile Filters and Prefabricated Drainage Geocomposites* (Bhatia SK and Suits LD (eds)). ASTM, Philadelphia, PA. ASTM STP 1281.

Williams ND and Abouzakhm MA (1989) Evaluation of geotextile/soil filtration characteristics using the hydraulic conductivity ratio analysis. *Geotextiles and Geomembranes* **8**: 1–26.

Young CW, Nummo TJ, Jasinski MP, Cogley DR and Capone SV (1982) *Clogging of Leachate Collection System used in Hazardous Waste Disposal Facilities*. Paper prepared for the US EPA, Research Triangle Park, NC. (As cited by Koerner and Koerner (1992).)

Ziani N, Lafleur J, Mlynarek J and Rollin AL (1995) Silt barriers to protect the environment. *Proceedings of Geosynthetics '95*, Nashville. Industrial Fabrics Association International, St Paul, MN, pp. 253–266.

Zitscher FF (1975) Recommendations for the use of plastics in soil and hydraulic engineering. *Die bautechnik* **52(12)**: 397–402.

Handbook of Geosynthetic Engineering
ISBN 978-0-7277-4175-2

http://dx.doi.org/10.1680/hge.41752.089

Chapter 4
Retaining walls

B. M. Das
Dean Emeritus, California State University, Sacramento, USA

Geosynthetics (geotextiles and geogrids) have been used extensively in the construction of retaining walls as a reinforcement material since their development. In most cases, granular materials are used as the backfill. This chapter presents the design considerations and stability analysis of retaining walls with geotextiles and geogrids as the reinforcement in granular backfill, and describes their stepwise general design procedure. Numerical examples illustrate and explain the design steps. Some simple questions are presented to assess understanding of the analysis and design concepts.

4.1. Introduction

Since the early 1970s, geotextiles have been used to reinforce soil in the construction of retaining walls in many parts of the world. In the early part of the 1980s, Netlon Ltd in the UK was the first to produce geogrids. In 1982, Tensar Corporation (now Tensar International Corporation) introduced geogrids in the USA. Since then, geogrids have been increasingly used as a soil-reinforcement material in the construction of retaining walls. This chapter provides general guidelines for designing retaining walls using geotextiles and geogrids as reinforcing materials in granular soil used as the backfill material. These walls are flexible compared with the rigid retaining walls constructed with reinforced concrete.

Figure 4.1 shows a schematic diagram of a geotextile-reinforced retaining wall. In most cases, a granular material is used as the backfill. In this type of retaining wall, the facing of the wall is formed by lapping the sheets, as shown with a lap length of l_1. When construction of the wall is finished, the exposed face of the wall must be covered; otherwise, the geotextile will deteriorate from exposure to ultraviolet light. *Bitumen emulsion* or *gunite* is sprayed on the wall face. A wire mesh anchored to the geotextile facing may be necessary to keep the coating on the face of the wall. Schematic diagrams of some typical retaining walls constructed with geogrid reinforcement are shown in Figure 4.2. Figure 4.2(a) shows a geogrid wrap-around wall. A geogrid-reinforced wall with gabion facing is shown in Figure 4.2(b). Figure 4.2(c) shows a vertical retaining wall with precast concrete panels as the facing.

4.2. Design considerations

4.2.1 Stability

At the present time, common practice in designing retaining walls with geosynthetic reinforcement is the use of limit equilibrium analysis. The analysis consists of two major parts:

(*a*) *internal stability*, which involves determining the tension and pull-out resistance in the reinforcing elements, the length of reinforcement, and the integrity of the facing elements
(*b*) *external stability*, which involves checking the overall stability of the stabilised mass as it relates to sliding, overturning, bearing capacity failure and deep-seated stability (Figure 4.3).

4.2.2 Lateral earth pressure

In order to conduct the stability check described above, the lateral earth pressure behind the retaining wall must be determined. The general guidelines for determining the horizontal and vertical earth pressures used in the design of retaining walls with geosynthetic reinforcement follow.

Figure 4.4 shows a retaining wall with a granular backfill having a unit weight of γ_1 and an effective friction angle of ϕ'_1. Below the base of the retaining wall, the *in situ* soil has been excavated and recompacted with the granular soil used for backfill. Below the backfill, the *in situ* soil has a unit weight of γ_2, an effective friction angle of ϕ'_2 and a cohesion of c'_2. A surcharge with an intensity of q per unit area lies at the top of the retaining wall. The wall has geosynthetic reinforcement ties at depths $z = 0$, S_V, $2S_V, \ldots,$ NS_V. The height of the wall is $NS_V = H$.

According to the Rankine active earth pressure theory:

$$\sigma'_a = \sigma'_v K_a - 2c'\sqrt{K_a} \tag{4.1}$$

where σ'_a is the Rankine active earth pressure at any depth z.

For dry granular soils with no surcharge at the top, $c' = 0$, $\sigma'_v = \gamma_1 z$ and $K_a = \tan^2(45 - \phi'_1/2)$. Thus

$$\sigma'_{a(1)} = \gamma_1 z K_a \tag{4.2}$$

Figure 4.1 A geotextile-reinforced retaining wall

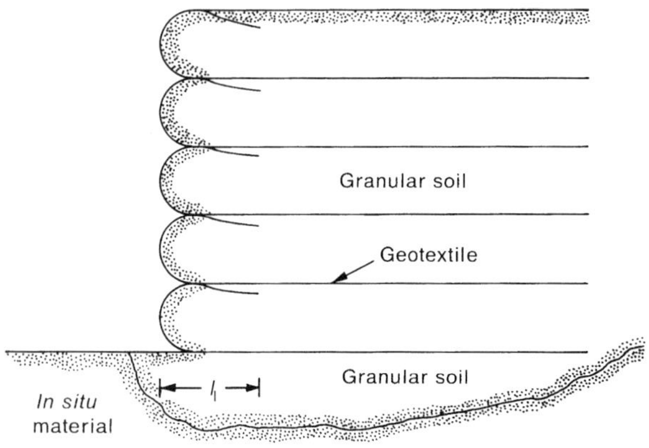

When a surcharge is added at the top, as shown in Figure 4.4(a),

$$\sigma'_v = \underbrace{\sigma'_{v(1)}}_{\text{due to soil only}} + \underbrace{\sigma'_{v(2)}}_{\text{due to the surcharge}} = \gamma_1 z \quad (4.3)$$

Figure 4.3 External stability checks: (a) sliding; (b) overturning; (c) bearing capacity; and (d) deep turning stability

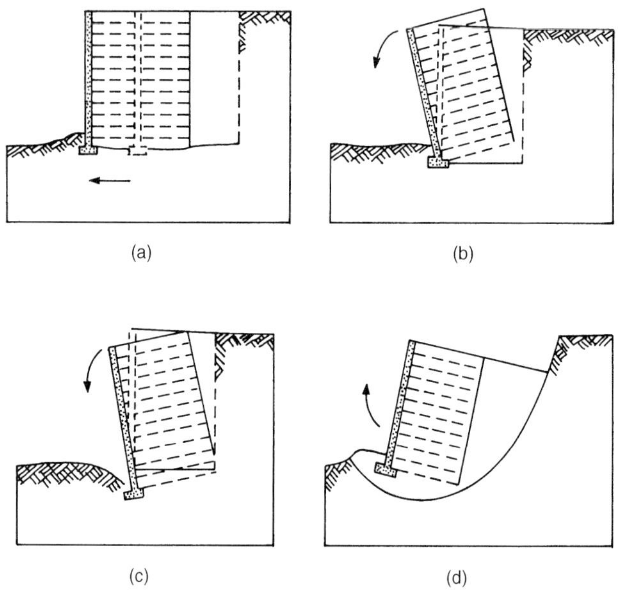

Figure 4.2 Typical schematic diagrams of retaining walls with geogrid reinforcement: (a) a geogrid wrap-around wall; (b) a wall with gabion facing; and (c) a concrete-panel-faced wall

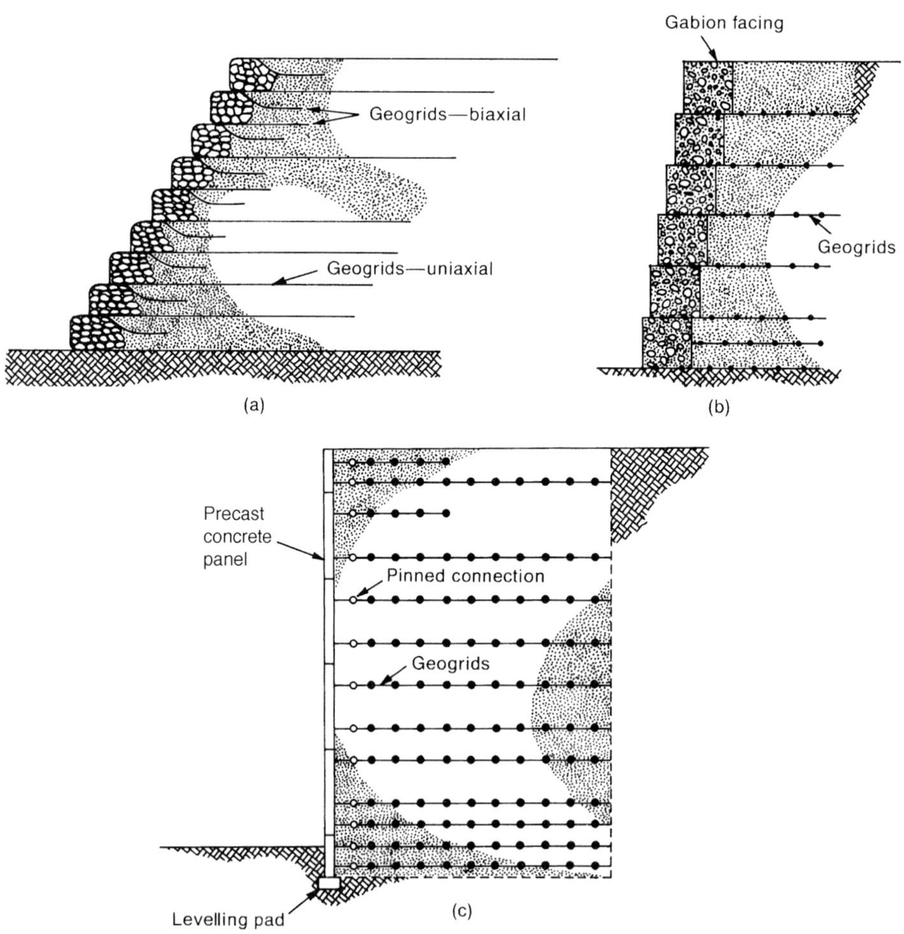

Figure 4.4 Analysis of a reinforced earth-retaining wall

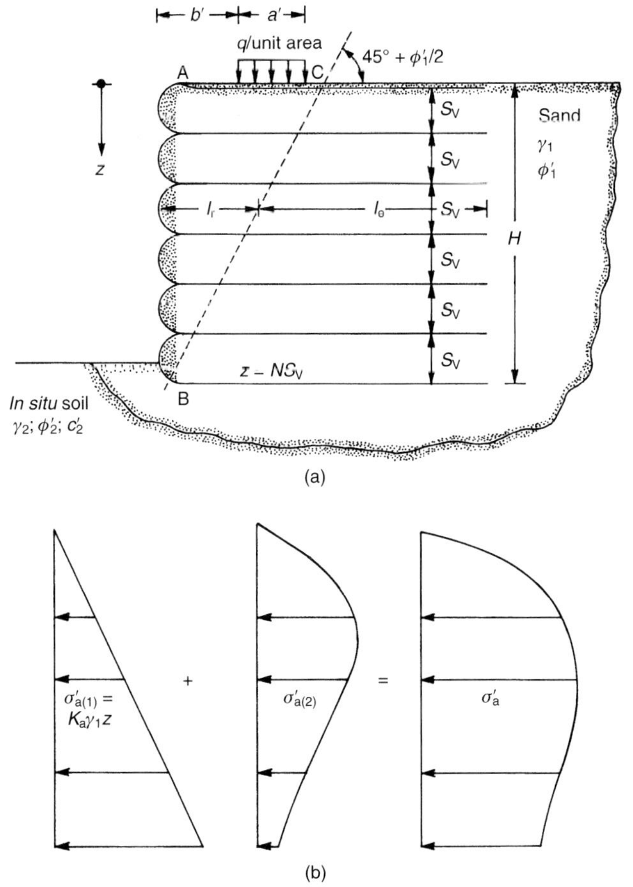

The magnitude of $\sigma'_{v(2)}$ can be calculated by using the 2 : 1 method of stress distribution, and is shown in Figure 4.5(a). According to Laba and Kennedy (1986):

$$\sigma'_{v(2)} = \frac{qa'}{a' + z} \quad (\text{for } z \leq 2b') \tag{4.4}$$

and

$$\sigma'_{v(2)} = \frac{qa'}{a' + \dfrac{z}{2} + b'} \quad (\text{for } z > 2b') \tag{4.5}$$

Also, when a surcharge is added at the top, the lateral pressure at any depth is

$$\sigma'_a = \underbrace{\sigma'_{a(1)}}_{= K_a\gamma_1 z \text{ due to soil only}} + \underbrace{\sigma'_{a(2)}}_{\text{due to the surcharge}} \tag{4.6}$$

According to Laba and Kennedy (1986), $\sigma'_{a(2)}$ may be expressed as (Figure 4.5(b)):

$$\sigma'_{a(2)} = M\left(\frac{2q}{\pi}\underbrace{(\beta - \sin\beta\cos 2\alpha)}_{\text{in radians}}\right) \tag{4.7}$$

Figure 4.5 (a) Notation for the relationship of $\sigma'_{v(2)}$ – Equations (4.4) and (4.5); and (b) notation for the relationship of $\sigma'_{a(2)}$ – Equations (4.7) and (4.8)

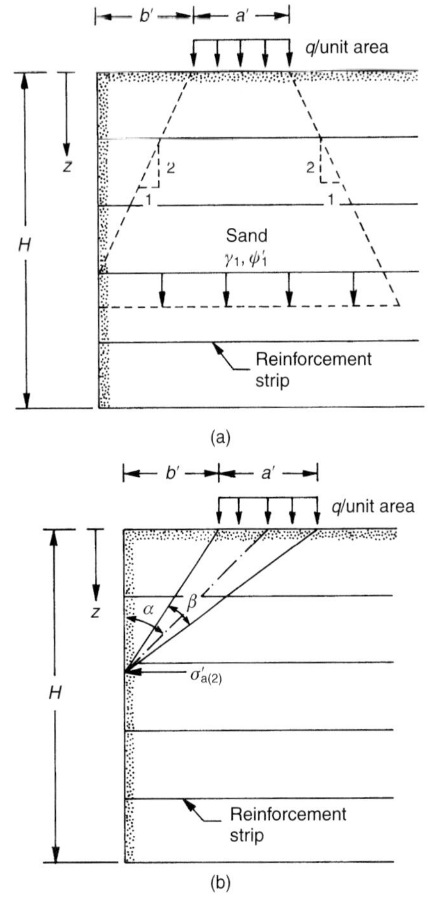

where

$$M = 1.4 - \frac{0.4b'}{0.14H} \geq 1 \tag{4.8}$$

The total active (lateral) pressure distribution on the retaining wall, calculated by using Equations 4.6, 4.7 and 4.8, is shown in Figure 4.4(b).

4.2.3 Tie force

Refer again to Figure 4.4. The tie force per unit length of the wall, T, developed at any depth z can be calculated as

$$T = \text{active earth pressure at depth } z \times S_V = \sigma'_a S_V \tag{4.9}$$

4.3. Design procedure for retaining walls with geotextile reinforcement

Referring to Figure 4.6, below is a step-by-step procedure for the design of retaining walls using a geotextile as the reinforcement in the granular backfill.

4.3.1 General

(*a*) Determine the height of the wall, H, and the properties of the granular backfill, such as unit weight (γ_1) and angle of friction (ϕ'_1).

Figure 4.6 Retaining wall with geotextile reinforcement

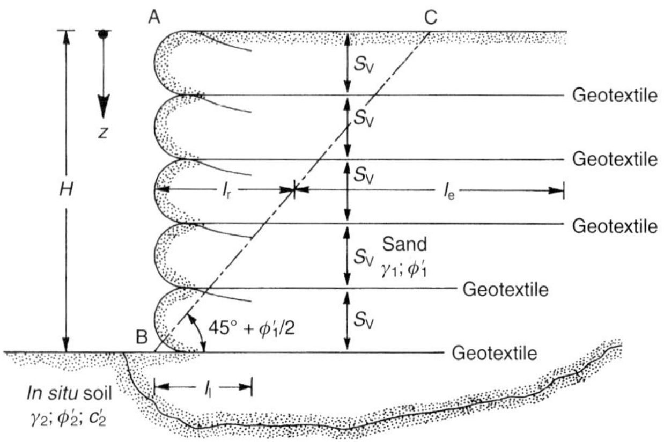

(*b*) Obtain the soil–geotextile interface friction angle (ϕ'_F).

(*c*) Obtain the *in situ* soil parameters, such as the unit weight (γ_2), effective friction angle (ϕ'_2) and cohesion (c'_2).

4.3.2 Internal stability

(*a*) Determine the active pressure distribution on the wall from

$$\sigma'_a = K_a \sigma'_v = K_a \gamma_1 z \tag{4.10}$$

where K_a is the Rankine active earth pressure coefficient $= \tan^2(45^\circ - \phi'_1/2)$, γ_1 is the unit weight of the granular backfill, and ϕ'_1 is the effective friction angle of the granular backfill.

(*b*) Select a geotextile fabric that has an allowable tensile strength of σ_G (kN/m). The allowable tensile strength for retaining wall construction may be expressed as (Koerner, 2005)

$$\sigma_G = \frac{\sigma_{ult}}{RF_{id} \times RF_{cr} \times RF_{cbd}} \tag{4.11}$$

where σ_{ult} is the ultimate tensile strength, RF_{id} is the reduction factor for installation damage, RF_{cr} is the reduction factor for creep, and RF_{cbd} is the reduction factor for chemical and biological degradation. The recommended values of the reduction factors are as follows:

RF_{id}: 1.1–2.0
RF_{cr}: 2.0–4.0
RF_{cbd}: 1.0–1.5

(*c*) Determine the vertical spacing of the layers at any depth z from

$$S_V = \frac{\sigma_G}{\sigma'_a \times FS_{(B)}} = \frac{\sigma_G}{\gamma_1 z K_a \times FS_{(B)}} \tag{4.12}$$

where $FS_{(B)}$ is the factor of safety against rupture of reinforcement. The magnitude of $FS_{(B)}$ is generally taken as 1.3–1.5. In all practical cases, S_V should not exceed 800 mm. For $H \leq 5$, the maximum value of S_V should be 500 mm.

(*d*) Geotextile layers at any depth z will fail by pull-out if the frictional resistance developed along their surfaces is less than the force to which the layers are being subjected. The *effective length* of the geotextile layers along which the frictional resistance is developed may be taken conservatively as the length that extends *beyond the limits of the Rankine active failure zone*, which is the zone ABC in Figure 4.6. Line BC in Figure 4.6 makes an angle of $(45^\circ + \phi'_1/2)$ with the horizontal. Now, the maximum friction force F_R per unit length of the wall that can be realised for a geotextile layer at depth z is

$$F_R = 2\sigma'_v l_e \tan \phi'_F \tag{4.13}$$

Thus, the factor of safety against pull-out, $FS_{(P)}$, at any depth z (from Equations 4.9 and 4.13) is

$$FS_{(P)} = \frac{F_R}{T} = \frac{2\sigma'_v l_e \tan \phi'_F}{\sigma'_a S_V} \tag{4.14}$$

or

$$l_e = \frac{S_V \sigma'_a \times FS_{(P)}}{2\sigma'_v \tan \phi'_F} \tag{4.15}$$

where

$$\sigma'_a = \gamma_1 z K_a$$

$$\sigma'_v = \gamma_1 z$$

So,

$$l_e = \frac{S_V K_a \times FS_{(P)}}{2 \tan \phi'_F} \tag{4.16}$$

The magnitude of $FS_{(P)}$ is generally taken as 1.3–1.5. If an experimental value of ϕ'_F is not available, it may be assumed to be about $\frac{2}{3}\phi'_1$. Now, determine the length of each layer of geotextile as

$$L = l_r + l_e \tag{4.17}$$

where

$$l_r = \frac{H - z}{\tan\left(45 + \frac{\phi'_1}{2}\right)} \tag{4.18}$$

So,

$$L = \frac{H - z}{\tan\left(45 + \frac{\phi'_1}{2}\right)} + \frac{S_V K_a \times FS_{(P)}}{2 \tan \phi'_F} \tag{4.19}$$

(*e*) Determine the lap length, l_1, from

$$l_1 = \frac{S_V \sigma'_a \times FS_{(P)}}{4\sigma'_v \tan \phi'_F} \quad (4.20)$$

The minimum lap length should be 1 m.

4.3.3 External stability

The external stability check includes checks for *overturning*, *sliding* and *bearing capacity failure*. They can be done as follows.

4.3.3.1 Check for overturning

(*a*) Referring to Figure 4.7, calculate the Rankine active force P_a per unit length of the wall as

$$P_a = \tfrac{1}{2}\gamma_1 H^2 K_a \quad (4.21)$$

where K_a is the Rankine lateral earth pressure coefficient $= \tan^2(45 - \phi'_1/2)$. The active force P_a acts as a distance of $z' = H/3$ measured from the bottom of the wall.

(*b*) Calculate the overturning moment, M_O, due to the active force as

$$M_O = P_a z' \quad (4.22)$$

(*c*) Calculate the resisting moment, M_R, due to the weight of the wall as

$$M_R = Wx \quad (4.23)$$

where $W = LH\gamma_1$ and x is the distance from the wall to the centre of gravity of soil mass (zone ABCD) considered.

Figure 4.7 External stability check

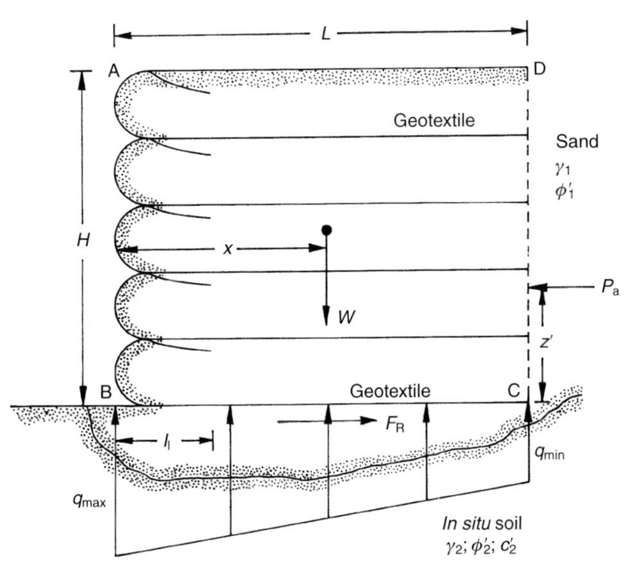

(*d*) The factor of safety against overturning can then be calculated as

$$FS_{(overturning)} = \frac{M_R}{M_O} \quad (4.24)$$

The factor of safety against overturning should be at least 3.

4.3.3.2 Check for sliding

(*a*) Calculate the horizontal driving force at the bottom of the wall, F_H, as

$$F_H = P_a \quad (4.25)$$

(*b*) Calculate the horizontal resisting force along BC (Figure 4.7) as

$$F_R = W \tan \phi'_F \quad (4.26)$$

(*c*) Check the factor of sliding against sliding as

$$FS_{(sliding)} = \frac{F_R}{F_H} \quad (4.27)$$

The minimum factor of safety against sliding should be 1.5.

4.3.3.3 Check for bearing capacity failure

The check against bearing capacity failure can be conducted using Meyerhof's effective area method (Meyerhof, 1953). The check can be done as follows:

(*a*) Calculate the eccentricity e due to the forces on the reinforced block, or

$$e = \frac{L}{2} - \frac{M_R - M_O}{W} \quad (4.28)$$

The eccentricity should be less than $L/6$.

(*b*) The stress q on the soil below the reinforced block can be given as

$$\left.\begin{matrix} q_{max} \\ q_{min} \end{matrix}\right\} = \frac{W}{L}\left(1 \pm \frac{6e}{L}\right) \quad (4.29)$$

(*c*) The ultimate bearing capacity q_u of the soil with the eccentric loading can be given as (Meyerhof, 1963)

$$q_u = c'_2 N_c + \tfrac{1}{2}\gamma_2 (L - 2e) N_\gamma \quad (4.30)$$

The variation of N_c and N_γ with ϕ'_2 was given by Prandtl (1921) and Vesic (1973), respectively, by the following relationships:

$$N_c = \left[\tan^2\left(45° + \frac{\phi'_2}{2}\right) e^{\pi \tan \phi_2} - 1\right] \cot \phi'_2 \quad (4.31)$$

and

$$N_\gamma = 2\left[\tan^2\left(45° + \frac{\phi'_2}{2}\right) e^{\pi \tan \phi'_2} - 1\right] \tan \phi'_2 \quad (4.32)$$

Table 4.1 gives the variation of N_c and N_γ with ϕ'_2.

Table 4.1 Variation of N_c and N_γ with ϕ'_2

ϕ'_2	N_c	N_γ	ϕ'_2	N_c	N_γ	ϕ'_2	N_c	N_γ
0	5.14	0.00	17	12.34	3.53	34	42.16	41.06
1	5.38	0.07	18	13.10	4.07	35	46.12	48.03
2	5.63	0.15	19	13.93	4.68	36	50.59	56.31
3	5.90	0.24	20	14.83	5.39	37	55.63	66.19
4	6.19	0.34	21	15.82	6.20	38	61.35	78.03
5	6.49	0.45	22	16.88	7.13	39	67.87	92.25
6	6.81	0.57	23	18.05	8.20	40	75.31	109.41
7	7.16	0.71	24	19.32	9.44	41	83.86	130.22
8	7.53	0.86	25	20.72	10.88	42	93.71	155.55
9	7.92	1.03	26	22.25	12.54	43	105.11	186.54
10	8.35	1.22	27	23.94	14.47	44	118.37	224.64
11	8.80	1.44	28	25.80	16.72	45	133.88	271.76
12	9.28	1.69	29	27.86	19.34	46	152.10	330.35
13	9.81	1.97	30	30.14	22.40	47	173.64	403.67
14	10.37	2.29	31	32.67	25.99	48	199.26	496.01
15	10.98	2.65	32	35.49	30.22	49	229.93	613.16
16	11.63	3.06	33	38.64	35.19	50	266.89	762.89

(*d*) Calculate the factor of safety against bearing capacity failure as:

$$FS_{(\text{bearing capacity})} = \frac{q_u}{q_{max}} \geq 3 \qquad (4.33)$$

Example 4.1

A geotextile-reinforced retaining wall is 6 m high: $\gamma_1 = 16\ \text{kN/m}^3$, $\phi'_1 = 32°$, $\gamma_2 = 17.2\ \text{kN/m}^3$, $\phi'_2 = 20°$, $c'_2 = 30\ \text{kN/m}^2$ and $\sigma_{ult} = 52.5\ \text{kN/m}$. For the geotextile, $RF_{id} = 1.4$, $RF_{cr} = 2$ and $RF_{cbd} = 1.25$. Determine S_V, L and l_1. Assume $FS_{(B)}$ and $FS_{(P)}$ to be 1.5, and $\phi'_F \approx \frac{2}{3}\phi'_1$.

Solution

$$K_a = \tan^2\left(45° - \frac{\phi'_1}{2}\right) = \tan^2(45° - 16°) = 0.307$$

From Equation 4.11,

$$\sigma_G = \frac{\sigma_{ult}}{RF_{id} \times RF_{cr} \times RF_{cbd}} = \frac{52.5}{1.4 \times 2 \times 1.25} = 15\ \text{kN/m}$$

Determination of S_V. From Equation 4.12:

$$S_V = \frac{\sigma_G}{\gamma_1 z K_a \times FS_{(B)}}$$

We will perform a few trials to find S_V.

At $z = 2$ m:

$$S_V = \frac{15}{16 \times 2 \times 0.307 \times 1.5} = 1.02\ \text{m}$$

At $z = 4$ m:

$$S_V = \frac{15}{16 \times 4 \times 0.307 \times 1.5} = 0.51\ \text{m}$$

At $z = 6$ m:

$$S_V = \frac{15}{16 \times 6 \times 0.307 \times 1.5} = 0.34\ \text{m}$$

For safety, use $S_V = 0.4$ m for $z = 0$–4 m and $S_V = 0.3$ m for $z = 4$–6 m (Figure 4.8).

Figure 4.8 Example 4.1

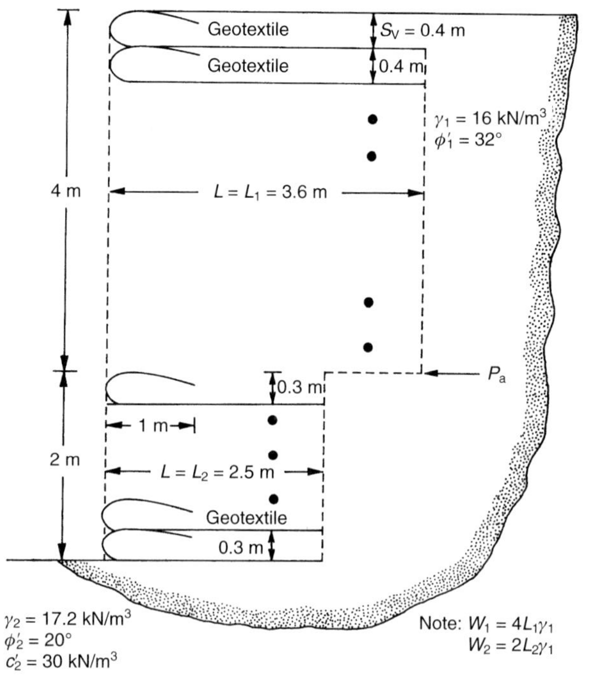

Determination of L. From Equation 4.19:

$$L = \frac{H - z}{\tan\left(45^\circ + \frac{\phi_1'}{2}\right)} + \frac{S_V K_a \times FS_{(P)}}{2\tan\phi_F'}$$

$$= \frac{6 - z}{\tan\left(45^\circ + \frac{32^\circ}{2}\right)} + \frac{S_V \times 0.307 \times 1.5}{2\tan(\frac{2}{3} \times 32^\circ)}$$

$$= 0.554 \times (6 - z) + 0.59S_v$$

Now the following table can be prepared:

z: m	S_V: m	L: m
0.4	0.4	3.56
2.0	0.4	2.45
4.0	0.4	1.35
6.0	0.3	0.18

For safety, use $L = 3.6$ m for $z \leq 4$ m and $L = 2.5$ m for $z = 4$–6 m.

Determination of l_1. From Equation 4.20:

$$l_1 = \frac{S_V \sigma_a' \times FS_{(P)}}{4\sigma_v' \tan\phi_F'} = \frac{S_V K_a FS_{(P)}}{4\tan\phi_F'}$$

At $z = 0.4$ m:

$$l_1 = \frac{0.4 \times 0.307 \times 1.5}{4\tan(\frac{2}{3} \times 32^\circ)} \approx 0.12\text{m}$$

Use $l_1 = 1$ m (see Figure 4.8).

Example 4.2

For the retaining wall shown in Figure 4.8, calculate the factor of safety against overturning, sliding and bearing capacity failure. Use the soil and geotextile parameters given in Example 4.1.

Solution

Factor of safety against overturning. From Equation 4.21:

$$P_a = \frac{1}{2}\gamma_1 H^2 \tan^2\left(45^\circ - \frac{\phi_1'}{2}\right)$$

$$= \frac{1}{2} \times 16 \times 6^2 \times \tan^2\left(45^\circ - \frac{32^\circ}{2}\right) \approx 88.5\,\text{kN/m}$$

$$z' = \frac{H}{3} = \frac{6}{3} = 2\,\text{m}$$

Taking the moment about the bottom of the wall:

$$M_O = P_a z' = 88.5 \times 2 = 177\,\text{kN m/m}$$

$$M_R = (4 \times L_1 \times \gamma_1)\frac{L_1}{2} + (2 \times L_2 \times \gamma_1)\frac{L_2}{2}$$

$$= (4 \times 3.6 \times 16)\frac{3.6}{2} + (2 \times 2.5 \times 16)\frac{2.5}{2}$$

$$= 514.7\,\text{Kn m/m}$$

$$FS_{(\text{overturning})} = \frac{514.7}{177} = 2.91 \qquad (\textit{Answer})$$

Factor of safety against sliding. From Equation 4.26:

$$F_R = (W_1 + W_2)\tan(\tfrac{2}{3}\phi_1')$$

$$= [(4 \times 3.6 \times 16) + (2 \times 2.5 \times 16)]\tan(\tfrac{2}{3} \times 32^\circ)$$

$$= 121.2\,\text{kN/m}$$

From Equation 4.27:

$$FS_{(\text{sliding})} = \frac{F_R}{F_H} = \frac{121.2}{88.5} \approx 1.37 \qquad (\textit{Answer})$$

Factor of safety against bearing capacity failure. From Equation 4.28:

$$e = \frac{L_2}{2} - \frac{M_R - M_O}{W_1 + W_2}$$

$$= \frac{2.5}{2} - \frac{514.7 - 177}{(4 \times 3.6 \times 16) + (2 \times 2.5 \times 16)} = 0.16\,\text{m}$$

From Equation 4.29:

$$q_{max} = \frac{W_1 + W_2}{L_2}\left(1 + \frac{6e}{L_2}\right)$$

$$= \left(\frac{(4 \times 3.6 \times 16) + (2 \times 2.5 \times 16)}{2.5}\right)\left(1 + \frac{6 \times 0.16}{2.5}\right)$$

$$= 171.8\,\text{kN/m}^2$$

From Equation 4.30:

$$q_u = c_2' N_c + \tfrac{1}{2}\gamma_2(L_2 - 2e)N_\gamma$$

For $\phi_2' = 20^\circ$, from Table 4.1, $N_c = 14.83$ and $N_\gamma = 5.39$:

$$q_u = (30 \times 14.83) + \tfrac{1}{2} \times 17.2 \times [2.5 - (2 \times 0.16)] \times 5.39$$

$$= 545.95\,\text{kN/m}^2$$

$$FS_{(\text{bearing capacity})} = \frac{545.95}{171.8} = 3.18 \qquad (\textit{Answer})$$

Comments

For practicality, since $FS_{(\text{overturning})}$ and $FS_{(\text{sliding})}$ are less than 3 and 1.5, respectively, it will be necessary to redesign the retaining wall. This can be done by making $L_1 = L_2 = 3.6$ m (see Figure 4.9). With this assumption:

$$M_O = 177\,\text{kN m/m}$$

Figure 4.9 Example 4.2

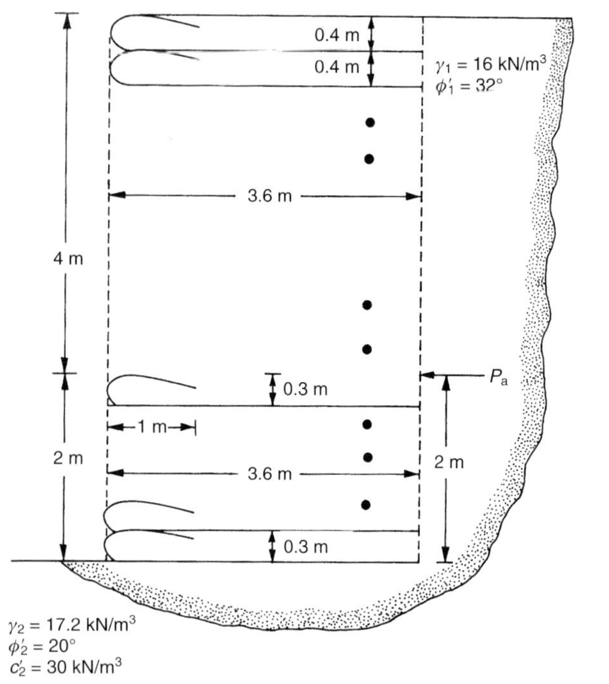

$$M_R = (6 \times 3.6 \times 16)\left(\frac{3.6}{2}\right) = 622\,\text{kN/m}$$

$$FS_{(\text{overturning})} = \frac{622}{177} = 3.5 \qquad (Answer)$$

$$F_R = (6 \times 3.6 \times 16)\tan(\tfrac{2}{3} \times 32^\circ) = 135\,\text{kN/m}$$

$$FS_{(\text{sliding})} = \frac{135}{88.5} = 1.5 \qquad (Answer)$$

$$e = \frac{3.6}{2} - \frac{622 - 177}{6 \times 3.6 \times 16} = 0.51 < \frac{L}{6}$$

$$q_{\max} = \left(\frac{6 \times 3.6 \times 16}{3.6}\right)\left(1 + \frac{6 \times 0.51}{3.6}\right) = 177.5\,\text{kN/m}^2$$

$$q_u = (30 \times 14.98) + \tfrac{1}{2} \times 17.2 \times [3.6 - (2 \times 0.51)] \times 5.39$$
$$= 569\,\text{kN/m}^2$$

$$FS_{(\text{bearing capacity})} = \frac{569}{177.6} = 3.2 \qquad (Answer)$$

4.4. Design procedure for retaining walls with geogrid reinforcement

The design of retaining walls with geogrid reinforcement, as shown in Figure 4.2(c), can be done using procedures similar to those described in Section 4.3, with the following modifications:

- Equation 4.12 should be replaced by

$$S_V = \frac{T_{\text{all}} C_r}{\gamma_1 z K_a \times FS_{(B)}} \qquad (4.34)$$

where

$$T_{\text{all}} = \text{allowable tensile strength}$$
$$= \frac{T_{\text{ult}}}{RF_{\text{id}} \times RF_{\text{cr}} \times RF_{\text{cbd}}} \qquad (4.35)$$

where T_{ult} is the ultimate tensile strength, RF_{id} is the reduction factor for installation damage (1.1–1.4), RF_{cr} is the reduction factor for creep (2.0–3.0), and RF_{cbd} is the reduction factor for chemical and biological degradation (1.1–1.5). C_r is the *coverage ratio*, which is the fractional plan area at any particular elevation that is actually occupied by the geogrid. For example, if there is a 0.3 m-wide space between each 1.2 m-wide piece of geogrid, the coverage ratio is

$$C_r = \frac{1.2\,\text{m}}{1.2\,\text{m} + 0.3\,\text{m}} = 0.8$$

- Equations 4.15 and 4.19 should be replaced, respectively, by

$$l_e = \frac{S_V K_a \times FS_{(P)}}{2 C_r C_i \tan \phi'_1} \qquad (4.36)$$

where C_i is the interaction coefficient and

$$L = \frac{H - z}{\tan\left(45 + \frac{\phi'_1}{2}\right)} + \frac{S_V K_a \times FS_{(P)}}{2 C_r C_i \tan \phi'_1} \qquad (4.37)$$

The interaction coefficient, C_i, can be determined experimentally in the laboratory. The following is an approximate range for C_i for various types of backfill:

Gravel, sandy gravel:	0.75–0.8
Well graded sand, gravelly sand:	0.7–0.75
Fine sand, silty sand:	0.55–0.6

Example 4.3

Consider a geogrid-reinforced retaining wall. Referring to Figure 4.2(c), given $H = 6$ m, $\gamma_1 = 16.5$ kN/m³, $\phi'_1 = 35^\circ$, $T_{\text{all}} = 45$ kN/m, $FS_{(B)} = 1.5$, $FS_{(P)} = 1.5$, $C_r = 0.8$ and $C_i = 0.75$, determine S_V and L for the design of the wall.

Solution

$$K_a = \tan^2\left(45 - \frac{\phi'_1}{2}\right) = \tan^2\left(45 - \frac{35}{2}\right) = 0.27$$

Determination of S_V. From Equation 4.34:

$$S_V = \frac{T_{\text{all}} C_r}{\gamma_1 z K_a \times FS_{(B)}} = \frac{45 \times 0.8}{16.5 \times z \times 0.27 \times 1.5} = \frac{5.39}{z}$$

At $z = 2$ m:

$$S_V = \frac{5.39}{2} = 2.7\,\text{m}$$

At $z=4$ m:

$$S_V = \frac{5.39}{4} = 1.35\,\text{m}$$

At $z=5$ m:

$$S_V = \frac{5.39}{5} = 1.08\,\text{m}$$

Use $S_V \approx 0.5$ m (which is the maximum: see Section 4.3.2).

Determination of L. From Equation (4.37):

$$L = \frac{H - z}{\tan\left(45 + \frac{\phi_1'}{2}\right)} + \frac{S_V K_a \times FS_{(P)}}{2C_r C_i \tan \phi_1'}$$

$$= \frac{6 - z}{\tan\left(45 + \frac{35}{2}\right)} + \frac{0.5\,\text{m} \times 0.27 \times 1.5}{2 \times 0.8 \times 0.75 \times \tan 35°}$$

At $z=1$ m:

$$L = 0.52 \times (6 - 1) + 0.241 = 2.841\,\text{m}$$

At $z=3$ m:

$$L = 0.52 \times (6 - 3) + 0.241 = 1.801\,\text{m}$$

At $z=5$ m:

$$L = 0.52 \times (6 - 5) + 0.241 = 0.761\,\text{m}$$

So, use $L=3$ m for $z=0$–6 m.

Note: $L=3$ may not be sufficient to satisfy the conditions of external stability, and may need to be increased. In most cases, $L=0.8H$ to $1.2H$ will be sufficient.

4.5. Self-evaluation questions

Use the following parameters in Figure 4.5 for questions 4.1 and 4.2:

Wall:	$H=8$ m
Soil:	$\gamma_1 = 17$ kN/m^3, $\phi_1' = 35°$
Reinforcement:	$S_V = 1$ m and $S_H = 1.5$ m
Surcharge:	$q = 70$ kN/m^2, $a' = 1.5$ m, $b' = 2$ m

4.1. The vertical stress σ_v' at a depth of $z=4$ m is nearly equal to:

(*a*) 120 kN/m^2
(*b*) 87 kN/m^2
(*c*) 75 kN/m^2
(*d*) 60 kN/m^2

4.2. The lateral pressure σ_a' at a depth $z=6$ m is nearly equal to:

(*a*) 22 kN/m^2
(*b*) 31 kN/m^2
(*c*) 46 kN/m^2
(*d*) 52 kN/m^2

In questions 4.3–4.9, consider a retaining wall with geotextile-reinforced granular backfill, as shown in Figure 4.7, with (H = 8 m):

Backfill soil:	$\gamma_1 = 17.8$ kN/m^3, $\phi_1' = 36°$
Geotextile:	$RF_{id} = 1.3$, $RF_{cr} = 2.4$, $RF_{cbd} = 1.25$, $FS_{(B)} = 1.5$, $FS_{(P)} = 1.5$

4.3. If σ_{ult} for the geotextile is 48 kN/m, the allowable tensile strength (σ_G) is approximately:

(*a*) 18 kN/m
(*b*) 14 kN/m
(*c*) 12 kN/m
(*d*) 10 kN/m

4.4. If $\sigma_G = 14$ kN/m, the maximum value of the reinforcement layer spacing (S_V) is approximately:

(*a*) 250 mm
(*b*) 300 mm
(*c*) 450 mm
(*d*) 500 mm

4.5. If $\sigma_G = 14$ kN/m, $S_V = 250$ mm and $\phi_F' = 24°$, the minimum length (L) of reinforcement at a depth $z=5$ m is approximately:

(*a*) 3.2 m
(*b*) 1.6 m
(*c*) 4.3 m
(*d*) 2.1 m

4.6. The active force P_a of the wall will be subjected to nearly:

(*a*) 180 kN/m
(*b*) 160 kN/m
(c) 150 kN/m
(*d*) 130 kN/m

4.7. If the length of the geotextile layer (L) is 6.5 m (for $z=0$–8 m), the factor of safety against overturning is nearly equal to:

(*a*) 9.4
(*b*) 8.3
(*c*) 7.2
(*d*) 5.1

4.8. If the length of the geotextile layer is 6 m (for $z=0$–8 m), the factor of safety against sliding is nearly equal to:

(*a*) 2.5
(*b*) 3.2
(*c*) 3.9
(*d*) 4.6

4.9. If the length of the geotextile layer (L) is 6.5 m (for $z=0$–8 m), the maximum stress (q_{max}) below the reinforced block is approximately equal to:

(*a*) 170 kN/m^2
(*b*) 190 kN/m^2
(*c*) 210 kN/m^2
(*d*) 230 kN/m^2

4.10. Consider a geogrid-reinforced retaining wall, as shown in Figure 4.2(c), with:

Soil: $\gamma_1 = 19$ kN/m^3, $\phi'_1 = 38°$
Geogrid: $C_r = 0.8$, $RF_{id} = 1.25$, $RF_{cr} = 2.5$, $RF_{cbd} = 1.25$, $T_{ult} = 120$ kN/m, $FS_{(B)} = 1.4$

The maximum spacing of geogrid layers (S_V) needed at a depth $z = 5$ m is nearly equal to:

(*a*) 1.1 m
(*b*) 0.78 m
(*c*) 0.65 m
(*d*) 0.5 m

Answers

4.1(*b*), 4.2(*b*), 4.3(*c*), 4.4(*a*), 4.5(*b*), 4.6(*c*), 4.7(*a*), 4.8(*a*), 4.9(*b*), 4.10(*b*).

4.6. Summary of main points

- Retaining walls with geotextiles and geogrids as the reinforcement in granular backfill are flexible retaining walls that are easy to construct, and can withstand earthquake forces without undergoing total collapse. Further field and laboratory studies are now underway in many countries to evaluate the exact nature of the lateral earth pressure distribution behind this type of retaining wall, which will lead to more economical designs.
- The design of a retaining wall with granular backfill, reinforced with geosynthetics can be divided into two parts: (*a*) internal stability checks, which are the checks for resistance to pull-out; and (*b*) break (or rupture) of reinforcement layers.
- The external stability check involves determination of the factors of safety against overturning, sliding and bearing capacity failure for the mass of reinforced granular backfill.
- The ranges for factors of safety recommended for the stability check are as follows:

Factor of safety against:	Range
Rupture/break of reinforcement, $FS_{(B)}$	1.3–1.5
Pull-out of reinforcement, $FS_{(P)}$	1.3–1.5
Overturning (minimum), $FS_{(overturning)}$	3
Sliding (minimum), $FS_{(sliding)}$	1.5
Bearing capacity (minimum) failure, $FS_{(bearing\ capacity)}$	3

- The maximum spacing of the reinforcement layers (S_V) should be about 500 m for $H \leq 5$–6 m and about 800 mm for $H > 6$ m.

4.7. Some common questions and their answers

Questions

1. Is there any limitation on the height of geosynthetic-reinforced earth-retaining walls?
2. What type of footing is required for geosynthetic-reinforced earth-retaining walls?
3. Compared with concrete walls, what are the major advantages of geosynthetic-reinforced earth-retaining walls?
4. What lateral pressure is considered for designing facing elements?
5. Can backfill be cohesive soils?
6. Is regular maintenance required for geosynthetic-reinforced earth-retaining walls?

Answers

1. No.

2. No footing of any special kind is required. The lowest geosynthetic layer can be placed directly on the foundation soil; however, the foundation soil should be compacted to some extent to limit settlement.

3. Geosynthetic-reinforced earth-retaining walls have a low cost–benefit ratio and low environmental impact. Additionally, being flexible structures, they resist seismic forces without major failure.

4. The lateral pressure is resisted mainly by geosynthetic reinforcement layers installed within the backfill. Facing elements are provided to prevent near-surface soil erosion; so, their design is mainly based on the pressures required for their survivability.

5. With cohesive backfills, geosynthetic reinforcement will not bring significant advantages. Moreover, cohesive backfills are impervious, thus allowing additional lateral pressure in the form of porewater pressure development. So, cohesive backfills should be avoided; otherwise, special arrangements will be required to drain out the water from the backfill.

6. Like other structures, repairs/maintenance may occasionally be required, but the maintenance cost is comparatively low.

REFERENCES

Koerner RB (2005) *Designing with Geosynthetics*, 5th edn. Prentice-Hall, Englewood Cliffs, NJ.
Laba JT and Kennedy JB (1986) Reinforced earth retaining wall analysis and design. *Canadian Geotechnical Journal* **23(3)**: 317–326.
Meyerhof GG (1953) The bearing capacity of foundations under eccentric and inclined loads. *Proceedings of the 3rd International Conference on Soil Mechanics and Foundation Engineering*, Zurich, pp. 440–445.

Meyerhof GG (1963) Some recent research on the bearing capacity of foundations. *Canadian Geotechnical Journal* **1**: 16–26.

Prandtl L (1921) Uber die eindringungsfestigkeit (härte) plastischer baustoffe und die festigkeit von schneiden. *Zeitschrift für angewandte Mathematik und Mechanik* **1(1)**: 15–20.

Vesic AS (1973) Analysis of ultimate load of shallow foundations. *Journal of the Soil Mechanics and Foundations Division, ASCE* **99(1)**: 45–73.

Handbook of Geosynthetic Engineering
ISBN 978-0-7277-4175-2

http://dx.doi.org/10.1680/hge.41752.101

Chapter 5
Embankments

E. M. Palmeira
Department of Civil and Environmental Engineering, University of Brasília, Brazil

The stabilisation of embankments on soft soils is one of the most traditional applications of geosynthetics in geotechnical engineering. In this chapter, the utilisation of such materials to increase the safety factors of embankments and to accelerate their settlements due to soft soil consolidation is examined. The combination of geosynthetic layers and piles is also addressed for cases where minimum levels of settlement are required. The content of the chapter focuses on the more practical design approaches available in the literature, but comments on the use of more sophisticated design tools are also presented. Relevant theoretical and practical aspects on the use of geosynthetics in embankments on soft soils are also addressed.

5.1. Introduction

Geosynthetics can be very attractive for works involving embankments built on soft foundation soils. Basically, layers of geosynthetics can serve as reinforcing materials or can accelerate the process of consolidation of the soft subgrade. The former function usually aims for a temporary increase in the safety factor of the embankment, and is associated with a faster rate of construction or the use of steeper slopes than would be possible without reinforcement. The latter function can also be associated with the need for a more stable embankment or staged construction, but also helps to accelerate the consolidation settlements. Figure 5.1 summarises the usual functions of geosynthetics in embankments on soft soils.

An additional benefit brought about by the presence of the reinforcement is to provide separation between good-quality fill materials and the fine-grained foundation soil, as shown in Figure 5.2(a). This is achieved when the reinforcement is also a filter for the foundation soil, as may be the case for non-woven geotextiles, or when a geosynthetic filter is used in conjunction with the reinforcement. The presence of the reinforcement also reduces the consumption of fill material, because it minimises or avoids local failure mechanisms caused by construction equipment during transport, spreading and compaction of the fill material (Figure 5.2(b)).

Quite a few works can be found in the literature dealing with the design of embankments on soft soils using geosynthetics. In this chapter, the main contributions from these works are presented and discussed, along with drainage aspects.

5.2. Geosynthetics as a basal reinforcement in embankments

5.2.1 Reinforcement roles and aspects to be considered in the analysis

The use of geosynthetic reinforcement can significantly increase the factor of safety of the embankment. Its use is particularly attractive for low ratios between foundation-soil thickness and embankment-base width (say, less than 0.7). For thick foundation soils, the contribution of the reinforcement can be less significant, consisting mainly of the provision of a rough boundary between the base of the embankment and the foundation soil. This boundary will only marginally increase the overall factor of safety, depending on the variation of undrained strength with depth in the subgrade.

Figure 5.3 shows the possible failure mechanisms that may occur in an embankment built on soft soil. Figure 5.3(a) shows the possibility of failure inside the embankment, which may occur for very steep embankment slopes on reasonably strong subgrades. This type of mechanism has to be predicted using a stability analysis, and it is not the most critical mechanism for situations of soft soil foundations. Figure 5.3(b) presents, schematically, a mechanism of expulsion of the soft soil laterally. This mechanism may occur for heavily reinforced embankments on thin foundation layers. The most critical situation is shown schematically in Figure 5.3(c), where the failure mechanism is characterised by a well-defined failure surface (or region) cutting the fill, the reinforcement and the soft soil. This mechanism can involve tensile failure of the reinforcement, or bond failure due to insufficient anchorage of the reinforcement.

Figure 5.1 Contributions from geosynthetics in embankments on soft soils

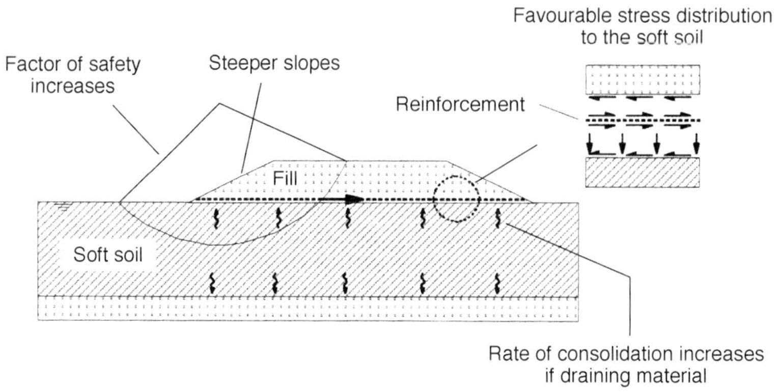

Figure 5.2 Beneficial effects of the reinforcement during construction: (a) separation and (b) reduction of local failures during construction

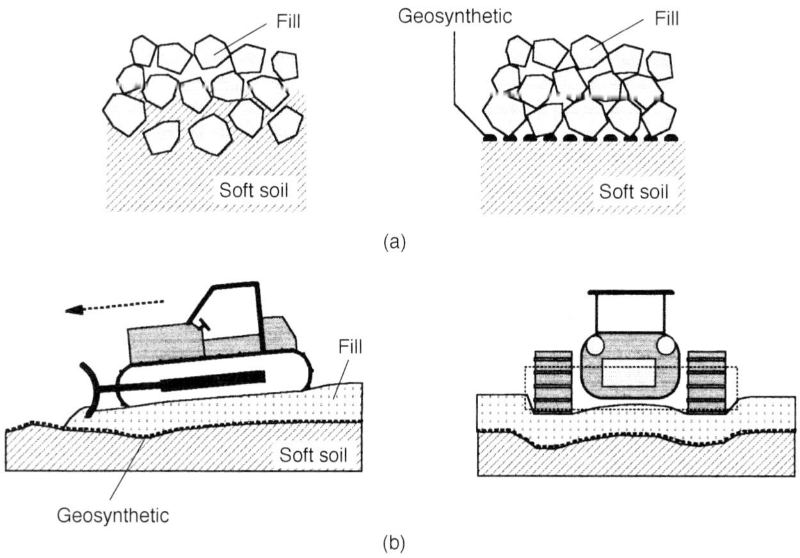

The usual practice for the design of reinforced embankments on soft soils is the use of limit equilibrium methods to estimate safety factors for the situations described above. Simple methods are usually preferred for preliminary routine analyses, such as the ones dealing with sliding wedges or circular failure surfaces. For the latter case, the Fellenius and the modified Bishop's methods are commonly used.

Figure 5.4 raises some important questions to be considered when analysing the stability of reinforced embankments on soft soils. The first question relates to the magnitude and inclination of the mobilised reinforcement force. The reinforcement layer is initially placed horizontally at the embankment base. As the embankment and the reinforcement deform, some inclination of the reinforcement force to the horizontal direction is expected. One can find works in the literature where the inclination of the reinforcement force can be taken into account in the analysis (Low *et al.*, 1990; Sabhahit *et al.*, 1994; Kaniraj, 1996). Usually, horizontal, tangential or bisectorial forces are considered. The consideration of this inclination in the stability analysis will affect the results of calculated safety factors. Large-scale direct shear tests on stiff granular materials (Palmeira and Milligan, 1989) suggest that, at peak strength, the contribution of the reinforcement deviation from its original direction to the stability is negligible. This is likely to be the case for stiff fill materials on soft foundations when the first sign of instability appears (e.g. cracks along the embankment surface). The influence of the inclination of the tensile reinforcement force will be significant under large-strain conditions, when usually the embankment would be considered to be compromised in operational terms. An additional difficulty, regarding the use of inclined reinforcement forces, is the impossibility, at the present state of knowledge, of using simple analysis to predict the force inclination to the horizontal during embankment loading. Besides, the consideration of a constant inclination of the reinforcement force would not be realistic or cinematically consistent. On the

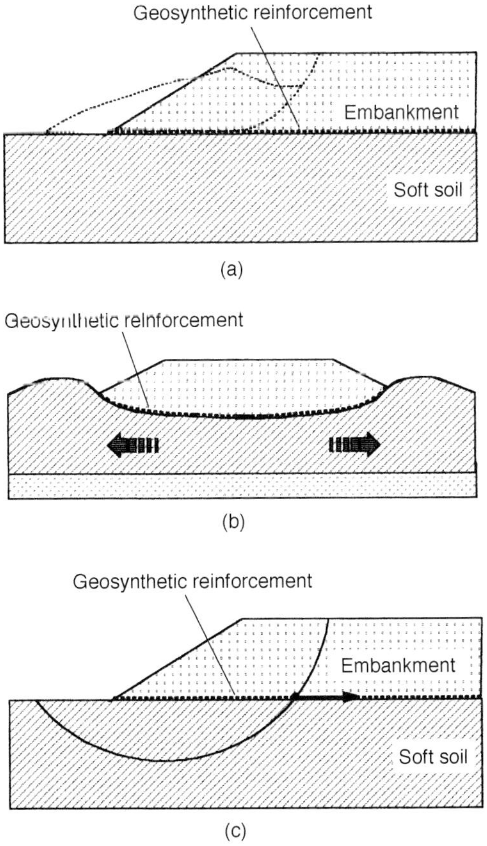

Figure 5.3 Instability mechanisms in an embankment on soft soil: (a) internal stability, (b) soil expulsion and (c) overall stability

other hand, the inclination of tangential and bisectorial forces to the horizontal will depend on the slip surface considered because, in both cases, the force inclination will depend on the inclination of the line tangent to the slip surface at the intersection with the reinforcement layer. At the present state of knowledge, the use of horizontal reinforcement forces in stability analyses of reinforced embankments is recommended, and the accuracy of this assumption has been observed by back-analyses of reinforced embankments led to failure, and by limit analyses (Michalowski, 1998; Palmeira *et al.*, 1998).

Depending on the polymer of which the geosynthetic reinforcement is composed, the mobilised tensile forces can be more or less affected by the rate of strain imposed during loading. This rate is a function of the speed of embankment lifting. Therefore, for those reinforcements significantly affected by the rate of strain, the tensile force–strain–time relationship has to be well known for an accurate prediction to be made of stability conditions.

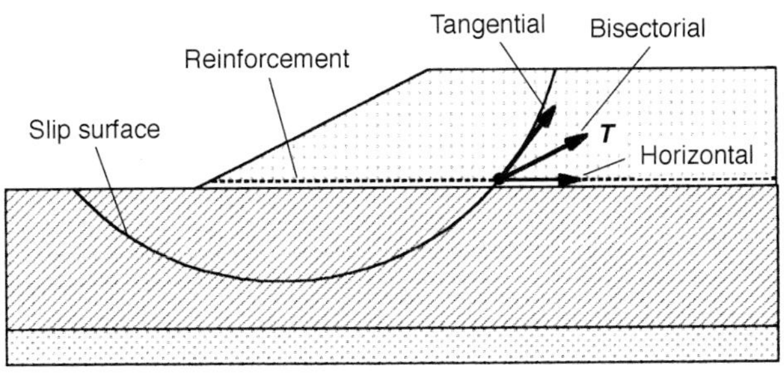

Figure 5.4 Orientation of the tensile force of the reinforcement

Another aspect to be considered in the stability analysis of reinforced or unreinforced embankments is the influence of the cohesion of the fill material on the results of the safety factors obtained. High fill cohesions or large slice inclinations to the horizontal inside the fill can cause negative normal forces at the base of slices in that region, which may cause misleading values of factors of safety. Discussions and ways of treating this type of problem can be found elsewhere (Whitman and Bailey, 1967; Chirapuntu and Duncan, 1975; Palmeira and Almeida, 1980). When the cohesive fill material is expected to crack throughout its entire height, it is wise to treat it simply as a surcharge on the foundation soil surface, with no contribution in terms of resisting the failure mechanism. For large deformations of granular fill materials, critical state strength parameters should be used for this material in the calculations.

Different methods of analysis or forms of definition of the safety factor may also affect the result obtained. For circular failure surfaces, depending on the method employed, the factor of safety may be defined from force equilibrium conditions or ratios between resisting and driving moments. In addition, for Bishop's method, for instance, the contribution from the reinforcement force can be considered in two ways: simply taking into account, in the analysis, the resisting moment of this force with respect to the centre of the circular surface; or taking the components of the reinforcement force (normal and tangential to the slice base) into the equilibrium equations for the slice. The latter approach explores the increment in frictional resistance caused by the increase in the normal force at the base of the slice, plus the component of the reinforcement force along the slice base direction. These two different ways of considering the effect of the reinforcement force described above may yield some difference between the results for the safety factor obtained in each case.

The analysis of reinforced embankments using traditional methods for slope-stability analysis consists of trying to find the critical failure surface for which the necessary force in the reinforcement is maximum. Usually, a target value for the safety factor is established, and the maximum necessary reinforcement force is determined by searching the critical failure surface.

5.2.2 Design approaches for reinforced embankments

5.2.2.1 Designing against expulsion of the foundation soil

Figure 5.5 shows an approach commonly used to verify the possibility of a saturated foundation soil expulsion under undrained conditions. The stability analysis consists of checking the stability of the foundation soil block below

Figure 5.5 Expulsion of the soft subgrade

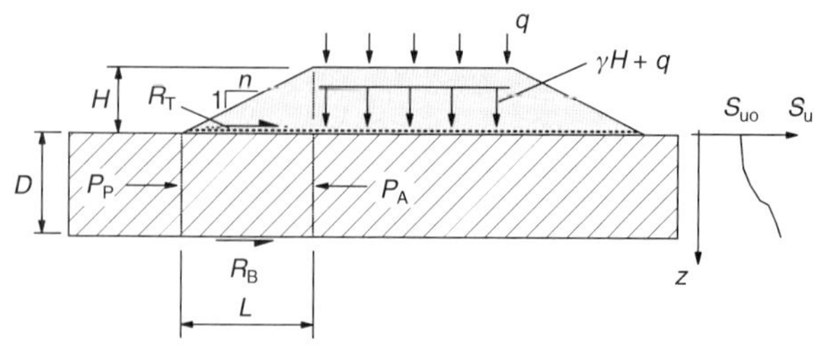

the embankment slope, as illustrated in Figure 5.5, which also shows the relevant dimensions and parameters to be considered in the calculations, such as the thickness of the soft soil D, the height of the embankment H, the width of the soft soil block L, the surcharge on the embankment q, slope of inclination of the embankment (n horizontal to 1 vertical), and the undrained strength of the soft soil S_u. The factor of safety against soil expulsion can be estimated as

$$F_e = \frac{P_P + R_B + R_T}{P_A} \tag{5.1}$$

where F_e is the safety factor against foundation soil expulsion, P_P is the passive reaction force against block movement, R_T is the force at the top of the soil block, R_B is the force at the base of the soil block, and P_A is the active thrust on the soil block.

The active and passive forces can be evaluated by means of earth pressures theories (e.g. Rankine's theory), while the forces at the base and top of the soil block can be estimated as a function of the undrained strength at the bottom of the foundation soil and adherence between the reinforcement layer and the surface of the foundation soil, respectively.

In a real situation, the failure mechanism is rather more complex than the above approach, because drainage is likely to occur at the top and bottom of the soft soil block, depending on the type of reinforcement used and on the type of soil below the soft layer. In this sense, the approach presented above is conservative. Rowe and Li (1999) discuss the effect of partial drainage of the soft soil on the stability of the embankment using the finite-element method.

5.2.2.2 Generalised failure of the embankment

In this case, several approaches for the stability analysis of the embankment can be considered. Below, some of these approaches are presented and discussed.

Combined failure surface (Jewell, 1987)

Jewell (1987) presents a methodology of stability analysis where the slip surface combines a circular arc in the foundation soil and a straight line inside the embankment, as shown in Figure 5.6. The effect of the embankment on the stability is represented by the active thrust acting on the vertical CD in Figure 5.6. For example, the active thrust can be estimated using the Rankine earth pressure theory. One advantage of this approach is that it allows the identification and analysis of a cracked cohesive embankment in a clearer and easier way. The depth of the crack will affect the value of the active thrust, as is the case in usual retaining wall problems with cohesive backfills. The factor of safety for the unreinforced embankment can be written as

$$F_o = \frac{M_r}{M_a} \tag{5.2}$$

where F_o is the factor of safety of the unreinforced embankment, M_r is the sum of the moments of the resisting forces with respect to the centre of the circular portion of the slip surface, and M_a is the sum of the moments of the driving forces with respect to the centre of the circular portion of the slip surface.

The procedure consists of the calculation of the factor of safety for several slip surfaces in order to obtain the overall minimum value for the factor of safety. For cases with no surcharge on the embankment platform, the critical circle is located on the vertical line passing through the middle of the embankment slope (Leshchinsky, 1987).

Figure 5.6 Combined failure mechanism
Adapted from Jewell (1987)

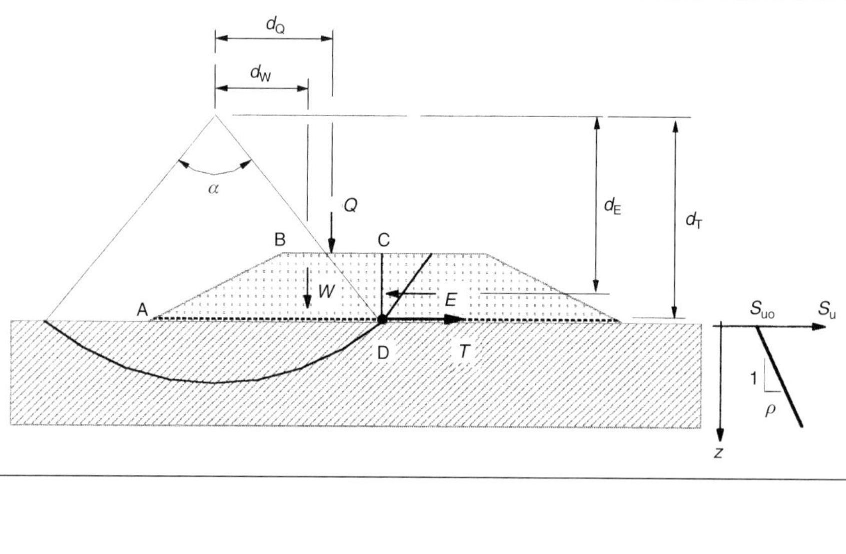

For the case of a reinforced embankment, the moment equilibrium equation for the sliding mass can be written as

$$\frac{M_r}{F_r} + Td_T = M_a \tag{5.3}$$

where F_r is the factor of safety for the reinforced embankment, T is the required tensile force in the reinforcement, and d_T is the arm of T with respect to the circle centre (see Figure 5.6).

Combining Equations 5.2 and 5.3 yields

$$T = \left(1 - \frac{F_o}{F_r}\right)\frac{M_a}{d_T} \tag{5.4}$$

Equation 5.4 is easier to use in the sense that if the value of F_r is established for the slip surface considered and the value of F_o for that surface is known, then the value of T can be directly obtained from Equation 5.4, knowing the values of d_T and M_a. The values of T for several surfaces have to be calculated in order to determine the maximum required tensile force of the reinforcement for the critical surface. It is important to point out that the critical slip circle is not necessarily the same in reinforced and unreinforced cases.

For a foundation soil with the undrained strength varying linearly with depth (see Figure 5.6), Equations (5.2) and (5.4) give

$$F_o = \frac{R^2\{\alpha[S_{uo} - \rho R\cos(\alpha/2)] + 2\rho R\sin(\alpha/2)\}}{Ed_E + Wd_W + Qd_Q} \tag{5.5}$$

and

$$T = \frac{F_r(Ed_E + Wd_W + Qd_Q) - R^2\{\alpha[S_{uo} - \rho R\cos(\alpha/2)] + 2\rho R\sin(\alpha/2)\}}{F_r d_T} \tag{5.6}$$

where R is the radius of the circular portion of the slip surface considered, α is the internal angle defining the circular portion of the slip surface, S_{uo} is the undrained strength at the foundation soil surface, ρ is the rate of increase in the undrained strength with depth, E is the active thrust in the embankment along the vertical CD, W is the weight of the wedge ABCD in the embankment, Q is the resultant force due to surcharge on the embankment surface (if any) along the length BC, and d_E, d_W and d_Q are the arms of the forces E, W and Q with respect to the centre of the circle, respectively (see Figure 5.6).

Low *et al.* (1990) approach

Low *et al.* (1990) presented a methodology for the design of reinforced embankments using circular failure surfaces. Mathematical expressions are presented for the determination of the relevant parameters of the problem, and a chart for the determination of the value of the tensile reinforcement force required is also provided. The geometry of the problem treated by Low *et al.* is presented in Figure 5.7. The maximum required tensile force in the reinforcement to guarantee a target safety factor F_r for the critical surface, for all the circles tangential to the horizontal line at the depth z, is given by

Figure 5.7 Low *et al.* approach
Adapted from Low *et al.* (1990), with permission from Elsevier

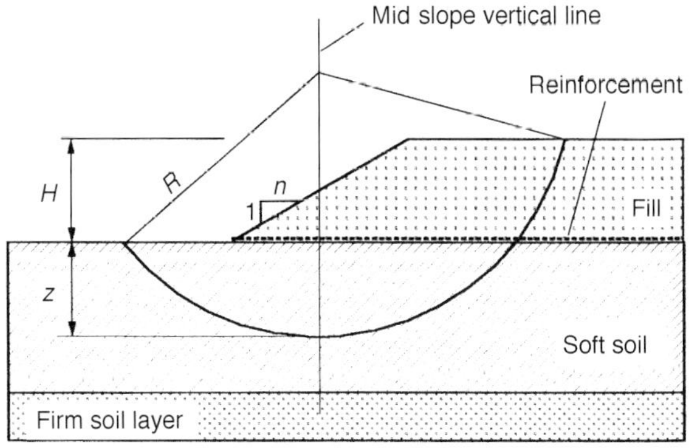

$$T = \left(1 - \frac{F_o}{F_r}\right)\frac{\gamma H^2}{I_R} \tag{5.7}$$

where F_o is the critical factor of safety for circular surfaces tangential to the same depth z in the unreinforced case, γ is the unit weight of the fill material, H is the embankment height, and I_R is the stability number, which can be obtained from Figure 5.8 as a function of z and the inclination of the slope of the embankment n. The depth z of the tangency of the failure surfaces must be varied along the foundation soil thickness for the determination of the overall maximum value of T.

The value of F_o at each depth considered for the analysis must be known in order to solve Equation 5.7 for different failure surfaces. Low (1989) presents a solution for the

Figure 5.8 Stability number for reinforced embankments – horizontal reinforcement force
Adapted from Low *et al.* (1990), with permission from Elsevier

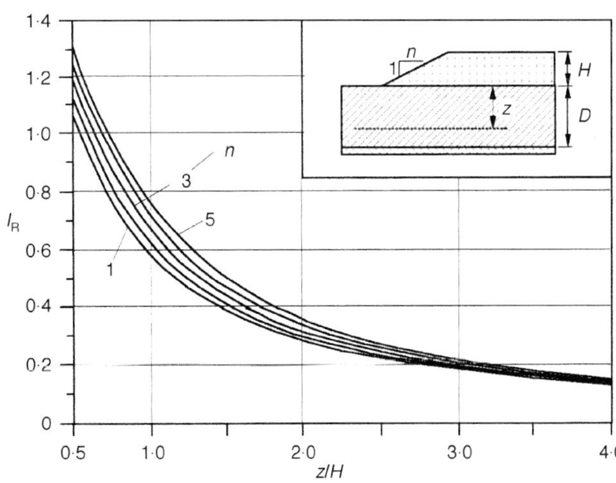

Figure 5.9 Stability numbers for unreinforced embankments
Adapted from Low (1989), with permission from ASCE

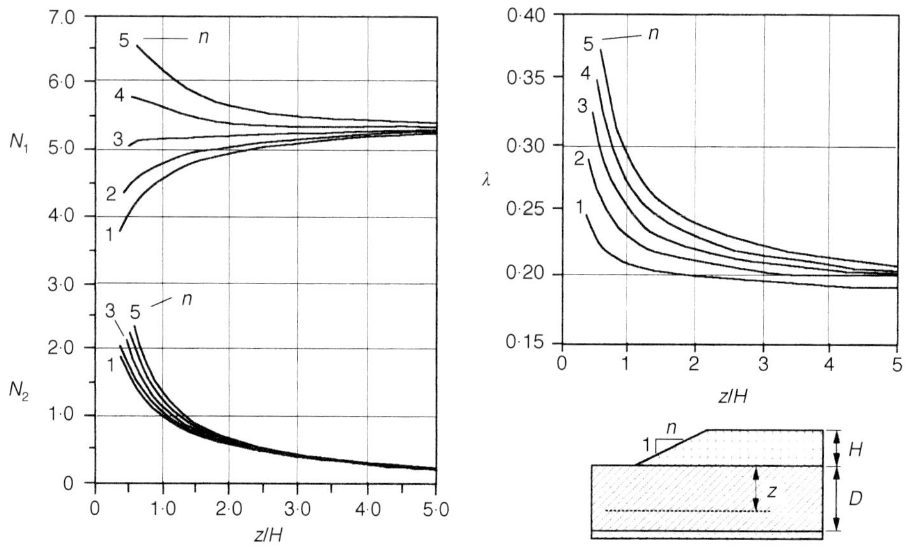

determination of the minimum factor of safety for the unreinforced case among all circles reaching an established depth z. In this case, the value of F_o is given by

$$F_o = N_1\frac{S_{ueq}}{\gamma H} + N_2\left(\frac{c}{\gamma H} + \lambda \tan\phi\right) \quad (5.8)$$

where N_1, N_2 and λ are the stability numbers obtained from Figure 5.9 as a function of the inclination of the embankment slope n, the height of the embankment H and the depth reached by the circle z. S_{ueq} is the equivalent undrained strength of the foundation along the depth z, γ is the unit weight of the fill, and c and ϕ are strength parameters of the fill material.

For the common condition of undrained strength profile, as for the one shown in Figure 5.10, the value of S_{ueq} can be obtained by the following expression:

$$S_{ueq} = 0.35S'_{uo} + 0.65S_{uz} + 0.35\left(\frac{z_c}{z}\right)^{1.1}\Delta S'_{uo} \quad (5.9)$$

where S'_{uo}, S_{uz}, $\Delta S'_{uo}$, z_c and z are defined in Figure 5.10.

The consideration of other forms of undrained strength profiles is presented in Low (1989).

The radius of the critical circle (minimum F_o) in the unreinforced case for a given depth z of tangency can be obtained using

$$R_o = \left[0.1303\frac{n^2+1}{(z/H)+0.5} + 1.5638\left(\frac{z}{H}+0.5\right)\right]H \geq z + H \quad (5.10)$$

where R_o is the radius of the critical circle tangential to the horizontal line at the depth z.

Figure 5.10 Common undrained strength profile of soft soil
Adapted from Low (1989), with permission from ASCE

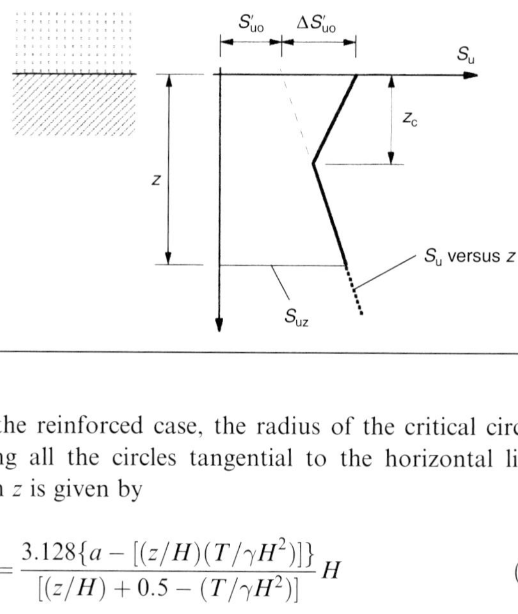

For the reinforced case, the radius of the critical circle R_r among all the circles tangential to the horizontal line at depth z is given by

$$R = \frac{3.128\{a - [(z/H)(T/\gamma H^2)]\}}{[(z/H) + 0.5 - (T/\gamma H^2)]}H \quad (5.11)$$

with

$$a = \frac{1}{2}\left(\frac{z}{H}+0.5\right)^2 + \frac{(n^2+1)}{24} \quad (5.12)$$

Kaniraj (1994) approach

Kaniraj (1994) presented a methodology similar to the one presented by Low *et al.* (1990). However, the presence of a drainage channel at the embankment toe, berm and cracks at the embankment surface (cohesive fills) can be considered in this case. The geometrical conditions for the problem are presented in Figure 5.11.

For the unreinforced case, the coordinates of the centre of the critical circle among all the circles tangential to a horizontal

Figure 5.11 Geometrical characteristics in the Kaniraj approach
Adapted from Kaniraj (1994), with permission from Elsevier

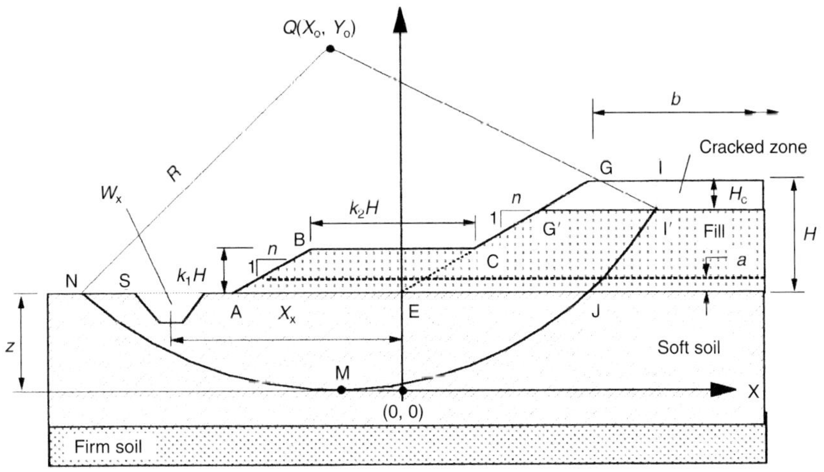

line at depth z are given by

$$\frac{X_o}{H} = \frac{n}{2} - k_1 k_2 + \frac{W_X}{\gamma H^2} \tag{5.13}$$

$$\frac{Y_o}{H} = 1.564\alpha_1 \tag{5.14}$$

where X_o and Y_o are the coordinates of the centre of the critical circle for depth z in the unreinforced case, k_1 is the ratio between the berm height and the embankment height (see Figure 5.11), k_2 is the ratio between the berm platform width and the embankment height, and W_X is the weight of soft soil removed during the excavation of the drainage channel.

The value of α_1 in Equation 5.14 is given by

$$\alpha_1 = \frac{\alpha_1'}{(z/H) + \beta - \dfrac{\beta^2}{2}} \tag{5.15}$$

with

$$\alpha_1' = \beta^2\left(1 - \frac{2}{3}\beta - \frac{z}{H}\right) + 2\beta\frac{z}{H} + \left(\frac{z}{H}\right)^2 + \frac{n^2}{12} + \mu - \frac{W_X}{\gamma H^2}\left[\frac{W_X}{\gamma H^2} + n + 2\left(\frac{X_X}{H} - k_1 k_2\right)\right] \tag{5.16}$$

$$\mu = k_1 k_2 (n + k_2)(1 - k_1) \tag{5.17}$$

$$\beta = 1 - \frac{H_c}{H} \tag{5.18}$$

where μ is a berm factor ($\mu = 0$ for embankments with no berm), and β is the ratio between the height of the non-cracked part of the embankment and the total embankment height (see Figure 5.11).

The minimum factor of safety for all circles tangential to the horizontal line at depth z in the unreinforced case is given by

$$F_o = S_f N_1' + S_e N_2' \tag{5.19}$$

with

$$S_f = \frac{S_u}{\gamma H} \tag{5.20}$$

$$S_e = \frac{c}{\gamma H} + \lambda \tan\phi \tag{5.21}$$

$$\lambda = 0.19 + \frac{0.02n}{z/H} \quad \text{for } z/H \geq 0.5 \tag{5.22}$$

where S_f and S_e are the normalised strength parameters for the soft soil and fill materials, respectively, S_u is the strength of the undrained soft soil, c and ϕ are shear strength parameters of the fill, and N_1' and N_2' are stability numbers given by

$$N_1' = 5.55\alpha_1^{0.47}\frac{(z/H)^{0.53}}{(z/H) + \beta - (\beta^2/2)} \tag{5.23}$$

$$N_2' = mN_1' \tag{5.24}$$

$$m = 0.5\left[\left(1 + \frac{\beta}{z/H}\right)^{0.53} - 1\right] \tag{5.25}$$

For the reinforced case, Kaniraj presents solutions for horizontal, tangential and bisectorial tensile reinforcement forces. For the horizontal reinforcement force, the coordinates of the centre of the critical circle for a depth of tangency z are given by Equation 5.13 (for X_o) and by

$$\left(\frac{Y_o}{H}\right)^{1.47} - 3.128\left(\frac{z+a}{H}\right)\left(\frac{Y_o}{H}\right)^{0.47} - 2.128\left(\frac{F_2}{F_1}\right) = 0 \tag{5.26}$$

with

$$F_1 = A_1 S_f + B_1 S_e \tag{5.27}$$

$$F_2 = F_2' - F_r \frac{z+a}{H}\left(\frac{z}{H} + \beta - \frac{\beta^2}{2}\right) \tag{5.28}$$

$$A_1 = 3.06\left(\frac{z}{H}\right)^{0.53} \tag{5.29}$$

$$B_1 = 1.53\left[\left(\frac{z}{H} + \beta\right)^{0.53} - \left(\frac{z}{H}\right)^{0.53}\right] \tag{5.30}$$

$$F_2' = F_r\left[\frac{1}{2}\left(\frac{z}{H}\right)^2 + \frac{\beta^2}{2} - \frac{\beta^3}{3} + \frac{n^2}{24} + \beta\frac{z}{H} - \frac{\beta^2}{2}\frac{z}{H} + \frac{\mu}{2} - \frac{W_X}{\gamma H^2}\left(\frac{X_X}{H} + \frac{n}{2} - k_1 k_2 + \frac{1}{2}\frac{W_X}{\gamma H^2}\right)\right] \tag{5.31}$$

where a is the elevation of the reinforcement layer with respect to the soft soil surface (see Figure 5.11) and F_r is the required safety factor for the reinforced embankment.

The tensile force of the reinforcement required for a factor of safety equal to F_r for the reinforced embankment is given by

$$T = \frac{\alpha_2}{L_a/H}\gamma H^2 \tag{5.32}$$

with

$$\alpha_2 = F_r\left(\frac{z}{H} + \beta - \frac{\beta^2}{2}\right)\frac{Y_o}{H} - F_2' - F_1\left(\frac{Y_o}{H}\right)^{1.47} \tag{5.33}$$

$$\frac{L_a}{H} = \frac{Y_o}{H} - \frac{z+a}{H} \tag{5.34}$$

For the reinforced case where the force in the reinforcement is considered tangential to the circle at the intersection between the failure surface and the reinforcement layer, the value of X_o is also given by Equation 5.13, while Y_o is given by

$$\frac{Y_o}{H} = 1.672\left(\frac{F_2'}{F_1}\right)^{0.68} \tag{5.35}$$

The required reinforcement force is again given by Equation 5.32, but with L_a given by

$$L_a = Y_o \tag{5.36}$$

After the calculations and determination of the critical circles, it is necessary to verify whether the geometrical conditions presented in Figure 5.11 have been satisfied. These conditions are expressed mathematically by the expressions below.

Condition 1. The circle centre must be located above, or at the bottom level, of the cracked zone in the fill. So,

$$\frac{Y_o}{H} \geq \beta + \frac{z}{H} \tag{5.37}$$

Condition 2. Both the berm and the drainage channel must be entirely inside the sliding mass bounded by the circular slip surface and the soil surface boundary. So,

$$\frac{Y_o}{H} \geq 0.5\left\{\frac{z}{H} + \frac{\left[\frac{n}{2} - k_1 k_2 + \left(W_X/\gamma H^2\right) + \left(X_S/H\right)\right]^2}{z/H}\right\} \tag{5.38}$$

where X_s is the distance along the horizontal between the extremity of the channel and the Y axis (distance between points S and E in Figure 5.11).

Condition 3. The extremity of the failure surface (point I′ in Figure 5.11) must lie on, or beneath, the embankment platform length. So,

$$nH \leq X_{I'} \leq nH + b \tag{5.39}$$

or

$$\frac{Y_o}{H} \geq \frac{1}{2}\left\{\frac{\left[(n/2) + k_1 k_2 - \left(W_X/\gamma H^2\right)\right]^2}{\beta + (z/H)} + \beta + \frac{z}{H}\right\} \tag{5.40}$$

and

$$\frac{Y_o}{H} \leq \frac{1}{2}\left\{\frac{\left[(n/2) + k_1 k_2 - \left(W_X/\gamma H^2\right) + (b/H)\right]^2}{\beta + (z/H)} + \beta + \frac{z}{H}\right\} \tag{5.41}$$

where $X_{I'}$ is the abscissa of point I′ and b is the width of the embankment platform (see Figure 5.11).

Analytical solution Jewell (1996)

Jewell (1996) presented an analytical solution considering the equilibrium of the block of soft soil below the embankment slope, as illustrated in Figure 5.12, in a similar way to the analysis of soft soil expulsion presented above. The equations to be used depend on the form of variation of the undrained strength of the soft soil with depth. So, for a foundation soil with uniform strength and limited depth,

$$F_o = \frac{S_u}{\gamma H}\left(\frac{8D + 2nH}{2D + k_a H}\right) \tag{5.42}$$

Figure 5.12 Jewell's stability analysis approach
Adapted from CIRIA SP123, Jewel (1996): http://www.ciria.org

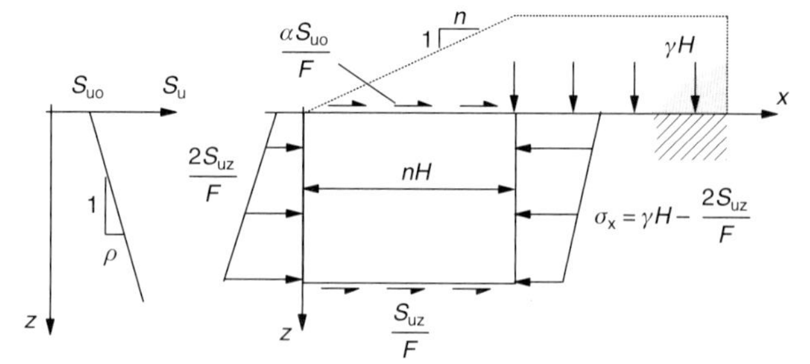

$$F_r = \frac{S_u}{\gamma H}\left(4 + (1+\alpha)\frac{nH}{D}\right) \qquad (5.43)$$

$$T_{max} = \gamma H^2\left(\frac{\alpha n D}{4D + (1+\alpha)nH} + \frac{k_a}{2}\right) \qquad (5.44)$$

For a foundation soil with strength increasing with depth,

$$F_r \text{ or } F_o = \frac{S_{uo}}{\gamma H}\left(4 + \frac{\rho n H}{S_{uo}} + 2\sqrt{\frac{2(1+\alpha)\rho n H}{S_{uo}}}\right) \qquad (5.45)$$

$$T_{max} = \gamma H^2\left(\frac{\alpha n S_{uo}}{F_r \gamma H} + \frac{k_a}{2}\right) \qquad (5.46)$$

$$z_{crit} = \sqrt{\frac{(1+\alpha)S_{uo} n H}{2\rho}} \qquad (5.47)$$

Jewell (1996) suggests the use of the expressions above under the condition

$$\frac{F_r \gamma H}{S_u} \geq 6 \qquad (5.48)$$

where F_o and F_r are the safety factors for unreinforced and reinforced embankments, respectively, T is the required reinforcement force, S_u is the undrained strength of the soft soil, n is the slope of the embankment, α is the ratio between the mobilised shear stress and the undrained strength at the subgrade surface (see Figure 5.12), H is the embankment height, k_a is the Rankine active earth pressure coefficient of the fill material, D is the thickness of the soft subgrade, γ is the unit weight of the fill, S_{uo} is the undrained strength at the surface of the subgrade, ρ is the rate of increase in the undrained strength with depth, and z_{crit} is the critical depth of the sliding block.

The value of α for the reinforced case is established by the designer ($0 < \alpha \leq 1$). For the unreinforced case, the value of α is given by the expression below, and the solution is obtained by trial and error in combination with Equation 5.45:

$$\alpha = -\frac{k_a F_o \gamma H_o}{2nS_{uo}} \qquad (5.49)$$

5.2.3 Choice of the reinforcement

After the determination of the maximum required reinforcement force to guarantee the stability of the embankment, the reinforcement must be chosen carefully. Several factors have to be taken into consideration, such as the embankment characteristics, the consequences of an embankment failure, deformation allowances, serviceability requirements, reinforcements available, etc. With regard to embankments on soft soils, the following factors may be of major concern when choosing the reinforcement

- tensile strength and stiffness
- soil–reinforcement bond characteristics
- creep characteristics
- geosynthetic resistance to mechanical damage
- durability.

With regard to tensile strength, the reinforcement chosen must adhere to the following condition (Jewell, 1996):

$$T_d = \frac{T_{ref}}{f} \geq T \qquad (5.50)$$

where T_d is the reinforcement design strength, T_{ref} is the reference tensile strength of the reinforcement obtained under appropriate testing techniques, taking into consideration the design life of the project (e.g. creep tests), f is the reduction factor for the tensile strength, and T is the required reinforcement force.

The value of f can be determined using the following expression (Jewell, 1996):

$$f = f_m f_{md} f_{env} \qquad (5.51)$$

where f_m is the reduction factor to account for reinforcement material uncertainties, f_{md} is the reduction factor to account for mechanical damage during handling and construction, and f_{env} is the reduction factor to account for strength losses caused by environmental factors (chemicals, biological factors, etc.).

The value of the required reinforcement stiffness J is still difficult to evaluate, mainly because routine design methods for reinforced embankments are based on limit equilibrium analyses. Depending on the serviceability requirements for the embankment, the reinforcement tensile strain is typically expected to be within 2–10%. Therefore, a crude way to estimate the required reinforcement stiffness is to divide the required reinforcement force by an expected strain within that range.

Jewell (1996) presents the expression below to estimate the reinforcement stiffness for the case of an embankment on a foundation soil with uniform strength and limited depth (for $\alpha \geq 0$):

$$J = \frac{G}{(1-\alpha)S_{ud}}\left[k_{ad}\gamma H^2\left(\frac{nH}{3D} + \frac{L_c}{2D}\right) + \alpha S_{ud} D\left(1 + \frac{nH}{D}\right)^2\right] \qquad (5.52)$$

where G is the shear modulus for the foundation soil, L_c is the width of the embankment crest, k_{ad} is the design active earth pressure coefficient of the fill material, and S_{ud} is the design undrained strength of the foundation soil.

For the use of Equation 5.52, the value of α is established by the designer ($0 < \alpha \leq 1$) and the value of S_{ud} is equal to S_u/F_r, with F_r given by Equation 5.43. The value of k_{ad} is obtained using the Rankine earth pressure theory and the design value for the friction angle of the fill material.

Large bond values between the soil and the reinforcement are also required for appropriate load transference between materials and to provide enough anchorage resistance for the reinforcement. Values for the adhesion and friction angle between the soil and the reinforcement can be evaluated using appropriate testing procedures (e.g. direct shear and pullout tests, as described in Chapter 2).

Reinforcement creep characteristics and durability may or may not be relevant, depending on the type of geosynthetic reinforcement considered and on how long the reinforcement will be required to guarantee the embankment stability. In most cases, the reinforcement is only required during embankment construction and for a short period afterwards, because the strength of the foundation soil increases with time due to consolidation. However, in cases where the reinforcement is required for long-term stability, its creep behaviour and durability as described in Chapter 1 must be carefully considered in the design.

5.2.4 Anchorage length of the reinforcement

The methodologies presented above allow for the estimation of the required reinforcement force in order to obtain a desired factor of safety for the reinforced embankment. In addition, the anchorage conditions of the reinforcement extremities must be evaluated in order to avoid slippage of the reinforcement along the anchorage length, as presented schematically in Figure 5.13. From this figure, the factor of safety against bond failure along the anchorage length can be estimated as

$$F_{\text{anch}} = \frac{L_{\text{anch}}(\alpha_{\text{fr}} S_{\text{uo}} + a_{\text{sr}}) + W \tan \delta_{\text{sr}}}{T} \tag{5.53}$$

where F_{anch} is the factor of safety against anchorage failure, L_{anch} is the anchorage length beyond the critical failure surface, α is the ratio between mobilised shear stress on the reinforcement and the undrained strength at the subgrade, a_{sr} is the adhesion between the fill material and the reinforcement, W is the fill weight above the anchorage length, δ_{sr} is the interface friction angle between the fill material and the reinforcement, and T is the required reinforcement tensile force. A similar approach must also be used to verify the anchorage condition along the length AI in Figure 5.13.

Figure 5.13 Reinforcement anchorage conditions

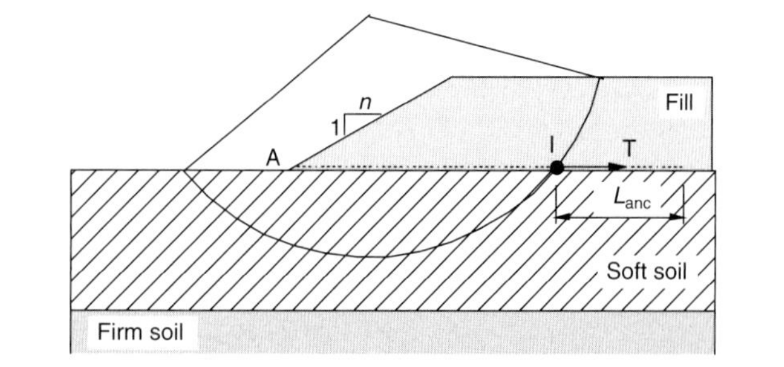

5.2.5 Additional remarks on analysis and design

It is well recognised that limit equilibrium methods have important limitations with regard to model failure mechanisms. Several factors are not taken into account in limit equilibrium analyses, such as stress–strain relationships for the materials involved, progressive failure and time-dependent factors. However, when used properly and with accurate input data, limit equilibrium stability analysis methods can be useful tools for the estimation of safety factors of reinforced embankments on soft soils. Palmeira *et al.* (1998) presented back-analyses of case histories of reinforced embankments on soft soils that led to failure (Volman *et al.*, 1977; Rowe and Soderman, 1984; Schaefer and Duncan, 1988; Delmas *et al.*, 1990; Loke *et al.*, 1994; Rowe *et al.*, 1995) using the modified Bishop method, the United States Army Corp of Engineers' method (USACE, 1970) and the method presented by Jewell (1996). These methods were used to predict the factor of safety at the failure height of the embankments in the field. Figure 5.14 presents some details of the case histories analysed. The data on the materials and reinforcements used were obtained from the original sources. Table 5.1 shows the predicted values of the safety factor for the case histories at failure height. It can be seen that, in most cases, the predicted values are satisfactorily close to unity. The largest deviations from unity were observed for case history 4 (Loke *et al.*, 1994). However, Bergado *et al.* (1994) reported values of predicted safety factors close to one for case history 4 when a site-dependent correction factor was employed for the undrained strength of the soft soil (obtained by vane tests). It is important to point out that an accurate prediction of the safety factor at the failure height for case history 6 (Schaefer and Duncan, 1988) was only possible with the knowledge of the load–strain–time relationship for the polymeric reinforcement used in that case history; the behaviour of this relationship is highly dependent on the strain rate imposed.

Different methods of analysis are based on different assumptions, so deviations between the results obtained are expected to occur. Figure 5.15 shows the variation in the required reinforcement force versus the fill cohesion, for a safety factor F_r of 1.3, in a hypothetical situation of an embankment on soft soil with a constant undrained strength throughout its depth (Silva, 1996). For cohesionless fill materials, the differences between predicted reinforcement forces are small, except in the case of the Fellenius method, the predictions from which are significantly more conservative. As the fill cohesion increases, it affects each method differently, depending on how each of them deals with the influence of the fill cohesion. The combined failure surface approach tends to be the most conservative for greater cohesion values because, beyond a certain value of fill cohesion, the embankment will be considered cracked along its entire height. Therefore, users should be aware of these types of variation, depending on the method employed and on the project characteristics (as mentioned earlier in this chapter). It is also important to check if the centre of the critical circle lies above the embankment platform for methods using circular failure surfaces. Sometimes, for very strong

Figure 5.14 Trial embankments that led to failure: (a) case history 1 (Volman *et al.*, 1977), (b) case history 2 (Rowe and Soderman, 1984), (c) case history 3 (Delmas *et al.*, 1990), (d) case history 4 (Loke *et al.*, 1994), (e) case history 5 (Rowe *et al.*, 1995) and (f) case history 6 (Schaefer and Duncan, 1988)
Adapted from Palmeira *et al.* (1998), with permission from Elsevier

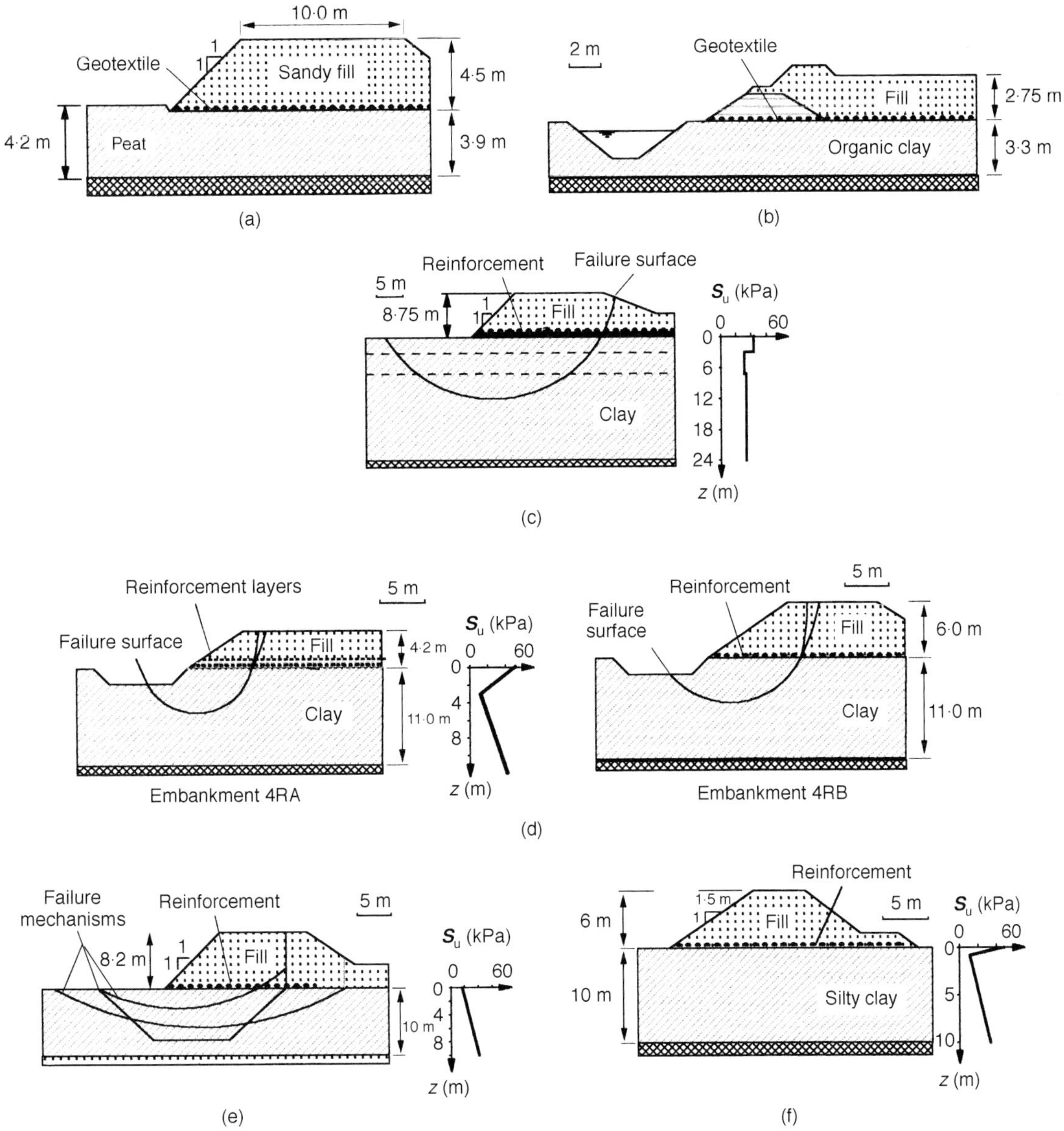

reinforcement layers or very weak soils, the circle's centre lies below the embankment platform, and this may lead to misleading values of safety factors or required reinforcement forces. This possibility is examined by Kaniraj (1994), with Equation 5.37, and by Low *et al.* (1990), with Equations 5.10 and 5.11.

Figure 5.16 presents some alternatives with regard to the installation of the geosynthetic reinforcement layer inside the embankment. Figure 5.16(a) shows the installation of the reinforcement layer inside the embankment, rather than along the interface between the fill material and the foundation soil. This solution is necessary when vertical drains are installed in the foundation, before reinforcement installation. It also favours a better reinforcement anchorage strength, particularly for geogrids, for which all the grid-bearing members will be buried inside the good-quality fill material. The elevation of the reinforcement layer, with respect to the foundation soil surface, is usually limited to 0.3–0.5 m, and often the geosynthetic lies on the drainage blanket of the embankment. This location of the reinforcement layer also minimises the possibility of mechanical damage to the geosynthetic in some situations, depending on the surface conditions of the foundation soil. When using a stability analysis method, such as Bishop's method, for a given slip circle large values of the reinforcement

Table 5.1 Comparisons between predicted and observed safety factors at failure for trial reinforced embankments
Data from Palmeira *et al.* (1998)

Case history	MBF*	MBM*	USACE	Jewell (1996)
1†	1.012	0.979	0.959	0.93
2	0.810	0.870	0.911	0.99
3	0.977	0.982	0.970	0.97
4 (4RA)	1.181/1.116‡	1.159	1.179	1.68
4 (4RB)	1.051/1.085‡	1.059	1.056	1.26
5	1.045	1.036	0.941	0.98
6 (6RA)	1.037	1.035	0.970	1.01
6 (6RB)	±	±	±	1.11

* Safety factors obtained using Bishop's method, based on the force equilibrium in the slices (MBF) and the moment of the reinforcement force with respect to the circle centre only (MBM).
† Reinforced embankment did not fail at final height.
‡ With and without a tension crack at the embankment surface.

Figure 5.15 Comparisons between values of the required reinforcement force predicted using different methods
Modified from Silva (1996)

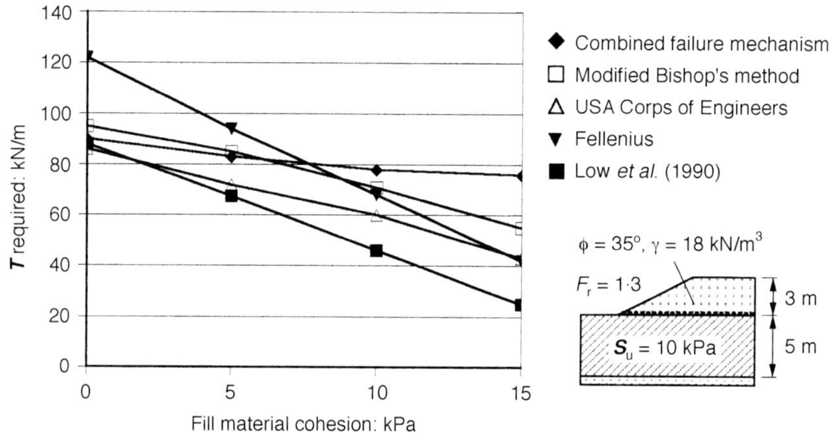

elevation will lead to the need for stronger reinforcement due to the reduction in the arm of the reinforcement force with respect to the centre of the slip circle.

Figure 5.16(b) shows the use of several layers of reinforcement along the embankment height. This concentration of reinforcement layers is usually restricted to the region close to the surface of the soft soil. In this case, combinations of different types of reinforcement (geotextiles and geogrids) with different functions (drainage and reinforcement) can be used. The use of the reinforcement can also be optimised by using reinforcement layers of different tensile strengths. The concentration of reinforcement layers at the base of the embankment creates a stiffer mass in that region, which tends to reduce differential settlement. In this case, the expulsion of the soft soil layer may be the most critical mechanism of destabilisation of the system. The same effect can be achieved with the use of geocells filled with the same, or better, quality fill material, as shown in Figure 5.16(c).

Figure 5.16 Geosynthetic installation: (a) reinforcement inside the embankment, (b) several reinforcement layers, (c) a geocell at the base of the embankment, (d) reinforcement with folded ends, (e) combination of reinforcement with berms and (f) reinforcement and piles

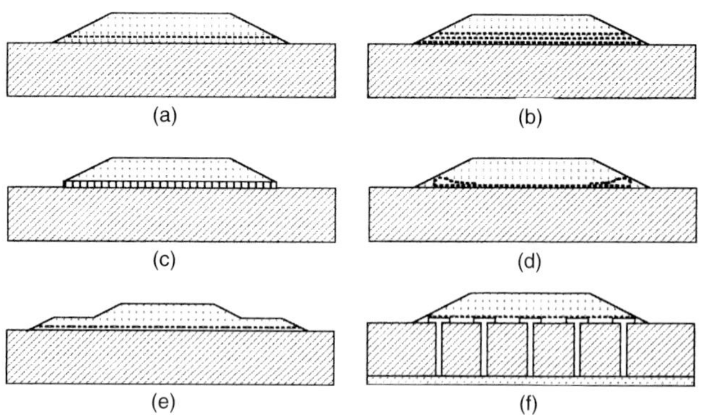

The use of folded reinforcement extremities to obtain a greater anchorage strength is shown schematically in Figure 5.16(d). Vertical longitudinal plates, to which the geosynthetic extremities can be fixed, can also be employed, although this leads to additional complications with regard to the construction, particularly for soft to very soft subgrades, and provides questionable overall benefits. An increase in the reinforcement anchorage and a reduction in the required reinforcement strength can also be achieved with the combined use of a reinforcement layer and berms, as illustrated in Figure 5.16(e).

When limits must be applied to the settlements of the embankment, the solution of combining geosynthetic layers with vertical piles with caps can be employed, as shown schematically in Figure 5.16(f), and this type of geosynthetic application is addressed later in this chapter. In this case, the presence of the geosynthetic layer provides a better distribution of the embankment weight to the pile caps in conjunction with fill arching in between caps. Additional information on and details of the performance of this type of application can be found in, for example, Gartung *et al.* (1996), Kempton *et al.* (1998), Cooper and Rose (1999), Collin *et al.* (2005) and Raithel *et al.* (2008).

The effect of the combined use of geosynthetic reinforcement and lateral berms, as discussed above, is shown in Figure 5.17 for a typical case of an embankment on soft soil (Silva and Palmeira, 1998). This figure shows the variation in the required reinforcement force, for an overall safety factor of 1.3, with the berm length, expressed as a fraction of the required berm length, obtained using the method proposed by Jakobson (1948) and for the case of an unreinforced embankment. Note that, in the case shown in Figure 5.17, the reinforcement layer lies only along the main embankment base. A considerable decrease in the required reinforcement force can be observed as the length of the berm increases.

Due care has to be taken during the installation of geosynthetic layers in the field in order to avoid, or minimise,

Figure 5.17 Required reduction in the reinforcement force due to the use of berms
Modified from Silva and Palmeira (1998)

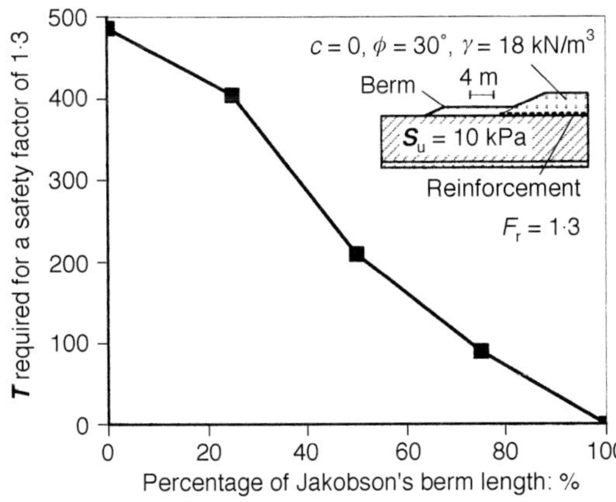

mechanical damage to the reinforcement during its installation and during the spreading and compaction of the fill material, particularly for very soft subgrades and/or very coarse fill materials (stones, rockfill, etc.). Palmeira (1998) presented a methodology to estimate geosynthetic tensile strains caused by loading conditions during embankment construction. Appropriate reduction factors must be applied to the tensile strength of the reinforcement in order to account for mechanical damage. The value of the reduction factor required to account for mechanical damage depends on the type of fill material, type of geosynthetic and the quality of the construction technique, and varies typically between 1.05 and 1.5 for normal design and construction conditions.

5.3. Geosynthetics for drainage in embankments

5.3.1 Introduction

Geosynthetics can also be used as drainage boundaries for the problem of embankments on soft soils. In this case, the primary role of the geosynthetic layer is to accelerate consolidation settlements that otherwise would take a long time to occur. The dissipation of excess pore pressures and the acceleration of the consolidation process also increase the strength of the subsoil, allowing for a faster rate of embankment lifting. Traditional sand blankets and vertical drains have been used for quite a long time, but the use of geosynthetic drainage systems in embankments on soft soils has increased markedly during the last three decades. In comparison to sand drains, the main advantages of geosynthetic drains are

- they are cost-effective in regions where granular material is scarce or its exploitation is prohibited by environmental constraints
- due to the industrial nature of their production, they can be manufactured to specified characteristics, and the quality can be carefully controlled
- they can be easily transported to remote construction areas
- they are easy to install, and usually only require simple and light construction equipment
- their installation is considerably faster than that of sand drains, which can lead to significant cost savings and a reduction in construction time. The installation of vertical geosynthetic drains is also a rather clean process compared with the usual vertical sand drain installation techniques, which tend to contaminate the drainage blanket at the soft soil surface with fines from the foundation soil.

In the sections below, design approaches for geosynthetic drainage systems to be used in embankments on soft soil are presented and discussed.

5.3.2 Geosynthetic drainage blanket at the base of the embankment

The simplest form of foundation soil drainage in embankments on soft soils consists of the installation of a drainage

Figure 5.18 A geosynthetic drainage blanket at the embankment base
Adapted from Giroud (1981)

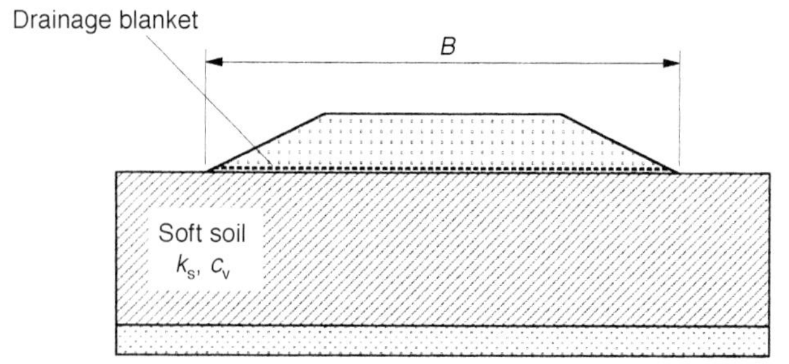

layer at the base of the embankment, as shown schematically in Figure 5.18. Giroud (1981) presented the following expression for the estimation of the required transmissivity of the drainage blanket:

$$\theta_{req} = k_p t_{GT} = \frac{B^2 k_s}{(c_v t_c)^{0.5}} \tag{5.54}$$

where θ_{req} is the required geosynthetic transmissivity, k_p is the in-plane coefficient of the permeability of the geosynthetic layer, t_{GT} is the thickness of the geosynthetic blanket, B is the width of the embankment base (see Figure 5.18), k_s is the coefficient of permeability of the foundation soil, c_v is the coefficient of consolidation of the foundation soil, and t_c is the time for the embankment construction.

It is important to note that the stress level on the geosynthetic has to be considered in order to choose a geosynthetic layer that will ensure a value of θ_{req}, calculated using Equation 5.54 under the vertical surcharge caused by the embankment.

5.3.3 Geosynthetic vertical drains

The use of vertical drains in conjunction with a horizontal drainage blanket to accelerate consolidation settlements can be considerably more efficient than using a drainage blanket alone. Several design methodologies are available for the design of vertical drains, and the reader is recommended to consult Magnan (1983) for a comprehensive and detailed study on vertical drains.

In contrast to traditional cylindrical sand drains, geosynthetic vertical drains, or band-shaped prefabricated drains, can have rectangular or circular cross-sectional areas, are delivered in rolls, and usually consist of a plastic core (or perforated tube) covered by a non-woven geotextile. Figure 5.19 shows common types of geosynthetic vertical drains. The non-woven geotextile cover must be an efficient filter for the surrounding soil, as per the filter criteria discussed in Chapter 3. The drain installation process consists of pushing one end of the drain into the soil until the required depth is reached, which is usually the bottom of the soft soil deposit or an appropriate permeable soil stratum, as shown in Figure 5.20. Special components must be used to attach the drain extremity and to drive it into the ground, and in order to keep it at the desired depth during casing retrieval (see Figure 5.20).

Figure 5.19 Common types of geosynthetic vertical drain
Reproduced from Palmeira (1996)

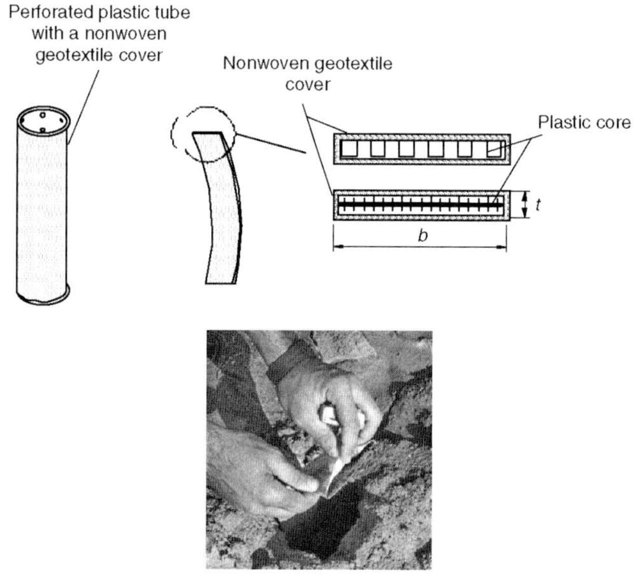

The required average consolidation ratio for the vertical drain system, when both vertical and radial drainage occur simultaneously in a saturated soft deposit, can be evaluated as (Carrillo, 1942)

$$U_r = \frac{U_{vr} - U_v}{1 - U_v} \tag{5.55}$$

Figure 5.20 Installation of a geosynthetic vertical drain
Reproduced from Palmeira (1996)

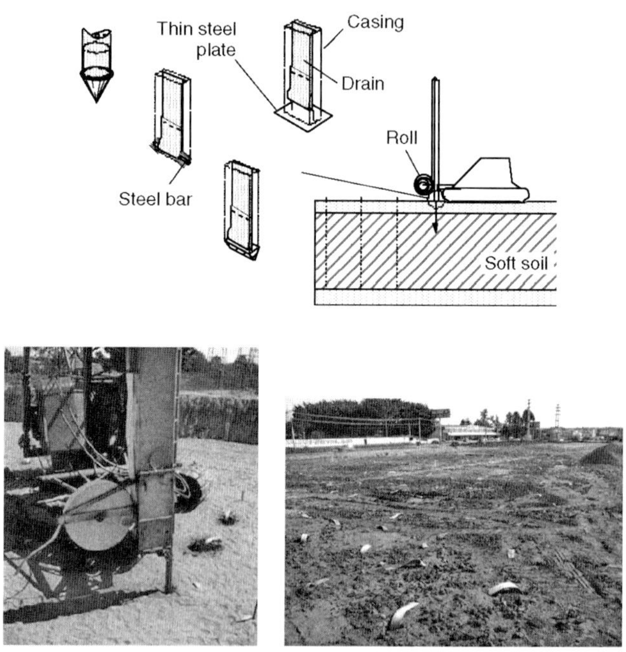

Figure 5.21 The average consolidation ratio versus the time factor for vertical drainage

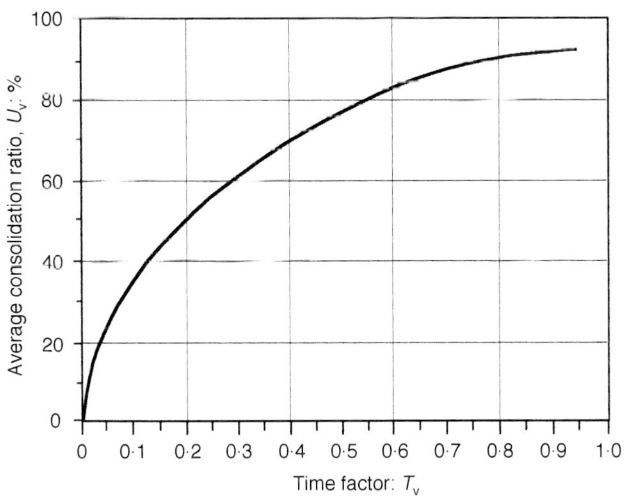

where U_r is the required average radial consolidation ratio from the vertical drains, U_{vr} is the target average consolidation ratio for the soft soil, taking into account the vertical and radial drainage, and U_v is the average consolidation ratio due to vertical drainage only.

The value of U_v occurs during the period of time considered and can be estimated using the traditional one-dimensional consolidation theory (Terzaghi, 1943; Lambe and Whitman, 1969) when tridimensional effects can be neglected, being a function of the time factor given by

$$T_v = \frac{c_v t}{H_d^2} \tag{5.56}$$

where T_v is the time factor, c_v is the coefficient of consolidation of the foundation soil, t is the time elapsed until U_v is reached, and H_d is the longest path followed by a water particle to the nearest drainage boundary along the vertical direction. The relationship between U_v and T_v is presented in Figure 5.21.

The design of vertical drains consists of the following steps: The designer establishes a target value of U_{vr} to be reached at a time t. The value of U_v at the time t is calculated using Equation 5.56 and Figure 5.21. Then, the required value of U_r is calculated using Equation 5.55. It is common for some designers to neglect the contribution of the vertical drainage at the time t ($U_v = 0$), which will lead to a more conservative design of the vertical drain system. By knowing the required value of U_r, the vertical drain system can be designed using one of the available design methods.

Hansbo (1979) presented the following solution for the value of the consolidation ratio due to radial drainage (Figure 5.22) of a saturated soil deposit for $n = D/d_w > 5$:

$$U_r = 1 - \exp(-8c_h t / D^2 \mu) \tag{5.57}$$

with

$$\mu = \ln\left(\frac{D}{d_s}\right) + \frac{k_h}{k_s}\ln\left(\frac{d_s}{d_w}\right) - 0.75 + \frac{2\pi H_d^2 k_h}{3q_w} \times \left(1 - \frac{k_h/k_s - 1}{(k_h/k_s)(D/d_s)^2}\right) \tag{5.58}$$

where c_h is the horizontal consolidation coefficient of the soft soil, D is the diameter of the dewatered soil cylinder around the vertical drain, t is the time of consolidation, d_s is the diameter of the disturbed zone around the vertical drain due to the drain installation, k_h is the undisturbed horizontal coefficient of permeability of the soft soil, k_s is the coefficient of permeability of the disturbed zone, d_w is the diameter of the drain, and q_w is the discharge capacity of the vertical drain.

The value of μ given by Equation 5.58 is an average value for the entire soft soil thickness (see Magnan, 1983).

Figure 5.22 Vertical drains and radial drainage

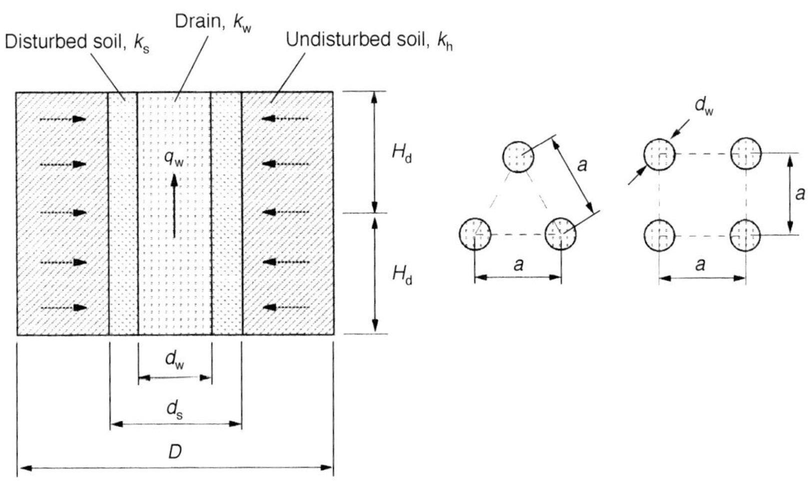

Hansbo (1979) points out that the value of d_s/d_w varies between 1.5 and 3, depending on the drain type and construction conditions. When a casing is used for drain installation, d_s/d_m varies typically between 2 and 3, where d_m is the casing diameter. van Impe (1989) suggests $k_s \approx k_h/5$ for preliminary estimates when accurate values of permeability coefficients are not available.

The drain discharge capacity q_w depends on the permeability coefficient of the drain, the drain dimensions, the hydraulic gradient in the drain and the stress level. Hansbo (1979) recommends that, when test results are not available, the value of q_w used should be less than 500 m^3/year. Holtz *et al.* (1989) suggest that a minimum value of q_w should lie between 100 and 150 m^3/year. Hansbo *et al.* (1981) and van Impe (1989) suggest a hydraulic gradient equal to one inside the drain, with the discharge capacity given by $A_w k_w$, where A_w is the cross-sectional area of the drain and k_w is the permeability coefficient of the drain.

Common band-shaped prefabricated drains are usually capable of presenting large values of q_w for the normal stress levels present in soft soil layers beneath embankments. However, some clogging of the drain filter is expected to occur during drain installation or with time, which will reduce the drain discharge capacity. Compression of the drain and damage to it caused during installation or by severe consolidation settlements can also reduce the value of q_w. Holtz (1987) recommends a minimum tensile strength for the drain of 5 kN/m and a strain at maximum tensile stress of 2–10%.

Equations 5.57 and 5.58 were initially derived for cylindrical vertical drains. Hansbo (1979) showed that a band-shaped drain behaves in a similar way to a cylindrical one with the same perimeter. Therefore, the value of d_w in Equation 5.58 should be replaced by the equivalent diameter of the band-shaped drain, given by

$$d_{eq} = \frac{2(b+t)}{\pi} \tag{5.59}$$

where d_{eq} is the equivalent drain diameter, and b and t are the width and thickness of the geosynthetic drain, respectively (see Figure 5.19).

The expressions for drain spacing are

$$a = \frac{D}{1.13} \quad \text{for drain installation in a square pattern} \tag{5.60}$$

or

$$a = \frac{D}{1.05} \quad \text{for drain installation in a triangular pattern} \tag{5.61}$$

The value of D used in Equation 5.60 or 5.61 must be determined from Equations 5.57 and 5.58 by trial and error.

5.4. Combined use of geosynthetic reinforcement and vertical drainage

The combination of geosynthetic reinforcement and prefabricated vertical drains can provide short-term stability of the embankment and the acceleration of consolidation settlements. Depending on the characteristics of the project and of the soils, it can also reduce the horizontal displacements of the soft foundation during and after embankment construction. This can be relevant when the embankment is constructed close to existing structures, like in the case of abutments approaching existing bridges. However, the analysis of the benefits of the combination of reinforcement and vertical drains in this case requires more sophisticated tools and input data for accurate estimates of displacements of the soft soil and their consequences on the neighbouring structure. Usually, numerical analysis (finite-element or finite-differences methods) are employed in such cases.

Macedo *et al.* (2009) carried out finite-element analyses on the influence of the presence of geosynthetic reinforcement and vertical drains on the effects of abutment construction on the behaviour of neighbour bridge piles. The type of problem analysed is shown schematically in Figure 5.23. Figures 5.24(a) and 5.24(b) show some results obtained from the study in terms of the distributions of the bending moments M along the pile length in situations with and without basal reinforcement and vertical drains for a pile distant 1 m from the embankment toe, 28 days after the end of embankment construction. It can be seen that for the case without vertical drains (see Figure 5.24(a)), the greater the reinforcement tensile stiffness J the smaller the bending moments in the pile. It should be pointed out that, in this case, the construction of the embankment without

Figure 5.23 Geosynthetic reinforcement with vertical drainage: (a) geometrical characteristics of the problem and (b) finite-element mesh
Adapted from Macedo *et al.* (2009)

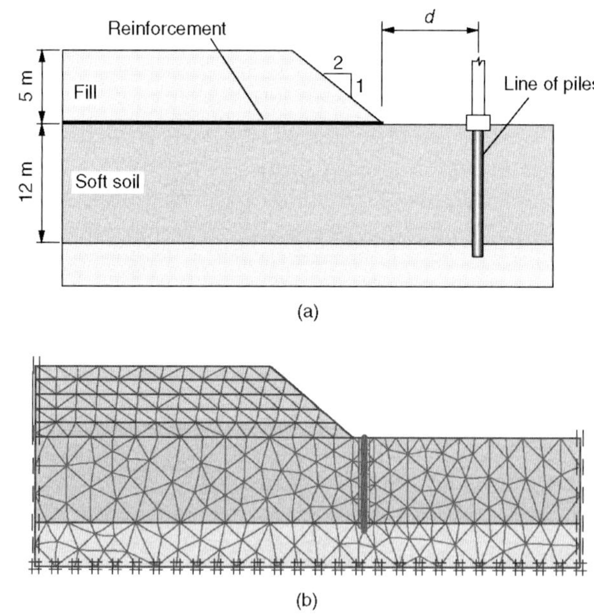

Figure 5.24 Distribution of bending moments along the pile length for $d = 1$ m and $t = 28$ days (FPT case): (a) without drains and (b) with drains

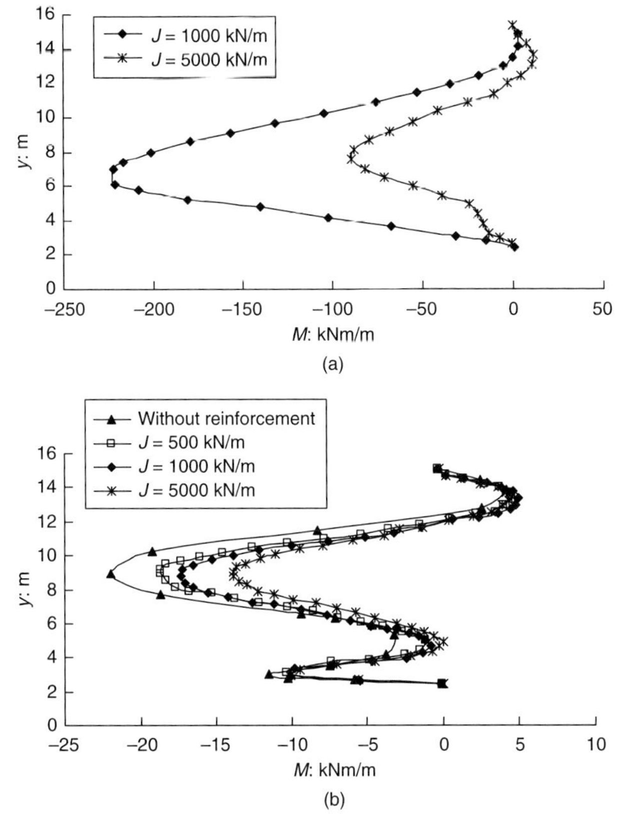

reinforcement would not be possible, because it would fail before its final height had been reached. However, the use of vertical drains would allow the construction of the unreinforced embankment, and Figure 5.24(b) presents the distributions of bending moments along the pile length for cases with and without basal reinforcement. The results shown in Figures 5.24(a) and 5.24(b) suggest that the combination of vertical drainage and basal reinforcement can reduce significantly the bending moments in the pile. The results obtained in similar analyses also showed the benefits of the use of geosynthetic reinforcement and vertical drainage in reducing the displacements of the soft foundation soil. It should be pointed out that, despite using sophisticated numerical tools, this type of analysis also has its limitations. Nevertheless, the results are encouraging in showing the benefits in using basal reinforcement and vertical drains to reduce the effects of embankment construction on the foundations of neighbouring structures. In real situations, the benefits brought by the use of geosynthetics in this type of problem are site specific, and should always be monitored with the use of proper geotechnical instrumentation. Fahel *et al.* (2000), Fahel and Palmeira (2002), Palmeira *et al.* (2003), and Araujo and Palmeira (2006), present some case histories of instrumented reinforced abutments on soft soils.

5.5. Piled embankments

5.5.1 Introduction

Piles under embankments are used when settlements must be minimised. Arching of the embankment soil in the region between pile caps yields to the transference of embankment loads to the piles, which can reduce significantly the stresses transmitted to the soft soil, embankment settlements and lateral displacements of the soft foundation material. The use of geosynthetics in piled embankments can decrease even further the stresses transmitted to the subgrade, and improve the distribution of loads among the piles. Two situations for the use of geosynthetics in such applications can be identified, as shown in Figures 5.25(a) and (b). Figure 5.25(a) shows the case in which the geosynthetic layer works as a membrane carrying part of the embankment load that otherwise would be transmitted to the foundation soil. Figure 5.25(b) shows the situation where more than one layer of geosynthetic reinforcement is used to form a

Figure 5.25 Uses of geosynthetics in piled embankments: (a) membrane effect, (b) load-transfer platform, (c) piles with caps and (d) caps and gravel layer *in situ*

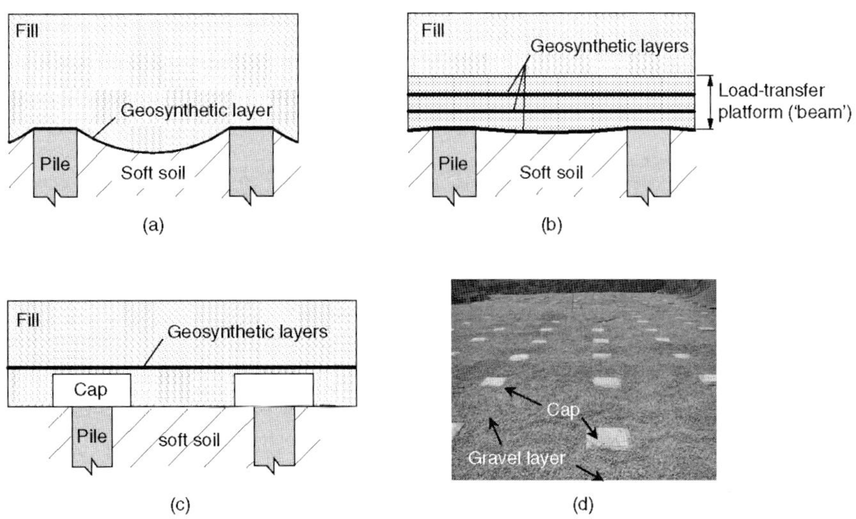

platform for a better transference of the embankment load to the piles and a greater reduction of settlements. Caps (typically square in plan) are often used on the pile top to reduce the embankment region subjected to arching and to increase load transference to the piles (Figure 5.25(c)).

5.5.2 Geosynthetics working as a membrane

5.5.2.1 Vertical stress transmitted to the soft soil

One of the key aspects for the design of a piled embankment is the vertical pressure that is transmitted to the soft layer in the region between the piles or caps. Several methods are available in the literature for applications where the geosynthetic layer works as a membrane (see Figure 5.25(a)) and their accuracy is subject to debate (Russell and Pierpoint, 1997; Love and Milligan, 2003) due to the complex state of stress in the embankment region undergoing arching and to the simplifying assumptions made by those methods. The determination of the stresses transmitted to the foundation soil is one of the key steps for the design of the geosynthetic layer as a membrane in piled embankments.

Based on soil arching studies by Terzaghi (1943), the following expression for the vertical stress at the level of the pile head (or cap surface) can be derived for a case without reinforcement (Russell and Pierpoint, 1997):

$$\frac{p_r}{\gamma H} = \frac{s^2 - a^2}{4HaK \tan\phi}\left[1 - \exp(-4aHK\tan\phi/(s^2 - a^2))\right] \tag{5.62}$$

where p_r is the vertical stress at the level of the pile head (or cap surface), s is the spacing between piles (centre-to-centre distance), γ is the embankment soil unit weight, H is the height of the embankment, K is the earth pressure coefficient, a is the cap (square) width, and ϕ is the embankment soil friction angle.

Hewlett and Randolph (1988) approached the arching mechanism in piled embankments by considering the limiting equilibrium of stresses in a hemispherical dome region of sand between pile caps. From results obtained in three-dimensional model tests conducted on sand, and using the theory of plasticity, they proposed the following equation to obtain the vertical stress between caps in a case without reinforcement:

At the crown of the soil arch:

$$\frac{p_r}{\gamma H} = \left(1 - \frac{a}{s}\right)^{2(K_p-1)}\left[1 - \frac{s}{\sqrt{2}H}\left(\frac{2K_p - 2}{2K_p - 3}\right)\right] + \frac{s-a}{\sqrt{2}H}\left(\frac{2K_p-2}{2K_p-3}\right) \tag{5.63}$$

The condition at the cap:

$$\frac{p_r}{\gamma H} = \frac{1}{\left(\frac{2K_p}{K_p+1}\right)\left[\left(1-\frac{a}{s}\right)^{(1-K_p)} - \left(1-\frac{a}{s}\right)\left(1+\frac{a}{s}K_p\right)\right] + \left(1-\frac{a^2}{s^2}\right)} \tag{5.64}$$

with

$$K_p = \frac{1+\sin\phi}{1-\sin\phi} \tag{5.65}$$

In a reinforced case, the value of p_r to be used for the reinforcement layer in the region between piles (or caps) would be the greatest between the values given by Equations 5.63 and 5.64.

The value of p_r is used in the design of the geosynthetic layer as a membrane and to estimate its tensile load. The approach used to do this depends on the design method chosen. Two of these methods are described in the following sections.

5.5.2.2 The British method (BS 8006)

Figure 5.26 presents the relevant information required for designing piled embankments with geosynthetic reinforcement according to BS 8006 (British Standards Institution (BSI), 1995). The maximum spacing (square pattern) between piles can be calculated using

$$s = \sqrt{\frac{Q}{\gamma H + q}} \tag{5.66}$$

where s is the spacing between piles, Q is the pile allowable load, γ is the unit weight of the fill material, H is the fill thickness above the reinforcement layer, and q is the surcharge along the fill surface.

The vertical stress on the pile cap can be obtained using (no partial safety factors included)

$$\frac{p'_c}{\sigma'_v} = \left(\frac{C_c a}{H}\right)^2 \tag{5.67}$$

with

$$\sigma'_v = \gamma H + q \tag{5.68}$$

where p'_c is the vertical stress on the pile cap, a is the cap width, and C_c is the soil arching coefficient, which can be estimated using:

Figure 5.26 Geometrical characteristics of the piled embankment

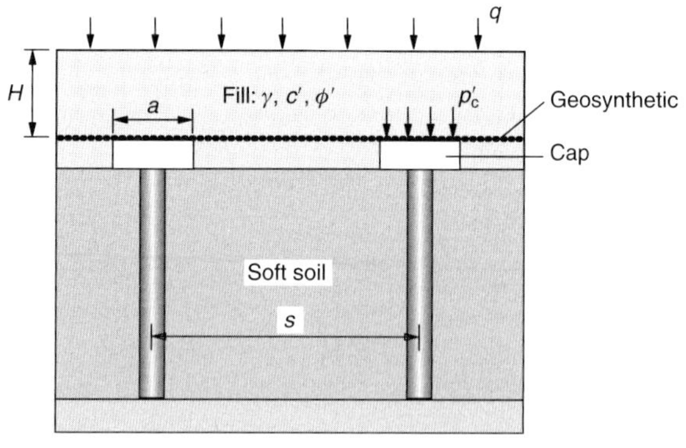

Figure 5.27 Contribution of the geosynthetic layer

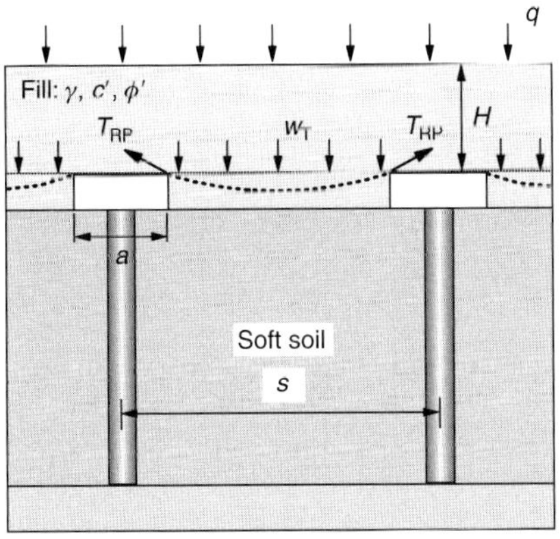

For end-bearing piles (unyielding):

$$C_c = 1.95\frac{H}{a} - 0.18 \quad (5.69)$$

For friction and other piles:

$$C_c = 1.50\frac{H}{a} - 0.07 \quad (5.70)$$

Regarding arching in the fill material, the following condition is required:

$$H \geq 0.7(s-a) \quad (5.71)$$

Figure 5.27 shows the deformed shape of the geosynthetic layer. The distributed vertical load w_T, acting on the reinforcement between adjacent piles, is given by

$$w_T = \frac{1.4s\gamma_e(s-a)}{s^2-a^2}\left(s^2 - a^2\frac{p'_c}{\sigma'_v}\right) \text{ if } H > 1.4(s-a) \quad (5.72)$$

or

$$w_T = \frac{s\gamma_e H}{s^2-a^2}\left(s^2 - a^2\frac{p'_c}{\sigma'_v}\right) \text{ if } 0.7(s-a) \leq H \leq 1.4(s-a) \quad (5.73)$$

with

$$\gamma_e = \gamma + q/H \quad (5.74)$$

Biaxial geogrids are preferred in applications in piled embankments. However, when two uniaxial geogrids are used, the transverse (lower) reinforcement layer will pick up more vertical load than the longitudinal (upper) reinforcement layer (Love and Milligan, 2003). Therefore, for a square piling layout (with square pile caps) when two layers of uniaxial reinforcement are used, Love and Milligan (2003) suggest that the value of w_T be given by

$$w_T = p_r s \quad \text{along the transverse direction} \quad (5.75)$$

and

$$w_T = p_r a \quad \text{along the longitudinal direction} \quad (5.76)$$

According to BS 8006 (BSI, 1995), the force and strain mobilised in the geosynthetic layer can be determined using

$$T_{RP} = \frac{w_T(s-a)}{2a}\sqrt{1+\frac{(s-a)^2}{16d^2}} \quad (5.77)$$

$$\varepsilon = \frac{8d^2}{3(s-a)^2} \quad (5.78)$$

where T_{RP} is the tensile force in the reinforcement, d is the maximum deflection of the reinforcement, and ε is the tensile strain in the reinforcement.

Combining Equations 5.77 and 5.78 yields

$$T_{RP} = \frac{w_T(s-a)}{2a}\sqrt{1+\frac{1}{6\varepsilon}} \quad (5.79)$$

Assuming a linear elastic behaviour of the geosynthetic reinforcement,

$$T_{RP} = J\varepsilon \quad (5.80)$$

where J is the reinforcement tensile stiffness.

Equations 5.79 and 5.80 must be solved by trial and error for the determination of the force and strain in the geosynthetic. A maximum tensile strain of 6% and further creep deformation less than 1% are recommended (BSI, 1995). The design strength (with the appropriate reduction factors having been applied) of the geosynthetic must not be smaller than the value of T_{RP}.

The tensile force in the reinforcement in the region of the slope of the embankment must be increased because of the action of the active thrust in the fill, as depicted in Figure 5.28. Thus, in this case, the value of T_{RP} is given by:

Along the transversal direction of the embankment:

$$T_d \geq T_{RP} + \Delta T_{RL} \quad (5.81)$$

Along the longitudinal direction of the embankment:

$$T_d \geq T_{RP} \quad (5.82)$$

Figure 5.28 Influence of the fill active thrust

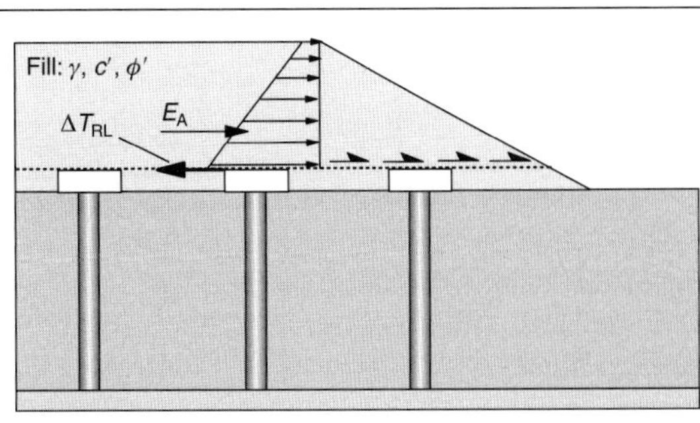

Figure 5.29 Support conditions underneath the embankment and definition of s
Adapted from Kempfert *et al.* (2004)

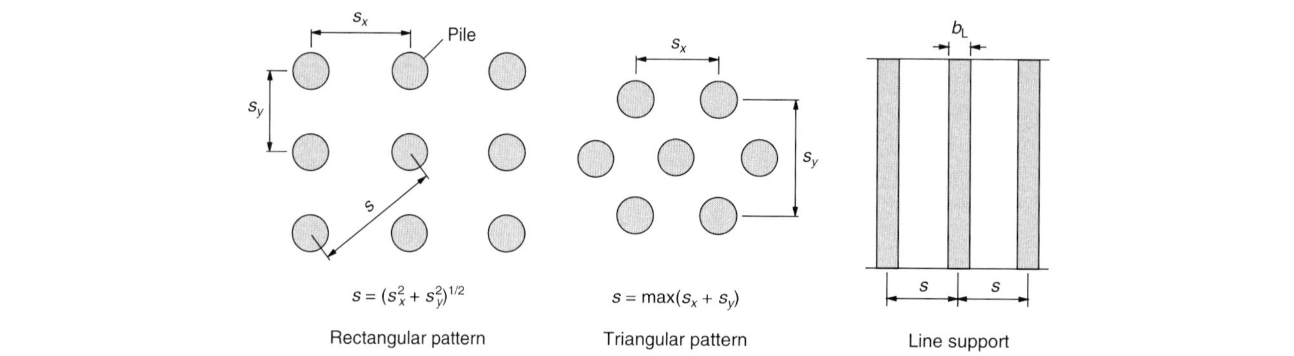

with

$$\Delta T_{\mathrm{RL}} = E_{\mathrm{A}} \tag{5.83}$$

where T_{d} is the design strength of the reinforcement, ΔT_{RL} is the increment of load in the reinforcement, and E_{A} is the active thrust in the fill.

Love and Milligan (2003) argue that the BS 8006 (BSI, 1995) method is based on two-dimensional pipeline theory, not the actual three-dimensional situation verified at the base of a piled embankment, and that the method does not satisfy simple vertical equilibrium and does not consider accurately the arching mechanism. However, it is still a design approach commonly used in several parts of the world. At present, it is being revised, and an updated version of the method is likely to be available soon.

5.5.2.3 The German method

Kempfert *et al.* (2004) present the possible support conditions provided by the piles underneath an embankment, as shown schematically in Figure 5.29. Piles are typically distributed in rectangular or triangular patterns.

Figure 5.30 shows the region of soil in the fill material undergoing arching, according to the German practice for piled embankments (Zaeske, 2001; Zaeske and Kempfert, 2002; Raithel *et al.*, 2008). From the study of the equilibrium of that region, the following equation for the vertical stress on the level of the soft soil layer can be derived (Kempfert *et al.*, 2004):

$$\sigma_{\mathrm{zo,k}} = \lambda_1^{\chi}\left(\gamma_{\mathrm{k}} + \frac{p_{\mathrm{k}}}{h}\right)\left\{h(\lambda_1 + h_{\mathrm{g}}^2\lambda_2)^{-\chi} + h_{\mathrm{g}}\left[\left(\lambda_1 + \frac{h_{\mathrm{g}}^2\lambda_2}{4}\right)^{-\chi} - (\lambda_1 + h_{\mathrm{g}}^2\lambda_2)^{-\chi}\right]\right\} \tag{5.84}$$

with

$$\chi = \frac{d(K_{\mathrm{crit}} - 1)}{\lambda_2 s} \tag{5.85}$$

$$K_{\mathrm{crit}} = \tan^2(45^{\circ} + \phi'_{\mathrm{k}}/2) \tag{5.86}$$

$$\lambda_1 = \tfrac{1}{8}(s - d)^2 \tag{5.87}$$

$$\lambda_2 = \frac{s^2 + 2ds - d^2}{2s^2} \tag{5.88}$$

where $\sigma_{\mathrm{zo,k}}$ is the vertical stress on the geosynthetic layer, p_{k} is the characteristic value of the live load, h is the embankment thickness, d is the pile diameter, ϕ_{k} is the characteristic value of the friction angle of the fill material, γ_{k} is the characteristic value of the unit weight of the fill material, and h_{g} is the vault height, given by

$$h_{\mathrm{g}} = s/2 \text{ if } h \geq s/2 \tag{5.89}$$

or

$$h_{\mathrm{g}} = h \text{ if } h < s/2 \tag{5.90}$$

Figure 5.30 Arching in the fill material
Adapted from Zaeske (2001)

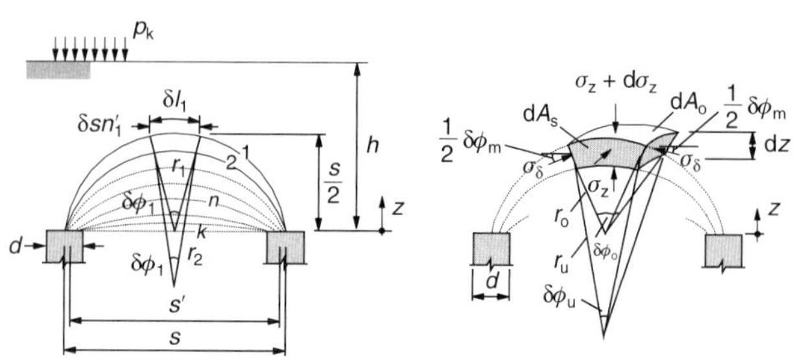

Figure 5.31 shows the normalised value of the vertical stress on the soft foundation layer for the particular case of an embankment with a soil friction angle of 30° (Kempfert *et al.*, 2004).

The theory of elastically embedded membranes (Zaeske, 2001) was used to predict the loads in the reinforcement. The maximum strain in the reinforcement occurs in the band bridging two neighbouring piles. Therefore, the maximum load in the reinforcement takes place within the width b_{Ers}, as shown schematically in Figure 5.32(a), and may be calculated based on a planar system (Figure

Figure 5.31 Vertical stress on the soft soil between piles for an embankment built with a cohesionless soil having a 30° friction angle
Adapted from Kempfert *et al.* (2004)

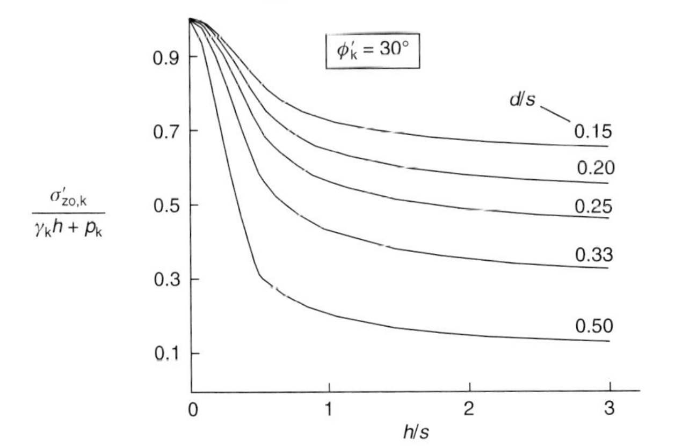

5.32(b)). When a biaxial geogrid is used as reinforcement, the geogrid must be analysed along both the x and the y directions. The reaction from the soft soil is estimated by considering the soil as a set of springs with a given characteristic value of the subgrade reaction modulus $k_{s,k}$.

The value of the vertical force on the reinforcement layer in each direction (x or y) depends on the pattern of distribution of the piles. For a rectangular distribution of piles (Figure 5.33) the values of the forces in each direction can be estimated as (EMPFEHLUNG, 2003)

$$F_{x,k} = A_{Lx}\sigma_{zo,k} \tag{5.91}$$

and

$$F_{y,k} = A_{Ly}\sigma_{zo,k} \tag{5.92}$$

with

$$A_{Lx} = \frac{1}{2}(s_x s_y) - \frac{d^2}{2}\left[\tan^{-1}\left(\frac{s_y}{s_x}\right)\right]\frac{\pi}{180} \tag{5.93}$$

$$A_{Ly} = \frac{1}{2}(s_x s_y) - \frac{d^2}{2}\left[\tan^{-1}\left(\frac{s_x}{s_y}\right)\right]\frac{\pi}{180} \tag{5.94}$$

Figure 5.32 Load transfer to the piles: (a) load transfer between piles and (b) simplified planar (two-dimensional) bearing system
Adapted from Zaeske (2001)

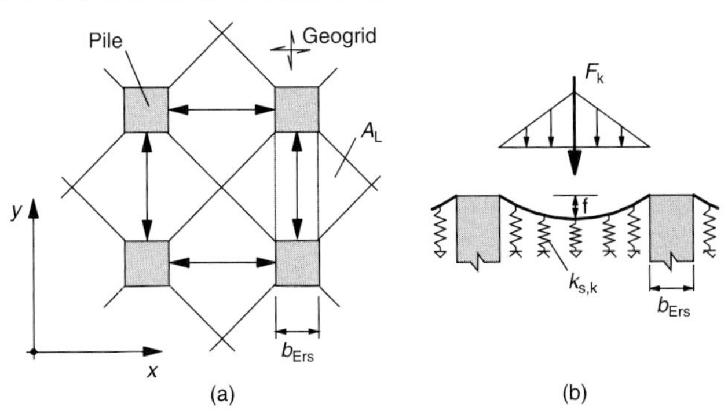

Figure 5.33 Calculation of $F_{x,k}$ and $F_{y,k}$ for piles in a rectangular pattern
Adapted from Kempfert *et al.* (2004)

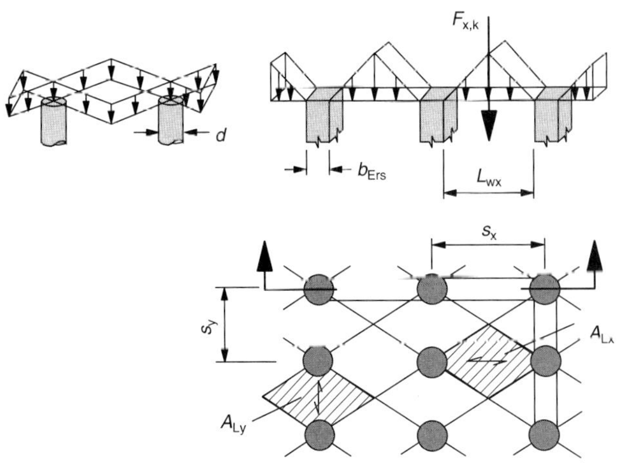

where s_x is the spacing between piles in the x direction, s_y is the spacing between piles in the y direction, and A_{Lx} and A_{Ly} are the loaded areas contributing to the calculation of the vertical force on the bands along the x and y directions, respectively.

For a triangular pattern of pile distribution (Figure 5.34), the vertical forces can be estimated as (Kempfert *et al.*, 2004)

$$F_{x,k} = \frac{j_x}{j_x + j_y}A_{Lxy}\sigma_{zo,k} \tag{5.95}$$

$$F_{y,k} = \frac{j_y}{j_x + j_y}A_{Lxy}\sigma_{zo,k} \tag{5.96}$$

with

$$A_{Lxy} = \frac{1}{2}s_x s_y - \frac{d^2}{4}\pi \tag{5.97}$$

Figure 5.34 Calculation of $F_{x,k}$ and $F_{y,k}$ for piles in a triangular pattern
Adapted from Kempfert *et al.* (2004)

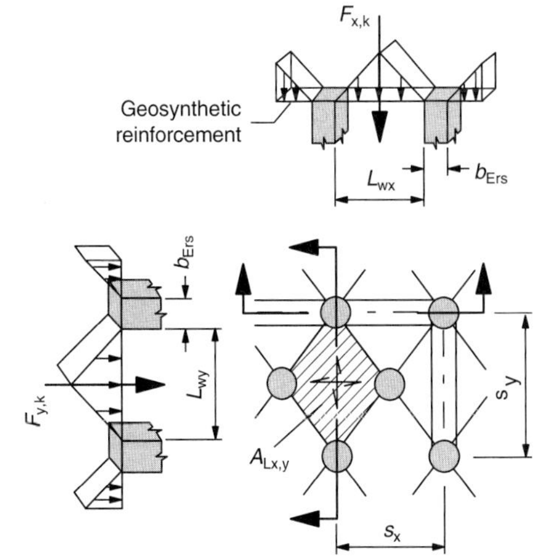

Figure 5.35 Maximum strain in the reinforcement
Adapted from EMPFEHLUNG (2003)

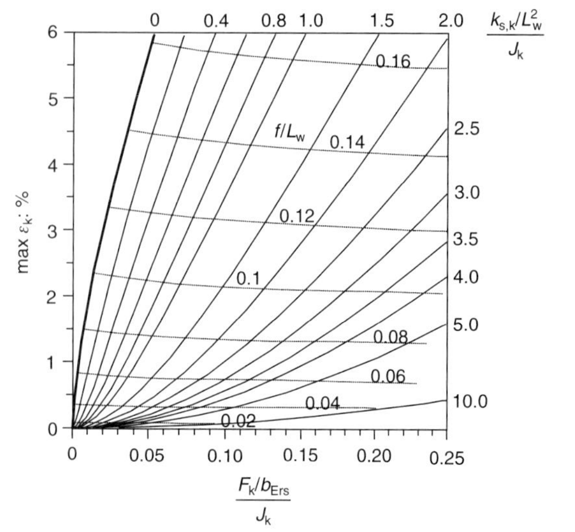

where A_{Lxy} is the loaded area contributing to the calculation of the vertical force, and j_x and j_y are the fractions of the loaded band areas along the x and y directions, respectively.

With the value of the total vertical force F_k, one can estimate the tensile strain in the reinforcement using the graph shown in Figure 5.35 (EMPFEHLUNG, 2003), where $k_{s,k}$ is the modulus of the subgrade reaction of the soft soil, L_w is the clear distance ($= s - d$), J_k is the tensile stiffness of the reinforcement, and b_{Ers} is the width of the support, given by

$$b_{Ers} = \frac{1}{2} d \sqrt{\pi} \qquad (5.98)$$

The force in the reinforcement is given by

$$T = \varepsilon_k J_k \qquad (5.99)$$

where ε_k is the tensile strain of the reinforcement.

The same approach regarding the increase in reinforcement forces, due to the active thrust from the embankment in its slope regions, applies (Equations 5.81 to 5.83).

Regarding the centre-to-centre distance of the piles s, the pile diameter d and the width of the pile caps, the following conditions must be observed (Kempfert *et al.*, 2004)

- In the case of static loads: $(s - d) \leq 3$ m or $(s - b_L) \leq 3$ m (see Figure 5.29).
- In the case of heavy live loads: $(s - d) \leq 2.5$ m or $(s - b_L) \leq 2.5$ m.
- $d/s \geq 0.15$ or $b_L/s \geq 0.15$.
- $(s - d) \leq 1.4(h - z)$.
- The ratio between the values of subgrade reaction modulus of the pile elements and that of the surrounding soft soil $k_{s,p}/k_{s,w} > 100$, to ensure full arching mobilisation, as assumed by the theory.

Figure 5.36 Reinforcement layer elevation z

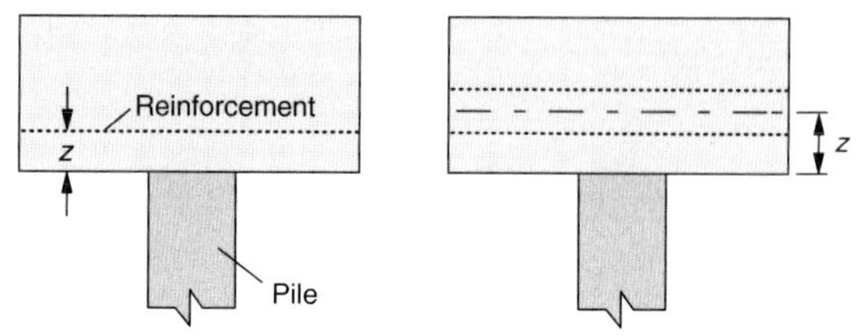

Normally, conventional pile systems fulfil this condition (Kempfert *et al.*, 2004).

With respect to the reinforcement layer, the following conditions must be observed (Kempfert *et al.*, 2004)

- The closer the distance between the reinforcement layer to the pile heads (or caps), the better it is to achieve maximum efficiency from the geosynthetic reinforcement. However, it is recommended to have a safe distance between the lowest reinforcement and the pile heads (or caps) to avoid damage of the reinforcement because of tearing at the edges of the pile heads or caps.
- Maximum of two reinforcement layers (Figure 5.36).
- $z \leq 0.15$ m for a single reinforcement layer (Figure 5.36).
- $z \leq 0.3$ m for two reinforcement layers (Figure 5.36).
- For two reinforcement layers the distance between them should be 0.15–0.30 m.
- For the reinforcement, a design value of tensile strength of ≥ 30 kN/m and of ultimate strain of $\leq 12\%$.
- Overlapping of reinforcement layers is generally allowed, but should only occur just above the pile (caps) and only in the secondary bearing direction. The overlapping length must be $\geq d$.

Cohesionless soils with internal friction angles ϕ_k greater than 30° should be used. The use of low-cohesion soils is permitted, but not preferred (Kempfert *et al.*, 2004).

5.5.3 Geosynthetics in load-transfer platforms

Collin *et al.* (2005) presents a design method for load-transfer platforms (see Figure 5.25(b)), using a beam theory based on the following conditions and assumptions

- The thickness (h, Figure 5.37) of the load transfer platform is not smaller than one half of the clear span between columns ($s - d$).
- A minimum of three layers of geosynthetic reinforcement is used to create the load-transfer platform.
- The minimum distance between reinforcement layers is 0.15 m.
- Selected (well-graded granular material) fill material is used in the load-transfer platform.
- The primary function of the reinforcement is to provide lateral confinement of the selected fill to facilitate soil arching within the height (thickness) of the load-transfer platform.

- The secondary function of the reinforcement is to support the wedge of soil below the arch.
- The vertical load from the embankment above the load-transfer platform is transferred to the columns below the platform.
- The initial reinforcement strain is limited to 5%.

The load supported by a column below the load-transfer platform is given by

$$Q = \pi \frac{D_e^2}{4}(\gamma H + q) \tag{5.100}$$

with

$$D_e = 1.05s \quad \text{for triangular spacing between columns} \tag{5.101}$$

or

$$D_e = 1.13s \quad \text{for square spacing between columns} \tag{5.102}$$

where Q is the load on the column, D_e is the diameter of the effective tributary area of the column, H is the embankment height, q is the live and dead load surcharge, γ is the unit weight of the embankment soil, and s is the spacing between columns.

The load carried by each reinforcement layer depends on the spacing between columns and the pattern of spacing (triangular or square). If the soft foundation soil is strong enough to support the first lift of fill material, the first layer of reinforcement is located 0.15–0.25 m above the soft soil surface. Each layer of reinforcement carries the load from the platform fill that is within the soil wedge below the arch. The fill load attributed to each reinforcement layer is that from the material located between that layer of reinforcement and the next reinforcement layer, as shown schematically in Figure 5.37.

The vertical stress on a given layer i of reinforcement can be calculated as

$$p_{Ti} = \frac{A_i + A_{i+1}}{2A_i}\gamma h_i \tag{5.103}$$

Figure 5.37 Geosynthetics in load-transfer platforms
Adapted from Collin *et al.* (2005)

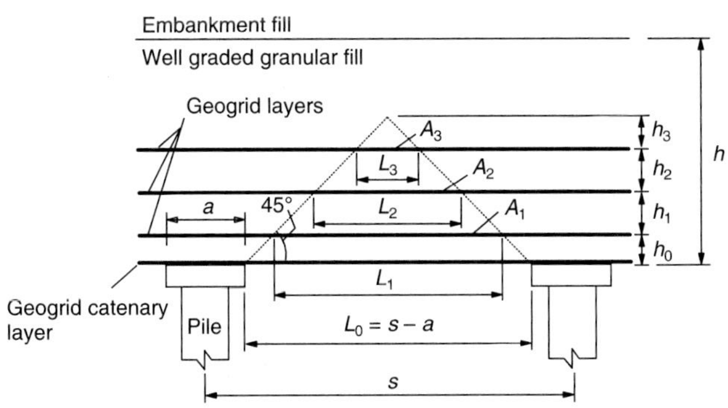

with

$$A_i = L_i^2 \quad \text{for square column spacing (see Figure 5.37)} \tag{5.104}$$

or

$$A_i = \frac{\sqrt{3}}{4}L_i^2 \quad \text{for triangular column spacing (see Figure 5.37)} \tag{5.105}$$

where p_{Ti} is the vertical stress on a reinforcement layer i, A_i is the area at reinforcement layer i (see Figure 5.37), L_i is the length of the side of the loaded area, and h_i is the thickness of soil between reinforcement layers (see Figure 5.37).

The tension membrane theory (Giroud *et al.*, 1990) was used to obtain the tensile load in the reinforcement, which can be calculated (Collin *et al.*, 2005) as

$$T_{rpi} = \frac{p_{Ti}\Omega D_i}{2} \tag{5.106}$$

with

$$D_i = 1.41L_i \quad \text{for square column spacing} \tag{5.107}$$

$$D_i = 0.867L_i \quad \text{for triangular column spacing} \tag{5.108}$$

where T_{rpi} is the tensile load in the reinforcement layer i, D_i is the design span for the tension membrane, and Ω is a dimensionless parameter that depends on the geosynthetic strain (Table 5.2).

The geogrid catenary layer (see Figure 5.37) is designed to support the load of the pyramid below the arch region. The vertical stress on this layer is given by (Collin *et al.*, 2005)

$$p_{TC} = \frac{\gamma h_p}{3} \quad \text{for square or triangular column spacing} \tag{5.109}$$

where p_{TC} is the vertical stress on the catenary layer, and h_p is the height of the pyramid below the arch.

The load in the catenary layer is given by

$$T_{rpC} = \frac{p_{TC}\Omega D}{2} \tag{5.110}$$

where D is the design span for the catenary tensioned membrane.

Table 5.2 Values of Ω
Data from Collin *et al.* (2005)

Reinforcement strain (%)	Ω
1	2.07
2	1.47
3	1.23
4	1.08
5	0.97

5.6. Summary of main points

- The design of reinforced embankments on soft soils requires accurate soil and reinforcement properties. Although some rather simple stability analysis approaches are useful tools for embankment design, good-quality data on the load–strain–time relationship for the reinforcement and accurate soil strength parameters are of utmost importance for good estimates of the safety factor of the embankment. The load–strain–time relationship of the reinforcement will be important to allow a proper choice of the value of reduction factors to be applied to the index strength of the reinforcement, in order to take into account strain rate effects. It is also important to know the rate of increase in the foundation soil strength with time due to consolidation, in order to obtain an optimised estimate of the required period of time for which the reinforcement action is important. This will also allow for a better choice of reduction factors for the reinforcement strength, and will lead to a more cost-effective design of the embankment.
- Some uncertainties regarding soil parameters may remain, even when laboratory or field tests are carefully conducted on the soils involved in the problem, particularly for the soft foundation soil. Other sources of uncertainties can be heterogeneities or anisotropy in the foundation soil not detected in the testing programme, or the use of a theoretical failure mechanism in the analysis which does not represent what is likely to occur in the field. A comprehensive testing programme is always recommended for design purposes. Even in less critical cases, in terms of stability requirements, the designed embankment should have a factor of safety greater than 1.3 in order to account for uncertainties in the analysis and in the input data used.
- With regard to the design of geosynthetic vertical drains, it is of fundamental importance to have an accurate estimate of the soil horizontal coefficient of consolidation. This may require some special laboratory or field tests, but their results are necessary for a good design of the vertical drainage system. The use of empirical relations, in general, is not recommended, due to the significant scatter between predicted and measured values of coefficients of consolidation in these cases.
- Regarding piled embankments, it should be pointed out that the limit equilibrium methods available can be considered as rather crude methods for the analysis of piled embankments in several cases, because of the complexity of this type of problem and the simplifying assumptions adopted in the methods. Therefore, in large projects and in critical or severe situations, the use of more sophisticated solutions and proper geotechnical instrumentation is advised. The monitoring of experimental instrumented embankments prior to the construction of large projects can also be very useful to improve design and construction methodologies, and to reduce costs in this type of geosynthetic application.
- When the design follows basic rules, considering the conditions mentioned above and sound engineering judgement, the use of geosynthetic reinforcement can provide a cost-effective and safe solution for the construction of embankments on soft soils.

5.7. Some common questions and their answers

Questions

1. Will the presence of the reinforcement be necessary for short-term stability conditions only?

2. Are the tensile strength and stiffness of the reinforcement the only pieces of information on the reinforcement relevant for design purposes?

3. What is an acceptable safety factor for a reinforced embankment on soft soil?

4. Are the design approaches presented in this chapter appropriate for embankments under seismic conditions?

5. Are the load conditions of the reinforcement at the end of embankment construction always the most severe ones?

6. What are the main factors that may cause vertical drains not to function properly?

Answers

1. Not always. The reinforcement action may be required for the entire lifetime of the project if the strength increase of the soft soil due to consolidation is not enough to maintain the stability of the embankment without the contribution from the reinforcement.

2. No. Obviously, the reinforcement tensile strength and stiffness are of fundamental importance in design, and works of embankments on soft soils usually require reinforcements with high values of tensile properties. However, other reinforcement characteristics and properties are also important, such as resistance against mechanical damage, level of interaction with the surrounding soils, and long-term behaviour and durability, depending on the duration of the required reinforcement action. Reinforcement degradation due to the environment may be relevant depending on the project lifetime, or if the reinforcement is in contact with chemically aggressive substances or materials. These and other factors highlight the importance of a proper choice of reduction factors and safety factors in the design of an embankment on soft soil.

3. As in most geotechnical engineering designs, the value of the safety factor will depend on the project characteristics, and on the consequences and repercussion of failures or poor performance. For less critical jobs, such as low embankments with negligible consequences in the case of failure, or in some temporary works, a safety factor as low as 1.2 may be acceptable. However, in more important projects, where failure can put lives at risk or cause significant material losses and delay of construction completion, safety factors of 1.5 or greater may be required. National standards in some countries may establish minimum values of safety factors. Special care must be taken in embankment construction close to existing structures, when deformations of the soft foundation soil may be a key issue in the design. It is also important to bear in mind that the scatter of test results of some key soil parameters for design, particularly for the soft soil (e.g. undrained strength and coefficient of consolidation), can be quite significant. In this context, probabilistic approaches are important design tools to establish the safety level of the embankment.

4. No. The design approaches presented in this chapter do not consider the action of accelerations due to seismic events. Under such conditions, more sophisticated design methods are required, such as computer programs for modelling slope stability, or finite-element programmes capable of taking into account the effects of seismic loads. The presence of the reinforcement will also certainly be beneficial under these conditions. Some of the seismic aspects described in Chapter 16 may be applicable in the design of embankments under seismic conditions.

5. Usually yes, but the reinforcement may be subjected to very severe conditions during the construction stages of the embankment; for instance, when the traffic of heavy construction equipment on thin fill layers overlying the geosynthetic reinforcement is allowed in poor or careless construction practices. This can overstress or cause mechanical damage to the reinforcement. In the case of piled embankments, the placement of the reinforcement layer in direct contact with the cap, associated to the cyclic loading due to the traffic of construction vehicles, can cause abrasion of the reinforcement, due to friction with the concrete cap, or tearing of the geosynthetic layer along the cap edges. Long-term stability of the embankment may also be compromised due to losses of reinforcement strength (reinforcement degradation) or reinforcement overstressing. Good construction practices and the proper choice of reduction factors will avoid these problems.

6. Usually, bad design or poor construction practices. Unfortunately, required design parameters for vertical drainage systems are often estimated from correlations, or just guessed, instead of being obtained from proper testing techniques. Bad construction practices or the use of large casings for drain installation may increase the disturbance of the soft soil beyond the level assumed in the design. Disturbance of soft soil can also compromise drainage efficiency in the case of closely spaced vertical drains (say, a drain spacing of less than 1.3 m), even in carefully conducted drain installations. Filter criteria must also be verified to avoid the possibility of clogging of the drain filter. Severe conditions for filter performance can be found in the case of vertical drains in highly organic soils, such as peat. The penetration of the vertical drain extremities in the top and bottom drainage layers must also be guaranteed, as this is always assumed in designs of vertical drains in thin to medium-thick soft soil layers. Care must be taken to avoid accidental closure of the edges of the surface drainage blanked by fine-grained fill materials (drainage blanked entirely buried in the embankment), because this will avoid or limit the exit of the drained water laterally, delaying the consolidation process.

REFERENCES

Araujo GLS and Palmeira EM (2006) Backanalysis of an instrumented reinforced abutment on soft soil. *Proceedings of the 8th International Conference on Geosynthetics*, vol. 3, Yokohama, pp. 1169–1172.

Bergado DT, Long PV, Lee CH, Loke KH and Werner G (1994) Performance of reinforced embankment on soft Bangkok clay with high-strength geotextile reinforcement. *Geotextiles and Geomembranes* **13**: 403–420.

BSI (1995) BS 8006: *Code of practice for strengthened–reinforced soils and other fills*. BSI, London.

Carrillo N (1942) Simple two- and three-dimensional cases in the theory of consolidation of soils. *Journal of Applied Mathematics and Physics* **21(1)**: 1–5.

Chirapuntu S and Duncan JM (1975) *The Role of Fill Strength in the Stability of Embankments on Soft Clay Foundations*. University of California, Berkeley, CA. Geotechnical Engineering Report No. TE 75-3.

Collin JG, Han J and Huang J (2005) Geosynthetic-reinforced column-support embankment design guidelines. *NAGS-GRI-19 Conference*, Las Vegas.

Cooper MR and Rose AN (1999) Stone column support for an embankment on deep alluvial soils. *Proceedings of the Institute of Civil Engineers, Geotechnical Engineering* **137**: 15–25.

Delmas P, Queyroi D, Quaresma M, Amand DS and Peuch A (1990) Failure of an experimental embankment on soft soil reinforced with geotextile: guiche. *Proceedings of the 4th International Conference on Geotextiles, Geomembranes and Related Products*, The Hague, pp. 1019–1025.

EMPFEHLUNG (2003) Bewehrte erdkörper auf punkt- oder linienförmigen traggliedern – entwurf Septermber 2003, Kapitel 6.9 für die Empfehlungen für Bewehrungen aus Geokunststoffen (EBGEO), Germany.

Fahel ARS and Palmeira EM (2002) Failure mechanism of a geogrid reinforced abutment on soft soil. *Proceedings of the 7th International Conference on Geosynthetics*, vol. 4, Nice, pp. 1565–1568.

Fahel ARS, Palmeira EM and Ortigao JAR (2000) Behaviour of geogrid reinforced abutments on soft soil in the BR 101-SC highway, Brazil. *GeoDenver 2000 – Advances in Transportation and Geoenvironmental Systems Using Geosynthetics*, vol. 1, ASCE, Denver, pp. 257–271. ASCE Geotechnical Special Publication No. 103.

Gartung E, Verspohl J, Alexiew D and Bergmair F (1996) Geogrid reinforced railway embankments on piles – monitoring. *Geosynthetics: Applications, Design and Construction – EUROGEO'96*, vol. 1, Maastricht, pp. 251–258.

Giroud JP (1981) Designing with geotextiles. *Materials of Construction* **14(82)**: 257–272.

Giroud JP, Bonaparte R, Beech JF and Gross BA (1990) Design of soil layer–geosynthetic systems overlying voids. *Geotextiles and Geomembranes* **9(1)**: 11–50.

Hansbo S (1979) Consolidation of clay by band-shaped prefabricated drains. *Ground Engineering* **12(5)**: 16–25.

Hansbo S, Jamiolkowski M and Kok L (1981) Consolidation by vertical drains. *Géotechnique* **31**: 45–66.

Hewlett WJ and Randolph MF (1988) Analysis of piled embankments. *Ground Engineering* Apr.: 12–18.

Holtz RD (1987) Preloading with prefabricated vertical strip drains. *Geotextiles and Geomembranes* **6(1–3)**: 109–131.

Holtz RD, Jamiolkowski M, Lancellotta R and Pedroni S (1989) Behaviour of bent prefabricated vertical drains. *Proceedings of the 12th International Conference on Soil Mechanics and Foundation Engineering*, Rio de Janeiro, pp. 1657–1660.

Jakobson B (1948) The design of embankments on soft soil. *Géotechnique* **1(1)**: 80–90.

Jewell RA (1987) *The Mechanics of Reinforced Embankments on Soft Soils*. University of Oxford, Oxford. OUEL Report No. 071/87.

Jewell RA (1996) *Soil Reinforcement with Geotextiles*. Construction Industry Research and Information Association, London. CIRIA Special Publication 123.

Kaniraj SR (1994) Rotational stability of unreinforced and reinforced embankments on soft soils. *Geotextiles and Geomemberanes* **13(11)**: 707–726.

Kaniraj SR (1996) Directional dependency of reinforcement force in reinforced embankments on soft soils. *Geotextiles and Geomembranes* **15(9)**: 507–519.

Kempfert H-G, Göbel C, Alexiew D and Heitz C (2004) German recommendations for reinforced embankments on pile-similar elements. *3rd European Geosynthetic Conference – EuroGeo 3*, vol. 1, Munich, pp. 279–284.

Kempton G, Russel D, Pierpoint ND and Jones CJFP (1998) Two- and three-dimensional numerical analysis on the performance of piled embankments. *Proceedings of the 6th International Conference on Geosynthetics*, vol. 2, Atlanta, pp. 767–772.

Lambe TW and Whitman RV (1969) *Soil Mechanics*. Wiley, New York.

Leshchinsky D (1987) Short-term stability of reinforced embankments over clay foundation. *Soils and Foundations* **27(3)**: 43–57.

Loke KH, Ganeshan V, Werner G and Bergado DT (1994) Composite behaviour of geotextile reinforced embankment on soft clay. *Proceedings of the 5th International Conference on Geotextiles, Geomembranes and Related Products*, vol. 1, Bangkok, pp. 25–28.

Love J and Milligan G (2003) Design methods for basally reinforced pile-supported embankments over soft ground. *Ground Engineering* Mar.: 39–43.

Low BK (1989) Stability analysis of embankments on soft ground. ASCE *Journal of Geotechnical Engineering* **115(2)**: 211–227.

Low BK, Wong HS, Lim C and Broms BB (1990) Slip circle analysis of reinforced embankments on soft ground. *Geotextiles and Geomembranes* **9(2)**: 165–181.

Macedo IL, Palmeira EM and Araujo GLS (2009) Effects of an abutment construction on soft soil on a neighbouring structure: Influence of different construction techniques using geosynthetics. *Soils and Rocks* **32(2)**: 71–82.

Magnan JP (1983) Theorie et pratique des drains verticaux. Tech. et Doc. e Lavoisier, Paris.

Michalowski RL (1998) Limit analysis in stability calculations of reinforced soil structures. *Geotextiles and Geomembranes* **16(6)**: 311–332.

Palmeira EM (1996) *Geosynthetics in Geotechnical and Geoenvironmental Engineering*. Graduate Programme of Geotechnics, University of Brasília. Publication No. GM 001-96 (in Portuguese).

Palmeira EM (1998) Geosynthetic reinforced unpaved roads on very soft soils: construction and maintenance effects. *Proceedings of the 6th International Conference on Geosynthetics*, vol. 2, Atlanta, pp. 885–890.

Palmeira EM and Almeida MSS (1980) *An Update of the Program BISPO for Slope Stability Analysis*. Institute of Highway Research, IPR/DNER, Brazil. Research Report No. 2.019-03.01-2/17/42 (in Portuguese).

Palmeira EM and Milligan GWE (1989) Large scale laboratory tests on reinforced sand. *Soils and Foundations* **29(1)**: 1–18.

Palmeira EM, Pereira JHF and Silva ARL (1998) Back analyses of geosynthetic reinforced embankments on soft soils. *Geotextiles and Geomembranes* **16(5)**: 274–292.

Palmeira EM, Fahel ARS and Campos LEP (2003) Geotextile reinforced abutments on soft foundations. In *Reinforced Soil Engineering* (Ling HI, Leshchinsky D and Tatsuoka F (eds)), vol. 1. Marcel Dekker, New York, pp. 221–237.

Raithel M, Kirchner A and Kempfert HG (2008) German recommendations for reinforced embankments on pile-similar elements. *4th Asian Regional Conference on Geosynthetics*, vol. 1, Shanghai, pp. 697–702.

Rowe RK and Li AL (1999) Reinforced embankments over soft foundations under undrained and partially drained conditions. *Geotextiles and Geomembranes* **17(3)**: 129–146.

Rowe RK and Soderman KL (1984) Comparison of predicted and observed behaviour of two test embankments. *Geotextiles and Geomembranes* **1(2)**: 143–160.

Rowe RK, Gnanendran CT, Landva AO and Valsangkar AJ (1995) Construction and performance of a full-scale geotextile reinforced test embankment, Sackville, New Brunswick. *Canadian Geotechnical Journal* **32**: 512–534.

Russell D and Pierpoint ND (1997) An assessment of design methods for piled embankments. *Ground Engineering* Nov.: 39–44.

Sabhahit N, Basudhar PK and Madhav MR (1994) Generalized stability analysis of reinforced embankments on soft clay. *Geotextiles and Geomembranes* **13(12)**: 765–780.

Schaefer VR and Duncan JM (1988) Finite element analysis of the St Alban test embankments. *Proceedings of the Symposium on Geosynthetics for Soil Improvement*, vol. 1. American Society of Civil Engineers (ASCE), Reston, VA. ASCE Geotechnical Special Publication No. 18, pp. 158–177.

Silva ARL (1996) The stability of geosynthetic reinforced embankments on soft soils. MSc thesis, University of Brasilia, Brazil (in Portuguese).

Silva ARL and Palmeira EM (1998) Stability of geosynthetic reinforced embankments on soft soils. *Proceedings of the 12th Brazilian Conference on Geotechnical Engineering*, Brasilia, pp. 1213–1220.

Terzaghi K (1943) *Theoretical Soil Mechanics*. Wiley, New York.

USACE (1970) *Engineering and Design Stability of Earth and Rock-fill Dams*. Department of the Army, United States Army Corps of Engineers, Washington, DC. Engineer Manual EM 1110-2-1902.

van Impe WF (1989) *Soil Improvement Techniques and their Evolution*. Balkema, Rotterdam.

Volman W, Krekt L and Risseeuw P (1977) Armature de traction en textile, un nouveau procédé pour améliorer la stabilité des grands remblais sur sols mous. *Proceedings Colloque International sur L'Emploi des Textiles en Geotechnique* **1**: 55–59.

Whitman RV and Bailey WA (1967) Use of computers for slope stability analysis. *Journal of the Soil Mechanics and Foundation Engineering Division*, ASCE **9(SM4)**: 475–498.

Zaeske D (2001) *Zur Wirkungsweise von unbewehrten und bewehrten mineralischen Tragschichten über pfahlartigen Gründungselementen. Schriftenreihe Geotechnik*. Universität Gh-Kassel. Helft 10.

Zaeske D and Kempfert HG (2002) *Berechnung und Wirkungsweise von unbewehrten und bewehrten mineralischen Tragschichten auf punkt und linienförmigen Traggliedern*. Bauingenieur, Band 77.

Handbook of Geosynthetic Engineering
ISBN 978-0-7277-4175-2

http://dx.doi.org/10.1680/hge.41752.129

Chapter 6
Shallow foundations

S. K. Shukla
Discipline of Civil Engineering, School of Engineering, Edith Cowan University, Perth, Australia

Geosynthetic-reinforced foundation soils are used to support several structures such as footings, pavements, embankments and retaining walls. This chapter deals with various aspects of geosynthetic-reinforced foundation soils subjected to loads from loaded shallow footings, including reinforcing mechanisms and functions, reinforcing patterns, modes of failure, model test results, methods of analysis for load-bearing capacity and settlement, the concepts of prestressing, geosynthetic profiles and strain analysis, and some field applications.

6.1. Introduction

The construction of shallow footings supported on geosynthetic-reinforced foundation soils has considerable potential as a cost-effective alternative to conventional deep foundations. In this technique, one or more layers of geosynthetic reinforcement (geotextile, geogrid, geocell or geocomposite) are placed inside a controlled granular fill beneath the footings in order to create a composite material with improved performance characteristics (Figure 6.1). The geosynthetic-reinforced foundation soils are also being used to support paved and unpaved roads, low embankments, railway tracks, oil drilling platforms, platforms for heavy industrial equipment, parking areas, closure covers for tailing dams, etc. Such reinforced foundation soils provide improved load-bearing capacity and reduced settlements by distributing the imposed loads over a wider area of weak subsoil. In the conventional construction techniques without the use of any reinforcement, a thick granular layer is needed, which may be costly or may not be possible, especially in sites that have a limited availability of good-quality granular materials. Moreover, the simplicity of the basic principles and the economic benefits over the conventional approaches make geosynthetic-reinforced foundation soil very attractive to designers. Also, the use of geosynthetics provides many other indirect benefits, as described in Section 1.7 of Chapter 1. This chapter presents several aspects of geosynthetic-reinforced foundation soils subjected to loads from loaded shallow footings of structures.

6.2. Reinforcing mechanisms and functions

Different concepts have been advanced to define the basic mechanism of reinforced soils. The effect of inclusion of relatively inextensible reinforcements (metals, fibre-reinforced plastics, etc., having a high modulus of deformation) in the soil can be explained using either an *induced-stresses concept* (Schlosser and Vidal, 1969) or an *induced deformations concept* (Bassett and Last, 1978). According to the induced-stresses concept, the tensile strength of the reinforcements and friction at the soil–reinforcement interfaces give an apparent cohesion to the reinforced soil system. The induced-deformations concept considers that the tensile reinforcements involve anisotropic restraint of the soil deformations. The behaviour of the soil, reinforced with extensible reinforcements, such as geosynthetics, does not fall within these concepts. The difference between the influences of inextensible and extensible reinforcements is significant in terms of the load–settlement behaviour of the reinforced soil system (Figure 6.2). The soil reinforced with extensible reinforcement (termed *ply-soil* by McGown and Andrawes (1977)) has greater extensibility and smaller losses of post-peak strength, compared with soil alone or to soil reinforced with inextensible reinforcement (termed *reinforced earth* by Vidal (1969)). However, some similarity between ply-soil and reinforced earth exists in that they both inhibit the development of internal tensile strains in the soil and develop tensile stresses.

Geosynthetics, in conjunction with foundation soils, may be considered to have five different roles that improve load-bearing capacity and settlement characteristics compared with conventional foundations (Shukla, 2004; Shukla *et al.*, 2009).

- Geosynthetics reduce the outward shear stresses transmitted from the overlying soil/fill to the top of the underlying foundation soil. This action of geosynthetics is known as the *shear stress reduction effect*. This effect results in a general-shear failure, rather than a local-shear failure (Figure 6.3(a)), thereby causing an increase in the load-bearing capacity of the foundation soil (Bourdeau *et al.*, 1982; Guido *et al.*, 1985; Love *et al.*, 1987; Espinoza, 1994; Espinoza and Bray, 1995;

Figure 6.1 A loaded footing resting on the geosynthetic-reinforced granular fill – soft soil system

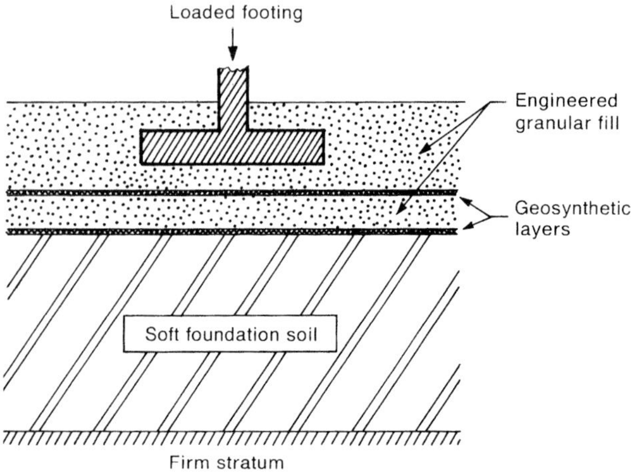

Figure 6.2 Postulated behaviour of a unit cell in plane-strain conditions with and without inclusions: (a) dense sand with inclusions and (b) loose sand with inclusions
Reproduced from McGown *et al.* (1978)

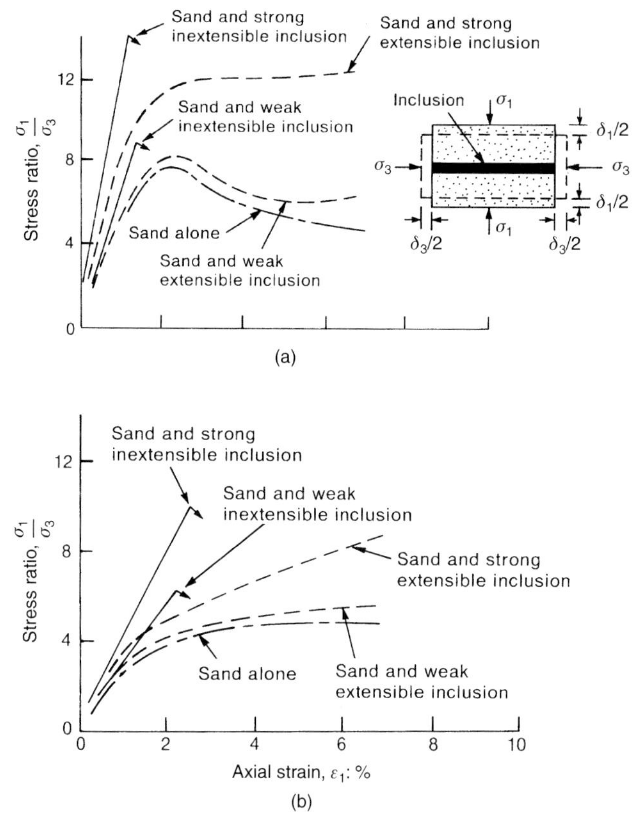

Adams and Collin, 1997). The reduction in the shear stress and the change in the failure mechanism are the primary benefits of the geosynthetic layer at small deformations.

- Geosynthetics redistribute the applied surface load by providing restraint of the granular fill if embedded in it, or by providing restraint of the granular fill and the soft foundation soil if placed at their interface, resulting in reduction of the applied stress (Figure 6.3(b)). This is referred to as the *slab effect* or *confinement effect* of geosynthetics (Bourdeau *et al.*, 1982; Giroud *et al.*, 1984; Madhav and Poorooshasb,

Figure 6.3 Influence of geotextile inclusion on a two-layer soil system: (a) change in the failure mode, (b) redistribution of the applied surface load and (c) the membrane effect
Reproduced from Bourdeau *et al.* (1982), Industrial Fabrics Association International; and Espinoza (1994), with permission from Elsevier

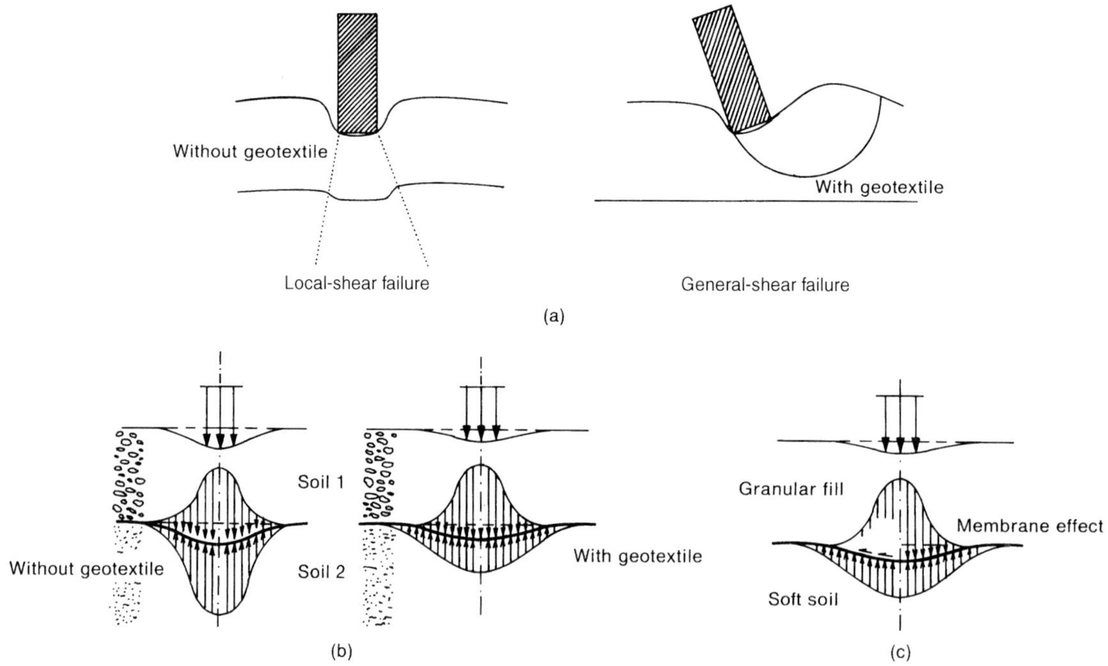

1989; Hausmann, 1990; Sellmeijer, 1990). The friction mobilised between the soil and the geosynthetic layer plays an important role in confining the soil.

- The deformed geosynthetic, sustaining normal and shear stresses, has a membrane force with a vertical component that resists applied loads; that is, deformed geosynthetics provide a vertical support to the overlying soil mass subjected to loading. This action of geosynthetics is popularly known as the *membrane effect* (Figure 6.3(c)) (Giroud and Noiray, 1981; Bourdeau *et al.*, 1982; Sellmeijer *et al.*, 1982; Love *et al.*, 1987; Madhav and Poorooshasb, 1988; Bourdeau, 1989; Sellmeijer, 1990, Shukla and Chandra, 1994a). Depending on the type of stresses – normal stress and shear stress – sustained by the geosynthetics during their action, the membrane support may be classified as 'normal stress membrane support' or 'interfacial shear stress membrane support', respectively (Espinoza and Bray, 1995). The edges of the geosynthetic layer need to be anchored in order to develop the membrane support contribution that results from normal stresses, whereas the membrane support contribution resulting from mobilised interfacial membrane shear stresses does not require any anchorage. The membrane effect of geosynthetics causes an increase in the load-bearing capacity of the foundation soil below the loaded area, with a downward loading on its surface at either side of the loaded area, thus reducing its heave potential. It should be noted that both geotextiles and geogrids can be effective in membrane action in the case of high deformation of reinforced foundation soils (Hass *et al.*, 1988).
- The use of geogrids has another benefit due to the interlocking of the soil through the apertures of the grid, which is known as the *anchoring effect* (Guido *et al.*, 1986). The transfer of stress from the soil to the geogrid reinforcement is made through bearing at the soil to the grid cross-bar interface.
- Geosynthetics (particularly, geotextiles, but perhaps also geogrids) improve the performance of the reinforced soil system by acting as a separator between the soft foundation soil and the granular fill. This influence is known as the *separation effect* of geosynthetics (Guido *et al.*, 1986; Nishida and Nishigata, 1994). The separation can be an important function compared with the above functions (which may collectively be called the *reinforcement function*) when the ratio of the applied stress (σ) on the subgrade soil to the shear strength (c_u) of the subgrade soil has a low value (less than 8), and it is essentially independent of the settlement of the reinforced soil system (Figure 6.4).

In general, the improved performance of a geosynthetic-reinforced foundation soil can be attributed to an increase in the shear strength of the foundation soil from the inclusion of the geosynthetic layer. The soil–geosynthetic system forms a composite material that inhibits development of the soil failure wedge beneath shallow spread footings.

Figure 6.4 Relationship between the separation and the reinforcement functions
Reproduced from Nishida and Nishigata (1994), Industrial Fabrics Association International

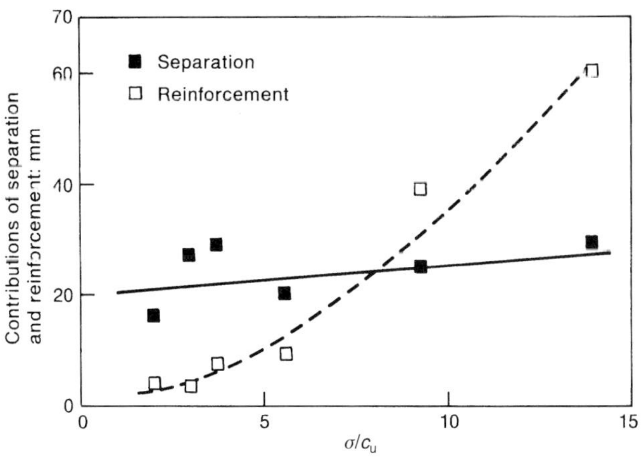

6.3. Reinforcing patterns

Analysis of strain fields suggests that the ideal reinforcing pattern below a shallow footing is as shown in Figure 6.5(a). The ideal pattern has reinforcement layers placed horizontally below the footing, which become progressively steeper the further from the footing (Bassett and Last, 1978). This means that the reinforcement should be placed in the direction of the major principal strain. This fact was stated by Hausmann (1990) in terms of stress. According to Hausmann, the tensile reinforcement is most effective if placed in the major principal plane in the direction of the minor principal stress, which in many practical geotechnical problems is horizontal, as shown in Figure 6.5(b).

6.4. Modes of failure

There are four possible modes of failure for geosynthetic-reinforced shallow foundations. These are as follows

Figure 6.5 Pattern of reinforcement beneath a footing: (a) the ideal reinforcement pattern and (b) a practical reinforcement pattern
(a) reproduced from Bassett and Last (1978), ASCE

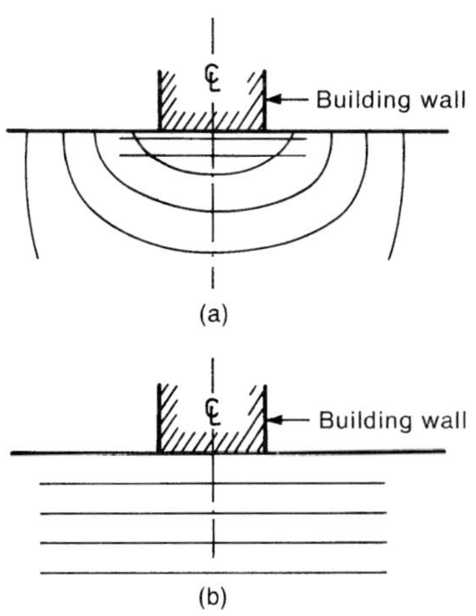

Figure 6.6 Possible modes of failure of geosynthetic-reinforced shallow foundations
Reproduced from Binquet and Lee (1975b), ASCE; and Koerner (1990), Pearson Education

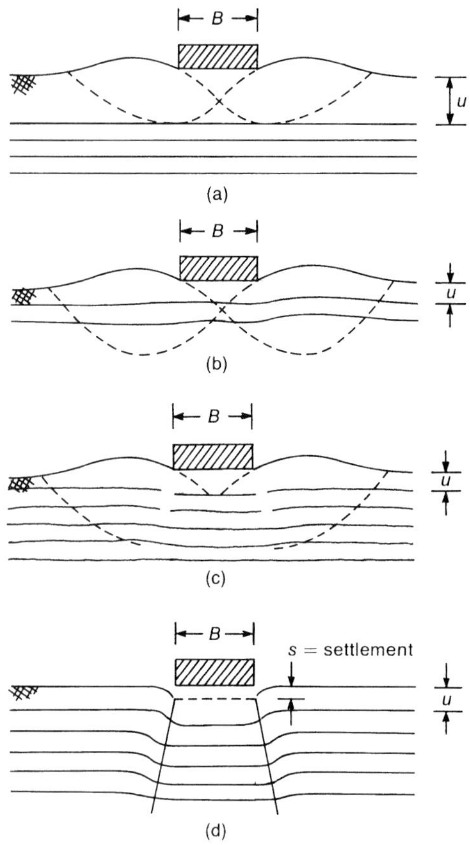

- *Bearing capacity failure of soil* above the uppermost geosynthetic layer (Figure 6.6(a)) – this type of failure is likely to occur if the depth of the uppermost layer of reinforcement (u) is greater than about two-thirds of the width of the footing (B) (i.e. $u/B > 0.67$), and if the reinforcement concentration in this layer is sufficiently large to form an effective lower boundary into which the shear zone will not penetrate. This class of bearing capacity problem corresponds to the bearing capacity of a shallow footing resting on the soil deposit overlying a strong rigid boundary.
- *Pull-out of the geosynthetic layer* (Figure 6.6(b)) – this type of failure is likely to occur for a shallow and light reinforcement ($u/B < 0.67$, and the number of reinforcement layers $N < 3$).
- *Breaking of the geosynthetic layer* (Figure 6.6(c)) – this type of failure is likely to occur with long, shallow and heavy reinforcement ($u/B < 0.67$, $N > 3$ or 4). The reinforcement layers always break approximately under the edge or towards the centre of the footing. The uppermost layer is most likely to break first, followed by the next deepest layer, and so on.
- *Creep failure of the geosynthetic layer* (Figure 6.6(d)) – this failure may occur due to long-term settlement caused by sustained surface loads and subsequent geosynthetic stress relaxation.

The first three modes of failure were first reported by Binquet and Lee (1975a,b) on the basis of observations made during laboratory model tests (on a footing resting on a sand layer reinforced with metallic reinforcements). The fourth mode of failure (creep failure) was discussed by Koerner (1990).

6.5. Model tests

A large number of model tests have been conducted in order to evaluate the beneficial effects of reinforcing the soils with geosynthetics, as related to the load-bearing capacity and the settlement characteristics of shallow foundations (Jarrett, 1984, 1986; Guido *et al.*, 1985; Milligan and Fannin, 1986; Love *et al.*, 1987; Sakti and Das, 1987; Koerner, 1990; Omar *et al.*, 1993; Khing *et al.*, 1994; Manjunath and Dewaikar, 1994; Ochiai *et al.*, 1994; Yetimoglu *et al.*, 1994; Adams and Collin, 1997; Krishnaswamy *et al.*, 2000; Dash *et al.*, 2001, 2003, 2007; Ghosh *et al.*, 2005; Patra *et al.*, 2006; Basudhar *et al.*, 2007; Lovisa *et al.*, 2010; Vinod and Bhaskar, 2010). Model tests have also been conducted on soil reinforced with a relatively inextensible reinforcement, such as metallic and fibre strips, and are reported in the literature (Binquet and Lee, 1975b; Akinmusuru and Akinbolade, 1981; Fragaszy and Lawton, 1984; Huang and Tatsuoka, 1988, 1990). Model test studies on foundation soils, reinforced with metallic and fibre strips, have elucidated many useful and basic facts on soil reinforcement. Geosynthetic-reinforced foundation soils and foundation soils reinforced with metallic and fibre strips show similar behaviour in many respects.

6.5.1 Reinforced granular soil

Guido *et al.* (1985) conducted laboratory model tests to study the bearing capacity of a square footing (side $B = 0.31$ m) resting on loose sand (relative density = 50%), reinforced with geotextiles varying in strength from 0.67 to 2.16 kN/m. The tests were performed in a square stiffened Plexiglass box of the dimensions shown in Figure 6.7(a). The square sheets of geotextile were placed concentrically under the square footing. The vertical load was applied to the footing by a hydraulic jack with a hand pump. The load was applied in small increments, and the resulting footing displacement was measured using two dial gauges, placed at opposite corners of the footing. For these tests, several parameters were varied – the depth below the footing of the first layer of geotextile, u; the vertical spacing of the layer of geotextile, h; the number of layers of the geotextile, N; the width of the square sheet of the geotextile, b; and the tensile strength of the geotextile, σ_G. For convenience in expressing and comparing test data, the results were presented in terms of a bearing capacity ratio (BCR), a term introduced by Binquet and Lee (1975a). This term is defined as follows:

$$\text{BCR} = \frac{q_{(\text{R})}}{q_\text{u}} \tag{6.1}$$

where q_u is the ultimate bearing capacity of the unreinforced soil, and $q_{(\text{R})}$ is the bearing capacity of the reinforced soil at a settlement corresponding to the settlement s_u at the ultimate bearing capacity q_u for the unreinforced soil.

Figure 6.7 (a) Geometry of the model, (b) load–settlement curves ($u/B = 0.5$, $h/B = 0.25$, $b/B = 2$), (c) BCR variation with N ($u/B = 0.5$, $h/B = 0.25$, $b/B = 2$), (d) BCR variation with the width ratio ($u/B = 0.5$, $h/B = 0.25$, $N = 2$) and (e) BCR variation with tensile strength ($u/B = 0.5$, $h/B = 0.25$, $b/B = 3$)
Reproduced from Guido *et al.* (1985), CRC press

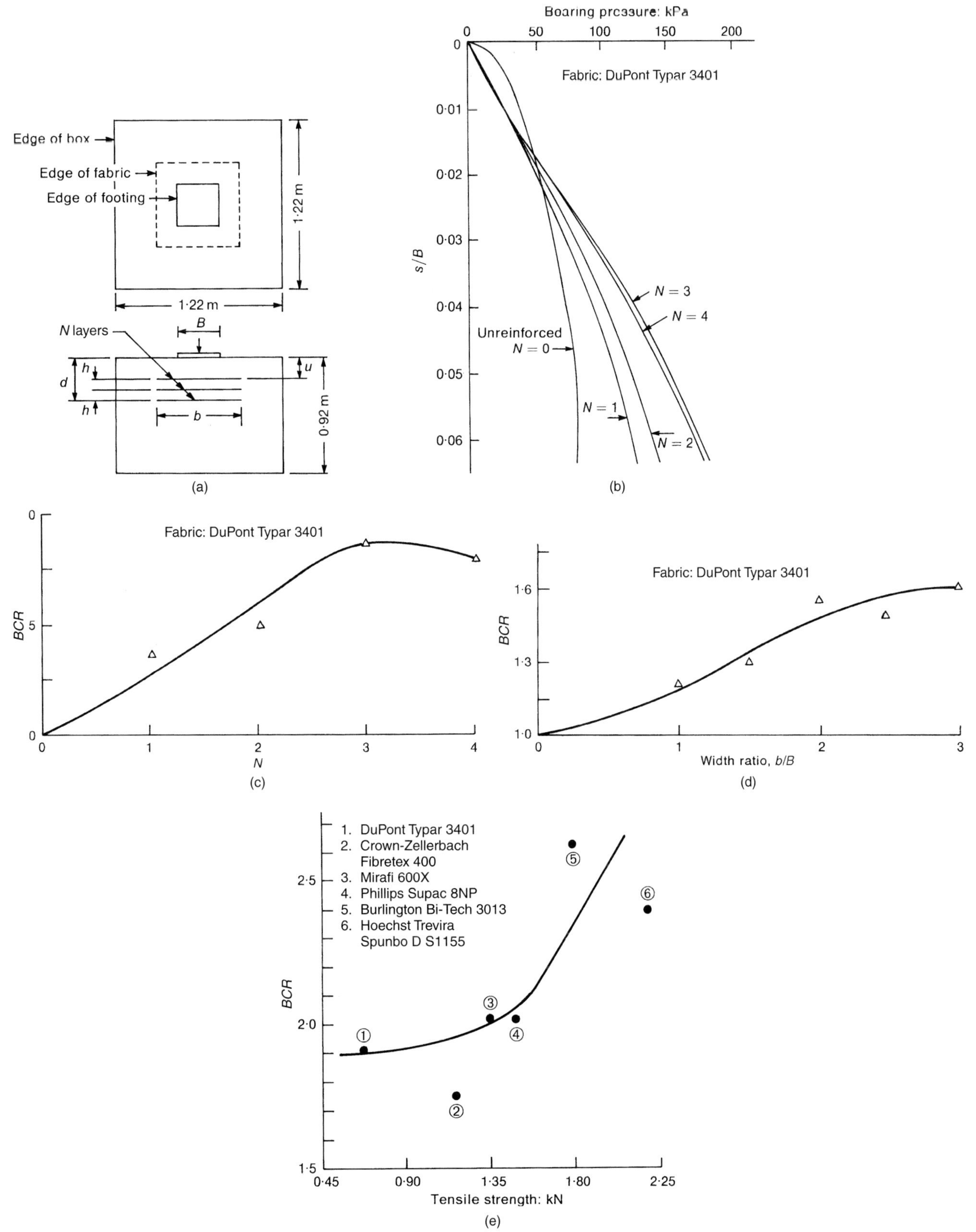

The typical load–settlement curves and variation in the BCR with some parameters are shown in Figures 6.7(b)–6.7(e). Based on the test results, the following generalised conclusions can be drawn

- All the parameters stated above have a substantial effect on the load-bearing capacity of the geotextile-reinforced foundation.
- When the geotextile layers are placed within a depth equal to the width of the foundation, they increase the load-bearing capacity of the foundation – but only after a measurable settlement has occurred. This result is logical because the geotextile layers have to deform before their reinforcing benefits can be realised.
- The presence of the geotextile layers changes the failure mode from local shear failure to general shear failure. The trends in variation for BCR have been reported to be independent of the soil type.

Small-scale laboratory model test results for the ultimate bearing capacity of strip and square footings, supported by sand reinforced with geogrid layers, as shown in Figure 6.8, have been presented by Omar *et al.* (1993). The general conclusions from the test observations are as follows

- For the development of the maximum bearing capacity ratio (BCR_{max}), the effective depth of the geogrid layer z is about $2B$ for strip footings and $1.4B$ for square footings.
- The maximum width of geogrid layers b_{max} required for mobilisation of the BCR_{max} is about $8B$ for strip footings and $4.5B$ for square footings.
- The maximum depth of placement u_{max} of the first layer of the geogrid should be less than about B for the geogrid to be effective.

Yetimoglu *et al.* (1994) investigated the bearing capacity of rectangular footings on geogrid-reinforced sand by performing laboratory model tests. From the test results, the following generalised conclusions can be drawn

- The bearing capacity of rectangular footings can be increased significantly by incorporating geogrid layers at strategic elevations in the foundation soil. However, the settlement at failure may not be affected significantly by the geogrid layer.
- For single-layer reinforced sand, the optimum embedment depth (the depth of the reinforcement layer at which the bearing capacity is highest) is approximately 0.3 times the footing width B. For multi-layer reinforced sand, the highest bearing capacity occurs at an embedment depth (for the first layer of reinforcement) of approximately $0.25B$. The optimum vertical spacing of the reinforcement layer is between $0.2B$ and $0.4B$.
- The bearing capacity of reinforced sand increases significantly with the size of the geogrid reinforcement and the number of reinforcement layers within a certain effective zone. The extent of the effective zone lies approximately within $1.5B$ from both the base and edges of the footing.

Figure 6.8 Strip and square footings supported by sand reinforced with layers of geogrid (q = load per unit area)
Reproduced, with permission, from Omar *et al.* (1993), © Canadian Science Publishing or its licensors

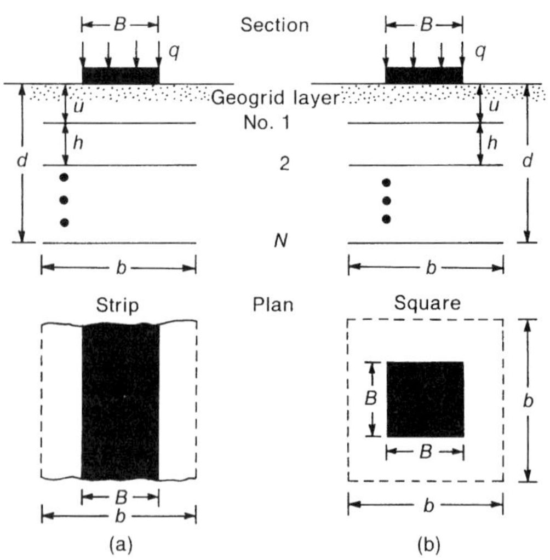

Ju *et al.* (1996) performed a series of bearing capacity tests on reinforced sand with strip footings. The sand was reinforced with a geonet of relatively weak tensile strength. The types of reinforcement used were one layer, multi-layer and mattress. Of the three reinforcing methods, the greatest ultimate bearing capacity was obtained from the multi-layer type, the optimum layer number was 4, and the ultimate bearing capacity ratio was 3.65.

A total of 34 large model load tests were conducted by Adams and Collin (1997) in order to evaluate the potential benefits of reinforcing the sand with geosynthetic layers below the shallow spread footings. The tests were performed in a reinforced concrete box 5.4 m wide by 6.9 m long by 6 m deep. One to three layers of the geogrid reinforcement, or one layer of geocell, were placed beneath the 0.30, 0.46, 0.61 and 0.91 m square footings. The depth of the reinforcement layers varied between 0.25 and 1.5 m. In the tests, precast steel-reinforced concrete footings were loaded with a hydraulic ram jacked against a reaction frame. The generalised conclusions from the tests are as follows

- The use of geosynthetic-reinforced soil foundations may increase the ultimate bearing capacity of shallow spread footings by a factor of 2.5.
- The maximum improvement in the bearing capacity at low strains ($s/B = 0.5\%$, where s is the settlement and B is the footing width) occurs when the top layer of reinforcement is within a depth of $0.25B$ from the bottom of the footing.
- For one layer of reinforcement, improvement in the bearing capacity occurs if the sand within the reinforced zone is compacted to a high relative density so that stress transfer to the reinforcement takes place before large soil strains occur.

Figure 6.9 Model footing test results: (a) BCR versus *N* and (b) BCR versus *u*/*B*
Reproduced from Ingold and Miller (1982), Industrial Fabrics Association International

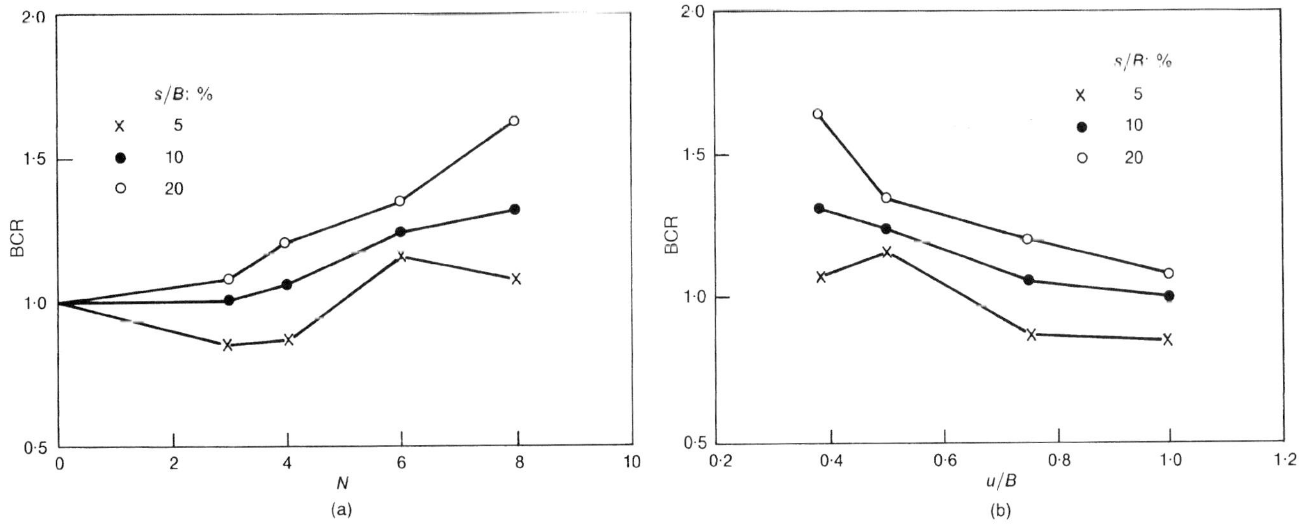

- When the first layer of reinforcement is placed 0.4*B* beneath the base of the footing, the spread footings on the reinforced soil foundation are likely to experience a general-shear plunging failure.

Small-scale laboratory model test results of the ultimate bearing capacity of a strip footing supported by sand, reinforced with multiple layers of geogrid, were presented by Shin and Das (2000). The tests were conducted with one type of sand compacted at two different relative densities and only one type of geogrid. The foundation depth was varied from zero to 0.75*B* (*B* is the footing width). The test results indicated that the BCR value determined from the surface footing tests would provide conservative estimates of the ultimate bearing capacity for footings at depths greater than zero.

Vinod and Bhaskar (2010) conducted laboratory model tests to investigate the short-term bearing capacity of a hand-knotted coir-reinforced sand bed. The optimum strength improvement was realised for a coir netting reinforcement length to footing width ratio of 3. A single layer of coir netting was most effective when placed at a depth between 0.25*B* and 0.4*B* below the base of the footing, where *B* is the width of the footing. For a multi-layer reinforced system, the optimum beneficial effect is obtained by reinforcing the zone of soil up to a depth of about 0.4*B* to 0.6*B* beneath the footing with three layers of coir netting.

6.5.2 Reinforced clay

One of the possibilities for increasing the ultimate bearing capacity of a shallow footing, supported by a saturated clay foundation under undrained conditions, is reinforcing it with geosynthetic layers. Ingold and Miller (1982) reported model test results conducted on geogrid-reinforced clay. The apparatus consisted of a rigid steel box 150 mm wide, 150 mm deep and 710 mm long, in which the clay foundation was loaded under undrained plane strain conditions using a rigid strip footing 50 mm wide. Figure 6.9 shows some model footing test results. It is noted from Figure 6.9(a) that the bearing capacity ratio, in general, increases with the number of reinforcing layers (*N*); however, at low settlement ratios (e.g. $s/B = 5\%$, where s is the footing settlement and B is the width of footing) and for a number of reinforcement layers less than 5, the reinforcement appears to weaken the foundation, as indicated by bearing capacity ratios less than unity. This tendency is seen in Figure 6.9(b), which shows that the BCR < 1 for a depth ratio $u/B > 0.65$ (u is the depth below the footing to the top of the reinforcement layer) and for a settlement ratio $s/B = 5\%$.

Sakti and Das (1987) reported some model test results on the bearing capacity of a strip footing on saturated reinforced clay. They used a heat-bonded non-woven geotextile as the reinforcement. From their tests, the following general conclusions can be drawn

- Beneficial effects of geotextile reinforcement are realised when the reinforcement is placed within a depth equal to the width of the footing.
- For maximum benefit, the first layer of geotextile should be placed at a depth of about 0.35 times the width of the footing.
- The minimum length of the reinforcing geotextile layer for maximum benefit is about four times the width of the footing.
- Geotextile reinforcements do not have much influence on the foundation settlement at ultimate load.

Figure 6.10 Model footing test results
Reproduced from Koerner (1990), Pearson Education

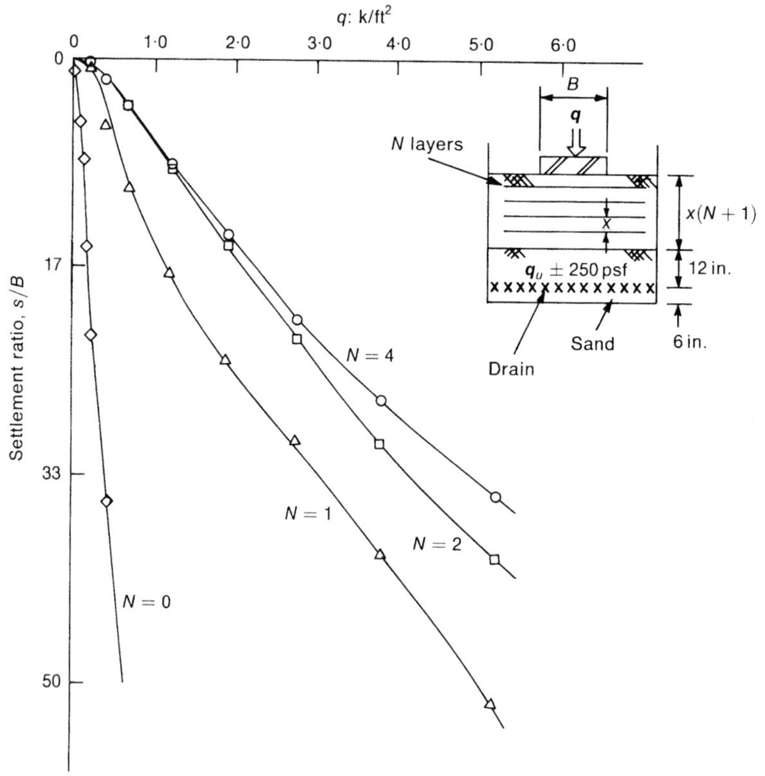

Koerner (1990) reported the results of model tests conducted at the Drexel University's Geosynthetic Research Institute. The loading tests were carried out on 6 in. round footings resting on soft saturated clay silt, at saturation above the plastic limit and reinforced with woven slit-film geotextile layers at 1.5 in. spacings (Figure 6.10). Some improvement in the load-bearing capacity is noted throughout, but the improvement is noteworthy only at large deformations.

Bearing capacity tests on model footings resting on clay subgrades, reinforced with horizontal layers of geogrids, were conducted by Mandal and Sah (1992). The test results show that the geogrid reinforcement increases the bearing capacity of subgrades, with improvements being observed at nearly all levels of deformation. The maximum percentage reduction in settlement with the use of geogrid reinforcement below the compacted and saturated clay is about 45%, and it occurs for the geogrid layer at a depth of $0.25B$ (B is the footing width) from the base of the square foundation.

6.5.3 Reinforced granular fill – soft foundation soil system

Common practice in the construction field is to clear and level the soft subgrade, spread the geosynthetic layer out on the surface (generally unstressed), and cover the geosynthetic layer with a suitable thickness of granular fill compacted by a certain standard procedure.

Jarrett (1984, 1986) carried out large-scale plane strain loading tests in the laboratory on a series of compacted gravel fills of thickness varying from 150 to 450 mm, constructed on a 1 m-deep peat subgrade (Figure 6.11(a)). The tests were carried out in a test pit 3.7 m long by 2.4 m wide and 2 m deep. Loading was applied to the compacted gravel surface through a 203 mm-wide beam spanning the full width of the test pit. In the reinforced cases, Tensar Geogrid Type SS2 was placed at the peat–gravel interface. The loading procedure adopted was an incremental static type. Figure 6.11(b) shows that, initially, no difference between the unreinforced and reinforced cases exist until some displacement occurs. However, the Tensar geogrids have a significant effect on the bearing capacity of compacted fill over a peat subgrade. The vertical movements shown in Figure 6.11(c) indicate the deformed profile of the geogrid under load. The central 'concave up-section' of the geogrid provides vertical support beneath the loaded beam. The magnitude of the vertical support is a function of the tension in the geogrid and its geometry. The 'concave down-section' that follows after the inflection point represents the lateral zone into which the vertical support forces are spread to the subgrade soil by the tension in the membrane.

Milligan and Fannin (1986) conducted model and full-scale loading tests of a geogrid-reinforced granular layer on a weak clay foundation. The results suggest a number of situations, such as less-stiff granular material and larger deformations, in which the geogrid can be more effective.

The effectiveness of geogrid reinforcement, placed at the base of a layer of granular fill resting on the surface of soft

Figure 6.11 (a) The test apparatus; (b) load–displacement curves and (c) reinforcement geometry after approximately 200 mm of central displacement
Reproduced from Jarrett (1986), Industrial Fabrics Association International

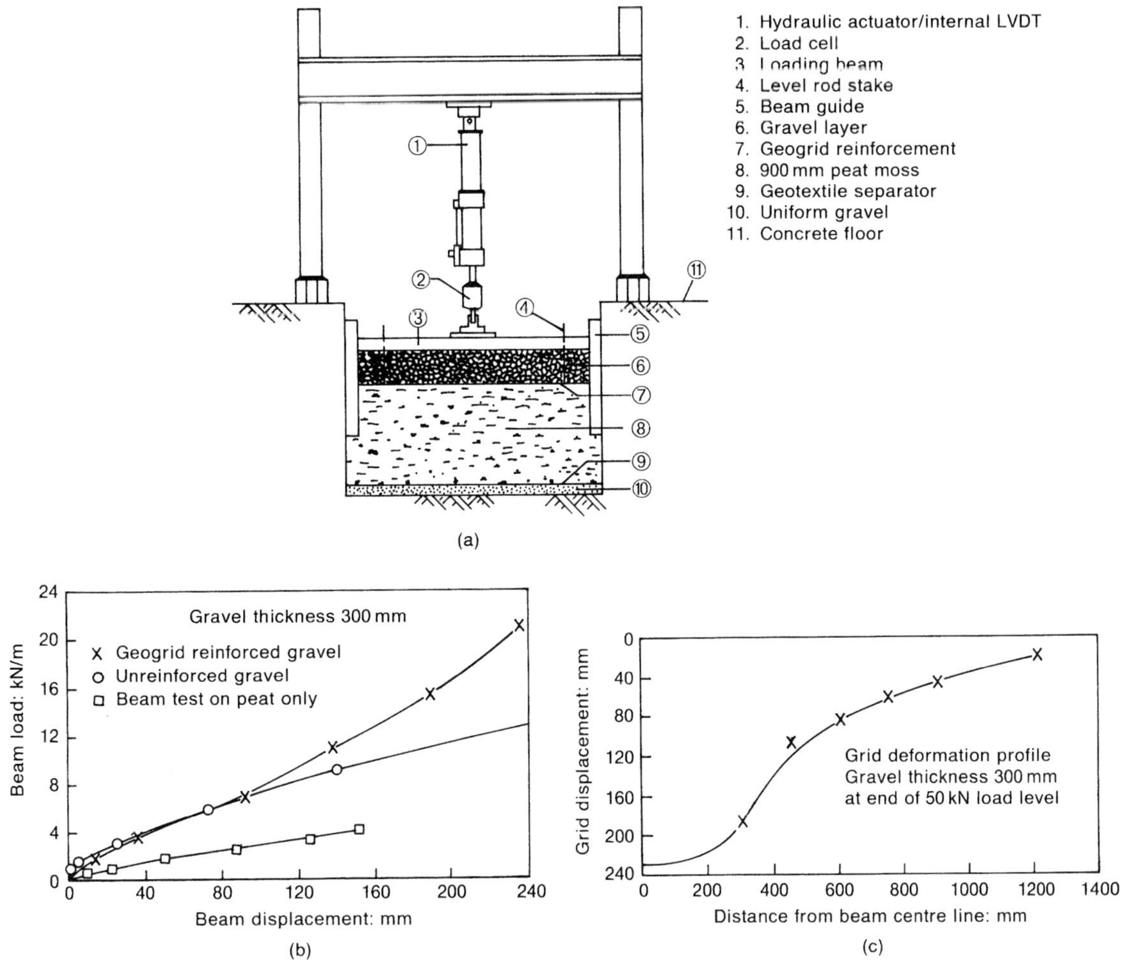

clay, was studied by Love *et al.* (1987) in the laboratory by conducting small-scale model tests. From the test results, the following conclusions can be drawn

- The geogrid reinforcement tends to reduce the shear stresses transmitted to the surface of the clay subgrade. The amount of reduction depends primarily on the strength of the clay, and the thickness and stiffness of the granular layer.
- The failure mechanisms in the clay are mobilised at quite small deformations of the fill, and large deformations are therefore not necessary for any benefits of the reinforcement to be felt. At large deformations, where these are permissible, additional benefit is obtained from the membrane action of the reinforcement.
- To have the desired effect, the reinforcement has to be stiff enough and strong enough to take the tension induced by the shear stresses from the granular layer above (and also the shear stresses from the clay beneath) without failing. Geogrids can be sufficiently stiff and strong to do so.
- There is a risk of soft clay being extruded through the grid, and the bond between the grid and granular layer being broken. When there is separation of the fill and clay, and the contribution of membrane forces are more significant than the reinforcing action, a geotextile would probably be more appropriate.

Kim and Cho (1988) reported a series of laboratory bearing capacity tests of a strip footing on a sand–clay layer with and without a reinforcing geotextile. The test results indicated that the contribution of a geotextile to the increase in the bearing capacity becomes high as the distance of the footing from the geotextile layer is reduced. It also becomes high as the footing depth and the footing settlement increase. The ratio of the sand layer depth to the footing width, which gives the greatest geotextile effect, falls between 0.5 and 1.0 for settlements where $s/B < 1.0$.

Khing *et al.* (1994) conducted a number of laboratory model tests to determine the ultimate and allowable bearing capacities of a surface strip footing, supported by a layer of strong sand underlain by a saturated weak clay deposit reinforced with a layer of geogrid reinforcement at the sand–clay interface. Based on the model test results, the following conclusions can be drawn

- The maximum benefit from the geogrid reinforcement, in increasing the ultimate bearing capacity, occurs when the thickness of the strong sand layer is about two-thirds of the width of the footing *B*.
- For the depth of a geogrid layer greater than, or equal to, about 1.5*B*, the contribution of the geogrid reinforcement to the bearing capacity improvement is negligible.
- The optimum width of the geogrid layer required to mobilise the maximum possible bearing capacity for a given sand–geogrid–clay combination is about 6*B*.

Manjunath and Dewaikar (1994) conducted laboratory model tests to determine the effect of a single layer of geosynthetic reinforcement (geotextile as well as geogrid) on the bearing capacity of shallow foundations. The tests were conducted with square footings resting on a compact sand layer overlying a soft clay subgrade. From the test results, the following conclusions can be drawn

- The ultimate bearing capacity of shallow footings on soft clays can be substantially improved by the inclusion of a reinforcing layer at a suitable location.
- A geotextile is more suitable than a geogrid, when footings are located on sand above a soft clay subgrade.
- The primary properties of the reinforcement material that affect the performance of footings on reinforced soil beds are their tensile strength, elastic modulus and aperture size.
- The size of the footing does not have any significant effect on the performance of the footings on reinforced soil beds.

Among the reinforcement practices for buildings, roads and embankments constructed on soft ground, the use of a

Figure 6.12 (a) Geocell mattress configuration, (b) a plan view of geocell mattress and (c) connection detail
Reproduced from Bush *et al.* (1990), with permission from Elsevier

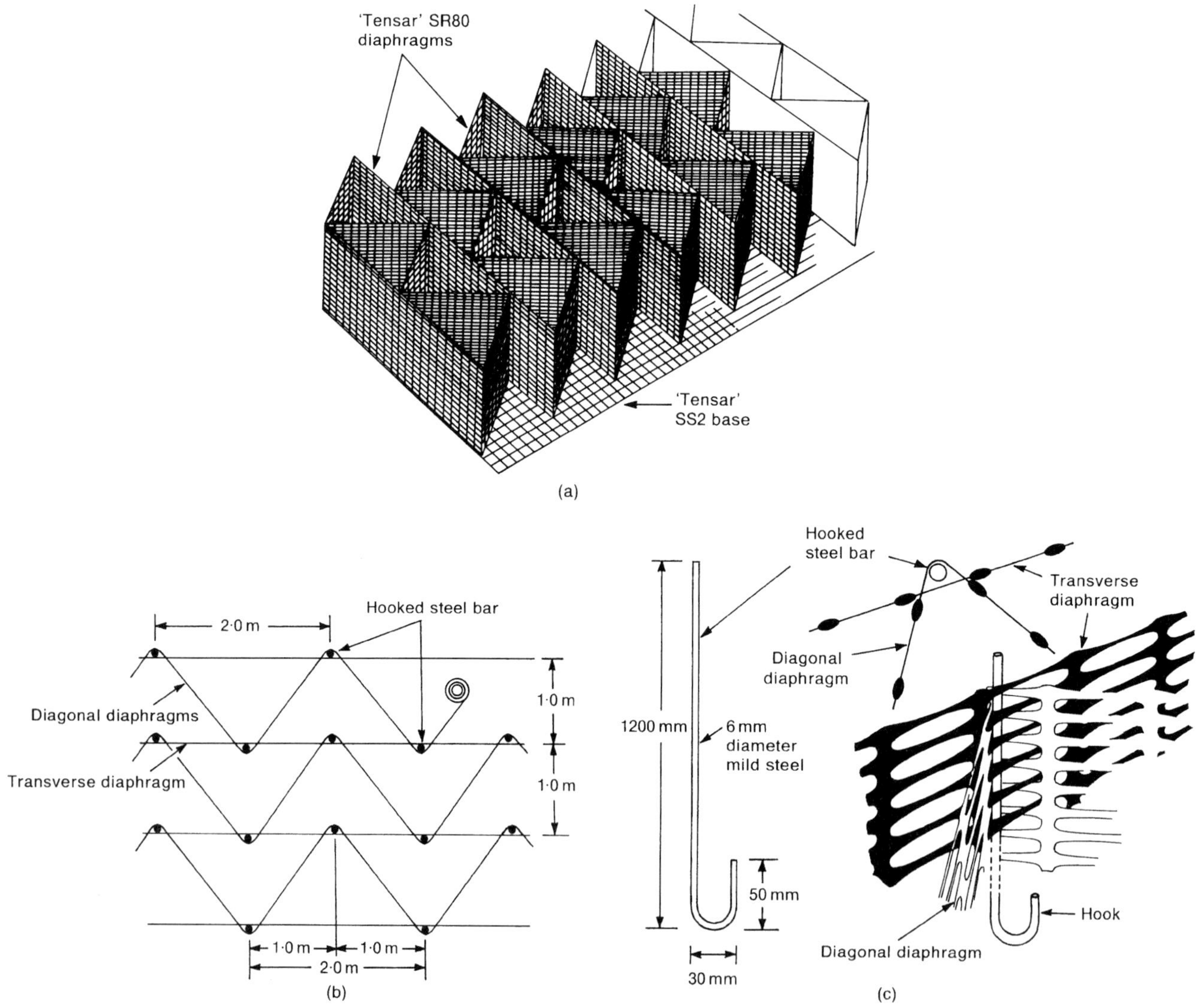

geocell foundation mattress is a unique method, in which the mattress is placed on the soft foundation soil of insufficient bearing capacity so as to withstand the weight of the superstructure. The geocell foundation mattress is a honey comb structure formed from a series of interlocking cells (Figure 6.12). These cells are fabricated directly on the soft foundation soil using uniaxial polymer geogrids in a vertical orientation connected to a biaxial base grid, and are then filled with granular material resulting in a structure usually 1 m deep. This arrangement forms a stiff platform that provides a working area for the contractor to push forward the construction of the geocell itself and subsequent structural load, and also forms a drainage blanket to assist in the consolidation of the underlying soft foundation soil. The incorporation of a geocell foundation mattress provides a relatively stiff foundation to the structure, and this maximises the bearing capacity of the underlying weak soil layer. The geocell mattress is self-contained, and, unlike constructions with horizontal layers of geotextiles, it needs no external anchorage beyond the base of the main structure. As a result of the flexible interaction with the supporting foundation soil underneath, even locally or unevenly applied vertical load propagates within the mattress and is transmitted widely to the supporting foundation soil (Ochiai *et al.*, 1994).

A series of large-scale static tests were undertaken by Bathurst and Jarrett (1988) to investigate the load–deformation behaviour of geocomposite mattresses (geocell or geoweb mattresses) constructed over a compressible peat subgrade, and to compare this behaviour with that of comparable unreinforced gravel bases and gravel bases reinforced with a single layer of geotextile or geogrid at the gravel–peat interface. In this investigation, the geoweb mattress reinforcement comprised non-perforated plastic strips welded together ultrasonically. The geocell reinforcement was constructed from strips of polymeric mesh (geogrid) attached by metal bodkins. The tests showed that the geocomposite mattresses significantly improved the load-bearing capacity of the gravel base layer in comparison with equivalent depths of unreinforced gravel bases. The stiffer geoweb construction gave a greater load-bearing capacity at a given rut depth than did the less stiff geocell construction. In addition, tests showed that the reinforcing effect due to the geocomposite construction of the geoweb was initiated at a lower rut depth than was due to the geocell structure. Comparisons between geoweb-reinforced gravel bases and unreinforced bases showed that the geoweb composites were equivalent to about twice the thickness of unreinforced gravel bases in their effectiveness.

Krishnaswamy *et al.* (2000) conducted laboratory model tests to quantify the improvement in the performance of embankments constructed on soft clays due to the provision of a geocell reinforcement layer at the base. The results of the tests have shown that the provision of a geocell layer at the base of the embankment improves the load-bearing capacity, and vertical and lateral deformations of the embankment. The tensile stiffness of the geogrid used to manufacture the geocell layer, and the aspect ratio (height-to diameter ratio) of the geocell pockets, have an important influence on the performance of geocell-supported embankments. Dash *et al.* (2001, 2003, 2007) investigated the performance of geocell-reinforced foundation soils with varying cell sizes, infill material properties and loading conditions. Their studies show that the effectiveness of the geocell reinforcement depends not only on the adequate load transmission to the fill material (via friction and interlocking), but also on the stiffness of the geocell material.

6.6. Load-bearing capacity analysis

Model tests have shown that the load-bearing capacity of a geosynthetic-reinforced foundation soil depends on several factors, such as the depth, length, number and stiffness of geosynthetic layers in the foundation soil. It is very difficult to make an exact analysis considering all the aspects of geosynthetics simultaneously. In view of the fact that load-bearing capacity considerations often govern the design of geosynthetic-reinforced soil systems for their use as foundations for shallow footings, embankments, unpaved roads, etc., several researchers carried out load-bearing capacity analyses considering the limited roles of geosynthetic reinforcements in improving the load-bearing capacity based on different sets of assumptions (Barenberg, 1980; Giroud and Noiray, 1981; Bourdeau *et al.*, 1982; Raumann, 1982; Sellmeijer *et al.*, 1982; Love *et al.*, 1987; Jewell, 1988; Bourdeau, 1989; Milligan *et al.*, 1989; Espinoza, 1994; Espinoza and Bray, 1995; Huang and Menq, 1997; Sharma *et al.*, 2009; Zhang *et al.*, 2010). Such semi-empirical methods of bearing capacity analysis do not consider the deformability of all components in analysis. Some of these methods are described in Chapter 7.

Finite element methods are commonly used to analyse reinforced soil systems considering strain compatibility requirements (Andrawes *et al.*, 1982; Love *et al.*, 1987; Rowe and Soderman, 1987; Abdel-Baki and Raymond, 1994; Yetimoglu *et al.*, 1994; Otani *et al.*, 1998). These methods provide valuable information regarding reinforced soil behaviour. Unfortunately, data preparation for finite element models is time consuming and, therefore, these models are not convenient for routine design calculations.

In carrying out load-bearing capacity analysis, there are two basic approaches of modelling for the interaction behaviour between soils and geosynthetics. One is that the soil and the geosynthetic are individually modelled and the other is that the geosynthetic layer and its surrounding soil are unified in the model. Both approaches are discussed in this chapter, with more emphasis on the former, because it is more useful for general practice.

6.6.1 Reinforced granular fill

Huang and Menq (1997) presented an analytical approach to the bearing capacity analysis of sandy ground reinforced with horizontal reinforcement layers. This analytical method was verified by Huang and Menq using the results of loading tests on a 101.6 mm-wide strip footing resting

Figure 6.13 The failure mechanism for reinforced sandy ground assumed by Binquet and Lee (1975b)

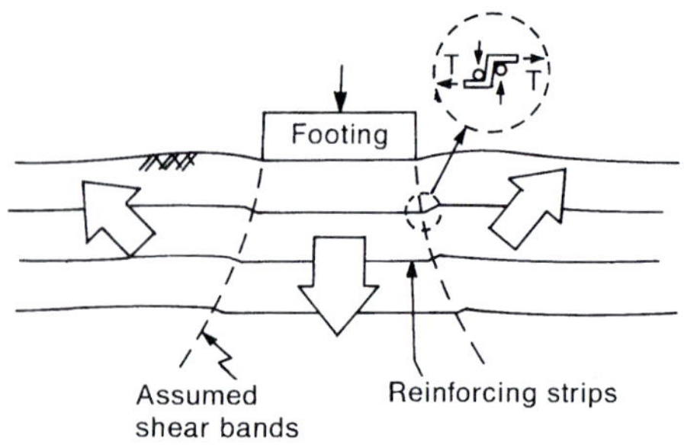

on geogrid-reinforced fine sand performed by Khing *et al.* (1992). The details of this method of analysis are described below.

Binquet and Lee (1975a,b) performed a pioneering study on the load-bearing capacity of footings resting on sandy ground reinforced with aluminium foil strips, and proposed a design method based on an assumed failure mechanism, as shown in Figure 6.13. According to this mechanism, the tensile force, developed in the vertically bending part of the reinforcement across the assumed shear band, increases the bearing capacity of the reinforced sandy ground. When the length of the reinforcement is short (e.g. equal to the width of footing (B)), as shown in Figure 6.14 (Huang and Tatsuoka, 1990), the model proposed by Binquet and Lee (1975b) is invalid. Schlosser *et al.* (1983) proposed a failure mechanism, shown in Figure 6.15, for the reinforced ground. Based on this failure mechanism, the bearing capacity of a strip footing resting on reinforced ground can be expressed as

$$q_{\mathrm{u(reinforced)}} = \gamma D_{\mathrm{R}} N_{\mathrm{q}} s_{\mathrm{q}} d_{\mathrm{q}} + 0.5(B + \Delta B)\gamma N_{\gamma} s_{\gamma} \qquad (6.2)$$

where $q_{\mathrm{u(reinforced)}}$ is the ultimate bearing capacity of the footing resting on reinforced ground, γ is the unit weight of sand, N_{q}, N_{γ} are bearing capacity factors, D_{R} is the depth of the reinforced zone from the ground surface, s_{q} and s_{γ} are shape factors, d_{q} is the depth factor ($= 1 + 0.35D_{\mathrm{R}}/B$), B is the width of the surface footing, ΔB is the increase in the footing width at the depth D_{R} due to the wide slab effect expressed by $2D_{\mathrm{R}} \tan\alpha$, and α is the load-spreading angle as shown in Figure 6.15.

Figure 6.14 The failure surface observed by Huang and Tatsuoka (1988, 1990)

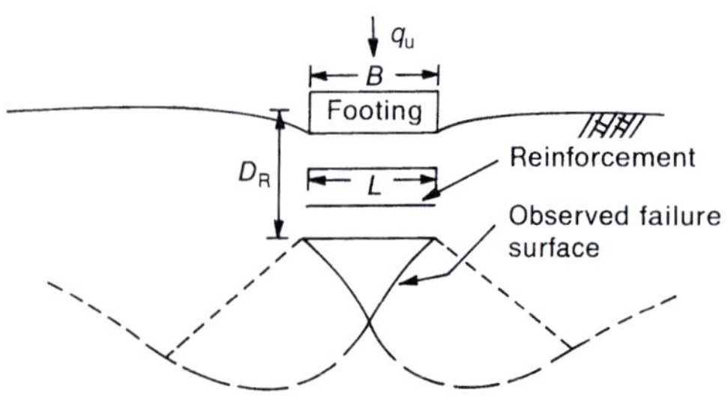

Figure 6.15 The failure mechanism of reinforced ground proposed by Schlosser *et al.* (1983)

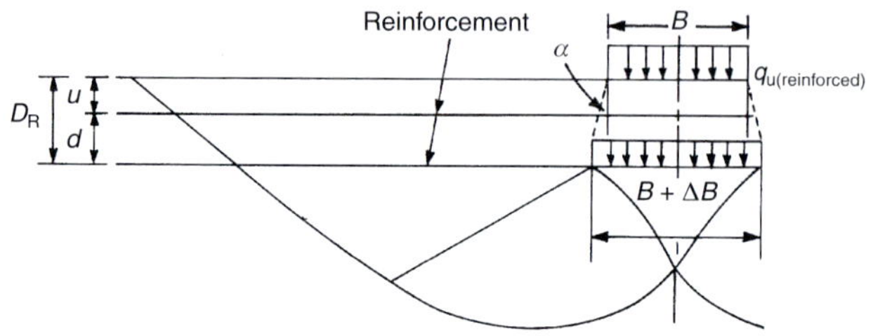

According to Equation 6.2, the following two mechanisms account for the increase in the bearing capacity of footings resting on densely reinforced sandy ground

- deep-footing mechanism
- wide-slab mechanism.

The deep-footing mechanism is applicable when a quasi-rigid zone is developed beneath the footing (Huang and Tatsuoka, 1988, 1990). The wide-slab mechanism is applicable only when a quasi-rigid earth slab below the footing extends beyond the width of the footing. For densely reinforced conditions (for either short or long strips), shear bands starting from the edges of the footing extend straight down approximately to the depth D_{R}, then form a wedge beneath the reinforced zone (Figure 6.16(a)). In this case, the bearing capacity of the reinforced ground is controlled by the strength of the zone, including the wedge denoted B in Figure 6.16(a).

For lightly reinforced conditions, the shear bands that start from the edges of the footing form a wedge within the reinforced zone, but the apex of the wedge is deeper than that for the unreinforced ground (Figure 6.16(b)). In this case, the bearing capacity of the reinforced ground is controlled by the strength of the block A immediately beneath the footing, and the failure may occur because of one of the following factors

- bond failure between the sand and the reinforcement
- an insufficient CR (covering ratio, which is the width of the reinforcing strip/centre-to-centre horizontal spacing of the reinforcing strips) of reinforcement

Figure 6.16 Failure modes of reinforced sand: (a) densely reinforced and (b) lightly reinforced
Reproduced from Huang and Tatsuoka (1988)

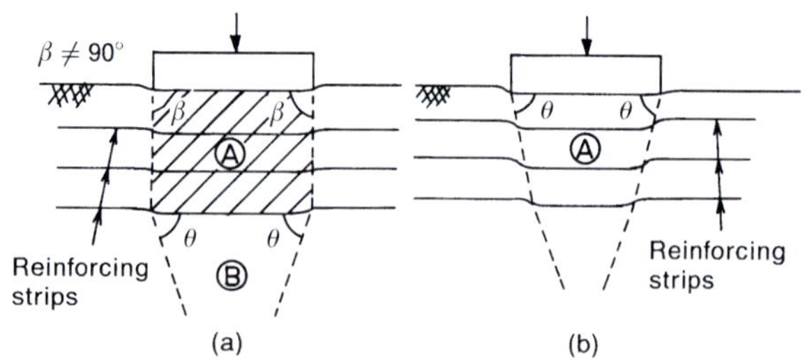

- rupture failure of the reinforcement (Huang and Tatsuoka, 1990).

For estimating the ultimate bearing capacity of a deep footing ($0 < D_f/B < 2.5$, where B is the width of the footing and D_f is the depth of the footing) placed on a homogeneous dry sand, the following equation has been suggested by Terzaghi (1943):

$$q_{u(unreinforced, D_f>0)} = \eta B\gamma N_\gamma + \gamma D_f N_q \tag{6.3}$$

where $q_{u(unreinforced, D_f>0)}$ is the ultimate bearing capacity for an unreinforced deep footing, $\eta = 0.5$ for a strip footing and $\eta = 0.4$ for a square footing. Based on Equation 6.3, a bearing capacity ratio for a deep footing, BCR_D, is defined as

$$BCR_D = \frac{q_{u(unreinforced, D_f>0)}}{q_{u(unreinforced, D_f=0)}} = 1 + \frac{1}{\eta}\frac{D_f}{B}\frac{N_q}{N_\gamma} \tag{6.4}$$

where $q_{u(unreinforced, D_f=0)}$ is the ultimate bearing capacity for a surface footing resting on unreinforced ground. The definitions of N_q and N_γ, suggested by Vesic (1973), are as follows:

$$N_q = e^{\pi \tan\phi}\tan^2\left(\frac{\pi}{4} + \frac{\phi}{2}\right) \tag{6.5}$$

$$N_\gamma = 2(N_q + 1)\tan\phi \tag{6.6}$$

where ϕ is the angle of internal friction.

A comparison of theoretical and measured BCR_D values studied by Huang and Menq (1997) infers that the value of the BCR_D is not susceptible to a change in the internal friction angle ϕ. This feature is important especially when loading tests from various sources are analysed in judging the 'deep-footing mechanism' of reinforced ground.

In the case of a deep-footing effect in reinforced sandy ground, Equation 6.4 can be used to estimate the theoretical value of the BCR_D, in which the term D_f should be replaced by D_R, which represents the depth of the reinforced zone. Developing this concept, Huang and Menq (1997) analysed various loading test results, including tests on geogrid-reinforced sandy ground, by calculating the BCR_D and comparing the measured value of the bearing capacity ratio, BCR_m, defined as

$$BCR_m = \frac{q_{u(reinforced)}}{q_{u(unreinforced, D_f=0)}} \tag{6.7}$$

where $q_{u(reinforced)}$ is the measured value of the ultimate bearing capacity for a surface footing placed on reinforced ground.

Equation 6.3 can be extended for reinforced ground based on the deep-footing and wide-slab mechanisms as

$$\begin{aligned} q_{u(reinforced)} &= \eta(B + \Delta B)\gamma N_\gamma + \gamma D_f N_q \\ &= q_{u(unreinforced, D_f=0)} + \eta\,\Delta B\,\gamma N_\gamma \end{aligned} \tag{6.8}$$

The last term in Equation 6.8 represents a component of bearing capacity contributed by the so-called wide-slab mechanism to the bearing capacity of reinforced ground, namely, $q_{u(slab)}$. Thus,

$$q_{u(slab)} = \eta\,\Delta B\,\gamma N_\gamma = \Delta B \frac{q_{u(unreinforced, D_f=0)}}{B} \tag{6.9}$$

Equation 6.8 can be rearranged as

$$\begin{aligned} q_{u(slab)} &= q_{u(reinforced)} - q_{u(unreinforced, D_f>0)} \\ &= q_{u(unreinforced, D_f=0)}(BCR_m - BCR_D) \end{aligned} \tag{6.10}$$

The tangent of the load-spreading angle from the vertical, $\tan\alpha$, can be obtained as

$$\tan\alpha = \frac{\Delta B}{2D_R} \tag{6.11}$$

Based on comparisons of measured and multiple-variable data regression for several model test results, the following relationship between the load-spreading angle, α, and the factors that control the scheme for reinforcement was presented by Huang and Menq:

$$\begin{aligned} \tan\alpha = {} & 0.680 - 2.071d/B + 0.743CR \\ & + 0.030L/B + 0.076N \end{aligned} \tag{6.12}$$

where d is the vertical spacing between two reinforcing layers, B is the footing width, L is the length of reinforcing layers and N is the total number of reinforcing layers. This relationship is valid under the following conditions: $\tan\alpha > 0$, $0.25 \leq d_f/B \leq 0.5$, $0.02 \leq CR \leq 1.0$, $1 < L/B \leq 10$ and $1 \leq N \leq 5$.

The agreement between the measured values for tests on geogrid-reinforced fine sand, using a 101.6 mm-wide strip footing performed by Khing *et al.* (1992), and the predicted values, using Equations 6.9–6.12, of the BCR of reinforced sandy ground is encouraging, especially when the contribution of N is eliminated in Equation 6.12 (Figure 6.17).

The bearing capacity of geosynthetic-reinforced granular soil was analysed using the finite element method by several authors (Andrawes *et al.*, 1982; Abdel-Baki and Raymond, 1994; Yetimoglu *et al.*, 1994). These studies have shown that the geosynthetic reinforcement has a major beneficial effect, increasing the bearing capacity of footings resting on geosynthetic-reinforced granular soil. The finite element method has been found useful in predicting the failure patterns of the model tests for the reinforced granular soil. The applicability of the results for any given situation will depend on the details of the specific finite elements and the constitutive models used in the analysis.

Dixit and Mandal (1993) applied a variational method to determine the bearing capacity of shallow strip footings loaded vertically and placed on a geosynthetic-reinforced sand layer. In this method, the shape of the failure surface

Figure 6.17 Comparison of predicted and measured values of the BCR for tests obtained by Khing *et al.* (1992)
Reproduced from Huang and Menq (1997), with permission from the ASCE

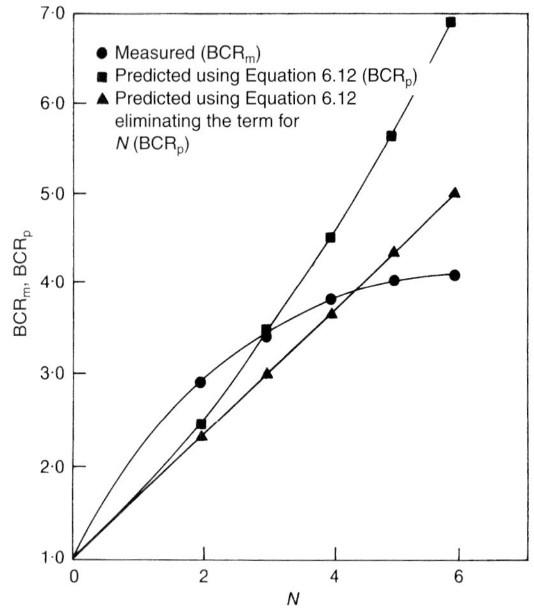

and the distribution of normal stress over it are determined using minimising theorems of variational calculus.

6.6.2 Reinforced clay

Ingold and Miller (1982) presented a method of analysis for the load-carrying capacity of geosynthetic-reinforced clay foundations in undrained conditions. This method uses the concept of a composite theory in which the effects of reinforcement are assumed to impart an equivalent undrained shear strength. Ideally, foundations can be designed using existing total stress theories, with the equivalent undrained shear strength in place of the shear strength of unreinforced clay. A comparison of model test results with the theoretical results based on the suggested simple design technique has shown sufficiently reasonable agreement.

6.6.3 Reinforced granular fill–soft foundation soil system

A bearing capacity analysis, presented by Espinoza and Bray (1995) for a single-layer geotextile-reinforced granular fill–soft foundation soil, is described here. The bearing capacity equation derived satisfies vertical and horizontal force equilibrium along the geotextile reinforcement, and incorporates two important membrane support contributions, namely normal stress membrane support and interfacial shear stress membrane support. The subgrade shear stress reduction effect of a geotextile is also included in the equation.

By considering the vertical force equilibrium of a differential geotextile element of unit area as shown in Figure 6.18(a), a general equilibrium equation is obtained:

$$q_{\mathrm{app}}(x) = q_{\mathrm{s}}(x) + q_{\mathrm{g}}(x) \tag{6.13}$$

Figure 6.18 Forces on a geotextile: (a) membrane contribution provided by the geotextile and (b) the vertical and horizontal force equilibrium of the deformed geotextile
Reproduced from Espinoza and Bray (1995), Industrial Fabrics Association International

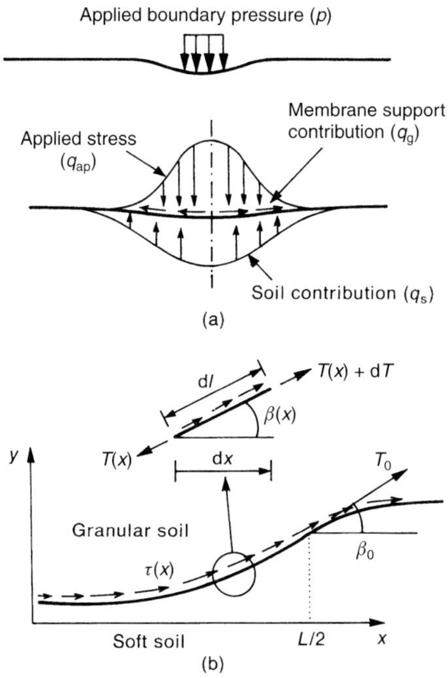

where $q_{\mathrm{app}}(x)$ is the force per unit area above the geotextile, $q_{\mathrm{s}}(x)$ is the vertical soil reaction per unit area, $q_{\mathrm{g}}(x)$ is the membrane support contribution per unit area, and x is the horizontal coordinate.

Assuming plane-strain conditions and considering the vertical and horizontal force equilibrium of the deformed geotextile (Figure 6.18(b)), it can be shown that (Espinoza, 1994)

$$q_{\mathrm{g}}(x) = T_{\mathrm{h}}(x)\frac{\mathrm{d}^2 y(x)}{\mathrm{d}x^2} \tag{6.14}$$

with

$$T_{\mathrm{h}} = T(x)\cos\beta(x) \tag{6.15}$$

$$T(x) = J\varepsilon(x) \tag{6.16}$$

$$\tan\beta(x) = \frac{\mathrm{d}y}{\mathrm{d}x} \tag{6.17}$$

where $y(x)$ is the vertical deflection of the geotextile, $\beta(x)$ is the angle that the deformed geotextile makes with the horizontal line at a distance x from the centreline, $T(x)$ is the geotextile tensile force, J is the geotextile stiffness modulus, $\varepsilon(x)$ is the geotextile strain, and $T_{\mathrm{h}}(x)$ is the horizontal component of the tensile force $T(x)$.

Espinoza (1994) defined the average membrane support contribution:

$$\bar{q}_{\mathrm{g}} = \frac{1}{L}\int_{-L/2}^{L/2} q_{\mathrm{g}}(x)\,\mathrm{d}x = \frac{1}{L}\int_{-L/2}^{L/2} T_{\mathrm{h}}(x)\frac{\mathrm{d}^2 y(x)}{\mathrm{d}x^2}\mathrm{d}x \tag{6.18}$$

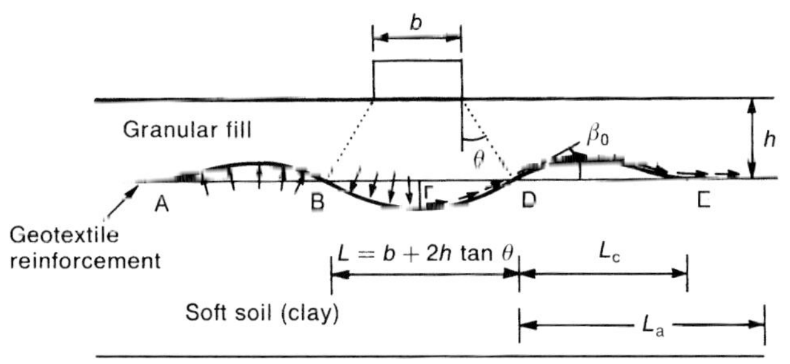

Figure 6.19 The failure mechanism
Reproduced from Espinoza and Bray (1995), Industrial Fabrics Association International

where L is the effective horizontal length of geotextile, defined by the segment joining the stationary points B and D shown in Figure 6.19. This equation satisfies the global vertical and horizontal force equilibrium.

The geotextile located outside the effective length (i.e. AB and DE in Figure 6.19) exerts a vertical pressure, $\bar{q}_{\mathrm{lat}}$, due to membrane support, thus reducing the heave potential of the subgrade soil. Considering an average surcharge load $(\bar{q}_{\mathrm{lat}} + \gamma h)$, the subgrade bearing capacity is given by

$$\bar{q}_{\mathrm{s}} = cN_{\mathrm{c}} + \gamma h + \bar{q}_{\mathrm{lat}} \tag{6.19}$$

where

$$\bar{q}_{\mathrm{lat}} = \frac{1}{L}\int_{L/2}^{L_{\mathrm{c}}+L/2} q_{\mathrm{g}}(x)\,\mathrm{d}x \tag{6.20}$$

and

$$N_{\mathrm{c}} = 1 + \frac{\pi}{2} + \alpha + \sin\alpha \tag{6.21}$$

where $\alpha = \cos^{-1}(\tau_{\mathrm{c}}/c_{\mathrm{u}})$, τ_{c} is the shear stress transmitted to the clay surface, c_{u} is the undrained shear strength of the clay, N_{c} is the bearing capacity factor, h is the thickness of the granular fill, γ is the unit weight of the fill, and L_{c} is the length of the geotextile preventing heave (Figure 6.19).

Equation 6.21 is based on the lower-bound plasticity theory for undrained loading on a semi-infinite saturated clay layer (Bolton, 1979). If the shear above the clay surface is zero (smooth footing), then $\alpha = \pi/2$, and N_{c} becomes $(\pi + 2)$, which is the classical bearing capacity factor for vertical loads on a rigid perfectly plastic material. An N_{c} factor larger than $(\pi + 2)$ may be used for rough footings that transmit inward shear to the clay.

The average vertical stress within the fill can be estimated using a load spreading angle, θ. The average pressure applied to the geotextile is given by

$$\bar{q}_{\mathrm{ap}} = \gamma h + \alpha_{\mathrm{b}} p \tag{6.22}$$

Table 6.1 Load spreading angle (note: h is expressed in cm)

Method	Spreading angle, θ: °	
	Without geotextile	With geotextile
Barenberg (1980)	$\tan^{-1}(0.3 + 5/h)$	$\tan^{-1}(0.6 + 5/h)$
Giroud and Noiray (1981)	$(\pi/4 - \phi/2)$	26.6–35.0
Raumann (1982)	28.8	33.0
Sellmeijer *et al.* (1982)	26.6–45.0	26.6–45.0
Love *et al.* (1987)	–	26.6–31.0

where $\alpha_{\mathrm{b}} = b/L$ (the width factor) and $L = b + 2h\tan\theta$. Table 6.1 shows different empirical values of the load spreading angle, θ, as reported in the literature.

Combining Equations 6.13, 6.18, 6.19 and 6.22, an average equilibrium equation is obtained as

$$\alpha_{\mathrm{b}} p = c_{\mathrm{u}} N_{\mathrm{c}} + \bar{q}_{\mathrm{t}} \tag{6.23}$$

with

$$\bar{q}_{\mathrm{t}} = \frac{2}{L}\int_0^{L/2} q_{\mathrm{g}}(x)\,\mathrm{d}x + \frac{1}{L}\int_{L/2}^{L_{\mathrm{c}}+L/2} q_{\mathrm{g}}(x) \tag{6.24}$$

where $\bar{q}_{\mathrm{t}}$ is the total membrane support contribution, which includes both normal stress membrane support (membrane contribution obtained from outside the effective length) and interfacial shear stress membrane support (membrane contribution obtained from within the effective length). Normal stress membrane support depends on proper anchorage outside the effective length. Interfacial shear stress membrane support depends on the applied load and the mobilised interface friction.

Assumptions regarding the geotextile strain distribution and deformation are necessary to numerically evaluate the integral expression given by Equation 6.24. An equation for the admissible surface pressure, p_{adm}, can be estimated as

$$p_{\mathrm{adm}} = \frac{c_{\mathrm{u}} N_{\mathrm{c}} + T_0 \sin\beta_0/L + 2\alpha_{\mathrm{r}}\gamma h\tan\psi_{\mathrm{m}}}{\alpha_{\mathrm{b}}(1 - 2\alpha_{\mathrm{r}}\tan\psi_{\mathrm{m}})} \tag{6.25}$$

where $\alpha_{\mathrm{r}} = r/L$ (the rutting factor), r is the rutting depth (Figure 6.19), T_0 is the tensile force in the geotextile layer at point D, β_0 is the inclination of the geotextile layer at point D, and ψ_{m} is the mobilised interface friction angle. The normal stress membrane support is reflected in the tensile force T_0, and the angle of deflection β_0 developed at the stationary points B and D in Figure 6.19. In many practical field situations, proper anchorage cannot be ensured at all times during construction (i.e. there is not enough anchorage length, L_{a}, or surcharge load, γ_{h}, or a combination of both). In such cases, $T_0 = 0$ should be used to estimate the admissible pressure. Even in cases where proper anchorage is provided (i.e. $T_0 > 0$), its effect will not be felt until large deformations are induced (i.e. $\beta_0 \gg 0$).

Figure 6.20 Mobilised shear
Reproduced from Espinoza and Bray (1995), Industrial Fabrics Association International

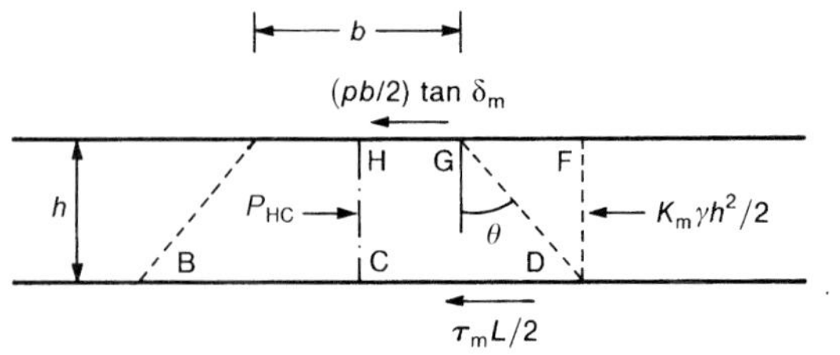

An expression can also be derived for the mobilised interface friction angle based on the strict equilibrium between the membrane and sliding block above it. An expression for this case, valid for the situation shown in Figure 6.20, is

$$\tan\psi_m = \frac{\alpha_h(K - K_{pm}) + M_c(\eta K - \tan\delta_m)}{1 + M_c + 2\alpha_r[\alpha_h(K - K_{pm}) - \eta K + \tan\delta_m]} \quad (6.26)$$

where K is an earth pressure coefficient, $K_{pm} = \tan^2(\pi/4 + \phi_m/2)$ (the mobilised passive earth pressure coefficient), ϕ_m is the mobilised soil friction angle, δ_m is the mobilised interface friction angle at the footing base, and $\alpha_h = h/L$ and $M_c = (c_u N_c + T_0 \sin\beta_0/L)\gamma h$ are dimensionless parameters.

Equations 6.25 and 6.26 have been used to predict admissible pressures for a small-scale model test set-up, and the results are compared with the footing pressures measured by Love *et al.* (1987) and Milligan *et al.* (1989) for a series of model tests with various granular fill thicknesses and subgrade strengths. Overall, the computed values of the admissible pressures compare favourably with those measured, and this finding provides support to the validity of the proposed Equations 6.25 and 6.26.

Ochiai *et al.* (1994) described a conventional approach for the assessment of the improvement of the bearing capacity due to placement of the geogrid mattress foundation. In this approach, a vertical load of intensity p and width B, applied to the mattress, is transmitted widely to the supporting foundation soil with the corresponding intensity p_m and width B_m (Figure 6.21). The ultimate bearing capacity q without the use of the mattress may be given by Terzaghi's equation, as follows:

$$q = cN_c + \tfrac{1}{2}\gamma B N_\gamma \quad (6.27)$$

where c is the cohesion and γ is the unit weight of the supporting foundation soil. On the other hand, the ultimate bearing capacity, q_m, with the use of a mattress, may be given as follows, assuming that the placement of the geogrid mattress has a surcharge effect on the bearing capacity of the supporting foundation:

$$q_m = cN_c + \gamma_m H N_q + \tfrac{1}{2}\gamma B_m N_\gamma \quad (6.28)$$

Figure 6.21 Effects of the use of a geogrid mattress
Reproduced from Ochiai *et al.* (1994), Industrial Fabrics Association International

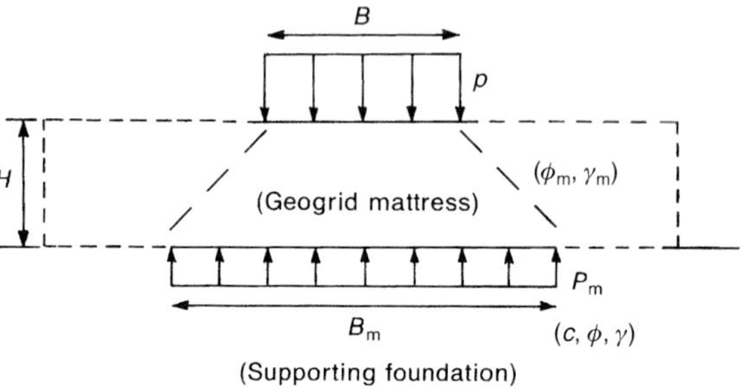

where m is the unit weight of the mattress and H is the thickness of the mattress. Therefore, the increase in the bearing capacity Δq due to the placement of the mattress can be given as

$$\Delta q = \gamma_m H N_q + \tfrac{1}{2}\gamma(B_m - B)N_\gamma \quad (6.29)$$

It is therefore found that the evaluation of the bearing capacity improvement requires the estimation of the width B_m. The experimental studies have revealed that the width of the supporting foundation soil over which the vertical stress is distributed becomes larger as the thickness of the geogrid mattress becomes greater, and as the vertical stiffness of the supporting foundation soil becomes lower. It was suggested, from a design point of view, that the width of the geogrid mattress should be at least large enough to accommodate the vertical stress distribution that takes place underneath the mattress.

Sharma *et al.* (2009) developed analytical solutions for estimating the ultimate bearing capacity of geogrid-reinforced foundations for both sand and silty clay soils by modifying Meyerhof and Hanna's (1978) solution to incorporate the reinforcement effect. The derivation is based on several simplifying assumptions; however, it has been reported that the predicted bearing capacity values from analytical solutions are in good agreement with some experimental results.

Zhang *et al.* (2010) proposed a calculation method for the bearing capacity of a soft subgrade treated with geocell reinforcement at the bottom of an embankment. This method considers that the 'vertical stress dispersion effect' and the 'membrane effect' are the main reinforcement mechanisms of the geocell when it is installed at the base of an embankment. It was reported that the calculated results were much closer to the experimental results. However, the tensile force in the geocell as considered in the derivation does not appear to exist in the field applications.

Several authors have analysed the geosynthetic-reinforced granular fill–soft soil system by finite element method (Love *et al.*, 1987; Koga *et al.*, 1988; Poran *et al.*, 1989; Floss and Gold, 1994; Otani *et al.*, 1998). The advantage of such an analysis is that the displacement distribution, and the stress distribution, can both be obtained in the

Figure 6.22 Effects of the geosynthetic reinforcement on the bearing capacity of the foundation: (a) $T = 80$ kN/m, (b) $T = 55$ kN/m, (c) $T = 35$ kN/m and (d) $T = 15$ kN/m
Reproduced from Otani *et al.* (1998), with permission from Elsevier

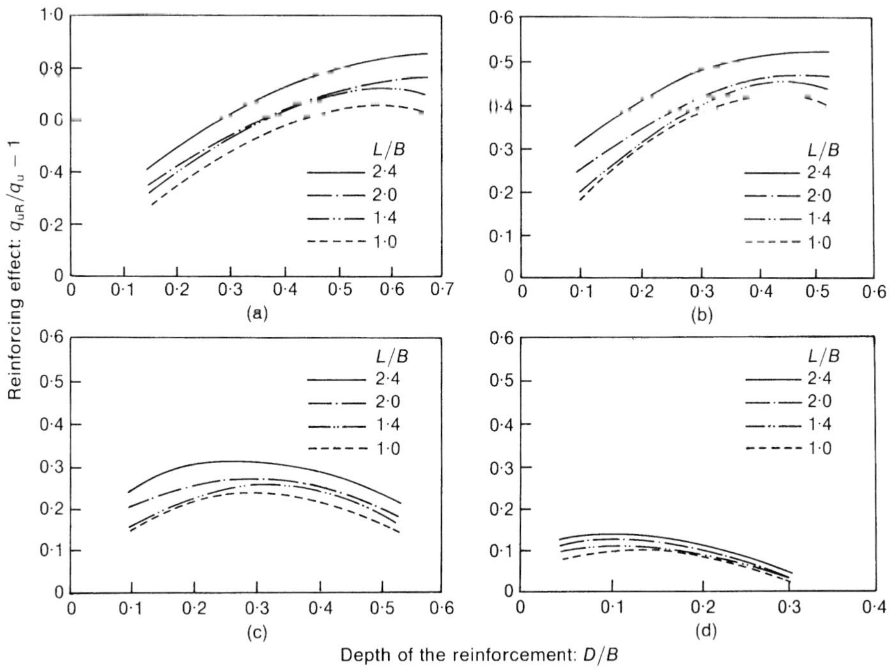

subsoil as well as in the soil–geosynthetic layer system. Nevertheless, it should be realised that the accuracy of the finite element results depends on the appropriate material properties used and the type of modelling adopted for the analysis. In the finite element analysis, the complete soil–geosynthetic layer system can be modelled using individual elements, such as bar elements for the geosynthetic layer, continuum elements for the soil and joint elements for the interface behaviour, or by using composite elements that comprise the soil–geosynthetic system as a whole. In the latter case, the properties of the composite element can be evaluated either experimentally or by a separate numerical analysis.

The bearing capacity analysis of a geosynthetic-reinforced cohesive foundation loaded by a flexible uniform strip footing was carried out by Otani *et al.* (1998) using a rigid plastic finite element formulation. This method is based on the upper bound theorem of the theory of plasticity, and the bearing capacity is obtained as a load factor at the ultimate limit state. The geosynthetic reinforcement and the surrounding sand layer (constructed around the geosynthetics in the cohesive ground for the purpose of increasing the friction between the geosynthetic and the adjacent soil) are modelled as a single composite material with an equivalent cohesion. The underlying soft ground is also assumed to be purely cohesive, and, hence, both the reinforced soil and the soft ground are modelled using the von Mises failure criterion. The method of analysis proposed was checked against the field measurements or the model test results. The analysis indicates that the bearing capacity of the geosynthetic-reinforced foundation increases as the depth and the length of the reinforcement are increased, but there is an optimum depth as well as an optimum number of geosynthetic layers, for which the maximum reinforcing effect is obtained. Figure 6.22 shows a simple design chart for the estimation of the bearing capacity of geosynthetic-reinforced foundations on soft ground. In this chart, L is the half length of the geosynthetic layer, B is the half width of the footing, D is the depth of the geosynthetic layer, T is the tensile strength of the geosynthetic layer, q_u is the ultimate bearing capacity of the unreinforced foundation soil, and q_{uR} is the ultimate bearing capacity of the reinforced foundation soil.

6.7. Settlement analysis

Model tests have shown that the inclusion of one or more geosynthetic layers to reinforce the granular base has been very effective in increasing the load-bearing capacity and reducing settlements of shallow footings resting on it. For analysing the reinforced soil systems, theoretical and experimental approaches have been used. Most of the theoretical work available is associated with the bearing capacity aspect.

Love *et al.* (1987) developed a finite element programme in which the subgrade is modelled as an elastic perfectly plastic material with limiting shear stress equal to undrained cohesion c_u, the granular fill material is modelled as an elastic-frictional material obeying the Matsuoka yield criterion (Matsuoka, 1976), and the geosynthetic reinforcement is modelled using three-noded line elements of appropriate stiffness that conform to the six-noded triangular soil elements on either side. The reinforcement is treated as perfectly

Figure 6.23 Comparison of finite element calculations with model test results (δ is the displacement, B is the width of loading and c_u is the undrained cohesion)
Reproduced from Love *et al.* (1987), © Canadian Science Publishing or its licensors

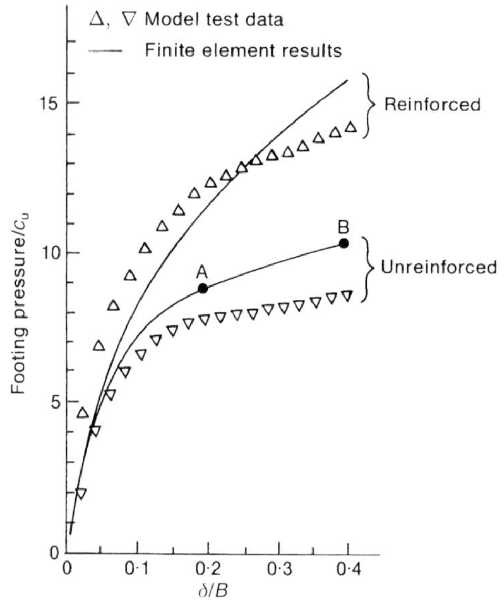

Figure 6.24 The vertical displacement profile of reinforced foundation soil: (a) a surface footing and (b) an embedded footing
Reproduced from Koga *et al.* (1988)

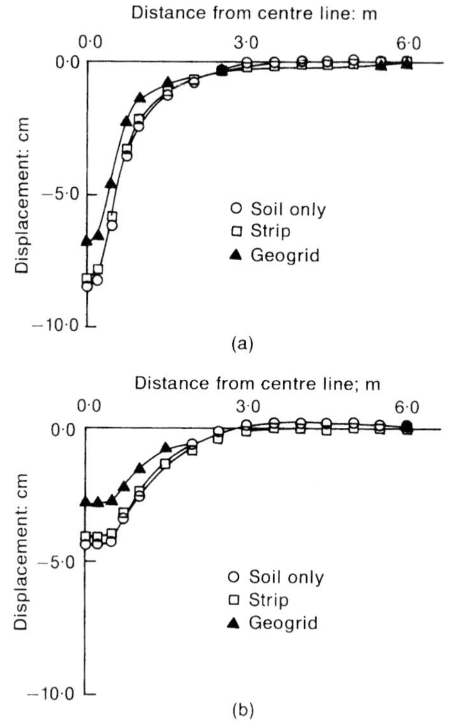

rough, so that any failure must occur in the soil elements adjacent to the reinforcement rather than at the interface. Yielding of the reinforcement was not considered, as none was observed in the model tests, and no compressive stress was allowed in the reinforcement. Figure 6.23 shows a comparison between the load–deflection curves from the finite element calculations and the model test results. In most model tests, the initial stiffness values are similar to the reinforced and unreinforced cases, and this result is also given by the finite element analysis.

Koga *et al.* (1988) carried out a finite element analysis for a soil reinforcement system of geogrids, with particular reference to an embankment on soft soil and a strip footing. In this analysis, individual elements for soil, reinforcement and interface behaviour were used. The vertical displacement along the horizontal surface has been observed, as shown in Figure 6.24(a), for the surface footing and, in Figure 6.24(b), for the embedded footing.

Poran *et al.* (1989) used finite element analysis for the evaluation of settlements of footings placed on geogrid-reinforced granular fill overlying a soft clay subgrade. The parametric results indicate the effects of geogrid reinforcement for the improvement of the load–deformation behaviour of such systems. The design procedure proposed is applicable for cases where the allowable footing settlements under working stress conditions are relatively small and cannot be analysed by simplified design methods.

The settlement characteristics of geosynthetic-reinforced foundation soil were studied by developing mechanical foundation models by Douglas (1987), Madhav and Poorooshasb (1988, 1989), Ghosh (1991), Ghosh and Madhav (1994), and Shukla and Chandra (1994a,b, 1995, 1996, 1998). Shukla and Chandra (1994a) presented a generalised mechanical model for the study of the time-dependent settlement behaviour of the geosynthetic-reinforced granular fill–soft soil system (Figure 6.25). In this model, the geosynthetic reinforcement and the granular fill are represented by the stretched rough elastic membrane and the Pasternak shear layer, respectively. The general assumptions are that the geosynthetic reinforcement is linearly elastic, rough enough to prevent slippage at the soil interface, and has no shear resistance. A perfectly rigid plastic friction model is adopted to represent the behaviour of the soil–geosynthetic interface in shear. The compressibility of the granular fill is represented by a layer of Winkler springs attached to the bottom of the Pasternak shear layer. The saturated soft foundation soil is idealised by the Terzaghi consolidation model, which has a dashpot and a spring. The spring represents the soil skeleton, and the dashpot simulates the dissipation of the excess porewater pressure. The spring constant is assumed to have a constant value with depth of the foundation soil and also with time. The equations governing the response of the model are given as

$$q = \overline{X_1}\frac{k_f k_s w}{k_s + k_f U} - [G_t H_t + \overline{X_2}(T_p + T)\cos\theta + \overline{X_1} G_b H_b]\frac{\partial^2 w}{\partial x^2} \quad (6.30)$$

Figure 6.25 The mechanical foundation model
Reproduced from Shukla and Chandra (1994a), with permission from Elsevier

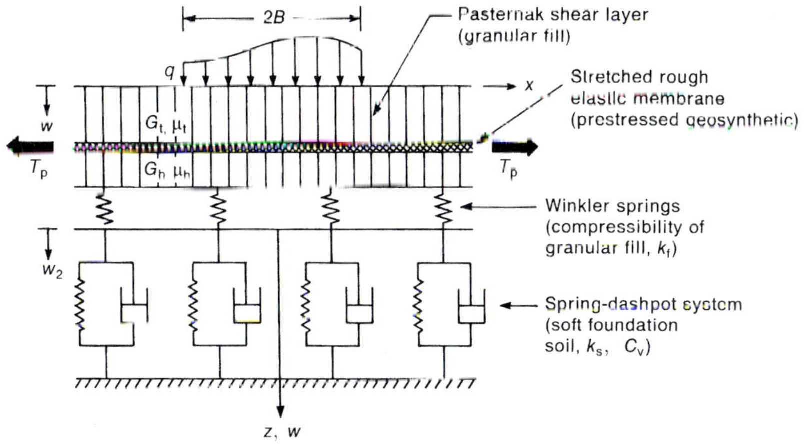

and

$$\frac{\partial T}{\partial x} = -\overline{X_3}\left(q + G_t H_t \frac{\partial^2 w}{\partial x^2}\right) - \overline{X_4}\left(\frac{k_f k_s w}{k_s + k_f U} - G_b H_b \frac{\partial^2 w}{\partial x^2}\right) \tag{6.31}$$

with

$$\overline{X_1} = \frac{1 + K_{0R}\tan^2\theta - (1 - K_{0R})\mu_b \tan\theta}{1 + K_{0R}\tan^2\theta + (1 - K_{0R})\mu_t \tan\theta} \tag{6.32a}$$

$$\overline{X_2} = \frac{1}{1 + K_{0R}\tan^2\theta + (1 - K_{0R})\mu_t \tan\theta} \tag{6.32b}$$

$$\overline{X_3} = \mu_t \cos\theta(1 + K_{0R}\tan^2\theta) - (1 - K_{0R})\sin\theta \tag{6.32c}$$

$$\overline{X_4} = \mu_b \cos\theta(1 + K_{0R}\tan^2\theta) + (1 - K_{0R})\sin\theta \tag{6.32d}$$

where q is the applied load intensity, $w(x, t)$ is the vertical surface displacement, $T(x, t)$ is the tensile force per unit length mobilised in the membrane, T_p is the pre-tension per unit length applied to the membrane, G_t and H_t are the shear modulus and thickness of the upper shear layer, respectively, G_b and H_b are the shear modulus and thickness of the lower shear layer, respectively, μ_t and μ_b are the interface friction coefficients at the top and bottom faces of the membrane, respectively, k_f is the modulus of the subgrade reaction of the granular fill, k_s is the modulus of the subgrade reaction of the soft foundation soil, K_{0R} is the coefficient of lateral stress at rest at an overconsolidation ratio (R) (which is defined here as the ratio of the maximum stress to which the granular fill is subjected through compaction to the existing stress under the working load), θ is the slope of the membrane, U is the average degree of consolidation of the soft foundation soil, C_v is the coefficient of consolidation, x is the distance measured from the centre of the loaded region along the x direction, B is the half width of loading, and t is any particular instant of time measured from the instant of loading.

The parameters of mechanical foundation models can be determined following the guidelines suggested by Selvadurai (1979) and Shukla and Chandra (1996). The parametric studies carried out by Shukla and Chandra (1994a) show the effects of various parameters on the settlement response of a geosynthetic-reinforced granular fill–soft soil system. Figure 6.26 shows the settlement profiles for a typical set of parameters at various stages of consolidation of the soft foundation soil. The trend of results obtained using the above generalised model is in good agreement with other reported works.

Yin (1997a) incorporated a deformation compatibility condition into the mechanical foundation model, and compared the results to two-dimensional finite element modelling results and the results from the mechanical foundation models suggested by Madhav and Poorooshasb (1988), Ghosh (1991), and Shukla and Chandra (1995). Yin (1997b) further improved the mechanical foundation model, incorporating a non-linear constitutive model for the granular fill and a non-linear spring model for the soft soil.

6.8. Prestressed geosynthetic-reinforced foundations

It is now well established that geosynthetics, particularly geotextiles, show their beneficial effects only after relatively large settlements (Andrawes *et al.*, 1982; Milligan and Love, 1984; Guido *et al.*, 1985; Rowe and Soderman, 1987; Madhav and Poorooshasb, 1988; Poorooshasb, 1989; Shukla and Chandra, 1994a,b; Yetimoglu, 1994; Lovisa *et al.*, 2010), which may not be a desirable feature for shallow footings, paved and unpaved roads, and embankments resting on geosynthetic-reinforced foundation soils. Andrawes *et al.* (1982) reported measured and predicted data from which they concluded that the influence of the geotextile on the load settlement behaviour of the strip footing on sand was very limited up to settlements equal to, approximately, 8% of the footing width. This suggests that up to that level of settlement, strains in the soil are insufficient to

Figure 6.26 Settlement profiles of a geosynthetic-reinforced granular fill–soft soil system for various degrees of consolidation of soft saturated foundation soil ($T_p^* = T_p/k_sB^2$, $\alpha = k_f/k_s$)
Reproduced from Shukla and Chandra (1994a), with permission from Elsevier

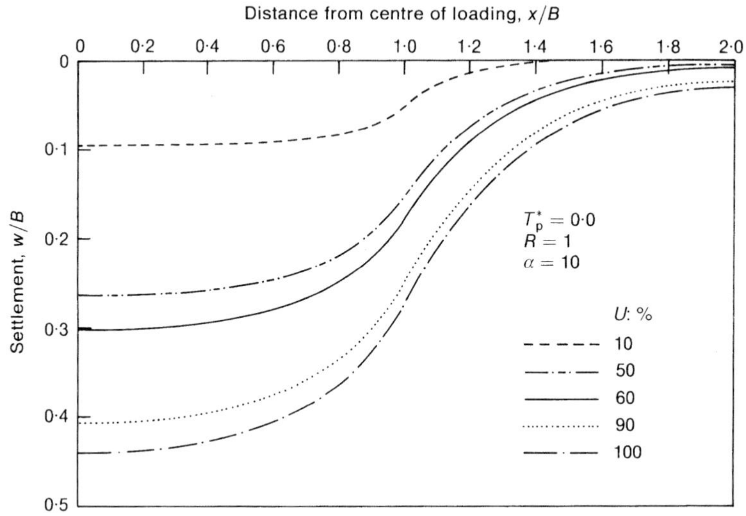

mobilise a significant tensile load in the geotextile. From large-scale laboratory tests, Milligan and Love (1984) showed that there was a marked improvement in the load-carrying capacity with a geogrid at high deformation and only a nominal beneficial effect at low deformation. Poorooshasb (1989) found that at a lower settlement level (less than 2.5 cm), the presence of geogrids had no effect at all. Hence, there is a need for a technique that can make geosynthetics more beneficial without the occurrence of large settlements. Prestressing the geosynthetics can be one of the techniques to achieve this goal (Shukla and Chandra, 1994a,b; Lovisa *et al.*, 2010).

The idea of prestressing the geosynthetics has been recognised by several workers in the past. Aboshi (1984) and Watary (1984) described a method to stabilise very soft clay, developed in Japan, called the 'rope sheet method', in which ropes are preloaded to 0.5–0.6 kN in order to increase their effectiveness. For the stabilisation of very soft clay using a geotextile, Broms (1987) suggested that the geotextile should be stretched as much as possible before the stabilising berms are placed along the perimeter of the geotextile sheet, in order to limit the penetration required to develop the necessary tension in the geotextile. Koerner (1990) expressed his view that a method of prestressing the geotextile would be a significant step forward in ground improvement. Hausmann (1990), while developing construction guidelines for geotextile applications in various geotechnical constructions, pointed out that simple procedures, such as pretensioning the geotextile, might enhance the reinforcement function in some applications.

Gorle and Thijs (1989) studied the effect of prestressing the geosynthetics in a two-layer model (soil–granular material) by conducting plate load tests. Special clamps were used for anchoring or prestressing the geosynthetics. Prestressing forces were varied between 15 and 8 kN/m. Releasing the anchorage of the prestressed geosynthetic resulted in a horizontal confining stress of (maximum) 20 kPa, which built up in the granular material. The results indicate that, for low-deformation soil–sand systems (California bearing ratio (CBR) of soil between 3 and 6%), the prestress in the geosynthetic limits the total settlement of the system and increases the cyclic bearing capacity, especially when the layer thickness is less than, or equal to, the radius of the loading surface. The prestressing of geosynthetics does not seem to be justified economically on the basis of the test results reported, considering the rather complicated prestressing procedure in full-scale applications.

Shukla and Chandra (1994b) studied the effect of prestressing the geosynthetic reinforcement on the settlement behaviour of a geosynthetic-reinforced granular fill–soft soil system by developing a new mechanical element, the 'stretched rough elastic membrane'. This study shows that the improvement in the settlement response increases with an increase in the prestress in the geosynthetic reinforcement within the loaded footing, and it is most significant at the centre of the loaded footing, resulting in reduction of the differential settlement of the footing (Figure 6.27).

Tatsuoka *et al.* (1997) described a new and unique construction method that aims at making reinforced soil structures, such as a geosynthetic-reinforced foundation soil to support a structure, very stiff and elastic. In this method, the prestress is introduced in the geosynthetic reinforcement by preloading the metallic tie rods that penetrate the reinforced soil and that are connected to top and bottom reaction blocks (Figure 6.28). Preloading the tie rods also causes a large reduction in the plastic deformation of the soil. This method will substantially reduce the settlement of the fill in

Figure 6.27 Settlement profiles – effect of prestressing for various load intensities ($W = w/B$; $G_t^* = G_t H_t / k_s B^2$, $G_b^* = G_b H_b / k_s B^2$ and $q^* = k_s B$)
Reproduced from Shukla and Chandra (1994b), with permission from Elsevier

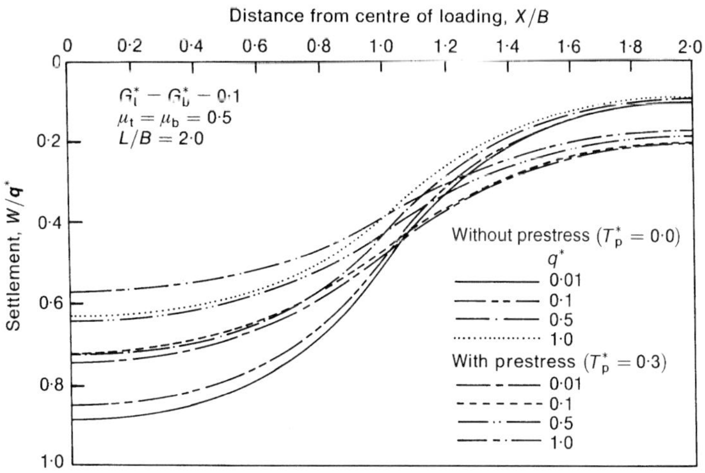

the construction of a heavy structure, thus making a pile foundation unnecessary. This method will also be effective in reducing the possible rocking motion of the structure during earthquakes.

Lovisa *et al.* (2010) investigated improvements in the settlement behaviour and load-bearing capacity of a prestressed geotextile-reinforced sand foundation using both experimental and numerical methods. The experimental programme, which involves small-scale plate load tests, was carried out using test facilities at the James Cook University. Clean, oven-dried beach sand (specific gravity = 2.62, minimum void ratio = 0.58, maximum void ratio = 0.84, effective grain size $D_{10} = 0.14$ mm, coefficient of uniformity $C_u = 2.71$, coefficient of curvature $C_c = 0.91$, peak friction angle = 39.5°, ultimate friction angle = 31°) at 50% relative density was used as a sand bed in the laboratory model tests. The sand was classified as poorly graded sand (SP), as per the Unified Soil Classification System. A single layer of woven geotextile (SG40/40) (thickness = 0.74 mm, mass per unit area = 186 g/m^2, tensile strength = 40 kN/m, elongation (MD – machine direction) = 15%, elongation (XD – cross-direction) = 8%, tearing strength (MD) = 400 N, tearing strength (XD) = 676 N, Burst strength (CBR plunger) = 4800 N, puncture resistance (drop cone) = 13 mm) was used to reinforce the sand bed. Loading tests were carried out on a circular rigid footing (thickness = 20 mm, diameter = 100 mm) fabricated from mild steel. Three different footing embedment depths of 0, 50 and 100 mm were investigated. The prestress applied was 2% of the allowable tensile strength of the geotextile (approximately 0.84 kN/m), and was distributed over three pulleys in both the x and y directions (Figure 6.29). The test tank was constructed from Perspex with the following dimensions: 0.8 m × 0.8 m in plan and 0.6 m high. Figure 6.30 shows the experimental apparatus. When the sand was poured into the model and no compaction was administered, a relative density of 18.3% was achieved. Additional compaction was applied using a 30 mm-diameter vibratory poker. For compacting soil layers near the footing and reinforcement locations, a 5 kg weight with a 200 mm-diameter base was strategically dropped from a height of 100 mm across the surface of the sand bed. It should be noted that at the prestressed geotextile level, the underlying sand was slightly mounded, so that when the prestress force was applied, the entire geotextile was in contact with the sand. The model footing was placed on the surface of the sand bed at a predetermined location in the test pit. A normal compressive load was applied in increments of 0.075 kN to the model footing by means of a mechanical jack supported against a reaction beam. Load increments were monitored using a load cell. An experimental set-up at

Figure 6.28 A heavy structure supported on a prestressed geosynthetic-reinforced fill
Reproduced from Tatsuoka *et al.* (1997), © Japanese Geotechnical Society

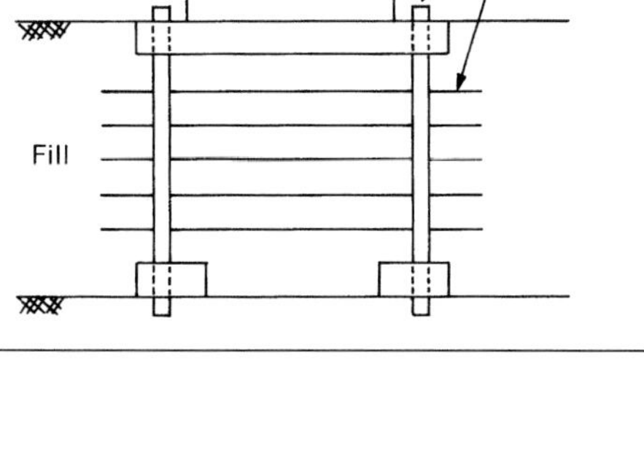

Figure 6.29 The experimental set-up: (a) placement of the prestressed geotextile layer and (b) the test tank with a prestressing arrangement
Reproduced from Lovisa *et al.* (2010), with permission from Elsevier

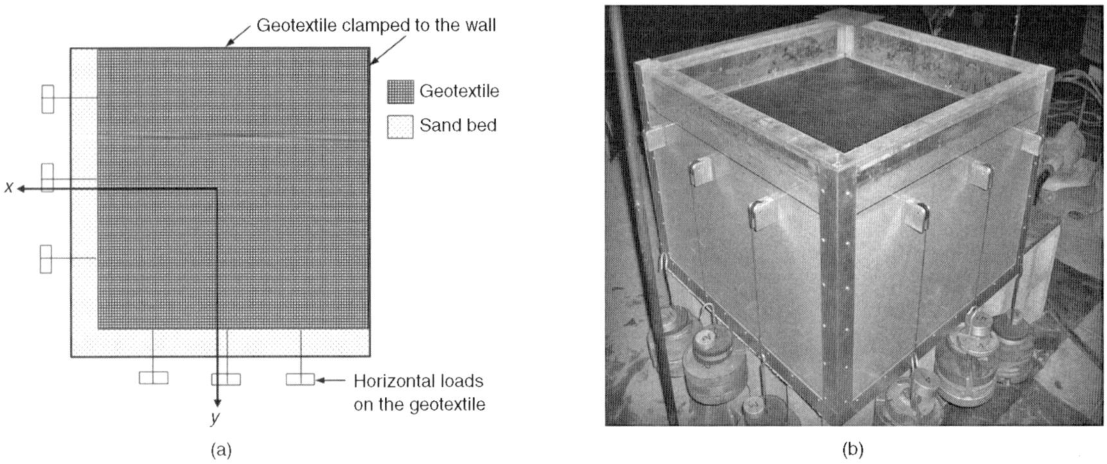

failure is shown in Figure 6.31. The settlement was measured using two dial gauges located at equal distances from the footing, and averaged to obtain the settlement at the centre of the footing. The test was concluded when the load capacity of the jack was reached, or when failure occurred within the soil as the predefined amount of displacement or tilt of the footing. The test tank was emptied and refilled for each test, to ensure standardised conditions throughout the investigation. Figure 6.32 shows pressure versus settlement curves

Figure 6.30 (a) Elevation and (b) plan of the physical model test arrangement
Reproduced from Lovisa *et al.* (2010), with permission from Elsevier

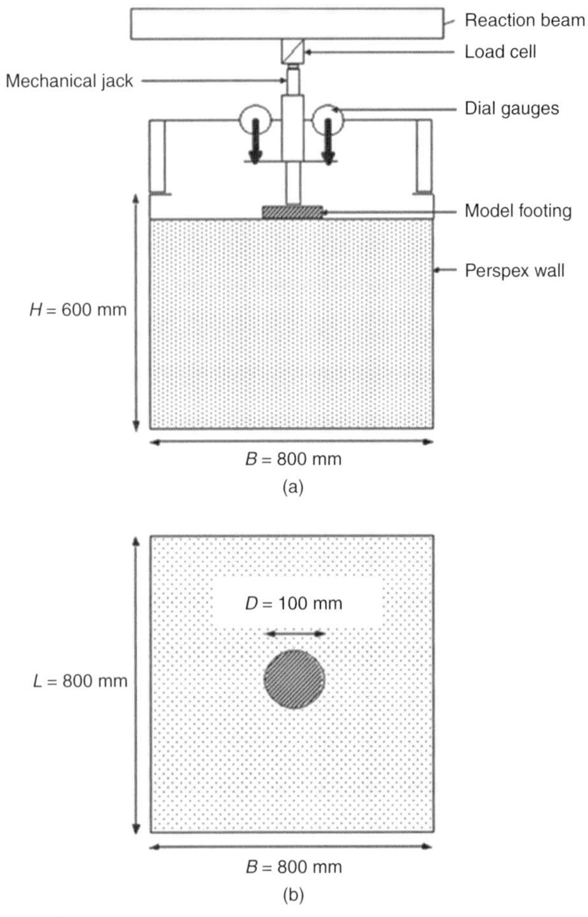

Figure 6.31 Photograph of the experimental set-up at failure (note tilting)
Reproduced from Lovisa *et al.* (2010), with permission from Elsevier

Figure 6.32 Pressure versus normalised settlement curves for the surface footing
Reproduced from Lovisa *et al.* (2010), with permission from Elsevier

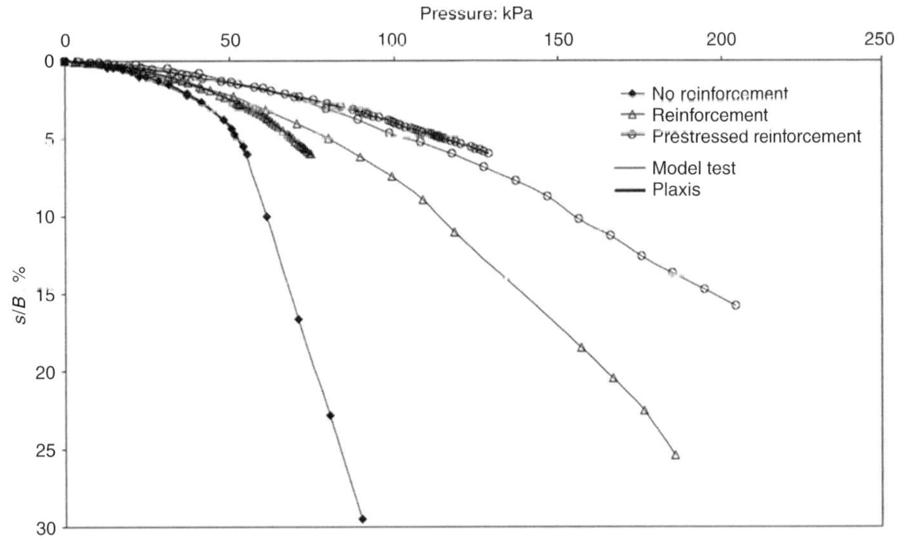

for the surface footing. Based on the test results obtained and the analysis conducted, the following conclusions can be drawn

- The addition of prestress to the geotextile reinforcement significantly improved the settlement response and the load-bearing capacity of the soil.
- The beneficial effects of geotextile reinforcement without prestress were insignificant beyond a footing embedment depth of 50 mm for low strains (i.e. small displacements). However, the addition of prestress to the geotextile reinforcement improved the settlement response and the load-bearing capacity for all footing depths.
- The ultimate bearing capacity of unreinforced sand foundations was in good agreement with the values obtained from the Meyerhof bearing capacity equations. The bearing capacity equation modification suggested in the literature to include reinforcement also produced theoretical results similar to those obtained from physical model tests.
- The relationship for the modulus of the subgrade reaction at a certain footing embedment depth suggested in the literature produced results that were in good agreement with those obtained from the physical model tests.
- Results obtained from finite element analysis using the program PLAXIS were generally in good agreement with those obtained from physical model tests. This validated the equation in the literature, relating the modulus of elasticity to the modulus of the subgrade reaction, where the parameter H was approximately 2.5–3 times the footing width.
- To achieve the benefits of prestressing the geosynthetic in field conditions, the geosynthetic should be pulled out following the laboratory method in principle, and anchored in trenches surrounding the area to be reinforced before placement of the granular fill over it. The pulling-out process may not be easy to simulate in field conditions, especially when a high prestress in the geosynthetic is required. However, it is expected that some suitable practical methods of prestressing the geosynthetic in field situations will be developed in the future.

It should be noted that the concept of prestressing the geosynthetic can also be used in other application areas of geosynthetics, especially those where strength/performance improvement is required without significant deflection/stretch of the geosynthetic layer. Shukla and Kumar (2008) have reported advantages of using a prestressed geosynthetic layer in the overall slope stability of a reinforced embankment on soft ground.

6.9. Geosynthetic deflection profiles and strain analysis

The geosynthetic layers in field applications may deflect under the loads and stresses due to surcharge, overburden, etc., when they are applied normal to their planes. The deflected shapes may be circular, parabolic or a combination thereof, depending on the ground conditions. For a geosynthetic layer spanning a cavity, a trench or any depression, a circular or parabolic shape, as shown in Figure 6.33, can be expected; however, the deflected shape of a geosynthetic layer resting on a foundation soil and subjected to a normal stress is not necessarily completely circular or parabolic. To simplify the mathematical derivations, it can be assumed to be a combination of both circular and parabolic curves, with the inflection points shown in Figure 6.34, which were observed in the settlement profiles reported in the literature (Jarrett, 1986; Miura *et al.*, 1990; Shukla and Chandra, 1995; Shukla, 1996).

Figure 6.33 The parabolic or circular shape of a deflected geosynthetic
Reproduced from Shukla and Sivakugan (2009)

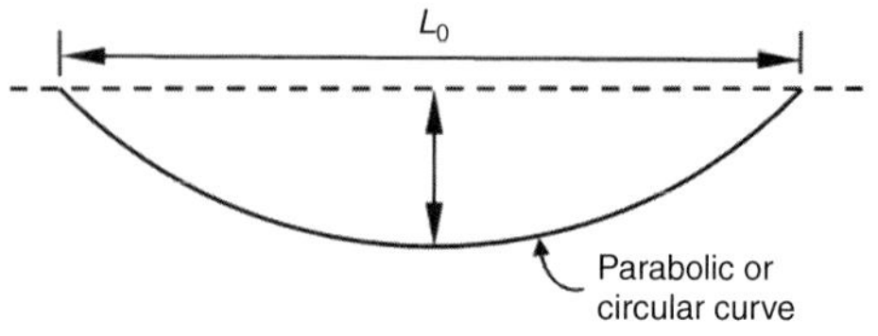

Considering the circular and parabolic concave shapes of the deflected geosynthetic with their ends at the same elevation (Figure 6.33), Giroud (1995) presented expressions for computing geosynthetic strain. Shukla (1996) observed that, in field applications, the two ends of the geosynthetic may be at different elevations, possibly due to inappropriate installation or as dictated by site conditions. For this field situation, an expression was presented by Shukla (1996) as an extension of the work of Giroud (1995). The expressions presented by Giroud (1995) and Shukla (1996) can only be applied to the situations where the deflected geosynthetic can be described by a concave parabola or a circular arc without any inflection point, as shown in Figure 6.33. For other possible deflected situations in field applications, such as that shown in Figure 6.34, where the deflected shape of the geosynthetic is described by two convex parabolas and a circular arc, Shukla and Sivakugan (2009) presented the derivation of an analytical expression for the geosynthetic strain (ε).

As shown in Figure 6.35, a horizontally installed geosynthetic layer of length L_0 within a soil mass elongates due to deflection caused by the applied normal load. If the shape of the deflected geosynthetic is a smooth curve in the cross-section considered, the elongation of the geosynthetic in the cross-section can be determined mathematically by assuming that the shape is a curve with a known equation such as a combination of parabolas (DGE and AFB) and a circle (BCD). The depth of the inflection points B and D on the deflected geosynthetic, measured from the initial elevation of the geosynthetic layer, is h. The length of the horizontal projection of the circular arc of the deflected geosynthetic is d, and r is the rut depth or the maximum deflection of the geosynthetic, which is different from the radius (R) of the circular arc BCD. It is assumed that no slippage occurs between the soil and the geosynthetic, and the geosynthetic strain is uniformly distributed. The expression for

Figure 6.34 A geosynthetic layer deflected under the applied normal stress in field applications
Reproduced from Shukla and Sivakugan (2009)

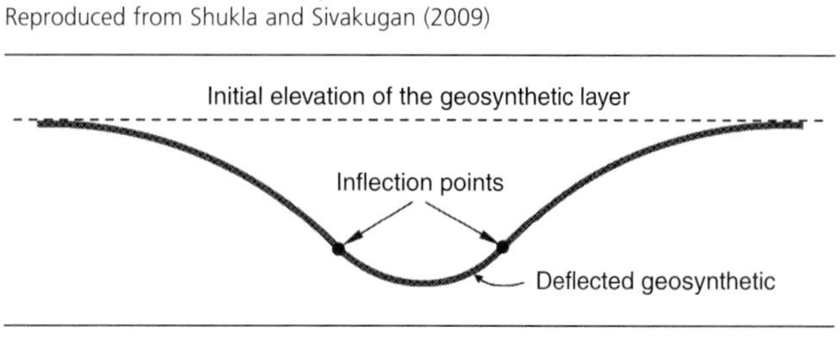

Figure 6.35 Shape of the deflected geosynthetic layer having a combination of circular and two parabolic curves
Reproduced from Shukla and Sivakugan (2009)

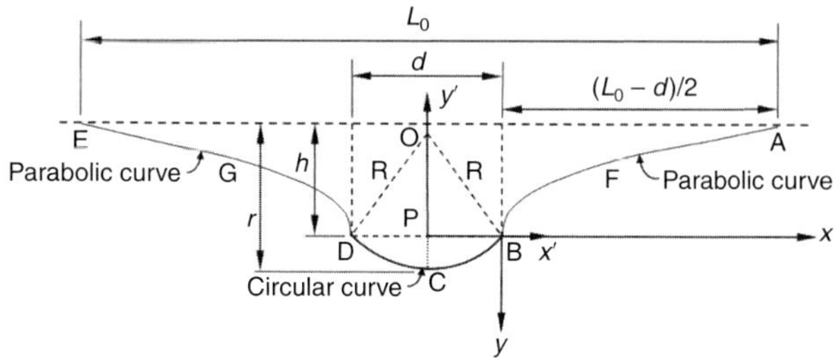

the geosynthetic strain $\varepsilon(=(L-L_0)/L_0)$ is presented as

$$\varepsilon = \frac{h}{L_0}\left\{\sqrt{1+\left(\frac{L_0-d}{h}\right)^2}+\frac{1}{\frac{L_0-d}{h}}\times \ln\left[\frac{L_0-d}{h}+\sqrt{1+\left(\frac{L_0-d}{h}\right)^2}\right]\right\}$$

$$+\frac{d}{L_0}\left(\frac{d}{4(r-h)}+\frac{r-h}{d}\right)\sin^{-1}\left(\frac{1}{\frac{d}{4(r-h)}+\frac{r-h}{d}}\right)-1 \tag{6.33}$$

Equation 6.33 is the general expression for determining the geosynthetic strain for the deflected shape of the geosynthetic with inflection points, as shown in Figure 6.35. This equation involves four geometric parameters: L_0, d, r and h. Although the ratio h/r has not been extensively reported in model studies, it has been estimated to be about 0.8 from the graphical results of Jarrett (1986). A suitable assumption of h/r will always result in appropriate values of strain for the generalised case; the confidence in the selection of h/r will improve with measurement of the same in experimental studies expected in the future.

It is important to note that consideration of a common tangent at the inflection point has not been considered in the derivation of Equation 6.33. In fact, the vertex of the parabolic curve has been assumed at the inflection point, for mathematical simplicity; therefore, the slope of the tangent to the parabola at the inflection point will never be equal to the slope of the tangent to the circular arc at this point. If this condition is considered by shifting the vertex to a suitable point, additional parameters will appear, and Equation 6.33 with the new condition will have more than four parameters with replacement of h, which is, practically, a more realistic parameter than new arbitrary parameters being introduced because of the condition of a common tangent. Since the geosynthetic remains embedded in the soil mass under the subjected loading, it can be expected that the slope of the tangents may not be exactly identical in most practical situations, and, therefore, the general

Figure 6.36 Shape of the deflected geosynthetic layer having a combination of two parabolic curves
Reproduced from Shukla and Sivakugan (2009)

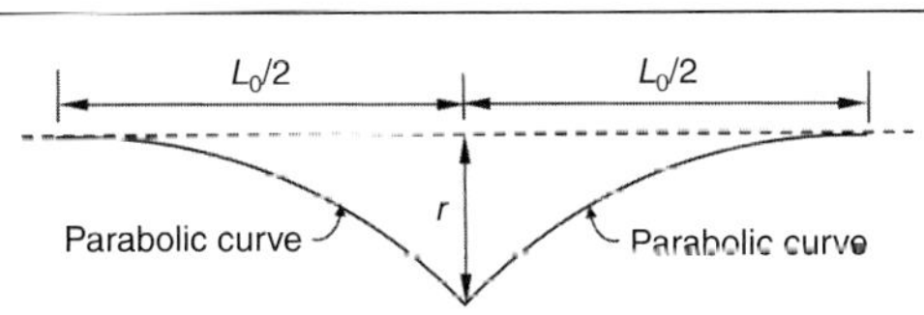

Equation 6.33 proposed herein is applicable to most field situations.

The general Equation 6.33 developed can be used to explain a few special cases discussed below.

Case 1: the geosynthetic layer is deformed into a circular arc.

For $d = L_0$, $h = 0$, the deflected shape of the curve is purely circular (Figure 6.33), with no parabolic segment present. Here, Equation 6.33 becomes

$$\varepsilon = \left(\frac{L_0}{4r} + \frac{r}{L_0}\right)\sin^{-1}\left(\frac{1}{\dfrac{L_0}{4r} + \dfrac{r}{L_0}}\right) - 1 \qquad (6.34)$$

Equation 6.34 is the same as that given by Giroud (1981), and, later, considering that the relative deflection r/L_0 is small, its approximation was presented by Giroud (1995) as

$$\varepsilon \approx \frac{8}{3}\left(\frac{r}{L_0}\right)^2 \qquad (6.35)$$

Case 2: the geosynthetic layer is deformed into a parabolic arc.

For $d = 0$, $h = r$, the deflected shape of the curve is a combination of two symmetrical convex parabolic curves (Figure 6.36), and Equation 6.33 becomes

$$\varepsilon = \sqrt{1 + \left(\frac{r}{L_0}\right)^2} + \left(\frac{r}{L_0}\right)^2 \times \ln\left[\left(\frac{L_0}{r}\right) + \sqrt{1 + \left(\frac{L_0}{r}\right)^2}\right] - 1 \qquad (6.36)$$

The deflected shape of the geosynthetic shown in Figure 6.36 may be possible when a line load is applied normal to the horizontally laid geosynthetic layer resting on very weak foundation soil.

Case 3: the geosynthetic layer is deformed into a profile consisting of circular and parabolic arcs, with $h/r = 0.8$ and $d/L_0 = 0.2$.

In the field situations, h/r and d/L_0 values may vary in some finite ranges; but for discussion of Equation 6.33, $h/r = 0.8$ and $d/L_0 = 0.2$ can be considered in the possible ranges. With these values, Equation 6.33 becomes

$$\varepsilon = \frac{4r}{5L_0}\left[\sqrt{1 + \left(\frac{L_0}{r}\right)^2} + \frac{r}{L_0}\ln\left\{\left(\frac{L_0}{r}\right) + \sqrt{1 + \left(\frac{L_0}{r}\right)^2}\right\}\right] + \frac{1}{5}\left[\frac{L_0}{4r} + \frac{r}{L_0}\right]\sin^{-1}\left[\frac{1}{\dfrac{L_0}{4r} + \dfrac{r}{L_0}}\right] \quad 1 \qquad (6.37)$$

Figure 6.37 shows the variation of geosynthetic strain (ε) with relative deflection (r/L_0) using Equations 6.34–6.37 and the equation for the parabolic shape of the deflected geosynthetic (Figure 6.33) given by Giroud (1995) as

$$\varepsilon = \frac{1}{2}\sqrt{1 + 16\left(\frac{r}{L_0}\right)^2} + \frac{1}{8}\left(\frac{r}{L_0}\right)\ln\left[4\left(\frac{r}{L_0}\right) + \sqrt{1 + 16\left(\frac{r}{L_0}\right)^2}\right] - 1 \qquad (6.38)$$

In Figure 6.37, it can be seen that the geosynthetic strain (ε) increases with an increase in the relative deflection (r/L_0), for all cases, irrespective of the shape of the deformed geosynthetics, which is expected intuitively. The two extreme cases corresponding to Equations 6.34 and 6.36 for the circular (Figure 6.33) and double convex parabolic curves (Figure 6.36) are the upper and lower limits, respectively, for the general Equation 6.33. Equation 6.37 ($h/r = 0.8$ and $d/L_0 = 0.2$), a special case of Equation 6.33, is therefore lying between the curves for Equations 6.34 and 6.36. At low strain levels, with r/L_0 less than 0.25, the approximate equation (Equation 6.35) predicts the geosynthetic strains very well, for circular and parabolic shapes as well as for the combined ones shown in Figures 6.35 and 6.36. For r/L_0 greater than approximately 0.25, the approximate Equation 6.35 predicts the geosynthetic strain values greater than the values calculated from any of the exact Equations 6.34, 6.36 or 6.38. Giroud (1995) noted that the approximate equation (Equation 6.35) can be used instead of his exact Equations 6.34 and 6.38 for low values of r/L_0 of up to 0.15. Figure 6.37 clearly shows that the approximation holds even for the situation where the deflected geosynthetics profile is approximated by parabolic and circular arcs with inflection points. In fact, the approximation holds for r/L_0 values as high as 0.25.

6.10. Field applications

Although geosynthetics have been widely used to reinforce many soil structures, the application of geosynthetics to increase the bearing capacity of shallow footings on soft foundation soil has been limited because the effect of geosynthetic reinforcement on the bearing capacity of foundation soil has not been verified quantitatively in the field, and also because geosynthetic reinforcement only shows beneficial effects after a relatively large settlement. This section deals with some applications of geosynthetics in different forms in foundation soil.

Figure 6.37 Variation in the geosynthetic strain with relative deflection
Reproduced from Shukla and Sivakugan (2009)

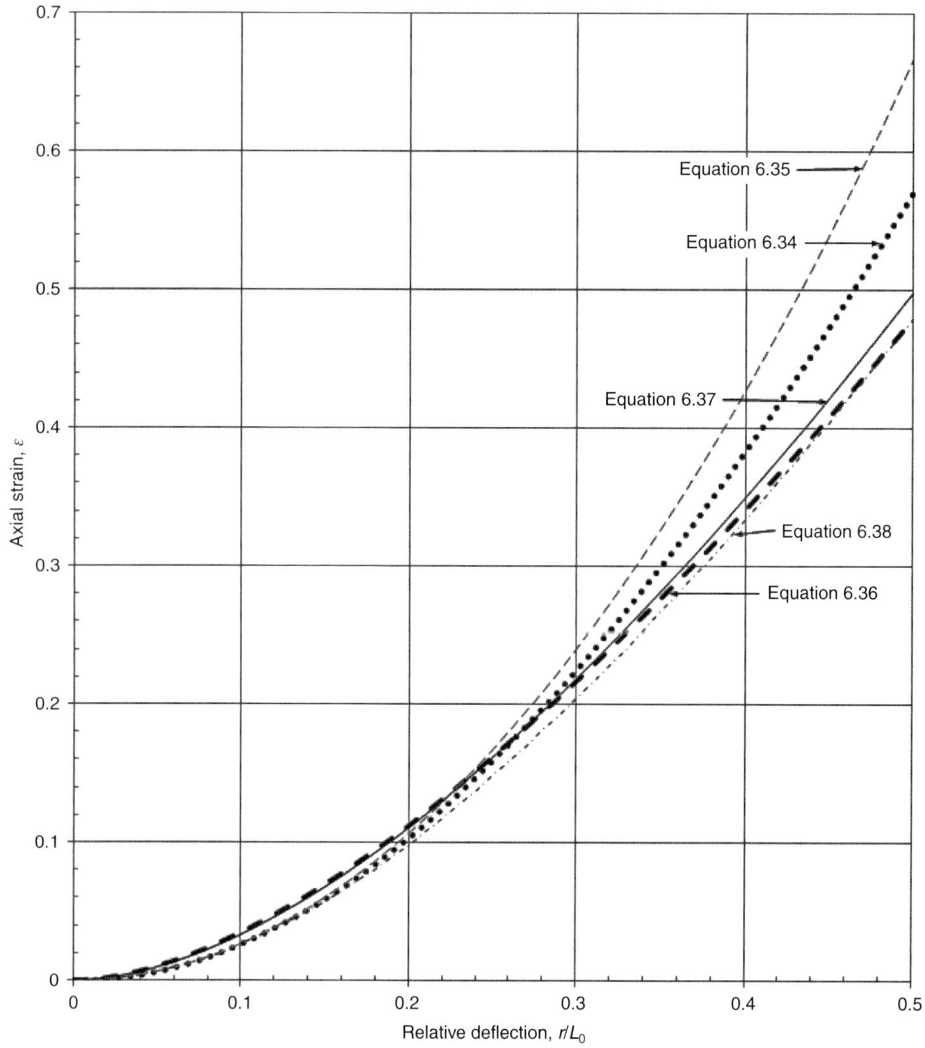

Wang *et al.* (1993) reported the use of a mat foundation made of aggregate, in geotextile bags, together with sand drains for supporting a gas-holder steel tank (volume = 1000 m^3, diameter = 15.5 m, height = 7.85 m, pressure at the base when filled with water = 112.8 kPa) at a site of muck clay, about 14 m thick, in China. The outline of the foundation treatment is shown in Figure 6.38. The geotextile was a knitted fabric of polystyrene, and its tensile strength was 16 kN/m. The sand drains have a diameter of 400 mm, intervals of 2.5 m and a depth of 16 m. The top of the sand drain is connected to a sand mat 200 mm thick. Above the sand drain is the mat, made of aggregate in geotextile bags. Based on the performance study, the following generalised conclusions were drawn

- Mat made of aggregate in geotextile bags can diffuse loads and prevent shallow soil failures more effectively than sand mats.
- Mat made of aggregate in geotextile bags can reduce settlement, control unequal settlement and improve the stability of a foundation.
- Mat made of aggregate in geotextile bags has the advantage of convenient construction and low cost.

Arman and Griffin (1993) reported the use of geogrid reinforcement for aircraft parking and taxiing areas in Europe for the US Air Force. The site was cleared of grass, and the sand was removed to a depth of 150 mm and stockpiled close to the site. The geogrid was placed in two layers: the first layer was installed 150 mm below grade, and the second layer was placed 75 mm below grade. The sand was then replaced over each layer of the grid and smoothed manually using ordinary garden rakes. The grid was lapped 600 mm longitudinally, and tied using plastic wire ties to hold it in place while the soil was being placed over it. The edges of the geogrids were held in place by tent stakes. Based on the performance study, the following conclusions were drawn

- Where soil conditions permit, a geogrid is a viable means of providing taxiways and parking stands, which can be constructed very quickly and inexpensively using unskilled labour.

Figure 6.38 Use of geotextile aggregate mat for supporting a gas-holder tank
Reproduced from Wang et al. (1993)

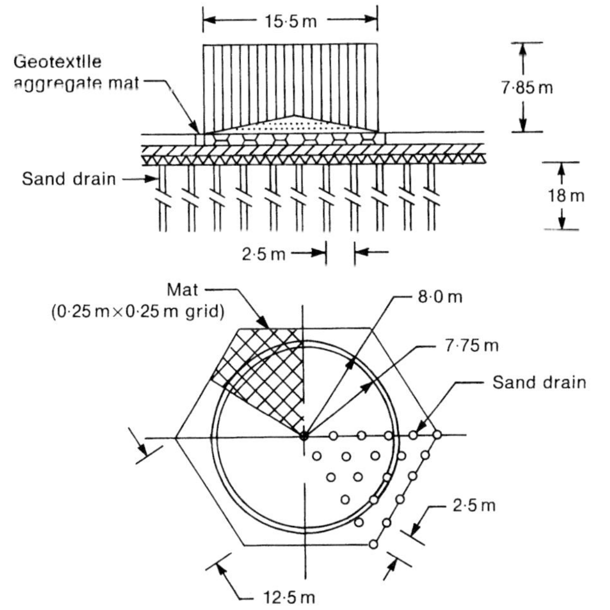

- Geogrids can be used to revive redundant aircraft facilities or to replace combat-damaged facilities much more quickly than ordinary construction methods.
- Since native soil is used as the final cover, the taxiways and parking areas can be made virtually undetectable, except perhaps by infrared detection devices.

Holtz and Massarsch (1993) reported the use of a multifilament woven polyester geotextile to carry horizontal forces in a bridge approach embankment supported on vertically driven relief piles (Figure 6.39) (construction carried out in Sweden in 1972). Three layers of geotextile were placed, with about 150 mm of compacted sand between each layer. The geotextile was handled very easily in the construction field: it was simply rolled out over the compacted subgrade fill. About 300 mm of overlap was used between each 1.7 m-wide strip. The performance study revealed that the geotextile used to reinforce the bridge approach embankment on the soft sensitive clays effectively reduced the horizontal movements and, thereby, probably prevented serious settlements and instability. The installation was significantly cheaper than a longer bridge structure.

Risseeuw and Voskamp (1993) reported the use of reinforcing geotextiles along with vertical drains to support an embankment in Hong Kong (Figure 6.40) (construction carried out in 1986 and 1987). Reinforcing geotextiles were used to assure the stability of the embankment built on top of the vertical drains during the consolidation process. It also allowed for fast construction of the embankment. The reinforcing geotextile was required to have an ultimate

Figure 6.39 Use of geotextile layers for supporting an embankment
Reproduced from Holtz and Massarsch (1993)

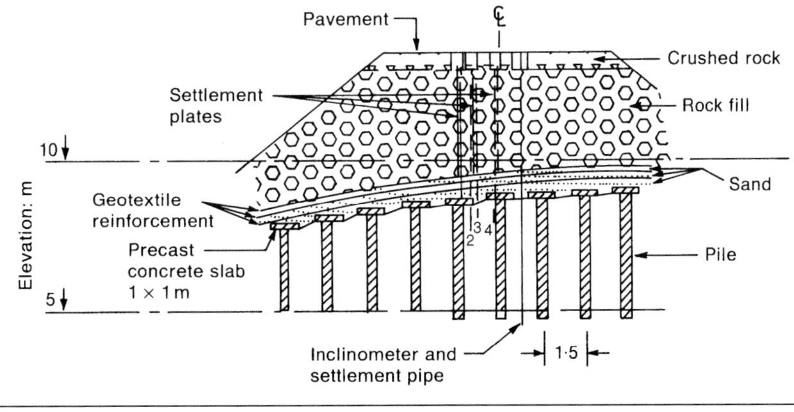

Figure 6.40 Use of geotextile layers for supporting an embankment
Reproduced from Risseeuw and Voskamp (1993)

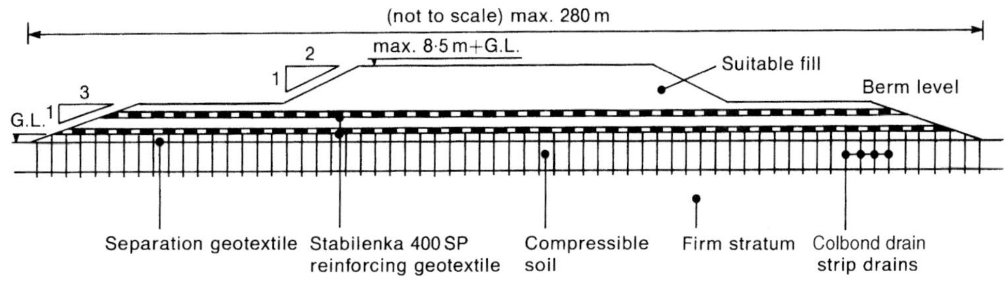

Figure 6.41 Use of a polymer geogrid in a street subgrade
Reproduced from Tsukada *et al.* (1993)

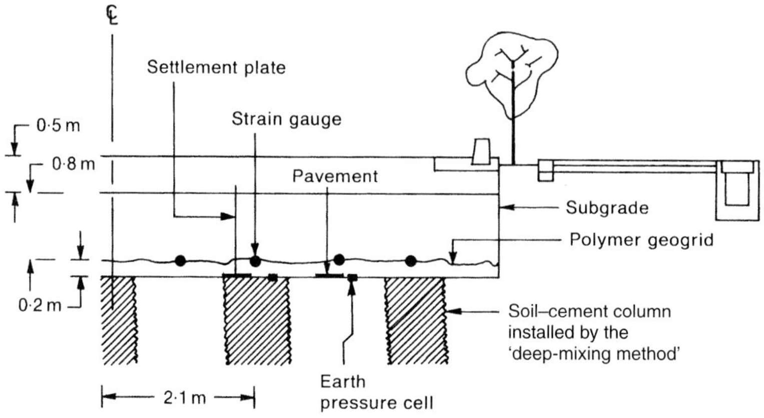

strength of 400 kN/m at a maximum elongation of 12%, while creep was limited to 2% in two years at a 133 kN/m force level.

Tsukada *et al.* (1993) reported the use of a polymer geogrid in the subgrade of a street along with soil–cement columns (Figure 6.41) (construction carried out in 1985). On the basis of the performance of the subgrade, the following conclusions were drawn

- The polymer geogrid is effective in preventing differential settlement between the soil–cement columns and the soil located between the columns.
- Subgrade rigidity increases with the placement of the polymer geogrid. As a result, vertical pressure transmitted to the subsoil between the columns reduces.
- The use of more layers of the polymer geogrid, or the use of a mattress-type geogrid, compares better to the use of one layer of geogrid in terms of the improvement of the subgrade rigidity.

Robertson and Gilchrist (1987) reported the selection of the geocell foundation mattress for Auchenhowie Road in Scotland as the most cost-effective and practical way of constructing a 4 m-high embankment over a drained lake bed, where the foundation soils comprised 4 m of soft, silty clay with an average undrained shear strength of 15 kPa overlying mudstone. The selection of the geocell mattress solution was made after an economic appraisal of both the 'excavation and replacement'. Other methods had been discounted because they were either impractical or would take too long to construct.

A geocell mattress formed part of a trial embankment on the Panci Toll Road Project, Bandung, Indonesia, and the preliminary results of the trial were reported by Oliver and Younger (1988). The performance of the geocell mattress, compared with horizontal layers of reinforcement, showed that the geocell mattress section had settled 33% less after four months under a 6.2 m-high embankment. Performance data show reduced differential settlements and reduced total settlements due to the load-spreading ability of the rigid geocell foundation mattress.

Cowland and Wong (1993) reported the use of geocell mattress foundations for two portions of an embankment constructed on very soft clays in Hong Kong. These

Figure 6.42 Construction method to stabilise very soft subgrades: (a) placement of the geofabric, (b) placing of stabilising berms, (c) placement of fill and (d) widening of berms
Reproduced from Broms (1987), with permission from Elsevier

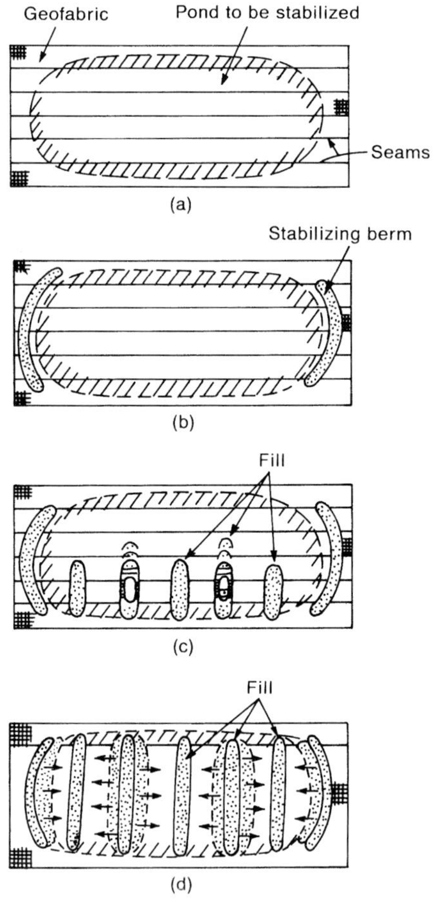

foundations essentially performed as plastic-reinforced rockfill rafts.

Broms (1987) described a method to stabilise very soft clay using a geofabric. The stabilising effect of the fabric is illustrated in Figure 6.42. This method was applied both in Malaysia and in Singapore with satisfactory results. A geotextile was used in Kuala Lumpur, Malaysia, to stabilise an 18 m-thick layer of very soft, silty clay in a settling pond associated with tin mining in the area, so that apartment buildings up to five storeys high could be constructed without excessive settlements. It was suggested that the fabric should be stretched as much as possible before stabilising berms are placed along the perimeter of the geofabric sheet, in order to limit the penetration required to develop the necessary tension in the fabric.

Toh *et al.* (1994) reported the use of a geotextile–bamboo fascine mattress foundation for filling over very soft deposits, such as slimes and peat soils, in Malaysia. The application was successful without any of the problems involved in mixing the fill with soft deposits, remoulding, mud waves, and general loss of control of the filling process. It is suggested that that the geotextile used should consist of the following properties: high extensibility and permeability; high resistance to bursting; and able to mitigate tear or puncture. Care should be exercised when sewing the geotextile sheets, to ensure a high level of seam efficiency.

6.11. Summary of main points

- Geosynthetic-reinforced foundations consist of one or more layers of geosynthetic reinforcement within the controlled soils/granular fills. Their performance depends not only on the properties of soil and geosynthetic reinforcement, but also on their interaction.
- A geosynthetic layer installed within the foundation soil in order to improve its load-carrying capacity and settlement characteristics can have several effects, such as the shear stress reduction effect, the confinement effect, the membrane effect, the anchoring effect and the separation effect.
- Geosynthetic-reinforced foundations can fail in four different modes: bearing capacity failure of soil above the uppermost geosynthetic reinforcement layer, geosynthetic pull-out, geosynthetic break and geosynthetic creep failure.
- For optimum effect, a geosynthetic layer should be placed at the soil–granular fill interface in the case of a thin layer of fill, otherwise near to the midpoint of a granular fill layer. Moreover, the zone of such placement should not involve high elastic tensile strains in the geosynthetic layer. Under these conditions, the geosynthetic layer can be highly effective in reinforcing the foundation soil and can thereby extend the life of structures.
- Small-scale model footing tests are always subjected to scale effects resulting in a larger value of the ultimate bearing capacity. In addition, these tests use full-scale geosynthetics. It would be helpful if geosynthetics could be modelled and scaled for compatibility.
- Though small-scale model footing tests and the mathematical modelling approaches described in this chapter have some shortcomings, they are valuable techniques in predicting trends in behaviour, in interpolating between results of full-scale trials and in understanding the mechanisms involved.
- Mathematical modelling is valuable in reducing the numbers of expensive full-scale trials that need to be carried out. Full-scale tests and field trials are, of course, essential to validate such models.
- The currently available methods of analysis for the load-bearing capacity and the settlement of geosynthetic-reinforced foundation soil do not consider explicitly the benefits from anchoring and the separation effects of geosynthetics, and, therefore, further studies in this direction are required.
- To make geosynthetics more effective at smaller deformations, they should be prestressed, and this area needs research, especially for the development of feasible and economical field methods for prestressing.
- The strain of the deflected geosynthetic, as per the field situations, can be estimated using analytical expressions proposed by researchers.

6.12. Some common questions and their answers

Questions

1. What is the practical reinforcement pattern beneath a footing?

2. What is the optimum vertical spacing of geosynthetic reinforcement layers?

3. What is the maximum depth of geosynthetic reinforcement layers?

4. Does a geogrid-reinforced foundation perform better than a geotextile-reinforced foundation?

5. How do geocell reinforcements improve the load-bearing capacity of foundation soils?

6. Do geonaturals (coir/jute meshes) as reinforcement layers within foundation soils improve their load-bearing capacity and reduce settlements?

7. Is numerical/finite element modelling of geosynthetic-reinforced foundation soils a complex/challenging task?

8. What is the major benefit from prestressing the geosynthetic for foundation applications?

9. Should a footing resting on a geosynthetic-reinforced foundation soil be analysed as a soil–structure interaction problem?

Answers

1. Reinforcements are installed horizontally in one or more parallel layers.

2. The experimental studies show that the optimum vertical spacing of geosynthetic reinforcement layers varies from $0.2B$ to $0.5B$, where B is the width/diameter of the footing.

3. The maximum depth of reinforcement layers varies from 1.0*B* to 2.0*B*, to achieve improvements in load-bearing capacity economically.

4. Yes, this has been observed in experimental studies. The main reason is that the tensile modulus of a geogrid is generally greater than that of a geotextile.

5. The geocell reinforcement arrests the lateral spreading of the infill soil by all-round confinement within its cells/pockets, and creates a stiffened mattress to support the footing with a redistribution of the footing load over a wider area, thus giving rise to an increase in the load-bearing capacity of the foundation soil.

6. Yes, but they result in short-term bearing capacity improvements and settlement reductions. Such reinforced soil beds can be used as reinforced platforms for heavy machinery, which are used for temporary works in engineering and other general projects.

7. Yes, the complexity of the modelling process is caused mainly by difficulty in modelling the true soil–geosynthetic interaction behaviour.

8. Increased load-carrying capacity and reduced total and differential settlements of the footing.

9. Yes.

REFERENCES

Abdel-Baki MS and Raymond GP (1994) Numerical analysis of geosynthetic reinforced soil slabs. *Proceedings of the International Conference on Geotextiles*, Geomembranes and Related Products, Singapore, pp. 317–320.

Aboshi H (1984) Soil improvement techniques in Japan. *Proceedings of the Seminar on Soil Improvement and Construction Techniques in Soft Ground*, Singapore, pp. 3–16.

Adams MT and Collin JG (1997) Large model spread footing load tests on geosynthetic reinforced soil foundations. *Journal of Geotechnical and Geoenvironmental Engineering* **123(1)**: 66–72.

Akinmusuru JO and Akinbolade JA (1981) Stability of loaded footings on reinforced soil. *Journal of Geotechnical Engineering*, ASCE **107(6)**: 819–827.

Andrawes KZ, McGown A, Wilson-Fahmy RF and Mashhour MM (1982) The finite element method of analysis applied to soil–geotextile systems. *Proceedings of the 2nd International Conference on Geotextiles*, Las Vegas, NV, pp. 695–700.

Arman A and Griffin PMJ (1993) Geogrid reinforcement for aircraft parking and taxiing facility areas spot locations, Europe. In *Geosynthetics Case Histories* (Raymond GP and Giroud JP (eds)). ISSMFE Technical Committee TC9, Geotextiles and Geosynthetics, London, pp. 198–199.

Barenberg EJ (1980) *Design Procedures for Soil–fabric–aggregate Systems with Mirafi 500X Fabric*. Department of Civil Engineering, University of Illinois at Urbana, Champaign, IL.

Bassett RH and Last NC (1978) Reinforcing earth below footings and embankments. *Proceedings of the Symposium on Earth Reinforcement*. ASCE, New York, pp. 202–231.

Basudhar PK, Saha S and Deb L (2007) Circular footings resting on geotextile-reinforced sand bed. *Geotextiles and Geomembranes* **25(6)**: 377–384.

Bathurst RJ and Jarrett PM (1988) Large-scale model tests of geocomposite mattresses over peat subgrades. *Transportation Research Record* **1188**: 28–36.

Binquet J and Lee KL (1975a) Bearing capacity tests on reinforced earth slabs. *Journal of Geotechnical Engineering Division*, ASCE **101(12)**: 1241–1255.

Binquet J and Lee KL (1975b) Bearing capacity analysis of reinforced earth slabs. *Journal of the Geotechnical Engineering Division*, ASCE **101(12)**: 1257–1276.

Bolton MD (1979) *A Guide to Soil Mechanics*. Macmillan Press, London.

Bourdeau PL (1989) Modeling of membrane action in a two-layer reinforced soil system. *Computers and Geotechnics* **7**: 19–36.

Bourdeau PL, Harr ME and Holtz RD (1982) Soil–fabric interaction – an analytical model. *Proceedings of the 2nd International Conference on Geotextiles*, Las Vegas, NV, pp. 387–391.

Broms BB (1987) Stabilization of very soft clay using geofabric. *Geotextiles and Geomembranes* **5**: 17–28.

Bush DI, Jenner CG and Bassett RH (1990) The design and construction of geocell foundation mattress supporting embankments over soft ground. *Geotextiles and Geomembranes* **9**: 83–98.

Cowland JW and Wong SCK (1993) Performance of a road embankment on soft clay supported on a geocell mattress foundation. *Geotextiles and Geomembranes* **12**: 687–705.

Dash SK, Krishnaswamy NR and Rajagopal K (2001) Bearing capacity of strip footings supported on geocell reinforced sand. *Geotextiles and Geomembranes* **19(4)**: 235–256.

Dash SK, Rajagopal K and Krishnaswamy NR (2007) Behaviour of geocell reinforced beds under strip loading. *Canadian Geotechnical Journal* **44**: 905–916.

Dash SK, Sireesh S and Sitharam TG (2003) Model studies on circular footing supported on geocell reinforced sand underlain by soft clay. *Geotextiles and Geomembranes* **21(4)**: 197–219.

Dixit RK and Mandal JN (1993) Bearing capacity of geosynthetic-reinforced soil using variational method. *Geotextiles and Geomembranes* **12**: 543–566.

Douglas RA (1987) Modelling geotextile behaviour in thin access road fills over peat subgrades. *Proceedings of the 40th Canadian Geotechnical Conference*, Regina, Saskatchewan, pp. 111–120.

Espinoza RD (1994) Soil–geotextile interaction: evaluation of membrane support. *Geotextiles and Geomembranes* **13**: 281–293.

Espinoza RD and Bray JD (1995) An integrated approach to evaluating single-layer reinforced soils. *Geosynthetics International* **2(4)**: 723–739.

Floss R and Gold G (1994) Causes for the improved bearing behaviour of the reinforced two-layer system. *Proceedings of the 5th International Conference on Geotextiles, Geomembranes and Related Products*, Singapore, pp. 147–150.
Fragaszy RJ and Lawton E (1984) Bearing capacity of reinforced sand sub-grades. *Journal of Geotechnical Engineering*, ASCE **110(10)**: 1500–1507.
Ghosh A, Ghosh A and Bera AK (2005) Bearing capacity of square footing on pond ash reinforced with jute-geotextile. *Geotextiles and Geomembranes* **23(2)**: 144–173.
Ghosh C (1991) *Modelling and analysis of reinforced foundation beds*. PhD thesis, Department of Civil Engineering, Indian Institute of Technology, Kanpur.
Ghosh C and Madhav MR (1994) Settlement response of a reinforced shallow earth bed. *Geotextiles and Geomembranes* **13(5)**: 643–656.
Giroud JP (1995) Determination of geosynthetic strain due to deflection. *Geosynthetics International* **2(3)**: 635–641.
Giroud JP and Noiray L (1981) Geotextile-reinforced unpaved road design. *Journal of the Geotechnical Division*, ASCE **107**: 1233–1254.
Giroud JP, Ah-Line C and Bonaparte R (1984) Design of unpaved roads and trafficked areas with geogrids. *Proceedings of the Symposium on Polymer Grid Reinforcement in Civil Engineering*, London, Paper No. 4.1.
Gorle D and Thijs M (1989) Geosynthetic-reinforced granular materials. *Proceedings of the 12th International Conference on Soil Mechanics and Foundation Engineering*, Rio de Janeiro, pp. 715–718.
Guido VA, Biesiadecki GL and Sullivan MJ (1985) Bearing capacity of a geotextile-reinforced foundation. *Proceedings of the 11th International Conference on Soil Mechanics and Foundation Engineering*, San Francisco, CA, pp. 1777–1780.
Guido VA, Dong KG and Sweeny A (1986) Comparison of geogrid and geotextile reinforced earth slabs. *Canadian Geotechnical Journal* **23(1)**: 435–440.
Hausmann MR (1990) *Engineering Principles of Ground Modification*. McGraw-Hill, New York.
Holtz RD and Massarsch KR (1993) Geotextile and relief piles for deep foundation improvement embankment near Goteborg, Sweden. In *Geosynthetics Case Histories* (Raymond GP and Giroud JP (eds)). ISSMFE Technical Committee TC9, Geotextiles and Geosynthetics, London, pp. 168–169.
Huang CC and Menq FY (1997) Deep-footing and wide-slab effects in reinforced sandy ground. *Journal of Geotechnical and Geoenvironmental Engineering* **123(1)**: 30–36.
Huang CC and Tatsuoka F (1988) Prediction of bearing capacity in level sandy ground reinforced with strip reinforcement. *Proceedings of the International Geotechnical Symposium on Theory and Practice of Earth Reinforcement*, Fukuoka, pp. 191–196.
Huang CC and Tatsuoka F (1990) Bearing capacity of reinforced horizontal sandy ground. *Geotextiles and Geomembranes* **9**: 51–82.
Ingold TS and Miller KS (1982) Analytical and laboratory investigations of reinforced clay. *Proceedings of the 2nd International Conference on Geotextiles*, Las Vegas, NV, pp. 587–592.
Jarrett PM (1984) Evaluation of geogrids for construction of roadways over muskeg. *Proceedings of the Symposium on Polymer Grid Reinforcement in Civil Engineering*, London, Paper No. 4.5.
Jarrett PM (1986) Load tests on geogrid reinforced gravel fills constructed on peat subgrades. *Proceedings of the 3rd International Conference on Geotextiles*, Vienna, pp. 87–89.
Jewell RA (1988) The mechanics of reinforced embankments on soft soils. *Geotextiles and Geomembranes* **7(4)**: 237–273.
Ju JW, Son SJ, Kim JY and Jung IG (1996) Bearing capacity of sand foundation reinforced by geonet. *Proceedings of the International Symposium on Earth Reinforcement*, Fukuoka, pp. 603–608.
Khing KH, Das BM, Puri VK, Cook EE and Yen SC (1992) Bearing capacity of two-closely spaced strip foundations on geogrid reinforced sand. *Proceedings of the Earth Reinforcement Practice*, Fukuoka, pp. 619–624.
Khing KH, Das BM, Puri VK, Yen SC and Cook EE (1994) Foundation on strong sand underlain by weak clay with geogrid at the interface. *Geotextiles and Geomembranes* **13**: 199–206.
Kim SI and Cho SD (1988) An experimental study on the contribution of geotextiles to bearing capacity of footings on weak clays. *Proceedings of the International Geotechnical Symposium on Theory and Practice of Earth Reinforcement*, Fukuoka, pp. 215–220.
Koerner RM (1990) *Designing with Geosynthetics*, 2nd edn. Prentice Hall, Upper Saddle River, NJ.
Koga K, Aramaki G and Valliappan S (1988) Finite element analysis of grid reinforcement. *Proceedings of the International Geotechnical Symposium on Theory and Practice of Earth Reinforcement*, Fukuoka, pp. 407–411.
Krishnaswamy NR, Rajagopal K and Latha GM (2000) Model studies on geocell supported embankments constructed over a soft clay foundation. *Geotechnical Testing Journal*, ASTM **23(1)**: 45–54.
Love JP, Burd HJ, Milligan GWE and Houlsby GT (1987) Analytical and model studies of reinforcement of a layer of granular fill on soft clay subgrade. *Canadian Geotechnical Journal* **24**: 611–622.
Lovisa J, Shukla SK and Sivakugan N (2010) Behaviour of prestressed geotextile-reinforced sand bed supporting a loaded circular footing. *Geotextiles and Geomembranes* **28(1)**: 23–32.
McGown A and Andrawes KZ (1977) The influence of non-woven fabric inclusions on the stress–strain behaviour of a soil mass. *Proceedings of the International Conference on the Use of Fabrics in Geotechnics*, Paris, pp. 161–166.
McGown A, Andrawes KZ and Al-Hasani MM (1978) Effect of inclusion properties on the behaviour of sand. *Géotechnique* **28(3)**: 327–346.
Madhav MR and Poorooshasb HB (1988) A new model for geosynthetic-reinforced soil. *Computers and Geotechnics* **6**: 277–290.

Madhav MR and Poorooshasb HB (1989) Modified Paternak model for reinforced soil. *Mathematical and Computational Modelling, an International Journal* **12**: 1505–1509.

Mandal JN and Sah HS (1992) Bearing capacity tests on geogrid-reinforced clay. *Geotextiles and Geomembranes* **11**: 327–333.

Manjunath VR and Dewaikar DM (1994) Model footing tests on geofabric reinforced granular fill overlying soft clay. *Proceedings of the 5th International Conference on Geotextiles, Geomembranes and Related Products*, Singapore, pp. 327–330.

Matsuoka H (1976) On the significance of the spatial mobilized plane. *Soils and Foundations* **16(1)**: 91–100.

Milligan GWE and Fannin J (1986) Model and full-scale tests of granular layers reinforced with a geogrid. *Proceedings of the 3rd International Conference on Geotextiles*, Vienna, pp. 61–65.

Milligan GWE and Love JP (1984) Model testing of geogrids under an aggregate layer in soft ground. *Proceedings of the Symposium on Polymer Grid Reinforcement in Civil Engineering*. Institution of Civil Engineers, London, Paper No. 4.2.

Milligan GWE, Jewel RA, Houlsby GT and Burd HJ (1989) A new approach to the design of unpaved roads – part 1. *Ground Engineering*, 25–29.

Miura N, Sakai A, Taesiri Y, Yamanouchi T and Yasuhara K (1990) Polymer grid reinforced pavement on soft clay grounds. *Geotextiles and Geomembranes* **9(1)**: 99–123.

Nishida K and Nishigata T (1994) The evaluation of separation function for geotextiles. *Proceedings of the 5th International Conference on Geotextiles, Geomembranes and Related Products*, Singapore, pp. 139–142.

Ochiai H, Tsukamoto Y, Hayashi S, Otani J and Ju JW (1994) Supporting capability of geogrid-mattress foundation. *Proceedings of the 5th International Conference on Geotextiles, Geomembranes and Related Products*, Singapore, pp. 321–326.

Oliver THL and Younger JS (1988) Embankment construction over soft ground using geogrid reinforcement techniques. *Proceedings of the Roads, Highways and Bridges Conference*, Hong Kong.

Omar MT, Das BM, Puri VK and Yen SC (1993) Ultimate bearing capacity of shallow foundations on sand with geogrid reinforcement. *Canadian Geotechnical Engineering* **30**: 545–549.

Otani J, Ochiai H and Yamamoto K (1998) Bearing capacity analysis of reinforced foundations on cohesive soil. *Geotextiles and Geomembranes* **16**: 195–206.

Patra CR, Das BM, Bhoi M and Shin EC (2006) Eccentrically loaded strip foundation on geogrids-reinforced sand. *Geotextiles and Geomembranes* **24(4)**: 254–259.

Poorooshasb HB (1989) Analysis of geosynthetic-reinforced soil using a simple transform function. *Computers and Geotechnics* **8**: 289–309.

Poran CJ, Herrmann LR and Romstad KM (1989) Finite element analysis of footings on geogrid-reinforced soil. *Proceedings of the Geosynthetics '89 Conference*, San Diego, CA, pp. 231–242.

Raumann G (1982) Geotextiles in unpaved roads: design considerations. *Proceedings of the 2nd International Conference on Geotextiles*, Las Vegas, NV, pp. 417–422.

Risseeuw P and Voskamp W (1993) Geotextile and vertical drains for deep foundation improvement, Lok Ma Chao, Hong Kong. In *Geosynthetics Case Histories* (Raymond GP and Giroud JP (eds)). ISSMFE Technical Committee TC9, Geotextiles and Geosynthetics, London, pp. 160–161.

Robertson J and Gilchrist AJT (1987) Design and construction of a reinforced embankment across soft lakebed deposits. *Proceedings of the International Conference on Foundations and Tunnels*, London, pp. 84–92.

Rowe RK and Soderman KL (1987) Stabilization of very soft soils using high strength geosynthetics: the role of finite element analyses. *Geotextiles and Geomembranes* **6**: 53–80.

Sakti JP and Das BM (1987) Model tests for strip foundation on clay reinforced with geotextile layers. *Transportation Research Record* **1153**: 40–45.

Schlosser F and Vidal H (1969) Reinforced earth. *Bulletin de Liaison des Laboratoires des Ponts et Chaussees*, No. 41.

Schlosser F, Jacobsen HM and Juran I (1983) Soil reinforcement. General report. *Proceedings of the 8th European Conference on Soil Mechechanics and Foundation Engineering*, Helsinki, pp. 83–103.

Sellmeijer JB (1990) Design of geotextile reinforced unpaved roads and parking areas. *Proceedings of the 4th International Conference on Geotextiles, Geomembranes and Related Products*, The Hague, pp. 177–182.

Sellmeijer JB, Kenter CJ and Van den Berg C (1982) Calculation method for fabric reinforced road. *Proceedings of the 2nd International Conference on Geotextiles*, Las Vegas, NV, pp. 393–398.

Selvadurai APS (1979) *Elastic Analysis of Soil–foundation Interaction*. Elsevier, Amsterdam.

Sharma R, Chen Q, Abu-Farsakh M and Yoon S (2009) Analytical modelling of geogrid reinforced soil foundation. *Geotextiles and Geomembranes* **27(1)**: 63–72.

Shin EC and Das BM (2000) Experimental study of bearing capacity of a strip foundation on geogrid-reinforced sand. *Geosynthetics International* **7(1)**: 59–71.

Shukla SK (1996) Discussion of 'Determination of geosynthetic strain due to deflection' by JP Giroud. *Geosynthetics International* **3(1)**: 141–144.

Shukla SK (2004) Discussion of 'Applications of geosynthetics for soil reinforcement' by M. I. M. Pinto. *Ground Improvement* **8(4)**: 179–181.

Shukla SK and Chandra S (1994a) A generalized mechanical model for geosynthetic-reinforced foundation soil. *Geotextiles and Geomembranes* **13**: 813–825.

Shukla SK and Chandra S (1994b) The effect of prestressing on the settlement characteristics of geosynthetic-reinforced soil. *Geotextiles and Geomembranes* **13**: 531–543.

Shukla SK and Chandra S (1995) Modelling of geosynthetic-reinforced engineered granular fill on soft soil. *Geosynthetics International* **2(3)**: 603–618.

Shukla SK and Chandra S (1996) A study on a new mechanical model for foundations and its elastic settlement response. *International Journal for Numerical and Analytical Methods in Geomechanics* **20**: 595–604.

Shukla SK and Chandra S (1998) Time-dependent analysis of axi-symmetrically loaded reinforced granular fill on soft subgrade. *Indian Geotechnical Journal* **28**: 195–213.

Shukla SK and Kumar R (2008) Overall slope stability of the prestressed geosynthetic-reinforced embankment on soft ground. *Geosynthetics International* **15(2)**: 165–171.

Shukla SK and Sivakugan N (2009) A general expression for geosynthetic strain due to deflection. *Geosynthetics International* **16(5)**: 402–407.

Shukla SK, Sivakugan N and Das BM (2009) Fundamental concepts of soil reinforcement – an overview. *International Journal of Geotechnical Engineering* **3(3)**: 329–342.

Tatsuoka F, Uchimura T and Tateyama M (1997) Preloaded and prestressed reinforced soil. *Soils and Foundations* **37(3)**: 79–94.

Terzaghi K (1943) *Theoretical Soil Mechanics*. Wiley, New York.

Toh CT, Chee SK, Lee CH and Wee SH (1994) Geotextile-bamboo fascine mattress for filling over very soft soils in Malaysia. *Geotextiles and Geomembranes* **13**: 357–369.

Tsukada Y, Isoda T and Yamanouchi T (1993) Geogrid subgrade reinforcement and deep foundation improvement, Yono City, Japan. In *Geosynthetics Case Histories* (Raymond GP and Giroud JP (eds)). ISSMFE Technical Committee TC9, Geotextiles and Geosynthetics, London, pp. 158–159.

Vesic AS (1973) Analysis of ultimate loads of shallow foundations. *Journal of Soil Mechanics and Foundation Division*, ASCE **99(1)**: 45–73.

Vidal H (1969) The principle of reinforced earth. *Highway Research Record* **282**: 1–16.

Vinod P and Bhaskar AB (2010) Contribution of hand-knotted coir netting to the bearing capacity of loose sand. *Geosynthetics International* **17(6)**: 443–449.

Wang TR, Ye ZX and Qi SM (1993) Geotextile–aggregate mat and vertical drains for deep foundation improvement, gas-holder at Hangzhou, China. In *Geosynthetics Case Histories* (Raymond GP and Giroud J (eds)). ISSMFE Technical Committee TC9, Geotextiles and Geosynthetics, London, pp. 166–167.

Watary Y (1984) Reclamation with clayey soils and method of earth spreading on the surface. *Proceedings of the Seminar on Soil Improvement and Construction Techniques in Soft Ground*, Singapore, pp. 103–119.

Yetimoglu T, Wu JTH and Saglamer A (1994) Bearing capacity of rectangular footings on geogrid-reinforced sand. *Journal of Geotechnical Engineering* **120(12)**: 2083–2099.

Yin JH (1997a) Modelling geosynthetic-reinforced granular fills over soft soil. *Geosynthetics International* **4**(2): 165–185.

Yin JH (1997b) A nonlinear model of geosynthetic-reinforced granular fill over soft soil. *Geosynthetics International* **4(5)**: 523–537.

Zhang L, Zhao M, Shi C and Zhao H (2010) Bearing capacity of geocell reinforcement in embankment engineering. *Geotextiles and Geomembranes* **28(5)**: 475–482.

Handbook of Geosynthetic Engineering
ISBN 978-0-7277-4175-2

http://dx.doi.org/10.1680/hge.41752.163

Chapter 7
Unpaved roads

P. L. Bourdeau* **and A. K. Ashmawy†**
*School of Civil Engineering, Purdue University, West Lafayette, USA
†School of Engineering, American University in Dubai, UAE

The functions of geosynthetics in the improvement of unpaved roads are reviewed, and the reinforcing mechanisms of geotextiles and geogrids are discussed. The fundamentals of the mechanical interaction between the subgrade, the geosynthetic layer and the granular base are described. Several design methods representative of the current state of practice are summarised, and their main features, as well as limitations, are discussed.

7.1. Introduction

Roadway construction was one of the earliest areas of application of geosynthetic materials. Geotextiles and geogrids have been successfully used for their ability to separate, filter, drain or reinforce soils in both paved and unpaved roads. Historical accounts of the development of geosynthetics are presented in Section 1.3. This chapter addresses the application of geosynthetics in unpaved roads (i.e. roadway structures that are not capped with concrete slabs or an asphaltic concrete wearing course). Examples of such designs are found in temporary access roads or tracks, forest roads, haul roads, etc. They generally consist of a layer of crushed stone or gravel fill laid directly on the subgrade. The aggregate layer serves as both a base course and a wearing course. It should be noted that, during their construction, all paved roads undergo a period of being 'unpaved', during which time they are subjected to the traffic of construction equipment.

Significant improvement can be expected to accrue from the use of geosynthetics in unpaved roads, built on soft to moderately soft fine-grained subgrades having a high water content. In such situations, improved performance (in terms of permanent deformation for a given number of axle loads) is obtained by placing geotextiles or geogrids at the interface between the subgrade and the aggregate layer. Conversely, for equivalent performance, successful application of geosynthetics would spare a significant volume of aggregates. Because geosynthetics can play multiple roles in unpaved roads, there are many factors involved in their selection, and different design solutions can be equally effective for a particular project. This explains why, in this type of application, a wide range of geosynthetics, including woven and non-woven geotextiles, geogrids and geocomposites, have been used with success.

The benefits obtained from the use of geosynthetics in unpaved roads can be observed not only with respect to structural performance and durability, but also with respect to construction and economy. These benefits are as follows

- On very soft subgrade soil, installation of a geotextile or a geogrid makes possible the construction of the aggregate layer without excessive loss of material. This separation role is often the major advantage of geosynthetics for construction on very soft soil.
- Compaction of the aggregate layer is made easier by the presence of a geosynthetic layer at the interface, especially when local heterogeneities (softer zones) of the subgrade are crossed. This results in better homogeneity of the gravel base layer and lesser spatial variability of its mechanical characteristics. The capacity of a geosynthetic to 'bridge' heterogeneities is a consequence of its reinforcing action.
- A geotextile placed at the interface between a fine-grained subgrade and a coarse-grained base course can minimise the contamination of the base course by fine particles pumped from the subgrade under repeated traffic loads. This ability of a geotextile to control the pumping of fines is related to its filtration capacity and opening size.
- The structural capacity of an unpaved road is improved by the reinforcing action of a geosynthetic when, under traffic load, the reinforcement placed at the interface contributes to a more efficient transfer of stresses from the base course to the subgrade. As a result, less rutting is experienced under repeated loading.
- A geotextile with a high hydraulic transmissivity can ensure that the contact zone between the subgrade and the base course will remain drained during periods of increased water content due to rainfall infiltration. Unpaved roads do not benefit from the surficial drainage that is otherwise provided by a pavement. Thus, the role of the underdrain, played by a

geosynthetic, can be critical to the performance of the system.

In the following sections, interaction between soil and geosynthetics in unpaved roads and design approaches are discussed. It is considered that the primary function of a geosynthetic is reinforcement. However, it should be noted that, in a number of cases, the relative importance of the geosynthetic's functions is difficult to assess, and these functions are not independent of each other. In such cases, the separation function or drainage might be the designer's primary focus.

7.2. Unpaved road reinforcement

In unpaved roads, the overall response of the reinforced soil mass and the resulting performance of the system depend on a number of factors

- the subgrade properties, including the groundwater conditions close to the surface
- the thickness and properties of the aggregate layer
- the location and properties of the geosynthetic used as reinforcement
- the loading conditions, including the magnitude and number of load applications.

Most research to date has focused on reinforcement under monotonic and static loading. Only relatively recently has more attention been given to repeated and dynamic loading due to traffic. While it is necessary first to understand the mechanisms of reinforcement under monotonic loading, it should be recognised that this is a mere simplification of the actual loading conditions for unpaved roads. In particular, the fact that vehicle traffic induces repeated and dynamic variations in stresses, including rotation of the principal stresses in the reinforced soil, makes the analysis much more complex than in the case of monotonic loading. This explains why there is still no solution available to address the problem on a rational basis and in its full complexity. Currently available analysis methods rely on a drastic simplification of the geometry and loading conditions, and practical design methods are combinations of analysis and empirical knowledge. A literature database on the subject has been summarised by Ashmawy and Bourdeau (1995).

7.2.1 Interactions under monotonic loading

In the case of monotonic loading, three main mechanisms of soil–reinforcement interaction have been identified: tensile membrane action, enhanced confinement and passive anchorage.

7.2.1.1 Membrane tension

Membrane tension is the major mechanism acting in reinforced unpaved roads laid on a compressible subgrade (Figure 7.1). The application of concentrated loads produces compression of the subgrade, together with deflections of the aggregate layer and of the geosynthetic placed at the interface. As a result of the geosynthetic deformation, tensile forces are developed and vertical support is provided to the aggregate by the membrane. Static equilibrium analysis of the forces shows that vertical stresses at the subgrade–geosynthetic interface are modified as compared with the unreinforced system. In particular, the peak value of the vertical stress induced by the applied load is significantly lower and the consecutive deformation is reduced. A number of analytical or numerical models have been proposed to address this mechanism in two-dimensional plane strain conditions (e.g. Bourdeau, 1989).

7.2.1.2 Enhanced confinement or lateral restraint

Enhanced confinement or lateral restraint develops in reinforced soil, where there is tendency for the soil to expand laterally under the effect of vertical compression. Because the reinforcement has a much greater tensile strength and stiffness than the soil, it can resist this tendency and restrain the lateral deformation of the soil. The stress transferred at the soil–reinforcement interface results in a confining pressure, within the soil, larger than it would be in the absence

Figure 7.1 Tensile membrane action in reinforced unpaved road: (a) without reinforcement and (b) with reinforcement

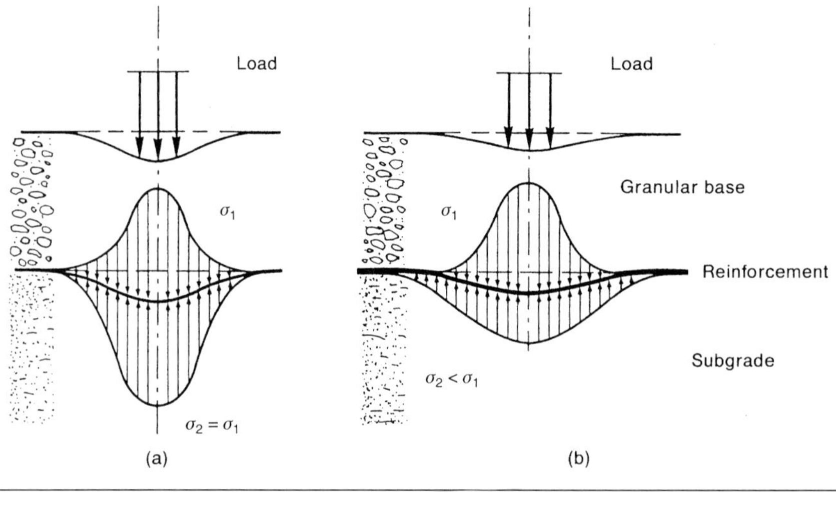

of reinforcement. This particular mechanism has been recognised as a major mode of reinforcement in multi-layered structures, such as mechanically stabilised walls. In unpaved road structures where, in general, a geosynthetic reinforcement is present only at the interface between the subgrade soil and the aggregate base course, there is also evidence of lateral restraint in both the aggregate and the subgrade. Deformation fields observed in unreinforced, as well as rein forced, two-layer systems (sand overlaying silt), subjected to approximately the same load increment, are shown in Figure 7.2 (Bourdeau *et al.*, 1991). Individual displacements of small lead-shot tracers were observed using x-ray imaging and are represented by their direction and magnitude in the unreinforced system (see Figure 7.2(a)), and a similar system reinforced at the interface with a geotextile (see Figure 7.2(b)). As the situation is symmetrical, only half of the model is shown. In the unreinforced case, significant lateral displacement occurs in the upper granular layer as well as in the silty subgrade. Ultimately, this would lead to bearing capacity failure by rotation and lateral sliding of the soil mass. In contrast, when the geotextile is present, most of the displacements taking place in the two layers are vertical, because the lateral displacements are restrained by the geotextile. As a result, the reinforced system is more resistant to bearing capacity failure, and the load plate settlement is significantly reduced. Lateral restraint of the aggregate base-course deformation has been identified as a major mechanism in the reinforcement of unpaved road models using geogrids (Milligan *et al.*, 1986; Love *et al.*, 1987), and is the basic concept of the design method proposed by Milligan *et al.* (1989) for this type of structure. On the basis of plate-load tests, Haliburton and Barron (1983) (see also Christopher *et al.*, 1984) observed that the optimum location of the reinforcement is at a depth of approximately one-third of the width of the loaded area. This observation emphasises the requirement for the reinforcement to intercept the bearing failure mechanisms, in order to develop lateral restraint effectively.

7.2.1.3 Passive anchorage

Passive anchorage, although not the primary mechanism of reinforcement in unpaved roads, is a necessary condition for the tensile membrane and enhanced confinement to be effective. Tensile forces developed in the geosynthetic under the loaded area need to be balanced by interface friction in the lateral anchorage zones. Numerical modelling indicates that the anchorage capacity can be a limiting factor on the effectiveness of the tensile membrane mode of reinforcement (Bourdeau, 1991).

7.2.2 Effect of repeated loading

Under repeated loading due to traffic, the three modes of interaction described in the preceding section are still the major mechanisms of reinforcement. In addition, two other mechanisms, specific to cyclic loading conditions, can develop.

Additional compaction of the aggregate base course is produced by repeated traffic load, and results in increasing stiffness and resistance of the granular material. Tests performed in the laboratory, after cyclic loads had been applied to reinforced soil, showed significant improvement in performance under monotonic load as compared to the values recorded prior to the cyclic loading. This was attributed to additional compaction (Madani *et al.*, 1979; Leflaive, 1985; Nimmesgern and Bush, 1991). In unpaved road structures, the membrane support and lateral restraint provided by the geosynthetic reinforcement allow this additional compaction of the upper layer to take place. The presence of reinforcement prevents the occurrence of bearing failure during the early cycles of loading.

Dynamic interlock is a mechanism specific to geogrid reinforcement (McGown *et al.*, 1990). Upon loading, aggregate particles become locked into the grid apertures and prevent the elastic part of the reinforcement tensile strain from being fully recovered during the unloading phase. As a result, the geogrid remains stressed and the lateral confinement of the aggregate layer is increased.

In contrast to the extensive literature published on the analytical or numerical modelling of monotonic loading conditions, only a few attempts have been made to model the behaviour of reinforced soil under repeated loading, and even fewer attempts have been made to specifically address reinforced unpaved roads. A review of these works has been presented by Ashmawy and Bourdeau (1995). In general, the predictions obtained from these models are not in close agreement with observations, for the following main reasons

- The interaction between the soil and the geosynthetic reinforcement is still not fully understood, including the fact that the different mechanisms of reinforcement are not independent but interfere with one another.
- The material response is not simulated accurately by the models, including the soil non-linear attributes, anisotropy induced by compaction, and the time-dependent behaviour of the geosynthetic.
- The effect of the applied load on the material and interface properties, in particular the dependence of these properties on the number of cycles and loading rate, is not well modelled.

Because of these difficulties, most knowledge accrued to date on the effect of repeated loading on reinforced soil is of an experimental nature, and this has resulted in empirical or semi-empirical design approaches being used.

Under cyclic loading, the strength of reinforced soil is usually expressed in terms of cycles to failure for a specific peak load, chosen as a fraction of the monotonic loading strength. An alternative criterion is to compare the cycles to failure for reinforced and unreinforced specimens subjected to the same peak load and load frequency. Triaxial tests performed on reinforced soil specimens indicate that the soil strength increases under cyclic loading when reinforcement is used (Bourdeau *et al.*, 1990; Ashmawy and Bourdeau, 1997,

Figure 7.2 X-ray observation of deformation fields in a two-dimensional model of an unpaved road under monotonic loading (load increment 94–125 kPa): (a) unreinforced system and (b) reinforced system
Reproduced from Bourdeau *et al.* (1991)

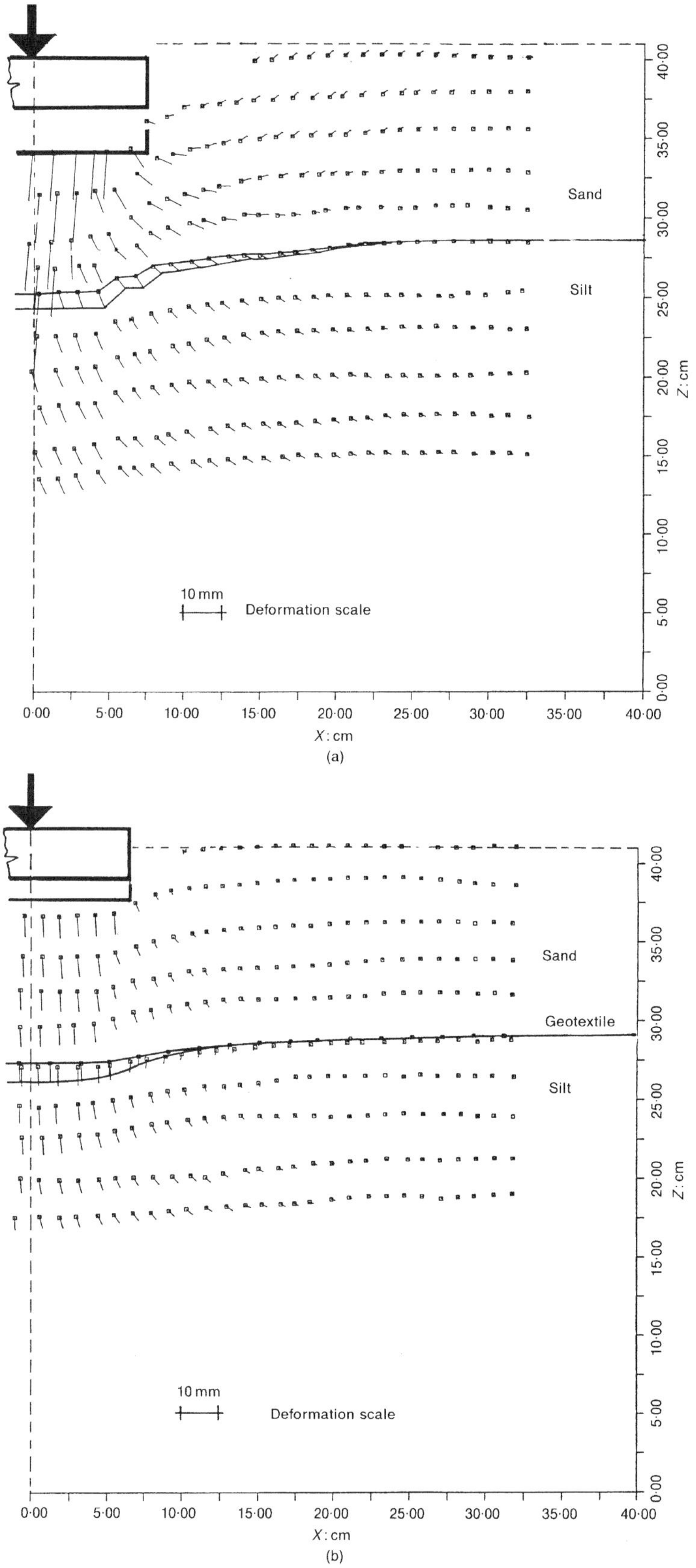

1998; Ashmawy *et al.*, 1999). According to Maher and Ho (1993), the number of cycles to failure is roughly proportional to the density of reinforcement. However, results obtained using triaxial specimens of soils, within which the reinforcement is distributed uniformly under the form of discs of geotextile or mixed fibres, should not be extrapolated without caution to unpaved road structures reinforced with localised reinforcement sheets and subjected to non-uniform boundary loads.

For roadway and railway structures, as well as embankments subjected to traffic loading, it was observed that permanent (i.e. non-elastic) deformations are reduced if geosynthetic reinforcement is used (Milligan *et al.*, 1986). It is important to note that appropriate selection of the type and location of the geosynthetic is necessary in order to achieve optimum improvement, and that more improvement is obtained if the subgrade is compressible. For these types of structure, it was observed that the use of geosynthetic reinforcement is barely beneficial on a firm subgrade, and the smaller the allowable deformation, the higher is the necessary reinforcement stiffness.

For unpaved road structures reinforced using geotextiles, Delmas *et al.* (1986) have identified the existence of a fatigue limit pressure, $p_{mP}(r_P)$, such that the rut depth r does not exceed the value of r_P when this limit pressure is applied for an infinite number of cycles. According to Delmas *et al.*, the rut depth r can be related to the number of load cycles N and the cyclic peak pressure p_m by the equation

$$\frac{p_m}{p_s(r)} = \frac{\alpha' \log_{10} N + \beta'}{\log_{10} N + \gamma'} \qquad (7.1)$$

where p_s is the static pressure corresponding to a rut depth r, and can be determined experimentally or using empirical relationships. The coefficients, α', β' and γ' vary in function with the geotextile type and the soil properties. Typical values are 0.6, 1.45 and 0.5 for α', β' and γ', respectively, in the absence of reinforcement. In reinforced structures, α' is typically <0.6.

Results obtained from large-scale tests, conducted on granular material resting on compressible subgrade (De Garidel and Javor, 1986; Chaddock, 1988; Chan *et al.*, 1989; Hirano *et al.*, 1990), indicate that the use of geogrid or geotextile reinforcements results in a reduction in the permanent deformation of the order of 25–50%, with smaller ranges of reduction with geotextiles. The number of cycles required to achieve a specific permanent deformation may increase by up to one order of magnitude if geosynthetic reinforcement is used, with typical ranges of five to seven times. Empirical formulae of the general form

$$S = a(\log_{10} N)^2 + b(\log_{10} N) + c \qquad (7.2)$$

have been proposed to compute the permanent deformation S as a function of the number of load cycles N for railway ballast reinforced with a geogrid on a compressible subgrade (Bathurst *et al.*, 1986) or fibre-reinforced triaxial specimens of sand (Leflaive, 1985). The generic form of Equation 7.2, and its applicability to different types of reinforced soil structure, suggest that it could quite likely also be fitted to unpaved road deformation records.

The magnitude of elastic rebound during cycles of loading and unloading on reinforced unpaved roads seems to be more sensitive to the subgrade compressibility than to the presence and location of a geosynthetic. Gourc (1983) observed that elastic rebound was controlled mostly by the tensile membrane effect. The presence of reinforcement may even decrease the amount of elastic rebound slightly, probably due to the progressive increase in the reinforced mass stiffness. During the first cycles of loading, the strength of the reinforcement is not fully mobilised. The reinforced soil stiffness increases with an increasing number of cycles, as interface friction, interlocking effects and enhanced compaction gradually take place. At later stages, the overall modulus may decrease as the soil approaches failure. Douglas (1991) suggested that the resilient modulus should be the main design criterion for unpaved roads, and introduced the parameter k^* to represent the ratio of the road-section stiffness to subgrade stiffness. While the stiffness of the road section increases with an increasing number of load cycles, k^* remains almost constant and, according to Douglas (1991), is a function of base thickness and depth of reinforcement.

7.2.3 Design for reinforcement

A number of methods have been proposed to address the structural design of reinforced unpaved roads. These can be classified as follows

- Methods based on a rational analyses of the reinforcement mechanism under static or monotonic loading. For bibliographic references on these methods, see, for example, the literature database presented by Ashmawy and Bourdeau (1995). Such models are not recommended for practical designs of unpaved roads because they do not account for the repetition of loads induced by traffic.
- Fully empirical methods based on observation of the performance of full-scale models or trial road sections. Examples of such methods are the procedures recommended by the French Geotextile Committee (1981) or the Swiss Society of Geotextile Professionals (1985). Both methods are based on extensive collections of field performance data. In the French method, a catalogue of more than 90 typical design situations are presented, corresponding to combinations of geometry, traffic and soil conditions. For each case, the minimal required geotextile characteristics are indicated in terms of tensile resistance and stiffness, tear strength, hydraulic conductivity and pore opening size. The Swiss method and its extension proposed by Jaeklin (1986) are the result of multivariate regression analyses. The resulting equations take into account the traffic intensity (based

on the load magnitude and characteristics, and the number of passages), rut depth, subgrade strength, type of fill aggregate, fill thickness, and number and type of geotextile layers. The output variables are the required tensile strength of the geotextile and the design failure strain. It is noted that these procedures, especially the French method, address at the same time the reinforcement function and other design requirements of the geosynthetic, such as filtration/separation, drainage and survivability of installation. This is due to the fact that the method results from observation of the overall performance of unpaved road sections. Like all empirical methods, the advantage of these procedures is in their simplicity of application, and their shortcoming is the limitation of their validity to particular materials and conditions.

- Rational methods based on analyses of the reinforcement mechanisms that account for traffic loading. Unavoidably, the formulation of such methods is possible only at the cost of drastic simplifications when representing the material properties, boundary conditions and traffic-loading features. The cyclic nature of the traffic is often modelled using empirical equations or adjustments that are combined with the rational static analysis. Despite such approximations, these methods seem to offer the most reliable and versatile approach. They have been used extensively by practitioners worldwide.

Three types of these methods are described more fully below.

7.2.3.1 Giroud and Noiray (1981)

This method is based on a static analysis of membrane support, combined with an empirical model to account for traffic loading. The static analysis is used to determine the savings Δh in aggregate fill thickness provided by the tensile membrane action of the reinforcement. Figures 7.3(a) to 7.3(c) show the load diffusion model, the assumed subgrade deformation mechanisms and the deformed geometry of the geotextile. Further hypotheses are made with respect to the design limit pressure on the subgrade. This pressure is assumed to be equal to the elastic limit bearing capacity q_e in the absence of geotextile, and to the ultimate bearing capacity q_{ult} in the presence of reinforcement.

According to Figure 7.3(a), these pressures are:

$$q_e = \pi c_u + \gamma h_0 \tag{7.3}$$

$$q_{ult} = (\pi + 2)c_u + \gamma h \tag{7.4}$$

where c_u is the undrained shear strength ($\phi_u = 0$) of the subgrade and γ is the aggregate unit weight. For geotextile reinforcement, the load diffusion angles α_0 and α, corresponding to the unreinforced and reinforced situations, respectively, are assumed both equal to $\tan^{-1}(0.6)$. This implies that the presence of the reinforcement does not modify significantly the load transmission mechanism through the aggregate layer. The vertical pressure at the bottom of the aggregate layer (see Figure 7.3(a)) is given by

$$p_0 = \frac{P}{2(B + 2h_0 \tan\alpha_0)(L + 2h_0 \tan\alpha_0)} + \gamma h_0 \tag{7.5}$$

Figure 7.3 (a) Load diffusion model, (b) kinematics of subgrade deformation and (c) deformation of the geotextile
Reproduced from Giroud and Noiray (1981), with permission from ASCE

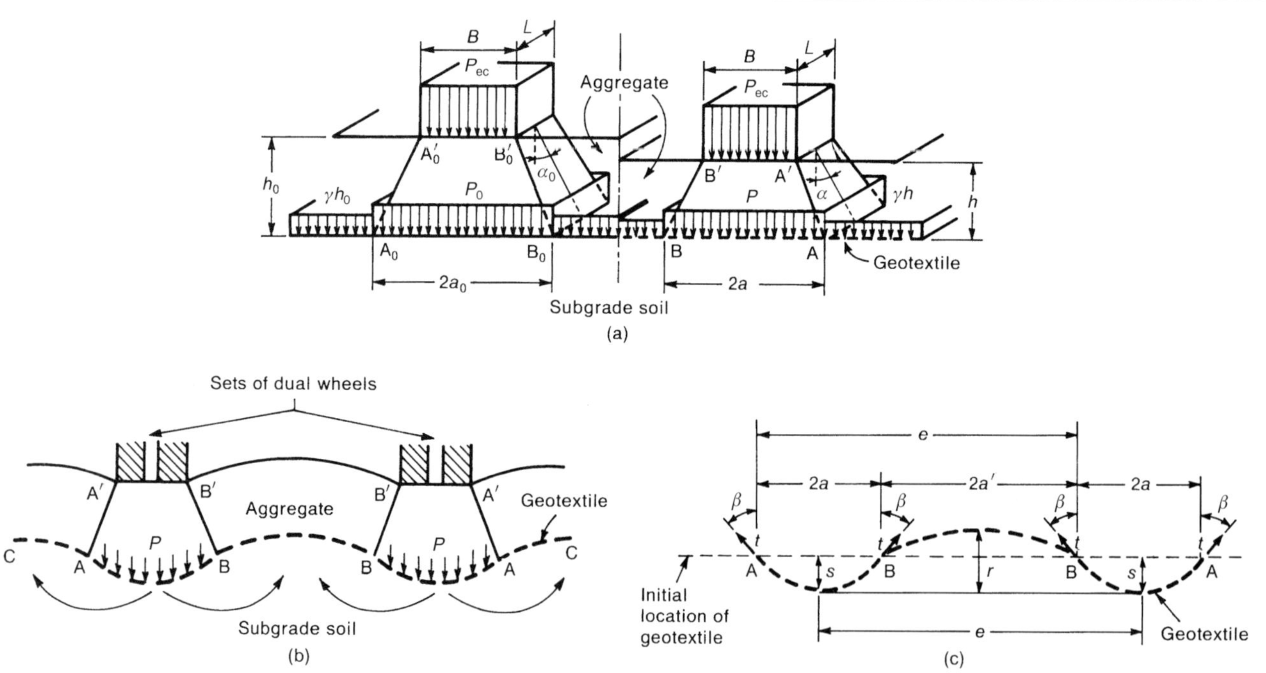

in the absence of geotextile, and

$$p = \frac{P}{2(B + 2h\tan\alpha)(L + 2h\tan\alpha)} + \gamma h \quad (7.6)$$

with geotextile, where P ($= 2LBp_{ec}$; p_{ec} is the *equivalent contact pressure*, which is assumed to be distributed uniformly between rectangle $L \times B$ and the aggregate layer) is the axle load, and other parameters are clearly shown in Figure 7.3.

At the limit state for the subgrade

$$p_0 = q_e \quad (7.7)$$

without geotextile, and

$$p = q_{ult} \quad (7.8)$$

with geotextile.

The corresponding two-dimensional deformation of the geotextile is assumed to be parabolic in shape, with fixed points located at A and B in Figure 7.3(c). For these geometric conditions and conservation of the volume of the undrained subgrade, the elastic tension of the membrane results in a release of vertical pressure of

$$p_g = \frac{E\varepsilon}{a\sqrt{1 + (a/2s)^2}} \quad (7.9)$$

where E and ε are the tensile modulus and tensile strain of the geotextile, respectively.

From Equations 7.3, 7.5 and 7.7, the limit-state condition without geotextile is expressed as

$$c_u = \frac{P}{2\pi(B + 2h_0\tan\alpha_0)(L + 2h_0\tan\alpha_0)} \quad (7.10)$$

In the presence of reinforcement, Equations 7.4, 7.6, 7.8 and 7.9 lead to

$$(\pi + 2)c_u = \frac{P}{2(B + 2h\tan\alpha)(L + 2h\tan\alpha)} + \frac{E\varepsilon}{a\sqrt{1 + (a/2s)^2}} \quad (7.11)$$

Solving Equation 7.10 for h_0 and Equation 7.11 for h allows us to determine the potential savings of aggregate thickness, due to the use of reinforcement, under static (or monotonic) loading:

$$\Delta h = h_0 - h \quad (7.12)$$

A further assumption, central to the method, is that the same benefit Δh can be obtained under repeated traffic loading, and this means that in the method the effect and analysis of the reinforcement is uncoupled from the cyclic nature of the loading. Under traffic loading, the required aggregate fill thickness h'_0, of the unpaved road without geotextile, is determined using an empirical method originally developed by Webster and Alford (1978) for a rut depth of $\gamma = 75$ mm and simplified by Giroud and Noiray (1981) to the form:

$$h'_0 = \frac{0.19\log_{10} N_s}{(\mathrm{CBR})^{0.63}} \quad (7.13)$$

where N_s is the number of passes with a standard axle load of 80 kN, and CBR is the California bearing ratio of the subgrade. Giroud and Noiray (1981) extended the above formula (Equation 7.13) to other values of axle load and rut depth using the following relationships:

$$N_s = N\left(\frac{P}{P_s}\right)^{3.95} \quad (7.14)$$

$$\log_{10} N = \log_{10} N_s - 2.34(r - 0.075) \quad (7.15)$$

where N_s is the number of passes of axle loads, $P_s = 80$ kN for a rut depth of 0.075 m, and N is the number of passes of axle loads P corresponding to r, the design value of rut depth.

They also introduced the subgrade undrained cohesion, using the empirical formula

$$c_u = 30 \times \mathrm{CBR} \quad (7.16)$$

where c_u is expressed in kilopascals (kPa).

Once h'_0 has been determined, the required aggregate thickness with geotextile reinforcement can be computed as

$$h = h'_0 - \Delta h \quad (7.17)$$

where Δh is given by Equation 7.12.

Giroud and Noiray (1981) also published a design chart for a particular set of parameters, as shown in Figure 7.4, in which p_c is the tyre inflation pressure. Similar charts developed for a broader range of input data can be found elsewhere (Holtz and Sivakugan, 1987).

Among the assumptions made by Giroud and Noiray (1981), the adoption of different limit-bearing pressures for the subgrade, in the presence or absence of geotextile, leads to results that may seem theoretically inconsistent. Because the computed performance of an unreinforced road is not similar to that of the same road reinforced with a zero-modulus geotextile, a significant portion of the reduction in aggregate thickness results merely from the fact that a geotextile is placed at the interface. However, it has been recognised in practice that even very low modulus geotextiles are beneficial because of their separation function. The mechanistic assumption introduced by Giroud and Noiray (1981) seems also to be supported by experimental results obtained on scale models by Bender and Barenberg (1978).

Figure 7.4 Design chart for standard axle load $P = 80$ kN and rut depth, $r = 0.3$ m
Reproduced from Giroud and Noiray (1981), as adapted by Koerner (1998), Pearson Education

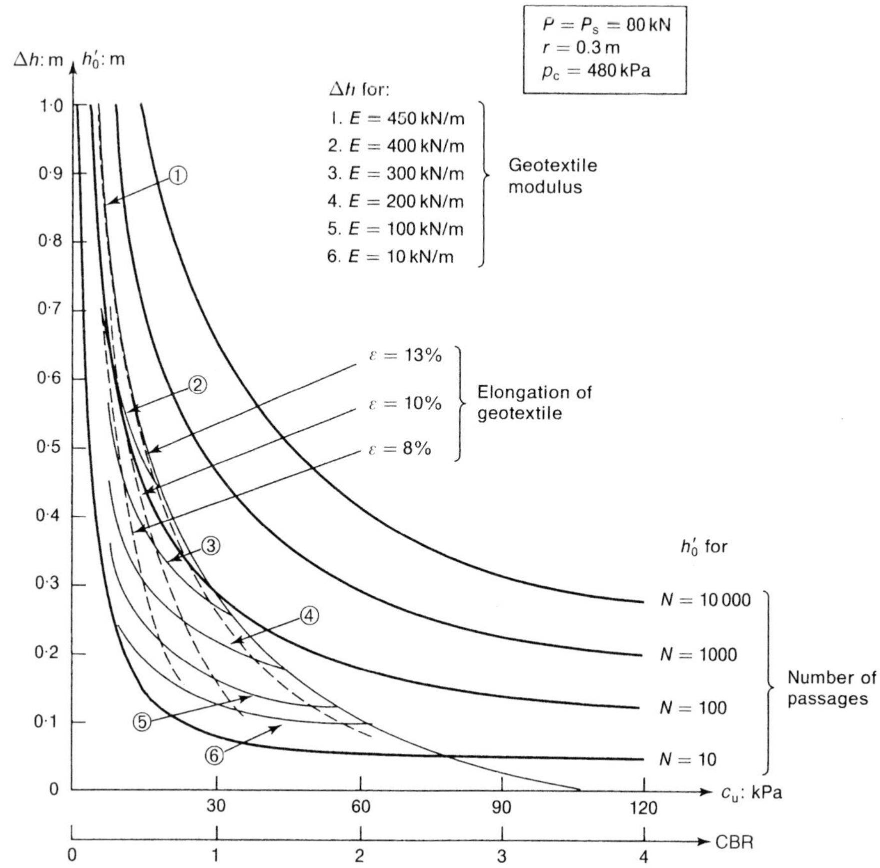

It should be noted that the Giroud and Noiray (1981) equations or design charts do not constitute a complete design procedure for the geotextile reinforcement. In addition to the determination of the aggregate thickness provided by the method, computations should be completed with verifications of the tensile resistance and lateral anchorage of the geotextile. These verifications can be performed for the tensile strain given by Equation 7.9 or the design chart shown in Figure 7.4.

7.2.3.2 Oxford method

Milligan *et al.* (1989) have proposed a method for the design of unreinforced and geotextile- or geogrid-reinforced unpaved roads. This method of analysis resulted from an extensive research programme conducted at Oxford University. The investigations consisted of experiments on models of unreinforced and reinforced roads under monotonic loading (Love, 1984) and cyclic loading (Fannin, 1987), as well as large-strain finite-element analysis for static loading (Burd, 1986). The results of these studies are summarised in Milligan *et al.* (1986) and Love *et al.* (1987).

The method of analysis considers first the plastic equilibrium of the granular base and clay subgrade under the effect of a static load. Originally developed for plane-strain conditions, the analysis can be extended to axisymmetric cases (Houlsby and Jewell, 1990). Conceptual models are presented in Figure 7.5(a) for the geometry of load diffusion, and in Figures 7.5(b) and 7.5(c) for the equilibrium of the base course wedge located beneath the loaded area in the unreinforced and reinforced case, respectively. In the unreinforced case, shear stresses induced by the load in the granular base develop along the interface between the base course and the subgrade, and therefore reduce the bearing capacity. If these shear stresses are sufficiently high, sliding may even occur at the interface. A non-dimensional form solution for the ultimate conditions at the interface and in the subgrade was devised:

$$N_{\mathrm{ca}} = 1 + \frac{\pi}{2} + \cos^{-1}\alpha_{\mathrm{a}}\sqrt{1-\alpha_{\mathrm{a}}^{2}} \tag{7.18}$$

Equation 7.18 represents the relationship between the bearing capacity factor of the clay subgrade in undrained conditions N_{ca} and the normalised shear stress at the interface α_a, with

$$N_{\mathrm{ca}} = (\sigma_{\mathrm{va}} - \sigma_{\mathrm{vo}})s_{\mathrm{u}} \tag{7.19}$$

$$\alpha_{\mathrm{a}} = \tau_{\mathrm{a}}/s_{\mathrm{u}} \tag{7.20}$$

Figure 7.5 Conceptual models for: (a) geometric load diffusion, (b) horizontal equilibrium of base course wedge without reinforcement and (c) horizontal equilibrium of base course wedge with reinforcement
Reproduced from Milligan *et al.* (1989), Emap

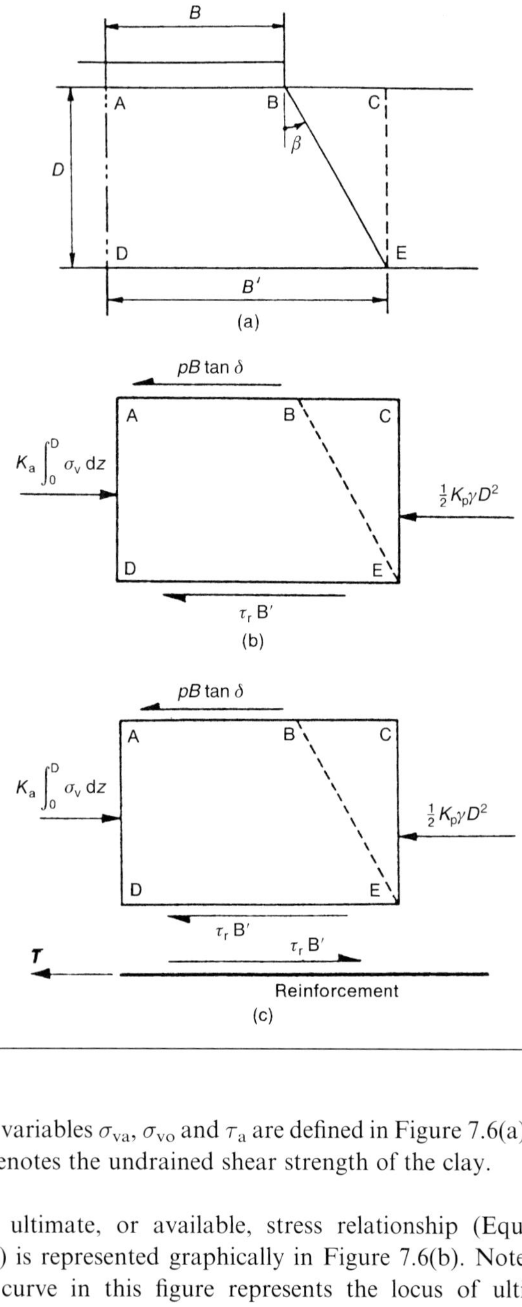

The variables σ_{va}, σ_{vo} and τ_a are defined in Figure 7.6(a), and s_u denotes the undrained shear strength of the clay.

The ultimate, or available, stress relationship (Equation 7.18) is represented graphically in Figure 7.6(b). Note that the curve in this figure represents the locus of ultimate stress conditions for the subgrade, including its bearing capacity and sliding along the interface. In the absence of reinforcement, the limit equilibrium of the granular base wedge (see Figure 7.5(b)) is expressed as

$$\alpha_r = \frac{1}{2}(K_a - K_p)\frac{\gamma D^2}{s_u B'} + N_{cr}\left[\frac{K_a}{\tan B'}\ln\left(\frac{B'}{B}\right) - \tan\delta\right] \tag{7.21}$$

where δ is the angle of soil–geosynthetic interface friction

$$\alpha_r = \tau_a - s_u \tag{7.22}$$

is the normalised required shear stress, and

$$N_{cr} = (\sigma_{vr} - \sigma_{vo})s_u \tag{7.23}$$

In Equation 7.23, σ_{vr} and σ_{vo} denote the overburden stress with and without reinforcement, respectively. The bearing capacity factors K_a and K_p are the Rankine earth pressure coefficients for active and passive states, respectively. In Figure 7.6(b), the linear relationship between α_r and N_{cr} is represented by a line such as GH. At the intersection of this line and the curve, point C, the required and the available combination of the normal and the shear stresses are compatible with the maximum bearing capacity factor.

The role attributed to the reinforcement is to carry the interface shear stress, thus allowing the subgrade bearing capacity to be maximised. If the shear stress developed at the base of the granular course is fully transferred to the geosynthetic, the clay subgrade is loaded only by normal vertical stresses and, in Figure 7.6(b), the ultimate state is represented by point E, where $\alpha_a = 0$ and $N_{ca} = (2 + \pi)$. In this case, the allowable surface applied pressure is

$$p = (2 + \pi)s_u\left(1 + \frac{D}{B}\tan\beta\right) \tag{7.24}$$

and the tensile force per unit length of road developed in the reinforcement is

$$T = \tau_r B' \tag{7.25}$$

A mandatory condition, for the analysis to be valid, is that full transfer of the shear stress can be achieved at the interface between the granular course and the reinforcement. Furthermore, the resulting elongation of the geosynthetic must be small enough so that no significant shear stress be transferred to the subgrade. This requires, for the geosynthetic used as reinforcement, both high interface interaction with the granular material and high tensile performance characteristics (modulus and strength). In addition to verifications of interface shear strength and tensile resistance of the reinforcement, it is necessary to check that the bearing capacity failure of the granular layer would not be the controlling factor.

Note that, in the above method, no account is taken of tensile membrane action of the reinforcement. The analysis rests entirely on the concept of the transfer of shear stress at the interface, and the consequent enhanced confinement of the granular base and enhanced bearing capacity of the subgrade. According to the authors of the method, the analysis, as well as the related experiment, demonstrate how unpaved road reinforcement can be effective without there necessarily being a large deflection of the reinforcement.

The Oxford method accounts for the repeated loading by applying an empirical correction to the results of the static analysis. The fatigue relationship recommended by the authors was proposed by De Groot *et al.* (1986), on the basis of full-scale tests of reinforced unpaved road sections

Figure 7.6 Bearing capacity model for the subgrade: (a) definition of variables and (b) normal and shear stress relationship
Reproduced from Milligan *et al.* (1989), Emap

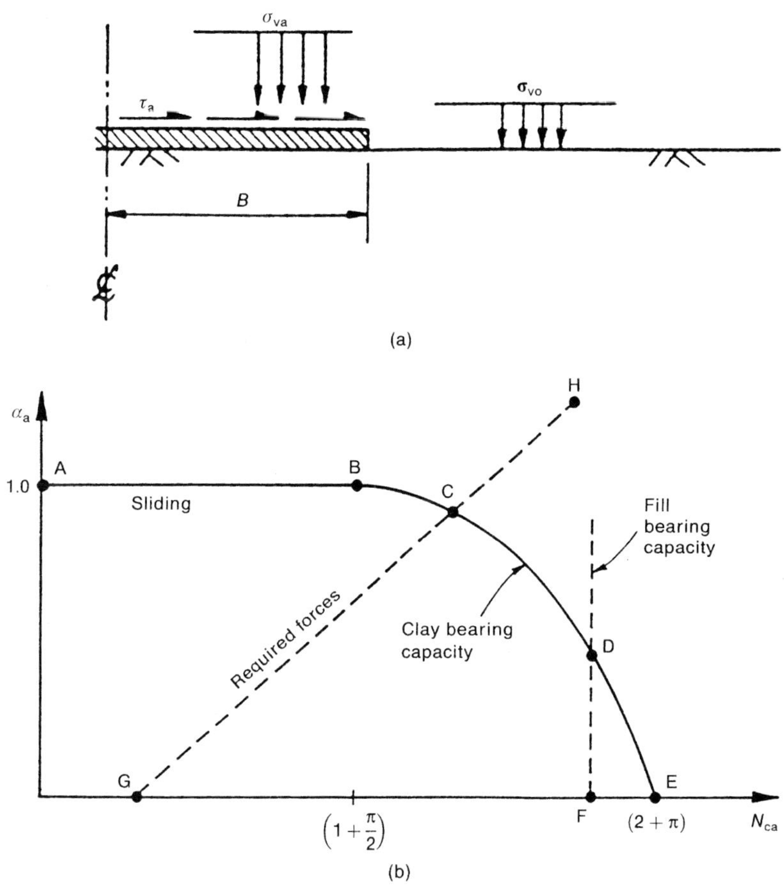

under traffic loading:

$$p_N = p_s/N^{0.16} \tag{7.26}$$

where p_s is the allowable static applied surface pressure, and p_N is the allowable pressure for N applications of the load.

Comparative computations were performed by Ashmawy and Bourdeau (1995) to investigate the consistency between different methods and their sensitivity to several design parameters. This study is summarised here. The methods considered were the Giroud and Noiray (1981) method, the Oxford method, as presented by Houlsby and Jewell (1990), and Jaeklin's (1986) empirical method.

The baseline values chosen for the parameters are shown in Table 7.1, and the results of the sensitivity analyses are summarised in Figure 7.7(a)–(d). Only one parameter is varied in each sensitivity analysis.

It is interesting to note that, in general, the empirically-based design by Jaeklin (1986) results in the least conservative values. As the analysis by Houlsby and Jewell (1990) is based on limit equilibrium considerations, the rut depth and the geosynthetic stiffness do not affect the design fill thickness. The method seems most suitable for small rut depths and high geotextile stiffnesses. Because the Giroud and Noiray (1981) method considers the membrane action of the reinforcement, the dependence of required fill thickness on the rut depth and geotextile stiffness are reflected in the results. The analysis indicates that the design fill thickness decreases with increasing rut depth, a consequence of the reduced performance requirements. For this case, only a 20% reduction in fill thickness resulted in a 400% increase in rut depth.

Table 7.1 Baseline case parameters used in comparative and sensitivity study
Data from Ashmawy and Bourdeau (1995)

Parameter	Value
Number of cycles	10 000
Rut depth	100 mm
Undrained shear strength (c_u)	40 kN/m^2
Geotextile stiffness	300 kN/m
Axle load	80 kN
Standard tyre pressure	480 kN/m^2
Equivalent contact pressure	340 kN/m^2

Figure 7.7 Comparison of design methods and the sensitivity study: (a) influence of rut depth, (b) influence of the number of load applications, (c) influence of the subgrade undrained shear strength and (d) influence of the reinforcement tensile modulus
Reproduced from Ashmawy and Bourdeau (1995)

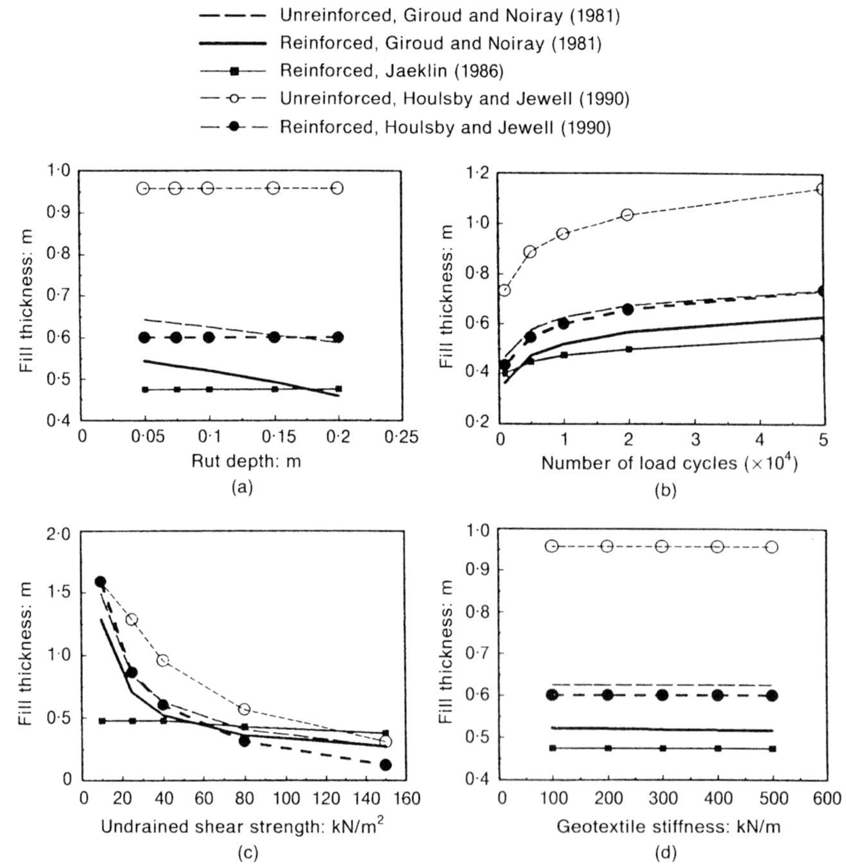

With all three methods, as the number of load cycles increases, so does the design fill thickness, the effect being more pronounced at fewer cycles. This phenomenon is related to the behaviour described earlier by Madani *et al.* (1979) and Delmas *et al.* (1986). For greater thicknesses, the applied pressure on the road surface drops below the fatigue limit pressure of the soil, and the number of cycles has little or no influence on the response of the structure. It is noted that an extrapolation was made for the Giroud and Noiray (1981) design for load applications in excess of 10 000 cycles. The geotextile stiffness seemed to have very little effect on the design fill thickness. This observation was made for the selected baseline parameters, but further analyses showed that the effect of geotextile modulus on required fill thickness becomes more pronounced at higher rut depths and lower subgrade strengths.

The required fill thickness for both reinforced and unreinforced cases decreases with increasing undrained subgrade strength c_u, except for the Jaeklin method. The amount of improvement introduced by the reinforcement can be related to the ratio between the reinforced and the unreinforced thickness. Figure 7.8 illustrates the variation in the thickness ratio with c_u for both the Giroud and Noiray (1981) and Houlsby and Jewell (1990) methods. A significant discrepancy is seen between these results, both from the standpoint of the relative improvement values, and the undrained strength of the subgrade for which optimal improvement was obtained.

Figure 7.8 Comparison of design methods and the sensitivity study – variation in the design fill thickness ratio with undrained shear strength of the subgrade
Reproduced from Ashmawy and Bourdeau (1995)

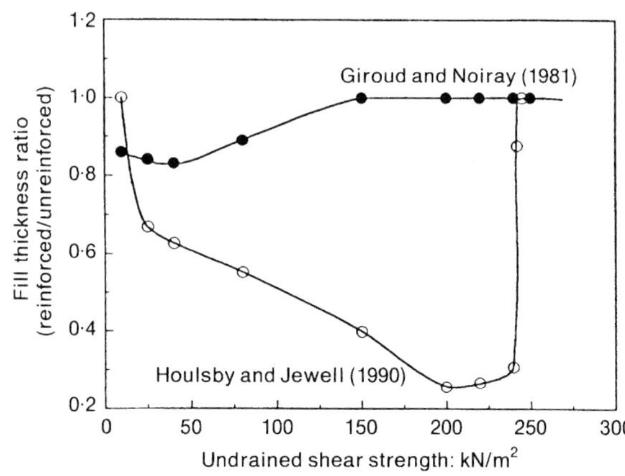

At low values of c_u the reinforcement has practically no effect, because the required fill thickness is very high (see Figure 7.7(c)). This is less pronounced in the Giroud and Noiray case because the analysis allows for the ultimate bearing capacity to develop in the reinforced design, while only the elastic limit is used in the unreinforced design. As the undrained strength of the subgrade increases, the benefit of using the geotextile also increases up to a certain limit. For higher values of c_u the design fill thickness decreases and the contribution of the reinforcement to the overall behaviour becomes minimal, as the subgrade can resist the applied loads. For the particular parameters given by the baseline case, and assuming a minimum acceptable fill thickness of 50 mm, the reinforcement is ineffective for $c_u \geq 250\ \mathrm{kN/m^2}$ for the Houlsby and Jewell method.

7.2.3.3 Giroud and Han (2004)

Presented as an evolution of an earlier analysis of geogrid-reinforced unpaved roads (Giroud *et al.*, 1985), this design method (Giroud and Han, 2004a,b) is, in fact, the result of a new, more comprehensive and rigorous formulation of the problem. More recent experimental and case studies (Fannin and Sigurdsson, 1996; Knapton and Austin, 1996; Wayne *et al.*, 1998; Gabr, 2001; Tingle and Webster, 2003) provided supporting data for the new method calibration and validation.

The geometry is axisymmetric, the load on the granular base course being uniformly distributed over a circular area as a function of the applied wheel load and tyre inflation pressure. The subgrade is considered to be a fully saturated fine-grained soil, with an undrained shear strength and compression modulus empirically related to its CBR coefficient.

Fundamental to the method is the notion, shared with the earlier methods of Giroud and Noiray (1981) and Giroud *et al.* (1985), that in presence of reinforcement the ultimate bearing capacity of the subgrade can be mobilised, while only local bearing capacity would be available in absence of reinforcement. This represents a major benefit of placing reinforcement at the subgrade–base interface: as the geogrid prevents local shear failure of the subgrade, application of the ultimate bearing pressure results in a smaller deformation than in the unreinforced situation.

Enhanced confinement of the granular base course is the other important mechanism of reinforcement addressed in the analysis. It is allowed for by effective interlocking between grains and grid apertures, and depends on the geogrid stiffness. Here, instead of relying on the overall tensile modulus of the geogrid to quantify this material property, the authors introduced a more pertinent parameter, the aperture stability modulus, as a measure of local confinement potential. The effect of this mechanism is to stiffen the granular base and broaden the load diffusion through the layer. This translates into a higher ratio of granular base to subgrade deformation moduli and an increase in the stress distribution angle at the interface, as compared with an unreinforced system.

Figure 7.9 Example of a design chart for a geogrid-reinforced unpaved road. *P*, wheel load; *p*, tyre pressure; CBR_{bc}, California bearing ratio of the granular base course; CBR_{sg}, California bearing ratio of the subgrade; *J*, geogrid aperture stability modulus; *s*, allowable rut depth; *N*, number of load cycles; *h*, required base course thickness

Reproduced from Giroud and Han (2004b), with permission from ASCE

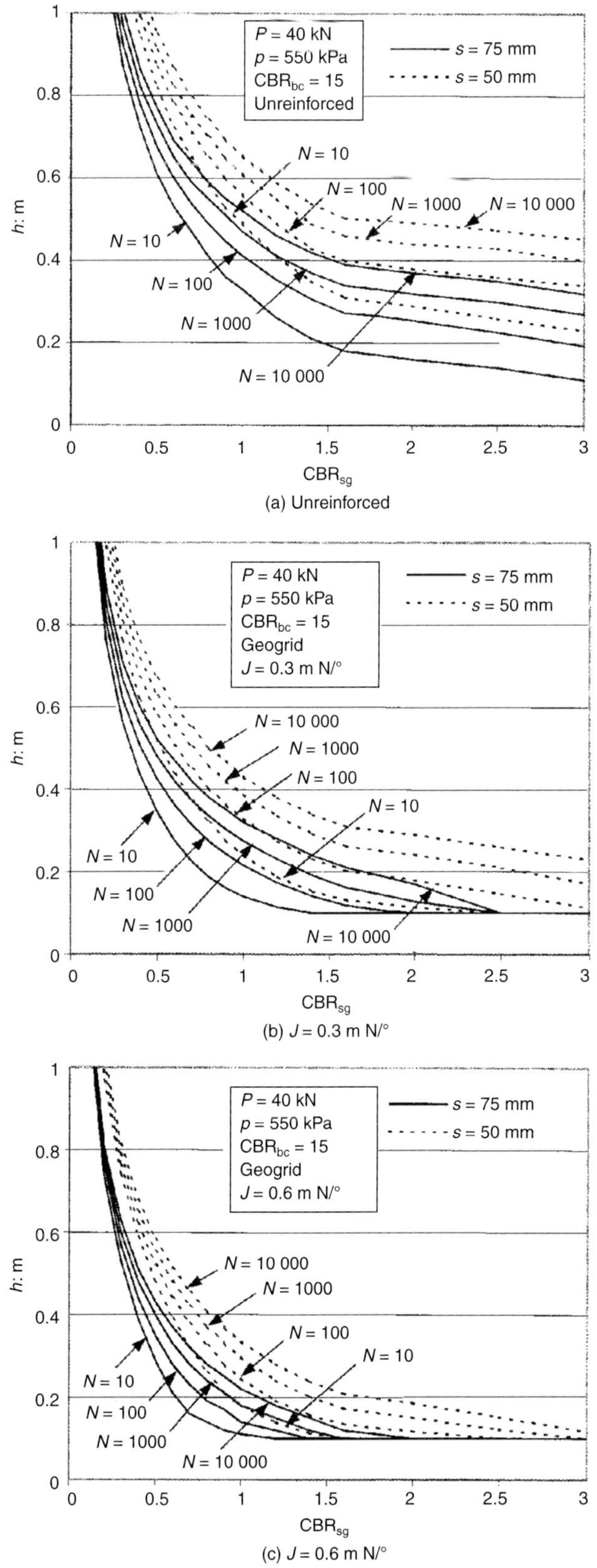

Another original attribute of the model is the integration of cyclic loading in the formulation. In previously presented methods, design equations were developed for only one static application of the load, and the results then modified using empirical formulae to account for the load repetition. In the Giroud and Han method, the effect of traffic is introduced directly in the model as a factor of degradation for some of the properties of the reinforced system. In particular, the stress distribution angle decreases with the repetition of load cycles and, in consequence, the enhanced confinement mechanism is less and less effective.

The analysis results in a compact design equation in which all the discussed factors are represented. However, iterations are required to solve the equation, because of its implicit form. A final form of the equation including calibrated coefficients is presented in Giroud and Han (2004b). Design charts, such as the one shown in Figure 7.9, can be produced. Comparison of the computed base thickness with measured values reported by Hammitt (1970) shows a good agreement.

It should be noted that the analysis does not account for the tensile membrane effect of the reinforcement, which is not the dominant mechanism with geogrids. The tensile force in the reinforcement and overall anchorage verification are not provided by this method. These features, as well as aperture size compatibility with the granular base grain size, are assumed to be adequate, and are not specifically addressed. Other limitations noted by the authors relate to the range of parameters that were represented in the supporting experimental data and the specific type of geogrid considered in the analysis.

7.3. Summary of main points

Placement of geotextiles and geogrids at the interface of a soft subgrade and a granular base course can significantly improve the bearing capacity and durability of an unpaved road. The main aspects of this type of application of geosynthetics are as follows

- The fundamental modes of reinforcement are tensile membrane action and enhanced lateral confinement of the base course. Material properties that are essential to these mechanisms are interface friction (for geotextiles) or grain–aperture interlock (for geogrids), tensile resistance and modulus for both types of geosynthetic, and high compressibility of the subgrade.
- Tensile membrane action occurs when there is a large deformation of the geosynthetic and a large deflection of the subgrade. Attenuation of traffic-induced vertical stress on the subgrade results from this mechanism.
- Enhanced confinement, by stiffening the granular base course, restrains the lateral deformation, facilitates the transfer of load and reduces shear stresses in the subgrade.
- Under cyclic loading due to traffic, degradation of base course characteristics and accumulation of permanent deformation in the subgrade are attenuated by the action of the geosynthetic.
- Design methods have been developed to quantify the effect of reinforcement on the required thickness of the granular base course, as compared with the necessary thickness in absence of reinforcement, for a given allowable rut depth and an expected number of wheel load cycles. These methods make use of rational analysis combined with empirical knowledge. However, due to our imperfect understanding of the complex interaction between the soil and the reinforcement under both monotonic and cyclic loading, and because of simplifications inherent in practical design, significant discrepancies can sometimes be seen between these methods.
- A primary condition for successful reinforcing of an unpaved road is that the selected geosynthetic is handled with professional care at the construction site, and is not significantly damaged during its installation and subsequent placement of the granular base course.

7.4. Some common questions and their answers

Questions

1. What types of geosynthetic are used in unpaved road reinforcing?

2. Do geosynthetics in unpaved roads have other functions, in addition to mechanical reinforcement?

3. What conditions or material properties are essential to successful reinforcing of unpaved roads?

4. What improvement can be expected from the use of geosynthetic reinforcement in unpaved roads?

Answers

1. Geotextiles and geogrids are generally used for this type of application.

2. Geotextiles can also act as separators/filters to prevent contamination of the granular base course by fines pumping from the subgrade under cyclic load. Geogrids act solely as reinforcement.

3. Important factors are: a high tensile modulus of the geosynthetic; a high compressibility of the subgrade; good interface friction between the geotextile and the granular base material; good interlocking between the geogrid apertures and the granular base material; stability of the geogrid apertures; resilience of the geosynthetic under cyclic loading; and survivability of the geosynthetic with regard to installation.

4. Reduced rutting under traffic load is expected, which can translate to either increased traffic capacity or savings in base course thickness.

REFERENCES

Ashmawy AK and Bourdeau PL (1995) Geosynthetic-reinforced soils under repeated loading: a review and comparative design study. *Geosynthetics International* **2(4)**: 643–678.

Ashmawy AK and Bourdeau PL (1997) Testing and analysis of geotextile-reinforced soil under cyclic loading. *Proceedings of the Geosynthetics '97 Conference*, Long Beach, CA, pp. 663–674.

Ashmawy AK and Bourdeau PL (1998) Effect of geotextile reinforcement on the stress–strain and volumetric behavior of sand. *Proceedings of the 6th International Conference on Geosynthetics*, Atlanta, GA, pp. 1079–1082.

Ashmawy AK, Bourdeau PL, Drnevich VP and Dysli M (1999) Cyclic response of geotextile-reinforced soil. *Soils and Foundations* **39(1)**: 43–52.

Bathurst RJ, Raymond GP and Jarrett PM (1986) Performance of geogrid-reinforced ballast railroad track support. *Proceedings of the 3rd International Conference on Geotextiles*, Vienna, pp. 43–48.

Bender DA and Barenberg EJ (1978) Design and behavior of soil–fabric–aggregate systems. *Transportation Research Record* **671**: 64–75. National Research Council, Washington, DC.

Bourdeau PL (1989) Modeling of membrane action in a two-layer reinforced soil system. *Computers and Geotechnics* **7(1 and 2)**: 19–36.

Bourdeau PL (1991) Membrane action in a two-layer soil system reinforced by geotextile. *Proceedings of Geosynthetics '91*, Atlanta, pp. 439–453. North American Geosynthetics Society.

Bourdeau PL, Miskin KK and Fuller JM (1990) Behavior of geotextile-reinforced soil under cyclic loading. *Proceedings of the 4th International Conference on Geotextiles*, The Hague, p. 251.

Bourdeau PL, Pardi L and Recordon E (1991) Observation of soil–reinforcement interaction by x-ray radiography. In *Performance of Reinforced Soil Structures* (British Geotechnical Society), Thomas Telford, London, pp. 347–352.

Burd HJ (1986) A large displacement analysis of a reinforced unpaved road. PhD thesis, Oxford University, Oxford.

Chaddock BCJ (1988) *Deformation of road foundations with geogrid reinforcement.* Department of Transportation, Transport and Road Research Laboratory (TRRL), Crowthorne. Research Report 140.

Chan F, Barksdale RD and Brown SF (1989) Aggregate base reinforcement of surfaced pavements. *Geotextiles and Geomembranes* **8(3)**: 165–189.

Christopher BR, Holtz RD and DiMaggio JA (1984) *Geotextile Engineering Manual.* US Department of Transportation, Federal Highway Administration, Washington, DC.

De Garidel R and Javor E (1986) Mechanical reinforcement of low-volume roads by geotextiles. *Proceedings of the 3rd International Conference on Geotextiles*, Vienna, pp. 147–152.

De Groot M, Janse E, Maagdenberg TAC and Van den Berg C (1986) Design method and guidelines for geotextile application in road construction. *Proceedings of the 3rd International Conference on Geotextiles*, Vienna, pp. 741–746.

Delmas Ph, Matichard Y, Gourc JP and Riondy G (1986) Unsurfaced roads reinforced by geotextiles. A seven year experiment. *Proceedings of the 3rd International Conference on Geotextiles*, Vienna, pp. 1015–1020.

Douglas RA (1991) Parametric study of the stiffness of unbound, geosynthetic-built roads on compressible subgrades. *Proceedings of the 44th Canadian Geotechnical Conference*, Calgary, pp. 80.1–80.7.

Fannin RJ (1987) Geogrid reinforcement of granular layers on soft clay – a study at model and full scale. PhD thesis, Oxford University, Oxford.

Fannin RJ and Sigurdsson O (1996) Field observations on stabilization of unpaved roads with geosynthetics. *Journal of Geotechnical Engineering*, ASCE **122(7)**: 544–553.

French Geotextile Committee (1981) *Recommendations pour l'Emploi des Géotextiles dans les Voies de Circulation Provisoires, les Voies à Faible Trafic et les Couches de Forme.* Comité Français des Géotextiles, Boulogne. (In French.)

Gabr M (2001) *Cyclic Place Loading Tests on Geogrid Reinforced Roads.* Tensar Earth Technologies, Inc., NC State University.

Giroud JP and Han J (2004a) Design method for geogrid-reinforced unpaved roads. I. Development of design method. *Journal of Geotechnical and Geoenvironmental Engineering, ASCE* **130(8)**: 775–786.

Giroud JP and Han J (2004b) Design method for geogrid-reinforced unpaved roads. II. Calibration and applications. *Journal of Geotechnical and Geoenvironmental Engineering, ASCE* **130(8)**: 787–797.

Giroud JP and Noiray L (1981) Geotextile-reinforced unpaved road design. *Journal of Geotechnical Engineering, ASCE* **107(9)**: 1233–1254.

Giroud JP, Ah-Line C and Bonaparte R (1985) Design of unpaved roads and trafficked areas with geogrids. In *Polymer Grid Reinforcement*, Thomas Telford, London, pp. 116–127.

Gourc JP (1983) Quelques aspects du comportement des géotextiles en mécanique des sols. Thèse de doctorat ès sciences, Université et Institut National Polytechnique de Grenoble.

Haliburton TA and Barron JV (1983) Optimum depth method for design of fabric reinforced unsurfaced roads. In *Engineering Fabrics in Transportation Construction*, Transportation Research Record. TRB No. 916, pp. 26–32.

Hammitt GM (1970) *Thickness Requirement for Unsurfaced Roads and Airfields, Bare Base Support.* US Army Corps of Engineers Waterways Experiment Station, Vicksburg, MI. Technical Report S-70-5.

Hirano I, Itoh M, Kawahara S, Shirasawa M and Shimizu H (1990) Test on trafficability of a low embankment on soft ground reinforced with geotextiles. *Proceedings of the 4th International Conference on Geotextiles, Geomembranes and Related Products*, The Hague, pp. 227–232.

Holtz RD and Sivakugan N (1987) Design charts for roads with geotextiles. *Geotextiles and Geomembranes* **5(3)**: 191–200.

Houlsby GT and Jewell RA (1990) Design of reinforced unpaved roads for small rut depths. *Proceedings of the 4th International Conference on Geotextiles, Geomembranes and Related Products*, The Hague, pp. 171–176.

Jaeklin FP (1986) Design of road base and geotextile by regression analysis from experienced data sources. *Proceedings of the 3rd International Conference on Geotextiles*, Vienna, pp. 87–92.

Knapton J and Austin RA (1996) Laboratory testing of reinforced unpaved roads. In *Earth Reinforcement* (Ochiai H, Yasufuku N and Omine K (eds)), Balkema, Rotterdam, pp. 615–618.

Koerner RM (1998) *Designing with Geosynthetics*, 4th edn. Prentice Hall, Upper Saddle River, NJ.

Leflaive E (1985) Sol renforcé par des fils continus: le Texsol. *Proceedings of the 11th International Conference on Soil Mechanics and Foundation Engineering*, vol. 3, San Francisco, pp. 1787–1790. (In French.)

Love JP (1984) Model testing of geogrids in unpaved roads. PhD thesis, Oxford University, Oxford.

Love JP, Burd HJ and Milligan GWE (1987) Analytical and model studies of reinforcement of granular fill on a soft clay subgrade. *Canadian Geotechnical Journal* **24(4)**: 611–622.

Madani C, Long NT and Legeay G (1979) Comportement dynamique de la terre armée à l'appareil triaxial. *Proceedings of the Colloque International sur le Renforcement des Sols*, Paris, pp. 83–88. (In French.)

Maher MH and Ho YC (1993) Behavior of fibre-reinforced cemented sand under static and cyclic loading. *Geotechnical Testing Journal, ASTM* **16(3)**: 330–338.

McGown A, Yeo KC and Yogarajah I (1990) Identification of dynamic interlock mechanism. *Proceedings of the International Reinforced Soil Conference*, University of Strathclyde, Glasgow. In *Performance of reinforced soil structures* (British Geotechnical Society), Thomas Telford, London, pp. 377–379.

Milligan GWE, Fanin RJ and Farrar DM (1986) Model and full-scale tests of granular layers reinforced with a geogrid. *Proceedings of the 3rd International Conference on Geotextiles*, Vienna, pp. 61–66.

Milligan GWE, Jewell RA, Houlsby GT and Burd HJ (1989) A new approach to the design of unpaved roads. *Ground Engineering* **22(3)**: 25–29 (Part 1); **22(11)**: 37–42 (Part 2).

Nimmesgern M and Bush D (1991) The effect of repeated traffic loading on geosynthetic reinforcement anchorage resistance. *Proceedings of the Geosynthetics '91*, Atlanta. North American Geosynthetics Society, pp. 665–672.

Swiss Society of Geotextiles Professionals (1985) *Le Manuel des Géotextiles*. Association Suisse des Professionels des Géotextiles, EMPA, St Gall. (In French and German.)

Tingle JS and Webster SL (2003) Review of Corps of Engineers design of geosynthetic reinforced unpaved roads. Presented at: TRB 82nd Meeting, Transportation Research Board, Washington, DC. Presentation and CD-ROM.

Wayne MH, Han J and Akins K (1998) The design of geosynthetic reinforced foundations. *Proceedings of Geo-Congress '98*, Geotechnical Special Publication No. 76, ASCE, pp. 1–18.

Webster SL and Alford SJ (1978) *Investigation of Construction Concepts for Pavement across Soft Ground.* US Army Corps of Engineers, Waterways Experiment Station, Vicksburg, MI. Technical Report S-78-6.

Handbook of Geosynthetic Engineering
ISBN 978-0-7277-4175-2

http://dx.doi.org/10.1680/hge.41752.179

Chapter 8
Paved roads

S. W. Perkins[*], R. R. Berg[†] and B. R. Christopher[‡]
[*]Department of Civil Engineering, Montana State University, USA
[†]Ryan R. Berg and Associates, Inc., Woodbury, Minnesota, USA
[‡]Christopher Consultants, Roswell, Georgia, USA

Geosynthetics are used in paved roads to provide reinforcement, separation, drainage and filtration. Common geosynthetics used include geotextiles, geogrids, geonets and geocomposites. Geosynthetics help control the development of common pavement distress features such as rutting and cracking. This is accomplished by structural reinforcement of the base aggregate layer, controlling infiltration into moisture-susceptible layers, and preventing contamination of structural layers by the migration of fines from underlying soils. These functions help preserve and in some cases enhance the structural integrity of the constructed pavement layers. In this chapter, the functions listed above are described and illustrated for paved roads. Guidelines for the selection and use of geosynthetics for paved road reinforcement are presented. These guidelines currently rely upon the use of a traffic benefit ratio or a base-course reduction ratio. These parameters are dependent on the geosynthetic used and the design specific roadway materials and layer thickness. The ratios are currently determined by constructing test sections incorporating these conditions. New methods incorporating mechanistic-empirical design principles are currently under development and are discussed.

8.1. Introduction

Paved roadways are designed and constructed to allow safe, efficient and economical transport of passenger and commercial vehicular traffic. To accomplish these objectives, distress of the pavement induced during construction and during the operational life of the roadway must be kept within acceptable limits. Design considerations and options necessary to limit pavement distress must be viewed in terms of roadway construction, operation and maintenance costs so that designs are economically optimised.

The use of geosynthetics in paved roadways presents several design options to the pavement engineer that allows for improved pavement performance in an economically efficient manner. Geosynthetics may be used in paved roadways for reinforcement, separation, filtration and drainage. Geosynthetics may be used as a construction expedient allowing for reduced construction time and cost, or as a design component intended to control distress induced by operational traffic. A paved roadway design incorporating geosynthetics may be economically efficient from the standpoint of reducing construction costs and/or operation and maintenance costs.

In this chapter, a distinction is made between geosynthetics when used for construction purposes and when used as a design component for the control of pavement distress under operational traffic. The four functions of reinforcement, separation, filtration and drainage can be utilised for both construction and operation applications; however, the mechanisms by which the geosynthetic provides these functions, the selection of the proper geosynthetic product and the design techniques used can be significantly different. Generally, if a geosynthetic is needed to stabilise the subgrade for construction purposes, the roadway is treated as an unpaved road and the geosynthetic is employed according to the guidelines and techniques discussed in Chapter 7. In this chapter, material is presented that defines the functions of reinforcement, separation, filtration and drainage for the control of pavement distress due to the application of vehicular traffic during the operational life of the pavement. The use of geosynthetics for reinforcement is addressed in more detail, with particular attention given to experimental evidence illustrating the mechanisms by which the geosynthetic provides reinforcement, and design guidelines and practices.

8.2. Distress features and their relationship to geosynthetics

Pavement distress can take on several forms, and is generally classified into categories attributed to structural deficiencies and those resulting in loss of function of the roadway. Structural deficiencies arise from the loss of mechanical properties that govern the load-carrying capacity of the roadway or by an increase in loading for which the roadway was not designed. In either event, structural deficiencies result in the loss of the ability of the roadway to carry vehicular loads for which it was designed. The loss of functional capacity of the roadway typically involves the development of conditions that result in discomfort to the roadway user, and include conditions such as an excessively rough riding surface, an excessively cracked riding surface, excessive rutting, potholes and asphalt bleeding. Functional problems may or may not be due to preceding structural deficiencies. In the event that functional problems are not related to preceding structural deficiencies, then the functional problem, if left uncorrected, will most likely lead to the development of a structural problem.

Roadway failure due to structural deficiencies can occur prematurely or expectedly at the end of the design life of the roadway. In either event, structural deficiencies have developed and may have been caused by a number of conditions. The development of permanent strain in the base and subgrade materials under repeated traffic loading can eventually result in excessive deformation or rutting of the roadway surface. In a grossly underdesigned road, this can happen relatively quickly. Even for a properly designed roadway, however, the accumulation of rutting can occur gradually over the life of the pavement, such that the pavement becomes inoperable only at the end of the design life of the pavement. In this case, geosynthetic reinforcement of the roadway system could be used to enhance the structural characteristics of the pavement system and to control rutting of the roadway.

In another case, the mixing of the subgrade with the base course, by either mechanical mixing or pumping of the subgrade soil, would lead to a deterioration in the mechanical properties of the base course layer. In this situation, use of a geosynthetic separator/filter would ensure the structural integrity of the base aggregate and the intended capacity of the roadway.

Drainage of water away from the pavement structure is a critical feature for maintaining the structural integrity of the roadway. Geosynthetics can be used to divert and carry water away from the pavement structure. These three situations where geosynthetics can be used to improve or maintain the structural capacity of a roadway are illustrated in Figure 8.1.

Figure 8.1 Situations where geosynthetics can be used for the prevention of roadway distress: (a) reinforcement, (b) separator/filter and (c) drainage

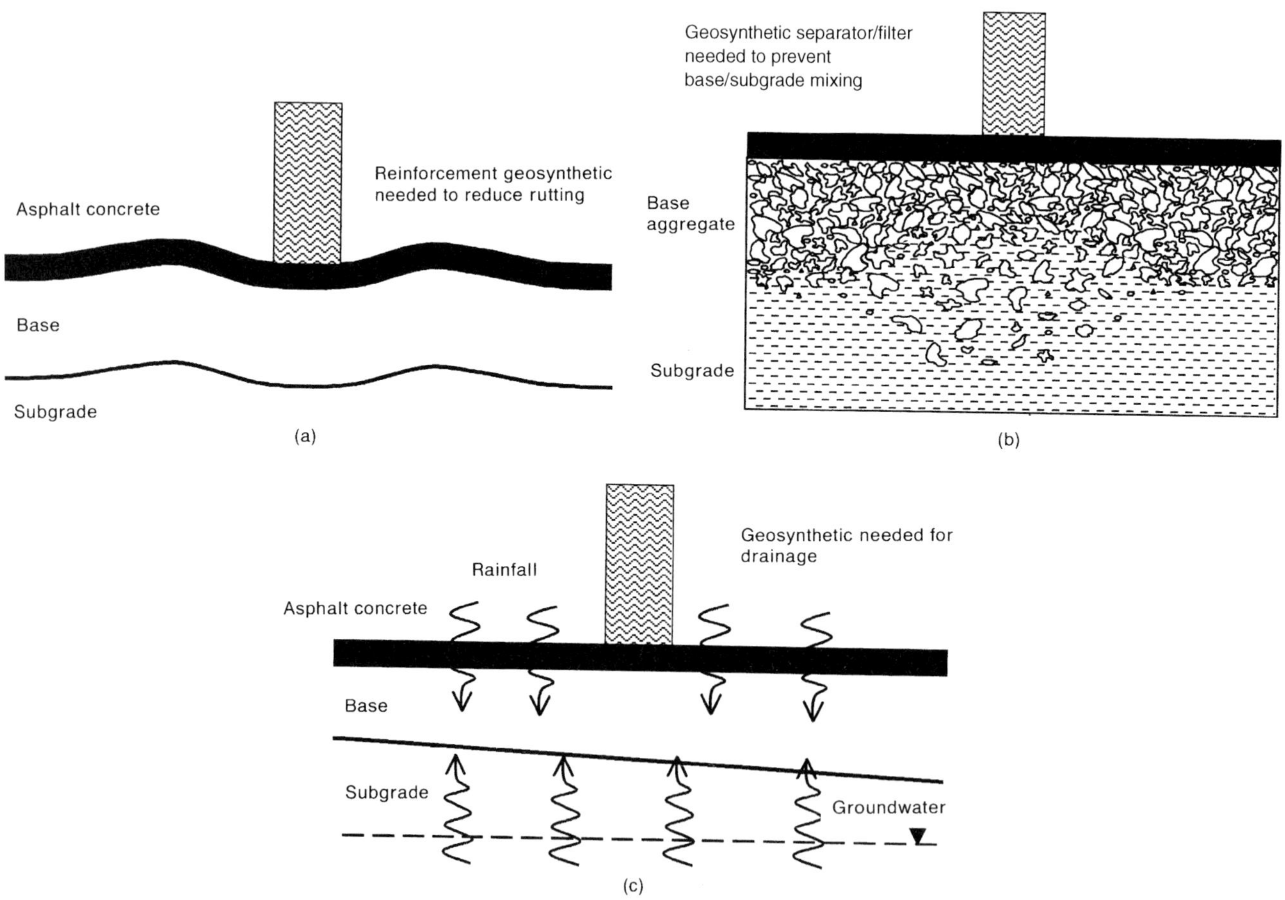

Geosynthetics are commonly used in the design of paved roadways in one of two ways to prevent roadway failure. In certain situations, geosynthetics can be used to alter the design cross-section of the roadway, generally by reducing the thickness of one or more of the pavement structural layers, such that a roadway of equal life results. Geosynthetics can also be added to the original design cross-section, such that the life of the roadway is extended while maintenance costs are decreased. The selection of a specific geosynthetic depends on the distress feature being addressed and the roadway conditions leading to the anticipated distress. Specific functions and benefits of geosynthetics are discussed in the section below.

8.3. Geosynthetic functions

The purpose of this section is to describe the basic functions and benefits of geosynthetics within the context of their use in paved roadways. These functions fall into four categories – namely, *reinforcement*, *separation*, *filtration* and *drainage* – and are described in general terms in this section. These 'primary' functions are often used to identify the specific application of the geosynthetic in the roadway system; however, in most applications a combination of these functions is actually required. The importance of these combinations will be reviewed at the end of this section.

8.3.1 Reinforcement

The function of reinforcement pertains to the ability of the geosynthetic to aid in supporting operational vehicular traffic loads. In Chapter 7, the tensioned-membrane reinforcement mechanism is described, where it is noted that relatively large deformations of the roadway surface are necessary to mobilise this particular reinforcement mechanism. Since a paved roadway is generally considered inoperable once a rut depth in excess of approximately 25 mm is reached, the tensioned-membrane reinforcement mechanism is not applicable for paved roadways. The principal mechanism responsible for reinforcement in paved roadways is one generally referred to as *base course lateral restraint*. This function was originally described by Bender and Barenberg (1978), and was later elaborated on by Kinney and Barenberg (1982) for geotextile-reinforced unpaved roads. This name may be somewhat misleading in that the function, as envisioned, incorporates mechanisms, in addition to preventing lateral movement of the base course aggregate. A more appropriate description might be to describe this reinforcement function, and its associated mechanisms, as one of a shear-resisting interface, as suggested in Perkins *et al.* (1998a).

The reinforcement function of a shear-resisting interface develops through shear interaction of the base course layer with the geosynthetic layer or layers contained in, or at the bottom of, the base aggregate (Figure 8.2), and consists of four separate reinforcement mechanisms. Vehicular loads applied to the roadway surface create a lateral spreading motion of the base course aggregate. Tensile lateral strains are created in the base below the applied load, as the material moves down and out, away from the load. Lateral movement of the base allows for vertical strain to develop, leading to a permanent rut in the wheel path.

Figure 8.2 Reinforcement mechanisms of a shear-resisting interface
Reproduced from Perkins (1999)

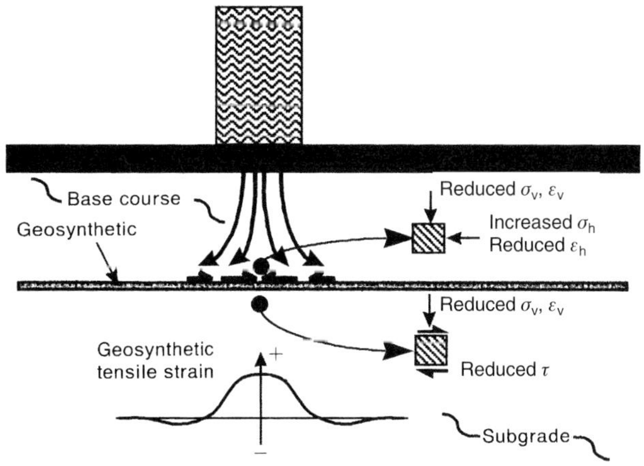

Placement of a geosynthetic layer or layers in the base course allows for shear interaction to develop between the aggregate and the geosynthetic as the base attempts to spread laterally. Shear load is transmitted from the base aggregate to the geosynthetic and places the geosynthetic in tension. The relatively high stiffness of the geosynthetic acts to retard the development of lateral tensile strain in the base adjacent to the geosynthetic. Lower lateral strain in the base results in less vertical deformation of the roadway surface. Hence, the first mechanism of reinforcement corresponds to direct prevention of lateral spreading of the base aggregate.

Shear stress developed between the base course aggregate and the geosynthetic provides an increase in lateral stress within the base. This increase in lateral confinement leads to an increase in the mean hydrostatic stress, which may tend to increase as additional traffic loads are applied through the dynamic interlock effect described by McGown *et al.* (1990). Granular materials generally exhibit an increase in the elastic modulus with increasing mean stress. The second reinforcement mechanism results from an increase in the stiffness of the base course aggregate when adequate interaction develops between the base and the geosynthetic. The increased stiffness of this layer results in lower vertical strains in the base. It would also be expected that an increase in the modulus of the base would result in lower dynamic, recoverable vertical deformations of the roadway surface, implying that fatigue of the asphalt concrete layer would be reduced.

The presence of a geosynthetic layer in the base can also lead to a change in the state of stress and strain in the subgrade. For layered systems, where a less stiff subgrade material lies beneath the base, an increase in the modulus of the base layer results in an improved, more broadly distributed vertical stress on the subgrade. In general, the vertical stress in the base and subgrade directly beneath the applied load should decrease as the base layer stiffness increases. The vertical stress on the subgrade will become more widely distributed, meaning that surface deformation will be reduced and more uniform. Hence, a third reinforcement mechanism results from an improved vertical stress distribution on the subgrade.

The fourth reinforcement mechanism results from a reduction in shear stress in the subgrade soil. It is expected that shear stress transmitted from the base course to the subgrade would decrease as shearing of the base transmits tensile load to the reinforcement. Less shear stress, coupled with less vertical stress, results in a less severe state of loading (Houlsby and Jewell, 1990), leading to lower vertical strain in the subgrade.

A prerequisite to realising the reinforcement mechanisms described above is the development of a strain distribution in the geosynthetic similar to that shown in Figure 8.2. Haas *et al.* (1988), Miura *et al.* (1990) and Perkins *et al.* (1998a,b) have presented data demonstrating such trends for paved roadways using geogrid reinforcement, while Perkins (1999) has shown this effect for a geotextile. Data from several of these studies are presented in a later section, to demonstrate these mechanisms.

Benefits of reinforcement on the design of flexible pavements are generally expressed in terms of an extension in the life of the pavement or an allowable reduction in the base course thickness. An extension in the life of the pavement is typically expressed in terms of the *traffic benefit ratio* (TBR), which is illustrated in Figure 8.3. The TBR is defined as the ratio of the number of traffic loads between otherwise identical reinforced and unreinforced pavements that can be applied to reach a particular permanent surface deformation of the pavement. The TBR indicates the additional amount of traffic loads that can be applied to a pavement when a geosynthetic is added, with all other pavement materials and geometry being equal.

Figure 8.3 Illustration of reinforcement benefit defined by the traffic benefit ratio (TBR)

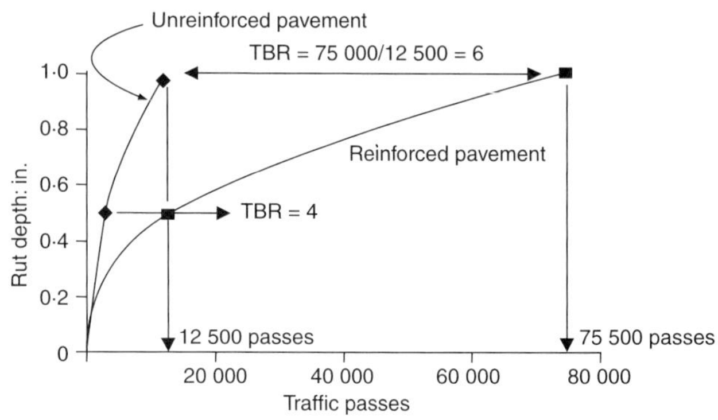

The benefit of reducing the base aggregate thickness is typically defined by the *base-course reduction ratio* (BRR), which is illustrated in Figure 8.4. The BRR defines the percentage reduction in the base course thickness of a reinforced pavement, such that equivalent life is obtained between the reinforced and the unreinforced pavements with the greater aggregate thickness. Design methods using the TBR and BRR are presented in Section 8.6.

8.3.2 Separation

In many situations, fines from the underlying subgrade can contaminate the base course layer of a roadway, and may occur during or after construction. Contamination of the base course layer leads to a reduction in strength, stiffness and drainage characteristics, promoting distress and early failure of the roadway. Yoder and Witczak (1975) state that fines as little as 20% by weight of the subgrade, mixed in with the base aggregate, will reduce the bearing capacity of the aggregate to that of the subgrade. Jorenby and Hicks (1986) conducted a laboratory study to examine the influence of added fines on the resilient modulus of base course aggregates. The study showed that up to 6% added fines could be tolerated before the aggregate stiffness reduced; however, a significant reduction in the resilient

Figure 8.4 Illustration of reinforcement benefit defined by the base-course reduction ratio (BRR)

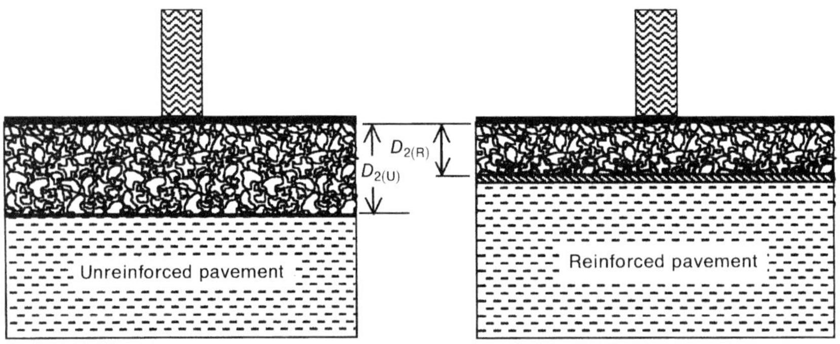

$$\mathrm{BRR} = \frac{D_{2(U)} - D_{2(R)}}{D_{2(U)}}$$

Figure 8.5 The geosynthetic separation function

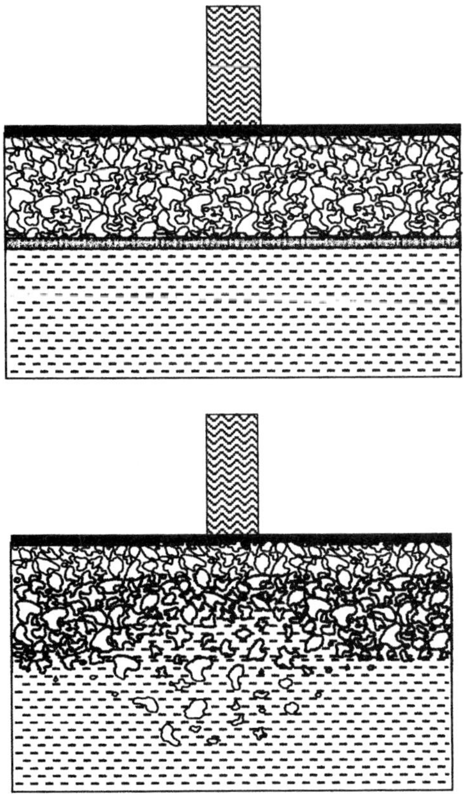

modulus occurred as the fines were increased above 6%. To maintain adequate drainage properties, only 2% fines can be allowed. Fines contamination also makes the base course layer more susceptible to frost heaving.

The function of separation refers to the ability of a geosynthetic to provide physical separation of subgrade and base or subbase aggregate materials, both during construction and during the operating life of the roadway, and is illustrated in Figure 8.5. Separation prevents mixing of the subgrade and aggregate where mixing is caused by some type of mechanical action. Mechanical actions causing mixing generally arise from physical forces imposed by construction or operating traffic, and may cause the aggregate to be pushed down into the soft subgrade and/or the subgrade to be squeezed up into the base aggregate. If the subgrade is weak at the time of construction, then the combination of a relatively thin initial base course lift, combined with heavy construction equipment, generally means that the potential for mixing is greatest during construction. Conversely, if the subgrade is relatively dry and strong during construction, yet there is an expectation that it will become wet and weaker during the life of the pavement, the potential for mixing is greatest during the operating life of the roadway. A properly designed geosynthetic separator allows the base course aggregate to remain 'clean', which preserves its strength and drainage characteristics.

Brorsson and Eriksson (1986) reported results from construction taking place on a frost-susceptible clayey silt subgrade that was too weak to walk on at the time of construction. Various geotextile products were used, with 900 mm of base eventually placed on top of the geotextiles. The sections were exhumed after 5–10 years, with very little mixing of the subgrade being observed. In addition, in many situations, the subgrade was seen to have consolidated and to have a much lower water content than during construction. Tsai *et al.* (1993) reported results from full-scale test roads where various geotextiles were seen to prevent base contamination observed in sections not containing a geotextile. Holtz (1996) exhumed geotextiles from paved roadways in 22 different sites in the state of Washington, USA, and showed that the geotextiles, in all situations, performed their intended separation function. In certain situations, evidence of geotextile damage, clogging and blinding was seen, but these conditions did not appear to influence the ability of the geotextile to carry out its intended function. Black and Holtz (1997) exhumed five different geotextiles from a full-scale test road where the geotextiles had been in place for five years. The geotextiles appeared to have aided in the consolidation of the subgrade. Some fines migration into the base layer was noted, but did not appear to influence the performance of the road. Bonaparte *et al.* (1988) exhumed geotextiles from existing unpaved roads, and found no evidence of contamination. Other studies, including those by Richardson and Behr (1990), Lawson (1992) and Tsai and Holtz (1997), have also shown positive results from geotextiles used for separation.

Several studies (Austin and Coleman, 1993; Fannin and Sigurdsson, 1996) have presented data showing that the inclusion of a geogrid limited the amount of fines contamination compared with a similar section without a geosynthetic, while in other situations, contamination of the base occurred as aggregate was pushed down or subgrade was squeezed up through the apertures of the geogrid. The ability of a geogrid to act as a separator appears to be related to the stress imposed by vehicular traffic at the subgrade–aggregate interface and the strength of the subgrade, along with the gradation of the base and the geogrid aperture size. If the geogrid is providing a reinforcement function, then there may be an added benefit in that the shear-resisting interface reinforcement function limits the shear and vertical stress in the subgrade, and thereby limits the potential for mechanical mixing. Additional information, needed to demonstrate and quantify these effects before designs, using geogrids as separators, can be safely developed.

For separation applications, unlike reinforcement applications, the strength and modulus of the geosynthetic is important only to ensure survivability of the material during construction and operation of the roadway. The addition of a geosynthetic separator ensures that the base course layer, in its entirety, will contribute, and continue to contribute, its intended structural support of vehicular loads; the geosynthetic separator itself is not seen to contribute structural support to the roadway. Holtz *et al.* (2008) have provided guidelines for situations where geotextiles are used for separation, filtration and drainage. Criteria for the survivability of geosynthetics for separation applications are provided in the AASHTO M288-96 specifications (AASHTO, 1997). These criteria consider the construction conditions of the subgrade, the contact pressure provided by the construction equipment, and the compacted base course thickness to be used. Based on the combination of these conditions, the survivability level of the geosynthetic is assessed. The survivability level is then expressed in terms of certain geosynthetic index properties. AASHTO M288-96 (AASHTO, 1997) also provides recommendations for filtration and drainage properties of geotextiles for use in roadways.

8.3.3 Filtration

Filtration refers to the ability of the geosynthetic to filter out fines contained in porewater as the water flows from the subgrade to the base. Flow is typically induced by porewater pressures generated in the subgrade as a result of traffic loading. Fines contained in the subgrade may become suspended in the porewater as a result of shearing action and can be carried into the base in the absence of a proper filter.

The key to this application lies in the ability of the geosynthetic to filter fines without becoming clogged. Snaith and Bell (1978) showed that the percentage of fines passing into the base was a function of the type of geotextile used. A heavy woven geotextile was found to be best in preventing the passage of fines. Bell *et al.* (1982) later showed that the material passing though the geotextile was a function of opening size. Larger O_{95} values resulted in greater contamination. Perkins and Brandon (1998) showed similar results. A strong correlation between the initial water content to the amount of subsequent pumping was also noted, indicating that the degree of saturation is an important consideration in determining the need for a geotextile filter. Interesting results were obtained in terms of the rate at which porewater pressure was dissipated. High rates of dissipation corresponded to greater levels of contamination. The majority of the contamination appeared to occur at points where large stones were in contact with the geotextile and created high stress concentrations. This result was also seen by Hoare (1982) and Lafleur *et al.* (1990), and indicates that thicker geotextiles will perform better due to their ability to reduce stress concentrations. Glynn and Cochrane (1987) expanded upon this concept, and showed that the small depressions created in the geotextile by the aggregate provided a location where water could pond, and where a clay slurry could be formed and pumped into the base layer.

Nishida and Nishigata (1994) also showed that the amount of fines pumped though woven geotextiles was proportional to the opening size of the material, but did not find a significant correlation with non-woven geotextiles. Rather, for non-woven geotextiles, the unit weight of the geotextile was inversely proportional to the amount of fines passed. This may also be related to the increase in thickness associated with heavier-weight geotextiles, which has been correlated to filtration performance by Giroud (1996). Alobaidi and Hoare (1994, 1996) showed that the amount of pumping was directly related to the amount of porewater pressure developed in the subgrade, which was, in turn, dependent on the cyclic stress and strain conditions developed at this point. Similar to other findings, an increase in the rate of porewater pressure dissipation was shown to cause more 'erosion' within the subgrade, and produced higher levels of contamination. Geotextiles with high permeability were seen to provide porewater pressure dissipation within a single load cycle, and showed more pumping than geotextiles with a greater thickness, which reduced the critical hydraulic gradient.

A full understanding of the need for a geotextile filter layer depends on the cyclic shear stress induced in the top of the subgrade due to vehicular loading, the excess porewater pressure developed in the subgrade and the tendency for detachment of soil particles that will then be free to move with the drainage water. With the need for a geotextile filter established, a geotextile having the characteristics necessary to prevent fine migration can be selected. Guidelines for the use of geotextiles for filtration are given in AASHTO M288-96 (AASHTO, 1997).

8.3.4 Drainage

Inadequate drainage in pavements is a principal cause of pavement distress. Adequate drainage of a pavement is believed to extend the life of a pavement system by up to two to three times that of a similar pavement having inadequate drainage (Cedergren, 1987). Drainage refers to the ability of the geosynthetic to intercept and drain water in the plane of the geosynthetic to the edge of the pavement, where it can be properly discharged through pavement edge drains. In addition, geosynthetics, specifically thick geocomposites, may serve as a capillary break to mitigate frost heaving in frost-susceptible soils (Christopher *et al.*, 1999).

The functions of the geosynthetic, required for drainage applications in pavements, are sufficient transmissivity, to carry water from the pavement system, and sufficient compressional stiffness, such that the cross-sectional area of the geosynthetic, through which water flows, is maintained. The use of a 100 mm-thick open-graded base layer has generally proven adequate to meet the requirement of the complete drainage of the pavement within a two-hour period. Based on this condition, Christopher *et al.* (1999) have estimated that a geosynthetic used for drainage would require a transmissivity of 0.00035–0.001 m/s.

The vertical stress imposed on the geosynthetic from traffic loads can be as great as 600 kPa, depending on the level of the geosynthetic within the pavement structure. Vertical stresses imposed during construction can often be more critical than those imposed from operational traffic. The repetitive nature of operational traffic and the likelihood for vertical compression of the geosynthetic due to creep must also be considered. These conditions are difficult to meet with non-woven geotextiles, and generally require that a geocomposite drainage layer be used.

Table 8.1 Application and associated functions of geosynthetics in roadway systems
Adapted from Holtz *et al.* (2008)

Application	Function(s)	Subgrade strength where most applicable	Qualifier
Separator	Separation Secondary: filtration[a]	90 kPa $\leq c_u \leq$ 240 kPa 3 $\leq$ CBR $\leq$ 8 30 MPa $\leq M_R \geq$ 80 MPa	Soils containing high fines (SC, CL, CH, ML, MH, SM, SC, GM, GC)
Stabilisation	Separation, filtration and some reinforcement (especially $CBR < 1$) Secondary: transmission	$c_u <$ 90 kPa CBR $<$ 3 $M_R <$ 30 MPa	Wet, saturated fine-grained soils (i.e. silt, clay and organic soils)
Base reinforcement	Reinforcement Secondary: separation	30 kPa $\leq c_u \leq$ 240 kPa 3 $\leq$ CBR $\leq$ 8 10 MPa $\leq M_R \geq$ 80 MPa	All subgrade conditions. Reinforcement located within 6–12 in. (150–300 mm) of pavement
Drainage	Transmission and filtration Secondary: separation	Not applicable	Poorly draining subgrade

CBR, California bearing ratio.
[a] Always evaluate filtration requirements.

8.3.5 Combined functions

The different applications of geosynthetics in roadway systems are often identified by their primary function, as described in the previous sections. However, in addition to the primary function, geosynthetics usually are required to perform one or more *secondary* functions. For example, for a geotextile filter to work, it must also provide separation. Geosynthetics used in drainage must also provide filtration to prevent clogging of the system or piping of the fines into the drain. For reinforcement geosynthetics, if separation is not provided, fines can migrate into and weaken the base-course aggregate, thus diminishing the effectiveness of the reinforcement function. The total contribution of the geosynthetic to a particular application is thus made up of both primary and secondary functions. Secondary functions can be equally important as the primary function, and in order to obtain optimum geosynthetic performance, both must be considered in the design computations and specifications (Holtz *et al.*, 2008). Table 8.1 provides a list of the primary and secondary functions associated with applications of geosynthetics in paved roads, along with the subgrade conditions where these applications are most effective.

As shown in Table 8.1, stabilisation of soft subgrade conditions to provide a working platform for construction of the pavement system is truly a multifunction application. As previously indicated, this application is primarily considered for construction, and uses the design methods that were reviewed in Chapter 7. While the stabilisation application is primarily used for initial construction, geosynthetics used in that application can also provide long-term benefits by maintaining the roadway design section and the base-course material integrity by preventing the aggregate from penetrating the subgrade. The separation function provided by geotextiles, geogrid/geotextile geocomposites or geogrids with appropriately designed filter aggregate prevents the migration of fines into base/sub-base materials, especially into open-graded bases, maintaining the support and drainage characteristics of the base over the life of a pavement system. By providing these multiple functions, the geosynthetic basically improves the reliability of the pavement system performance, especially during seasonally weak subgrade conditions, and thus should ultimately increase the life of the roadway (Perkins *et al.*, 2010).

8.4. History and experimental evidence for base reinforcement

The use of geosynthetics for base reinforcement has been examined over a period of approximately 20 years, with a number of studies contributing to the body of knowledge. Research began in the early 1980s, and tended to follow in the footsteps of work performed on geosynthetic-reinforced unpaved roads. Geotextiles used for reinforcement were first examined by Brown *et al.* (1982) and Ruddock *et al.* (1982). Work with geogrid reinforcement followed in the late 1980s (Barker, 1987; Haas *et al.*, 1988; Barksdale *et al.*, 1989). Perkins and Ismeik (1997a) have provided a review of these studies, and have summarised and discussed design procedures and numerical modelling efforts in a companion paper (Perkins and Ismeik, 1997b).

Since paved roads are considered to be inoperable once large surface deformations are seen, most studies have focused on paved road performance and improvements offered by geosynthetics for pavement deformation less than 25 mm. A number of studies have demonstrated that the service life of the pavement, as defined by the number of load repetitions carried by the pavement to reach a particular permanent surface deformation, can be increased by a factor ranging from just over 1 to in excess of 100 by the inclusion of a geosynthetic in the base aggregate layer (Haas *et al.*, 1988; Barksdale *et al.*, 1989; Miura *et al.*, 1990; Webster, 1993;

Figure 8.6 Permanent surface deformation versus load cycle for a geogrid-reinforced test section compared with a similar unreinforced test section
Data taken from Perkins (1999)

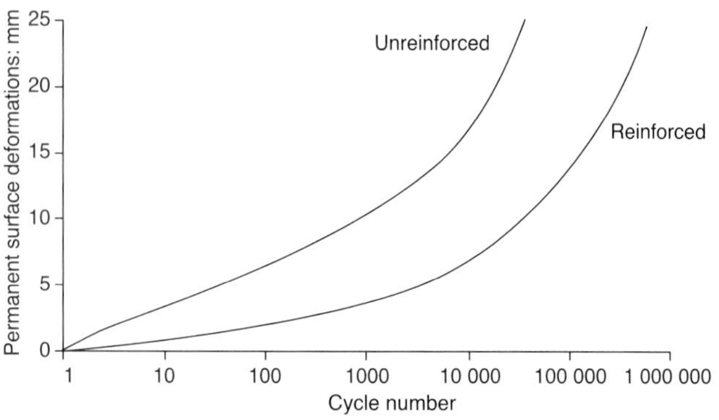

Cancelli *et al.*, 1996; Collin *et al.*, 1996; Moghaddas-Nejad and Small, 1996; Kinney *et al.*, 1998a,b; Cancelli and Montanelli, 1999; Perkins, 1999). Typical results from Perkins (1999) are illustrated in Figure 8.6, where a test section containing a geogrid is compared with a similar unreinforced test section. These test sections used approximately 75 mm of asphalt concrete over 300 mm of aggregate base placed on a weak subgrade, having a California bearing ratio (CBR) of approximately 1.5. The geogrid in this case was placed at the subgrade–base aggregate interface. Additionally, studies have shown that the base aggregate thickness of a reinforced section can be reduced by values ranging from 22% to 50%, such that equal service life results (Haas *et al.*, 1988; Anderson and Killeavy, 1989; Webster, 1993; Cancelli *et al.*, 1996). Improvement has been seen for all levels of rut depth below that corresponding to an inoperable condition (25 mm). The importance of variables such as the subgrade strength, the base course thickness and type, the geosynthetic type, the location and number of layers, and the magnitude of the traffic load, have all been shown to impact reinforcement performance improvement. Berg *et al.* (2000) have given a comprehensive review of available studies, where the variables involved in the studies and the

Figure 8.7 Permanent radial strain in the bottom of the base versus radial distance at 40 000 load cycles
Reproduced from Perkins (1999)

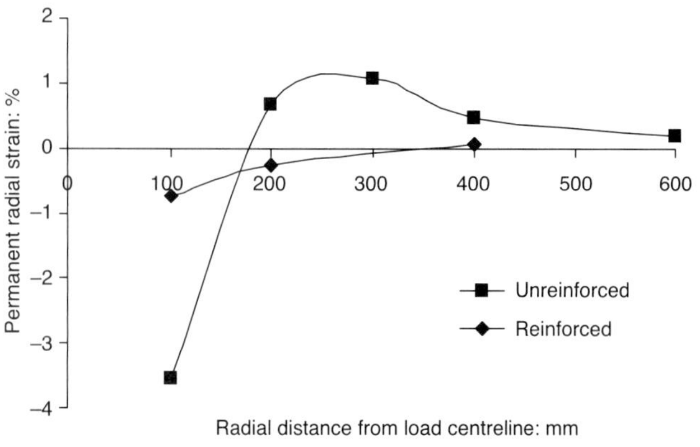

Figure 8.8 Dynamic vertical stress in the top of the subgrade versus radial distance at 40 000 load cycles
Reproduced from Perkins (1999)

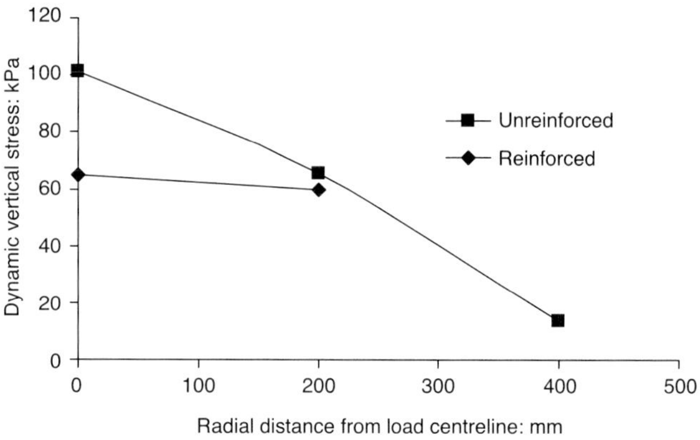

type and magnitude of the benefit derived by the geosynthetic has been summarised. The results from this report are outlined in Section 8.5.

Many of the reinforcement mechanisms previously described in Section 8.3.1 have been demonstrated experimentally. As described in Section 8.3.1, several studies have demonstrated a strain distribution in the geosynthetic that corresponds to that sketched in Figure 8.2, indicating that the geosynthetic is engaged to prevent lateral spreading of the base course aggregate. Perkins (1999) has shown that radial strain in the bottom of the base is considerably reduced by the presence of reinforcement, as illustrated in Figure 8.7 for the two test sections given in Figure 8.6. Haas *et al.* (1988) and Perkins (1999) have shown that vertical stress on the subgrade was less when reinforcement was present. The results from Perkins (1999) are given in Figure 8.8, where it is seen that vertical stress below the load centreline is reduced by approximately 35% for this particular set of pavement design variables. Perkins (1999) has shown results similar to that seen in Figure 8.7 for radial strain in the top of the subgrade, indicating that shear in the top of the subgrade is reduced by the addition of reinforcement. Perkins (1999) has also shown that an increase in the aggregate thickness results in mechanical improvements to the system that are similar to the reinforcement mechanisms illustrated above when a geosynthetic reinforcement layer is added to the pavement. These results show that, under the proper set of design conditions, reinforcement provides a significant structural enhancement to the system and can be used effectively to limit both permanent and dynamic deformation of the pavement system.

8.5. Summary of critical design variables for base reinforcement

A compilation of the studies summarised in Section 8.4 by Berg *et al.* (2000) has allowed for general conclusions to be drawn pertaining to the variables that are important in the use of geosynthetics for base reinforcement and the conditions under which geosynthetic materials should be used for this application. Table 8.2 provides a list of the variables

Table 8.2 Variables and ranges of values over which reinforcement is effective
Adapted from Berg *et al.* (2000)

Pavement component	Variable	Range from test studies	Condition where reinforcement appears to provide most benefit
Geosynthetic	Structure	Extruded, knitted and woven geogrids, woven and non-woven geotextiles, geogrid–geotextile composites	See Table 8.3
	Modulus at 2% and/or 5% strain	100–750 kN/m	Higher modulus improves potential for performance
	Location	Geogrid	Moderate load (≤80 kN axle load): bottom of thin bases (≤250 mm), middle for thick bases (>300 mm) Heavy load (>80 kN axle load): bottom for thin bases (≤300 mm), middle for thick bases (>350 mm)
		Geotextile	Between base and subgrade
		Geogrid–geotextile composite	Bottom of open-graded base
	Surface	Slick to rough	Rough
	Geogrid aperture	15–64 mm	$>D_{50}$ of adjacent aggregate
	Aperture stiffness	Rigid to flexible	Rigid
Subgrade condition	Soil type	SP, SM, CL, CH, ML, MH, Pt	No relation noted
	Strength	CBR = 0.5–27	CBR < 8 (resilient modulus, M_R < 80 MPa)
Base	Thickness	40–640 mm	≤250 mm for moderate loads
	Gradation	Well to poorly graded	Well graded
	Angularity	Angular to subrounded	Angular
Sub-base	Thickness	0–300 mm	No sub-base
	Angularity	Angular to rounded	Angular
Pavement	Type	Asphalt, concrete	Asphalt
	Thickness	25–180 mm	75 mm
Design	Pavement loading	200–1800 kPa	Should not be used on underdesigned pavements
Construction	Pre-rutting	None to pre-rutted	Unknown

that are believed to control the effectiveness of geosynthetics in this application. Table 8.2 also provides a range of these parameters as used in studies reported in the literature and an estimate of the range of values over which reinforcement is effective. Table 8.3 provides a summary of the conditions for which various geosynthetic products should be used. Tables 8.2 and 8.3 should serve as a guide when deciding whether geosynthetic reinforcement is applicable for a particular design situation, and are used in the design approach discussed in Section 8.6.

8.6. Design solutions and approaches for base reinforcement

Perkins and Ismeik (1997b) and Berg *et al.* (2000) have summarised techniques available for the design of geosynthetic-reinforced flexible pavements. Proposed design methods are either based on empirical and analytical considerations or analytical models modified by experimental data. Empirical methods are generally limited to a specific set of conditions associated with the experiments used to calibrate the method, and can be difficult to apply to design situations having a different set of design conditions. Analytical models generally fall within the category of finite-element models suitable for research purposes, but, at times, are often difficult to apply in practice.

Berg *et al.* (2000) proposed a design approach that relies upon the assessment of reinforcement benefit as defined by the TBR or BRR, as introduced in Section 8.3.1 and in Figures 8.3 and 8.4. The reinforcement benefit may be defined by empirical techniques or analytical solutions validated by experimental data. The reinforcement benefit defined in this manner is then used to modify an existing unreinforced pavement design.

Berg *et al.* (2000) proposed a design procedure given by the steps listed below

- Step 1: initial assessment of applicability of the technology.

Table 8.3 Geosynthetic selection guide
Adapted from Berg *et al.* (2000)

Roadway design conditions		Geosynthetic type					
Subgrade	Base/sub-base thickness: mm[a]	Geotextile		Geogrid[b]		Geotextile–geogrid composite	
		Non-woven	Woven	Extruded	Knitted or woven	Open-graded base[c]	Well-graded base
Low (CBR < 3) ($M_R < 30$ MPa)	150–300	●[d]	●	●	□	●	○[e]
	>300	●[d]	●[d]	◆	◆	◆	○[e]
Firm to very stiff ($3 \leq$ CBR ≤ 8) (30 MPa $\leq M_R \leq$ 80 MPa)	150–300	○[f]	◆	●	□	●	○[e]
	>300	○[f]	○[f]	◆[g]	□	□	○[e]
Firmer (CBR > 8) ($M_R > 80$ MPa)	150–300	○	○	◆	□	□	○[e]
	>300	○	○	○	○	○	○[e]

Key: ● usually applicable; ◆ applicable for some conditions; ○ usually not applicable; □ insufficient information at this time.
Notes:
[a] Total base or sub-base thickness with geosynthetic reinforcement. Reinforcement may be placed at the bottom of the base or sub-base, or within the base for greater (>300 mm) thicknesses. Thicknesses less than 150 mm not recommended for construction over soft subgrade. Placement of less than 150 mm over a geosynthetic not recommended.
[b] For an open-graded base or thin bases over wet, fine-grained subgrades, a separation geotextile should be considered with geogrid reinforcement.
[c] Potential assumes the base is placed directly on the subgrade. A sub-base may also provide filtration.
[d] Reinforcement usually applicable, but typically addressed as a subgrade stabilisation application.
[e] The geotextile component of the composite is not likely to be required for filtration with a well-graded base course; therefore, composite reinforcement is usually not applicable.
[f] Separation and filtration application; reinforcement is usually not applicable.
[g] Usually applicable when placed up in the base course aggregate. Usually not applicable when placed at the bottom of the base-course aggregate.

- Step 2: design of the unreinforced pavement.
- Step 3: definition of the qualitative benefits of reinforcement for the project.
- Step 4: definition of the quantitative benefits of reinforcement (TBR or BRR).
- Step 5: design of the reinforced pavement using the benefits defined in step 4.
- Step 6: analysis of life-cycle costs.
- Step 7: development of a project specification.
- Step 8: development of construction drawings and bid documents.
- Step 9: construction of the roadway.

Step 1 involves assessing the project-related variables given in Tables 8.2 and 8.3, and making a judgement on whether the project conditions are favourable or unfavourable for reinforcement to be effective, and what types of reinforcement products (as defined in Table 8.3) are appropriate for the project. Step 2 involves the design of a conventional unreinforced typical pavement design cross-section or a series of cross-sections, if appropriate, for the project. Any acceptable design procedure can be used for this step. Step 3 involves an assessment of the qualitative benefits that will be derived by the addition of the reinforcement. The two main benefits that should be assessed are whether the geosynthetic will be used for an extension of the life of the pavement (i.e. the application of additional vehicle passes), a reduction of the base aggregate thickness, or a combination of the two. Berg *et al.* (2000) have listed additional secondary benefits that should also be considered.

Step 4 is the most critical step in the design process and requires the greatest amount of judgement. This step requires the definition of the value, or values, of benefit (TBR and/or BRR) that will be used in the design of the reinforced pavement. Several approaches are available for the definition of these benefit values. Berg *et al.* (2000) have presented an empirical technique where these values are determined by a careful comparison of project design conditions, as defined in previous steps, to conditions present in studies reported in the literature. The majority of these studies have been summarised in Berg *et al.* (2000) in a form that allows direct comparison to known project conditions. In the absence of suitable comparison studies, an experimental demonstration method involving the construction of reinforced and unreinforced pavement test sections has been suggested and described in Berg *et al.* (2000), and may be used for the definition of benefit for the project conditions. Perkins (2001) presents a spreadsheet program (available at http://www.mdt.state.mt.us/departments/researchmgmt/grfb/grfb.html) that calculates the reinforcement benefit as a function of critical design input parameters. The program is based on a finite-element model that has been validated by full-scale test section. The reasonableness of benefit values should be carefully evaluated so that the reliability of the pavement is not undermined.

Step 5 involves the direct application of the TBR or BRR to modify the unreinforced pavement design defined in step 2. The TBR can be directly used to define an increased number of vehicle passes that can be applied to the pavement, while the BRR can be used to define a reduced base aggregate thickness so that equal life results. Within the context of an AASHTO pavement design approach, it is possible to calculate a BRR knowing a TBR, and vice versa for the specific project design conditions; however, this approach has not been experimentally or analytically validated.

With the unreinforced and reinforced pavement designs defined, a life-cycle cost analysis should be performed to assess the economic benefit of reinforcement. This step will dictate whether it is economically beneficial to use the geosynthetic reinforcement. The remaining steps involve the development of project specifications, construction drawings, bid documents and plans for construction monitoring. Berg *et al.* (2000) have presented a draft specification that may be adopted for this application.

8.7. Summary of main points

- Geosynthetics help control distress and damage in paved roads by providing reinforcement, separation, drainage and filtration.
- Geosynthetics are economically efficient from the standpoint of reducing construction costs, maintenance and/or life-cycle costs.
- Reinforcement is provided by the confinement of the base aggregate layer, as construction and operational traffic tends to cause lateral spreading of aggregate below the wheel path.
- Separation is provided by keeping good-quality aggregate layers physically separate from underlying fine-grained soils.
- Drainage geosynthetics provide for the transport of infiltrating water out of the pavement layers, and prevent this water from saturating pavement layers and seeping into underlying subgrade soils.
- Filtration prevents fines contained in underlying soils from migrating into overlying structural aggregate layers.
- Reinforcement is effective generally for thin pavements resting on weak subgrades. Table 8.1 provides guidelines for variables and ranges of values over which reinforcement is effective.
- Table 8.2 provides guidelines for the selection of geosynthetics and considers the subgrade strength/stiffness and base/sub-base thickness.
- Guidelines for the design of paved roads reinforced with geosynthetics currently use the traffic benefit ratio or the base-course reduction ratio to define the associated benefit from the geosynthetic. These ratios are currently defined from test sections incorporating similar conditions to the pavement design and are used to modify the unreinforced cross-section or service life.
- Mechanistic-empirical design methods for reinforced pavements are currently under development, and are expected to be incorporated into national/international design guides in the near future.

8.8. Some common questions and their answers

Questions

1. What are the different interaction mechanisms for geogrids and geotextiles in base reinforcement applications?
2. Does geogrid reinforcement work better than geotextile reinforcement?
3. Which type of geogrid works best – woven geogrids, extruded geogrids or welded geogrids?
4. Is creep of the geosynthetic important for base reinforcement applications?
5. What is the purpose (i.e. functions) of geosynthetics used in a roadway constructed on a soft subgrade?
6. Is reinforcement the only function that is important for pavement support?
7. Where do I obtain the engineering properties of geosynthetics for design?
8. Can I reduce the pavement design thickness when a geosynthetic is used for base reinforcement?
9. In specifying a geosynthetic, should a specific product 'or equal' be specified?
10. What are the property requirements for a geotextile used in stabilisation applications?
11. What are the property requirements for a geogrid used in base reinforcement?
12. When a geosynthetic is used for stabilisation of soft soil to provide a working platform for construction, is there any benefit given to the pavement support?
13. What benefit do geosynthetics used in base reinforcement provide?
14. Do geosynthetic separators provide any benefit for subgrade soils with CBR > 3 ($M_r > 30$ MPa)?
15. Don't geotextile filters used in roadway drainage clog?
16. Can any geotextile be used as a separator?
17. Do geogrids provide separation?

Answers

1. Geogrids provide interaction through interlock, with some friction along longitudinal elements, while geotextiles provide interaction through friction.

2. Not necessarily. If the geogrid does not provide separation, contamination of the aggregate layer may occur and can mitigate the effect of geogrid reinforcement.

3. Test sections are required to determine which type of geogrid will work best for the aggregate material that will be used, the pavement section geometry, loadings and the subgrade conditions.

4. Static creep is not an issue due to low strains in this application. However, creep under dynamic load should be evaluated in determining the modulus of the geosynthetic. Note that, in some cases, the modulus may increase under cyclic loading.

5. This application is a form of mechanical stabilisation, and can include the functions of separation, reinforcement, filtration and drainage to some level (especially in a wet, saturated subgrade).

6. If the subgrade contains fine-grained soil, and open-graded aggregate is to be used (i.e. for drainage), then the separation function will also be important.

7. While manufacturers routinely test their products, and several publications list property values, it is highly recommended that some independent tests be performed to check the property values for design as well as quality assurance. The pavement reinforcement traffic benefit ratio (TBR) and base-course reduction ratio (BRR) design values are currently defined from test sections incorporating similar conditions to the pavement design.

8. Yes, based on the BRR determined from field trials and numerical modelling. However, the greatest life-cycle cost benefit is usually in the increased life of the pavement attributed to the reinforcement expressed as the TBR.

9. It is strongly recommended that the geosynthetics be specified based on the properties required to perform the intended functions and, if products are known to meet those property requirements, they could be mentioned. However, the words 'or equal' should not be used, although the words 'or equivalent' may be used. An equivalent product would be one that meets or exceeds the listed required properties.

10. Strength to survive installation/construction, opening size and permeability/permittivity.

11. Strength to survive installation/construction, dynamic modulus, cyclic pull-out resistance, aperture stability and aperture size.

12. There can be some improvement in the subgrade support condition, and an equivalent modulus can be given to that improvement through field testing to determine the support value.

13. The benefits of base reinforcement are expressed in terms of ability to carry increased traffic over the life of the pavement for the reinforced section versus the unreinforced section, expressed as the TBR, or the percentage of reduction in the base-course thickness of a reinforced pavement for an equivalent life to that required for the unreinforced pavements, expressed as the BRR.

14. In subgrade soils that are seasonally weak (e.g. from frost heave) or for high-fine-content soils that are susceptible to pumping, a geotextile separator may be of benefit in preventing migration of fines at a much higher CBR or M_r value.

15. Not if they are appropriately designed to be compatible with the soils that they are required to filter.

16. The geotextile should be selected based on the properties required to survive installation/construction, as well as opening characteristics that are compatible with the subgrade soil so that piping and clogging will not occur. Furthermore, the geotextile should have sufficient permittivity to allow for free flow of water from the subgrade into the aggregate and from the aggregate into the subgrade.

17. Geogrids can also prevent aggregate penetration into the subgrade; however, the geogrid does not prevent intrusion of subgrade soils up into the base/sub-base course. Either the aggregate must prevent migration of fines (the aggregate must meet the filtration criteria for the subgrade) or a separation geotextile should be used in combination with the geogrid (optimally a geocomposite).

REFERENCES

AASHTO (1997) Specification M288-96 on geotextiles. *Standard Specification for Transportation Materials and Methods of Sampling and Testing*, 18th edn. Federal Highway Administration, Washington, DC.

Alobaidi I and Hoare D (1994) Factors affecting the pumping of fines at the subgrade–subbase interface of highway pavements: a laboratory study. *Geosynthetics International* **1(2)**: 221–259.

Alobaidi I and Hoare D (1996) The development of pore water pressure at the subgrade–subbase interface of a highway pavement and its effect on pumping of fines. *Geotextiles and Geomembranes* **14(2)**: 111–135.

Anderson P and Killeavy M (1989) Geotextiles and geogrids: cost effective alternate materials for pavement design and construction. *Proceedings of the Conference Geosynthetics '89*, vol. 2, San Diego, CA, pp. 353–360.

Austin DN and Coleman DM (1993) Field evaluation of geosynthetic-reinforced haul roads over soft foundation soils. *Proceedings of the Conference Geosynthetics '93*, vol. 1, Vancouver, British Columbia, pp. 65–80.

Barker WR (1987) *Open-graded Bases for Airfield Pavements*. USAE Waterways Experiment Station, Vicksburg, MS. Technical Report GL-87-16.

Barksdale RD, Brown SF and Chan F (1989) *Potential Benefits of Geosynthetics in Flexible Pavement Systems. National Cooperative Highway Research Program Report No. 315*. Transportation Research Board, National Research Council, Washington, DC.

Bell AL, McCullough LM and Snaith MS (1982) An experimental investigation of subbase protection using geotextile. *Proceedings of the 2nd International Conference on Geotextiles*, vol. 2, Las Vegas, NV, pp. 435–440.

Bender DA and Barenberg EJ (1978) Design and behavior of soil–fabric–aggregate systems. *Transportation Research Record* **671**: 64–75.

Berg RR, Christopher BR and Perkins SW (2000) *Geosynthetic Reinforcement of the Aggregate Base/Subbase Courses of Pavement Structures*. Geosynthetic Materials Association, Roseville, MN.

Black PJ and Holtz RD (1997) *Performance of Geotextile Separators: Bucoda Test Site – Phase II*. Washington State Department of Transportation Report WA-RD 440.1, 210 pp.

Bonaparte R, Ah-Line C and Charron R (1988) Survivability and durability of a non-woven geotextile. *Geosynthetics for Soil Improvement. Geotechnical Special Publication No. 18*, American Society of Civil Engineers, Reston, VA, pp. 68–91.

Brorsson I and Eriksson L (1986) Long-term properties of geotextiles and their function as a separator in road construction. *Proceedings of the 3rd International Conference on Geotextiles*, vol. 1, Vienna, pp. 93–98.

Brown SF, Jones CPD and Brodrick BV (1982) Use of non-woven fabrics in permanent road pavements. *Proceedings of the Institution of Civil Engineers*, London, *Part 2* **73**: 541–563.

Cancelli A and Montanelli F (1999) In-ground test for geosynthetic reinforced flexible paved roads. *Proceedings of the Conference Geosynthetics '99*, vol. 2, Boston, MA, pp. 863–879.

Cancelli A, Montanelli F, Rimoldi P and Zhao A (1996) Full-scale laboratory testing on geosynthetics reinforced paved roads. *Proceedings of the International Symposium on Earth Reinforcement, Fukuoka/Kyushuer*, Balkema, Rotterdam, pp. 573–578.

Cedergren HR (1987) *Drainage of Highway and Airfield Pavements*. Krieger, Malabar, FL.

Christopher BR, Hayden SA and Zhao A (1999) Roadway base and subgrade geocomposite drainage layers. In *Testing and performance of geosynthetics in subsurface drainage* (Baldwin JS and Suits LD (eds)), American Society for Testing and Materials, West Conshohocken, PA.

Collin JG, Kinney TC and Fu X (1996) Full-scale highway load test of flexible pavement systems with geogrid reinforced base courses. *Geosynthetics Intentional* **3(4)**: 537–549.

Fannin RJ and Sigurdsson O (1996) Field observations on stabilization of unpaved roads with geosynthetics. *Journal of the Geotechnical Engineering*, ASCE **122(7)**: 544–553.

Giroud JP (1996) Granular filters and geotextile filters. *Proceedings of Geofilters '96, Montreal* (Lafleur J and Rollin AR (eds)), pp. 565–680.

Glynn DT and Cochrane SR (1987) The behavior of geotextiles as separating membranes on glacial till subgrades. *Proceedings of Geosynthetics '87*, vol. 1, New Orleans, LA, pp. 26–37.

Haas R, Wall J and Carroll RG (1988) Geogrid reinforcement of granular bases in flexible pavements. *Transportation Research Record* **1188**: 19–27.

Hoare DJ (1982) A laboratory study into pumping clay through geotextiles under dynamic loading. *Proceedings of the 2nd International Conference on Geotextiles*, vol. 2, Las Vegas, NV, pp. 423–428.

Holtz RD (1996) *Performance of Geotextile Separators*. Washington State Department of Transportation, Washington, DC. Report WA-RD 321.2.

Holtz RD, Christopher BR and Berg RR (2008) *Geosynthetic Design and Construction Guidelines*. Federal Highway Administration, Washington, DC. Publication No. FHWA HI-07-092

Houlsby GT and Jewell RA (1990) Design of reinforced unpaved roads for small rut depths. *Proceedings of the 4th International Conference on Geotextiles, Geomembranes and Related Products*, The Hague, pp. 171–176.

Jorenby BN and Hicks RG (1986) Base course contamination limits. *Transportation Research Record* **1095**: 86–101.

Kinney TC and Barenberg EJ (1982) The strengthening effect of geotextiles on soil–geotextile–aggregate systems. *Proceedings of the 2nd International Conference on Geotextiles*, vol. 2, Las Vegas, NV, pp. 347–352.

Kinney TC, Abbott J and Schuler J (1998a) *Benefits of Using Geogrids for Base Reinforcement with Regard to Rutting*. Transportation Research Board, Washington, DC. Paper Preprint 981472.

Kinney TC, Kleinhans-Stone D and Schuler J (1998b) Using geogrids for base reinforcement as measured by a falling weight deflectometer in a full-scale laboratory loading. *Transportation Research Board*, Paper Preprint 981471, presented at TRB, Washington, DC.

Lafleur J, Rollin AL and Mlynarek J (1990) Clogging of geotextiles under pumping loads. *Proceedings of the 4th International Conference Geotextiles, Geomembranes and Related Products*, vol. 1, The Hague, pp. 189–192.

Lawson CR (1992) Some examples of separation geotextiles under road pavements. *Proceedings of the Institution of Civil Engineers, Transport* **95**: 197–200.

McGown A, Yeo KC and Yogarajah I (1990) Identification of a dynamic interlock mechanism. Performance of reinforced soil structures. *Proceedings of the International Reinforced Soil Conference*, Glasgow, pp. 377–379.

Miura N, Sakai A, Taesiri Y, Yamanouchi T and Yasuhara K (1990) Polymer grid reinforced pavement on soft clay grounds. *Geotextiles and Geomembranes* **9**: 99–123.

Moghaddas-Nejad F and Small JC (1996) Effect of geogrid reinforcement in model track tests on pavements. *Journal of Transportation Engineering, ASCE* **122(6)**: 468–474.

Nishida K and Nishigata T (1994) The evaluation of separation function of geotextiles. *Proceedings of the 5th International Conference on Geotextiles, Geomembranes and Related Products*, vol. 1, Singapore, pp. 139–142.

Perkins KE and Brandon TL (1998) *Performance of Soil–geotextile Systems in Dynamic Loading Tests*. Transportation Research Board, Washington, DC. Paper Preprint 980992.

Perkins SW (1999) Mechanical response of geosynthetic-reinforced flexible pavements. *Geosynthetics International* **6(5)**: 347–382.

Perkins SW (2001) *Mechanistic-empirical Modelling and Design Development of Geosynthetic Reinforced Flexible Pavements*. US Department of Transportation, Federal Highway Administration, Washington DC. Report No. FHWA/MT-01-002/99160-1.

Perkins SW and Ismeik M (1997a) A synthesis and evaluation of geosynthetic-reinforced base layers in flexible pavements: part I. *Geosynthetics International* **4(6)**: 549–604.

Perkins SW and Ismeik M (1997b) A synthesis and evaluation of geosynthetic-reinforced base layers in flexible pavements: part II. *Geosynthetics International* **4(6)**: 605–621.

Perkins SW, Ismeik M and Fogelsong ML (1998a) Mechanical response of a geosynthetic-reinforced pavement system to cyclic loading. *Proceedings of the 5th*

International Conference on the Bearing Capacity of Roads and Airfields, vol. 3, Trondheim, pp. 1503–1512.

Perkins SW, Ismeik M, Fogelsong ML, Wang Y and Cuelho EV (1998b) Geosynthetic-reinforced pavements: overview and preliminary results. *Proceedings of the 6th International Conference on Geosynthetics*, vol. 2, Atlanta, GA, pp. 951–958.

Perkins SW, Perkins SW, Christopher BR *et al.* (2010) Geosynthetics in pavement reinforcement applications. *Proceedings of the 9th International Conference on Geosynthetics*, Guaruja, pp. 165–192.

Richardson GN and Behr LH (1990) Survivability of subgrade separators. *Geotechnical Fabrics Report*, May/Jun: 22–26.

Ruddock EC, Potter JF and McAvoy MR (1982) A full-scale experiment on granular and bituminous road pavements laid on fabrics. *Proceedings of the 2nd International Conference on Geotextiles*, vol. 2, Las Vegas, NV, pp. 365–370.

Snaith MS and Bell AL (1978) The filtration behaviour of construction fabrics under conditions of dynamic loading. *Gétechnique* **28(4)**: 466–468.

Tsai WS and Holtz RD (1997) Laboratory model tests to evaluate geotextile separators in-service. *Proceedings of the Conference Geosynthetics '97*, vol. 2, Long Beach, CA, pp. 633–646.

Tsai WS, Savage MB, Holtz RD, Christopher BR and Allen T (1993) Evaluation of geotextiles as separators in a full-scale road test. *Proceedings of the Conference Geosynthetics '93*, vol. 1, Vancouver, pp. 35–48.

Webster SL (1993) *Geogrid Reinforced Base Courses For Flexible Pavements For Light Aircraft, Test Section Construction, Behavior Under Traffic, Laboratory Tests, and Design Criteria.* USAE Waterways Experiment Station, Vicksburg, MS. Technical Report GL-93-6.

Yoder EJ and Witczak MW (1975) *Principles of Pavement Design*, 2nd edn. Wiley, Chichester.

Handbook of Geosynthetic Engineering
ISBN 978-0-7277-4175-2

http://dx.doi.org/10.1680/hge.41752.193

Chapter 9
Railway tracks

S. A. Tan* **and S. K. Shukla†**
*Department of Civil Engineering, the National University of Singapore, Singapore
†Discipline of Civil Engineering, School of Engineering, Edith Cowan University, Perth, Australia

Many factors influence the safe and efficient operation of railway tracks throughout the world. The most important tasks for railway track engineers are the design, installation and maintenance of a highly stable track network that will reliably carry goods and passengers with safety and high speed. The use of geosynthetics in construction has improved the functions of railway tracks. This chapter introduces the components of conventional track structures and their functions, and describes the properties, design and installation of geosynthetics for the stabilisation and drainage of railway tracks, along with some case histories.

9.1. Introduction

Railway track engineers are required to understand the basics of railway track substructure design and performance under repeated wheel loading. The key feature of the multi-layer substructure system is to reduce both lateral and vertical stresses and strains in the weaker subgrade layer to such an extent that the subgrade soils do not experience progressive failures from cumulative cyclic shearing. Apart from the accumulated strains from repeated wheel loads, railway track substructure suffers degradation and progressive weakening due to the effects of water coupled with repeated wheel loadings. Without geosynthetics, the unprotected subgrade of cohesive soils will weaken considerably when loaded repeatedly in fully saturated and soaked condition, causing subsequent mud pumping and sub-ballast contamination. This will inevitably cause excessive track settlements and serviceability failure of the track system. Geosynthetics have been very successful in preventing premature track substructure failure and extending significantly the service life of track systems built on relatively poor cohesive subgrade soils (Shukla and Yin, 2006). This chapter outlines how properly designed and installed geosynthetics can provide good protection of the subgrade/sub-ballast/ballast layers by serving the principal functions of separation, filtration, lateral confinement and lateral drainage, so that each sub-layer component remains stable throughout its service life. To obtain the desired functions of geosynthetics in track substructure, it is often the case that appropriate geocomposite products are needed for different subgrade soils and drainage conditions.

9.2. Track components and substructure

The purpose of a railway track structure is to provide safe and economical train transportation. This requires the track to serve as a stable guideway with appropriate vertical and horizontal alignment. To achieve this role, each component of the system must perform its specific functions satisfactorily in response to the traffic loads and environmental factors imposed on the system.

Figure 9.1 shows the main components of ballasted track structures. These may be grouped into two main categories: superstructure and substructure. The superstructure consists of the rails, the fastening system and the sleepers (ties). The substructure consists of the subgrade, the sub-ballast and the ballast. Thus, the superstructure and substructure are separated by the sleeper–ballast interface.

9.2.1 Subgrade

The subgrade is the platform upon which the track structure is constructed. Its main function is to provide a stable foundation for the sub-ballast and ballast layers. The influence of the traffic-induced stresses extends downwards as much as 5 m below the bottom of the sleepers. This is considerably beyond the depth of the ballast and sub-ballast. Hence, the subgrade is a very important substructure component that has a significant influence on track performance and maintenance. For example, subgrade is a major component of the superstructure support resiliency, and so it contributes substantially to the elastic deflection of the rail under wheel loading. In addition, the magnitude of the subgrade stiffness is believed to influence ballast, rail and sleeper deterioration. Subgrade compression is also a source of rail differential settlement.

The subgrade may be divided into two categories: natural ground (formation) and placed soil (fill). Anything other than soils existing locally is generally uneconomical to use

Figure 9.1 Components of railway track structure
Reproduced from Selig and Waters (1994)

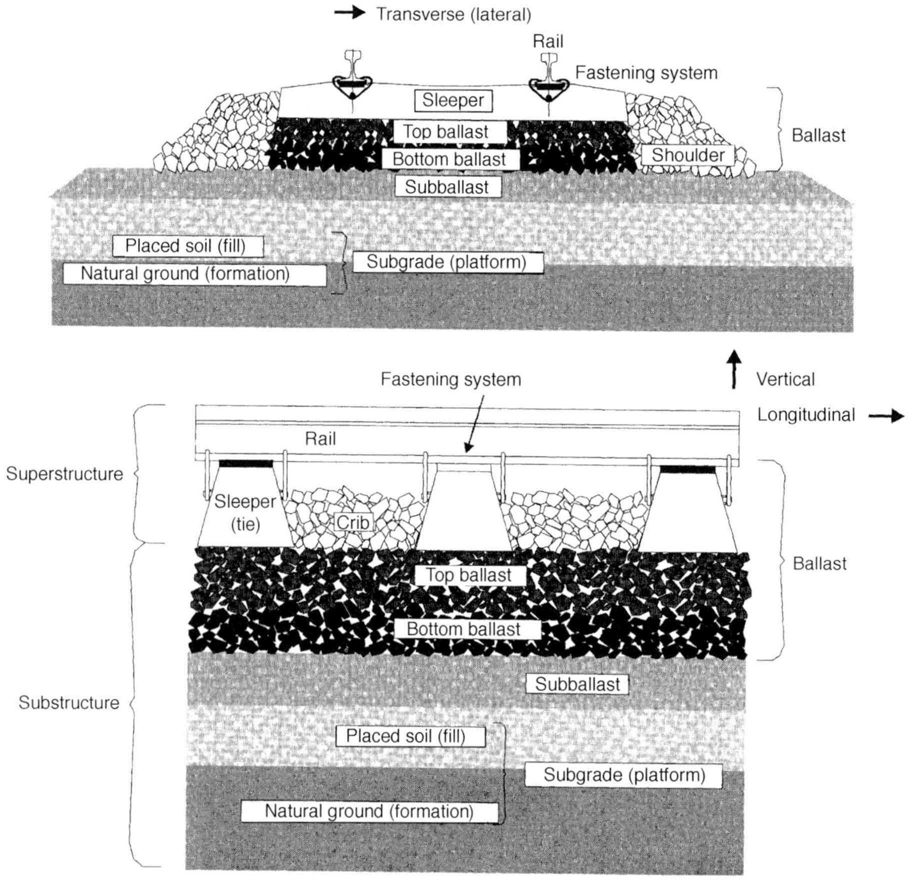

for the subgrade. Existing ground will be used without disturbance as much as possible. However, techniques are available to improve soil formations in place if they are unfavourable. These ground improvement techniques include grouting, compaction and admixture stabilisation with cement, lime, bitumen and/or fly ash combinations. Often, some of the formation must be removed to construct the track at its required elevation. Placed fill is used either to replace the upper portion of unsuitable existing ground or to raise the platform to the required elevation for the rest of the track structure.

To serve as a stable platform, the following subgrade failure modes must be avoided

- excessive progressive settlement from repeated traffic loading
- consolidation settlement and massive shear failure under the combined weights of the train, track structure and earth loads above it
- progressive shear failure (excessive heave near shoulder areas) from repeated wheel loading
- significant volume change (swelling and shrinking) from moisture change
- frost heave and thaw softening
- subgrade attrition.

In addition to its other functions, the subgrade must provide a suitable base for construction of the sub-ballast and ballast.

9.2.2 Sub-ballast

The layer between the ballast and the subgrade is the sub-ballast. It fulfils two functions that are also listed in the following section for ballast. These are

- to reduce the traffic-induced stresses at the bottom of the ballast layer to a tolerable level for the subgrade
- to extend the frost protection of the subgrade.

In fulfilling these functions, the sub-ballast reduces the otherwise required greater thickness of the more expensive ballast material. However, the sub-ballast has other important functions that cannot be fulfilled by the ballast. These are

- to prevent inter-penetration of the subgrade by ballast stones (separation function)
- to prevent upward migration of fine material emanating from the subgrade (separation function)
- to prevent subgrade attrition by the ballast, which, in the presence of water, leads to the formation of mud pumping and, hence, prevents this source of pumping

- to shed water (i.e. intercept water coming from the ballast and direct it away from the subgrade to ditches at the sides of the track)
- to permit drainage of water that might be flowing upward from the subgrade.

All these functions are very important for satisfactory track performance. Hence, in the absence of a sub-ballast layer, high maintenance can be expected unless these functions are fulfilled in some other manner. The last three functions form a subset of the sub-ballast functions, which represent what is sometimes known as a *blanket layer*.

The most common and most suitable sub-ballast materials are broadly-graded naturally occurring or processed sand–gravel mixtures, or broadly-graded crushed natural aggregates or slags. They must have durable particles and satisfy the filter/separation requirements for the ballast and subgrade. These requirements are discussed in later sections. Some of the functions of the sub-ballast may be provided by

- cement, lime or asphalt-stabilised local soils
- asphalt concrete layers
- geosynthetics, such as geomembranes, geogrids and geotextiles (filter fabrics).

9.2.3 Ballast

Ballast is the selected crushed granular material placed as the top layer of the substructure in which the sleepers are embedded.

Traditionally, angular, crushed, hard stones and rocks, uniformly graded, free of dust and dirt, and not prone to cementing action, have been considered good ballast materials. However, at present, no universal agreement exists concerning the proper specifications for the index characteristics of ballast material, such as size, shape, hardness, abrasion resistance and composition, that provide the best track performance. This is a complex subject that is still being researched. Availability and economic considerations have been the prime factors considered in the selection of ballast materials. Thus, a wide variety of materials have been used for ballast, such as crushed granite, basalt, limestone, slag and gravel.

The ballast performs many functions. The most important ones are

- to resist vertical (including uplift), lateral and longitudinal forces, applied to the sleepers to retain the track in its required position
- to provide some of the resiliency and energy absorption for the track
- to provide large voids for storage of fouling material in the ballast, and movement of particles through the ballast
- to facilitate maintenance surfacing and lining operations (to adjust track geometry) by the ability to rearrange ballast particles with tamping
- to provide immediate drainage of water falling onto the track
- to reduce pressures from the sleeper-bearing area to acceptable stress levels for the underlying material.

It should be noted that although increasing the ballast layer thickness will reduce the average stress, high-contact stresses from the ballast particles will require durable material in the layer supporting the ballast.

Other functions of the ballast are

- to alleviate frost problems by not being frost susceptible and by providing an insulating layer to protect the underlying layers
- to inhibit vegetation growth by providing a cover layer that is not suitable for vegetation
- to absorb airborne noise
- to provide adequate electrical resistance between rails
- to facilitate redesign/reconstruction of track.

As shown in Figure 9.1, the ballast may be subdivided into the following four zones:

- crib – the material between the sleepers
- shoulder – the material beyond the sleeper ends down to the bottom of the ballast layer
- top ballast – the upper portion of the supporting ballast layer, which is disturbed by tamping
- bottom ballast – the lower portion of the supporting ballast layer, which is not disturbed by tamping and which is generally the more fouled portion.

In addition, the term ‘boxing’ may be used to designate all the ballast around the sleeper that is above the bottom of the sleeper (i.e. the upper shoulders and the cribs).

The mechanical properties of ballast result from a combination of the physical properties of the individual ballast material and its in-place physical state. The physical state can be defined by the in-place density, while the physical properties of the material can be described by various indices, such as particle size, shape, angularity, hardness, surface texture and durability. The in-place unit weight of ballast is a result of compaction processes. Maintenance tamping usually creates the initial unit weight. Subsequent compaction results from train traffic combined with environmental factors.

In service, the ballast gradation changes as a result of

- mechanical particle degradation during construction and maintenance work, and under traffic loading
- chemical and mechanical weathering degradation from environmental changes
- the migration of fine particles from the surface and the underlying layers.

Figure 9.2 Fouled ballast from subgrade mud pumping

The migration of fine particles causes the ballast to become fouled and lose its open-graded characteristics, so that the ability of ballast to perform its important functions decreases and ultimately may be lost. An example of fouled ballast from the mud pumping of the subgrade is shown in Figure 9.2.

Selig and Waters (1994) recommended that the degree of ballast fouling (contamination) might be quantitatively represented from the gradation curve by the weight of fines. This would always be greater or equal to the percentage of fines. Gradations were obtained for samples of ballast taken from a wide variety of track sites in the USA. Based on these data, representative gradations ranging from clean to highly-fouled conditions were developed. A fouling index (FI) was defined:

$$\mathrm{FI} = P_4 + P_{200} \tag{9.1}$$

where P_4 is the percentage passing the 4.75 mm (#4, ASTM standard sieve size) sieve, and P_{200} is the percentage passing the 0.075 mm (#200, ASTM standard sieve size) sieve. Categories of fouling are given in Table 9.1.

9.3. Functions of geosynthetics

It has long been recognised that subgrade mud-pumping and load-bearing capacity failure beneath railway tracks are problems that can be handled by the use of geotextiles, geogrids and/or geomembranes at the ballast–subgrade interface (Koerner, 2005; Shukla and Yin, 2006). The design difficulty lies in which type of geosynthetic is most suitable.

Table 9.1 Fouling index
Reproduced from Selig and Waters (1994)

Fouling category	Fouling index (FI)	Ballast hydraulic conductivity, *k*: mm/s
Clean	<1	25–50
Moderately clean	1–9	2.5–25
Moderately fouled	10–19	1.5–2.5
Fouled	20–39	0.005–1.5
Highly fouled	>39	<0.005

The function of a geosynthetic beneath a railway track is fundamentally different from that beneath an unpaved road (described in Chapter 7) or a paved road (described in Chapter 8). The essential differences for railway track are

- the ballast used to support the sleeper is very coarse, uniform and angular
- the regular repeated loading from the axles can set up resonant oscillations in the subgrade, making wet subgrade with fine soils very susceptible to mud pumping
- the rail-track system produces long-distance waves of both positive and negative pressure in the ground ahead of the train itself.

There are four principal functions that can be provided when a properly designed and installed geosynthetic is placed within the track structure. These are

- separation, in new railway tracks, between *in situ* soil and new ballast
- separation, in rehabilitated railway tracks, between old contaminated ballast and new clean ballast
- filtration of soil porewater rising from the soil beneath the geosynthetic, due to rising water conditions, or the dynamic pumping action of the wheel loadings, across the plane of the geosynthetic
- lateral confinement-type reinforcement in order to contain the overlying ballast stone
- lateral drainage of water entering from above or below the geosynthetic within its plane, leading to side drainage ditches.

9.3.1 Separation

This is the key function of the geotextile in most railway track applications. The geotextile acts as a barrier preventing the intermixing of fine subgrade materials from contaminating clean ballast. For this application, the geotextile either replaces the sub-ballast or assists it in the separation function.

9.3.2 Filtration

While performing as a separator, at the same time the geotextile must act as a filter, allowing water to pass freely across the plane of the geotextile. Water should be 'pumped' from a wet subgrade during train passby, and fines from the subgrade should be filtered out and retained in the subgrade. If the filtration design criteria for soil retention are not met, subgrade mud can still be pumped up into the ballast from the subgrade through the geotextile.

Selig and Waters (1994) reported a failure of geotextile to prevent mud contamination of ballast three years after rehabilitation works, as shown in Figure 9.3. It was observed

Figure 9.3 Geotextile filtration (soil retention) failure where mud pumping still occurs through geotextile under train dynamic loadings

that the test site, with geotextile alone, is not able to prevent mud pumping, but another test site in the same location had a 50 mm sand protective layer below the geotextile, and mud pumping was effectively prevented.

Leuttich *et al.* (1992) proposed that the soil retention criteria for a geotextile filter on a subgrade with fine soils, and design under bidirectional dynamic flow conditions, should be as per the guidelines given in Table 9.2.

9.3.3 Confinement/reinforcement

A geotextile acting as a separator also tends to confine the supporting ballast beneath load-bearing members of the track. The confined ballast is better able to retain a degree of reinforcement to the trackbed. In addition to the prevention of contamination of sub-ballast and ballast, the geotextile can also have a role as a stress absorber at the subgrade level. A strong geotextile at this level can absorb stress and reduce the imposed loads on the subgrade. By reducing lateral shear stresses in the subgrade, the geotextile may help to increase the overall load-bearing capacity.

9.3.4 Drainage

A properly designed and installed geotextile allows water entering the plane of the fabric to be transmitted laterally away from the areas of loading. Water from precipitation and pumping action can be carried through the plane of the fabric to the edge of the track to adjoining ditches. Excess pore pressures from wet subgrade pumping are relieved, and ballast contamination is minimised.

Raymond (1982, 1986, 1993a,b) and Raymond and Bathurst (1990), for Canadian Railways, have shown that the basic functional requirements of geotextiles placed below clean ballast in track construction and rehabilitation are as follows

- to drain water from the trackbed on a long-term basis, both laterally and by gravity along the plane of the geotextile, without a build-up of excessive porewater pressures (drainage)
- to withstand abrasive forces of moving aggregate caused by tamping, compacting processes generated during initial construction and during subsequent cyclic maintenance, and by frequent passage of trains (survivability – abrasion)
- to filter and hold back soil particles while allowing the passage of water (filtration)
- to separate two dissimilar soil types, sizes and gradings that would readily mix under the influence of repeated loading (separation)
- to have the ability to elongate around protruding large angular particles without rupture or puncture (survivability – puncture).

Irrespective of the difficulty of identifying a single geotextile function (since multifunction application is involved), the acceptance of geotextiles by railway track companies is high and increasing, especially in the USA, Canada and Europe.

9.4. Properties of geosynthetics

During the early years when geotextiles were first being tested for railway tracks in the USA and Canada, the specified physical properties were those being promoted by

Table 9.2 Soil retention criteria for dynamic flow conditions

From soil property tests	Recommended opening size
More than 30% clay ($D_{30} < 0.002$ mm), non-dispersive soil (DHR < 0.5)	$O_{95} < 10D_{50}$, $O_{95} < D_{90}$, $O_{95} < 0.1$ mm
More than 30% clay ($D_{30} < 0.002$ mm), dispersive soil (DHR > 0.5)	Use 50–150 mm of fine sand as a protective layer, and design geotextile as the filter for the sand
Less than 30% clay ($D_{30} < 0.002$ mm) and more than 50% fines ($D_{50} < 0.0075$ mm), plastic soil (PI > 5)	Check DHR value, and use either of the above two criteria based on the DHR value
Less than 30% clay ($D_{30} < 0.002$ mm) and more than 50% fines ($D_{50} < 0.0075$ mm) Non-plastic soil (PI < 5)	$O_{95} < D_{50}$

DHR, double-hydrometer ratio for the soil test; O_{95}, geotextile opening size; PI, plastic index.

Table 9.3 Properties of geotextiles recommended by AREA

Test method	Regular	Heavy	Extra-heavy
Puncture resistance: N ASTM D-4833-88	500	675	900
Abrasion resistance: N ASTM D-3884 (Taber test at 1000 rev.; 1 kg load/wheel)	675	810	1080
Grab strength: N ASTM D-4632	900	1080	1440
Elongation: % ASTM D-4632	50	50	50
Trapezoidal tear strength: N ASTM D-4533	450	540	720
Cross-plane permeability: cm/s ASTM D-4491	0.2	0.2	0.2
Permittivity: l/s ASTM D-4491	0.5	0.4	0.3
In-plane transmissivity: $m^2/min \times 10^{-4}$ ASTM D-4716	2	4	6
AOS, US standard sieve: μm ASTM D-4751	70	70	70

manufacturers, and some manufacturers stressed only the physical properties provided by their particular products. This was confusing for users.

The American Railway Engineering Association (AREA, 1985) has now developed and published a standard specification for the use of geotextiles in railway track stabilisation. The specification recommends minimum physical property values for three categories of non-woven geotextiles: regular, heavy and extra-heavy. Selections of one of these, while based on subgrade conditions, are somewhat subjective. Therefore, many designers use the heavy and extra-heavy geotextiles, as the cost of geotextiles is low compared with the overall cost of the track rehabilitation work being done at the time of installation.

The selected geotextile must meet the following four durability criteria

- It must be tough to withstand the stresses of the installation process. Properties required are:
  - tensile strength
  - burst strength
  - grab strength
  - tear strength
  - resistance to ultraviolet light degradation for two weeks' exposure with negligible strength loss.
- It must be strong enough to withstand static and dynamic loads, high pore pressures, and severe abrasive action to which it is subjected during its useful life. Properties required are:
  - puncture resistance
  - abrasion resistance
  - elongation at failure.
- It must be resistant to excessive clogging or blinding, allowing water to pass freely across and within the plane of the geotextile, while at the same time filtering out and retaining fines in the subgrade. Properties required are:
  - cross-plane permeability (permittivity)
  - in-plane permeability (transmissivity)
  - apparent opening size (AOS).
- It must be resistant to rot, and attacks by insects and rodents. It must be resistant to chemicals, such as acids and alkalis, and to the spillage of diesel fuel.

Table 9.3 shows the index properties recommended by AREA for average roll values that should be considered when specifying geotextiles for railway tracks. Table 9.4 shows the properties of geotextiles recommended by Indian Railways, as presented by Yog *et al.* (1989). This was a tentative specification at that time.

Woven fabrics, while having excellent tensile strength, provide poor abrasion resistance and low in-plane permeability.

Table 9.4 Indian Railways' specifications for geotextiles in rail-track foundation

Parameter	Specification	Testing method
1. Composition	Polypropylene/polyester	IS 667:1981
2. Mode of manufacture	Needle-punched non-woven	Visual observation
3. Denier	4–10	IS 10014:1981, fibre test
4. Thickness	3.00 mm and above	IS 7702:1975, pressure of 2 kPa
5. Weight	400 g/m^2 and above	IS 1964:1970
6. Tensile strength	Minimum of 60 kg	IS 1969:1985, cut strip of 200 × 50 mm
7. Elongation at break	40–100%	IS 1969:1985, cut strip of 200 × 50 mm
8. Pore size	Maximum of 120 μm	BS 3321:1969
9. AOS	40–75 μm	By dry sieving
10. Roll width	Single roll preferred, jointed seam strength must be 90% of geotextile strength	
11. Roll length	As per site requirement to suit work with minimum joints	

Further, woven products have little or no ability to transmit water within their plane. Therefore, the most common choice by the railway track designers in the USA is thick needle-punched, non-woven geotextiles, having a mass per unit area of 400–680 g/m^2. Other considerations include the following

- *Chemical resistance*. Most geotextiles in railways are manufactured from polypropylene or polyester polymers. Both materials are highly resistant to rot, mildew, insects, rodents, and commonly encountered chemicals and diesel fuel. However, it should be noted that polyester is seriously damaged by high alkalinity. Polyester yarns have been observed to totally disintegrate in pond-liner applications in the USA.
- *Abrasion resistance*. In terms of the effective life of a railway geotextile, resistance to severe abrasion within the ballasted track is of critical importance. Van Dine *et al.* (1982) first reported an assessment of a geotextile abrasion resistance. Raymond and Bathurst (1990), who recommended the laboratory test using the Taber Abrasor (ASTM D3884), extended this work. Test data have shown that abrasion resistance is a function of opening size. Geotextiles, having low opening size, would be more abrasive-resistant due to the fact that the maximum size particle able to penetrate the geotextile is smaller. The degree of needle-punching that a non-woven needle-punched geotextile receives during its manufacture determines the amount of interlock between fibres and, therefore, influences the geotextile abrasion resistance. Raymond and Bathurst (1990) suggest that a minimum of 80 penetrations/cm^2 should be adequate.

Based on extensive laboratory tests on both unused and exhumed geotextiles from railway track installations in Canada, the following recommendations for geotextiles were used in railway rehabilitation works

- type of geosynthetic – needle-punched non-woven geotextiles, with 80 penetrations/cm^2 or greater
- fibre size – 0.7 tex or less
- fibre strength – 0.4 g/tex or more
- fibre polymer – polyester
- yarn length – 100 mm or greater
- filtration opening size – 75 μm or less
- in-plane coefficient of permeability – 0.005 cm/s or more
- elongation – 60% or more according to ASTM D1682
- colour – must not cause snow blindness
- abrasion resistance – for 1050 g/m^2 or greater, the geotextile must withstand 200 kPa on a 102 mm burst sample after 5000 revolutions of H-18 stones, each loaded with 1000 g using a rotary platform double-head Taber Abraser (ASTM D3884)
- width and length without seaming – specified by the client
- mass – 1050 g/m^2 or greater for track rehabilitation without the use of capping sand.

9.5. Design procedure

A review of the geosynthetic literature on railway track applications shows that they are somewhat inconsistent. Railway track specifications seem to favour relatively heavy non-woven needled-punched geotextiles because of their high flexibility and in-plane (transmissivity) characteristics. The logic behind high flexibility is apparent, since geotextiles must deform around relatively large ballast stone, and not fail or form potential slip plane. In-plane drainage itself is not a dominant function, because any geotextile that acts as an effective separator and filter would preserve the integrity of the drainage of the ballast. Koerner (2005) recommends the following design procedure

(*a*) Design the geotextile as a separator – this function is always required. Burst strength, grab strength, puncture resistance and impact resistance should be considered.

(*b*) Design the geotextile as a filter – this function is also usually required. The general requirements of adequate permeability, soil retention and long-term soil-to-geotextile flow equilibrium are needed, as in all filtration design. Note, however, that railway loads are dynamic; thus, porewater pressures must be rapidly dissipated. For this reason, high permittivity is required.

(*c*) Consider geotextile flexibility if the cross-section is raised above the adjacent subgrade. Here, a very flexible geotextile is an advantage in laterally confining the ballast stone in its proper location. Quantification of this type of lateral confinement is, however, very subjective.

(*d*) Consider the depth of the geotextile beneath the bottom of the tie. The very high dynamic load of the railway, acting on the ballast, imparts accelerations to the stone that are gradually diminished with depth. If the geotextile location is not deep enough, it will suffer from abrasion at the points of contact with the ballast. Raymond (1982, 1986, 1993a,b) and Raymond and Bathurst (1990) have evaluated a number of exhumed geotextiles beneath Canadian and US railway tracks, and found that many are pockmarked with abrasion holes. In fact, there are so many cases that they have quantified the situation. It is seen that the major damage occurs within 250 mm of the tie, but no damage is noticeable deeper than 350 mm. From these data, it can be safely concluded that the minimum depth for geotextile placement is 350 mm for abrasion protection. If this depth is excessive, a highly abrasion-resistant geotextile must be used. An example of abrasion damage to a geotextile due to inadequate ballast thickness is shown in an exhumed geotextile in Figure 9.4.

(*e*) The last step is to consider geotextile survivability during installation. To compact ballast under ties, the railway track industry uses a series of vibrating steel prongs forced into the ballast. Considering both the forces exerted and the vibratory action, high geotextile puncture resistance is required. Hence, it is necessary

Figure 9.4 Abrasion failure of a geotextile placed too close to the track structures
Reproduced from Raymond (1982), Industrial Fabrics Association International

to keep the geotextile deep or to use a special high-puncture-resistant geotextile.

9.6. Installation of geosynthetics

The acceptance and use of geotextiles for track stabilisation is now common practice in the USA, Canada and Europe. Geotextiles are also being used in high-maintenance locations, such as turnouts, rail crossings, switches and highway crossings. One of the most important areas served by geotextiles is beneath the mainline track for the stabilisation of marginal or poor subgrade, which can suffer from severe mud pumping and subsidence. These conditions often require the issuance of 'slow orders'. Trains become delayed, maintenance costs increase and there is greater possibility of costly derailments. All these add up to potential injury to passengers, inconvenience to customers and loss of operating revenue.

For optimum performance, a geotextile must be installed properly. The geotextile can be installed under existing tracks in a number of ways, but is usually placed in conjunction with undercutting, ploughing or sledding operations, as described by Walls and Newby (1993). In some instances, track sections are removed by crane during rehabilitation of the track bed, with the geotextile being installed at the same time. The following important points must be considered

- The surface, over which the geotextile is being placed, should be prepared and contoured to remove debris and road-bed irregularities, with cross-fall gradients to facilitate the drainage of water from the track centreline to adjacent ditches and drains.
- When joining geotextiles, an overlap of at least 0.5 m is recommended.
- The geotextile should be placed so that water entering the geotextile can drain away from the track.
- It is essential that the geotextile be placed at least 305 mm below the bottom surface of the cross-ties. This is to prevent damage from normal tamping operations, as shown in Figure 9.5.

Figure 9.5 Installation of a geotextile and sledding of sub-ballast

9.7. Case histories in railway track stabilisation

9.7.1 Experience from Canada and the USA

Walls and Newby (1993) reported a railway track rehabilitation in Alabama. They describe the first US application of a geotextile for separation and a geogrid for reinforcement of the track bed. In 1976, a 2000 m-long section of track was relocated by about 365 m east of its present location. Most of the new track was situated in deep, wide cuts through inter-bedded sand and weak clay layers at an elevation of about 7.5 m below the original groundwater level. Owing to the excessive deformation and fouled ballast, track realignment and resurfacing was required every 2–4 weeks. Furthermore, train speed was reduced to 8 km/h through this short section, in order to prevent derailments. In May 1983, rehabilitation work was done to address the low bearing capacity of the subgrade, to prevent soil contamination of the ballast, and to prevent dissipation of the high pore-water pressures caused by cyclic train loading. Tests at the Royal Military College of Canada showed that the number of cyclic loadings required to cause a 50 mm permanent rut could be increased by a factor of ten with geogrid-reinforced ballast over weak subgrades. The design involved removing the 300 mm of fouled ballast, placing a 380 g/m^2 non-woven needle-punched geotextile, followed by a geogrid of Tensar SS2, followed by 300 mm of clean ballast. After an initial observation of three months, in which the reinforced track structure performed satisfactorily, it was decided to increase train speed to 80 km/h. Subsequently, the track was reported to have been in service for four years without any problems, with only routine track maintenance required.

Raymond (1993a,b) reported the use of geotextiles for railway-switch and grade-crossing rehabilitation in Canada. The use of geotextiles to mitigate mud-pumping problems, and to maintain and drain areas, was of immense value. To assess the effect of abrasion with installation depth, data were obtained at several sites with needle-punched resin-treated geotextiles, all having a mass of between 450 and 510 g/m^2. After excavation, the estimated damage of the worst 300 mm × 300 mm (generally below the intersection of the rail and tie) was established by measuring the percentage of worn-out areas. The results ranged from 0.3% at a depth of 350 mm to 4.1% at a depth of 175 mm. Below 250 mm, the amount and rate of change of the damage was small. The results suggest that a minimum depth of 250 mm of ballast is needed before any ballast tamping operations, where, in practice, a 300 mm depth is preferred. For geotextiles used directly on the undercut subgrade surface, a 1000 g/m^2 resin-bonded non-woven needle-punched geotextile, made from fibres with a linear density of less than 0.7 tex and a tenacity greater than 4 mN/tex, is recommended.

9.7.2 European experience

The experience of the European railways was summarised by Van Santvoort (1994) as follows

- There are several alternative methods for using geotextiles in the track bed, with or without protective layers above or below, including adjacent granular filter layers directly in contact with the subgrade, and placed in the ballast without extra protection. Each method requires very specific properties of the geotextile.
- There is very little data on the use of woven geotextiles to provide a filter layer in the trackbed.
- Laboratory tests on silty clay subgrade (95% finer than 60 μm), where there is a special overlying protective gravel layer without filter properties, show that no geotextile tested can prevent the passage of silty clay slurry.
- In all known cases where a geotextile has been used in a track in combination with an adjacent protective layer, it is observed that the success was due partly to the filtering properties of that layer. In the absence of these filtering properties, the track will have a very short life on fine cohesive subgrades unless the soil contains a substantial sand content.
- In the experience of most railways, the use of a geotextile without any overlying or underlying protective layers in the trackbed, with a fine cohesive subgrade in wet conditions, will have only a very short track life. The use of a geotextile without an overlying protective layer can cause problems in subsequent ballast cleaning.
- It is recommended that the largest AOS for any geotextile used in the track should not be greater than 60 μm.
- A synthesis of European experience indicates that, provided the above rules are met, satisfactory results are obtained with needle-punched geotextiles with a minimum grade of 350 g/m^2, or heat-bonded geotextiles with a minimum grade of 250 g/m^2.
- In cases where the subsoil is fine-grained and poorly drained, it is advocated that a thin layer of sand should be used beneath the geotextile.
- The difference of the rate of passage of slurry in dynamic tests shows a slight advantage for needle-punched geotextiles over the thermally-bonded kind. This advantage is counterbalanced by the superior mechanical integrity of thermally-bonded geotextiles over the needle-punched material. When used with protective sand layers, or when the subgrade is a mixed soil containing similar proportions of sand, silt and clay, the different types of geotextile referred to above have equal merit.

9.7.3 Indian experience

Yog *et al.* (1989) reported on the use of geotextiles in rail-track foundations in India. Results of two trials were reported – one having a problem of mud pumping and the other with a problem of weak load-bearing capacity. Trial site 1 is between Sandila and Balamau stations, where 270 m of track bed consists of 50 mm of sand, a layer of 400 g/m^2 non-woven needle-punched geotextile, another 50 mm of sand and 300 mm of ballast, and was built in December 1985. After $3\frac{1}{2}$ years, it was observed that the mud pumping had been checked, and no fabric damage had occurred. Trial site 2 is between Sandila and Rahimabad stations, where 100 m of section was built in March 1987. The subgrade continued to deform, and penetration of 98 cm was observed by February 1989. The conclusion is that a lightweight, flexible geotextile is effective to prevent mud pumping, but cannot prevent shear failure of a weak subgrade, although it may help to reduce the vertical load intensity on the subgrade marginally.

Jain and Azeem (1998) reported the tremendous increase in the traffic level, speed and axle loads on Indian railways. Most of this increase is on core routes of about 11 500 km. As of March 1997, 750 km of track was under permanent speed restriction due to weak subgrade. In addition, temporary speed restrictions were imposed during monsoons for about 500 km. Trials on the use of low-strength, low-modulus geotextiles and geomeshes for subgrade stabilisation on railway tracks were conducted. Geotextiles were placed under a ballast cushion of 250 mm, sandwiched between two nominal sand layers, each 50 mm thick. The observations revealed that upward migration of fines was checked by the geotextile; however, the subgrade continued to deform, and the geotextile eventually ruptured under the outer rail seat. Geomeshes were tried, and they were also unable to prevent shear failure near the rail seats. Finally, geogrid reinforcement was used, and it was demonstrated that a high-strength, high-modulus biaxial geogrid would prevent bearing failure and reduce the sub-ballast thickness by about 40%. Thus, a combination of a geotextile and a geogrid will be the best option for tackling the twin problems of mud pumping and weak subgrade for high-load-capacity railway tracks.

9.7.4 Australian experience

Indraratna *et al.* (2010) conducted a field trial on an instrumented track at Bulli, New South Wales (NSW), Australia, with the specific aims of studying the benefits of a geocomposite installed at the ballast–capping interface, and to evaluate the performance of moderately graded recycled ballast in comparison with traditionally very uniform fresh ballast. The track was constructed between two turnouts at Bulli along the New South Coast. The total length of the instrumented track section was 60 m, and it was divided into four sections, each of 15 m length. Fresh and recycled ballast was used at sections 1 and 4, respectively, without inclusion of a geocomposite layer, while sections 2 and 3 were built by placing a geocomposite layer at the base of the fresh and recycled ballast, respectively.

The settlement pegs, displacement transducers and pressure cells were installed at specific locations. Concrete sleepers were used in the test track. The overall trackbed thickness was 450 mm, including a ballast layer of 300 mm and a capping layer of 150 mm in thickness. The particle size, gradation and other index properties of fresh ballast used were in accordance with the technical specification of the Rail Infrastructure Corporation of NSW (RIC), which represents sharp angular coarse aggregates of crushed volcanic basalt. Recycled ballast was collected from spoil stockpiles of a recycled plant commissioned by RIC at its Chullora yard near Sydney. The finest fraction of less than 9.5 mm was removed by screening. The capping material comprised a sand–gravel mixture. A bi-oriented geogrid was placed over the non-woven polypropylene geotextile, to serve as the geocomposites layer, which was installed at the ballast–capping interface.

The field trial highlighted the performance of different layers of the track exposed to repeated wheel loads. As expected, the performance of layers of the track differed in adjacent sections where the material properties were different. Under the edge of the sleepers, the compression of both fresh and recycled ballast was higher, due to reduced lateral constraints. The recycled ballast produced less settlements and strains compared with the fresh ballast. This was due to the moderately graded composition of the recycled ballast used here, in contrast to the very uniform fresh ballast. The test results demonstrated the potential benefits of using a geocomposite in track to reduce vertical and lateral deformations. The geocomposite reduced the lateral strains of fresh ballast significantly, thus minimising the lateral spread of ballast, thereby eliminating the need for additional layers of crib and shoulder ballast during maintenance. Also, recycled ballast, when used with a geocomposite layer, was found to perform as well as fresh ballast without a geocomposite, with obvious implications for reduced maintenance costs. This study also showed that the inclusion of a geosynthetic layer at the ballast-capping interface might also serve as an alternative method of increasing the internal confining pressure.

The measurement of *in situ* stresses indicated that the maximum vertical cyclic stress decreased significantly with the increase in depth, while the maximum horizontal cyclic stress decreased only marginally with depth. The geosynthetic layer increased both the friction interlock and the confining pressure. In rail-track environments, the confining pressure is of major concern. Therefore, the need to increase track confinement (. . . by incorporating a geosynthetic layer within the ballast layer itself) was further justified to improve track stability. Another important finding of this study was that wheel imperfections such as wheel-flats can increase the vertical maximum cyclic stress significantly, causing further ballast degradation.

9.7.5 Brazilian experience

Fernandes *et al.* (2008) presented a study on the use of geosynthetic reinforcement to reduce the consumption of good-quality sub-ballast material in a railway track in the state of Minas Gerais, Brazil. Six test sections on the railway were instrumented and monitored for a period of two years. Non-woven geotextile and a geogrid were used as reinforcement in different positions in the sub-ballast. Six experimental railway test sections, each 25 m long, were constructed as part of the research programme. These sections were localised in Line 2 of the Variante Capitao Eduardo, Vitoria–Minas railway, between kilometres 40 and 42. This railway is intensely trafficked, with approximately 16 compositions (two locomotives of 160 t plus 100 wagons of approximately 100 t each: 300 kN axle load) daily. The steel sleepers were of type UIC865, with dimensions of 2.2 m × 0.3 m × 0.02 m. The rail type was TR 68.

The results of *in situ* tests and strain measurements showed that the presence of the geosynthetic reinforcement reduced the compressibility of the system. Reduced breakage of the ballast material and greater abrasion resistance were observed in the test sections with geosynthetics. This can be attributed to the greater lateral confinement provided by the reinforcement layer. Similar track performance was observed for the two types of geosynthetic tested. However, significant mechanical damage was observed in the geotextile used after 600 days in service. This suggests that the long-term performance of the geotextile might be severely compromised by increasing mechanical damage, and that the use of a thicker, more durable geotextile would have been more appropriate. The results obtained validated the use of an alternative, low-cost, sub-ballast material in combination with geosynthetic reinforcement. This solution can be attractive for the mining industry in situations where reasonably good-quality mining wastes are plentiful, and conventional track construction materials are scarce or expensive.

9.8. Geosynthetic drains for track drainage applications

9.8.1 Sources of water

Sources of water entering track substructure are (Figure 9.6)

- precipitation
- surface flow
- subsurface seepage.

Excess water may create a saturated state in ballast and sub-ballast, and cause significant increases in track maintenance costs because of problems, such as the following

Figure 9.6 Sources of water entering track substructure

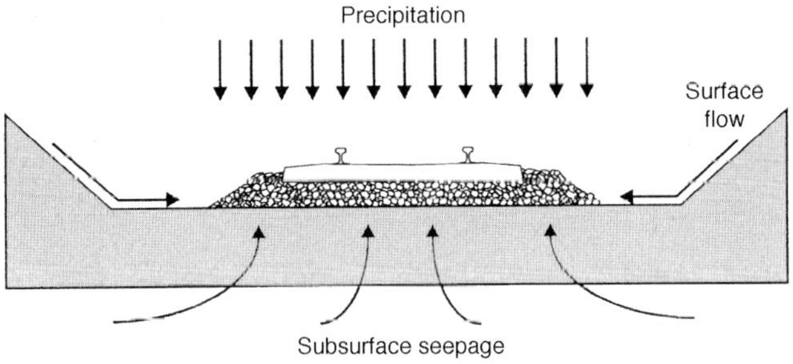

- excess porewater pressure increase under cyclic load, which causes an increase in plastic strain accumulation, a decrease in stiffness and a decrease in strength
- loss of strength due to an increase in the water content
- subgrade attrition and slurry formation from ballast action
- hydraulic pumping of fine soils from the subgrade up into the sub-ballast and ballast
- volume change from swelling of expansive soils, if any
- frost heave/thaw softening
- ballast degradation from slurry abrasion, chemical action and freezing of water
- sleeper attrition from slurry abrasion.

Clearly, adequate drainage to prevent or minimise these problems has a major influence on maintenance costs. Because each source of water requires different drainage methods, the sources must be identified in order to determine effective drainage solutions.

9.8.2 Track drainage requirements

- The first requirement is to keep the ballast clean, to allow lateral drainage.
- The second requirement is the design of a cross-fall gradient in the subgrade and sub-ballast for lateral drainage, as shown in Figure 9.7.
- The third requirement is to provide a means of carrying away the water that comes out of the substructure. This will require the use of edge drains, as shown in Figure 9.7. Two conditions must be avoided: one is the bathtub effect, as shown in Figure 9.8, and the other is fouled shoulders, resulting in a bathtub condition.

Figure 9.7 Sub-ballast and subgrade with cross-fall sloped into side drains

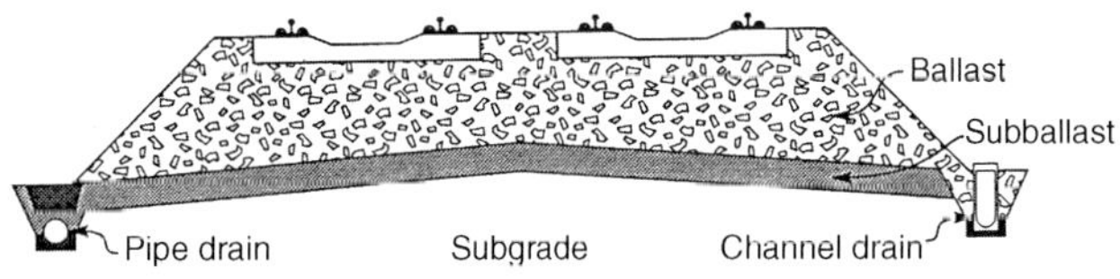

9.8.3 Side drains

Side drains are located on one or both sides of a track, parallel to its route until an outfall is reached. These are common to all railway track drain installations, and can be designed to intercept and carry away surface water, as well as seepage from the ballast, sub-ballast and subgrade. Off-take drains may be provided as an intermediate outlet for the side-drain system, to limit the drain length and to provide a shorter distance to a natural drainage course that would be available parallel to the track.

The most effective side drain is the side of an embankment, provided it is near the toe of the ballast, and the top of the embankment is sloped to shed water. However, embankment shoulder protection is required to control erosion.

The next most effective side drains are open ditches. They must have the capacity to carry away water from the substructure, as well as adjacent surface runoff. Ditches must be sloped steeply enough to prevent sedimentation, but not to cause erosion of the ditches. If velocities higher than the soil erosion limits are anticipated, the ditch may be protected from erosion by concrete or geosynthetic lining. The advantages of ditch drains are that they are economical to construct and can handle large stormwater flows. However, the track geometry and the surrounding ground topography restrict ditch drain geometry.

Imbert *et al.* (1996) reported on 20 years of successful French experience with the use of a bituminous geomembrane for water-proofing subgrade under railway ballast. For new track construction, 60 000 m^2 of bituminous geomembrane had been used to renovate track at Gare du Nord station in Paris, and to protect against rainfall infiltration in the gypsum-bearing subgrade. Also, these geomembranes were used to waterproof side earth ditches in accessible locations, where these products were easily transported and laid with manual labour.

Figure 9.8 The bathtub condition for ballast and sub-ballast

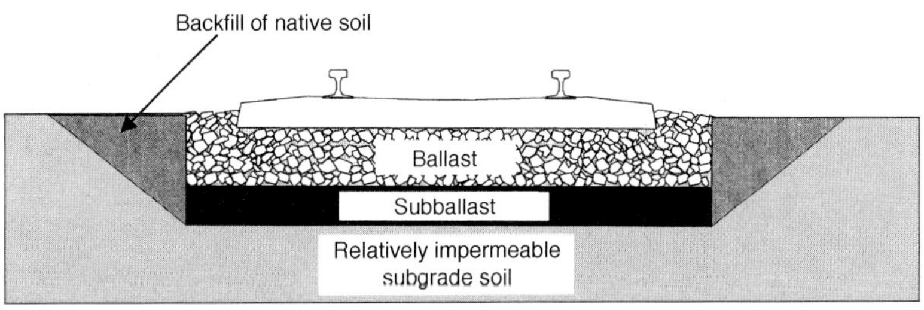

Figure 9.9 Examples of French drains as deep side drains

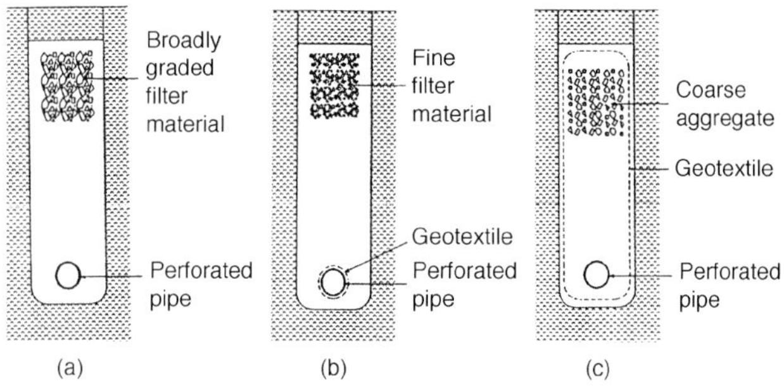

Ditch drains are not effective for removing subgrade water, either because they are not deep enough or because they are lined. A deep side drain is then required to provide a sufficient hydraulic gradient to allow water seepage, as well as to keep the phreatic surface well below the top of the subgrade. Examples of deep drains, known as French drains, are shown in Figure 9.9. These drains consist of a geotextile filter wrapped around a coarse aggregate, surrounding a perforated pipe at the base of the drain system. The drain system shown in Figure 9.9(b) is not recommended, as the filter around the perforated pipe may fail over time due to clogging of the geotextile. The best of the three designs is the system shown in Figure 9.9(c), where there is no possibility of clogging around the perforated pipe.

The wraparound geotextile must satisfy the separation requirement that

$$\text{AOS} < D_{85} \text{ of protected soil} \quad (9.2)$$

where AOS is the apparent opening size of the geotextile and D_{85} is the diameter for which 85% of soil particles are finer (mm).

The gradation of the granular envelope around the perforated pipe must be related to the hole size, as recommended by Cedegren (1977).

For slotted pipes:

$$D_{85} \text{ (size of filter material size)} > 1.2\text{–}2 \text{ times the slot width} \quad (9.3)$$

For circular holes:

$$D_{85} \text{ (size of filter material)} > \text{hole diameter} \quad (9.4)$$

Geocomposite edge drains (Figure 9.10) have been developed as an alternative to granular trench drains. They consist of a flat plastic core wrapped in a geotextile sleeve/filter. These are called fin drains (or prefabricated vertical drains, PVDs), and are designed for easy installation in either a slot or a trench dug in the soil along the edge of a track or pavement. This can be done in a continuous process without

Figure 9.10 A geocomposite edge drain (fin drain)

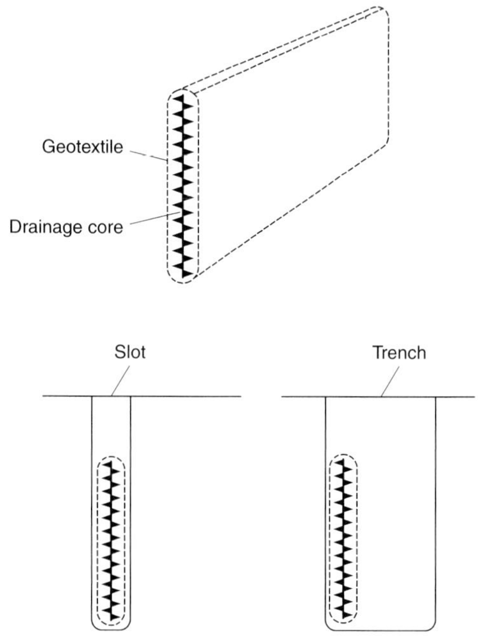

the need for much backfilling. It should be noted that there are two major shortcomings with PVDs. Firstly, the tensile strengths of the core and the geotextile filter are normally not compatible and sometimes differ a lot. As a result, one component, either the core or the filter, will break first. Once one component breaks, the PVD will no longer function properly. Secondly, as the filter is fitted loosely to the core, the filter will indent into the drainage channels of the core under earth pressure. As a result, the discharge capacity of the PVD will be reduced (Broms *et al.*, 1994; Liu and Chu, 2009). To overcome these two problems, a new type of PVD, the so-called integrated PVD, in which the core and filter are bonded together by heat melting, was introduced by Liu and Chu (2009). The integrated PVD offers a number of advantages over the separable PVD. Both the tensile strength and the discharge capacity of the integrated PVD are higher than those of the separable drain of the same material. Additionally, the integrated PVD has a stronger resistance to bending, uses less filter material and is more suitable for mass production. Therefore, integrated PVDs can be efficient edge drains in railway tracks and other drainage projects.

Fin drains act primarily as a rapid groundwater collector. The water discharge capacity is not intended to be equivalent to a long-distance carrier drain such as side ditches. Where a large volume of water is involved, or the distance between water discharge points is great, the fin drain discharges into a lower carrier, which can be a closed jointed traditional carrier drain pipe. An example in which a fin drain can be incorporated into a track drainage scheme is shown in Figure 9.11.

Selig and Waters (1994) reported that, in 1984, fin drains were installed at three locations on the Southern Railway

Figure 9.11 An example of the installation of a fin drain

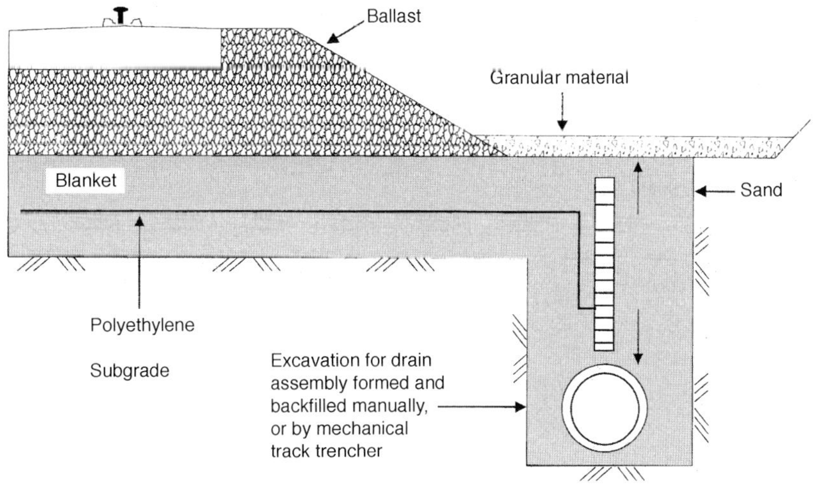

tracks in which wet subgrades were threatening to contaminate the ballast. An inspection of the fin drains carried out two years after installation indicated that they were functioning properly.

9.8.4 Drainage of subgrade seepage

Geocomposite side drains may also be used to collect water seeping upward from the subgrade, as shown in Figure 9.12. These drains must be deep enough to adequately lower the phreatic surface. The required depth of the drain depends on the permeability of the soil, including the effects of fissures and pervious seams. Depths of at least 1–2 m or more may be required. Trench-type drains, or other internal drains, may also be installed on the slopes of cuts to help in the lowering of the phreatic surface.

Cross-drains may also be used to supplement side drains. Cross-drains are sometimes installed beneath track where lateral drains are insufficient to control the groundwater beneath the centreline of the track. However, the use of cross-drains for general groundwater control is impractical. In high-permeability soils (i.e. sands and gravels), cross-drains are rarely required. In fine-grained soils, the permeability is so low that cross-drains must be closely spaced to be effective, which will be uneconomical. A better solution is to raise the track and place it on a blanket of granular material, or to excavate a portion of the subgrade and replace it with granular material. In these ways, the lateral drains can control the flow. In wide flat areas, cross-drains, as shown in Figure 9.13, may be necessary.

Figure 9.12 Lowering of the phreatic surface (groundwater level) with side drains

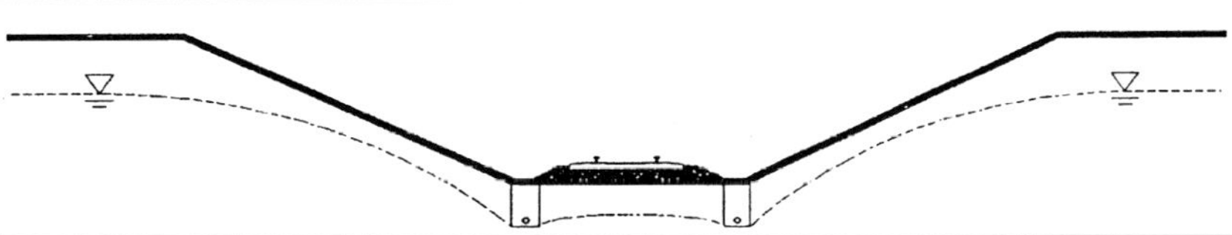

9.9. Summary of main points

This chapter gives a comprehensive picture of the nature of the track substructure system and its functions, and explains the role of geosynthetics in track substructure stabilisation and rehabilitation works. The main points are summarised as follows

- For stable tracks, it is crucial that the substructure system remains stable under repeated wheel loads during its service life. In particular, under high water-level conditions, mud pumping of subgrade fines into the sub-ballast will cause ballast fouling and contamination, leading to its failure as a load-dissipating layer. This will cause failure of subgrade soils and excessive settlements, leading to failure of the track system as a whole.
- The principal functions of geosynthetics in railway track applications are separation, filtration, lateral confinement/reinforcement and drainage. Geosynthetics protect the individual layers of subgrade/sub-ballast/ballast from mud contamination, degradation and weakening over time under repeated wheel loadings.
- The physical properties of geosynthetics should be examined thoroughly in order to aid in the selection of the correct geosynthetics for different track conditions. The list of key properties of geosynthetics needed to perform adequately in track substructure stabilisation should include durability, chemical resistance and abrasion resistance. Consideration of these characteristics will ensure the successful design of track

Figure 9.13 An example of the installation of cross-drains

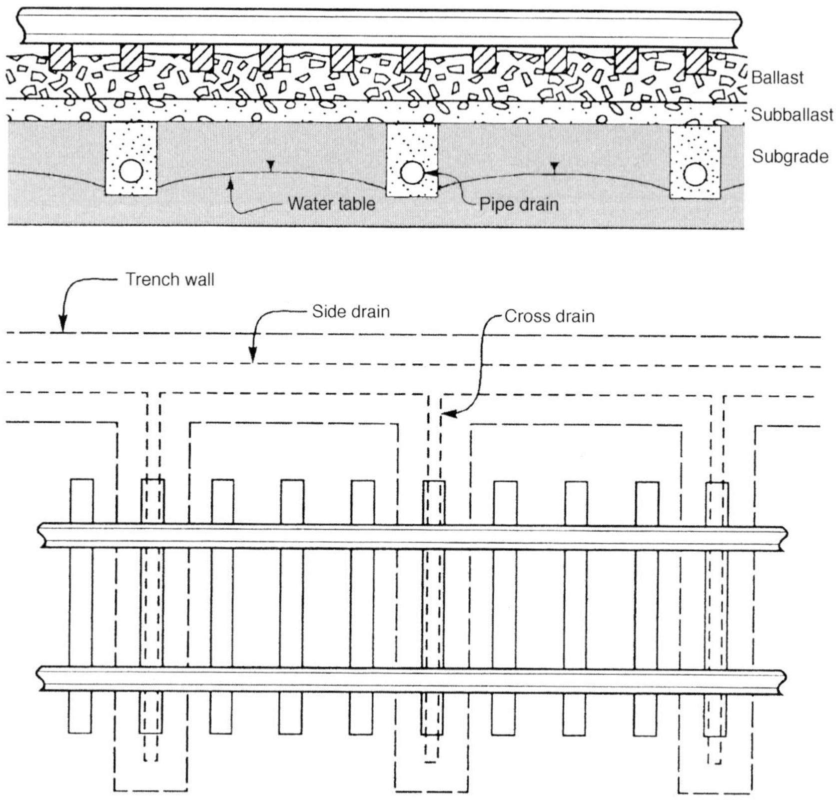

substructure with geosynthetics. It is also essential to take the necessary precautions for good installation practices, to prevent premature geosynthetic damage.

- Site drainage plays a significant role in track stability. Keeping subgrade soils as dry as possible is therefore a definite step towards protecting the track substructure system. The use of geotextile-wrapped trench drains and geocomposite drains provides rapid and cost-effective solutions for the requirements of subsurface and side drainage in railway tracks.
- The successful cases and studies of track stabilisation in Canadian, US, European, Indian, Australian and Brazilian experience show that correctly designed and installed geosynthetics produce very good expected performance and stability. However, it is important to note that geosynthetics are not an answer to all track-related problems, but, when used properly, they can help to maintain track integrity – and when this is accomplished, they enhance safety, reduce maintenance costs and improve operating revenues.
- Further research is required for a proper understanding of the beneficial effects of geosynthetics in railway tracks, with particular reference to the development of analysis and design methods.

9.10. Some common questions and their answers

Questions

1. How do applications of geosynthetics for railway tracks differ from geosynthetic applications for highway pavements?

2. What is the evidence that geosynthetics are needed for track substructure stabilisation?

3. How do geosynthetics help with subgrade problems?

4. How do geosynthetics help with sub-ballast stability?

5. How do we select appropriate geosynthetics for track substructure works?

Answers

1. The common problem in unprotected railway track substructure is mud pumping of fine subgrade soils, causing sub-ballast fouling and degradation. The function of geosynthetics at the subgrade–sub-ballast interface is different from the highway pavement situation. The essential differences are: (*a*) ballast and sub-ballast stones are very coarse, uniform and angular; (*b*) repeated railway wheel loads may set up resonant loadings in the soils; and (*c*) the railway track system feeds a long wave of both positive and negative pressure into the ground ahead of the train itself. These factors lead to aggressive mud-pumping actions when subgrade soils have a significant fines (silts and clays) fraction that can migrate upwards and cause ballast fouling.

2. The accurate prediction of the railway track maintenance cycle needs an understanding of the complex mechanisms of track deterioration. Geosynthetics can play a significant role in protecting and enhancing the performance of the different layers of the track substructure. In general, it has been observed in laboratory and field trials that track substructure is stabilised, and performs very well when appropriate types of geosynthetics are used, compared with the unprotected track system.

3. The key to track stability is protection of the subgrade from water saturation, softening and degradation from repeated wheel loads. To achieve this, thick geotextiles are needed as a separation layer between the subgrade and the ballast or sub-ballast, to prevent the intermixing of fine soils and contamination of coarse ballast. In addition to the separation function, the geosynthetic layer at the subgrade is needed to drain away rainfall infiltration in ballast from soft subgrade soils. Thus, a low-permeability geotextile or geomembrane is needed to seal the subgrade from above, while the filtration function is needed to prevent the upward migration of fines into the ballast from below. These two conflicting functions can be met using geocomposites.

4. The key function of sub-ballast is drainage and stress-absorbing capacity to dissipate stresses from wheel loads to acceptably low values at the subgrade layer. To achieve this, the ballast must remain clean and uncontaminated so that it has high frictional strength and stiffness during its service life. Laboratory and field measurements have shown that a strong geotextile or geogrid can provide lateral basal confinement, and reduce shear stresses in the subgrade soils below. Cohesive soils under shear loadings in wet conditions have low-bearing strength. Keeping these soils dry and reducing their shear loadings are the benefits of using high-strength geosynthetics with reinforcement and drainage functions to protect the weak subgrade soils.

5. From the discussions presented in this chapter, it is obvious that we do not have a 'one size fits all' situation in railway track substructure stabilisation. The appropriate choice of geotextiles and geogrids, with suitable separation, filtration, reinforcement and drainage capabilities, is needed for different site applications. Thus, the use of geocomposites is the way forward in very adverse site applications. For example:

- A site with a low risk of water ingress from the underlying water table would need only the basal reinforcement function and geomembrane sealing from rainfall infiltration. In this case, a geocomposite, consisting of a geomembrane sandwiched between two protective lightweight (400 g/m^2) non-woven geotextiles, may be adequate.
- Use of a single-layer geotextile for separation and filtration only is adequate for a site with a good subgrade with low fines content, natural drainage and low rainfall conditions.
- In extremely difficult conditions of high rainfall, a high water table and a poor subgrade, a multi-layer geocomposite is highly recommended. The use of a reinforced bituminous geomembrane liner system was shown to prevent water from coming into contact with the surface of a highly plastic soft clay subgrade on the Burlington Northern Santa Fe railway, in the USA. The product is a combination of non-woven geotextile, fibre-glass reinforcement and bituminous impregnation, which results in long life expectancy and very high resistance to stresses and ageing.

REFERENCES

AREA (1985) *American Railway Engineering Association Manual*. AREA, Lanham, MA.

Broms BB, Chu J and Choa V (1994) Measuring the discharge capacity of band drains by a new drain tester. *Proceedings of the 5th International Conference on Geotextiles, Geomembranes and Related Products*, Singapore, pp. 803–806.

Cedegren HR (1977) *Seepage, Drainage and Flownets*. Wiley, New York.

Fernandes G, Palmeira EM and Gomes RC (2008) Performance of geosynthetic reinforced alternative sub ballast material in a railway track. *Geosynthetics International* **15(5)**: 311–321.

Imbert B, Breul B and Herment R (1996) More than 20 years of experience in using a bituminous geomembrane beneath French railway ballast. In *Geosynthetics: Applications, Design and Construction* (De Groot MB, Den Hoedt G and Termaat RJ (eds)), Balkema, Rotterdam, pp. 283–286.

Indraratna B, Nimbalkar S, Christie D, Rujikiatkamjorn C and Vinod J (2010) Field assessment of the performance of a ballasted rail track with and without geosynthetics. *Journal of Geotechnical and Geoenvironmental Engineering* **136(7)**: 907–917.

Jain VK and Azeem A (1998) Rail transport support upgradation – potential evaluation of innovative geosynthetics. *Proceedings of the 6th International Conference on Geosynthetics*, Atlanta, GA. Industrial Fabrics Association International, Roseville, MN, pp. 1089–1092.

Koerner RM (2005) *Designing with Geosynthetics*, 5th edn. Prentice-Hall, Englewood Cliffs, NJ.

Leuttich SM, Giroud JP and Bachus RC (1992) Geotextile filter design guide. *Geotextiles and Geomembranes* **11(4)**: 19–34.

Liu HL and Chu J (2009) A new type of prefabricated vertical drain with improved properties. *Geotextiles and Geomembranes* **27(2)**: 152–155.

Raymond GP (1982) Geotextiles for railroad bed rehabilitation. *Proceedings of the 2nd International Conference on Geotextiles*, Las Vegas, NV, pp. 479–484.

Raymond GP (1986) Performance assessment of a railway turnout geotextile. *Canadian Geotechnical Journal* **23(4)**: 472–480.

Raymond GP (1993a) Geotextile for railroad branch line upgrading. *Proceedings on Geosynthetics Case Histories*. Technical Committee 9, International Society for Soil Mechanics and Foundation Engineering (ISSMFE), London, pp. 122–123.

Raymond GP (1993b) Geotextiles for railway switch and grade crossing rehabilitation maritime provinces, Canada. *Proceedings on Geosynthetics Case Histories*. Technical Committee 9, ISSMFE, London, pp. 124–125.

Raymond GP and Bathurst RJ (1990) Tests results on exhumed railway track geotextiles. *Proceedings of 4th International Conference on Geotextiles, Geomembranes and Related Products*, The Hague, pp. 197–202.

Selig ET and Waters JM (1994) *Track Geotechnology and Substructure Management*. Thomas Telford, London.

Shukla SK and Yin JH (2006) *Fundamentals of Geosynthetic Engineering*. Taylor and Francis, London.

Van Dine D, Williams SE and Raymond GP (1982) An evaluation of abrasion tests for geotextiles. *Proceedings of the 2nd International Conference on Geotextiles*, Las Vegas, NV, pp. 811–816.

Van Santvoort GPTM (1994) The use of geotextiles in railway construction in the Netherlands. In *Geotextiles and Geomembranes in Civil Engineering* (Van Santvoort GPTM (ed.)), Balkema, Rotterdam, pp. 378–387.

Walls JC and Newby JE (1993) Geosynthetics for railroad track rehabilitation. *Proceedings on Geosynthetics Case Histories*, Technical Committee 9, ISSMFE, London, pp. 126–127.

Yog AK, Krishna B and Azeem A (1989) Geotextile in rail track foundation. *Proceedings of the International Workshop on Geotextiles*, Bangalore, pp. 97–102.

Handbook of Geosynthetic Engineering
ISBN 978-0-7277-4175-2

http://dx.doi.org/10.1680/hge.41752.209

Chapter 10
Slopes – erosion control

T. S. Ingold* **and M.-H. Li†‡**
*Consulting Geotechnical Engineer, St Albans, UK
†Department of Landscape Architecture and Urban Planning, Texas A&M University, College Station, USA
‡Texas Transportation Institute, College Station, USA

Geosynthetics have become very important materials for slope erosion control since the late 1980s. Due to stringent water-quality regulations, many geosynthetic manufacturers have emerged and produced numerous effective products that have become useful additions to the traditional agricultural erosion control methods. Civil engineers are increasingly adopting a design approach using commercial products because of availability of materials, approved performance and ease-of-use during the design process. This chapter begins with an overview of soil erosion, erosion mechanics and the types of erosion control systems, followed by updated design approaches and performance testing methods. Ageing and cost-effectiveness of geosynthetic erosion control products are also covered in this chapter.

10.1. Introduction

Wind, water and gravity are the prime agents of erosion. These elements combine in a variety of ways to perform the dual role of abrading and simultaneously transporting both soil and rock. There are five distinct categories of erosion

- glacial erosion, in which the active agents are ice sheets and glaciers
- marine erosion, in which the sea is energised by wind and gravity to produce waves, tides and currents
- river erosion, in which the agents are corrosion, hydraulic lifting, scouring, cavitation and abrasion (river erosion also triggers mass movement in the form of landslips)
- Aeolian erosion, in which overland winds blow away fine cohesionless soil particles and wind corrosion (sand-blasting)
- rain erosion, in which corrosion, rain splash, rain wash and sheet wash attack the land mass, so causing surface erosion.

The principles involved in mitigating the effects of these processes involve sealing, strengthening, or in some other way retaining the rock and soil particles by a barrier, or barriers, of durable protective material that can harmlessly dissipate the erosive energy of the attacking agent. This chapter is concerned only with the control of surface erosion of slopes caused by wind and rain, and, therefore, provides no further consideration of other forms of erosion, save for river erosion and mass movement in as much as they interact with rain erosion. A study on erosion control methods, using synthetic products, is presented, along with methods based on agronomic systems. Recent studies on the ageing and longevity of erosion control products, and also cost-effectiveness, are added for the sake of better understanding and completeness of the subject.

10.2. Interaction of rain and river erosion

Rain erosion can act on a land surface of any degree of slope; however, the severity of rain erosion increases with increasing slope steepness and slope length. In turn, the steepness of the slope will be controlled by river erosion, as propounded in the theory of cyclic slope evolution. This includes the notion that erosion commences by the rapid downcutting of streams and rivers into the landmass, leading to the formation of steep-sided valleys. When the river bed reaches a mature profile, downcutting and vertical erosion give way to lateral erosion, which slackens the valley slopes. Finally, lateral erosion leads to the development of a low-lying flat land surface with sluggish meandering rivers. This is considered to define a base level below which erosion effectively ceases.

The significance of this to the control of surface rain erosion is that river erosion and its attendant effects on slope stability should be considered as primary mechanisms. In short, the establishment of an adequate rain erosion control system is worthless if it is applied to a slope that is unstable or likely to be rendered unstable within the perceived service life of the rain erosion control system. Before embarking on the detailed design of such a system, it is prudent to check slope stability and make provision for any necessary remedial works. In broad terms, two aspects need to be considered

- oversteepening of river banks, leading to slope instability
- the effects of groundwater regimes on overall slope stability.

The first of these considerations aims at guarding against river or canal bank erosion associated with localised oversteepening of the bank. Left unchecked, this can lead to the development of retrogressive landslips that work their way up slope, and so threaten overall slope stability. The second consideration relates to overall slope stability. An approximate assessment of the factor of safety is given by infinite slope analysis, where for a long planar slope of cohesionless soil with a soil strength represented by ϕ, the internal angle of shearing resistance, the factor of safety, F, for a slope of $1:n$ (1 vertical to n horizontal), is given by

$$F = n \tan \phi \quad \text{(for a dry slope)} \tag{10.1a}$$

$$F = \tfrac{1}{2} n \tan \phi \quad \text{(for a wet slope)} \tag{10.1b}$$

Equation 10.1a applies to a slope with a depressed groundwater table, while Equation 10.1b applies to a slope subject to parallel seepage with the phreatic surface coincident with the ground surface. In considering surface erosion control systems, it should be remembered that these can reduce surface water runoff by increasing surface water infiltration. As a result of this, the groundwater level may rise. This can lead to a reduction in slope stability, as indicated in Equation 10.1b. Where plants are being used to control surface erosion, these can be selected with root systems that act as reinforcement to the soil, and so mitigate any deleterious effects associated with an induced rise in groundwater level.

10.3. Mechanics of surface erosion

The susceptibility of soil to wind or rain erosion is quantified by its erodibility. For rain erosion this can be conveniently expressed in units of grams of soil loss per millimetre of rainfall. The ability of wind or rain to cause erosion is quantified by its erosivity. The most suitable expression of the erosivity of rainfall is an index based on the kinetic energy of the rain. Thus, the erosivity of rainstorm is a function of its intensity and duration, and of the mass, diameter and velocity of the raindrops. The effect of these variables is reflected in Table 10.1, which shows that the kinetic energy per unit area per unit time can vary by more than three orders of magnitude between drizzle and rainfall from a cloudburst.

Table 10.1 Rainfall kinetic energy related to rainfall intensity and droplet diameter

Rainfall form	Intensity: mm/h	Diameter: mm	Kinetic energy: J/m^2/h
Drizzle	<1	0.9	2
Light	1	1.2	10
Moderate	4	1.6	50
Heavy	15	2.1	350
Excessive	40	2.4	1000
Cloudburst	100	2.9	3000
Cloudburst	100	4.0	4000
Cloudburst	100	6.0	4500

The mechanism of soil loss by rain erosion is a combination of soil detachment and subsequent transportation. The two prime agents in this are raindrop impact and overland flow. On impacting with an unprotected soil surface, raindrops loosen the soil particles, with the resulting splash of particle-laden water causing an incremental movement of the suspended particles down slope. If the rainfall intensity exceeds the current permeability of the bare soil, then surplus water will run down the slope as overland flow transporting both soil particles detached by raindrop impact and particles loosened by the overland flow itself. The ability of overland flow to transport soil particles is approximately an order of magnitude higher than that of rain splash (Morgan, 1986); consequently, raindrop impact may be regarded as the primary agent of detachment, and overland flow as the primary agent of transport. An exception to this is when overland, or sheet flow, becomes canalised into rill flow. In this case, rill flow may prove to be two orders of magnitude greater than that of sheet flow. The magnitude of soil loss resulting from rain erosion will be a function of other variables, including slope inclination and length, and, of course, the degree of protection afforded to the bare soil by vegetative cover or an erosion control system. Rates of erosion, expressed as mass of soil loss, in tonnes per hectare per year, will vary from place to place, and an indication of this is given in Table 10.2 (Morgan, 1986).

Table 10.2 Rates of erosion in selected countries (t/ha/year)

Country	Natural	Bare soil
USA	0.03–3.0	4–90
China	<2.0	280–360
Nigeria	0.5–1.0	3–150
India	0.5–1.0	10–20
Belgium	0.1–0.5	7–85

10.4. Classification of erosion control systems

The applicability of rain erosion control methods vary according to the classification of the site, which can be considered under the following three main headings

- urban
- cultivated
- pasture.

Erosion control methods can be classified into three broad categories, with the applicability of the method depending on the classification of the site

- agronomic

Figure 10.1 A classification system for engineered agronomic erosion control systems

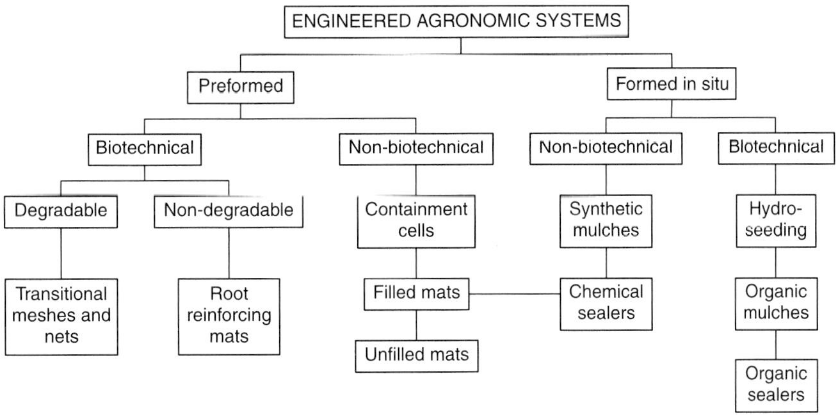

- soil management
- mechanical.

Agronomic or biological methods make use of vegetative or other forms of protective cover to mitigate erosion. Soil and land management is concerned with the ways in which the land is used, and the ways of preparing the soil to promote dense vegetative growth and to improve its structure so that it is more resistant to erosion; for example, the use of conservation tillage. Mechanical methods depend on manipulating the surface topography on a macro scale; for example, by the construction of terraces or silt fences to control the kinetic energy of overland flow. Soil management, land management and mechanical methods fall largely in the domain of the soil conservationist having long-term dealings with cultivated sites and pasture lands.

The domain of immediate interest to civil engineers is that of the urban site or, more specifically, the neo-urban site, where the process of construction may have destroyed natural protective cover. In this situation, the need is for erosion control systems that can be designed to give a required level of performance and installed in a modicum of time. Such systems fall within the agronomic classification, which can be subclassified as follows

- agricultural
- forestation
- engineered.

Agricultural methods involve protection of bare soil, either by live plant cover afforded by maturing crops, or by mulching between growing seasons using traditional mulches derived from crop residues such as straw, maize stalks or standing stubble. To further mitigate erosion and to maintain a sufficiently protective stand of crops in the growing season, it may be necessary to employ crop rotation. Forestation, afforestation and agroforestry, which is forestry incorporated with farming, all provide a suitable method for reducing runoff and therefore erosion. Engineered agronomic methods are more recent and have expanded substantially over the last two decades with the development of synthetic materials and improved chemical binders. A classification system for the various engineered agronomic systems is given in Figure 10.1. The main subclassification, made in Figure 10.1, is between preformed and formed *in situ* systems. Both of these can be further subcategorised as biotechnical or non-biotechnical.

Preformed biotechnical systems use nets, meshes or mats to give short-term protection while vegetative cover becomes established. Where adequate long-term protection is afforded by vegetative cover alone, biodegradable nets and meshes, usually formed from natural fibres, can be used to provide short-term protection. Where vegetative cover alone is inadequate (… where high-velocity overland flow of long duration might occur) then synthetic root-reinforcing mats might be employed. Formed *in situ* biotechnical systems use residual organic matter, such as straw and organic sealers, to provide short- to intermediate-term protection. Hydro-seeding with adhesives can provide short-term protection, with long-term protection being afforded as the vegetative cover develops.

Preformed non-biotechnical systems do not rely upon vegetative cover to give long-term protection. Instead, the soil is covered with unfilled mats, which subsequently tend to fill with soil initially lost further up slope, or mats which are filled with soil at the time of laying. A variation on this is to fill the mats with bound material such as bitumen-stabilised soil. A much thicker protective cover can be obtained using cellular soil containment cells, which, without vegetative cover, tend to form a series of micro terraces down the slope. Formed *in situ*, non-biotechnical systems form low-permeability soil cover by either chemically sealing the soil surface or covering it with a synthetic mulch.

10.5. Design approach

10.5.1 Use of the universal soil loss equation

The objective of any erosion control system is to limit soil loss to an acceptable level during the service life of the system. An acceptable soil loss for agricultural purposes is of the order of 10 t/ha/year (4 t/acre/year). This rate of soil loss would be tolerable in most types of civil engineering works, including highways, where a prime consideration is the time taken for siltation of the drainage systems. The sediment yield, produced by rain erosion, from soil stripped bare by the construction process can be approximated from the universal soil loss equation (USLE) (Wischmeier and Smith, 1958) or from more sophisticated versions of the USLE, such as that due to Morgan *et al.* (1985). All take a similar form and all include a crop factor, as shown below:

$$A = RKLSCP \tag{10.2}$$

where A is the mean annual soil loss (mass/areas), R is the rainfall erosivity index, K is the soil erodibility index, L is the slope length factor, S is the slope steepness factor, C is the crop factor and P is the conservation practice factor.

The conservation practice factor is taken as unity unless the site is terraced. The crop factor, C, represents the ratio of soil loss under a given crop to that for bare soil. Where vegetative cover can be established to give long-term erosion control, regionally published values of the crop factor can be employed. Fifield *et al.* (1989) quote crop factors of 0.01 for vegetated organic mulches, soil sealants and synthetic mats. Morgan (1986) quotes values in the range 0.004–0.01 for cultivated grass and 0.001–0.002 for forest and woodland. Evenly cultivated grasses will generally withstand overland flow velocities of 1 m/s for periods up to 48 hours (Whitehead, 1976). Where these limits might be exceeded, consideration should be given to using root-reinforcing mats, or mats filled with bound soil or aggregate. The use of non-biotechnical solutions (see Figure 10.1) might also be envisaged in the event that the predicted soil loss could not be reduced to an acceptable value using vegetation. A similar design approach can be adopted for wind erosion using the soil loss equation by Woodruff and Siddoway (1965).

The above observations and quoted crop factors for rain erosion only apply to established biotechnical systems; for example, in the long term. Like all other geotechnical problems, consideration must be given to both the long-term and the short-term conditions. In the short-term, pre-formed biotechnical systems rely solely on the protection afforded by the erosion control substrate, be this net, mesh or mat. Since in the short term no vegetation has been established, it is more appropriate to refer to a yield factor rather than a crop factor. The basic definition remains unchanged in as much as the yield factor is defined as the ratio of soil loss from soil protected by an erosion control system to soil loss from bare soil. The yield factor can assume a much larger value than the crop factor, which applies to the long-term condition where vegetative cover has been established.

Some five months after the trial installation of biotechnical systems, Fifield *et al.* (1988) quoted yield factors as high as 0.3, whereas about one year later, and presumably due to the establishment of denser vegetative cover, a value of 0.01 was recommended. A corollary to this is that, unless vegetative cover is established during clement weather, an assessment should be made of both short-term and long-term soil losses to check that the short-term loss is acceptable. Unfortunately, there appears to be little information on yield factors for the short-term condition.

10.5.2 Use of product evaluation results

Another common design approach popular with US civil and highway engineers is using performance evaluation results of the engineered agronomic systems (i.e. erosion control products). The most widely used example of this is the Approved Product List published by the Texas Department of Transportation (TxDOT, 2011). According to the TxDOT Maintenance Division, the Approved Product List has become a design reference by at least 22 other US state departments of transportation because of its thoroughness and broad product listing (Li *et al.*, 2009). Many US municipalities are also referencing the Approved Product List in their design manuals.

In the late 1980s, more stringent stormwater quality regulations appeared and more highway construction projects were being built; subsequently, the erosion control industry began to expand. The concern naturally focused on how well products perform to control erosion on highway slopes. In order to select the best-performing products in design, TxDOT launched a testing programme in 1991 to evaluate the performance of erosion control products marketed by manufacturers for highway projects. These products included both preformed and formed *in situ* systems. In the USA, common preformed systems include rolled erosion control products (RECPs) such as blankets and mats; and formed *in situ* systems are spray-on products with or without chemical binders. Spray-on products with chemical binders such as guar tackifier are called bonded fibre matrix (BFM). Simple spray-on products (no binders) are termed cellulose fibre mulches. This performance evaluation is conducted at the Hydraulics, Sedimentation, and Erosion Control Laboratory of the Texas Transportation Institute.

The Approved Product List is a performance-based list that catalogues erosion control products having exceeded TxDOT's minimum performance standards (TxDOT, 2000) (Table 10.3). Each product is evaluated to determine its ability to minimise sediment loss and establish vegetation. Once a product passes the minimum standards in both categories, the product is included in the list. Because each product has been evaluated under a specific slope and soil condition, the design approach for the end user is a simple selection process. Products for protecting slopes are classified as follows

- blankets (i.e. RECPs) for 1 : 3 (or flatter) on clay
- blankets for 1 : 3 (or flatter) on sand
- blankets for 1 : 3 (or steeper) on clay
- blankets for 1 : 3 (or steeper) on sand

Table 10.3 TxDOT minimum performance standards for the Approved Product List

Condition	Sediment loss on sand: kg/10 m^2	Sediment loss on clay: kg/10 m^2	Vegetation density: % cover compared with a control tray
50% (1 : 2) slope	<308.8	<3.86	>50
3% (1 : 3) slope	<138.9	<3.86	>50

- mulches for 1 : 4 (or flatter) slope on clay
- mulches for 1 : 4 (or flatter) slope on sand.

10.6. Study of short-term yield factors

A series of tests has been carried out to make direct measurement of short-term yield factors for a variety of soils and five different erosion control products (Ingold and Thomson, 1986). The five products were as follows

- A coarse-woven jute net with a mass of 500 g/m^2, a water absorbency of 485%, a nominal opening size of 11 mm × 18 mm and an open area ratio of 60%.
- A fine-woven jute net with a mass of 320 g/m^2, a water absorbency of 210%, a nominal opening size of 5 mm × 5 mm and an open area ratio of 60%.
- A coarse-woven coir fibre net with a mass of 480 g/m^2, a water absorbency of 65%, a nominal opening size of 15 mm × 15 mm and an open area ratio of 65%.
- A heat-bonded polyamide mat, 9 mm thick, with a mass per unit area of 260 g/m^2 and a water retention of 120%.
- A wood-wool mulch sandwiched between two layers of lightweight polypropylene mesh to form a mat with a mass of 360 g/m^2. The outer meshes have a nominal opening size of 25 mm × 37 mm and an open area close to 100%.

The data from these tests have subsequently been analysed with a view to determining the mechanisms that control the yield factor. Measured soil losses from the bare control plots and plots protected by the various erosion control systems allowed short-term yield factors to be calculated directly. Typical values for a sandy loam on a 1 : 2 slope are reproduced in Figure 10.2, which shows that the highest yield factor was obtained for the fine jute net. Figure 10.3 shows the mean measured erodibilities for each system. In this example, all the system erodibilities are less than that of the bare soil, which reflects the fact that all the systems are affording some degree of protection. What is worthy of note is that the fine jute net, which has the highest yield factor, does not have the highest system erodibility. Figure 10.4 gives the mean measured runoff for the various systems, and shows that the runoff for the fine jute net was the largest for all the systems and larger than that for the bare control plot. Since the sediment yield is the product of erodibility and runoff, it follows that the high yield for the fine jute net was dominated by its high runoff.

It is useful to formulate this notion in terms of a simple mathematical expression involving the yield factor (YF), the runoff factor (RF) and the erodibility factor (EF). The runoff factor is defined as the ratio of the runoff from a system-protected plot to the runoff from the unprotected control plot. Similarly, the erodibility factor is the ratio of erodibilities for the protected and unprotected plots. Combining these factors leads to

$$YF = RF \times EF \tag{10.3}$$

Figure 10.2 Yield factors for a sandy loam soil

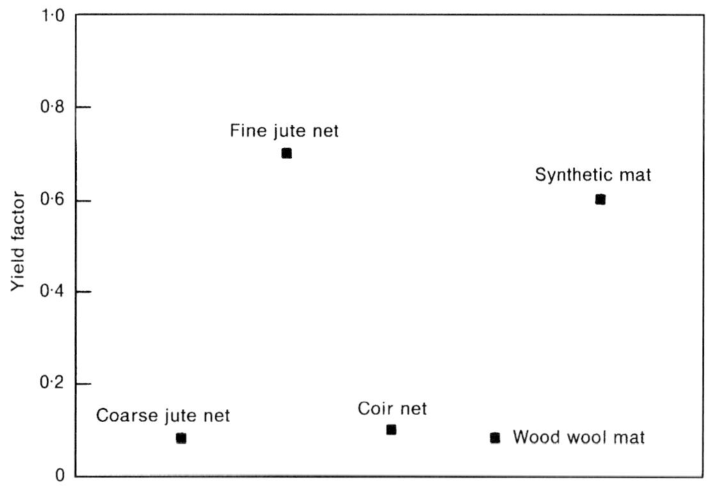

Figure 10.3 Measured erodibilities for a sandy loam

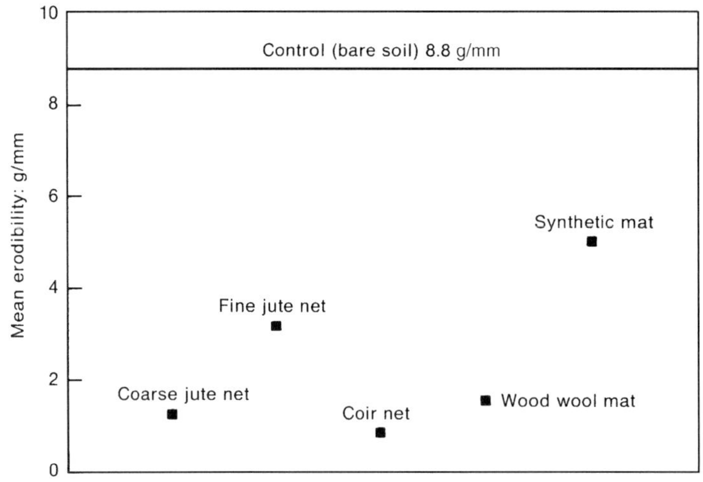

Figure 10.4 Measured runoffs for a sandy loam

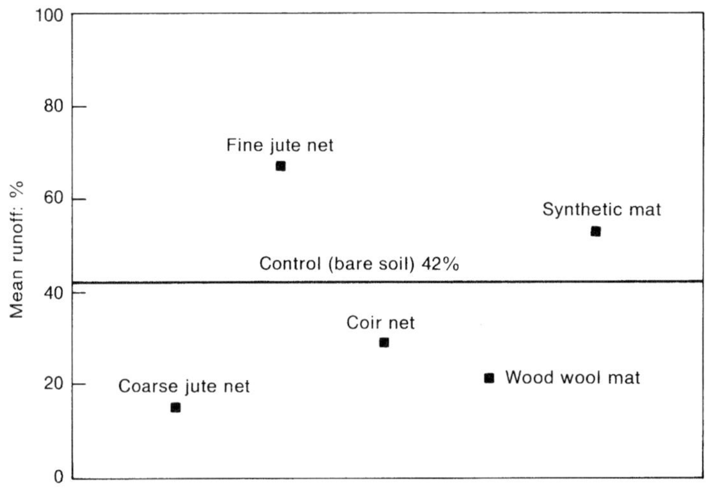

If a system reduces the sediment yield by exercising a balanced control over both runoff and erodibility, then RF and EF would be equal to one another. In this case, Equation 10.4 follows from Equation 10.3:

$$\mathrm{RF} = \mathrm{EF} = \sqrt{\mathrm{YF}} \tag{10.4}$$

The erodibility and runoff factors can be normalised by dividing by $\sqrt{\mathrm{YF}}$, which allows Equation 10.3 to be expressed in the normalised form

$$(\mathrm{RF}/\sqrt{\mathrm{YF}}) \times (\mathrm{EF}/\sqrt{\mathrm{YF}}) = 1 \tag{10.5}$$

Equation 10.5 defines a unique relationship between normalised erodibility and runoff factors, which is plotted as the solid curve in Figure 10.5.

Test data for all five erosion control products and a variety of soil types are superimposed on Figure 10.5 to allow comparison with the theoretical curve. Statistically, the agreement is reasonable, with a coefficient of correlation of 0.951 between the theoretical curve and the test data. The solid diagonal line in Figure 10.5 defines a balanced system. Test data points falling above this line identify soil loss that is dominated by high erodibility; for example, the erosion control system is efficient in controlling runoff, but inefficient in controlling erodibility. The reverse is the case for data points falling below the diagonal line. The normalised plot in Figure 10.5 gives no indication of the absolute values of the yield factors.

This is remedied in Figure 10.6, which is a plot of erodibility factors against runoff factors for a range of yield factors derived from Equation 10.3. The test data for the natural fibre nets are generally bounded by yield factors of 0.1–0.5, and all show a good control of erodibility, with the higher values of yield factor being caused by an apparent inefficiency in controlling runoff. The wood wool mat renders yield factors as low as 0.08, with all values being consistently below 0.2. That the test data are all close to the diagonal line of equality indicates that the wool mat exercises a balanced control over both erodibility and runoff factors. The test

Figure 10.5 Normalised runoff factors plotted against erodibility factors

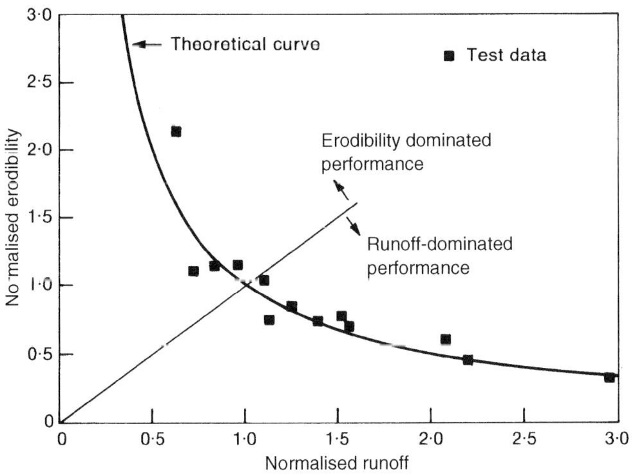

Figure 10.7 System water retention plotted against soil loss by rain splash

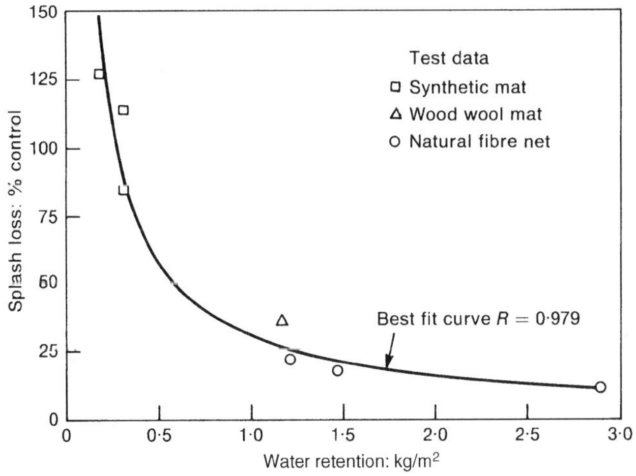

data for the synthetic mats return yield factors in the range 0.5–1.1. Although performance is balanced, there is little control over the runoff factor, and an erodibility factor that can be larger than the bare control plot. This may be influenced by sediment loss from soil infilling the mat. Some light is shed on this apparent anomaly by analysing the data published by Rickson (1988) for splash cup tests carried out on a clay loam protected by various systems, including natural fibre nets, wood wool mats, and both filled and unfilled synthetic mats. The test data have been interpreted in Figure 10.7 as a plot of the water-retention capacity of each system against soil loss by rain splash.

The data fall into two distinct clusters. The first pertains to the natural fibre and wood-wool system, which reduces loss to between 11.5 and 35.6% of the bare soil control. The second cluster relates to synthetic mats, where soil loss by splash varied from 84.6% for unfilled mats up to 127% for filled mats. This mechanism may explain the high erodibility factors for synthetic mats shown in Figure 10.6.

Figure 10.6 Runoff factors plotted against erodibility factors

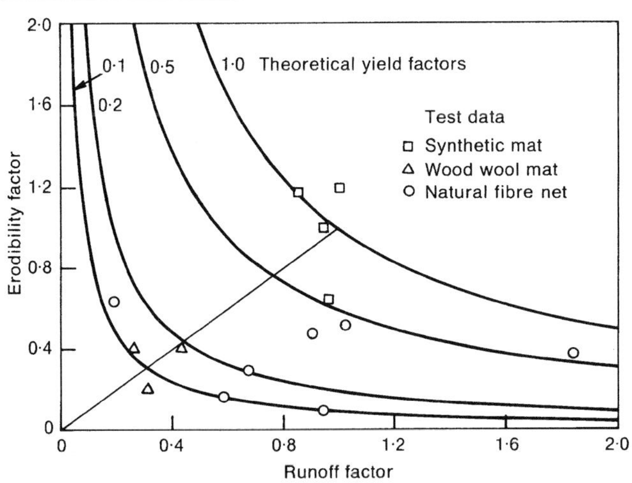

10.7. Erosion control performance testing of geosynthetics

10.7.1 Testing methods

Performance testing of erosion control geosynthetics started in the late 1980s and followed the research approach in agriculture of empirical field experiments (e.g. Fifield *et al.*, 1988). The need to evaluate the performance of erosion control products rapidly elevated because of the increasing availability of products. This led to more civil engineers adopting and specifying commercially available products, in addition to the traditional agricultural methods. In response to the research need, testing programmes from academics and the private industry emerged. For example, the Hydraulics, Sedimentation, and Erosion Control Laboratory, mentioned in Section 10.5.2, was built early on in 1991. Its original testing protocol has been adopted and modified by several laboratories, and has become the framework of present erosion and sediment control performance testing standards.

Performance testing of an erosion control geosynthetic product typically refers to its ability to reduce sediment yield and/or allow vegetation establishment. In essence, performance testing is a form of soil erosion research. According to Mutchler *et al.* (1994, p. 11), 'soil erosion research must be based on experimental results of some form'. Although soil erosion has been modelled in three different types, that is empirical, conceptual and physically based (Nearing *et al.*, 1994), the empirical approach is the primary method in evaluating product performance.

Methods of three different scales have been used to evaluate erosion control product performance, including full, large and small (or bench) scales. In terms of generating erosive force (i.e. raindrops), artificial simulators have been the primary means. Very few rely on natural rains. Among the three scales, large-scale testing, whether indoors or outdoors, is the most widely accepted method. Robeson *et al.* (2008) reviewed five testing facilities that the erosion control manufacturers regularly used in the USA. They summarised key testing parameters. Highlights of desirable parameters include

- a drop-size distribution near that of natural rains with a median diameter between 2 and 3 mm
- drop impact velocities near those of natural rains
- simulated rainfall intensities between 25 and 180 mm/h
- raindrop characteristics and rainfall intensity applied uniformly over the entire study plot by the rainfall simulator
- repeatability of the rainfall simulator
- a minimum length of test plots of greater than 5 m
- soil types to include sand and clay
- the gradient of the test slope to range from 0 to 100%
- preparation of the soil surface for special interests, such as compaction, soil moisture, ambient soil temperature and surface roughness.

As mentioned, test methods at different scales have been used for geosynthetic performance evaluation. Comparison of results from different-scale testing methods became evidently needed, and has been the focus of several studies. Lipscomb *et al.* (2005) compared performance data collected from full- and bench-scale testing, and could not find meaningful correlation between them. Despite of the desire of using bench-scale testing for cost saving and efficiency, current consensus remains that the use of large- or full-scale testing is a reliable means for performance evaluation, whereas bench-scale testing still needs significant improvement.

Vegetation establishment is the other dimension that measures performance of an erosion control product. No matter how well a product protects slopes from erosion, the goal is to grow vegetation for ultimate protection. The ability of a product to encourage or inhibit grass seed germination affects vegetation establishment. For this performance testing, a product is installed on slope surfaces (small or large scale) along with seeds of native or adapted grass species. For comparison, a control set is installed with seeds only on bare soils. Approximately 30–90 days are given for growth prior to the measurement of vegetation density. The difference in vegetation percentage cover between the target product and the control determines the performance of the product in vegetation establishment. Table 10.3 shows the TxDOT minimum standards in evaluating the performance of a product, including sediment loss and vegetation density.

10.7.2 Results from various field and laboratory tests

Dudeck *et al.* (1967) have reported trials on a 1 : 3 cutting slope in a silty sandy loam. Four simulated storms were applied to the slope with rainfall intensities of between 64 and 127 mm/h associated with storm durations of 1.4 hours and 0.3 hour, respectively. Results are given in Figure 10.8, which shows that the performance of jute, closely followed by a wood-wool mat, outstripped other systems. The results of other parts of the trial, associated with rates of grass growth, were not so favourable to jute,

Figure 10.8 Relative efficiency

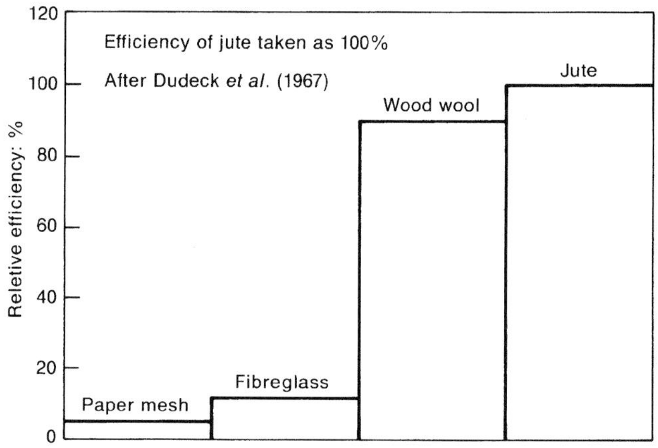

which, at 26 days after planting, showed only 37% cover compared with the 67% achieved by the wood-wool mat.

Under more severe conditions, the performance of some natural fibre products may be lacking. Aspects of this are reported by Kay (1978), who carried out rainfall trials on a sandy loam with a rainfall intensity of 150 mm/h applied for up to six hours. The measured soil losses from 1 : 5 and 1 : 2 slopes for bare soil and soil protected by a jute net are given in Figures 10.9(a) and 10.9(b), respectively. These results have been converted to yield factors, which are plotted against storm duration in Figure 10.10. A feature of this plot is that, for a given slope and rainfall intensity, the yield factor is time dependent until an equilibrium value is obtained at long storm durations, in this example at about 5–6 hours. From this, it can be seen that for short storm durations the yield factors are very low, with values of 0.005 and 0.025 for slopes of 1 : 5 and 1 : 2, respectively. However, for long storm durations the yield factors deteriorate until equilibrium is reached, and the steady-state values rise to 0.05 and 0.20 for 1 : 5 and 1 : 2 slopes, respectively. This deterioration is due to the jute progressively losing its ability to control runoff.

This is reflected in the work by Cancelli *et al.* (1990), who carried out laboratory trials to separate out performance based on rain splash and performance based on runoff. In both cases, a silty fine sand at a 1 : 2 slope was employed. To assess soil loss created by rain splash, a storm intensity of 75 mm/h was employed. Since the soil was of high permeability, some 56% of the rainfall was dissipated by surface infiltration, so reducing the effective intensity to 33 mm/h. Under these conditions, jute returned by far the highest efficiency, with a soil loss of approximately 3 g per litre of rainfall. This excellent performance is compared with that of other products in Figure 10.11. To assess soil loss associated with runoff, the slope was tested by trickling water over the surface at a rate of 6.6 litres per minute per metre of width. This flow rate was calculated to model a storm intensity of 75 mm/h, allowing for 56% infiltration, over a slope length of 15 m. The results of this trial are summarised in Figure 10.12, which indicates that jute returned by far the

Figure 10.9 Jute performance on: (a) a 1 : 5 slope; and (b) a 1 : 2 slope

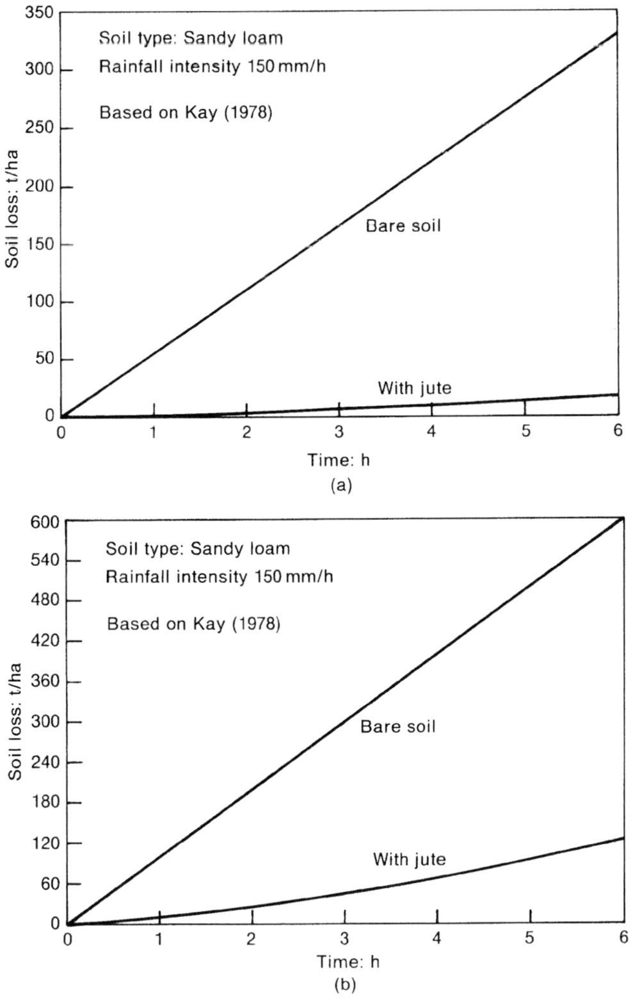

worst result, which was even worse than that for the bare soil control sample. Thus, the work of Kay (1978) and Cancelli *et al.* (1990) indicates the need to more fully assess and improve the performance of some natural fibre systems with respect to the control of runoff.

Figure 10.10 Variation in yield factors with storm duration

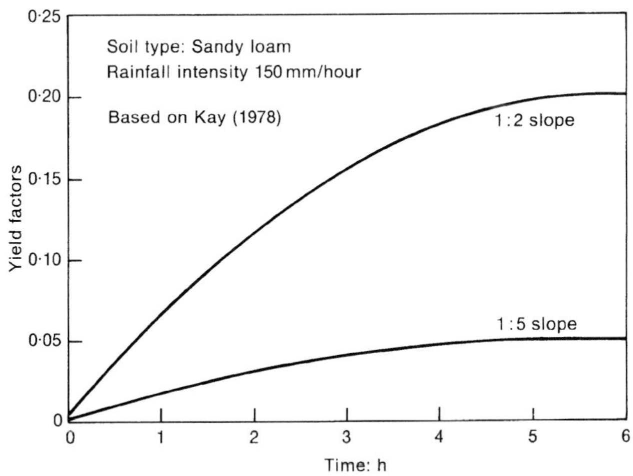

Figure 10.11 Erodibility by product type based on rain splash

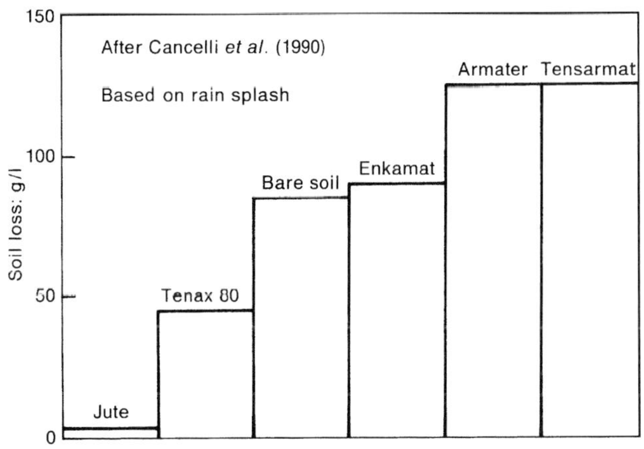

10.7.3 Ageing of erosion control geosynthetics

Geosynthetics degrade over time. Products such as RECPs installed in the field to control erosion are degraded by numerous mechanisms, including oxidation, hydrolysis, radioactivity, and physical, chemical and biological actions. According to Theisen (2005), the durability of an erosion control material is directly related to its functional longevity. Microbial activity, ultraviolet (UV) light degradation and chemical degradation can result in material breakdown that reduces durability. Among all the environmental factors, UV light and elevated temperatures are believed to be the most critical. The damage is due to nanometre waves of UV radiation penetrating into the polymer which, along with elevated temperatures, causes chain scission of polymers (Lord and Halse, 1989; Koerner *et al.*, 2005). High cyclic temperatures induce stress changes and adversely alter the stress–strain characteristics of the geotextile, reducing its strength due to wear and tear in the fibres. Long-term elevated temperatures can also reduce the strength of geosynthetics in most cases.

During rainfall, stormwater runoff can exert forces on the RECPs by wetting them and thus causing property changes. While water promotes the growth of vegetation in the soil to

Figure 10.12 Erodibility by product type based on runoff

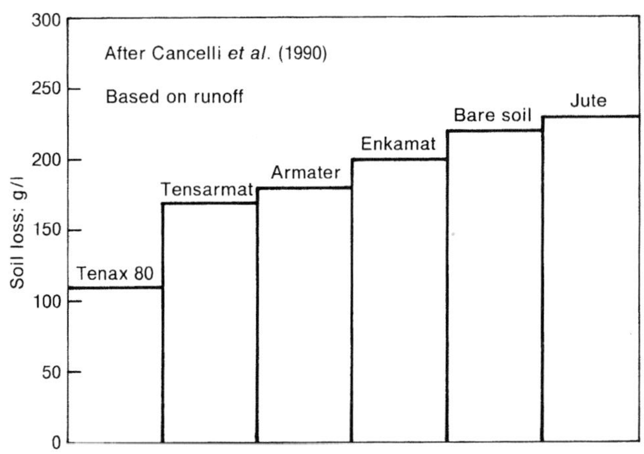

form permanent vegetative protection, it often combines with the raised temperatures and UV light to cause photo-oxidation of the polymers. Meanwhile, human factors such as mowing operations or vehicles running off pavement, not necessarily environmental factors, are common causes that damage or weaken RECPs installed on slopes.

Li and Khanna (2008) conducted laboratory testing to study the ageing effect on seven channel erosion control RECPs that were made of natural, synthetic or composite materials, similar to the raw materials composing the RECPs for slope erosion control. They compared used (installed in channels of a field laboratory for three years) and unused (stored in a limited climate control laboratory also for three years) products on ten physical properties, including mass per unit area, thickness, tensile strength, stiffness, light penetration, resiliency, water absorption, swell, specific gravity and smoulder resistance. They found that tensile strength was the only property that showed consistent, significant changes. All RECPs lost notable tensile strength after a three-year field exposure. Used materials of the natural type showed a 68% strength loss, while those of the composite and synthetic types had a smaller strength loss in the range 37–61%. Li and Khanna's (2008) results confirm Koerner's (2005) statement claiming that tensile strength is the single most important property for geotextiles (geosynthetics). In essence, tensile strength is a reliable property for evaluating ageing effects on RECPs.

10.8. Lifetime cost and the performance of erosion control products

Life cycle cost-effectiveness of construction materials and products is important to civil engineers. Geosynthetics used for slope erosion control are no exception. To date, one can find ample literature documenting costs for erosion (and sediment) control devices (e.g. Goldman *et al.*, 1986; Mayo *et al.*, 1993; Bhatia *et al.*, 2002), or studies on the water-quality performance of erosion (and sediment) control measures (e.g. Schueler, 1987; Mayo *et al.*, 1993; Strecker *et al.*, 2001; Lake, 2004). However, few studies cover both cost and performance in a single study. The California Department of Transportation (Caltrans, 2000) and Bhatia *et al.* (2002) included cost and performance in their studies of RECPs; however, the cost data reported in both studies were the initial installation cost (labour and material) – no long-term cost/benefit was included.

Woodhouse and Leo (2008) included the claimed longevity of products in analysing cost versus effectiveness of eight erosion control products (four RECPs and four mulches with chemical binders). They collected sediment loss data of products installed on 1:2, 93 m^2 slopes. They also collected and compared the unit price of these products and their claimed longevity. They found that the unit cost correlated with soil loss negatively, but with longevity positively. They suggested that the key to selecting a cost-effective product is to determine the length of time that the product is needed for protection, and then use a product with matching longevity.

A recent study examining lifetime cost and the performance of slope erosion control products was conducted by Li *et al.* (2009). This study included 58 products (48 RECPs and ten mulches), all of which were approved for highway use and listed on the TxDOT Approved Product List. These products consisted of natural, synthetic and composite materials. They quantified the lifetime performance of the products using: (*a*) the erosion control performance data, (*b*) product cost data collected from surveys and (*c*) predicted product longevity based on material composition. They also examined the relationship between product price, soil loss performance and predicted longevity. Li *et al.* (2009)

Figure 10.13 The price index corresponding to predicted product longevity based on material composition
Reproduced from Li *et al.* (2009)

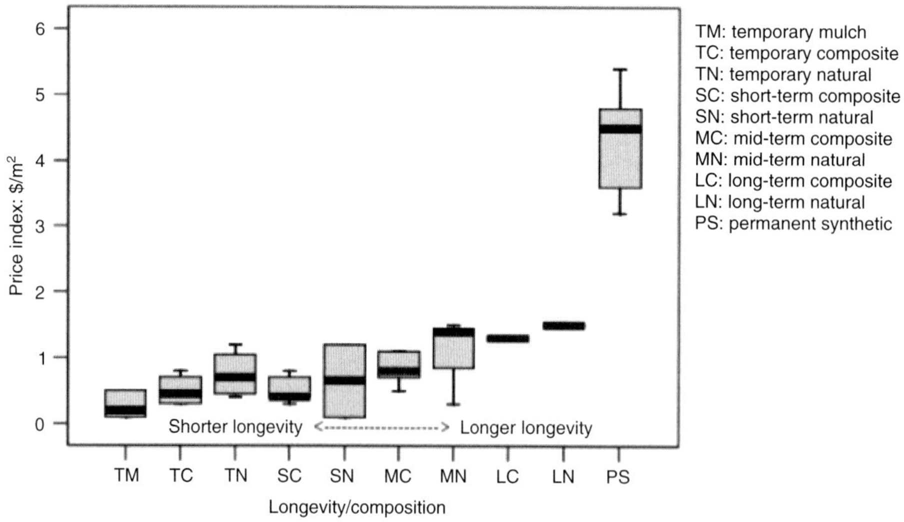

Figure 10.14 The soil loss index corresponding to predicted product longevity based on material composition: (a) clay slope and (b) sand slope
Reproduced from Li *et al.* (2009)

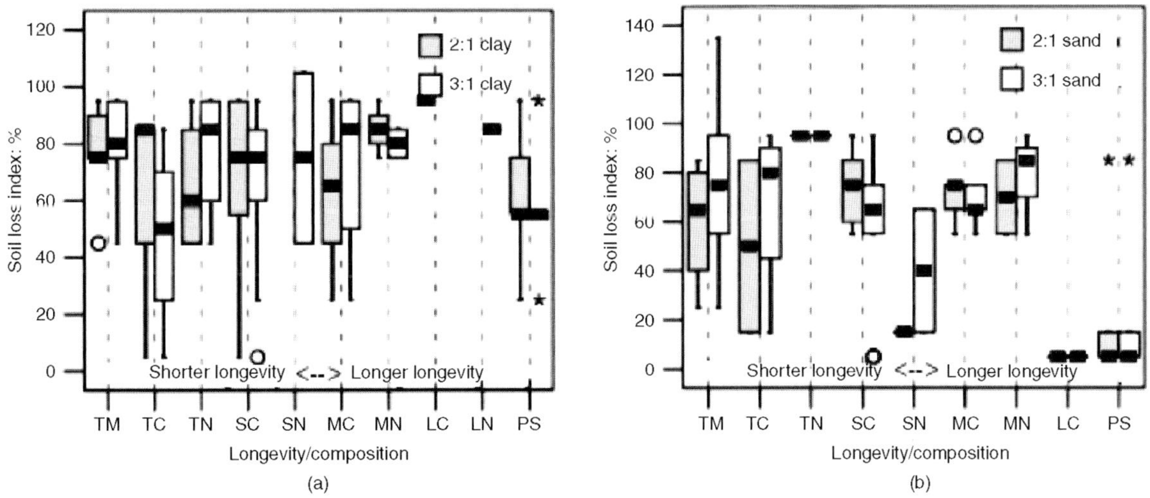

concluded the following

- The product price was highly correlated with the predicted product longevity. Figure 10.13 presents the box-and-whisker plot of the price index versus longevity, showing that the product price increased with the predicted longevity.
- The predicted longevity did not correlate with the initial soil loss performance of the products. Figure 10.14 shows no obvious trend between the soil loss index and predicted longevity.
- The lifetime product performance correlated well with the product price (Figure 10.15), which suggests that the marketed product prices well represent the lifetime worth of the products.

10.9. Summary of main points

The current knowledge and design approach of erosion control evolved from agriculture research and practices. Until recently, the universal soil loss equation (USLE), developed from decades of empirical agriculture experiments and later revised for broader soil loss modelling, has been the primary tool in designing erosion control systems. However, a simpler and more reliable design approach has emerged in recent decades, mainly based on the performance evaluation results of erosion control products as a design tool. By selecting products that have been evaluated and approved, one can estimate soil loss to a certain level of accuracy. Today, civil engineers have a wide range of choices for slope erosion control, including traditional agricultural methods as well as commercially developed geosynthetic products. Laboratory testing has been the primary method in evaluating product performance. The currently recognised testing protocol continues to be large-scale testing, whereas bench-scale testing remains questionable and is in need of improvement. Other important points from the literature cited in this chapter include the following

- Soil loss in the short-term can exceed the total long-term soil loss prediction. Initial tests using a rainfall simulator indicate that natural fibre erosion control products, such as coarse jute nets, coir nets and wood-wool mats, might be expected to return lower yield factors, and, therefore, a greater degree of erosion control, than their synthetic counterparts.
- Rolled erosion control products (RECPs) deteriorate over time due to environmental factors such as UV light and temperature. The physical properties of RECPs installed in the field are constantly deteriorating during the ageing process. This ageing effect can be most reliably measured by tensile strength.
- Erosion control product prices increase with product longevity. Based on the correlation between the lifetime product performance and the product price, marketed product prices represent the lifetime worth of products well.

10.10. Some common questions and their answers

Questions

1. What are the causes of soil erosion?

2. How do you interpret the crop factor and the yield factor?

3. Why are more and more engineers specifying geosynthetic erosion control products?

4. Are geosynthetic products temporary or permanent applications?

5. What is the difference between erosion and sediment controls?

6. There are so many erosion control products out there. How do I select the best product for my situation?

7. Why is large-scale testing mostly used for the performance evaluation of erosion control products?

Figure 10.15 Correlation between price and lifetime product performance of erosion control geosynthetics: (a) 1 : 3 clay slope, (b) 1 : 2 clay slope, (c) 1 : 3 sand slope and (d) 1 : 2 sand slope
Reproduced from Li *et al.* (2009)

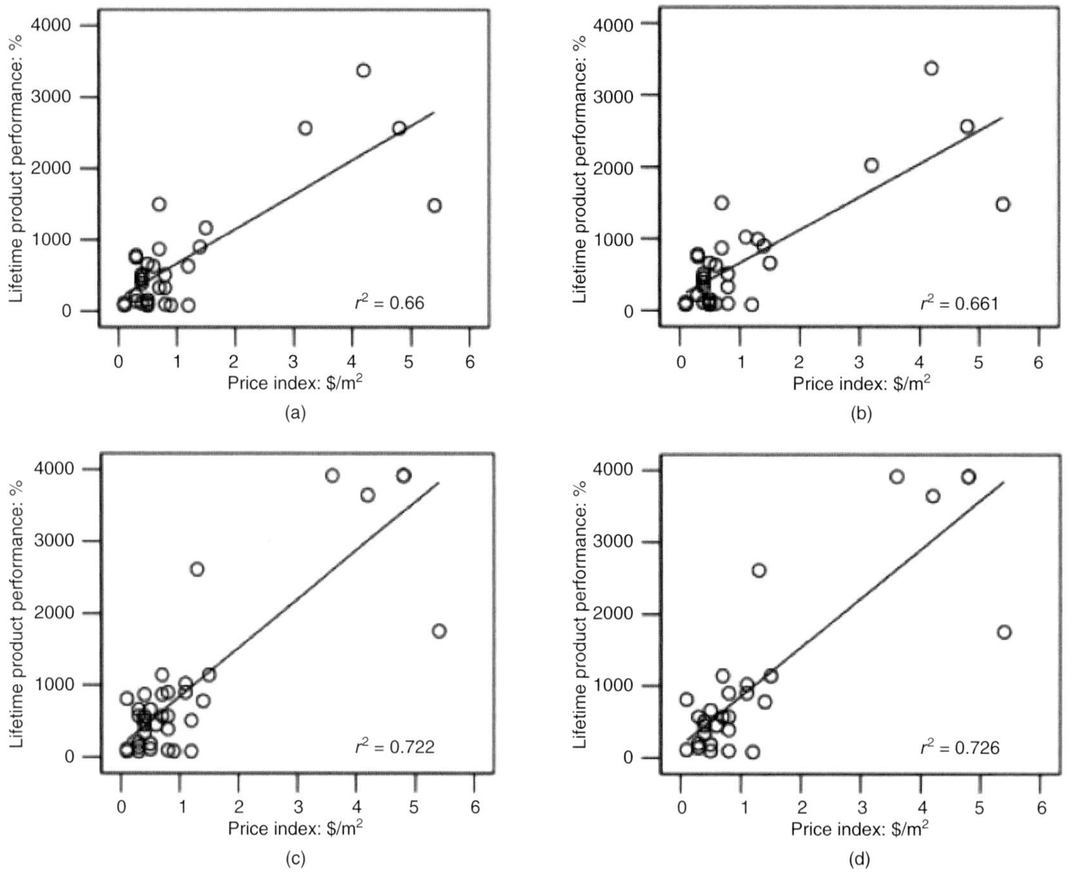

Answers

1. Wind and water are the main agents of soil erosion. The process begins with the detachment of soil particles, followed by sediment transport either by wind or water, and finally deposition somewhere else. Large raindrops (4 mm or greater in diameter) can reach a terminal velocity of greater than 30 km/h when they hit the ground surface. Such impact can easily detach soil particles, which are then readily transported.

2. The crop factor gives a measure of the protection afforded by vegetative cover. The crop factor would typically be in the range 0.001–0.01, which relates to protection efficiencies in the range 99.0–99.9%. This only applies in the long term, where adequate vegetative cover is fully established. In the short term, before vegetative cover is established, biotechnical systems rely solely on the protection afforded by the erosion control substrate be this net, mesh or mat. Clearly, the crop factor does not apply in the short term. Instead of the crop factor, use may be made of a corresponding short-term parameter defined as the yield factor. Large-scale tests have returned yield factors in the range 0.08–1.12. A yield factor greater than unity signifies a soil loss greater than that for the bare soil.

3. This increase can be attributed to three factors, including the availability of products for various site conditions, product performance approved by reliable academic research laboratories, and ease-of-use during the design process. The design decision process of erosion control for temporary protection can be accomplished by simply selecting products that have been approved for specific conditions such as slopes, soil type and project duration. A well-known example is the Approved Product List published by the Texas Department of Transportation. For quality control, the physical properties of all products on the Approved Product List are re-examined every few years to ensure that products are manufactured consistently.

Despite the fact that commercial products have gained popularity, designers should not overlook other traditional means of erosion control originating from agricultural practices. Measures such as soil roughening, use of straw as mulch and crimped straws are also proven, effective techniques. The concerns using these techniques include whether materials are locally available and their consistent quality.

4. Most slope erosion control products are for temporary purposes. The reason is that the magnitude of shear stress that would occur on slopes is small (compared with concentrated flow channels) and the ultimate goal for erosion control is to establish vegetation. Hence, the role of products is to protect bare soils, provide conditions for seed germination, and allow grass to mature (by anchoring and supporting the new growth). Another type of erosion control that is beyond the scope of this chapter but worth mentioning is channel application. Because of high shear stresses from concentrated flows exerting on channels, most channel erosion control products are intended for permanent protection. It should be noted that all synthetic materials degrade over time. The notion of permanent protection is that the loss in strength of product due to degradation is complemented by vegetation. Ultimately, roots and products form a complete system that stabilises channels permanently.

5. Although often used interchangeably, erosion and sediment controls are two different processes, and require different sets of tools and techniques to mitigate. Erosion control is any measure that protects the soil surface from soil particle detachment (by wind or rain). These often include mats, blankets, hydraulic mulches, etc. Sediment control methods are passive systems that capture or intercept eroded sediments through filtering or settling. Devices, such as in-channel wattles and check structures, that could intercept and slow down sediment-laden flows are sediment controls. A holistic approach is to use both erosion and sediment controls for complete environmental protection.

6. Erosion control products that have been evaluated and approved by reliable, objective laboratories deserve consideration. The Approved Product List, published by the Texas Department of Transportation, that lists high-performing erosion control products is one example. To make the most accurate selection, the designer should determine the duration of protection a site may need in order to select products with similar longevity. Such design practice can yield cost-effective erosion control solutions. Typically, products made of natural materials are less expensive and have a short lifespan, while synthetic products last the longest and cost more. Composite products that include a combination of natural and synthetic materials are typically in between.

Designers may also find manufacturer-published crop factors useful. For erosion control products, the crop factor is calculated as the ratio of soil loss from a product-protected plot to soil loss from an unprotected (control) plot in large-scale testing. Small crop factors represent high-performing products. Designers should certainly beware of how and where a crop factor for a product is measured.

7. The main reason is the cost of testing and validity for field application. Typically, full-scale testing is conducted outdoors, and requires high cost and much labour. Testing outdoors can also be quite challenging due to uncontrollable conditions such as natural rainfall and other factors outside of human control. Research has proven a correlation between tests conducted full scale and large-scale testing. However, no correlation between bench-scale testing and large/full-scale testing has been identified.

REFERENCES

Bhatia SK, Smith JL, Lake D and Walowsky D (2002) A technical and economic evaluation of geosynthetic rolled erosion control products in highway drainage channels. *Geosynthetics International* **9(2)**: 125–148.

Caltrans (2000) *District 7 Erosion Control Pilot Study*. California Department of Transportation, Sacramento, CA.

Cancelli A, Monti R and Rimoldi P (1990) Comparative study of geosynthetics for erosion control. *Proceedings of the 4th International Conference on Geotextiles and Geomembranes*, The Hague.

Dudeck AE, Swanson NP and Dedrick AR (1967) *Mulch Performance on Steep Construction Slopes*. Rural and Urban Roads.

Fifield JS, Malnor LK and Dezman LE (1989) Effectiveness of erosion control products on steep slopes to control sediment. *Proceedings of the 20th International Erosion Control Association Conference*, Vancouver.

Fifield JS, Malnor LK, Ritcher B and Dezman LE (1988) Field testing erosion control products to control sediment. *Proceedings of the 19th International Erosion Control Association Conference*, New Orleans, LA.

Goldman SJ, Jackson K and Bursztynsky TA (1986) *Erosion and Sediment Control Handbook*. McGraw-Hill, New York.

Ingold TS and Thomson JC (1986) Results of current research of synthetic and natural fibre erosion control systems. *Proceedings of the 17th International Erosion Control Association Conference*, Dallas, TX.

Kay BL (1978) *Mulches for Erosion Control and Plant Establishment on Distressed Sites*. University of California, Davis, CA. Agronomy Progress Report No. 87.

Koerner RM (2005) *Designing with Geosynthetics*, 5th edn. Prentice Hall, Upper Saddle River, NJ.

Koerner RM, Koerner GR and Hsuan YG (2005) Lifetime prediction of exposed geo-membranes. *Erosion of Soils and Scour of Foundations – Geo-frontiers 2005*, American Society of Civil Engineers, Reston, VA.

Lake Jr DW (2004) Cost analysis of erosion and sediment control practices. In *New York State Standards and Specifications for Erosion and Sediment Control*, New York State Department of Environmental Conservation, New York, Section 10.

Li MH and Khanna S (2008) Aging of rolled erosion control products for channel erosion control. *Geosynthetics International* **15(4)**: 224–231.

Li MH, Yi YJ and McFalls J (2009) Lifetime cost and performance of slope erosion control products for construction. *Geosynthetics International* **16(3)**: 139–146.

Lord AEL and Halse YH (1989) Polymer durability – the materials aspects. In *Durability and Aging of Geosynthetics* (Koerner RM (ed.)), Elsevier, St Louis, MI, pp. 293–323.

Lipscomb CM, Hudson V and Thornton CI (2005) National Cooperative Highway Research Program Project 20-07 (162) report – guidelines for the correlation of test results from bench-scale and large-scale testing on rolled erosion control products. Colorado State University, Fort Collins, CO (unpublished).

Mayo L, Lehman D, Olinger L, Donovan B and Mangarella P (1993) *Urban BMP Cost and Effectiveness Summary Data for 6217(g) Guidance: Erosion and Sediment Control During Construction*. Environmental Protection Agency, Assessment and Watershed Protection Division, Washington, DC.

Morgan RPC (1986) *Soil Erosion and Conservation*. Longman, Harlow.

Morgan RPC, Morgan DDV and Finney HJ (1985) A predictive model for the assessment of soil erosion risk. *Journal of Agricultural Engineering Research* **30**: 245–253.

Mutchler CK, Murphree CE and McGregor KC (1994) Laboratory and field plots for erosion research. In *Soil Erosion Research Methods* (Lal R (ed.)), 2nd edn. Soil and Water Conservation Society and St Lucie Press, Boca Ratan, FL, pp. 11–37.

Nearing MA, Lane LJ and Lopes VL (1994) Modeling soil erosion. In *Soil Erosion Research Methods* (Lal R (ed.)), 2nd edn. Soil and Water Conservation Society and St Lucie Press, Boca Ratan, FL, pp. 127–156.

Rickson RJ (1988) Geotextile applications in steep land agriculture. *Proceedings of the International Conference on Steep Land Agriculture*.

Robeson MD, Theisen MS and Spittle K (2008) Evaluation of large-scale rainfall testing facilities. *Proceedings of the 39th International Erosion Control Association Conference*, Orlando, FL.

Schueler TR (1987) *Controlling Urban Runoff: A Practical Manual for Planning and Designing Urban BMPs*. Department of Environmental Programs, Metropolitan Washington Council of Governments, Washington, DC.

Strecker EW, Quigley MM, Urbonas BR, Jones JE and Clary JK (2001) Determining urban storm water BMP effectiveness. *Journal of Water Resources Planning and Management* **127(3)**: 144–149.

Theisen MS (2005) Science combined with art makes for cost effective rolled erosion control product design practices. *Erosion of Soils and Scour of Foundations – Geofrontiers 2005*, American Society of Civil Engineers, Reston, VA.

TxDOT (2000) *Final Performance Analysis Through 1999 Evaluation Cycles–Slope Protection Products, Flexible Channel Liner Products & Hydraulic Mulches*. Texas Department of Transportation, Maintenance Division, Austin, TX.

TxDOT (2011) Product Evaluation Program. http://www.dot.state.tx.us/business/doing_business/product_evaluation/default.htm (accessed 18/07/2011).

Whitehead E (1976) *A Guide to the Use of Grass in Hydraulic Engineering Practice*. Construction Industry Research and Information Association, London.

Wischmeier WH and Smith DD (1958) Rainfall energy and its relationship to soil losses. *Transcript of American Geophysics* **39(2)**: 285–291.

Woodhouse ER and Leo T (2008) Temporary slope protection: cost vs. effectiveness. *Proceedings of the 39th International Erosion Control Association Conference*, Orlando.

Woodruff NP and Siddoway FH (1965) A wind erosion equation. *Proceedings of the Soil Science Society of America* **29**: 602–608.

Handbook of Geosynthetic Engineering
ISBN 978-0-7277-4175-2

http://dx.doi.org/10.1680/hge.41752.223

Chapter 11
Slopes – stabilisation

S. K. Shukla*, N. Sivakugan† and B. M. Das‡
*Discipline of Civil Engineering, School of Engineering, Edith Cowan University, Perth, Australia
†Discipline of Civil and Environmental Engineering, School of Engineering and Physical Sciences, James Cook University, Townsville, Australia
‡College of Engineering and Computer Science, California State University, California, USA

The use of geosynthetic reinforcements for slope stabilisation allows a reduction in earthworks by changing the geometry, and even allows the use of soils with average mechanical properties. Geosynthetic-reinforced slopes are generally compacted fill embankments that incorporate geosynthetic layers as tensile reinforcement to enhance stability. The reinforcement holds together the soil mass from both sides of the failure surface, thus increasing the factor of safety of the existing slope. Analytical, numerical and experimental research work, and many case studies on geosynthetic-reinforced slopes, have been reported in the literature. This chapter presents a comprehensive overview of the application of geosynthetics in stabilising soil slopes, including the suitability of geosynthetics, reinforcing mechanisms, methods of slope stability analysis and design, model studies, and typical slope stabilisation methods, with some specific construction guidelines.

11.1. Introduction

Slopes can be natural or man-made (e.g. excavations, embankments). Several natural and man-made factors contribute to slope instability (Shukla, 1997). Many of the problems associated with the stability of natural slopes (also referred to as hillsides) are radically different from those of man-made slopes, mainly in terms of the nature of the geomaterials involved, the environmental conditions, location of the groundwater level and stress history. In man-made slopes, there are also essential differences between cuts and embankments. The embankments can be built with relatively selected and well-controlled materials; in cuts, however, this possibility does not exist. Several methods are available to increase the stability of unstable slopes (Broms and Wong, 1990; Abramson *et al.*, 2002). These methods can be adopted singly or in combination. The choice depends primarily on the cost and the consequence of slope failure. The more commonly used slope stabilisation methods can be classified as follows (Broms and Wong, 1990)

- geometric methods, in which the geometry of the slope is changed
- hydrological methods, in which the groundwater table is lowered or the water content of the soil is reduced
- chemical and mechanical methods, in which the shear strength of the sliding soil mass is increased or the external force causing the slope failure is reduced.

Geometrical methods include slope flattening, removal of part of the soil or load from the top of the slope, construction of pressure berms at the toe, terracing, replacement of slipped material by free-draining material, and recompaction of slip debris. Hydrological methods include the installation of surface and subsurface drains, inverted filters, and thermal methods. Chemical and mechanical methods include grouting, construction of restraining structures (such as concrete gravity or cantilever walls), gabion structures, crib walls, embankment piles, lime and cement columns, ground anchors, soil nailing and root piles, earth reinforcement, and the planting of grasses and shrubs.

Reinforcing the steep slopes of embankments or earth walls by the installation of tensile-resistant components is a very old construction method. Tree branches have been used to stabilise slopes since ancient times. In modern times, Henri Vidal (1966, 1969), a French architect and engineer, is credited with developing a soil-reinforcing technique to the stage where it could be economically applied to large civil engineering structures, including natural slopes, cut slopes or slopes of embankments.

The advent of geosynthetic-reinforcement materials has brought a new dimension of efficiency to the design and construction of reinforced slopes, retaining walls, etc., due to their corrosion resistance and long-term stability. Today, geosynthetics offer a welcome additional technology for low-cost slope stabilisation (Hausmann, 1990; Shukla *et al.*, 2011). They may be used to

- prevent deep-seated failure by 'tieback' action
- contain surface soils in combination with soil nailings
- protect slope surfaces against erosion
- control sediment transport by wind and water.

Reinforced slopes are basically compacted fill embankments that incorporate geosynthetic tensile reinforcement arranged in horizontal planes. The tensile reinforcement holds the soil mass together across any critical failure surface, to ensure stability of the slope. Facing treatments ranging from vegetation to armour systems are applied to prevent ravelling and sloughing of the face.

This chapter deals with several aspects of slopes stabilised with geosynthetics as a major component, such as types and orientations of geosynthetics, modes of failure, review of methods of slope stability analysis, model tests, and stabilisation methods in practice.

11.2. Types and orientations of geosynthetics

Geotextiles, both woven and non-woven, and geogrids are being increasingly used for reinforcing steep slopes. Geotextiles, especially non-woven, exhibit considerable strain before breaking. Also, a non-woven geotextile has significantly lower stiffness than the ground. Hence, the deformation of a geotextile-reinforced soil slope is dominated not by the geotextile but by the soil slope. Due to the low stiffness of non-woven geotextiles, relatively low stresses are induced in them. Their function, however, is to provide adequate deformability and to redistribute the forces from areas of high stresses to areas of low stresses, thus avoiding the crushing of the soil. Further, non-woven geotextiles facilitate better drainage and help to prevent the build-up of porewater pressures that can reduce the shear strength. It has been observed that both woven geotextiles and geogrids bond well with granular soils, and their contribution to slope stability is likely to be limited by their material properties (e.g. stiffness and strength) rather than by the bond stresses developed with the soil (Jewell, 1985).

The tensile reinforcement is most effective when placed in the direction of tensile normal strains, ideally in the direction and along the line of action of the major principal tensile strain (Ingold, 1982). Figure 11.1(a) shows an ideal reinforcement layout within an embankment slope, where the reinforcement layers are placed along the major principal strain contours. As can be seen, although horizontal layers of reinforcement would be correctly aligned under the crest of the slope, they would have inappropriate inclinations under the batter, especially at the toe, making construction difficult. Even though an idealised reinforcement layout might be determined, it would be impractical if it took the form of that shown in Figure 11.1(a). Consequently, for practical reasons, geosynthetic layers are usually placed in horizontal layers within the slope, as shown in Figure 11.1(b) (Ingold, 1982; Broms and Wong, 1986; Koerner and Robins, 1986; Shukla *et al.*, 2011).

Figure 11.1 Reinforcement orientations: (a) ideal and (b) practical
Adapted from Ingold (1982), Industrial Fabrics Association International; and Shukla *et al.* (2011), © J. Ross Publishing, Inc.

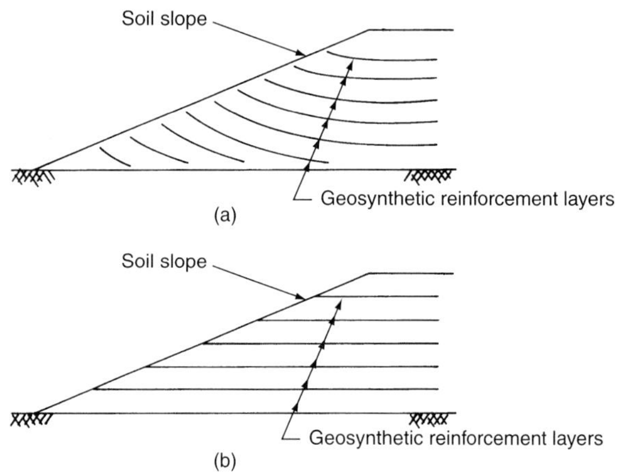

11.3. Function of reinforcement against slope failure

Figure 11.2(a) shows the active zone of the soil slope where instability will occur and the restraint zone in which the soil will remain stable. The required function of any reinforcing system would be to maintain the integrity of the active zone and effectively anchor this to the restraint zone in order to maintain the overall integrity of the soil slope. This may be achieved by the introduction of a series of horizontal reinforcements or restraining members, as shown in Figure 11.2(b). This arrangement of reinforcement can lead to three prime modes of failure, namely tensile failure of the reinforcement, pull-out from the restraint zone or pull-out from the active zone. Using horizontal reinforcement, it would be difficult to guard against the latter mode of failure. This may be due to the difficulty in obtaining adequate bond lengths. This can be illustrated by reference to Figure 11.2(b), which shows a bond length ac for the entire active zone. This bond length may be adequate to generate the required restoring force for the active zone as a rigid mass; however, the active zone itself contains an infinite number of prospective failure surfaces. Many of these may be close to the face of the batter, as typified by the dashed line in Figure 11.2(b), where the bond length would be reduced to the length ab, and, as such, be inadequate to

Figure 11.2 (a) An unreinforced slope and (b) a reinforced slope
Adapted from Ingold (1982), Industrial Fabrics Association International; and Shukla *et al.* (2011), © J. Ross Publishing, Inc.

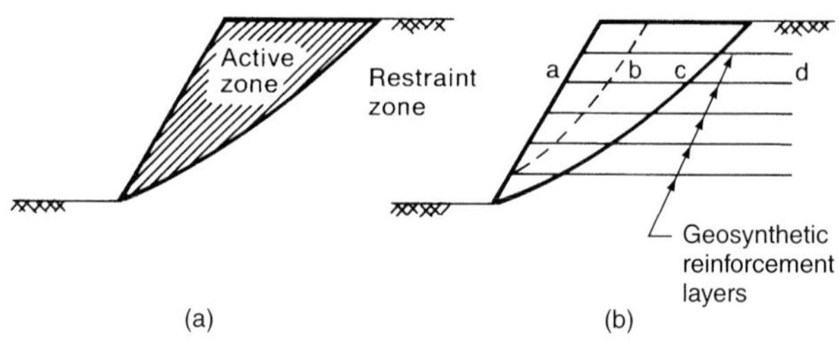

Figure 11.3 Soil slope with encapsulating reinforcement
Adapted from Ingold (1982), Industrial Fabrics Association International

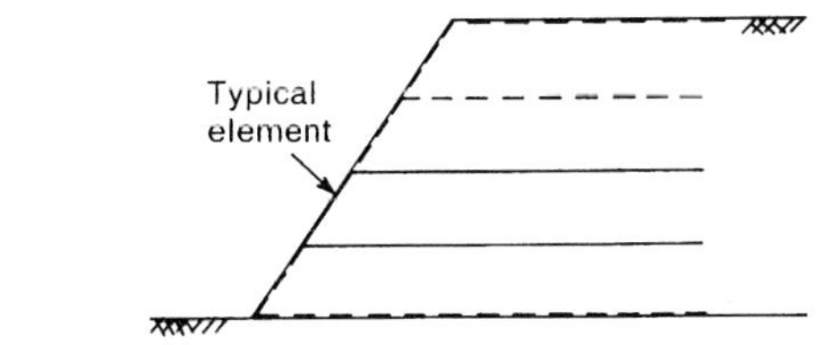

restrain the more superficial slips. This reaffirms the soundness of using encapsulating reinforcement or facing elements, where a positive restraining effect can be administered at the very surface of the slope by the application of normal stresses (Figure 11.3).

Reinforcements within a slope can also contribute to stability in two ways. First, reinforcement directly improves the shear resistance, enabling the reinforced soil to resist the increased shear loading caused by the steep face. Second, the reinforced zone, which extends to the slope face, acts to hold the unreinforced soil mass of the interior in equilibrium without overstressing the underlying foundation soils or exceeding the available shear resistance to outward shearing across the surface of the reinforcement layers (Jewell, 1985).

The stability of geotextile-reinforced slopes has been studied by researchers, based on centrifuge modelling and digital image analysis (Zornberg *et al.*, 1998a; Zornberg and Arriga, 2003; Viswanadham and Mahajan, 2007; Hu *et al.*, 2010; Shukla *et al.*, 2011). The studies show that failure in the slope models is characterised by well-defined shear surfaces through the toe of the slope, which is in good agreement with current design methods for reinforced slopes based on limit equilibrium, which is described in detail in Section 11.4. The maximum strain in the reinforcement does not occur near the toe of the reinforced slope: it is approximately at mid-height of the reinforced slopes, at the point on the critical failure surface directly below the crest of the slope. The occurrence of a particular type of failure is found to depend on the tensile stress–strain characteristics of a geotextile layer. With the presence of geotextile layer, the failure surface shifts away from the face, with an increase in slope inclination, and the reinforcement tension distribution is a function of the slope inclination of reinforced slope.

11.4. Stability analysis of reinforced slopes

From stability considerations, a given or proposed slope should meet the safety requirements; that is, the soil mass under given loads should have an adequate safety factor with respect to shear failure, and the deformation of the soil mass under the given loads should not exceed certain tolerable limits. The analyses are generally made for the worst conditions, which seldom occur in the design life. Methods, originally developed for analysing unreinforced slopes, have been extended to analyse reinforced slopes, taking into account the presence of reinforcements. There are basically four methods for analysing the stability of geosynthetic-reinforced soil slopes

- the limit equilibrium method
- the limit analysis method
- the slip line method
- the finite element method.

11.4.1 Limit equilibrium method

Limit equilibrium methods have been used traditionally to analyse the stability of slopes with and without reinforcements. Several limit equilibrium methods have been used in different studies (Ingold, 1982; Murray, 1982; Leshchinsky and Reinschmidt, 1985; Leshchinsky and Volk, 1985, 1986; Schmertmann *et al.*, 1987; Leshchinsky and Boedeker, 1989; Jewell, 1990; Wright and Duncan, 1991; Mandal and Labhane, 1992; Srbulov, 2001). In these methods of analysis, it is considered that the failure occurs along an assumed or a known failure surface. At failure, the shear strength is fully mobilised all the way along the failure surface, and the overall slope and each part of it are in static equilibrium. The mobilised shear strength required to maintain a condition of limiting equilibrium is compared with the available shear strength, giving the average factor of safety along the failure surface as below:

$$\text{FS} = \frac{\text{shear strength available}}{\text{shear strength required for stability}} \tag{11.1}$$

The shear strength of the soil is normally estimated using the Mohr–Coulomb strength criterion. The allowable tensile strength of geosynthetic layers is taken into account while calculating the available shear strength. Several slip surfaces are considered, and the most critical one is identified; the corresponding (smallest) factor of safety is then taken to be the factor of safety of the slope. It should be greater than 1.3. The problem is generally considered in two dimensions; that is, conditions of plane strain are used. A two-dimensional plane strain analysis is found to give a conservative result compared with a three-dimensional analysis (dish-shaped surface).

For an assumed circular arc failure surface within the shallow slope (inclination $\beta \leq 45°$) reinforced with horizontal geosynthetic layers (Figure 11.4), the factor of safety, in terms of the shear strength parameters of soil and the allowable tensile strength of the geosynthetic, can be obtained as below, following the method of slices, commonly used for slope stability analysis of unreinforced soil slopes:

$$\text{FS} = \frac{\text{moment of shear strength of soil and allowable tensile strength of geosynthetic layer along failure arc}}{\text{moment of weight of failure mass}}$$

$$= \frac{\sum_{i=1}^{n}(N_i \tan\phi + c\Delta l_i)R + \sum_{j=1}^{m} T_j y_j}{\sum_{i=1}^{n}(w_i \sin\theta_i)R} \tag{11.2}$$

Figure 11.4 Details of method of slices for circular slip analysis

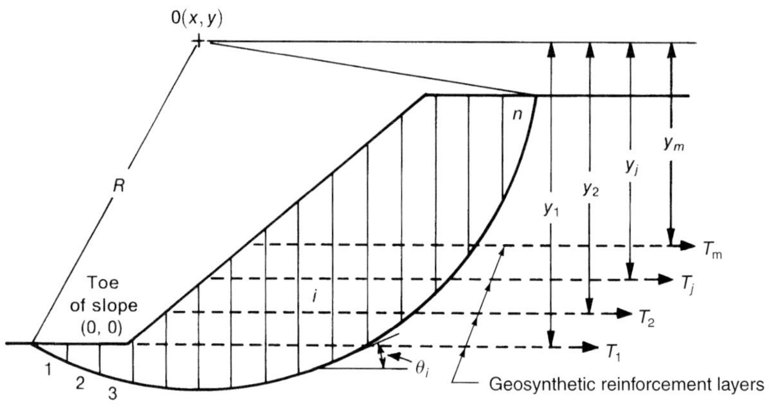

where w_i is the weight of the ith slice; θ_i is the angle made by the tangent, to the failure arc at the centre of the ith slice with the horizontal; $N_i = w_i \cos\theta_i$, Δl_i is the arc length of the ith slice; R is the radius of the circular failure arc; c and ϕ are shear strength parameters, cohesion and the angle of shearing resistance (total or effective depending upon field situations), respectively; T_j is the allowable geosynthetic tensile strength for the jth layer; y_j is the moment arm for the jth geosynthetic layer; n is the number of slices; and m is the number of geosynthetic layers.

The stability of steep reinforced slopes (inclination $\beta > 45°$) can be analysed by the tie-back wedge analysis approach used for vertical walls, as described in Chapter 4.

Leshchinsky and Reinschmidt (1985) presented a method to analyse geotextile-reinforced slopes based on limit equilibrium and variational extremisation. The results indicate that the potential failure surfaces are either planar or log-spiral, and the use of a strong geotextile results in a deep-seated failure. The presence of a geotextile layer increases the compressive stress over the critical slip surface, and it also decreases the tensile stress that tends to develop at the crest.

Zornberg *et al.* (1998b) evaluated the limit equilibrium methods with respect to their ability to predict the failure of geosynthetic-reinforced slope models tested in a geotechnical centrifuge. The evaluation indicates that limit equilibrium should consider the horizontal orientation of reinforcement forces, that significant contribution to stability is provided by the overlapping reinforcement layers, and that different rigorous limit equilibrium methodologies provide equally good results. Very good agreement was obtained between the g levels at failure obtained experimentally and those predicted by limit equilibrium. Equally good agreement was obtained between experimental and predicted locations of the failure surfaces.

Srbulov (2001) analysed the stability of geogrid-reinforced steep slopes using a method based on limit equilibrium. The method of analysis takes into account strains along the boundaries of rigid wedges, in addition to the forces considered by classical methods of limit equilibrium. The method also accounts for potential progressive failure in brittle soil–geogrid composites; however, the solution by this method remains approximate due to various assumptions. This extended method is simple to use and requires less input data than complete stress–strain methods. However, it inherits the disadvantage of conventional limit equilibrium methods: that a sliding mechanism must be assumed to govern slope stability.

Limit equilibrium methods do not furnish any information on soil deformations. Nevertheless, these methods have been very useful in solving slope stability problems, and need less computational effort. By adopting suitable factors of safety, the choice of which is largely governed by experience, the deformation can be limited. It is required to consider separate factors of safety for the soil and geosynthetic layers because their deformational characteristics are different.

11.4.2 Limit analysis method

Limit analysis is a universal method for a realistic solution to slope stability problems (Sawicki and Lesniewska, 1989; Michalowski and Zhao, 1993, 1994, 1995; Zhao, 1996; Jiang and Magnan, 1997; Michalowski, 1997; Porbaha and Lesniewska, 1999; Ausilio *et al.*, 2000; Porbaha *et al.*, 2000; Askari *et al.*, 2010). It is based on plasticity theory. This method can be applied to slopes (and also to other structures) of arbitrary geometry, complicated loading conditions and homogeneous, as well as heterogeneous, plastic materials. Using the limit theorems, it is possible to bracket the collapse load even if it cannot be determined exactly. In the lower-bound approach, we determine whether there exists an equilibrium stress field that is in equilibrium with the applied load, and with which the plastic yield condition is not violated anywhere in the slope. If such a stress field exists, it can be ascertained that the applied load is less than the limit load, and no plastic failure will occur in the slope. In the upper-bound approach, we search for a kinematically admissible velocity field; we then calculate the corresponding internal and external plastic power dissipations.

If the external power dissipation is higher than the internal one, the load can be said to be greater than the limit load. In this way, the limit load can be defined as the load under which there exists a statically admissible stress field, yet a free plastic flow can occur. An efficient and accurate numerical technique, such as the finite-element method, is vital to make limit analysis applicable to complicated problems of slope stability.

Zhao (1996) presented a limit analysis of the geosynthetic-reinforced soil slope using a kinematic solution of the plasticity theory. The approach presented is based on the upper-bound theorem of plasticity. The total energy dissipation during the incipient plastic failure process was assumed to be equal to the sum of the energy dissipation in the soil and in the geosynthetic reinforcement. The geosynthetic reinforcement was assumed to dissipate energy during incipient collapse only in the tensile mode. The soil was also assumed to be uniform and homogeneous. In the analysis, both translational and rotational rigid block collapse mechanisms were considered. It was pointed out that a rotational failure mechanism always yielded lower stability factor values for load-free geosynthetic-reinforced slopes. Comparisons of the limit analysis method with the limit equilibrium and the slip line methods indicated reasonably similar results.

Porbaha *et al.* (2000) applied a kinematic approach based on the framework of limit analysis for the stability analysis of reinforced vertical and sloping model walls, with cohesive backfill that were brought to failure under self-weight in a geotechnical centrifuge. The prototype equivalent heights predicted by the analyses are within the distress range; that is, the development of tension crack and the collapse of the retaining walls occurred during centrifuge tests.

Ausilio *et al.* (2000) analysed the seismic stability of slopes reinforced with geosynthetics within the framework of the pseudo-static approach. Calculations were conducted by applying the kinematic theorem of limit analysis. Different failure modes were considered, and, for each mode, analytical expressions were derived that may enable one to readily calculate the reinforcement force required to prevent failure and the yield acceleration of slopes subjected to earthquake loading. Several results were presented in order to illustrate the influence of seismic forces on slope stability. A suitable procedure based on the assessment of earthquake-induced permanent displacement was proposed for the design of reinforced slopes in seismically active areas.

11.4.3 Slip-line method

The slip-line method is based on the derived failure criterion describing the failure of a homogenised geosynthetic-reinforced soil composite and the application of the method of stress characteristics (Anthoine, 1989; de Buhan *et al.*, 1989). It differs from limit equilibrium and limit analysis methods in that the soil and the reinforcement are first homogenised, and the stability analysis assumes an anisotropic continuum, rather than soil and reinforcement as two separate constituents. The derivation of the failure criterion for a geosynthetic-reinforced soil composite was presented by Michalowski and Zhao (1995). The limit loads on geosynthetic-reinforced soil slopes can be calculated using the slip-line method as described by Zhao (1996). This approach is expected to have a wider application in the analysis of slopes with less conventional reinforcements such as continuous filaments or for fibre-reinforced soil slopes.

11.4.4 Finite-element method

The finite-element method of analysis is generally based on a quasi-elastic continuum mechanics approach in which stresses and strains are calculated. Since geosynthetic-reinforced soil slopes exhibit large deformations during the stage construction process, it is appropriate to adopt a non-linear soil model for the stress–strain analysis with a suitable failure criterion (e.g. the Mohr–Coulomb criterion). Such models of varying degrees of complexity have been developed. They require additional parameters, but these can usually be furnished by the standard triaxial test if shear and volumetric strain measurements can be carried out with sufficient accuracy. The geosynthetics are also required to be modelled by an appropriate constitutive model. More details on this method can be found in the work of Rowe and Soderman (1985), Almeida *et al.* (1986), Ali and Tee (1990), Chalaturnyk *et al.* (1990), and Porbaha and Kobayashi (1997). The finite-element studies have shown that significant reductions in the shearing, horizontal and vertical strains within the slope occur because of the presence of the reinforcement. The finite-element method is certainly the most comprehensive approach to analyse the stability of soil slopes subjected to seismic loadings. However, its use usually requires high numerical costs and accurate measurements of the properties of soil and geosynthetic reinforcement, which are often difficult to achieve.

11.5. Design aspects of reinforced slopes

Geosynthetic-reinforced slopes are designed to provide the following three basic modes of stability (Simac, 1992; Shukla and Yin, 2006)

- internal stability (Figure 11.5(a))
- global stability (Figure 11.5(b))
- surficial stability (Figure 11.5(c)).

The factor of safety must be adequate for both the short-term and long-term conditions, and for all possible modes of slope failure, similar to those for unreinforced slopes.

Internal stability is influenced by the quantity, strength, length and vertical spacing of the primary reinforcement elements. As the system moves towards failure, the soil deforms, creating tension in the reinforcement across the failure plane. A limit equilibrium analysis can be carried out to determine the magnitude of reinforcement tension necessary to maintain equilibrium with a minimum factor of safety of 1.3.

Figure 11.5 Modes of slope failure: (a) internal stability, (b) global stability and (c) surficial stability
Adapted from Simac (1992), Industrial Association International; and Shukla and Yin (2006), © Taylor & Francis Group

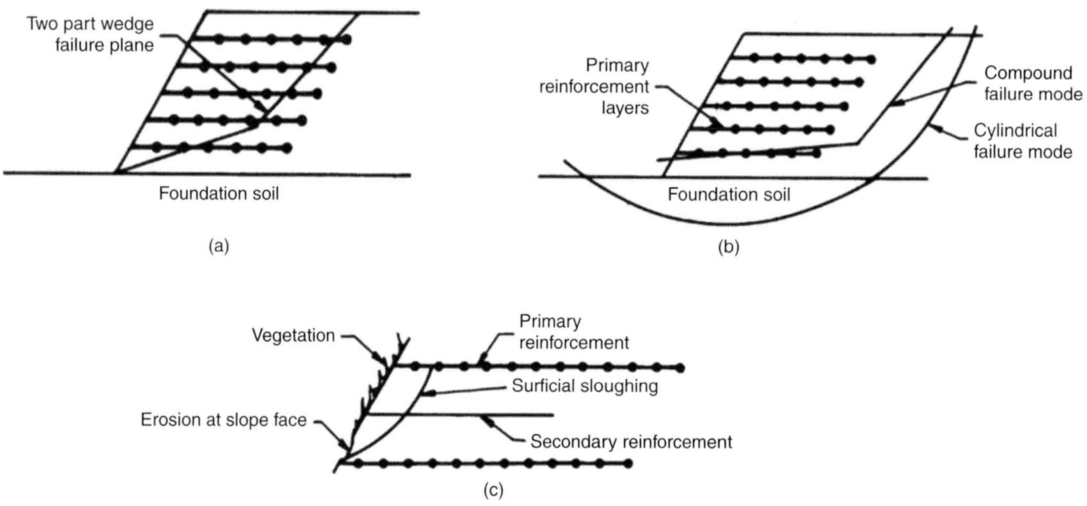

Conventional slope stability analysis of potential failures outside the reinforcement zone is designated *global stability*. The global stability of the entire reinforced-soil mass is usually controlled by the foundation upon which it rests. The influence of the foundation soil strength, groundwater conditions, soil stratigraphy, imposed surcharge loadings and slope geometry must be analysed to ensure satisfactory performance. A compound failure mode should also be analysed, where the potential failure surface passes partially through the reinforced mass and the soil that is being retained by the reinforced slope.

Surficial stability determines the secondary reinforcement requirements of the reinforced slope to preclude surficial sloughing of the slope face during and after construction. Depending on the magnitude of erosive forces imposed on the slope face, erosion control measures can range from temporary to permanent armoured systems. The available experience suggests a maximum of 60 cm vertical spacing between the secondary reinforcement layers, recommended to be a minimum of 120 cm in length. The secondary reinforcement can also be recommended at each compaction lift, at about 22–30 cm intervals, in the case of some cohesionless soils.

Field and numerical model test results indicate that the limit equilibrium approach to reinforced slope design provides a suitable but conservative design approach (Christopher *et al.*, 1990). A step-by-step design procedure incorporating the circular arc approach may be found in Christopher and Leshchinsky (1991). Simplified charts for the design of geosynthetic-reinforced slopes have been proposed by many research workers (Jewell and Woods, 1984; Jewell *et al.*, 1984; Christopher and Holtz, 1985). These charts can be used to evaluate the preliminary stability of geosynthetic-reinforced slopes before more thorough design procedures are performed. Critical reinforced slopes, as well as permanent slopes (having a design life of greater than 1–3 years), should be designed using comprehensive slope stability analyses. The factor of safety against slope stability should be taken from the critical surface requiring the maximum reinforcement.

The major steps for the design of a reinforced slope can be given as follows (Shukla and Yin, 2006)

- *Step 1*: define the geometry of the slope (slope height, H; slope angle, β).
- *Step 2*: define the loading conditions (surcharge load, temporary live load and dynamic load).
- *Step 3*: determine the engineering properties (permeability, shear strength and consolidation parameters) of the foundation soils and the slope soils.
- *Step 4*: locate the groundwater table. For slope and landslide repair projects, identify the cause of instability and locate the previous failure surface.
- *Step 5*: determine the properties (gradation and plasticity index, compaction characteristics, shear strength parameters, and chemical composition that may affect the durability of geosynthetic reinforcement) of available fill.
- *Step 6*: determine the geosynthetic properties (strength and modulus, and soil–geosynthetic interface friction) and assign limits to tolerable geosynthetic deformations. The maximum geosynthetic strain can be 2–10%. Drainage, constructability (survivability) and environmental requirements should be considered in the selection of the geosynthetics.
- *Step 7*: determine the factor of safety of the unreinforced slope and determine the geosynthetic reinforcement requirement (vertical spacing and length) based on the internal stability analysis.

- *Step 8*: check the factors of safety against external stability, including sliding, load-bearing failure, foundation settlement, deep-seated slope failure and dynamic stability, as carried out for conventional retaining wall designs, assuming that the geosynthetic-reinforced soil mass acts as a rigid body in spite of the fact that it is really quite flexible. The minimum values of the factors of safety against sliding, load-bearing capacity failure, deep-seated failure and dynamic loading are generally taken to be 1.5, 2.0, 1.3 and 1.1, respectively.
- *Step 9*: check the requirements for surface and subsurface water control, where surface water runoff and drainage are critical for maintaining slope stability.

The design details and concepts for some popular stabilisation methods are described in the form of case studies in Section 11.7, along with application guidelines. It should be noted that reinforced slopes essentially are mechanically stabilised earth (MSE) structures with similar properties and design criteria as those of vertical-faced, gravity MSE walls, such as reinforced earth or a concrete segmental unit combined with geogrid reinforcement. Therefore, for convenience in designs, one can also consider reinforced slopes as the gravity earth-retaining structures with a sloped face, for which the details are described in Chapter 4.

Reinforced-slope design can ideally be carried out using a conventional slope stability computer program modified to account for the stabilising effect of the reinforcements. However, to facilitate the complete design of geosynthetic-reinforced slopes, many types of software are commercially available, though some are limited to specific soil and reinforcement conditions. Leshchinsky (1997) mentioned the software called 'Reslope'. For a given problem, including the ultimate strength of the reinforcement layers, the software yields the optimal length and spacing of the geosynthetic layers. This layout satisfies the specified various factors of safety input by the user. Specific details of design practices in Europe, Japan and the USA are summarised in the review papers by Christopher *et al.* (2005), Jones (2005), and Otani and Ochiai (2005) respectively.

11.6. Model tests

This section presents the behaviour of geosynthetic-reinforced slopes, subjected to loading in the vicinity of the crest, as observed in model tests conducted by several researchers (Selvadurai and Gnanendran, 1989; Resl, 1990; Das *et al.*, 1996; Yoo, 2001; Mittal *et al.*, 2009; Choudhary *et al.*, 2010).

Selvadurai and Gnanendran (1989) conducted model tests in a reinforced concrete test tank with the following dimensions: 1500 mm in length, 880 mm in width and 1200 mm in depth. The rigid strip foundation was modelled by a steel box section measuring 104 mm × 870 mm in plan area. The longitudinal sides of the tank were fitted with highly polished stainless steel sheets to reduce the friction between the soil and the sides of the test tank, and to induce a near plane strain state in the tested soil mass. The constant rate of movement of the footing was controlled by a worm gear–actuator assembly driven by an electric motor. The reaction frame was anchored to the reinforced concrete floor, independent of the test container. The sand was used as a fill material for the entire experimental investigation. In all the experiments, the bulk density of the mortar sand in its compacted state was maintained at 17.6 kN/m^3. The approximate angle of internal friction (ϕ) for the sand was estimated to be 43°. The length of the Tensar geogrid for each depth was such that it was present from the boundary of the tank to the sloped fill surface. The dimensions of the typical test configuration are shown in Figure 11.6(a). The load–displacement relationships observed for a rigid footing, located on reinforced and unreinforced slopes, are shown in Figure 11.6(b). The location of the failure surface was estimated at the termination of each experiment. The influence of the depth of embedment of the geogrid reinforcement (Tensar SS2 type) on the failure path is summarised in Figure 11.7. These observations indicate that the failure paths exhibit a general slope failure pattern when $u/B < 1$. When $u/B > 1$, the failure path is significantly altered, and the failure occurs at the soil–geogrid interface. For geogrid depths where $u/B > 1$, it would appear that the plane of the geogrid acts as a plane of weakness (i.e. the plane of failure occurs just above the geogrid). The location of failure paths derived from additional tests involving Tensar geogrids SS0, SS1 and AR1 showed characteristics similar to those illustrated in Figure 11.7.

The dimensions of the typical test configuration are shown in Figure 11.6. Based on the experimental studies, the following generalised conclusions can be drawn.

- The load-carrying capacity of a footing on a sloped fill structure can be improved in excess of 50% by incorporating the geogrid reinforcement.
- When considering the ultimate bearing capacity, the optimum depth for the geogrid reinforcement is between 0.5 and 0.9 times the width of the foundation.
- The initial stiffness of the footing (defined as the slope of the load–displacement curve during initial loading) can be increased in excess of 25% by incorporating a geogrid reinforcement layer at a depth between 0.5 and 0.9 times the width of the foundation.
- The primary properties of a geogrid that govern its effectiveness in improving the load-carrying capacity of the sloped fill are identified as the aperture size, the modulus of elasticity and the tensile strength.
- The location of the geogrid layer at a depth greater than twice the width of the footing does not lead to any improvement in either the load-carrying capacity or the stiffness characteristics of the footing on a sloped fill.

Resl (1990) conducted laboratory model tests in order to understand the reinforcing mechanism of non-woven geotextiles used for reinforcing steep slopes. Figure 11.8

Figure 11.6 Model tests: (a) typical test configuration for a rigid footing on a reinforced slope and (b) load-displacement relationships for a rigid footing located on reinforced and unreinforced slopes
Adapted from Selvadurai and Gnanendran (1989), © Canadian Science Publishing or its licensors

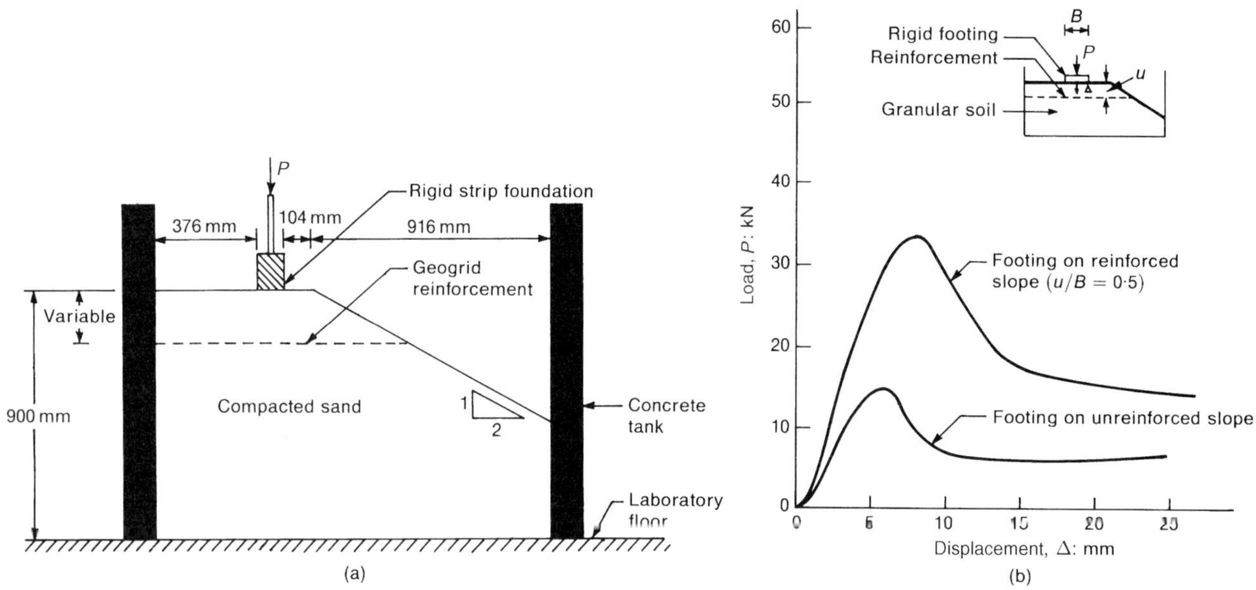

illustrates the test configuration. Sand (angle of shearing resistance $\phi = 32°$, cohesion $c = 0$) was used as a fill material. The digging out of the geotextile samples after the tests showed that failure occurred due to rupture of the geotextile, and not due to slippage or pull-out.

Figure 11.7 Influence of the depth of location of the reinforcement on the development of the failure planes within the soil slope: (a) shallow embedment of geogrid reinforcement and (b) deep embedment of geogrid reinforcement
Adapted from Selvadurai and Gnanendran (1989), © Canadian Science Publishing or its licensors

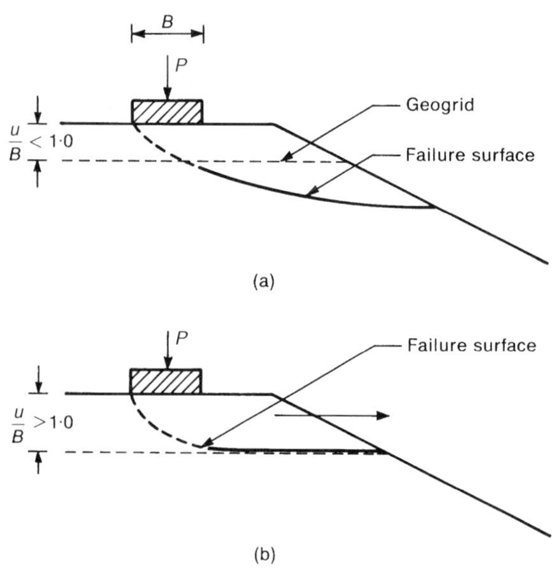

Das *et al.* (1996) presented the results of a number of bearing capacity tests for a model strip foundation resting on a biaxial geogrid-reinforced clay slope. The geometric parameters of the test model are shown in Figure 11.9. Based on the study, the following conclusions can be drawn

- Other conditions remaining the same, the first layer of the geogrid should be located at a depth of $0.4B$ (B = width of the footing) below the foundation for the maximum increase in the ultimate bearing capacity derived for reinforcement.
- The maximum depth of reinforcement, which contributes to the bearing capacity improvement, is about $1.72B$.

Yoo (2001) investigated the behaviour of a strip footing on a geogrid-reinforced soil slope using laboratory model tests, as well as finite-element analyses. The study shows that the failure zones tend to become wider and deeper than that for the unreinforced slope. Neglecting such a trend may lead to an unconservative estimation of the required anchorage length of geogrid reinforcements, especially for those at large depths. Significant benefit of geogrid reinforcements in terms of footing performance can be derived by inclusion of short geogrid layers directly below the footing.

Choudhary *et al.* (2010) conducted laboratory model tests to investigate the bearing capacity behaviour of a strip footing resting on the top of a geogrid-reinforced fly ash slope. A series of model footing tests covering a wide range of boundary conditions, including unreinforced cases, were conducted by varying parameters such as the location and depth of

Figure 11.8 Model test configuration
Adapted from Resl (1990), Industrial Fabrics Association International

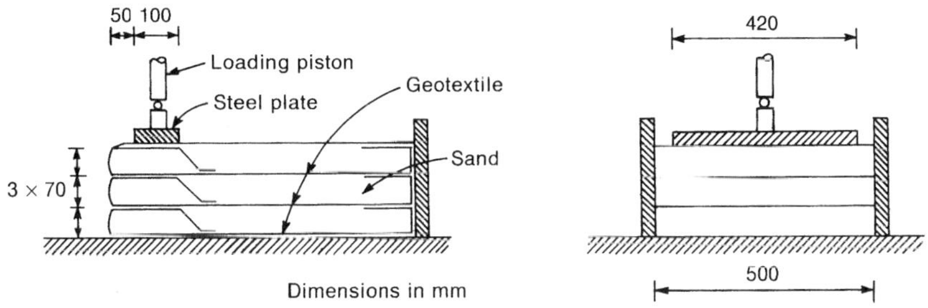

Figure 11.9 Geometric parameters for a surface strip foundation on a geogrid-reinforced clay slope
Adapted from Das *et al.* (1996)

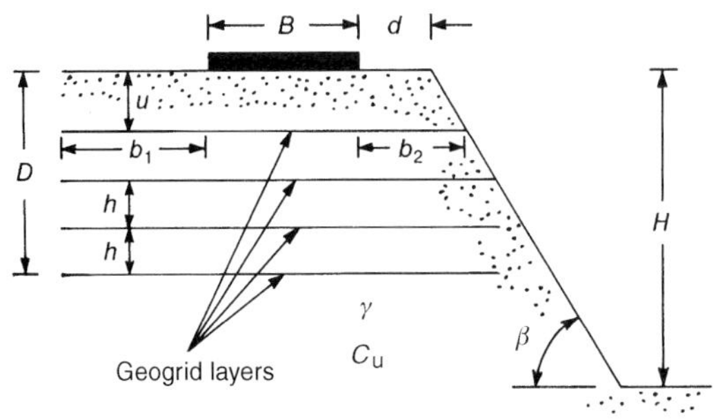

embedment of a single geogrid layer, the number of geogrid layers, the location of the footing relative to the slope crest, slope angles, and width of the footing. The results of the investigation indicate that both the pressure–settlement behaviour and the ultimate bearing capacity of a footing resting on the top of a fly ash slope can be enhanced by the presence of reinforcing layers. The efficiency of the fly ash–geogrid system increases with an increasing number of geogrid layers and the edge distance of the footing from the slope. The location of a single geogrid layer at a depth greater than 2.5 times the footing width does not improve the load-carrying capacity significantly. The bearing capacity ratio initially increases with an increase in the embedment ratio up to 1.0, and thereafter it decreases irrespective of the edge distance and slope angle. The load-carrying capacity of a footing located on top of a reinforced fly ash slope increases with an increase in the number of reinforcing layers for all the edge distances and slope angles adopted in the study. However, the increase is significant for up to four reinforcing layers, and thereafter the increase is only marginal.

11.7. Stabilisation methods in practice

The slope stabilisation methods generally reduce driving forces and increase resisting forces, or both. In many situations, the conventional repair of failed embankments and cut slopes involves the removal of the slide debris and reinstatement of the slope with free-draining granular materials. This method can be expensive because of the cost of excavation, granular materials and haulage. In an alternative repair procedure, geosynthetics (geotextiles or geogrids) are used as reinforcement so that the slide debris itself can be reused, making the stabilisation work economical. The use of geosynthetics in stabilising slopes provides a wide range of design advantages, as mentioned below (Simac, 1992; Shukla *et al.*, 2011)

- reduction in the land requirement to facilitate a change in grade
- provision of additional usable area at the toe or crest of slope
- use of available on-site soil to balance earthwork quantities
- elimination of the import costs of select fill or export costs of unsuitable fill
- meeting steep changes in grade, without the expense of retaining walls
- elimination of concrete face treatments, when not required for surficial stability or erosion control
- provision of a natural vegetated face treatment for environmentally sensitive areas
- provision of noise abatement for high-traffic areas and minimisation of vandalism
- offer of a design that is easily adjustable for surcharge loadings from buildings and vehicles.

The construction of reinforced slopes may highlight some of the above advantages in the following applications

- repair of failed slopes
- construction of new embankments
- widening of existing embankments
- construction of alternatives to retaining walls.

Many slope stabilisation methods in the form of case studies have been reported in the literature. This section deals with selected methods to stabilise slopes using geosynthetics, in various forms, along with construction guidelines.

11.7.1 *Geofabric-wrapped drain* method

A *geofabric-wrapped drain* (GWD) method (Broms and Wong, 1986) was used successfully in Singapore to stabilise a steep slope in residual soil and weathered rock. By this method, the stability of existing stable soil slopes can be increased, failed slopes can be stabilised, or new steep slopes or high embankments can be constructed without exceeding the bearing capacity of the soil. In these applications, the function of the geotextile, both as a tensile reinforcement and as a filter, is utilised.

In this method, geotextile-wrapped drains consisting of granular materials are installed along the slopes, as shown in Figure 11.10(a). The drains reduce the porewater pressure within the slopes during the rainy season and, thereby, the shear strength is increased. The geotextile layer acts as a filter around the drains, which prevents the migration of soil (internal erosion) within the slope into the drains. It also reinforces the soil along potential sliding zones or planes.

One additional advantage with this method is that the temporary decrease in the stability of the slope is only marginal during the construction of the deep trenches required for the drains. Here, only a limited width of the slope is affected. When concrete gravity or cantilever walls are used, the stability of the slope is generally reduced considerably during construction.

Figure 11.10 (a) Slope stabilisation using geofabric-wrapped drains, (b) a flow net showing steady state seepage and (c) computation of design tensile reinforcement to be provided by a geotextile
Adapted from Broms and Wong (1986), John Wiley & Sons, Inc.

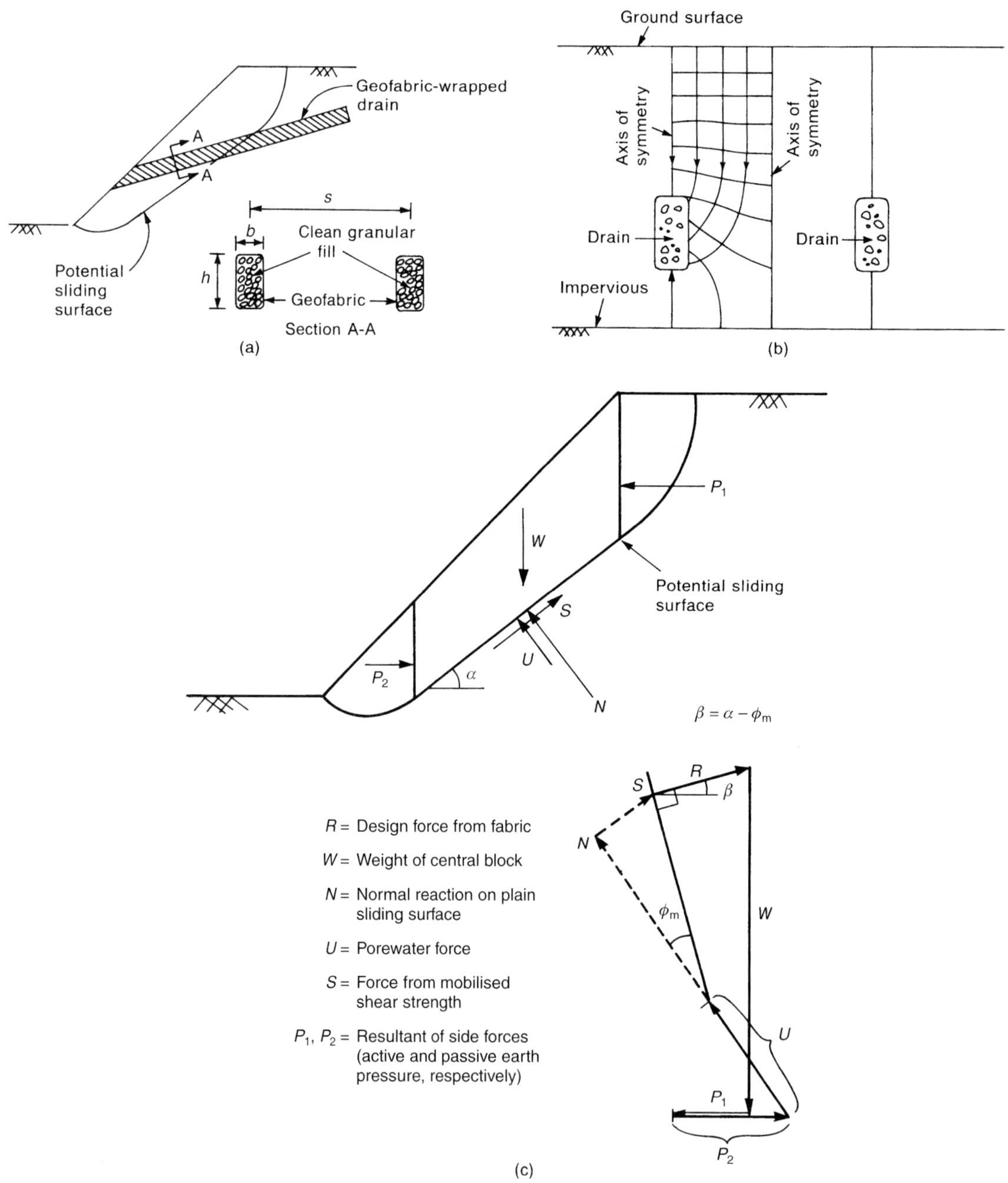

11.7.1.1 Spacing of drains

The required spacing of the drains wrapped in geotextile, as well as the dimensions of the drains, depends on the porewater pressures in the slope, which can be evaluated by means of a flow net. An example of a flow net that corresponds to steady-state seepage after prolonged rainstorm is shown in Figure 11.10(b). The granular material in the drains is considered to be infinitely pervious in relation to the slope material. The porewater pressure in the slope is reduced considerably by the drains, both above and between the drains, as can be seen from the flow net. For general situations, 0.5 m-wide and 1.0 m-high drains, spaced 3.0 m apart, would be reasonable. The flow lines above the drains will then be almost vertical, and the excess porewater pressures will be low.

11.7.1.2 Depth of drains

The drains should be located deep enough so that they intersect potential slip surfaces in the soil. The required depth of the drains depends on the difficulties of excavating trenches along the slopes. The maximum depth is about 4 m. For slopes in residual soils or weathered rocks, this depth is usually sufficient because most slope failures in these materials are shallow, having a maximum depth of the failure surface of less than 3–4 m.

11.7.1.3 Required tensile strength

The required tensile strength of the geotextile can be calculated by considering the force polygon for the sliding soil mass above possible sliding surfaces in the soil (Figure 11.10(c)). The sliding surface is often located at the contact between the completely weathered and the underlying, partially weathered, material.

11.7.1.4 Orientation

For a planar sliding surface, the orientation of the geotextile-wrapped drains should be perpendicular to the resultant of the normal reaction force and the force that corresponds to the mobilised shear strength along the potential failure surface, as shown in Figure 11.10(c), in order to utilise the geotextile effectively.

11.7.1.5 Required number of layers

The required number of layers (N) of the geotextile in each drain can be determined as follows:

$$N = \frac{F_s Rs}{aT} \tag{11.3}$$

where R is the force per unit width (kN/m) to be resisted by the geotextile, s is the drain spacing (m), T is the tensile strength per unit width (kN/m) of the geotextile, a is the effective perimeter of the drain (m) and F_s is the factor of safety.

11.7.1.6 Deformation

The geotextiles available in the market generally require an elongation of 14 50% before the ultimate tensile strength of the geotextile is mobilised. The strain required to mobilise the ultimate strength is much less for woven geotextiles than for non-woven geotextiles. Only woven geotextiles should therefore be used. In view of the large strain required at failure, a factor of safety of at least 3 should be used in the design.

The transfer length L that is required to transfer the load in the geotextile to the surrounding soil can be calculated as follows:

$$L = \frac{Rs}{2(hK\sigma'_v + b\sigma'_v)\tan\phi'_a} \tag{11.4}$$

where σ'_v is the vertical effective stress at mid-height (centre) of the drains, K is the lateral earth pressure coefficient for the compacted granular material in the drains, h is the height of the drains, b is the width of the drains, and ϕ'_a is the friction angle between the geotextile and the soil.

The deformation δ of the geotextile to mobilise the required tensile force can be calculated from the following equation:

$$\delta = L\frac{e}{100} \tag{11.5}$$

where e is the percentage of elongation needed to mobilise the required tensile resistance of the geotextile.

11.7.1.7 Compaction

During the construction of the granular fill drains, it is important to compact the fill carefully. The compaction will increase the lateral earth pressure and therefore the friction between the geotextile and the soil, resulting in a reduced transfer length L. For a well-compacted fill, a value of K equal to at least 1.0 can be used in the calculation of the transfer length. The lateral earth pressure is highly dependent on the degree of compaction of the granular fill.

A second important point, with respect to compaction of the granular fill drains, is that the compaction should be done in the downhill direction in order to pretension the geotextile. In this way, the elongation of the geotextile, which is necessary to mobilise the required tensile force as well as the required displacement of the slope, will be reduced.

This method was adopted for the stabilisation of a landslide on the campus of the Nanyang Technological Institute (NTI) in Singapore. The landslide occurred in early 1984, during a period of heavy rainfall, on the NTI campus in the western part of Singapore. One student dormitory, Block E, was located at the toe of the slope. Two other dormitories were at the crest. An existing rubble wall, which had been constructed along the whole length of the slope, with a height varying from 1.70 to 3.50 m, failed during the landslide. The average height of the slope was about 7 m. The inclination of the slope was 37° prior to the failure. A scupper drain at the toe of the rubble wall was damaged and closed up as a result of the movement of the slope. The ground immediately in front of the displaced rubble wall heaved about 200 mm. The whole sliding mass continued to move at a slow rate

Figure 11.11 Stabilisation scheme – the NTI Block E slide
Adapted from Broms and Wong (1986), John Wiley & Sons, Inc.

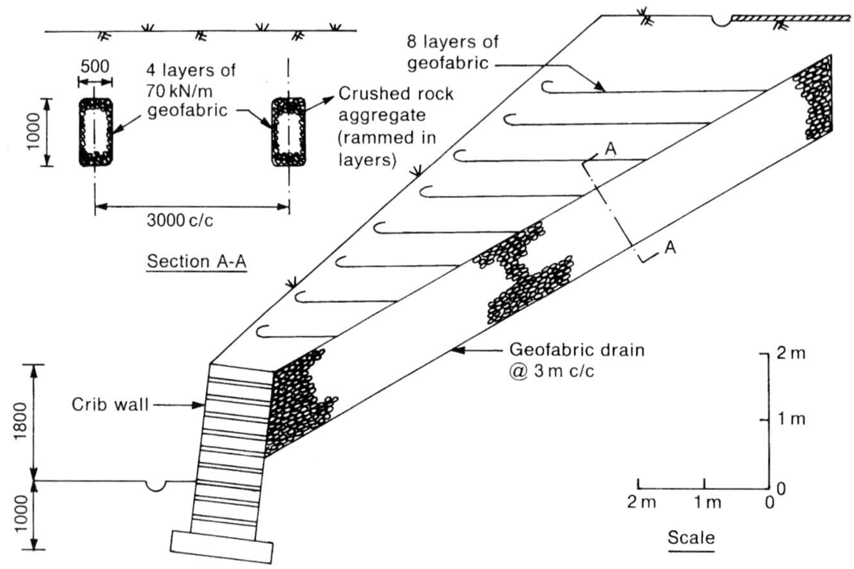

during the rest of 1984. Large cracks appeared on the displaced rubble wall. The total displacement of the wall was approximately 700 mm at the end of 1984. The toe had moved about 300 mm. The slope was composed of residual soil, and highly and completely weathered sedimentary rocks.

The remedial stabilisation works at the Block E slope consisted of the installation of eight fabric-wrapped crushed rock drains (Figure 11.11). The drains, 0.5 m wide and 1.0 m high, were spaced 3.0 m apart. Based on the sliding surface and a residual friction angle of 18°, the required tensile force of the geofabric was 85 kN per metre of the slope or 255 kN per drain. For each drain, four layers of 3.4 m-wide polyester fabric, with an ultimate tensile strength of 70 kN/m (238 kN per layer) at 14% elongation, were used. The fabric was wrapped around the two sides and the bottom of each drain. The drains were connected to the crib wall at the lower end of the slope for drainage. The crib wall was filled with crushed rock to allow discharge of the water from the transverse drains. Horizontal layers of the fabric were also placed in the slope between the ground surface and the transverse drains, to increase the stability of the slope with respect to shallow slides above the transverse drains. The far end of each fabric strip was anchored in the crushed rock drain. Another layer of the fabric was placed along the drains between the horizontal strips as a filter to prevent the soil above from being washed (eroded) into the drains. No further movements of the slope were observed after the installation of the drains.

11.7.2 Anchored geosynthetic system method

Koerner (1984), and Koerner and Robins (1986), proposed an *anchored geosynthetic system* (AGS) method, which is also known as the *anchored spider netting* method. It is an *in situ* slope stabilisation method in which a geosynthetic material (geotextile, geogrid or geonet) or other porous material is placed directly on the unstable or questionable slope, and anchored to it with long steel rod nails at discretely reinforced nodes, 1–2 m apart. These nails must be long enough to penetrate up to, and beyond, the actual or potential failure surface. Figure 11.12 shows the idealised cross-section of a slope stabilised by this method, along with its conceptualisation. When the rods are properly fastened, they begin pulling the surface netting into the soil, placing the net in tension and the contained soil in compression. When suitably deployed, this method offers a number of advantages in arresting slope failures

- the steel rods, in penetrating the failure surface, aid stability
- the stress caused by netting at the ground surface aids stability
- the surface netting stress mobilises normal stress at the base of the failure surface, which aids stability
- the entire system causes soil densification, which increases the shear strength parameters of soil.

It is important to recognise that the mechanism by which an AGS stabilises a slope is different from that of reinforced earth or soil nailing. Both reinforced-earth ties and soil nails are passive systems that rely on soil strains to mobilise pull-out, bending and shear resistances of the inclusions. By contrast, the anchors in an AGS are actively tensioned during installation. The increase in stability of the slope thus does not rely on soil movement to mobilise the soil–anchor interaction, but, rather, the increased stresses on potential failure surfaces, imparted by the tensioned fabric, increase the stability of the slopes (Ghiassian *et al.*, 1996).

In recognition of the above, the analysis of a slope stabilised by anchored geosynthetics follows a traditional limiting

Figure 11.12 (a) Idealised cross-section of anchored spider netting used to stabilise a soil slope, (b) a free-body diagram of netting, (c) a free-body diagram of contained soil and (d) a free-body diagram of anchor
Adapted from Koerner (1984); Koerner and Robins (1986), Industrial Fabrics Association International

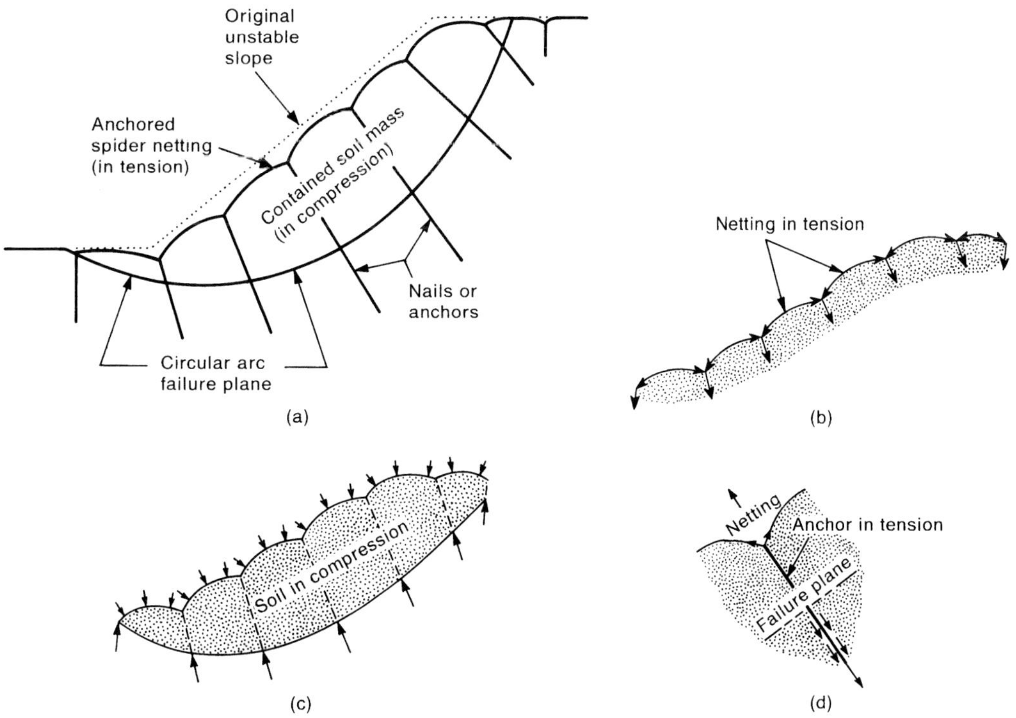

equilibrium stability analysis for slopes. The effects of the AGS are considered as additional forces acting on a potential failure surface. Bending and shearing resistances of the anchors are disregarded for several reasons. First, the bending and shearing resistances of the anchors are not mobilised. Second, the spacing of anchors is typically greater than the spacing in soil nailing and, thus, a coherent soil mass may not develop. A slope may fail by erosion and flow of soil around anchors; therefore, the bending and shearing resistances of the anchors may be irrelevant. Third, a variety of materials could be used for the anchors in an AGS, including cables with duck-billed anchors, which have essentially no bending resistance. Finally, the assumption is conservative.

A case history has been described where a 4.5 m-high clayey silt slope at a uniform slope angle of 25° has been stabilised. The slope was in an active state of failure. The slope was hand cleared of vegetation, and graded so as not to have any concave depressions. A knitted geonet made of bitumen-coated nylon was used as the netting. The anchors were 13 mm-diameter steel rods in 1.2 m-long sections, which were threaded into one another during installation. Since the existing failure zone was a shallow slope failure, the rods were only 2.4 m long. This length easily penetrated the failure plane as it drew the net into the surface soil. Two adjacent widths of netting were used on this slope (each being 5.6 m wide), along with a total of 73 steel rod anchors. Upon completion of the anchored spider netting, the slope was seeded with a rapid-growing rye grass and mulched. The grass grew within two weeks and completely hid the netting. The slope was reported to be in a stable condition.

11.7.3 Methods based on the construction of reinforced soil structures

Many case histories have reported the stabilisation of slopes by constructing geosynthetic-reinforced slopes (reinforced soil structures (RSSs) with an inclination $\beta \leq 70°$) or geosynthetic-reinforced retaining walls (reinforced soil structures with an inclination $\beta > 70°$). In many stabilisation projects, the failed soil was used as backfill material to make them economical. Chapter 4 provides the analysis and design of geosynthetic-reinforced retaining walls.

Dixon (1993) reported the geogrid-reinforced soil repair of a slope failure in clay on the North Circular Road in London in the UK. Figure 11.13 shows the cross-section of the repair of the slip failure. The slope was a cut slope (side slope = 2H : 1V, maximum height = 8 m) formed in London Clay in 1975. Some seven years after construction, slip failures began to occur along a 500 m length of cutting, causing damage to fence lines and spillage onto the carriageway. The main earthworks began in September 1985, with the excavation and removal from the site of a 35 m-long strip of slipped soil. Excavation extended beyond the failure plane, with benched steps cut into the undisturbed clay. To control any seepage, a 300 mm-thick granular drainage layer was

Figure 11.13 Cross-section of the repair of the slip failure
Adapted from Dixon (1993)

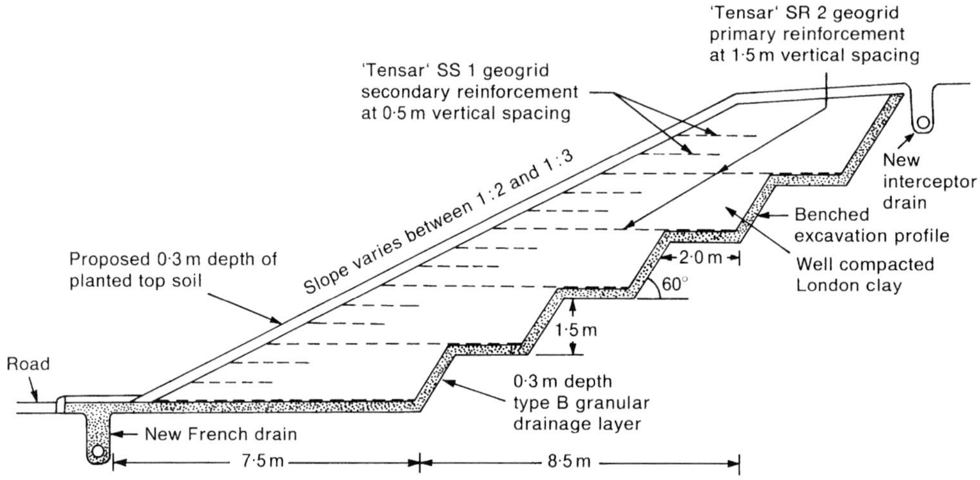

included over the excavated surface on the north side, where the forest slopes towards the cutting. The general sequence then adopted was to reinstate the first strip using fill excavated from an adjacent strip, thereby minimising double handling. The second strip was then reinstated using the fill excavated from a third strip, and so on. Fill was tipped from a dump-truck and placed using a bulldozer, and compacted to a maximum layer depth of 200 mm using a vibrating roller towed by the bulldozer. The 2.0 m-wide Tensar SS1 secondary reinforcements were obtained by cutting the standard 4.0 m-wide rolls into half on-site with a disc cutter. Tensar SR2 rolls were cut to the required length, and were laid perpendicular to the slope alignment. Adjacent rolls were butt jointed. The slope face was overfilled and trimmed in the conventional manner. The earthworks were completed in early February 1986. No special site equipment or expertise was required for the installation, which was carried out using conventional plant and labour. The average construction time per 35 m-long strip was about three days (typical strip quantities: excavation 1200 m^3, fill 800 m^3, gravel drain 350 m^3, geogrids 2000 m^3). After construction, no discernible movements were noted, and the grass cover on the slope was reported to be in good condition with a pleasing appearance.

Ali (1993) presented the design, construction and performance of the rehabilitation work for a 16 m-high slope failure in Malaysia. The slope consisted of residual soil deposits more than 10 m in thickness overlying a granite bedrock formation. The slope failure was most likely induced by infiltration due to heavy rain. The rehabilitation scheme consisted of two parts: a lower (70°) steep reinforced soil structure, 7 m high; and an upper (25°) gentle unreinforced slope. The reinforced soil structure acted as a gravity retaining system. The main fill material for this structure was compacted sand. Geogrids were used as the soil reinforcement. In order to facilitate turfing on the outer face, a thin layer of compacted cohesive soil was provided on the outer face, as shown in Figure 11.14. The design of the reinforced soil structure was based on a limit equilibrium analysis involving the use of a two-part wedge failure surface. A systematic search was made until the most critical wedge failure surface was identified. The reinforcements were accounted for by including a mobilised tensile force in the equations of equilibrium at every point at which a trial wedge surface intersected a reinforcement member. Internal and external stability checks were made using professional computer programs. The safety factor against global failure was found to be more than 1.5.

The construction follows the same procedure as that adopted for a compacted fill slope. Initially, the slope failure debris was removed, and excavation was then carried out to the required founding level. Before placing and compacting sand fill, horizontal subsoil-drainage pipes were positioned at the base to drain out of the reinforced soil. The drainage system consisted of 10 cm corrugated PVC pipes wrapped with geotextiles. Sand fill was then placed and compacted in layers, each 150 mm thick. As the level of fill was built up progressively, the geogrid reinforcement was placed at

Figure 11.14 Rehabilitation scheme for slope failure in Malaysia
Adapted from Ali (1993), with permission from Elsevier

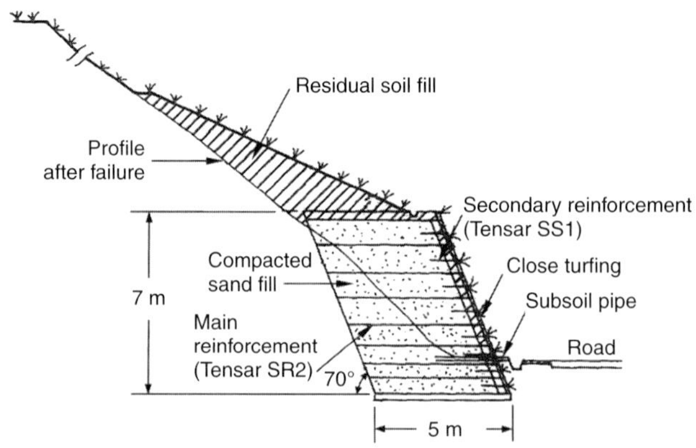

the required levels by stretching it tightly, and by pulling and fixing it down with timber pegs at the ends.

The performance of the slope was monitored with standpipe piezometers, pressure cells, inclinometer tubings, surface settlement markers and vibrating wire strain gauges from the beginning of construction up to a few months after construction. The tensile forces at different locations in the geogrids increased approximately linearly with the increase in fill height up to the end of construction. Shortly after the end of construction, the tensile force reached a maximum value, and it then remained constant with time. At each level of the geogrid layer, the tensile force increased progressively from the outer face to a maximum value and then decreased to zero at the other end of the geogrid in the upper levels, although at the lowest level the trend was less distinct. In the upper unreinforced slope, the horizontal movement taken for 10 months starting at the end of construction was found to be 30 mm, and the corresponding movement in the reinforced soil structure was only 6 mm or less. Furthermore, about 70% of the movement occurred within the first month after construction. The rate of settlement of the reinforced soil structure decreased progressively with time, and, after a few months, the rate was insignificant. The total settlement 10 months after construction was 14 mm. The horizontal earth pressure increased significantly within the five-month period, and the variation with depth was non-linear. The coefficient of the lateral earth pressure was found to be 0.19 and 0.23 at the time of completion and five months later, respectively. All these performance-monitoring strategies indicated that the slope stabilisation scheme performed satisfactorily.

Poh and Broms (1995) presented the application of an inexpensive slope stabilisation scheme that uses woven geotextile with old rubber tyres to arrest the deterioration of a hill slope adjacent to a 100 m-high microwave transmission tower on the Indonesian island of Batnam (Figure 11.15). Woven geofabric with a wrap and a weft tensile strength of 80 kN/m was used as reinforcement to resist the lateral earth pressure, and as a drainage medium to improve the drainage. Rubber tyres filled with granite aggregate and quarry waste were used as facing units only; that is, they were not designed to sustain lateral loads. The internal stability was checked with respect to the local tensile failure of the geotextile, and the adhesion between the geotextile and the soil. In assessing the overall stability of the wall, three modes of failure were investigated: (*a*) the failure of the wall by sliding as a rigid body; (*b*) the bearing failure of the foundation; and (*c*) the stability of the entire slope. A total of 10–12 unskilled labourers were needed for approximately 50 days to complete the 54 m-long wall using only simple hand tools. Slightly less than one 5.5 m × 100 m roll of the geotextile and 1400 old tyres were used. The total cost was less than 40% of the estimated cost of a conventional retaining wall. The initial performance evaluation indicated no visible movement of the wall.

Figure 11.15 Typical cross-section of a rubber tyre wall
Adapted from Poh and Broms (1995), with permission from ASCE

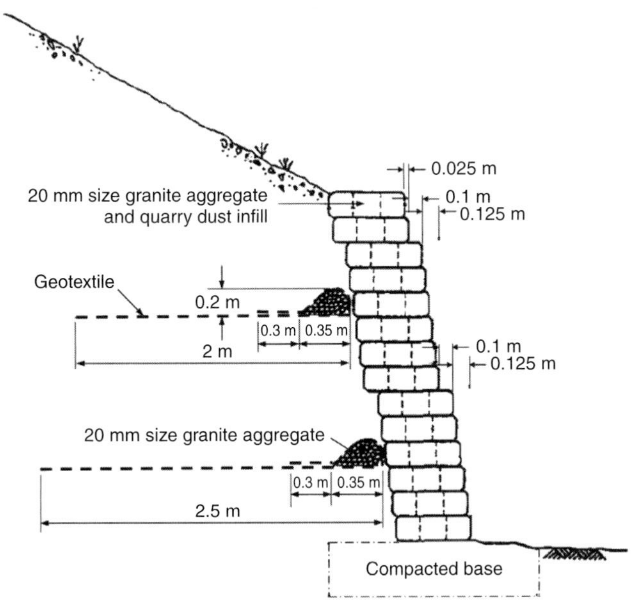

Rimoldi and Jaecklin (1996) summarised the construction methods for green-faced reinforced soil walls and steep slopes in four main schemes

- Straight reinforcement (Figure 11.16(a)) – this type of reinforcement, made of geosynthetics only, is mainly used for shallow slopes ($\beta < 50°$). Generally, the face is left exposed, or covered by a geomat or biomat, to prevent erosion. Therefore, the reinforcing geosynthetic is installed just at the face, without any wraparound. The installation is very easy. The reinforcement is laid down horizontally and straight, then the soil is spread and compacted to the required height, smoothing the face with a vibrating table or with the bucket of a back-hoe.
- Reinforcement wrapped around the face (Figure 11.16(b)) – in this scheme, the geosynthetic is

Figure 11.16 The construction schemes for green-faced structures: (a) straight reinforcement, (b) wraparound reinforcement, (c) a mixed scheme and (d) face blocks plus straight reinforcement
Adapted from Rimoldi and Jaecklin (1996)

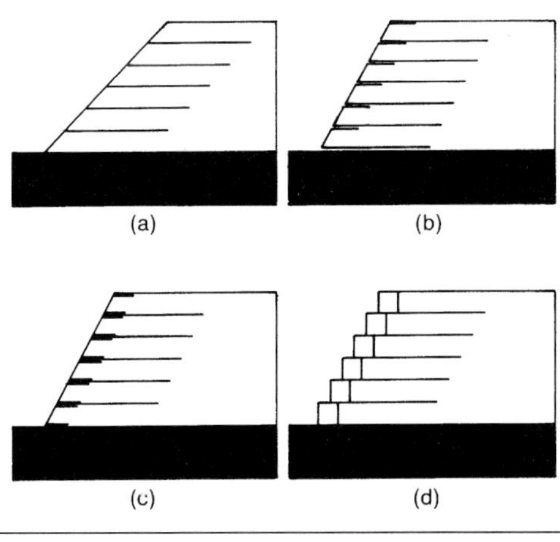

used both for reinforcement of the fill and for face protection from soil washing and progressive erosion, by wrapping it around the face of the slope. This 'wraparound' technique has been the most widely used construction method in Europe. The 'wraparound' installation procedure can be used with or without formworks. The use of formworks is suggested particularly when it is necessary to have a smooth and uniform face finishing.

The simplest construction method with the wraparound technique is without any formwork – it consists of placing a geogrid layer, in laying down, spreading and compacting the fill soil, in smoothing and levelling the face of the slope at the desired angle with a vibrating table or with the bucket of a back-hoe, and then wrapping the geogrid around the face and fixing it with a 'U' staple. This method provides fast construction and affords good results if it is not necessary to obtain a perfectly smoothed face. In fact, 'bulging' of the face often occurs, with an unpleasant aesthetic effect. Wraparound can also be made using movable formworks, or a straight steel mesh or steel mesh shaped as an 'L' or a 'C'. There may also be other suitable means. If steel meshes are used, they are left in place after the construction is terminated, which saves a lot of time and, hence, allows a very fast construction rate – a typical team of four to five workers, well equipped and with enough experience, can install about 50 m^2 of wall face in one working day, and, in particular situations, 100 m^2 of face in one day can also be achieved. The reinforcing geosynthetics can be connected to the steel meshes, but usually the two elements are independent.

- Mixed scheme – straight reinforcement plus another geosynthetic wrapped around the face in a 'C' shape (Figure 11.16(c)). In this scheme, the two functions of reinforcement and face protection are performed by two different geosynthetics. The reinforcing geosynthetic has high tensile strength and modulus, while the other one for face protection is lighter, and is engineered to support the growing vegetation and to retain the soil, preventing washout and erosion.
- Front blocks tied back by straight reinforcement (Figure 11.16(d)) – in this scheme, a front block is used both to support the facing during construction and in order to provide the final face finishing. Blocks are usually made of compacted soil, encased in containers, made either of gabion baskets or of a geosynthetic wrapped all around. Blocks are mechanically connected to straight reinforcing geosynthetics. This 'front blocks method' has the advantage of not being dependent on weather situations. The prefabrication does not disturb any traffic and can be near the site. A standard excavator is used to place the face blocks quickly, and the same excavator is also used for backfilling. No hydroseeding is needed because the grass seeds are already included inside the face blocks, and grass starts growing immediately.

Over-steep geogrid-reinforced slopes are usually associated with vegetation, and the facing of the slope is wrapped around by the geogrids or sometimes the facing is temporarily supported by a steel mesh, allowing vegetation to grow through the mesh apertures. Hard facing is also in use with geogrid-reinforced soil walls. Hard facing, as opposed to soft facing, refers to large precast concrete panels or small modular concrete blocks (MCBs). The blocks may have some kind of keys or inserts, which provide a mechanical interlock with the layer above. MCBs provide flexibility with respect to the layout of the curves and corners. They can tolerate larger differential settlements than conventional structures. MCBs are manufactured from concrete, and are produced in different sizes, textures and colours, and, therefore, they provide a varied choice to the engineer (Figure 11.17). Typically, all the blocks shown in Figure 11.17 are 250–450 mm long, 250–500 mm wide and 150–200 mm high. The mass of each block varies, typically, from 25 to 48 kg. The MCBs are laid dry (i.e. without mortar), and the geogrid reinforcements are placed between the block courses and connected by means of insert keys or pins, or by only the frictional interface between the courses. The footings for geogrid-reinforced modular concrete block wall systems (GRMCBWSs) can be constructed from granular compacted materials or from cast-in-place concrete. The walls are usually constructed with stepped facing, resulting in a batter ranging between 5° and 20°. The overall shape is equivalent to a steep slope, as opposed to a vertical wall, and, therefore, analysis can be carried out using steep slope procedures. One advantage of GRMCBWSs is the simplicity of installation because the blocks are easily transportable. It is estimated that 30–40 m^2 of wall can be erected by four persons over an eight-hour working day. As for the cost comparison, it is estimated that walls exceeding 1.0 m in

Figure 11.17 Examples of MCB units used in the UK: (a) porcupine, (b) keystone and (c) geoblock
Adapted from Dikran and Rimoldi (1996)

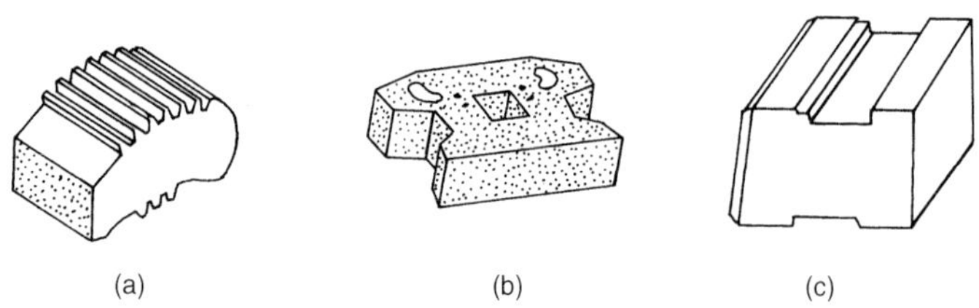

Figure 11.18 Geogrid-reinforced modular concrete block wall: Sevenoaks School, Kent
Adapted from Dikran and Rimoldi (1996)

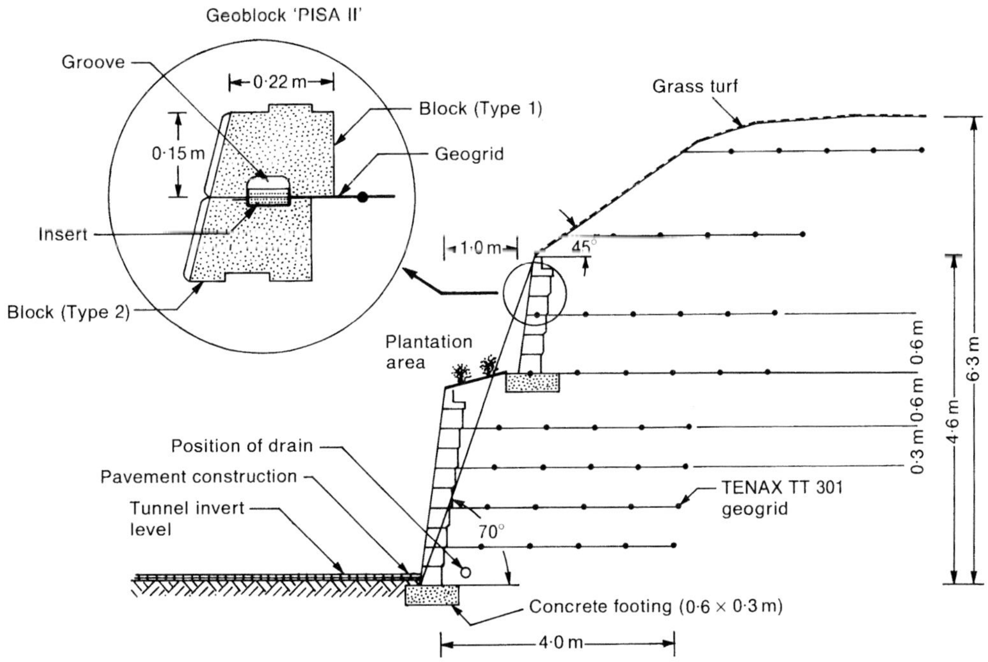

height typically offer a 25–35% cost saving over conventional cast-in-place concrete retaining walls.

Dikran and Rimoldi (1996) described a case study in which a GRMCBWS was successfully used for facing steep cuttings for the approaches of a tunnel under the main A225 Tonbridge Road in Sevenoaks, Kent, in the UK. Two walls were designed and built at each end of the tunnel, with a total length of 120 m, and heights varying from 6.3 to 10 m. The geometry of the walls was chosen in a manner that gave a pleasant aesthetic view and provided adequate stability (Figure 11.18). The blocks used in the project were of the Geoblock type (Figure 11.17), and the construction sequence was as follows

(*a*) The foundation was prepared and the footings were cast using mass concrete.
(*b*) The first course of the blocks was placed along the desired building line using standard rib units of type 1 (Figure 11.18).
(*c*) The second course was built using the insert blocks of type 2 (Figure 11.18), where the first layer of Tenax TT301 SAMP geogrid was required.
(*d*) The inserts were placed in the groove in the top of the block, with the narrow end of the finger pointing towards the face of the wall.
(*e*) Tenax TT301 SAMP geogrids were cut to the required lengths and placed over the inserts, so that every aperture in the grid was located over a finger of the inserts. The next row of blocks was placed over the insert and geogrid to hold the grid in place. The geogrid was then pulled from the back of the wall, so that the transverse rib of the grid was pulled back across the end of the fingers of the inserts.
(*f*) Excavated silty sand materials were used for the geogrid-reinforced fill. The compaction was carried out in 150 mm layers, using a plate compactor in areas within 1 m from the face of the wall and a vibrating roller with a mass per metre width of 1300 kg for the remainder of the length of the reinforcement.
(*g*) When the fill was placed and compacted up to the level of the next area of ground, the geogrid was then laid down on the top of the fill, and the construction continued as in points (*b*)–(*f*) above. The grass turf was placed over the back slopes.

A number of case histories on slope stabilisation using geosynthetics have been reported in some geosynthetic-related books (Raymond and Giroud, 1993; Koerner, 2005; Shukla and Yin, 2006). Recently, Porri *et al.* (2010) presented a study on slope stabilisation, with woven geogrids fabricated with polyester yarns. The slope was located on the San Bernardo hill on the east side of the city of Salta, Argentina. In February 2008, due to intense rainfall (70 mm in one hour) over several days, there were two landslides that hit the slopes and part of the asphalt surface in the access road to the San Bernardo hill. Due to the landslide, the material was deposited on the road. Repair with a conventional concrete wall was discarded from the beginning, because of its high cost, irreversible environmental impact, the limited space available for construction, and problems with the foundation and the time for construction. Dismissal of the traditional option led to the use of a non-conventional bagged stone wall alternative – the Terramesh System – and

Figure 11.19 (a) Final design of the slope intervention and (b) detail of the intermediate part (retaining wall 2) of the slope
Adapted from Porri *et al.* (2010), International Geosynthetics Society

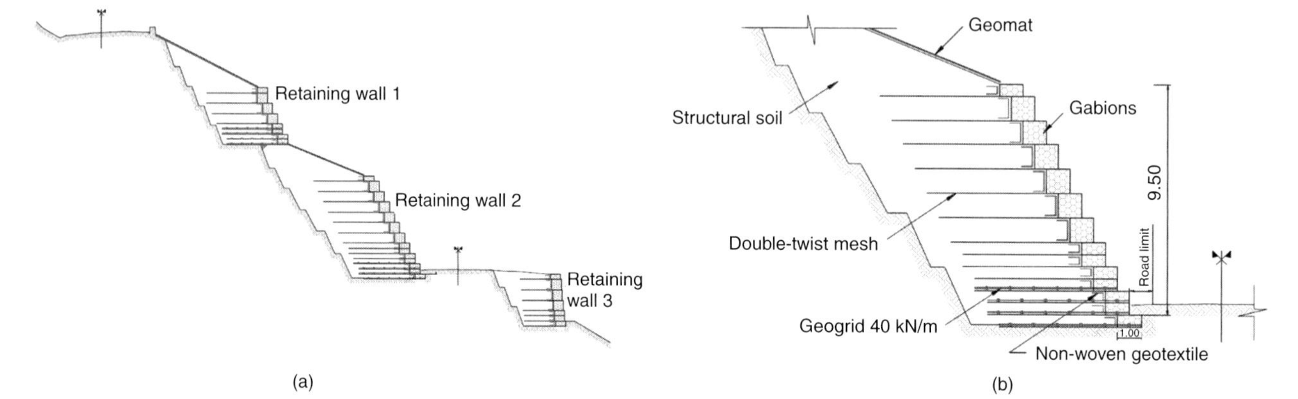

additional reinforcement of woven geogrids fabricated with polyester yarns, of 40–60 kN/m ultimate tensile strength. The analysis of slope stability was performed using the Macstars software, which checks the stability of reinforced soils; that is, structures that provide the slope stability using reinforcement units that are able to absorb the tensile stress. The selected alternative was built in a very limited time with the use of 70% of the material involved in the landslide, thereby reducing costs and direct environmental effects. Figure 11.19 shows the final stabilised slope and the details of its intermediate part.

11.8. Summary of main points

- The use of geosynthetic reinforcement for slope stabilisation allows a steeper slope and hence a reduction in the earthwork by changing the geometry, and even allows the use of soils with average mechanical properties.
- Geosynthetic reinforcement layers should ideally be placed in the direction and along the line of action of the major principal tensile strain within the soil slope; however, for practical reasons, reinforcement layers are usually placed in horizontal layers within the slope.
- Among the available methods of analysing the stability of geosynthetic-reinforced slopes, limit equilibrium methods are the most popular. Essentially, in each method, a failure mechanism is assumed, and some of the limit equilibrium requirements are satisfied. Most of the limit equilibrium methods, with their unrealistic oriented slip surfaces, are not correct from the viewpoint of the mathematical theory of plasticity, and they do not furnish any information on soil deformations. Other methods of slope stability analysis, described in this chapter, are attractive, but they are really only suited to research studies.
- Geosynthetic-reinforced slopes are designed to provide the following three basic modes of stability: internal stability, global stability and surficial stability. The factors of safety must be adequate for both the short-term and long-term conditions, and for all possible modes of slope failure, similar to those for unreinforced slopes.
- Laboratory model tests help to understand the behaviour of geosynthetic-reinforced slopes. However, such tests on a large scale under simulating field situations can be a more effective means of understanding the realistic behaviour of geosynthetic-reinforced slopes. Centrifuge model tests are an alternative, better option for experimental studies.
- There are several slope stabilisation methods such as the geofabric-wrapped drain (GWD) method, the anchored geosynthetic system (AGS) method and several reinforced soil structure (RSS) methods, which should be selected based on the specific field situations, available funds, and the importance of the stabilised slope sites.

11.9. Some common questions and their answers

Questions

1. Normally, woven geotextiles, geogrids and some geocomposites are used in reinforcement applications. Are there any specific advantages to using non-woven geotextiles in stabilising slopes?

2. How do geosynthetic reinforcements play an effective role in stabilising soil slopes?

3. Why should the geosynthetic reinforcement be placed ideally along the major principal tensile strain direction? Shouldn't geosynthetics be used where the major principal strains are compressive?

4. Is it possible to standardise the slope stabilisation methods?

5. Is there any height limitation for stabilising a soil slope with geosynthetic reinforcements?

6. How does the role of a geosynthetic in slope stabilisation differ from its application for slope erosion control?

Answers

1. It's true that woven geotextiles, geogrids and some geocomposites are the commonly used reinforcements; however, the function of the non-woven geotextile as reinforcement is to provide adequate deformability and to redistribute the forces from areas of high stresses to areas of low stresses, thus avoiding the crushing of the soil. Further, the non-woven geotextiles facilitate better drainage and help to prevent the build-up of porewater pressures that can reduce the shear strength.

2. A soil slope has two zones: the active zone of the soil slope, where instability may occur; and the restraint zone, in which the soil will remain stable. The geosynthetic layers installed horizontally within the soil slopes maintain the integrity of the active zone, and effectively anchor this to the restraint zone in order to maintain the overall integrity of the soil slope.

3. Most geosynthetics have negligible compressive strength compared with their tensile strength. Soil is very weak in tension, so inclusion of geosynthetic reinforcement layers in the direction of the major principal tensile strain provides tensile strength to the reinforced soil mass. Geosynthetics are not installed in the direction of compressive strains.

4. Standardisation of the slope stabilisation methods is not possible due to the variety of site situations observed under field conditions, so there are variations in the stabilisation practices adopted in different countries.

5. The case histories presented in this chapter and also in other published work show that a slope of any height can be stabilised with geosynthetic reinforcements.

6. A geosynthetic installed to stabilise a slope relies on its tensile strength to prevent slope failures/movements (sliding, rotation, etc.); they usually function as reinforcement. A geosynthetic layer installed for slope erosion control protects the soil slope surface from water and wind erosion while accelerating vegetative development; thus, it mainly performs filtration and screening functions, and requires little tensile strength.

REFERENCES

Abramson LW, Lee TS, Sharma S and Boyee G (2002) *Slope Stability and Stabilization Methods*. Wiley, New York.

Ali FH (1993) Field behaviour of a geogrid-reinforced slope. *Geotextiles and Geomembranes* **12(1)**: 53–72.

Ali FH and Tee HE (1990) Reinforced slopes: field behaviour and prediction. *Proceedings of the 4th International Conference on Geotextiles, Geomembranes and Related Products*, The Hague, pp. 17–20.

Almeida MSS, Britto AM and Parry RHG (1986) Numerical modeling of a centrifuged embankment on soft clay. *Canadian Geotechnical Journal* **23**: 103–114.

Anthoine A (1989) Mixed modeling of reinforced soils within the framework of the yield design theory. *Computers and Geotechnics* **7(1–2)**: 67–82.

Askari F, Farzaneh O and Mohamadzadeh H (2010) Three dimensional stability analysis of reinforced slopes. *Proceedings of the 17th International Conference on Soil Mechanics and Geotechnical Engineering*, Alexandria, pp. 1674–1677.

Ausilio E, Conte E and Dente G (2000) Seismic stability analysis of reinforced slopes. *Soil Dynamics and Earthquake Engineering* **19**: 159–172.

Broms BB and Wong IH (1986) Stabilization of slopes in residual soils with geofabric. *Proceedings of the 3rd International Conference on Geotextiles*, Vienna, pp. 295–300.

Broms BB and Wong KS (1990) Landslides. In *Foundation Engineering Handbook* (Fang HY (ed.)), Van Nostrand Reinhold, New York.

Chalaturnyk RJ, Scott JD, Chan DHK and Richards EA (1990) Stresses and deformations in a reinforced soil slope. *Canadian Geotechnical Journal* **27**: 224–232.

Choudhary AK, Jha JN and Gill KS (2010) Laboratory investigation of bearing capacity behaviour of strip footing on reinforced fly ash slope. *Geotextiles and Geomembranes* **28**: 393–402.

Christopher BR and Holtz RD (1985) *Geotextile Engineering Manual*. Federal Highway Administration, Washington, DC. Report No. FHWA-TS-86/203.

Christopher BR and Leschinsky D (1991) Design of geosynthetically reinforced slopes. *Proceedings of the Geotechnical Engineering Congress*, Boulder, CO, pp. 988–1005.

Christopher BR, Gill SA, Giroud JP *et al.* (1990) *Reinforced Soil Structures*, vol. I. *Design and Construction*. Federal Highway Administration, Washington, DC. Report FHWA-RO-89-043.

Christopher BR, Leshchinsky D and Stulgis R (2005) Geosynthetic-reinforced soil walls and slopes: US perspective. *Proceedings of the 18th GRI Conference*, Austin, TX, GSP 141.

Das BM, Omar MT and Singh G (1996) Strip foundation on geogrid-reinforced clay. *Proceedings of the 1st European Geosynthetics Conference – Eurogeo 1*, Maastricht, pp. 419–426.

de Buhan P, Mangiavcchi R, Nova R, Pellegrini G and Salencon J (1989) Yield design of reinforced earth walls by homogenization method. *Géotechnique* **39(2)**: 189–201.

Dikran SS and Rimoldi P (1996) Hard facing for steep reinforced slopes: a case history from the UK. *Proceedings of the 1st European Geosynthetics Conference – Eurogeo 1*, Maastricht, pp. 131–136.

Dixon JH (1993) Geogrid reinforced soil repair of a slope failure in clay, North Circular Road, London, United Kingdom. In *Geosynthetics Case Histories* (Raymond GP and Giroud JP (eds)), Technical Committee TC9, International Society for Soil Mechanics and Foundation Engineering, London, pp. 168–169.

Ghiassian H, Hryciw RD and Gray DH (1996) Laboratory testing apparatus for slopes stabilized by anchored geosynthetics. *Goetechnical Testing Journal* **19(1)**: 65–73.

Hausmann MR (1990) *Engineering Principles of Ground Modification*. McGraw-Hill, Singapore.

Hu Y, Zhang G, Zhang JM and Lee CF (2010) Centrifuge modeling of geotextile-reinforced cohesive slopes. *Geotextiles and Geomembranes* **28**: 12–22.

Ingold TS (1982) An analytical study of geosynthetic reinforced embankments. *Proceedings of the 2nd*

International Conference on Geotextiles, Las Vegas, NV, pp. 683–688.

Jewell RA (1985) Material properties for the design of geotextile reinforced slopes. *Geotextiles and Geomembranes* **2(2)**: 83–109.

Jewell RA (1990) Revised design charts for steep reinforced slopes. In *Reinforced Embankments, Theory and Practice* (Shercliff DA (ed.)), Thomas Telford, London, pp. 1–30.

Jewell RA and Woods RI (1984) Simplified design charts for steep reinforced slopes. *Proceedings of the Symposium on Reinforced Soil*, University of Mississipi, Oxford, MS.

Jewell RA, Paine N and Woods RI (1984) Design methods for steep reinforced embankments. *Proceedings of the Symposium of Polymer Grid Reinforcement*, Institute of Civil Engineering, London, pp. 18–30.

Jiang GL and Magnan JP (1997) Stability analysis of embankments: comparison of limit analysis with method of slices. *Géotechnique* **47(4)**: 857–872.

Jones CJFP (2005) Geosynthetic-reinforced soil walls and slopes: European perspective. *Proceedings of the 18th GRI Conference*, Austin, TX, GSP 141.

Koerner RM (1984) In-situ soil stabilization using anchored nets. *Proceedings of the Conference on Low Cost and Energy Saving Construction Methods*, Rio de Janeiro, pp. 465–478.

Koerner RM (2005) *Designing with Geosynthetics*, 5th edn. Prentice Hall, Upper Saddle River, NJ.

Koerner RM and Robins JC (1986) *In-situ* stabilization of soil slopes using nailed geosynthetics. *Proceedings of the 3rd International Conference on Geotextiles*, Vienna, pp. 395–400.

Leshchinsky D (1997) Software to facilitate design of geosynthetic-reinforced steep slopes. *Geotechnical Fabrics Report*, Jan/Feb: 40–46.

Leshchinsky D and Boedeker RH (1989) Geosynthetic reinforced soil structures. *Journal of Geotechnical Engineering* **115(10)**: 1459–1478.

Leshchinsky D and Reinschmidt AJ (1985) Stability of membrane reinforced slopes. *Journal of Geotechnical Engineering* **111(11)**: 1285–1300.

Leshchinsky D and Volk JC (1985) Stability charts for geotextile reinforced walls. *Transportation Research Report* **1131**: 5–16.

Leshchinsky D and Volk JC (1986) Predictive equation for the stability of geostextile reinforced earth structure. *Proceedings of the 3rd International Conference on Geotextiles*, Vienna, pp. 383–388.

Mandal JN and Labhane L (1992) A procedure for the design and analysis of geosynthetic reinforced soil slopes. *Geotechnical and Geological Engineering* **10(4)**: 291–319.

Michalowski RL (1997) Stability of uniformly reinforced slopes. *Journal of Geotechnical and Geoenvironmental Engineering* **123(6)**: 546–556.

Michalowski RL and Zhao A (1993) Failure criteria for homogenized reinforced soils and application in limit analysis of slopes. *Proceedings of Geosynthetics '93*, Vancouver, pp. 443–453.

Michalowski RL and Zhao A (1994) The effect of reinforcement length and distribution on safety of slopes. *Proceedings of the 5th International Conference on Geotextiles, Geomembranes and Related Products*, Singapore, pp. 495–498.

Michalowski RL and Zhao A (1995) Continuum versus structural approach to stability of reinforced soil structures. *Journal of Geotechnical Engineering Division, ASCE* **121**: 152–162.

Mittal S, Shah MY and Verma NK (2009) Experimental study of footings on reinforced earth slopes. *International Journal of Geotechnical Engineering* **3(2)**: 251–260.

Murray R (1982) Fabric reinforcement of embankments and cuttings. *Proceedings of the 2nd International Conference on Geotextiles*, Las Vegas, NV, pp. 707–713.

Otani J and Ochiai H (2005) Geosynthetic-reinforced soil walls and slopes: Japanese perspective. *Proceedings of the 18th GRI Conference*, Austin, TX, GSP 141.

Poh SH and Broms BB (1995) Slope stabilization using old rubber tires and geotextiles. *Journal of Performance of Constructed Facilities* **9(1)**: 76–79.

Porbaha A and Kobayashi M (1997) Finite element analysis of centrifuge model tests. *Proceedings of the 6th International Symposium on Numerical Models in Geomechanics – NUMOG VI*, Montreal, pp. 257–262.

Porbaha A and Lesniewska D (1999) Stability analysis of reinforced soil structures using rigid plastic approach. *Proceedings of the 11th Pan American Conference on Soil Mechanics and Geotechnical Engineering*, Iguassu Falls.

Porbaha A, Zhao A, Kobayashi M and Ishida T (2000) Upper bound estimate of scaled reinforced soil retaining walls. *Geotextiles and Geomembranes* **18**: 403–413.

Porri PA, Rivelli FR and Benedetti DT (2010) Slope stabilization with woven geogrids in San Bernardo Hill. *Proceedings of the 9th International Conference on Geosynthetics*, vol. 4, Guaruja, pp. 1777–1780.

Raymond GP and Giroud JP (1993) *Geosynthetics Case Histories*. ISSMFE Technical Committee TC9, International Society for Soil Mechanics and Foundation Engineering, London.

Resl S (1990) Soil-reinforcing mechanisms of nonwoven geotextiles. *Proceedings of the 4th International Conference on Geotextiles, Geomembranes and Related Products*, The Hague, pp. 93–96.

Rimoldi P and Jaecklin F (1996) Green faced reinforced soil walls and steep slopes: the state-of-the-art in Europe. *Proceedings of the 1st European Geosynthetics Conference – Eurogeo 1*, Maastricht, pp. 361–380.

Rowe RK and Soderman KL (1985) An approximate method for estimating the stability of geotextile reinforced embankments. *Canadian Geotechnical Journal* **22(3)**: 392–398.

Sawicki A and Lesniewska D (1989) Limit analysis of cohesive slopes reinforced with geotextiles. *Computers and Geotechnics* **7**: 53–66.

Schmertmann GR, Chourery-Curtis VE, Johnson RD and Bonaparte R (1987) Design charts for geogrid reinforced soil slopes. *Proceedings of Geosynthetics '87*, New Orleans, LA, pp. 108–120.

Selvadurai APS and Gnanendran CT (1989) An experimental study of a footing located on a sloped fill: influence of a soil reinforcement layer. *Canadian Geotechnical Journal* **26**: 467–473.

Shukla SK (1997) A study on causes of landslides in Arunachal Pradesh. *Proceedings of the Indian Geotechnical Conference*, Vadodara, pp. 613–616.

Shukla SK and Yin JH (2006) *Fundamentals of Geosynthetic Engineering*. Taylor and Francis, London.

Shukla SK, Sivakugan N and Das BM (2011) A state-of-the-art review of geosynthetic-reinforced slopes. *International Journal of Geotechnical Engineering* **5(1)**. 17–32.

Simac MR (1992) Reinforced slopes: a proven geotechnical innovation. *Geotechnical Fabrics Report*, May/Jun: 13–25.

Srbulov M (2001) Analyses of stability of geogrid reinforced steep slopes and retaining walls. *Computers and Geotechnics* **28(4)**: 255–268.

Vidal H (1966) La terre armée. *Annales de L'Institut Technique du Bâtiment et des Travaux Publics* **19(223–234)**: 888–939.

Vidal H (1969) The principle of reinforced earth. *Highway Research Record* **282**: 1–16.

Viswanadham BVS and Mahajan RR (2007) Centrifuge model tests on geotextile-reinforced slopes. *Geosynthetics International* **14(6)**: 365–379.

Wright SG and Duncan JM (1991) Limit equilibrium stability analyses for reinforced slopes. *Transportation Research Report* **1330**: 40–46.

Yoo C (2001) Laboratory investigation of bearing capacity behaviour of strip footing on geogrid-reinforced sand slope. *Geotextiles and Geomembranes* **19(5)**: 279–298.

Zhao A (1996) Limit analysis of geosynthetic-reinforced slopes. *Geosynthetics International* **3(6)**: 721–740.

Zornberg JG and Arriaga F (2003) Strain distribution within geosynthetic-reinforced slopes. *Journal of Geotechnical and Geoenvironmental Engineering* **129(1)**: 32–45.

Zornberg JG, Sitar N and Mitchell JK (1998a) Performance of geosynthetic-reinforced slopes at failure. *Journal of Geotechnical and Geoenvironmental Engineering* **124(8)**: 670–683.

Zornberg JG, Sitar N and Mitchell JK (1998b) Limit equilibrium as basis for design of geosynthetic reinforced slopes. *Journal of Geotechnical and Geoenvironmental Engineering* **124(8)**: 684–698.

Handbook of Geosynthetic Engineering
ISBN 978-0-7277-4175-2

http://dx.doi.org/10.1680/hgce.41752.245

Chapter 12
Landfills

H. Zanzinger*, E. Gartung† and W. P. Hornsey‡
*SKZ – TeConA GmbH, Wuerzburg, Germany
†LGA, Geotechnical Institute, Nuremberg, Germany
‡Geofabrics Australasia Pty Ltd, Gold Coast, Australia

Geosynthetics form an integral part of any modern landfill in its base lining and capping systems. As the awareness of the potential environmental hazards of traditional unlined 'rubbish dumps' has grown, the need for engineered containment systems has increased. Today, most, if not all, engineered landfill liner systems incorporate one or more layers of geosynthetics in order to optimise the performance of the liner system and limit the potential impact on the surrounding environment. This chapter introduces the concept of engineered waste containment and discusses the various geosynthetic materials used in the construction of the liner systems.

12.1. Introduction

Solid residues from production processes, as well as from daily life, are reused or recycled wherever possible, in order to reduce the amount of waste that has to be treated and finally disposed of. During the last few years, in some countries, legal, economical and educational efforts have led to a significant reduction in the generation of waste. So, for example, in Germany, the predictions of the landfill space required for the coming decades are currently being revised. In some cases, the design of landfill facilities is delayed or can even be given up completely due to the decreasing amount of refuse. In spite of this development, there is still, and will be, a great demand for solid-waste landfills in most parts of the world. The design and construction of landfills remains a major challenge to civil engineers.

Waste material may contain substances that can be harmful to the environment. It is therefore mandatory to handle and store waste in such a way that any contamination of the ground, as well as of the groundwater, is prevented. The primary engineering outcome in designing, constructing and operating solid-waste landfills is to provide efficient barriers against contamination. As water is the most important transporting agent for pollutants, the infiltration of water into, and the extraction of water out of, the solid-waste body must be controlled by reliable engineered methods. Landfill liners and caps are the most significant engineered elements of landfill structures. The basal and cap liner systems are crucial elements with respect to landfill integrity and stability.

Most lining and capping systems consist of a hydraulic barrier of some form – compacted natural clay, a geosynthetic clay liner, a geomembrane or a combination of these. In order to optimise the performance of the liner and cover system, a drainage layer (e.g. gravel, sand or geonet) should be incorporated in the design. The drainage layer effectively reduces the head on the liner, and ultimately reduces the contaminant movement through the liner system. There is a close relationship between the sealing and drainage elements of the basal and capping liners. Drainage systems must maintain minimum gradients to facilitate gravitational flow. Therefore, to some extent, the dewatering systems dictate the geometry of the surfaces of the sealing layers. Another important consideration is that leachate collection pipes should be placed in such a way that unavoidable penetrations through the sealing layers do not reduce the efficiency of the liners. The sealing and drainage elements form integral parts of barrier systems and have to be designed accordingly, as they influence each other during, and after, construction. Obviously, the placement of drainage gravel above geomembranes must be executed with the greatest care to avoid perforating the liner. In looking at the close interrelationship between the sealing and dewatering elements, this chapter deals with liners and covers, together with some aspects of dewatering systems.

The physical and biological properties of the solid waste, as well as the availability of construction material, are important parameters in the design of liners and covers. The geological, hydrological and climatic conditions at the landfill site are also major factors. There are various possibilities for the construction of a safe technical barrier for solid-waste landfills. As the interactions between the waste material, the natural climatic influences and the liners or covers are complex, the performance of landfill liners and covers cannot be quantified appropriately using simple analytical formulae. During the last few years, considerable progress has been achieved in scientific research, but to date there are no encompassing rational computational design models for landfill liners and covers. Many geotechnical questions related to the performance of landfills are still unanswered (Van Impe, 1995).

However, landfill construction cannot be delayed until all the relevant scientific problems have been solved. Landfills are needed now and have to be built. Therefore, the current design fundamentals vary from country to country, and in many cases between individual states and regions within countries, depending on the regulatory body responsible for the landfill. The design fundamentals are established on the basis of a combination of experience, engineering judgement and analytical procedures.

With the aim of achieving a suitable degree of safety with respect to the environment, nationally or regionally responsible authorities have issued minimum requirements and some basic rules for the design of landfill liner and capping systems. As stated previously, these rules differ from one country to another, and sometimes even within one country.

12.2. Multi-barrier concept

Traditionally, rubbish dumps had inadequate or no liner systems to prevent the migration of contaminants, which proved unsatisfactory from an environmental perspective. The modern landfill structure essentially consists of a large containment, with the solid-waste body inside (Figure 12.1). The migration of harmful substances is prevented by several barriers.

The geology of the site is important in this respect. The ground should have a low hydraulic conductivity and a high capacity for the adsorption of toxic material, such as heavy metals, etc. It must be sufficiently stable and should not undergo excessive settlements under the load of the landfill body, as settlement may overstress the liner and thereby limit its effectiveness.

The base and side liners should be highly engineered barriers designed to cut off the migration path of the contaminants, and these engineered structural components must be installed under a strict quality-control regimen.

After termination of landfill operations, a capping system must be installed. This covers the waste body, prevents the ingress of surface water and the emission of gas, odours and dust from the waste, and it facilitates landscaping.

Special attention is paid to the properties and the placement of the waste material. The waste body is considered a barrier by itself. In Germany, the refuse must be in such a condition that the stability of the landfill is ensured, and there is little or no tendency for harmful material to be dissolved and transported with seeping water. In addition, the deformation due to settlements should be predictable and small. Therefore, the integrity of the cover would not be impaired in the long term. However, this is not the case in many other countries, where co-disposal methodologies are used; this process allows the addition of liquors, and recycling of leachate is employed to speed up the biological degradation of the waste material. This methodology has the potential to reduce the stability and increase the settlement of the waste body; these factors should be considered during the design process.

In summary, the landfill structure forms a multi-barrier system (Stief, 1986). Each of the barriers has to meet certain technical minimum requirements, independent of the performance of other barriers.

12.3. Landfill categories

In agreement with the European Landfill Directive (European Community, 1999), in Germany, three categories of solid-waste landfill are distinguished with respect to the deposited waste material (Technische Anleitung Siedlungsabfall (TASi), 1993). The chemical composition of the constituents of the waste is the governing criterion for the assignment of the landfill category. The delivered waste material is analysed by means of tests at the entrance of the landfill site to make sure that the acceptance criteria are met. The control of the waste material is quite efficient, and so the contents of the landfills that are currently in operation are known. However, more landfills were placed in the past than at present, and there is now the problem

Figure 12.1 The barriers of landfills

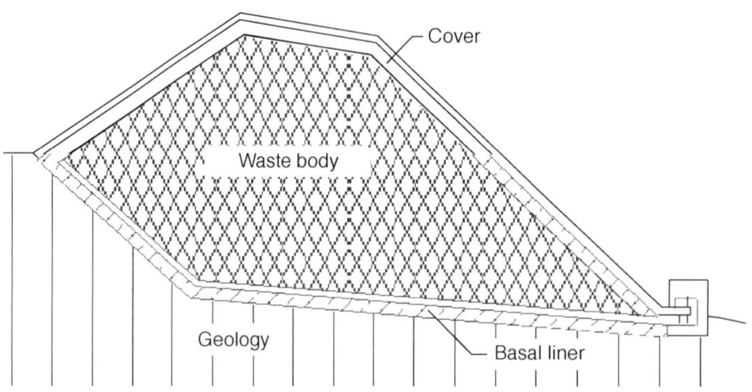

of old landfills that do not comply with the current German technical standards. Their environmental impact has to be evaluated and, in many cases, improvements by engineering/technical means are necessary (e.g. they have to be provided with covers).

Inert refuse with the lowest potential for containing harmful substances, such as mineral waste and construction material or demolition debris, are assigned to landfill category I, according to the German regulations. Except for the general requirements of sufficient bearing capacity and predictable, and not excessive, settlement, the geological conditions at a category I site do not have to meet any technical minimum standards. The German regulations recommend compacted clay liners at the bottom, as well as at the cap, and drainage systems. Geosynthetic clay liners and drainage geocomposites are frequently used as economical alternatives in the design of category I landfills (Gartung and Zanzinger, 1998; Zanzinger, 2000). Following the philosophy that the waste body is itself an important barrier against the contamination of the environment, strict criteria have to be met by the solid waste that is assigned to category I landfills.

The assignment criteria for the waste material are also selected with respect to the performance of the landfill structure. Bodies of waste that do not contain degradable material will not exhibit major deformations in the long term, and the mechanical properties of the refuse can be determined according to the usual soil mechanics practice. The design rules of the standard landfill are based on this type of waste material. However, at the present time, many municipal solid-waste landfills are depositing residues that do not meet the maximum organic carbon content criterion. They still contain a lot of organic material that will undergo degradation processes for a long time. This aspect is important for the capping systems. The cap either has to be sufficiently flexible to tolerate large deformations without losing its integrity, or it should not be put in place until the major deformations associated with the degradation processes have ceased.

Landfill category II comprises the majority of solid-waste landfills, which contain residues from incinerators, as well as typical municipal waste and similar materials, which may contain dangerous substances. Landfills in this category are provided with a liner at the bottom and a cover at the top, these consisting of geomembranes and compacted clay liners.

Waste that contains harmful substances exceeding the criteria for municipal waste landfills has to be disposed of in hazardous-waste landfills. Those are constructed according to the same basic principles as category II landfills. However, the thickness of the mineral component of the basal composite liner must be twice the thickness of liners in municipal-waste landfills, and a few other details differ slightly. In the following section, mainly liners and covers of category II landfills are discussed.

Some countries, such as South Africa, have set up national standards, which provide minimum requirements for landfill liner systems based not on only the type of waste disposed, but also on the climatic conditions associated with the site. In South Africa, where the climatic conditions vary from semi-arid to subtropical, the changes in rainfall intensity (ranging from 50 to 3000 mm/year) and groundwater conditions are acknowledged to have an effect on the design (DWAF, 1998). Table 12.1 shows the general classifications, and Figure 12.2 shows the minimum liner requirements for low-hazard waste in wet or alternately dry regions, and hazardous and co-disposal sites.

12.4. Basal lining systems

12.4.1 Functional layers

The landfill containment is sealed at the bottom by a basal lining system composed of several layers, each one serving a particular purpose. The example shown in Figure 12.3 comprises a seal, a protector and a drainage blanket.

In order to provide a continuous system of low permeability, the seal is placed directly above the subsoil without a drainage layer in between. The main seal may consist of a single or of a double liner, and the liner itself could consist of an impervious mono-layer or of a composite. The example in Figure 12.3 shows a composite liner, consisting of a compacted clay liner and a geomembrane.

As the geomembrane is rather thin and is sensitive to mechanical damage, a protective layer is needed above it. This layer can be a geosynthetic product, a soil, or a composite of both.

In order to prevent any build up of leachate head above the sealing layers, a drainage blanket is incorporated in the basal lining system. The build up of leachate above the liner creates two problems for the liner system, namely

- increased hydraulic pressure on the liner system, forcing leachate and contaminants through the liner at a faster rate
- increased temperature at the liner, which will increase the rate of degradation of the geomembrane and cause the clay liner to shrink, thus increasing the permeability of the system.

Finally, it may be necessary to place a transition or filter between the drainage blanket and the waste body to prevent clogging, and to maintain the long-term performance of the drainage system.

12.4.2 Concept of the composite liner

Extensive research by August *et al.* (1992) during the 1980s has led to the conclusion that a composite liner of the type shown in Figure 12.3 is the most efficient seal against the migration of harmful components of the leachate. Accordingly, the German regulations require such a composite liner at the bottom of municipal solid-waste landfills and hazardous-waste landfills as a standard.

Table 12.1 South African landfill characterisation
Data from DWAF (1998)

Minimum requirements	Classification									
	G General waste								H Hazardous waste	
	C Communal landfill		S Small landfill		M Medium landfill		L Large landfill		H:h Hazard rating 3 and 4	H:H Hazard rating 1–4
	B^-	B^+	B^-	B^+	B^-	B^+	B^-	B^+		
Apply as appropriate	R	R	R	R	R	R	R	R	R	R
Classify proposed site	R	R	R	R	R	R	R	R	R	R
Notify IAPs of the necessity for and intention to develop a landfill	R	R	R	R	R	R	R	R	R	R
Liaise with IAPs	R	R	R	R	R	R	R	R	R	R
Eliminate areas with fatal flaws	R	R	R	R	R	R	R	R	R	R
Identify candidate landfill sites	R	R	R	R	R	R	R	R	R	R
Buffer zone (m)	200	200	400	400	F	F	F	F	F	F
Minimum unsaturated zone (m)	2	2	2	F	F	F	F	F	F	F
Rank sites as indicated	F	F	R	R	R	R	R	R	R	R
Present ranked sites to IAPs	F	F	R	R	R	R	R	R	R	R
Site feasibility study	F	F	R	R	R	R	R	R	R	R
Site description	R	R	R	R	R	R	R	R	R	R
Complete permit application form	R	R	R	R	R	R	R	R	R	R

B^-, no significant leachate produced; B^+, significant leachate produced; F, special consideration to be given by the expert or departmental representative; IAP, Integrated and Affected Parties; N, not required; R, required; H:h, hazardous waste landfill that can receive waste with a hazard rating of 3 and 4 (low to moderate hazard); H:H, hazardous waste landfill that can receive waste with a hazard rating of 1 and 2 (high to extreme hazard).

The polymer component acts as a cut-off for the flow of liquid. Due to its non-polar molecular structure, it prevents the diffusion of polar substances and, therefore, is an absolute barrier against heavy metal cations. Non-polar molecules of hydrocarbons or chlorinated hydrocarbons that may permeate through the geomembrane are retarded at the surface of the compacted clay liner due to its strongly polar molecular structure. The effect is a decrease in the concentration gradient across the geomembrane and, consequently, a reduction in the rate of permeation. Therefore, it is primarily the interface between the geomembrane and the clay liner that allows the system to act as an efficient barrier against the movement of contaminants, such as hydrocarbons, provided both components of the sealing system are in intimate contact.

In order to achieve the excellent performance of the composite liner that has been demonstrated in small-scale laboratory testing, in real-world applications, high-quality standards have to be met by the properties of the geomembrane, by the material used for the mineral sealing layer and by the workmanship during construction. Manufacturing quality assurance (MQA) and construction quality assurance (CQA) are of the utmost importance in the construction of composite liners.

Traditionally, engineers have assessed the suitability of the clay used in both compacted clay liners (CCLs) and geosynthetic clay liners (GCLs) on the basis of its permeability in deionised and deaired water. However, this is a dangerous basis to work on, particularly when dealing with hazardous waste. Recent work on the compatibility of various GCLs, incorporating sodium bentonite of different physical and chemical composition, showed a marked difference in performance when permeated with high pH liquor (Benson *et al.*, 2008). The performance of the clay component of a liner (both CCLs and GCLs) should, therefore, be assessed on the basis of the chemical compatibility of the clay with the liquor it is intended to contain.

The composite liner described here has an excellent performance as a barrier against contaminant emission. But, like any other system, it has also certain limitations (Gartung, 1992), mainly with respect to its mechanical behaviour. For example, the frictional resistance at the surface of smooth geomembranes is limited, which may result in slope stability problems.

12.4.3 Alternative liners

In situations where it is not easy, or impossible, to construct the composite liner (e.g. at very steep slopes), or where the waste material undergoes substantial exothermic processes and the temperature development would be detrimental to polymer geosynthetics, or if other limits of the system are approached, alternative bottom liners can be used. Alternatives should be

Figure 12.2 Minimum base liner configuration: (a) general waste, large landfill in a dry region, (b) general waste, large landfill in wet region and (c) hazardous waste, rating 1–4
Reproduced from DWAF (1998)

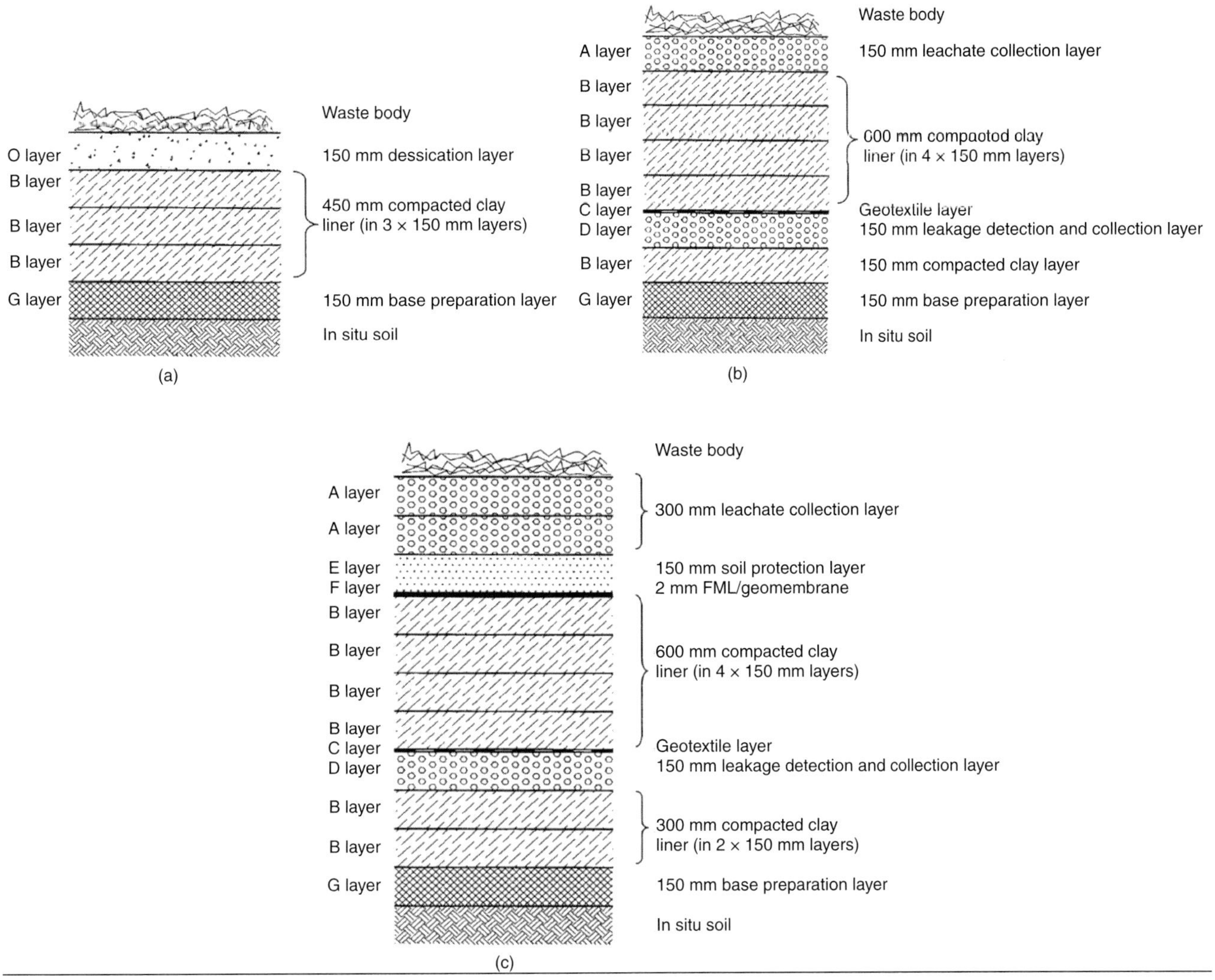

equivalent to the standard systems in their performance. In order to achieve the composite effect described in the previous section, alternative sealing systems should also consist of at least two elements: one that retains polar and another one that retains non-polar potential contaminants.

In Germany, before the development of the composite lining technique, most basal liners were constructed as a single CCL. A recent, very comprehensive investigation into the properties of a 12–15-year-old CCL below a municipal-waste landfill, with a functioning drainage system, led to the conclusion that the mineral liner had performed very satisfactorily. Extensive testing of samples across the entire thickness of the liner, applying soil mechanics, geochemical, mineralogical and microbiological methods, revealed that there were no traces of contaminant migration (Gartung *et al.*, 1996).

For permanent landfill structures, basal liners consisting of geomembranes without supplementary mineral liners were seldom employed in Europe. Single-sheet geomembrane liners without mineral sealing layers are used for the temporary storage of waste. For permanent structures, they are normally considered inappropriate in view of their sensitivity to mechanical damage and lack of redundancy with respect to their long-term performance.

Multilayer sealing systems comprising a CCL and more than one geomembrane are not very common in Europe. It is difficult to install a high-quality, well-compacted clay layer above a geomembrane. Thus the use of an additional geomembrane in a basal liner system does not necessarily lead to a substantial increase in safety against leachate migration.

In the USA, double composite liners are favoured. Below the primary geomembrane, and in combination with a GCL or a CCL, there is a geosynthetic leak-detection layer and then the secondary geomembrane, again as part of a composite liner including a CCL (Koerner, 1993). In Europe, the

Figure 12.3 The basal lining system for a municipal-waste landfill

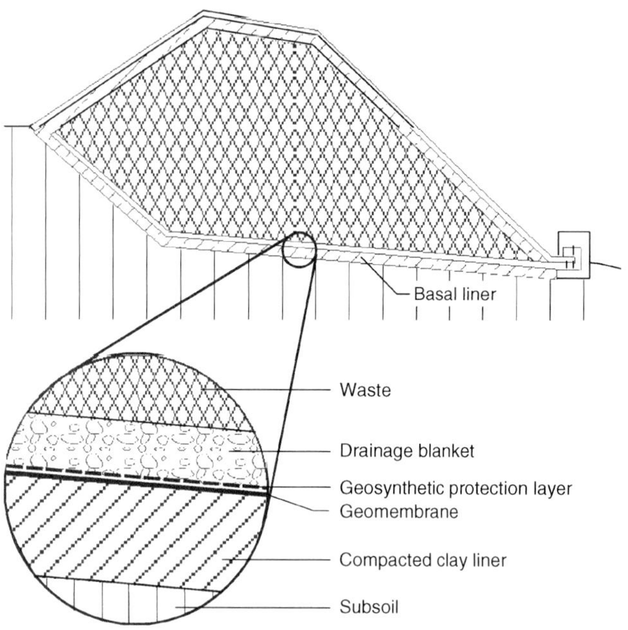

philosophy of leak detection at the base of waste deposits has been followed in only a few special cases, and the design concept of double-liner systems is not very well established. Some mineral double liners, with either gravel or geosynthetic leak-detection blankets, have been constructed, but at present alternative landfill bottom liners play only a minor role in central Europe. The composite liner consisting of a geomembrane and a mineral layer of low hydraulic conductivity, as shown in Figure 12.3, is the most common standard solution.

12.5. Components of the composite liner

12.5.1 Compacted clay liner

The characteristic property of the mineral sealing layer is its low hydraulic conductivity, and in Germany this is required to be less than 5×10^{-10} m/s. The soil must contain at least 20% clay-sized particles, half of which must be clay minerals. The clay must be compacted when wet, with an optimum water content of at least 95% Proctor density. The suitability of the selected soil must be proven in laboratory tests. In test fields, the contractor has to demonstrate his proficiency and the adequacy of his equipment prior to the start of construction.

The required minimum thickness of the CCL for composite liners is 0.5 m in Switzerland, and 0.75 m for category II landfills in Germany. For hazardous-waste landfills, the German instructions specify a minimum thickness of 1.5 m, placed in lifts of 0.25 m each.

As, in some parts of Europe, it is difficult to obtain enough natural clay to meet the specifications for a certified CCL, alternative mineral seals have been developed. These are based on blending granular soil with a small amount of bentonite. Horn (1989) called such mineral seals 'Bentokies', and they have been used to a great extent in landfill construction in southern Bavaria. Other blended mineral sealing materials are the patented DYWIDAG Mineralgemisch, developed by Finsterwalder and Mann (1990), and Chemoton, a mixed-in plant sealing material that uses waterglass and chemicals to obtain extremely low permeability and a very high resistance to aggressive chemicals (Lauf and Müllner, 1993). The components of such mineral sealing materials are mixed in plant, transported to the site, and placed and compacted with modern construction equipment. Often their performance is superior to that of sealing layers composed of compacted natural clays, because the components are specially selected, processed and blended under controlled working conditions.

12.5.2 Geosynthetic clay liner

GCLs were first developed in the 1980s in Germany as an alternative to CCLs. Most GCLs consist of a thin layer of sodium bentonite clay secured between two geotextiles, the primary function of which is to limit the migration of liquids. The geotextile component is essentially the packaging that allows the placement of a uniform layer of processed bentonite (generally ~10 mm thick) to serve as the barrier. While the geotextile component does add some benefit in terms of improving the shear strength of the clay when a needle-punched GCL is used, it remains of secondary importance when considering a GCL as a barrier (Gates *et al.*, 2009). Due to the relatively pure clay (>70% montmorillonite content) used in GCLs, maximum hydraulic conductivities of 3×10^{-11} to 5×10^{-11} m/s are the industry standard.

GCLs are used to either replace or supplement poor-quality CCLs, the advantages for designers, installers and facility owners being as follows

- factory production provides for a consistent, uniform, high-quality liner
- needle punching allows liner application on steeper slopes
- simple installation allows for rapid construction
- the liner has a high deformation capability
- the reduced thickness increases landfill storage space.

The clay component of a GCL is the basis of its performance as a liner, and a minimum clay quality should be specified when using GCLs in containment applications. Table 12.2 provides a guide to the key criteria that should be met by the clay component of the GCL, and these criteria should be incorporated in any GCL specification. This will ensure that a high-quality clay is used. However, when used in conjunction with aggressive leachates (i.e. pH extremes), project-specific compatibility testing is recommended.

GCLs should be installed in such a way as to limit the following

- repetitive wetting and drying cycles
- cation exchange
- root penetration.

Table 12.2 The key criteria that should be met by the clay component of a GCL
Data from Gates *et al.* (2009)

Property	Range or value
Montmorillonite content	>70%
Octahedral magnesium content	0.4–0.8 atoms per unit cell 2–4% by weight
Octahedral iron content	0.36–0.6 atoms per unit cell 3–5% by weight
Layer charge	<0.85 e^- per unit cell
Layer charge location	Predominantly (>50%) octahedral sheet
Cation exchange capacity	<110 meq per 100 g (cmol per kg)
Exchange cation	Sodium (Wyoming type)

12.5.3 Geomembrane

The function of the geomembrane in the basal liner system of a solid-waste landfill is to retain leachate, a liquid that may be composed of many different substances, some of which can be harmful. In most cases, the composition of the leachate cannot be predicted with a sufficient certainty, and it may vary with time. The only means of assessing the properties of a leachate is chemical analyses carried out on a great number of samples taken at many different landfill sites.

Geomembranes used for landfill liners should be impervious to all the components found in the leachate and also to those that might occur. The geomembrane should also be able to resist any chemical or biological attack in the landfill milieu, without losing its functional properties. Furthermore, geomembranes must be strong enough mechanically to survive transport, handling, placement and subsequent construction activities. The deformation behaviour of the geomembrane must lie within acceptable limits, be compatible with the deformations of the other components of the landfill structure, and be predictable. In addition, the strength and the interface frictional resistance must be able to meet the stability requirements of the landfill structure.

In summary, the physical, mechanical and endurance properties of the geomembrane have to be such that the membrane meets a number of requirements. Koerner (1998) lists 20 test methods for the determination of the parameters that describe the relevant properties of geomembranes. Some standards for determining the parameters have been mentioned in Section 1.8 of Chapter 1. All important material parameters must be specified in order to ensure that the geomembrane is suitable as a landfill liner.

There is a wide variety of materials and processing techniques for the production of geomembranes, as discussed in Section 1.5 of Chapter 1. Manufacturers are capable of meeting specific requirements in different ways, focusing on the subjects that seem most important in a particular case. Optimising plastics to meet one criterion (e.g. chemical durability) can lead to deficiencies in other properties (e.g. the deformation characteristics). If all possible variations could be exercised in practice, there would be countless types of different geomembranes, and the designer, the contractor and the regulator would hardly be able to evaluate the suitability of a product for a particular application unless all the parameters could be tested each time. This would be prohibitive in terms of cost and delays. The practising civil engineer would need the help of a polymer expert if all the characteristic material parameters of a geomembrane had to be selected for each specific project. For a successful application of geomembranes in landfill construction, it is more appropriate to limit the variation of geomembrane properties by standardisation.

Most European countries have requirements for landfill geomembranes. In Germany, an approval system has been laid down by legal action. The Federal instructions (TA-Abfall, 1991; TASi, 1993) specify that only approved geomembranes shall be used in landfill construction. The approval criteria were agreed upon by experts representing the geomembrane manufacturers, the testing and research institutions, the designers, and the regulators.

The procedure for the approval of geomembranes for landfill applications is executed by the Federal Institution for Material Research and Testing (Bundesanstalt für Materialforschung und-prüfung (BAM), 1992). On the basis of more than 20 years of experience with polyethylene in civil engineering applications and comparative testing of different geomembrane materials in BAM's laboratories for polymers (August *et al.*, 1984), it has been decided that only polyethylene should be used in the sealing systems of landfills.

The geomembranes approved in Germany exhibit excellent chemical resistance and sealing performance against a wide variety of substances that could be encountered in the leachate of municipal- or hazardous-waste landfills. Polyethylene, with a density between 0.932 and 0.942 g/cm^3, commonly called high-density polyethylene (HDPE), is used. It must have a carbon black content of 1.8–2.6%, and meet a number of strict requirements with respect to its physical and chemical properties.

By using suitability tests, it has to be demonstrated that the geomembranes meet certain general physical requirements, specific physical requirements, requirements under combined physical–chemical action, and chemical and biological requirements. Altogether, 22 tests have to be performed on the geomembrane, including some rather time-consuming long-term examinations. The results of the suitability tests are documented and used as reference parameters in the manufacturing quality control. The allowable deviations from the reference values are very small. Corbet and Peters (1996) have laid out the German and US experience on geomembranes.

In order to warrant a sufficient robustness of the geomembrane in handling, the specified minimum thickness of approved geomembranes is 2.5 mm. This thickness also happens to be very satisfactory with respect to the sealing function. However, HDPE geomembranes of 2.5 mm thickness are not very flexible.

The minimum width of the geomembrane roll is 5 m, in order to minimise the amount of field seaming needed to create large waterproof sheets. The size and weight of the geomembrane rolls are specified to ensure that they can be transported to the site and placed without major handling problems.

The approval documents also contain specifications on the quality assurance system of the geomembrane manufacturer. The manufacturer must follow the quality assurance procedures. In addition, the production is supervised by an external inspector, and some of the most critical material parameters are checked for conformance. Emphasis is placed on the homogeneity of the product with respect to the carbon black content, carbon black distribution, geometry, thickness, straight edges, surface properties and permeability.

Due to the approval system, the variability of the geomembranes used in the German landfill construction is limited, and the design engineer can rely on the properties declared in the product documents.

12.5.4 Protective layer for the geomembrane

The basal lining system includes a drainage blanket above the geomembrane liner. The blanket consists of very coarse gravel or crushed rock, typically of grain diameter 16–32 mm, but it can be as large as 75 mm, depending on the clogging potential of the leachate and rock source. Positioned below the waste body, which is tens of metres thick, and moving construction equipment, the coarse grains exert considerable point loads on to the basal sealing layers. In order to avoid perforations of the geomembrane, some form of puncture protection is needed.

There is no international agreement on the allowable strain or damage to geomembranes, and therefore the type or extent of protection required has not been standardised. There are two extremes in terms of design philosophy

- providing installation damage protection only – possibly an under-design
- limiting strains to 0.25% – possibly over-conservative.

Comparative testing in different German geosynthetic laboratories, and subsequent discussions by experts, led to preliminary criteria for geosynthetic protective layers (GDA, 1997). A loading test was developed, to which the geomembrane, covered by the protective layer and the drainage material, has to be submitted (Gallagher *et al.*, 1999). After 1000 hours under the specified load, the geomembrane is inspected visually. No scratches, grooves, indentations or holes are tolerated at the surface of the geomembrane. A metal sheet with plastic deformation properties placed below the geomembrane records the deformations of the membrane material. The maximum allowable local strain at the lower surface of an HDPE geomembrane after the test is 0.25%. This value was derived from the maximum allowable design strain for the geomembrane of 3% due to settlements or other influences, to which the value of local strain of 0.25% is superimposed. The 3% strain limit is imposed to eliminate long-term stress cracking of the HDPE; if geomembranes made from other materials (e.g. polypropylene (PP), low-density polyethylene (LDPE)) are used, then greater strains may be allowable.

Due to these strict mechanical performance requirements, protective layers in German landfills consist of heavy geosynthetic products or geotextiles in combination with mineral material. A typical example is a needle-punched non-woven HDPE of 1200 g/m plus 100–150 mm of sand or crushed stone of grain size 0–8 mm.

There is some concern about the durability of geosynthetic protectors. If the same criteria apply for the durability of protecting geosynthetic layers as for the geomembranes themselves, it becomes necessary to combine mineral and polymer components in protective layers, because most of the fine fibres of polymer products do not pass the severe incubation tests with highly oxidising chemicals.

German geosynthetics manufacturers have developed special composites of geotextiles and sand to serve as protective layers above geomembranes at the base of landfills. If the geotextile components of these composites degrade under the influence of leachate in the long run, the mineral components are left and still perform their function. Saathoff and Sehrbrock (1994) have reported a method of filling the voids of a drainage geocomposite with sand at the site. Kirschner and Kreit (1994) described geotextile mats that are filled on site with a sand bentonite slurry, and Müller-Rochholz and Asser (1994) reported on flat sand-filled cushions, manufactured industrially and placed by hand at the site. The system described by Müller-Rochholz and Asser has been developed further, and now sand-filled mats are available that are transported to the site on rolls and installed very conveniently by unrolling.

The evaluation of the performance of protective layers for geomembranes has been a topic of scientific research. Brummermann (1997) suggested new testing methods and new equipment, leading to more consistent results than the technique presently used. For American landfills, Wilson-Fahmy *et al.* (1995) developed a design concept for puncture protection based primarily on theoretical studies and laboratory tests.

Regarding the performance of protectors, there is little or no experience from the field in Europe. In large-scale model tests, Zanzinger (1999) examined some German protectors by exposure to a uniformly distributed load of 800 kPa for more than 1000 hours. He observed considerably greater deformations of the geomembrane in the large-scale tests than in his previous standard tests. According to this, the

German approach may be less conservative than it appears at first glance. Most likely, under actual field conditions, the geomembranes at the base of large landfills will experience strains in excess of 0.25%, in spite of the heavy protecting layers.

The field of liner protection and allowable geomembrane deformation requires a great deal of research before a consensus can be achieved on the correct design methodology to adopt.

12.5.5 Drainage systems

Drainage blankets are incorporated in all landfill liner systems. A drainage blanket allows the leachate generated within the landfill to be collected, treated and disposed of. No landfill liner system should be constructed without a drainage layer forming part of the system. The drainage blanket prevents the build-up of leachate above the liner system, and thus reduces the temperature on the liner and the amount of leachate that will pass through the liner system. Both these issues will affect the longevity of the liner system and the contamination potential of the site.

Traditionally, drainage systems have consisted of 20–75 mm coarse gravel laid in 300–600 mm thick layers over the base and side slopes of the landfill. Recently, geocomposite drainage nets have been developed to replace gravel drainage layers on side slopes, where the construction of gravel blanket drainage layers is difficult and expensive. These products are light in weight and are simple to install, and have relatively high in-plane flow rates; however, factors such as biological clogging and crush resistance need to be accommodated in the design.

Biaxial geonets (Figure 12.4(a)) are used primarily in capping applications where low confining pressures do not restrict the flow capacity. Triaxial geonets (Figure 12.4(b)) are used in high-load applications, where geotextile intrusion and strand crushing can limit the flow capacity.

12.6. Construction of liners

12.6.1 Preparations

The construction materials of the composite basal liner, clay soils and geomembranes differ greatly in their material properties. While the mineral component can follow any three-dimensional geometrical feature, as long as it can be shaped by earth moving and compaction equipment, geomembranes are planar elements. Therefore, the bottom of the landfill has to be designed such that the geomembrane can be spread evenly without distortions. In the ideal situation, the surface of the mineral liner consists of planes that intersect only at straight lines.

As all landfills differ in their geometry, it is necessary to prepare individual detailed drawings for the placement of the geomembrane. The drawings show the size, shape and designated number of the individual geomembrane sheets and the seams. Often, the geometry of the landfill requires triangular or other specially tailored pieces. Information on the geometrical details, and the predetermined sequence of placement of the individual sheets, is passed from the designer to the geomembrane construction personnel using drawings.

12.6.2 General aspects of installation

The placement of the basal liner of a solid-waste landfill is very important and requires very delicate construction work. It involves the operation of heavy earth-moving equipment, as well as the handling of sensitive geosynthetic products. Soil layers of up to 1 m thick, with a mass of tonnes per square metre, and geomembranes of 2.5 mm thickness are installed together. The installation of the basal liner must be executed without faults because any existing defects that are not noticed before the start of waste-filling operations would be practically irreparable. Defects in the basal liner cannot be tolerated.

The construction personnel must seek to achieve high-quality installation, and must be sufficiently skilled to

Figure 12.4 Geonets: (a) biaxial and (b) triaxial

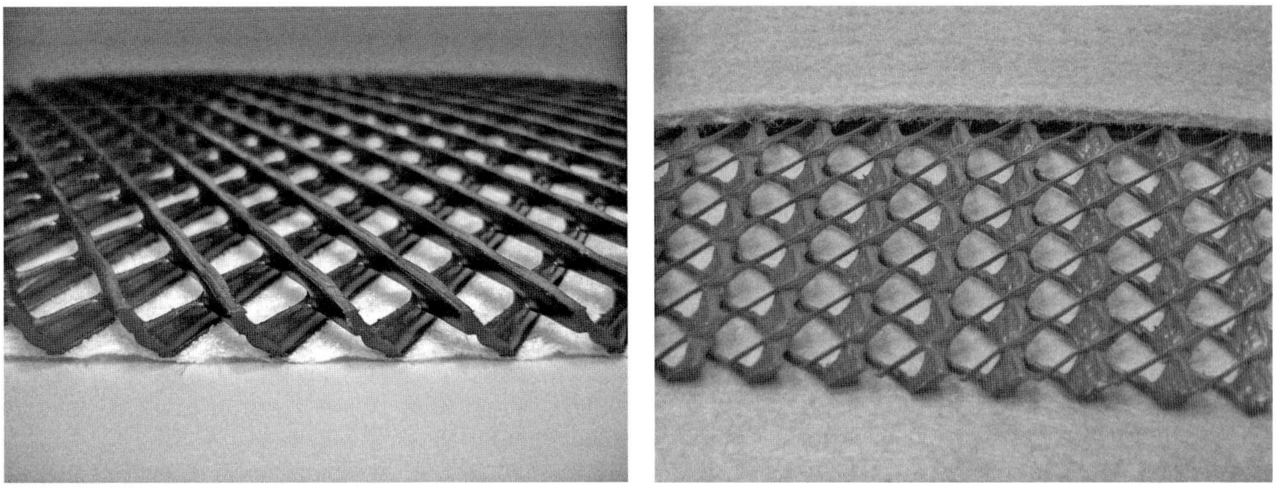

(a) (b)

achieve the same efficiency of the impervious composite liner under the difficult conditions in the field as have been obtained in small-scale models, under ideal conditions, by scientists in the laboratory. These requirements demand thorough preparation, scrutiny of the execution of the construction work and perfect timing of the various construction operations. The contractor doing the earthwork and the installer of the geomembrane must work in synchronisation under the coordination of the construction manager.

It must be taken into account that all construction operations at a landfill site are sensitive to weather conditions. Obviously, the placement of a clay liner is impossible during heavy rain, snowfall or frost, and partly finished clay blankets must be protected from the water and against desiccation due to dry wind and sunshine when the construction work is interrupted for weekends, owing to bad weather or for any other reason. For such temporary protection, thin plastic membranes are used.

The installation of geomembranes also requires favourable weather. It cannot be done in the rain. The minimum temperature for seaming polyethylene sheets is 5°C. Sufficient time has to be allocated for the placement of geomembranes to cope with unavoidable delays due to unfavourable weather, which may occur frequently in many parts of Europe.

The manufacturer of the geomembrane must establish his own instructions for handling and installation. If the manufacturer does not execute the construction work, he must subcontract it to a specialist. The approval document of the geomembrane manufacturer lists the firms authorised to place and seam the specified geomembranes. The construction personnel must be qualified and experienced, and the technicians must be certified welders. The seaming methods to be applied are specified in the approval documents. Strict rules are to be followed for the execution of the construction work.

12.6.3 Placement of the geomembrane

Great attention is paid to the preparation of the surface of the CCL. In order to obtain the intimate contact needed for the required composite sealing effect, the clay surface must be planar, smooth and free from stones, gravel or any other objects. The surface layer of the mineral liner may not contain single grains of more than 10 mm diameter. Such grains must be fully embedded in the clay matrix. No mineral grains with sharp edges are allowed at the surface. Deviations from the theoretical planar surface should not exceed 20 mm over a distance of 4 m. The ruts produced by the compaction equipment may not be deeper than 5 mm. The clay liner must retain its compaction water content, and no desiccation cracks are allowed.

In order to obtain the required composite effect of the geomembrane and the CCL, the geomembrane has to be placed without any voids between it and the clay surface. Therefore, ideally, the geomembrane, when spread, should not exhibit any waves. This is very difficult to achieve in practice. In particular, under clear and sunny conditions, the black polyethylene membrane heats up due to its high coefficient of thermal expansion, and the formation of waves in the geomembrane cannot be avoided under such conditions. However, at night, when the sun has set and the air temperature drops, the geomembrane will contract and the waves will disappear. This physical effect is used systematically in order to achieve the desired intimate contact between the geomembrane and the clay liner. Schicketanz (1992) has developed great expertise in the technology for the placement of geomembranes, which follows the daily rhythm of temperatures at the construction site.

The rhythm of the temperature differences governs the construction sequence. The clay liner is prepared to almost finish at least one day in advance of the placement of the geomembrane. Very early in the morning, the final surfacing work on the clay layer is carried out. Subsequently, the geomembrane is placed and seamed. As mentioned before, the geomembrane will form waves during the day, as the temperature is increasing, and it is very important that the welders are experienced in their work so they can seam the geomembrane sheets together without creating any pockets in spite of the waves that invariably exist. The seaming operation must be finished before the evening. Then the geomembrane is covered with a non-woven geotextile or other protector and, at specially selected locations, soil material is placed in such a way that, as the geomembrane contracts at night, it reaches the desired position. By using this technique, the geomembrane is stretched, and intimate contact with the clay surface is obtained. The geomembrane undergoes a certain amount of prestressing, but the resultant tensile forces are small and of no concern because they are reduced by relaxation of the polymer material over time.

The described operation is somewhat complicated in practice because the base of the landfill does not consist of a single continuous plane. The bottom liner system has to accommodate the gradients of the dewatering layer and the pipes. Hence, the surface of the bottom liner is composed of a sequence of roof-shaped planes inclined by at least 3% towards the leachate collection pipes, and by 1% in their longitudinal direction. Shallow grooves have to be prepared for the construction of the bedding layers for the pipes. As the geomembrane has to follow this profile, the placement of ballast is necessary to prevent uplifting of the geomembrane during its contraction at night. Obviously, such details require a lot of manual work and great skill. Their preparation is time consuming, and these details are the most vulnerable spots of the bottom liner system.

The geomembrane sheets are seamed together by dual hot wedge fusion. Automatic seaming machines are used that control and record the important parameters – advance rate, temperature, contact pressure and the length of advancement. If any irregularities occur, they are noted, and can be traced back and located precisely on the basis of the printed welding records. Once finished, each air channel of the dual seam is tested by inflation. Each channel must withstand an air pressure of 5 bar for 10 minutes without noticeable loss of pressure.

In areas where difficult access does not allow the automatic hot wedge welding machine to be used (e.g. sump bottoms, pipe penetrations or patches), the extrusion fillet method is applied. This requires a great deal of skill to obtain the same quality as attained with the automatic dual hot wedge fusion technique.

12.6.4 Quality assurance

The placement of the bottom liner system is executed under the construction quality control (CQC) programme of the contractor and the construction quality assurance (CQA) of an external inspector. All personnel responsible for quality management must be experienced in construction with geosynthetics.

Only geomembranes without any visible flaws are accepted. Experience shows that the quality of geomembranes manufactured under a quality assurance system, such as the one approved in Germany, is generally very good. If a geomembrane roll has to be rejected upon delivery to the construction site, the objections are usually due to damage that occurred during loading, transport or unloading. Sometimes the action of unloading the bulky, heavy geomembrane rolls from their shipping containers is very tricky, and personnel need to be trained to handle geomembrane rolls successfully and without any damage.

The quality inspectors have to check whether the seaming equipment is suitable for the job, and whether it functions properly. In particular, the generator producing the electric power has to meet the demands of the welding operations if uniform seams are to be produced. Every day, at the beginning and at the end of the seaming work, the controlling parameters of the seaming machine must be determined by making a test strip. The geometry is checked, and the seam is examined visually and by means of peel tests. Once the welding parameters have been established for the day, the work proceeds at a constant rate. If the weather conditions change, adjustments have to be made on the basis of new test strips. The advance rate, temperature and pressure are recorded automatically by the seaming machine, and occasional checks are made by the inspector.

Experience shows that a reliable execution of the CQA programme is of utmost importance for the geomembrane. Even though the education of the installers is generally very good, and although the construction personnel are aware of the importance of their work, mistakes do occur.

Fortunately, seam defects are often detected in time and can be corrected without delay, provided the CQA personnel are on site continuously. Sometimes, owners of landfills do not recognise the need for the external inspector to be present on site for the entire period of geomembrane installation, and they request only occasional visits. However, the chances are that the costs of having the external supervisor on site constantly will be more than compensated for by the savings made with regard to the need for corrections and for delays as a result of deficiencies noticed at a later time.

12.7. Leachate collection and removal

12.7.1 Drainage blanket and filters

As long as the landfill is being operated, there is no cover to prevent the infiltration of rain water and snow melt. Together with the placement moisture content of the solid waste, these liquids generate leachate, which must be collected at the bottom of the landfill and removed for proper treatment. The leachate collection system comprises a coarse-grained mineral drainage blanket, a transition at the top of the drainage blanket, leachate collection pipes, and access shafts or tunnels.

In municipal-waste deposits, which contain plastics, paper and other typical domestic refuse, no special filter is installed in the bottom dewatering systems. Experience shows that untreated domestic waste forms a filtering transition zone by itself. On the other hand, it has been observed in model tests (Ramke and Brune, 1990) that filters above drainage blankets percolated by very active biological leachate from municipal waste were completely encrusted when conditions were optimal for microbial growth.

Kossendey *et al.* (1996) studied the effect of microbial life on the long-term performance of geotextile filters at the base of landfills, specifically in large-scale permeation tests with well-defined hydraulic and biological boundary conditions. They found that the rate and the extent of bio-clogging depends on the living conditions of the micro-organisms. The amount of nutrition contained in the leachate is the governing factor. The species of micro-organisms cannot be controlled. There are always enough different types of microbes present in the environment to initiate the development of a mixed population of various bacteria, fungi, algae and other microbes. While these are growing, they occupy some of the void space in the waste above the filter of the geotextile and of the drainage gravel below, and the hydraulic conductivity of the system decreases. When the supply of nutrition is reduced, the amount of biological matter contained in the geotextile filter decreases, and the hydraulic conductivity recovers.

The study by Kossendey *et al.* reveals that considerable bioactivity develops even under poor nourishment conditions, such as may be expected at the base of hazardous-waste landfills or landfills for residues from incinerators. However, in such cases, the growth rate is slow, and most likely the hydraulic conductivity remaining after a long permeation time is sufficient with respect to the filtration function of the tested non-woven geotextiles. This result is quite important with regard to the modern landfills that will be operating in Europe in the future – landfills that will mainly contain residues from incinerators. A filter is needed in this type of landfill. The filter can be composed of either mineral granular material or, preferably, suitable geotextiles. Geotextile filters have the advantages of easier handling and installation, and lower mass and volume.

12.7.2 Leachate collection pipes and access shafts

As mentioned previously, the bottom of the landfill has a roof-shaped profile shape, and perforated pipes are installed at spaces of 30 m or less. The pipes are made of HDPE or PP. There have been many structural failures of rigid clay and concrete pipes, which were used more than 15 years ago. So these materials are totally excluded from landfill construction now, and only polymers are accepted. Structural analyses have to be carried out for flexible pipes installed at the bottom of a landfill, and this is the subject of research. Zanzinger and Gartung (1997) reported the results of finite-element computations and large-scale model tests, and these indicate that pipe deformations, rather than stability problems such as buckling, are the decisive design criteria.

Pipes must be accessible for cleaning, maintenance, camera inspection, measurements and leachate sampling. Pipes lead to access shafts or, in some cases, to tunnels where these necessary operations can be executed. The shafts should be placed outside the landfill body and be manufactured from a polymer material. If they are made of reinforced concrete, their external surfaces have to be lined with geomembranes. The same applies to concrete tunnels below landfills. If vertical shafts are placed within the waste body, they are submitted to the internal deformations of the waste, lateral pressures, vertical frictional forces, elevated temperatures, chemical attack, gases, etc. As a rule, vertical shafts in landfills should be avoided. In cases where this is impossible, design recommendations can be drawn from field observations, measurements and theoretical studies (Gartung *et al.*, 1993).

Vertical shaft structures in landfills must be founded above the basal liner in order to avoid leaks that would most likely occur if the shafts penetrate through the bottom sealing.

12.7.3 Consequences for the basal seal

The various components of the leachate collection and removal system of a solid-waste landfill require a great deal of specialist skill and intelligent engineering. The placement of the drainage blanket must be done very delicately, because the geomembrane liner must not be damaged. The design and installation of pipes above the bottom liner involve many technical details; for example, the bedding of the pipe, the penetration of pipes through the seals, and the connections of pipes and shafts or tunnels. The construction of shaft or tunnel structures is a major engineering assignment.

All technical details of the leachate collection and removal system are of great importance with regard to the safety of the entire landfill structure. If they are not built correctly, leaks develop at these details and all the efforts at constructing an efficient liner are greatly impaired. The bottom liner and the leachate collection and removal system act together as the basal barrier of the landfill.

12.8. Cover system

12.8.1 General

When the filling process of a solid-waste landfill, or a large part of it, is completed, the surface of the waste body has to be covered by a cap. The cover system must prevent the infiltration of rain water, and the emission of odours, dust and gas, and it has to facilitate landscaping and the growth of vegetation.

The main components of the cover of a landfill are a regulating soil layer immediately above the waste body, a gas venting system, the sealing layers, a drainage system and the restoration profile. Depending on the requirements of the particular landfill category, these layers vary to some extent. A capping system with geosynthetics is shown in Figure 12.5.

The properties and the behaviour of the waste influence the performance of the cap, and must be taken into account in the design and construction. For waste bodies that contain mineral solids that do not undergo chemical or biological reactions, no major long-term settlements are expected. This applies to landfills that mainly contain ashes from incinerators, and should apply to hazardous wastes as well.

For landfills without long-term differential settlements, the placement of the cover can be carried out as soon as the design height is reached.

Common municipal waste landfills are essentially bioreactors, where degradation processes take place in the waste body, associated with significant volume changes and gas production. The surfaces of this type of landfill usually undergo large settlements for quite some time. It is also likely that substantial settlement differences occur locally, which sometimes cannot be followed by mineral seals without the development of leaks. As the bioreactors need a certain amount of water to continue the degradation processes, some leakage is probably of no concern. It makes sense to provide municipal waste landfills with CCLs or GCLs, and mineral layers of low hydraulic conductivity, as interim covers. Later, these interim covers become part of the final capping system, which contains a geomembrane as the main seal. The geomembrane should be placed when most of the anticipated differential settlements have occurred. To determine the right time for this action, the deformation of the interim cover surface should be monitored.

12.8.2 Regulating the soil and gas venting layer

Solid wastes are not suitable for finish profiling; a regulating soil layer is required for this purpose. If gas is generated in the landfill body, a gas venting system has to be installed below the cover. The functions of the regulating layer and the gas venting layer may be combined by using sufficiently pervious soil. Geosynthetic composites may be used as gas collecting sheets if the permeability of the regulating soil

Figure 12.5 An alternative capping system using geosynthetic components

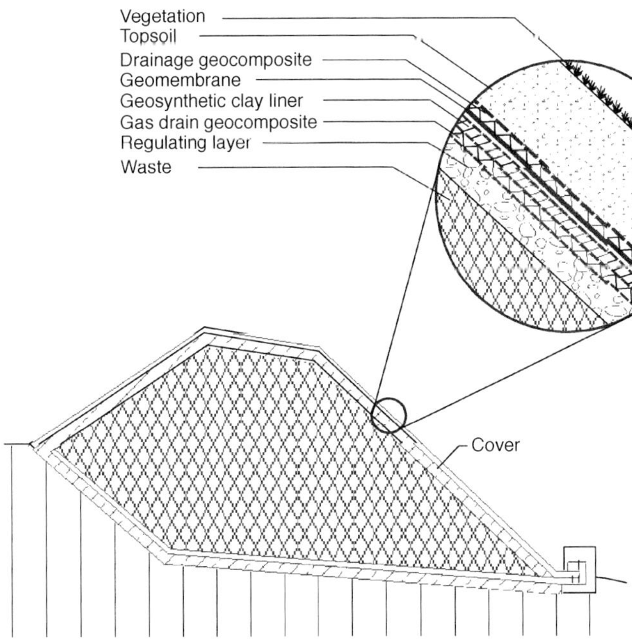

layer cannot be relied upon. The collected gas has to be conducted out of the landfill by pipes and should be submitted to caloric use, or it has to be burnt.

12.8.3 Mineral sealing layer

For covers on deposits of inert mineral residues, and for interim sealing layers of municipal waste landfills, a compacted clay layer placed in two lifts, each 0.25 m thick with a hydraulic conductivity of no more than 5×10^{-9} m/s, is commonly used.

Mineral liners in capping systems are exposed to fluctuations in their water content. Under central European climatic conditions, the evapo-transpiration rate is relatively high during the growing season, from about April until late September, while at the same time precipitation may be low. During this time, the water content of the clay layer is reduced. In autumn and winter, evapo-transpiration is decreasing, precipitation may be high and the mineral liner is rewetted. Observations by Melchior (1993) at large test fields indicate that, under certain unfavourable boundary conditions, desiccation causes the formation of micro-fissures and cracks in the cohesive cover soil during summer. These defects do not heal, and are utilised by plants for root paths. The detrimental effect of the desiccation due to thermal gradients is increased further by the suction of the roots. As a result, within two to three climatic cycles, the mineral liner experiences fissuring to a considerable extent, the overall hydraulic conductivity increases and the sealing function is impeded.

12.8.4 Geosynthetic clay liners

Alternatively, a GCL can be installed. The question of the equivalency of GCLs and CCLs has been discussed by Koerner and Daniel (1995) and Stief (1995), among others. The properties, testing methods and quality assurance aspects of GCLs have been compiled by Gartung and Zanzinger (1998).

Practical experience shows that GCLs, as part of a capping system, have some advantages over CCLs. Handling and installation are much easier, less time is needed for placement, waste storage space can be saved due to the lesser thickness, and the quality of the manufactured geosynthetic product shows less variability than that of the natural clay soils.

On the other hand, it has to be kept in mind that, due to the low thickness and mass of bentonite, GCLs are extremely sensitive to damage during and after construction, and thus great care must be taken in the construction when using GCLs. The design of a geosynthetic barrier containing GCLs is necessary to consider that the desiccation of the GCLs must be avoided, for a number of reasons. For example, dried out sodium bentonite will exchange cations with the adjacent soil and become calcium bentonite, which has much poorer swelling properties than sodium bentonite. In some projects in Germany it was found that, under certain conditions and with inadequate protection of the GCLs against desiccation, the barrier function of the GCLs was lost (Gartung and Zanzinger, 1998).

The chemical composition and thickness of cover materials must be considered during the design phase, as these factors can impact on the performance of the GCL. Buckley *et al.* (2010) have shown that the cover material can have a marked influence on the performance of a GCL used in a capping liner. They suggested some measures to limit the detrimental effects of cation exchange, desiccation and root penetration on GCL performance, including the following

- in landfill caps, avoid GCL cover soils that have high calcium or magnesium contents
- undertake a mineralogical analysis of the cover and subgrade soils before they are considered for use in construction
- prior to construction, measure soil-leachate soluble and exchangeable cations of the cover and subgrade soils
- in thin landfill caps or where there is a high potential for root penetration, use a plastic-laminated GCL product and a butyl tape jointing system
- ensure hydration of the subgrade soil with fresh water to enable prehydration of the GCL
- limit vegetation above the GCL to species having non-aggressive root systems.

12.8.5 Geomembranes

Fissuring of and the growth of roots in the mineral seals of landfill capping systems can be prevented by the placement of a geomembrane. A geomembrane functions as a barrier against root penetration as well as against moisture migration. The final sealing layers of landfill cover systems should consist of a combination of CCLs or GCLs with geomembranes. As the seal at the top of the landfill is not acted on by chemicals, the synergistic composite effect of polymers and clay soils, which facilitates the retention of polar as well as non-polar substances at the bottom of the landfill, is not effective in the capping system. Therefore, at the cover, the two components do not really act as a composite, but rather as a double liner.

Even though the geomembranes in covers are not exposed to a corrosive chemical environment, in Germany the same types of geomembrane are used for caps as for basal liners. They are made of HDPE, their thickness is 2.5 mm, and only approved geomembranes are used in landfill caps. The advantages are high robustness and reliable quality. However, their limited flexibility is of some disadvantage. The installation of thinner geomembranes or of softer polymers, such as very low density polyethylene (VLDPE), would be more favourable with respect to the anticipated deformations of the landfill surface.

The construction requirements and installation techniques are essentially the same for the geomembranes of the cover as those for the bottom liner. The seaming technique, and all the details of CQC and CQA described in Section 12.6 for basal liners also, apply to covers.

The surface of the landfill, or of the regulating layer, has to be modelled to a shape that allows planar geomembranes to be spread without distortion. This design requirement is especially important when HDPE membranes of 2.5 mm thickness are used. It is impossible to place these on three-dimensionally curved surfaces having small diameters of curvature.

Usually, landfills are hills with sloping surfaces, and thus slope stability is a very important issue in the design and construction of landfill covers. Often, it is not possible to mobilise enough shear resistance to achieve stability on smooth geomembrane surfaces. In such cases, geomembranes having specially structured rough surfaces are used in the cover construction. These structured geomembranes undergo the same stringent suitability tests as the smooth geomembranes used for the basal liner. Particular attention is paid to their long-term tensile strength and stress cracking resistance. In order to avoid tensile forces in the geomembrane, the mobilised friction at the lower surface of the geomembrane should be greater than that at the upper surface. If the slope-stability analysis leads to the conclusion that sufficient safety in the balance of forces can only be reached by additional reinforcing elements, geogrids are placed above the sealing layers of the capping system.

12.8.6 Dewatering of cover systems

A small amount of precipitation runs off from the landfill surface directly. Most of it evaporates or is stored in the topsoil layer, where it is available for plant growth. The remaining water percolates through the topsoil and reaches the sealing barrier. In order to avoid the build-up of water pressure on the seal, a dewatering layer is installed in the capping system above the sealing layer.

The minimum gradient of the cap drainage layer is 5% in German landfills. For long slopes, the maximum inclination should not be steeper than 1 in 3 (vertical to horizontal) for practical reasons with respect to landscaping and maintenance work. Of course, the maximum permissible slope angle is determined on the basis of the slope-stability analysis, which must be done for the most critical section of each landfill individually.

There are a number of, mostly, older landfills with slopes of 1 in 2.5 (vertical to horizontal), or locally even of 1 in 2. When these steep slopes have to be provided with a dewatering layer above the sealing element of the capping system, it is often necessary to install a geogrid reinforcement, because the frictional resistance at the interfaces of the various layers of the capping system is insufficient for the required slope stability.

The standard profile of landfill covers according to the German instructions contains a layer of granular soil having a minimum hydraulic conductivity of 1×10^{-3} m/s and a thickness of at least 0.3 m. Geosynthetic drainage layers are often used instead. In both cases, the design engineer has to undertake hydrological and hydraulic studies in order to determine whether additional relief drainage pipes must be incorporated in the dewatering system. Design analyses are also needed to specify the final gradients, select the granular drainage material or drainage geocomposites, and prepare the project documents, including drawings of details, specifications and the quality assurance plan. The principal design criteria and analytical methods are described by Ramke (1995), and are included in the technical recommendations of the German Geotechnical Society (GDA, 1997).

The criterion of minimum hydraulic conductivity allows fractions of clean coarse sand or fine gravel to be used for the drainage layer. However, sometimes the same coarse gravel is used for the cover drain as for the leachate collection system. If such a coarse granular drainage layer is placed onto a geomembrane, it is necessary to protect the geomembrane against puncturing. Geosynthetic drains are usually sufficiently robust and act as cushions, and thus no special layer is needed for puncture protection at the bottom of the restoration profile layer when geocomposite drains are used.

12.8.7 Drainage geocomposites

Drainage geocomposites have some advantages over granular drains. The mass of construction material to be handled is much smaller and, as their thickness is low, the waste-storage volume can be maximised. In addition, their placement is fast and simple. The quality of industrially manufactured geocomposites is more uniform than that of natural soils used for drainage layers, and in some parts of the world it may be difficult to find suitable granular drainage materials. Finally, synthetic drains often turn out to be cheaper than granular drainage layers. For these reasons, drainage geocomposites are being used increasingly in landfill capping systems.

Drainage geocomposites have to meet mechanical and hydraulic requirements. Their internal shear strength and the shear resistance at the interfaces with soils or geomembranes have to be in accordance with the demands for slope stability. The specific shear parameters have to be determined experimentally by means of laboratory tests on samples of 300×300 mm. It is necessary to conduct the tests on both dry and water-saturated drainage layers, because the shear parameters can be significantly different under the two conditions.

As most geosynthetic drains are compressed under loads, their hydraulic conduction capacity depends on the load applied perpendicular to the plane of the geocomposite. The hydraulic flow capacity has to be measured in laboratory tests under the pertinent stress conditions to determine adequate hydraulic gradients. As some geosynthetic drainage products are sensitive to shear forces, and may even collapse structurally at a certain shear stress level, it is necessary to test them under a combined normal and shear force. Creep may also be of concern, and has to be considered. Zanzinger (2000) has summarised the aspects of testing drainage geocomposites. He recommends determining the in-plane water flow capacity of the geocomposite in contact with the adjacent soil, or with foam layers, in order to include the effects of the intrusion of soil particles, which depends on the structure of the geosynthetic, on the grain-size distribution and on other properties, such as stiffness and the consistency of the soil, as well as the state of stress.

Most of the drainage geocomposites used in Europe consist of three layers – an upper and a lower non-woven filter layer, with a geospacer in between. Geonets, which are common in the USA, are used less frequently in Europe. The different drainage geocomposites can vary considerably in their mechanical and hydraulic properties, and also in their drainage performance according to project-specific conditions. Therefore, their suitability for the function of cover dewatering at a particular site should be determined by means of laboratory tests and design calculations.

Regarding the boundary conditions at a site, it may be advisable to carry out large-scale test constructions, particularly if new products are applied, if steep slopes have to be covered, or if very heavy construction equipment will be used. In the past, some drainage geocomposites suffered severe damage during installation because their structural strength was marginal. Such products should not be used. The robustness of the geocomposite, and the size of the delivered unit, whether it is completely prefabricated or has to be constructed on site, are important selection criteria. Seaming and connections with drainage pipes also have to be considered.

The placement of drainage geocomposites, which are essentially very thin and sensitive sheets, has to be executed under a CQC programme of the same type as applied to the installation of geomembranes.

12.9. Summary of main points

- The components of liners and covers consist of mineral materials and of geosynthetic products in combination. Geomembranes, drainage geocomposites, geotextiles, geosynthetic clay liners and geogrids are generally accepted as reliable components in the construction of landfills.
- Correct design and installation of liner systems is critical to provide long-term containment within any landfill. While the use of geosynthetics in these applications is relatively new (<50 years), our understanding of the requirements is growing rapidly. This has prompted the development of new geosynthetics to meet the demand.
- A 'black box' design method is not suitable for containment applications where leachate constituents can change over time and from site to site. It is therefore important to carry out site-specific compatibility testing of all components of the liner system, particularly where pH extremes (<4 and >10) are encountered.
- The design of the various components of any liner system is the first step in the development of the landfill, but without correct installation and construction quality assurance, the project runs a high risk of not achieving the design goals.
- This chapter has detailed the world's best practice in terms of liner system installation, aimed at minimising leaks through the system and maintaining its integrity for the design life of the structure.
- Construction of a liner system should be carried out under a strict quality assurance programme, both from a product quality and a product performance

perspective; a proper installation will provide a form of surety.

12.10. Some common questions and their answers

Questions

1. Do modern liner systems leak?
2. Why is there no international standard for liner systems?
3. How can a thin geosynthetic clay liner (GCL) (<10 mm thick) replace a compacted clay layer (CCL) that is 600–900 mm thick?
4. Will HDPE geomembranes from different suppliers behave in the same way?
5. Will geosynthetic clay liners from different suppliers behave in the same way?
6. Why is the limitation of wrinkles desirable in geomembranes?
7. Why is it important to cover a compacted clay liner (CCL) immediately after compaction is completed?
8. Can manufacturers' quality assurance testing replace a construction quality assurance testing?

Answers

1. It is virtually impossible to design and install a liner system which does not leak, and the cost of developing a 100% water-tight system could be cost prohibitive. The key is to assess the risks associated with any leaks, and then design and construct a system which would minimise the impact on the surrounding environment while balancing the risk and cost implications.

2. Due to the variability in waste types, climatic conditions, technological level and wealth of the various countries of the world, it is unlikely that any international consensus on minimum standards is likely to be achieved. Where no minimum requirements or standards exist, regulators and designers should consider all existing knowledge and adopt practices that best suit their application.

3. The GCL is able to replace the CCL due to the purity of the clay contained within the GCL (i.e. a montmorillonite content >70% and 80% of particles <0.5 μm in size).

4. Not necessarily. Manufacturers will have different resin suppliers, each of which has their own recipe. These different base resins can have an effect on certain performance aspects of the geomembrane.

5. Not necessarily. Each manufacturer will have its own clay source, the quality of which will vary. It should be noted that manufacturers' specifications are based on testing using deaired and deionised water, in conjunction with liquors of high or low pH, high salinity, etc.; the performance can vary significantly from the published data.

6. For two key reasons: (a) wrinkles prevent intimate contact with clay subgrades, thus allowing leaks to transmit across a large surface area of the containment facility and increasing leakage through the system; and (b) wrinkles or folds cause stress concentrations in the geomembrane, leading to environmental stress cracking that increases leakage rates through the liner system.

7. If the CCL is left exposed, moisture is lost due to evaporation, causing cracking of the CCL; these cracks will never become completely sealed without reworking. The cracks form preferential flow paths, resulting in a porous layer rather than a very low permeability layer.

8. No. Random third-party testing of materials supplied is required to certify product conformance.

REFERENCES

August H, Tatzky R, Patsuska G and Win T (1984) Untersuchung des Permeationsverhaltens von handelsüblichen Kunststoffdichtungsbahnen als Deponiebasisabdichtung gegenüber Sickerwasser, organischen Lösungsmitteln und deren wäßrige Lösungen. *Forschungsbericht Nr. 103 02 208, Abfallwirtschaft. UFOPLAN des Bundesministers des Innern, ImAuftrage des UBA*, Bundesanstaltfür Materialforschung und-prüfung (BAM), Berlin (In German).

August H, Tatzky-Gerth R, Preuschmann R and Jakob I (1992) Permeationsverhalten von Kombinationsdichtungen bei Deponien und Altlasten gegenüber wassergefährdenden Stoffen. *Forschungs- und Entwicklungsvorhaben 102 03 412*, BAM, Berlin (In German).

Benson CH, Wang X, Gassner FW and Foo DCF (2008) Hydraulic conductivity of two geosynthetic clay liners permeated with an Alumina residue leachate. *Proceedings of the First Pan American Geosynthetics Conference & Exhibition*, Cancun, pp. 94–101.

Brummermann K (1997) Schutzschichten für Kunststoffdichtungsbahnen in Deponiebasisdichtungen–Prüfung und Bewertungihrer Wirksamkeit. Dissertation, Institut für Grundbau, Bodenmechanik und Energiewasserbau, University of Hanover, Heft 46 (In German).

Buckley JL, Gates WP, Gibbs D and Gassner F (2010) Forensic examination of field GCL performance in landfill capping and mining containment applications. *Proceedings of the 3rd International Symposium on Geosynthetic Clay Liners*, Wartzburg.

Bundesanstalt für Materialforschung und-prüfung (BAM) (1992) Richtliniefür die Zulassung von Kunststoffdichtungsbahnen als Bestandteil einer Kombinationsdichtung für Siedlungs-und Sonderabfalldeponien sowie für Abdichtungen von Altlasten. BAM, Berlin, Germany (In German).

Corbet SP and Peters M (1996) First Germany/USA geomembrane workshop. *Geotextiles and Geomembranes* **14**(12): 647–726.

DWAF (1998) *Minimum Requirements for Waste Disposal by Landfill*, 2nd edn. Department of Water Affairs and Forestry, South Africa.

European Community (1999) Council directive 1999/31/EC of 26 April 1999 on the landfill of waste. *Official Journal of the European Communities* **L182**.

Finsterwalder K and Mann U (1990) Stofftransport durch mineralische Abdichtungen. Neuzeitliche. *Deponietechnik, Jessberger (Herausg)*, Balkema, Rotterdam (In German).

Gallagher EM, Darbyshire W and Warwick RG (1999) Performance testing of landfill geoprotectors: background,

critique, development and current UK practice. *Geosynthetics International* **6**: 283–301.
Gartung E (1992) Anwendungsgrenzen der Kombinationsdichtung im Deponiebau. *Veröffentlichungen des LGA-Grundbauinstituts*, Nürnberg, Heft 65, pp. 85–113 (In German).
Gartung E and Zanzinger H (1998) Engineering properties and use of geosynthetic clay liners. In *Geotechnical Engineering of Landfills* (Dixon N, Murray EJ and Jones DRV (eds)), Thomas Telford, London, pp. 131–149.
Gartung E, Prühs H and Nowack F (1993) Measurements on vertical shafts in landfills. *Proceedings of the Sardinia 93, 4th International Landfill Symposium*, pp. 461–468.
Gartung E, Müllner B, Heimerl H and Kohler E (1996) Die mineralische Basisabdichtung der Siedlungsabfalldeponie Aurach nach mehr als zwölfjährigem Betrieb. *Veröffentlichungen des LGA-Grundbauinstituts*, Nürnberg, Heft 75 (In German).
Gates WP, Hornsey WP and Buckley JL (2009) Geosynthetic clay liners – is the key component being overlooked? *Proceedings of the Geoafrica 2009 Conference*, Cape Town.
GDA (1997) *Empfehlungen des Arbeitskreises Geotechnik der Deponien und Altlasten*. VerlagErnst&Sohn, Berlin (In German).
Horn A (1989) Mineralische Deponie Flächendichtungen aus gemischtkörnigen Böden. *Bautechnik 69*, Heft 9, pp. 311 ff. (In German).
Kirschner R and Kreit V (1994) Innovative protective mattresses for landfill geomembranes. *Proceedings of the 5th International Conference on Geotextiles, Geomembranes and Related Products*, Singapore, pp. 1015–1018.
Koerner RM (1993) Geomembrane liners. In *Geotechnical Practice for Waste Disposal* (Daniel DE (ed.)), Chapman & Hall, London.
Koerner RM (1998) *Designing with Geosynthetics*. Prentice Hall, Englewood Cliffs, NJ.
Koerner RM and Daniel DE (1995) A suggested methodology for assessing the equivalency of geosynthetic clay liners to compacted clay liners. In *Geosynthetic Clay Liners* (Koerner RM, Gartung E and Zanzinger H (eds)), Balkema, Rotterdam, pp. 73–98.
Kossendey T, Gartung E and Schmidt S (1996) Microbiological influences on the long-term performance of geotextile filters. *Proceedings of the Geofilters*, Montreal. Balkema, Rotterdam, pp. 115–124.
Lauf G and Müllner B (1993) A new barrier material with high chemical resistance. *Proceedings of the Sardinia 93, 4th International Landfill Symposium*, pp. 499–505.
Melchior S (1993) Wasserhaushalt und Wirksamkeit mehrschichtiger Abdeck-systeme für Deponien und Altlasten. *Hamburger bodenkundliche Arbeiten*, Band 22 (In German).
Müller-Rochholz J and Asser JD (1994) Sandfilled geosynthetics for the protection of landfill liners. *Proceedings of the 5th International Conference on Geotextiles, Geomembranes and Related Product*, Singapore, pp. 1011–1014.
Ramke HG (1995) Oberflächenentwässerung von Deponien–Ansätze zur hydraulischen Berechnung. *Veröffentlichungen des LGA-Grundbauinstituts*, Nürnberg, Heft 75, pp. 131–165 (In German).
Ramke HG and Brune M (1990) Untersuchungen zur Funktionsfähigkeit von Entwässerungsschichten in Deponiebasisabdichtungen. *Abschlußbericht BMfT*, FKZ 14504573 (In German).
Saathoff F and Sehrbrock U (1994) Indicators for selection of protection layers for geomembranes. *Proceedings of the 5th International Conference on Geotextiles, Geomembranes and Related Product*, Singapore, pp. 1019–1022.
Schicketanz R (1992) Wirkungsweise der Kombinationsdichtung und Anfor-derungen an die mineralische Oberfläche. *Müll und Abfall* **5/92** (In German).
Stief K (1986) Das Multibarrierenkonzept als Grundlage von Planung, Bau, Betrieb und Nachsorge von Deponien. *Müll und Abfall* **1/86** (In German).
Stief K (1995) On the equivalency of liner systems – the state of discussions in Germany. In *Geosynthetic Clay Liners* (Koerner RM, Gartung E and Zanzinger H (eds)), Balkema, Rotterdam, pp. 3–15.
TA Abfall (1991) *Zweite allgemeine Verwaltungsvorschrift zum Abfallgesetz, Teil 1: Technische Anleitung zur Lagerung, chemisch/physikalischen, biolo- gischen Behandlung, Verbrennung und Ablagerung von besonders überwachungsbedürftigen Abfällenvom 12. März 1991*. Carl Heymanns-Verlag, Köln, GMBL, 42. Jahrg., No. 8, pp. 139 ff. (In German).
TASi (Technische Anleitung Siedlungsabfall) (1993) Technische Anleitung zur Verwertung, Behandlung und sonstigen Entsorgung von Siedlungsabfällen. *Bundesanzeiger* (In German).
Van Impe WF (1995) ISSMFE Policy and the challenges of environmental geotechnics. *Proceedings of the Luso-Brasilian Seminar on Environmental Geotechnics*, Lisbon, Portugal.
Wilson-Fahmy RF, Narejo D and Koerner RM (1995) *Puncture Protection of Geomembranes*. Geosynthetic Research Institute, Philadelphia, PA.
Zanzinger H (1999) Efficiency of geosynthetic protection layers – investigation of the performance in a large-scale model test. *Geosynthetics International* **6(4)**: 303–317.
Zanzinger H (2000) Reduction factors for the long-term water flow capacity of drainage-geocomposites. *Filters and Drainage in Geotechnical and Environmental Engineering. Proceedings of the 3rd International Conference, Geofilters* 2000, Rotterdam, pp. 245–249.
Zanzinger H and Gartung E (1997) Model tests of drainage pipes under heavy load. *Proceedings of the 14th International Conference on Soil Mechanics and Foundation Engineering*, Rotterdam, pp. 1869–1872.

Handbook of Geosynthetic Engineering
ISBN 978-0-7277-4175-2

http://dx.doi.org/10.1680/hgc.41752.263

Chapter 13
Earth dams

D. N. Singh*, S. K. Shukla† and D. N. Arnepalli‡
*Department of Civil Engineering, Indian Institute of Technology Bombay, Mumbai, India
†Discipline of Civil Engineering, School of Engineering, Edith Cowan University, Perth, Australia
‡Department of Civil Engineering, Indian Institute of Technology Madras, Chennai, India

Earth dams are normally constructed using locally available soils and rocks. They provide a sustainable source of water to meet agricultural, municipal and industrial demands, and often play a major role in flood protection and control. As a result of ageing, material deterioration, internal erosion, piping and desiccation of upstream surface of the dam, the seepage losses through the dam section increase dramatically and may cause structural instability of the dam structure. In this chapter an attempt is made to provide an overview of the application of both conventional and geosynthetic materials to minimise the seepage losses and to improve the overall performance. The chapter also highlights the role of the ageing mechanism and the durability of individual geosynthetic components with regard to accomplishing the required overall performance of the earth dam.

13.1. Introduction

Earth dams are massive structures for impounding water and are normally constructed using locally available soils and rocks. Figure 13.1 illustrates the various parts of an earth dam. One of the principal advantages of earth dams is that their construction is very economical compared to the construction costs of concrete dams. Although the design of an earth dam is a complex art, with each situation being different from any other, the basic steps involved in the design are quite easy to follow. These are

(*a*) a thorough exploration of the foundation and abutments, and an evaluation of the quantities and characteristics of all construction materials available within a reasonable distance from the site
(*b*) the selection of a possible trial design
(*c*) an analysis of the safety of the trial design
(*d*) the modification of the design in order to meet stability requirements
(*e*) the preparation of a detailed cost estimation
(*f*) the final selection of a design that seems to offer the best combination of economy, safety and convenience of construction.

Although a conventional design incorporates the above steps to a great extent, some recent developments in embankment and dam construction have imposed several challenges to achieving perfection and an economical cross-section, in terms of both time and money. In recent times the use of geosynthetics, in conjunction with the conventional earth dam construction materials, has increased, and has brought with it a challenging task with regard to the Civil Engineering practice. The use of geosynthetics in an earth dam affects the construction procedure and stability of the dam. In fact, efficient use of geosynthetics requires special attention. The properties of geosynthetics must be evaluated (see Section 1.6) based on specific criteria and functional requirements, such as effectiveness as a barrier, a filter, a drainage medium, protection, reinforcement and a separator. In addition, the long-term performance of an earth dam as a whole is dependent on the service life of individual components, in particular the geosynthetic systems, of the dam. Hence it is essential to estimate the service life of these components under simulated realistic field conditions by identifying the critical ageing mechanism.

In recent years, geosynthetics have also been very widely and successfully used in the rehabilitation of older dams. Primarily, the requirements for these materials to work in an efficient manner are imperviousness, flexibility, mechanical strength, frost and heat resistance, availability (by industrial production) and easy workability. These requirements are met by several polymer materials, as described in Chapter 1.

With the above background, in this chapter various factors related to the use of geosynthetics in earth dam construction are discussed and conventional earth dam construction practices using impervious membranes made from conventional manufactured materials, such as concrete, steel and asphaltic concrete, are briefly reviewed. Some case studies are included to highlight the construction process and the efficiency of installation of the geosynthetic materials.

Figure 13.1 Parts of an earth dam. (Note: not all elements shown would be incorporated in any one dam)
Reproduced from Sowers and Sally (1962)

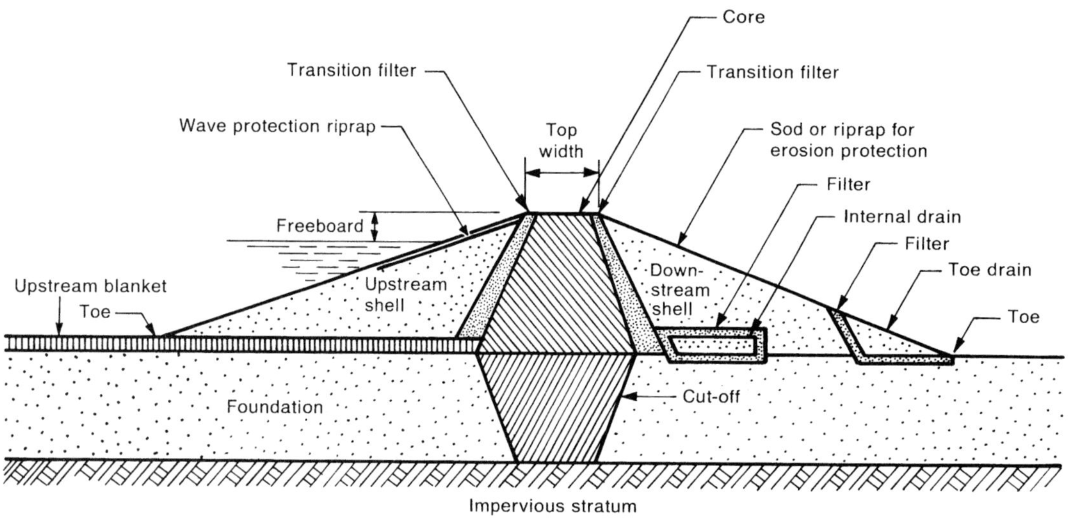

13.2. Use of conventional materials

For earth dams up to a height of at least 150 ft or so, founded on reasonably incompressible materials, there is a lot of available experience to indicate clearly that the fluid barrier or membrane can be constructed satisfactorily as a single monolithic concrete slab, without expansion joints. The reinforcing steel should be the same in each direction and should compose at least 0.5% of the cross-sectional area of the slab. When a single reinforced slab is used, some leakage always occurs through the inevitable hairline cracks or leakage may occur due to construction inadequacies that are seen soon after the filling of the reservoir. Consequently, it is necessary to provide a drain behind the slab on an earth dam to drain out the seepage water from the embankment.

Reinforced concrete slabs have been used in the McKay Dam, Oregon, as the water barrier on the upstream slope (Walker, 1958). This dam was built in 1925. It is 160 ft high and built of gravels and cobbles in thin compacted layers (Figure 13.2). Even though the crest length is 2700 ft, there are no expansion joints in the concrete slab. It has been reported that no maintenance of any kind has been required on the slab since construction. However, over the years, a few isolated hairline cracks have been observed.

Figure 13.2 The McKay Dam, Oregon
Reproduced from Walker (1958), material in the public domain

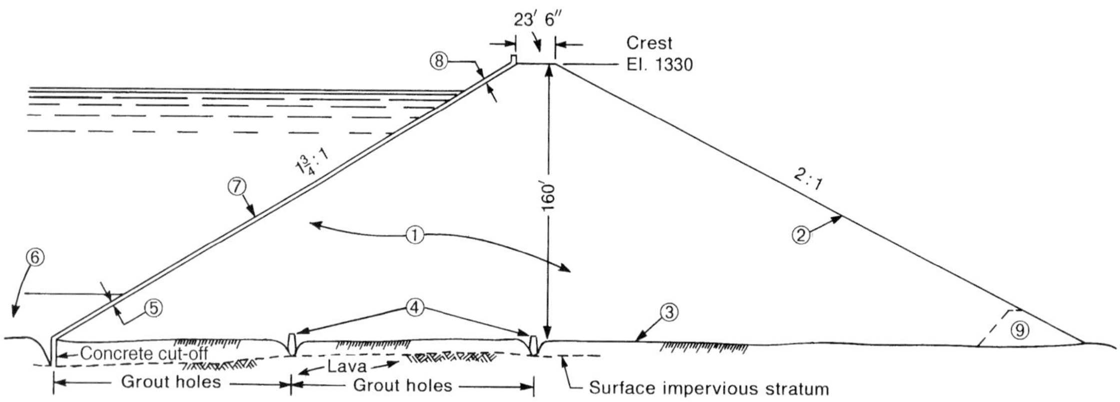

1. Sand and gravel rolled in 8 in. layers by a traction engine
2. Large cobbles and boulders in embankment material deposited on downstream face
3. Stripped to compact gravel and boulders, cemented gravel, or solid rock surface
4. Concrete cut-off walls
5. 12.5 in. concrete paving
6. Blanket of earth material extending 180 ft upstream
7. Continuous reinforced concrete slab
8. 8 in. concrete
9. Rock fill

In another situation, an embankment section and upstream concrete slab similar to that of the McKay Dam was used for the Don Martin Dam in northern Mexico which is 100 ft high and has a crest length of 4000 ft. Its performance has been as satisfactory as that of the McKay Dam. A few cracks have been reported in the slab due to differential settlement at the right-hand end of the dam where it connects with a massive concrete retaining wall at the spillway. Although the cracks were filled with asphalt 14 years after the construction of the dam, no further distress was reported (Sherard *et al.*, 1963).

Welded steel plates can be used at any site where reinforced concrete might be considered. Although steel is somewhat more expensive, it has the following two advantages over concrete

- it is completely watertight and may not require additional drainage systems behind the steel fluid barrier, whereas fine cracks in the concrete will inevitably allow some leakage
- it is more flexible and better able to cope with differential settlements without rupture.

Experience of the use of steel plate membranes in earth dams indicates that a steel facing has approximately the same life-span as that of a reinforced concrete facing. In addition, in no case has the corrosion of the steel plate led to serious problems. The only limitation is that the plate cannot be placed directly on soil containing an appreciable percentage of clay or silt.

Steel plates have been used in the El Vado Dam, New Mexico, where they have a maximum height of 175 ft and a crest length of 1200 ft. This dam has a rolled fill embankment, consisting primarily of gravel. The steel plate, having a thickness of 0.25 in., is underlain directly by the gravel fill embankment. A typical section is shown in Figure 13.3. The dam has been in continuous operation and the steel plate is still in excellent condition, except for requiring some minor inexpensive repair work from time to time. However, some wrinkling, buckling and tearing of the steel plate has been caused on the left-hand side of the dam by the movement of the abutment. In some other cases, it has been observed that the steel plate showed minor rust pitting and leakage has developed through the opening of the welds (Sherard *et al.*, 1963).

Asphaltic concrete is used to protect the upstream slopes of small dams against erosion and it sometimes acts as a water barrier. Professionals have stated that such membranes offer many advantages over other materials. The main advantage of asphaltic concrete is that its cost is less than that of either concrete or steel. These membranes are more flexible than reinforced concrete slabs and thus are better able to undergo differential settlements without cracking. At the same time, the construction is very rapid. The most important feature of asphaltic concrete membranes is that leaks, which develop under certain circumstances, are self-sealing. As far as maintenance is concerned, the portion of the asphaltic concrete above the reservoir level is easier to repair than either steel or concrete. However, the primary disadvantage is that the material is relatively soft and can be damaged more easily than either concrete or steel by falling rock, sabotage, or other activities of man or nature.

The asphaltic concrete used generally consists of very well-graded aggregate from a maximum size of about 1 in. to fine sand sizes and contains approximately 10% by weight of gravel rock dust (filler). Pure asphalt binder of 8–10% by weight of aggregate is used. The material is mixed and compacted when hot. The principal properties, such as the permeability, rate of ageing and resistance to alternate freezing and thawing, depend on the degree of compaction and the resultant air content in the compacted asphaltic concrete, which in turn is governed by its pore-size distribution characteristics. In practice, an air content of 2–3% appears to be optimum.

The successful performance of the Ghrib Dam, Algeria, built in 1935, over years of extreme temperature conditions, provides strong support for the contention that asphaltic concrete is suitable for use in impervious membranes on embankment dams. A typical cross-section of the Ghrib Dam is shown in Figure 13.4. A 12 cm-thick asphaltic concrete slab is sandwiched between a 12 cm lower layer of

Figure 13.3 The El Vado Dam, New Mexico
Reproduced from Segar (1935), The McGraw-Hill Companies

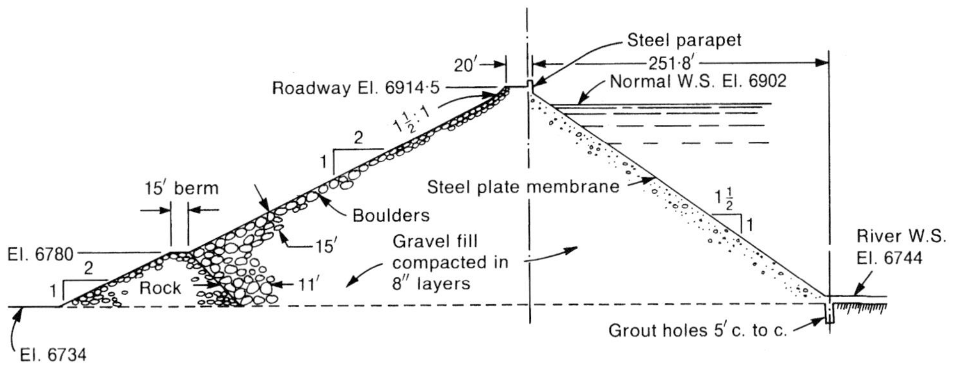

Figure 13.4 The Ghrib Dam, Algeria, with details of asphaltic concrete upstream membrane
Reproduced from Thevenin (1958)

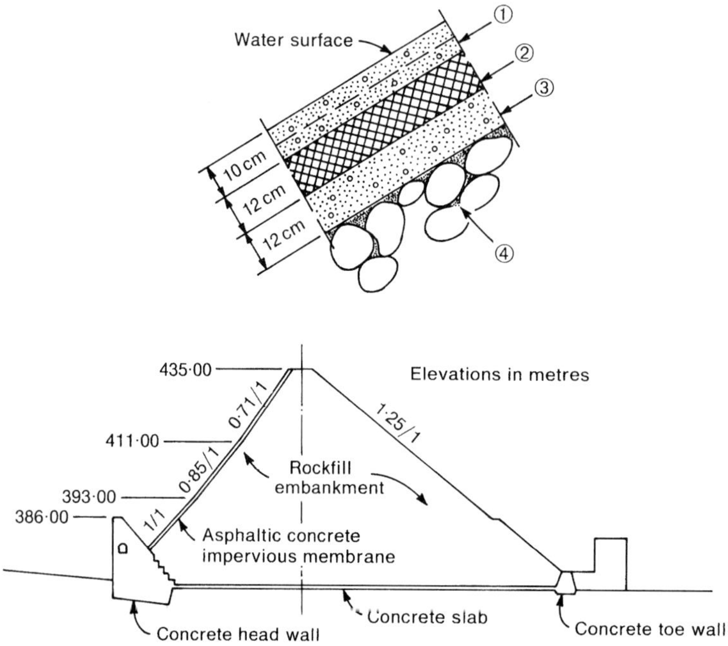

1. Porous concrete reinforced with 5 mm galvanised steel wires in 90 mm grid (removed completely in 1953)
2. Asphaltic concrete in two equal layers
3. Porous concrete with open drains running to gallery at upstream toe
4. Voids in rockfill embankment surface filled with sand–cement mortar

porous concrete which acts as a drain, and a 10 cm outer layer composed of porous concrete reinforced with a galvanised steel wire mesh which provides both thermal and physical protection. The underlying porous concrete drainage layer was placed for the purpose of collecting and controlling any water that escapes through micro-cracks in the asphaltic concrete. It contains interior open drains that run vertically down the slope at intervals, discharging into the reinforced concrete gallery that runs along the upstream toe of the dam. The locations at which the water enters the gallery provide some indication of the position of leaks in the membrane. In 1939, when the water level reached a new height, due to the rupture of the asphaltic concrete caused by differential settlement of the supporting layer, leakage appeared in the control gallery. The leakage gradually decreased and, after several months, virtually stopped, providing evidence of the remarkable self-repairing feature of asphaltic concrete (Sherard *et al.*, 1963).

13.3. Use of geosynthetics

Apart from the conventional membranes used in earth dam construction, as discussed in the previous section, geosynthetics (mainly geomembranes and geocomposites) are now popularly being used in the construction of new earth dams and the rehabilitation of existing water-impounding structures, due to their excellent long-term stability and performance, in order to achieve specific purposes. Moreover, properly designed and efficiently installed geosynthetics in earth dams contribute to an increase in safety, which corresponds to a positive environmental impact on dam structures.

The reasons why geosynthetics are used extensively are

- in earth dams they may serve the function of drainage, filtration, separation, reinforcement, protection and/or water barrier
- they are soft and flexible and can therefore endure some of the elastoplastic deformation that results from subsidence, expansion, landslide and seepage of soil
- they possess a certain amount of mechanical strength, which is favourable with regard to the selection of the dam construction materials
- the permeability of geomembranes is much lower than that of clay or concrete.

In view of the above-mentioned merits, Degoutte (1987) has described practical examples of the use of geotextiles in low-height earth dams as filters and as protection material. Christopher and Fischer (1992) have reviewed the current state of the practice in designing with geotextiles by highlighting the commonly accepted design criteria. Bouazza *et al.* (2006) have reported on the unsaturated hydraulic characteristics of geotextiles and their profound impact on the overall performance of water-impounding structures that are constructed with earth. Although the majority of the non-woven geotextiles exhibit very small suction and air-entry values, a tiny change in the suction value results in a drastic change in their hydraulic conductivity. Thus the inclusion of geotextiles in unsaturated earth dams, for purposes of drainage, separation or filtration, can potentially impede the flow of water and may result in redistribution of the moisture profiles within the dam, which in turn affects the overall stability of the system.

A review of the application of geosynthetics in earth dams, for different functions, is presented in the following sections. Case studies of different types of dam in which geosynthetics have been used are also presented.

13.3.1 Geosynthetics as a fluid barrier

According to the available records, geomembranes were first used as a waterproofing element or fluid barrier in a dam in 1959 at the Contrada Sabetta Dam in Italy (Cazzuffi, 1987). This rockfill dam, which is 32.5 m high, has performed successfully. Two sheets of polyisobutylene geomembrane of 2 mm thickness, protected with concrete slabs, were installed on the upstream face of the dam during the initial construction. In most cases, geomembranes are externally protected against a variety of environmental agents, such as wind, waves, floating debris, ice, vandalism, temperature variations and ultraviolet radiation, by the superposition of a protective layer, such as concrete slabs, precast concrete elements, or geosynthetic-reinforced gunite. The cross-section of the Contrada Sabetta Dam is shown in Figure 13.5. After 20 years of service, samples of the polyisobutylene geomembrane were collected from the barrier, and tested for their physical and chemical properties. Based on the observed data, it has been concluded that the barrier membrane was intact and its performance was still good (Cazzuffi, 1999). However, polyisobutylene membranes are no longer used as a fluid barrier, not because of poor performance, but primarily because modern geomembranes made of thermoplastics are easier to sew (Cazzuffi *et al.*, 2010).

In another application, a 0.9 mm-thick polyvinyl chloride (PVC) geomembrane was used in the Dobsina earth fill dam (10 m high, crest length 204 m) in Slovakia. The PVC geomembrane was placed on the 1 in 2.5 (vertical to horizontal) upstream face and then protected with prefabricated concrete slabs (Cazzuffi *et al.*, 2010). In 1967, the Miel earth fill dam (15 m high, crest length 130 m) in France was constructed with 1 mm-thick butyl rubber geomembrane as a fluid barrier on its upstream slope. A PVC geomembrane was covered with 0.20 m-thick sand and gravel, which was in turn overlain by 0.70 m-thick rip rap (ICOLD, 1993a).

For a geosynthetic to act as a fluid barrier, the procedure based on the use of metallic ribs may be adopted. The procedure allows continuous fastening along vertical lines, and horizontal prestressing, which eliminates sagging due to self-weight. The geosynthetic acts as the main drainage system by means of small holes along the side of the ribs. The geotextile provides protection against puncturing, resulting from the existing coarse surface on the upstream face. It is also designed to provide local drainage, so as to eliminate the water between the body of the dam and the

Figure 13.5 The Contrada Sabetta rockfill dam, Italy
Reproduced from Cazzuffi (1987), Global Trade Media; and ICOLD (1991), © ICOLD-CIGB

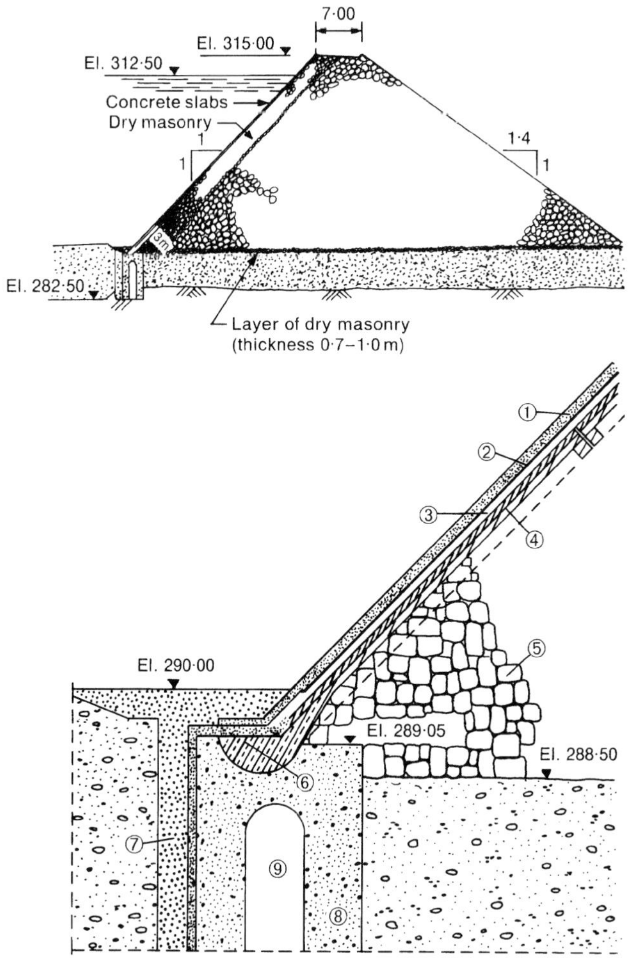

1. Concrete slab (0.20 m thick)
2. One sheet of bituminous paper-felt and two sheets of polyisobutylene geomembrane (2.0 mm thick) and bituminous adhesive
3. Porous cement concrete (2.0 m thick)
4. Reinforced concrete slabs (0.25 m thick)
5. Dry masonry (thickness 2.00–3.00 m)
6. Joint between plinth and upstream facing
7. Plastic concrete diaphragm wall
8. Concrete plinth
9. Inspection and drainage gallery

geomembrane. Mechanical fastening of the geocomposite to the face of the dam is done by means of metal plates and anchors, which are tested at 2 MPa pressure. In order to avoid seepage of reservoir water into the upstream face of an earth dam, the geomembrane is attached all along its perimeter with a watertight seal, which consists of an AISI 304 steel profile anchored to the dam face with threaded rods, washers and bolts every 15 cm. The surface of the concrete, onto which the profiles press the geocomposite, is prepared with a layer of epoxy mortar. There is a synthetic rubber strip between the steel profile and the geocomposite. Normally, flat and C-shaped profiles, 80 × 8 mm and 70 × 36 × 4 mm, respectively, and 2000 mm long, are used. Testing confirms the watertightness of the system. Daily measurements of water loss showed a maximum total loss of 2.7 l/s; only 14% of the total loss has been attributed to the PVC geomembrane. However, regular inspections should be carried out on site, whenever the reservoir is drained. The benefits associated with this type of construction are (Cazzuffi, 1987)

- total waterproofing
- new and efficient upstream-face drainage system
- prefabrication
- low cost and low installation time
- easy and practically cost-free maintenance
- long-term durability, etc.

In another example, a geomembrane composite was used at cofferdams in the Shuikou hydroelectric power station, in Fujian province, Nanping, China. The cofferdams on the upper and lower reaches were constructed using stone rip rap underwater, with a fill of sand–gravel in the middle above the water level. The upper part, in the centre of the cofferdam body, is a composite geomembrane, and the lower part is a concrete seepage-prevention wall. The construction of the Shuikou cofferdam shows that both plastic concrete and composite geomembrane possess good properties for the prevention of seepage, and can adapt to the deformation of the cofferdam body and foundation. Measurements have shown that the unit seepage capacity is 1.5 m^3/day, which satisfies the design seepage-prevention requirement.

At the Zhushou reservoir, in Sichuan province, Zhushou, China, the original design was a clay core stone ballast. It was noticed that the soil used for the core had a considerable gravel content, which resulted in a very high permeability (as high as 10^{-5} m/s). As such, it was decided to lay a geomembrane composite layer at the upper side of the gravely clay core wall, combining the composite geomembrane with the gravely clay core wall to prevent seepage. It has been observed that, as the composite geomembrane is combined with a gravely clay core wall to prevent seepage, the phreatic line lowers in height, leading to a decrease in seepage (Tao *et al.*, 1994).

Geomembranes have also been used for the waterproofing of earth dam cores. For example, the Bilancino Dam in Italy consists of a 42 m-high zoned embankment, including a 15 m-high cofferdam (Figure 13.6). The upstream face of the cofferdam has been waterproofed using a PVC geomembrane (thickness, $t_{GM} = 1.2$ mm), with a polypropylene (PP) non-woven geotextile (mass per unit area, $\mu = 350$ g/m^2) used as mechanical protection against any possible damage due to puncturing by the rockfill underneath (Baldovin, 1993).

Fayoux and Potie (2006) have described the application of reinforced PVC geomembranes as a waterproofing material to control the seepage losses from irrigation water storage reservoirs at the Canary Islands of Morocco. It has been reported that the price of a stored cubic metre of water is two to three times cheaper in geomembrane-waterproofed reservoirs than in conventional concrete-lined reservoirs. Although the PVC geomembrane had exhibited a small

Figure 13.6 Bilancino zoned embankment dam, Italy
Reproduced from Baldovin (1993), Taylor & Francis Group

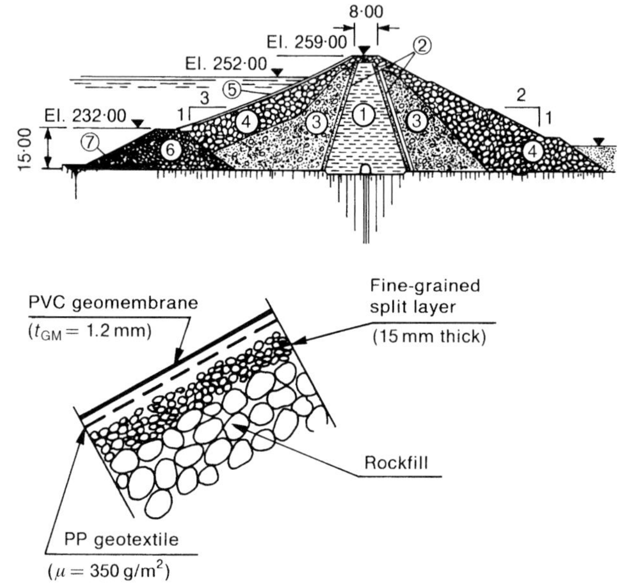

1. Clayey silt core
2. Filters
3. Transitions
4. Rockfill
5. Rip rap
6. Cofferdam (with upstream face 1:2 and downstream face 1:1.5
7. PVC geomembrane (t_{GM} = 1.2 mm) and PP geotextile (μ = 350 g/m^2)

variation in its plasticity behaviour due to its exposure to ultraviolet radiation for almost 22 years, the hydraulic characteristics of the membrane remained satisfactory.

To protect the Three Gorges Dam project, during its construction, a rockfill cofferdam of height 88.5 m was constructed. The body of the cofferdam dam was formed by dumping decomposed granite sand into deep water. To control the seepage through the cofferdam, two vertical concrete diaphragms, along with a composite geomembrane, were constructed in the middle of the cofferdam. The overall stability of the cofferdam was ensured by monitoring the tensile strain in the geomembrane, using large strain gauges, over three years. It was observed that the average tensile strain of the geomembrane in the cofferdam was about 11%, which is below the breaking stain of the geomembrane (Bao *et al.*, 2006). In addition, Kaichang *et al.* (2009) undertook extensive laboratory experiments to assess the mechanical and hydraulic characteristics of the composite geomembrane used in the cofferdam of the Three Gorges Dam. Similarly, Shu *et al.* (2010) have described the application of a composite geomembrane to control the seepage at the Xixiayuan reservoir on the Yellow River. To prevent seepage from the reservoir, the upstream face of the Xixiayuan earth–rock dam (20.2 m high, crest length 2609 m) was provided with a composite geomembrane.

Erlingsson and Hauksson (2009) have reported a seismic stability analysis of a proposed geomembrane-faced rockfill dam. The dam site is situated in the south Iceland seismic zone and has experienced two major earthquakes with a peak ground acceleration of 0.84*g* in 2000 and 2008. The stability analysis was undertaken to ensure that the proposed rockfill dam would be able to withstand this earthquake loading without sustaining major damage or undergoing total collapse. The largest relative deformations due to the applied earthquake loads along the geomembrane interface were found to be 60 cm. These deformations were mainly localised, and hence the total stability of the proposed dam would not be endangered.

Girard *et al.* (1990) and Briancon *et al.* (2002) have evaluated the interface friction angle of the various geosynthetic lining systems that are commonly used in hydraulic Civil Engineering structures, and hence their stability can be assessed. For this purpose, large box shear tests were carried out on an inclined failure plane. The tests were done under a very low normal stress, and under both dry and wet hydraulic conditions that may prevail in the field. The study highlights the importance of taking into account the effect of moisture when designing geosynthetic lining systems.

Despite their efficient function as fluid barriers, geomembrane liners may not be considered to be 100% impervious, as they are susceptible to damage during manufacture, transport and installation, and over their service lifetime. The defects are usually tears, flaws, punctures and seam imperfections, which ultimately diminish the effectiveness of the geomembrane as a fluid barrier. Several researchers have proposed numerous analytical equations to quantify the leakage rates thorough these defects (Giroud and Bonaparte, 1989; Giroud, 1997; Touze-Foltz and Giroud, 2003; Rowe, 2005). In addition, Weber and Zornberg (2005) have evaluated the leakage rates through such defects in a geomembrane under a very high hydraulic gradient.

Although geomembranes have been used extensively as efficient fluid barriers to minimise the seepage rates in earth, masonry and concrete dams, their serviceability and overall performance need to be assessed. With this in view, researchers have evaluated the ageing mechanism of different geosynthetic systems used in the construction of water-impounding structures (Koerner and Hsuan, 2003; Rowe *et al.*, 2004; Brachman *et al.*, 2008; Hsuan *et al.*, 2008; Lodi *et al.*, 2008; Rowe *et al.*, 2010a,b). Koerner and Hsuan (2003) have also evaluated and compared the plasticiser-to-resin and water-to-plasticiser bonding strengths. Knowledge of these bond strengths is essential, as a reasonable amount of plasticiser is added to the resin during the formulation of the PVC geomembranes in order to provide the membrane with sufficient flexibility throughout its service life.

13.3.2 Geosynthetics as a drainage channel

The first reported application of a geosynthetic as a chimney drain was at the 11 m-high Brugnens earth dam in France, constructed in 1973. There, a thick polyester (PET), needle-punched non-woven geotextile was used as a drain (Giroud, 1992). More recently, geosynthetics have been used for drainage at La Parade homogeneous earthfill dam in France (Navassartian *et al.*, 1993). In France since 1987, geocomposite shaft drains (including a non-woven geotextile draining core between two PP non-woven geotextile filters) have been used instead of granular drains in the construction of a number of homogeneous earthfill dams, about 10 m high. The geocomposite drains were set down gradually with alternate earth layers (Figure 13.7).

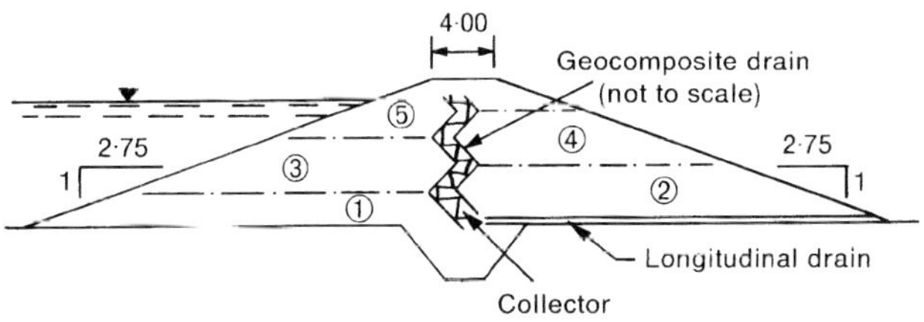

1. First compacted layer
2. Second compacted layer
3. Third compacted layer
4. Fourth compacted layer
5. Fifth compacted layer

Figure 13.7 La Parade homogeneous earthfill dam, France
Reproduced from Navassartian *et al.* (1993)

The chimney drain concept can also be used for rehabilitation purposes. In the case of embankment dams, which exhibit seepage through their downstream slope, the construction of a drainage system in the downstream zone is required. A geocomposite drain, placed on the entire downstream slope, or on only the lower portion of it, and covered with backfill, also performs well. The geocomposite drain must be connected to the new toe of the dam by outlet pipes or by a drainage blanket. This technique has been used at the 13 m high Reeves Lake Dam in the USA, which was repaired in 1990 by placing a geocomposite drain (including a polyethylene (PE) geonet core between two PP thermobonded non-woven geotextile filters) on the downstream slope (Wilson, 1992).

The Ben Boyd Dam in Australia (Figure 13.8) was designed as a water reservoir; its height is approximately 30 m. At the time of the design, detailed soil testing revealed that the foundation soils have dispersive characteristics (McDonald *et al.*, 1981). As the locally available filter sand did not fully meet the retention criteria for the foundation soil, a geotextile (which was effective where the sand was deficient) was included, in addition to the sand, at the underside of the downstream drainage blanket.

13.3.3 Geosynthetics as a filter

Geosynthetics, especially geotextiles, are also used as filters, at various locations, within dams

- beneath upstream and downstream armouring
- between toe drains and *in situ* material
- either side of chimney and blanket drains
- either side of core material, as transitions
- both in and beneath shell material, as consolidation expedients.

Figure 13.8 Details of the downstream drainage blanket at the Ben Boyd Dam, Australia, 1977
Reproduced from McDonald *et al.* (1981)

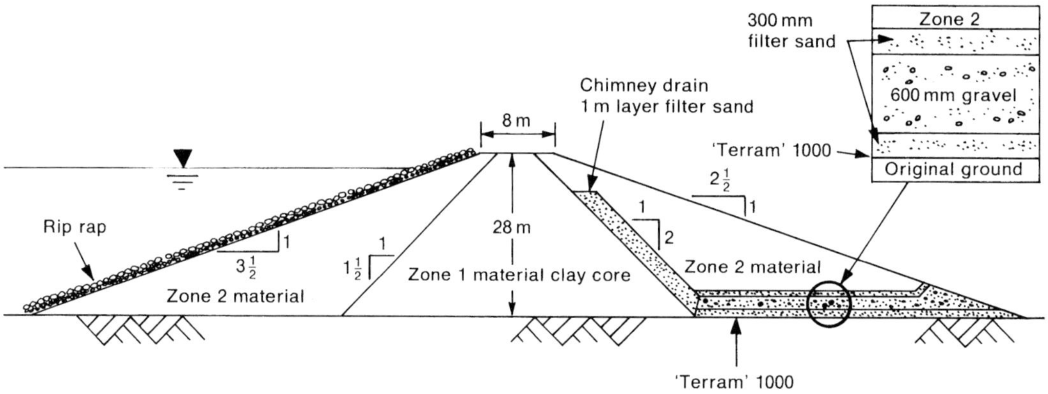

An interesting use of geotextiles is as a replacement for or as supplementary to granular filters. However, geotextiles do not act in the same manner as granular filters, mainly due to the fact that the stability of the interfaces between different soils under hydraulic flow involves complex mechanisms (Gourc and Faure, 1990; Artières *et al.*, 2009).

The potential uses of geotextiles in dams may be divided into two groups: construction expedients and permanent uses. Construction expedients include uses such as in temporary haul roads and in the drainage of fill material to speed up consolidation. The permanent uses of geotextiles should be at non-critical locations within the dam body, where maintenance is relatively easy. For example, they can be used beneath thin coverings of embankment protection (rip rap) on the upstream or downstream face, or in the toe drain. In exceptional cases, where much higher risks are acceptable and the required life of the dam is relatively short, the use of geotextiles as primary filters within smaller dams may be considered. It is a good practice to confine the use of geotextiles within dams to expedite construction, or as an adjunct to chimney and blanket drains, where their purpose is primarily to hold back core material and thereby, prevent contamination of the main granular filter. Geotextiles should not be seen as an alternative to a granular filter in dams (Legge *et al.*, 1994).

In 1970, a continuous-filament non-woven geotextile was used, for the first time, in a large earth dam (Valcros Dam in France) to perform a major function as a filter (Giroud *et al.*, 1977; Artières *et al.*, 2009). The geotextile was used to act as a filter on the upstream slope between the rock and the earth fill and on the downstream slope around the main drains. It was reported by Delmas *et al.* (1994) that after 22 years the efficiency of the continuous-filament needle-punched geotextile was the same as its efficiency when placed originally.

Geotextiles were used in association with granular filters in the Hans Strijdom Dam, South Africa (Figure 13.9). Problems were encountered in obtaining filter sand in sufficient quantity for the construction of this 57 m high, zoned rockfill dam and a PET non-woven geotextile filter ($\mu = 340\ g/m^2$) was placed between the core material and

Figure 13.9 The Hans Strijdom zoned rockfill dam, South Africa
Reproduced from Hollingwarth and Druyts (1982)

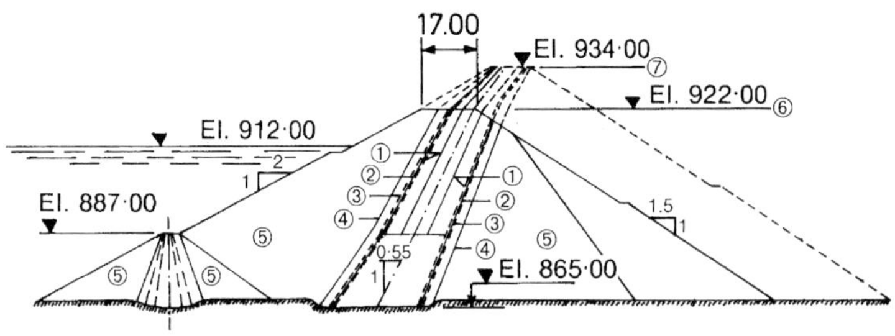

1. PET geotextile filter (μ = 340 g/m^2)
2. Sand filter
3. Gravel filter
4. Selected rockfill transition
5. Rockfill
6. Attained crest level
7. Final foreseen crest level

Figure 13.10 The New Esna Barrage – the lower part of this dam was built underwater, in a fast-running stream. The four positions where geotextile filters are used are shown by the dashed lines, and indicated by numbers in squares. The primary function was retention
Reproduced from Sembenelli and Sembenelli (1994), Industrial Fabrics Association International

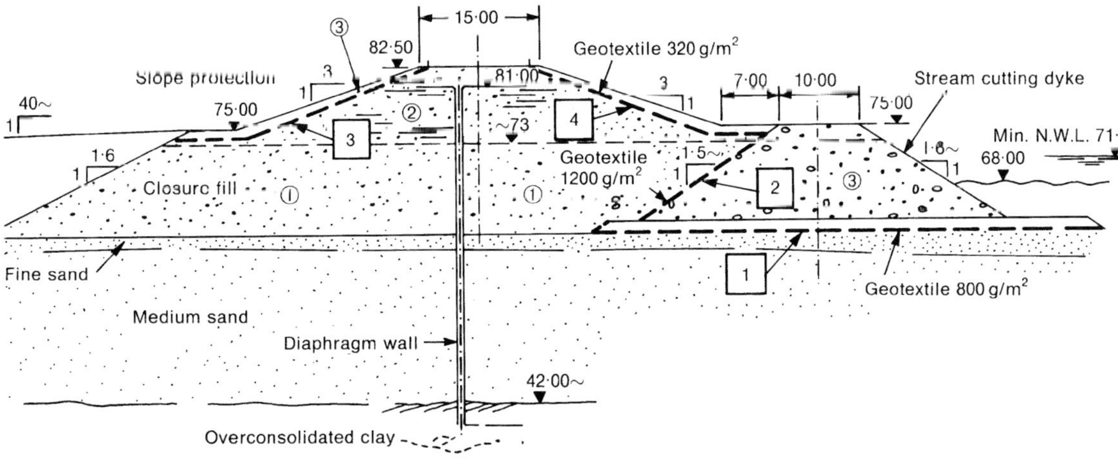

1. Pit run El Deir gravel and sand <100 mm dumped in water
2. Pit run El Deir gravel and sand <100 mm placed in 0.5 m lifts compacted with vibrating roller
3. Rockfill 50–100 mm excluding quarried or crushed materials

the sand filter. This reduced the thickness of the sand filter to 1 m on the upstream and downstream sides of the core zone (Hollingwarth and Druyts, 1982).

The New Esna Barrage Dam on the Nile (Figure 13.10) was a replacement for the Esna Barrage Dam, built in 1908, which had been suffering from scouring and age-related damage. In the new design, the hydraulic fill was substituted by a well-graded, processed, alluvial material that came from a nearby wadi, and geotextiles were used extensively, as filter and erosion preventers. In the design, geotextiles were used at the following four specific locations

- between the fine, alluvial sand of the Nile and the uniform cobbles of the stream cutting dyke, forming the free-draining toe of the dam
- between the well-graded sand and gravel, dumped in the water to create the body of the embankment, and the toe cobbles
- under the rounded rip rap protection specified on the upstream slope
- under the granular slope protection specified on the downstream side of the dam.

In such a mechanism, the main aim of using geotextiles is to prevent the migration of the river bed sand and the fines of the embankment dam to the toe (Sembenelli and Sembenelli, 1994).

Luettich *et al.* (1992) attempted to address the common filter design problems and existing solutions by using a comprehensive systematic approach, which may be helpful to practising engineers and designers. Although the design procedures at that time often included only permeability and retention criteria, the authors highlighted the importance of incorporating several other concepts, such as identifying and prioritising the primary function(s) of a material for a given application, quantifying the internal stability of the retained fill, survivability and durability. Artières *et al.* (2009) have presented a history of filter design concepts, with an emphasis on the major breakthroughs, improvements and evaluation of new filtration systems with optimised properties. In addition, the present authors have introduced a new monitoring solution, based on optical fibres, to detect early signs of malfunctioning, such as leaks or instability, which can be used to improve the service life and long-term performance of earth dam structures.

13.3.4 Geosynthetics as a protective layer

In many embankment dams, where a geomembrane is used as a barrier to fluid, a thick geotextile must be placed on one or both sides of the geomembrane itself, in order to protect it from potential damage by adjacent materials, typically the granular layer underneath and the external cover layer. In this technique, two different layers of geotextile are laid, both having a protection function. The lower geotextile, bonded in the factory to the geomembrane, can be glued to the original bituminous concrete facing, while the upper geotextile is independently placed between the PVC geomembrane and the cast-in-place concrete cover layer. This technique has been employed at the Codole Dam in France, as shown in Figure 13.11.

13.3.5 Geosynthetics as a reinforcement

The first dam in which geosynthetics were used for reinforcement purposes was the Maraval Dam in France. This 8 m

Figure 13.11 The Codole rockfill dam, France
Reproduced from ICOLD (1991), © ICOLD-CIGB

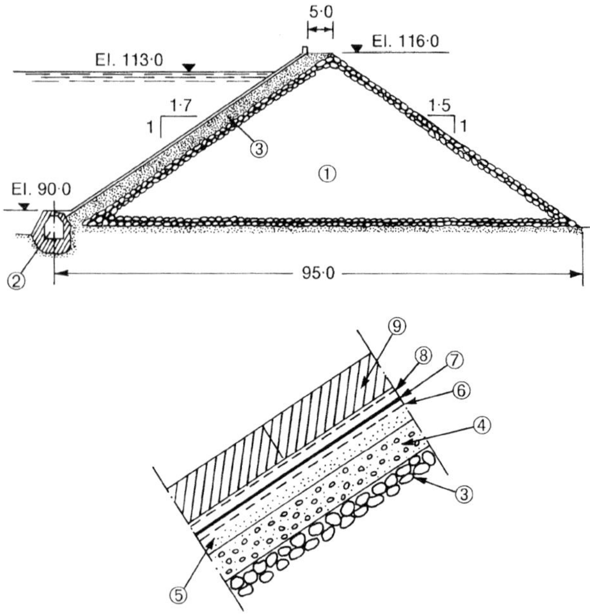

1. Rockfill (up to 1.00 m size)
2. Inspection and drainage gallery
3. Sand and gravel layer (2.00 m thick, 25–120 mm grain size)
4. Gravel layer (0.15 m thick, 25–50 mm grain size)
5. Cold premix layer (50 mm thick, 6–12 mm grain size)
6. Geotextile (μ = 400 g/m^2) bonded to geomembrane
7. PVC geomembrane (t_{GM} = 2.0 mm)
8. Geotextile (μ = 400 g/m^2)
9. Concrete slabs (0.14 m thick, 4.5 × 5.0 m^2 size)

high earth dam was constructed in 1976. As shown in Figure 13.12, the dam has a sloping upstream face lined with a 4.8 mm-thick bituminous geomembrane. A vertical downstream face has been obtained by constructing a multi-layered geotextile–soil mass, reinforced with a high-strength PET woven geotextile ($\mu = 750$ g/m^2, $T = 210$ kN/m). Due to the vertical downstream face, the spillway is short and therefore was not very expensive to construct. This is beneficial for such small dams where spillways usually represent a large fraction of the total construction cost.

Another interesting feature of this construction is that the dam was overtopped three times during construction, with

Figure 13.12 The Maraval earth dam, France
Reproduced from Kern (1977)

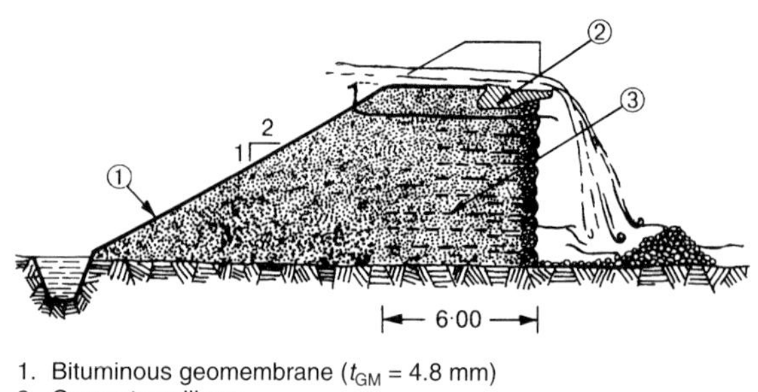

1. Bituminous geomembrane (t_{GM} = 4.8 mm)
2. Concrete spillway
3. Earth mass reinforced with PET geotextiles (μ = 350 g/m^2)

virtually no damage (Kern, 1977). However, due to exposure of the geotextile on the vertical downstream face, this technique has not been developed for the construction of small dams. Efforts have been made to use more aesthetically attractive metallic facing systems for dams of a low or moderate height (up to a maximum of 22.5 m) (ICOLD, 1993a).

Geosynthetics have been used for reinforcement at the Davis Creek Dam in the USA. This dam, constructed in 1990, is 33 m high and its upper part has a steep geogrid reinforcement on the downstream slope. Two types of geogrid were used: six layers (each 5 m long) of one type of high-density polyethylene (HDPE) were used to provide adequate deep-seated stability; and 19 layers (each 2 m long) of a lighter HDPE were used to give adequate near-surface stability. Of course, particular attention was given to establishing vegetation on the downstream slope. To achieve this, hydro-seed was covered with natural fibres (of coconuts) and irrigated for a long time (Engemoen and Hensley, 1989). A typical cross-section of such an arrangement is shown in Figure 13.13.

For rehabilitation purposes, such as heightening a dam itself to increase the storage capacity or heightening the free board, the use of a geosynthetic is ideal, and it is recommended that much more research be conducted in this area.

Finally, in seismically active regions, it is a good practice to adopt the use of geosynthetics in order to reinforce the bituminous concrete layer incorporated in embankment dam revetments. Usually, PET woven geotextiles ($\mu = 250$ g/m^2, $T = 50$ kN/m), which are able to resist thermal shocks due

Figure 13.13 The Davis Creek earth dam, USA
Reproduced from Engemoen and Hensley (1989)

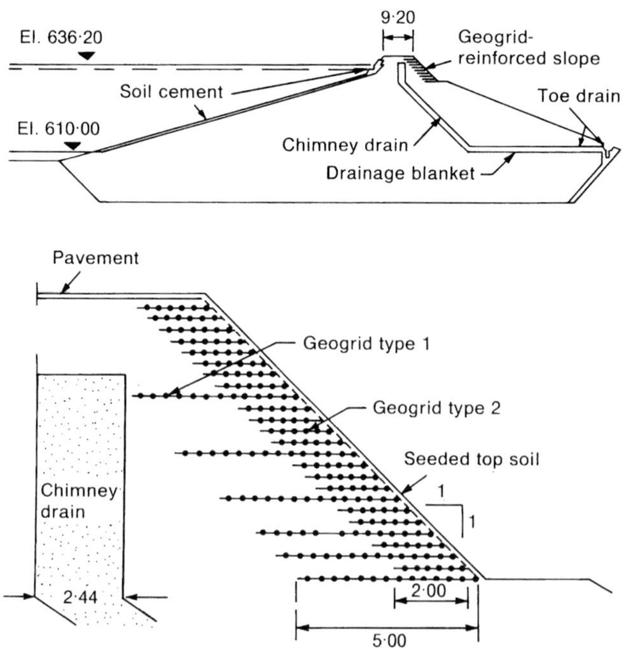

to contact with bitumen at about 160°C, are selected for this purpose, as suggested by Cazzuffi (1988).

13.3.6 Geosynthetics as an erosion control layer

Geosynthetics are also being used to control surface erosion in a number of embankment dams, both in new constructions and for rehabilitation purposes. Surface erosion may occur either due to atmospheric agents (mostly rain water) or by overtopping of the dam.

In the case of erosion caused by rain water, the entire downstream facing and the upper portion of the upstream facing, which is in direct contact with the stored water, is protected using typical techniques adopted for river bank revetments (e.g. a rip rap, in which geotextiles perform as a filter; or (see ICOLD, 1993b) soil–cement blankets, concrete slabs, bituminous concrete layers and so on, in which geosynthetics can be incorporated to provide separation or even reinforcement). The products most commonly used to control surface erosion due to atmospheric agents are geomats and geocells. Sometimes biotechnical mats are also adopted, particularly when biodegradation of the geotextile is desirable, such as when they are used in a temporary role during vegetation growth. This is common practice for solving the problems induced by erosion due to rainfall and the consequent runoff, as described in Chapter 10.

The most challenging application of geosynthetics is related to the protection against overtopping, which represents a crucial aspect of dam engineering. Many failures of embankment dams have been induced in the past by overtopping. Documented case histories show that the phenomena induced by overtopping are influenced by several factors, such as the valley morphology, the dam type and size, the physical and mechanical characteristics of the construction soils, the hydraulic regimen of the reservoir, and so on. The soil grain sizes have an important influence on the failure mechanism (Croce, 1989). Figure 13.14 illustrates a typical failure mechanism induced by overtopping, according to physical models of embankment dams constructed with coarse-grained materials and without protection on the downstream face.

Downstream slope protection was provided in the 12 m high Moochalabra Dam in Australia, which was constructed in 1971. Metallic grids and meshes were adopted to reinforce the downstream face for a time longer than the construction period (Johnson, 1973). The dam has already been overtopped several times, without sustaining any substantial damage, as shown in Figure 13.15.

For the rehabilitation of small dams, articulate concrete blocks linked by cables and resting on a geotextile were used in 1991 in the USA, in order to protect the crest and the downstream slope against overtopping (Wooten *et al.*, 1990). In this case, woven geotextiles were adopted mainly to perform as a filter material. However, the opening size of the geotextile was selected not only to satisfy the filter criteria, but also to allow penetration by grass roots. In fact, the articulate blocks were covered by a grassed topsoil layer in order to give it a natural appearance and additional anchorage for articulate concrete blocks, as shown in Figure 13.16. This indicates that geosynthetics, in association with earth materials and vegetation, form a stable solution to resisting overtopping phenomena in earth embankment dams.

Figure 13.14 A typical failure mechanism induced by overtopping: (a) first phase, (b) second phase and (c) third phase
Reproduced from Croce (1989)

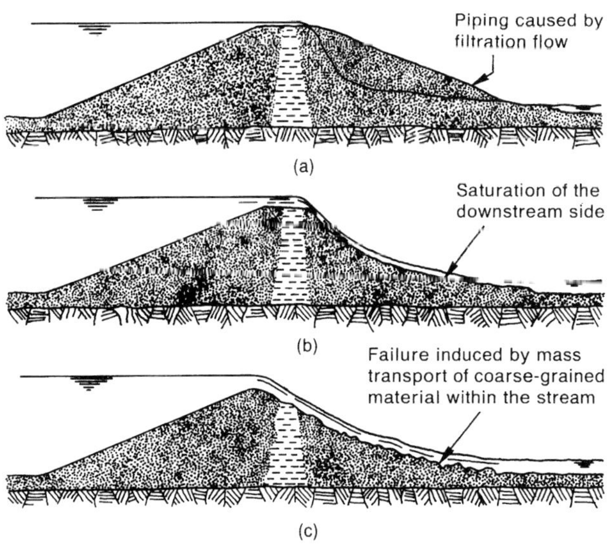

The use of gabions and mattresses on the upstream or downstream faces of earth dams has been well established for more than 30 years. Nowadays, gabions and mattresses are used widely, because they function as reinforced and flexible structures that can resist both tension and compression. These units may be used to form a protective lining for dams with an impermeable core and this can prove to be the most versatile, safe and, often, most economical solution. Gabions and mattresses are also successfully used in auxiliary structures, such as a lining to spillways, outlet channels, sedimentation basins, temporary works, the protection of fill embankments, etc. In these applications, the ease and rapidity of installation are the determining factors for the selection of the type of structure.

There are many examples where this technique has been used to protect against surface erosion. In the Grenada Dam in the USA, a revetment with 12 in.-thick PVC-coated gabions

Figure 13.15 The Moochalabra zoned rockfill dam, Australia
Reproduced from Johnson (1973)

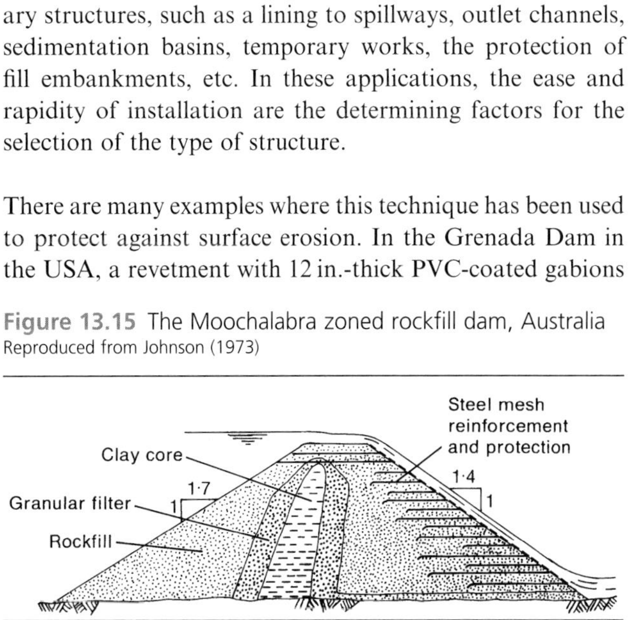

Figure 13.16 Detail of the articulate concrete blocks, with a geotextile filter and a grassed topsoil cover, used to protect the downstream face against overtopping at the Blue Ridge Parkways dam (8.5–12 m high) in the USA
Reproduced from Wooten *et al.* (1990), ASCE

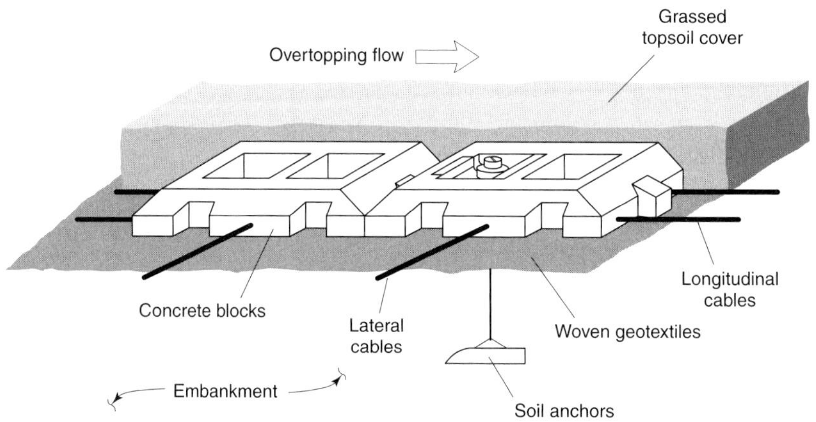

has been used. This dam was constructed to control the flooding of the Yazoo River basin. The gabion protection replaced a previously existing rip rap revetment that was completely destroyed during an extraordinary flood in 1983.

The upstream face of the Paduli Dam in Italy, across the Enza River, was completely made up of and sealed with sand–asphaltic–mastic-grouted mattresses. On completion of works, the surface was coated with bituminous aluminium paint to improve its resistance to sunlight.

13.4. River bed and bank protection

Geofabric earth pillows and flexible mattresses are being increasingly used for river bed and bank protection. In 1984, this technique was used at the Houzhou embankment, Jianli County, Fuyang City, China (Bao *et al.*, 1994). The geofabric used was made of PP. Each geofabric bag, filled with 6–7 t of soil (sand), was 10 m long, 0.80 m wide and 0.90 m high. The cushions were 70 m long and 10 m wide, with both sides reinforced with a 10–12 mm nylon rope. In addition, at each end of the cushions there was an extra 5–12 m of nylon rope, available for use as connectors to the piles and tail pillows on the shore.

While using flexible mattresses, care must be paid when they are laid on the bed slopes. Care must also be taken against the plain slides, drifts or uplifts that arise due to the action of the stream flow. Submerged mattresses must be properly fixed on the slope and partly embedded in a trench that is specially dug at a given spot and then rock ballasted. The ballast used can be made of concrete blocks or cement soil blocks (Bao *et al.*, 1994). Figure 13.17 shows a typical section of a geofabric bank-protection scheme.

Figure 13.17 A typical geofabric bank-protection scheme
Reproduced from Bao *et al.* (1994); Industrial Fabrics Association International

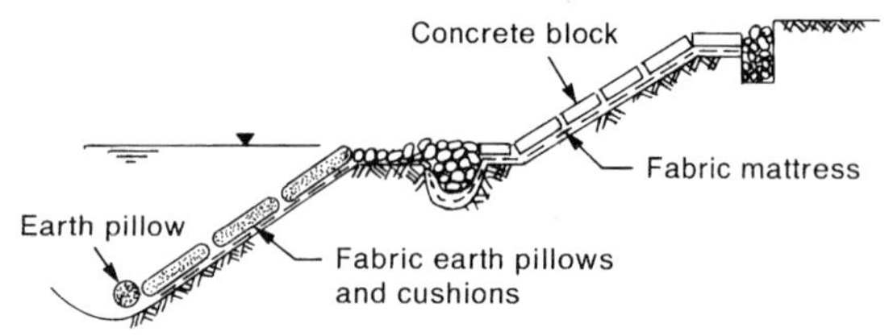

Haselsteiner *et al.* (2008) have reported the various construction methods and concepts used in the design of overflow dikes encompassing geosynthetic protection systems. The influence of the active embedded length of the reinforcement layers, and the assumed hydraulic loads on the overall slope stability of reinforced soils and internal erosion processes, have been evaluated using laboratory model test results.

Every year, a good number of small-sized earth dams (less than 15 m high) are severely damaged or collapse totally when they are subjected to a flood that exceeds the drainage capacity of a spillway. In order to overcome this problem, Matsushima *et al.* (2008) have proposed an innovative stabilisation technique that involves the use of geosynthetic soil bags with extended tails or reinforcement elements. The efficacy of the proposed stabilisation methodology against overtopping has been evaluated by conducting hydraulic overflow-induced failure tests using a 3.5 m high, 2.3 m wide model and different flood levels. Koffler *et al.* (2008) have also briefly highlighted the role of geosynthetics in protection against erosion of river and coastal banks, and marine and hydraulic structures.

13.5. Design considerations

In general, the design procedure is guided by the International Commission on Large Dams (ICOLD). A desktop analysis may be undertaken based on the available guidelines. Once a desktop analysis has been completed and suitable geosynthetics identified, they are subjected to soil–geosynthetic compatibility testing before making a final selection, which includes consideration of the minimum strength and deformation requirements of the geosynthetics. These parameters need to take account of both the short-term loading expected during installation and construction, as well as post-construction loads and deformations. While overall embankment settlement may be small, local stresses and strains may be large due to differential settlements or shrinkage of the soil. It is for this reason that a geosynthetic needs to maintain its restraining characteristics, even after local concentration of stresses and strains has occurred. There may be a substantial change in the pore size of the geotextiles due to elongation (Legge, 1986). However, the main concern is the extent to which the pores of woven tape and staple fibre products elongate when the fabric is placed under tension. It should be noted that all dams must be designed on the understanding that there is a significant risk that the core will crack and the possibility of internal erosion of the core has to be allowed for in the filter design (McKenna, 1989). Luettich *et al.* (1992) proposed a filter design methodology consisting of the following nine steps.

- Step 1: Define the application filter requirements – identify the drainage material and define the retention versus permeability trade-off.
- Step 2: Define the boundary conditions – evaluate the confining stress and define the flow conditions.
- Step 3: Determine the soil retention requirements – determine the steady-state flow or dynamic flow conditions, define the soil particle-size distribution and Atterberg limits, define the soil dispersion potential and the soil density conditions and determine the maximum allowable geotextile opening size O_{95}.
- Step 4: Determine the geotextile permeability requirements – define the soil hydraulic conductivity, define the hydraulic gradient for the application and determine the minimum allowable geotextile permeability.
- Step 5: Determine the anti-clogging requirements.
- Step 6: Determine the survivability requirements.
- Step 7: Determine the durability requirements.
- Step 8: Miscellaneous design considerations – the geotextile structure, intrusion of the geotextile into the drainage layer, extrusion of fine-grained soil through the geotextile when subjected to high confining pressures, abrasion of the geotextile due to dynamic action, intimate contact of the soil and geotextile, biological and biochemical clogging factors and safety factors.
- Step 9: Select a geotextile filter – ensure that the geotextile has the properties required in Steps 3 to 8 and, if necessary, verify through testing.

The above methodology can be applied for the use of geosynthetics as a filter in dams.

13.6. Summary of main points

- Geosynthetics play several functions (fluid barrier, filtration, drainage, protection, reinforcement) in earth dams, depending on their location within the dam structure and the purpose of their installation.
- Proper use of geosynthetics minimises seepage losses, and improves the overall stability and performance of the dam.
- The case histories presented in this chapter report the comprehensive construction details of important earth and rockfill dam sections, and the limitations and difficulties associated with conventional Civil Engineering materials.
- The use of geosynthetics is associated with a reduction in the amount of natural earth materials that need to be obtained and placed on the dam sites. This is a positive environmental impact of using geosynthetics.
- The construction of earth dams is still an art and an empirical process. As such, engineers must learn from their own experience and must familiarise themselves with the recent practices and advances in the area of earth dam construction before executing a project.

13.7. Some common questions and their answers

Questions

1. How do I select the suitable geomembrane for use as a fluid barrier?

2. Is it essential to cover or protect the geomembrane fluid barrier?

3. How do you evaluate the serviceability and long-term performance of an individual geosynthetic component, and its impact on the overall performance of an earth dam?

4. Why is a drainage layer essential underneath conventional fluid-barrier systems?

5. How are geocomposite drains effective in the rehabilitation of existing earth dams?

Answers

1. The use of polymer geomembranes as a fluid-barrier material in dams, landfill and containment applications has gained in popularity in recent years due to their low permeability, good chemical resistance and excellent durability. The selection of a suitable geomembrane (type of polymer) for a fluid barrier is primarily based on its chemical compatibility, but is also dependent on the ease of installation and seaming. In fact, flexible PVC geomembranes are preferred over rigid PE geomembranes because it is very difficult to install and seam PE geomembranes, especially around small complex geometry.

2. Research has shown that covered bituminous and PVC geomembranes are highly durable and have good long-term performance. Sharp objects or stones that may fall on the geomembrane during the installation and service life are responsible for indentation or puncture of the geomembrane. Hence it is advisable to cover or protect geomembranes with a suitable protection layer (e.g. a thick geotextile) for enhanced serviceability.

3. The majority of the geosynthetic components are made up of polymers, and therefore their physical and chemical properties are expected to deteriorate upon exposure to field conditions, mainly due to ageing phenomena. Thus it is essential to estimate the service life and long-term performance of a geosynthetic material under simulated field conditions. To this end, accelerated ageing tests are commonly used to assess the degradation mechanism, and the service life (long-term performance) is estimated by means of mathematical modelling. Based on the predicted serviceability and long-term performance of the individual geosynthetic component, the overall performance of the earth dam may be evaluated using a suitable numerical tool.

4. As mentioned in Section 13.2, when a single reinforced concrete slab or asphalt slab is used as a fluid barrier, some leakage always occurs through the inevitable hairline cracks, or due to construction inadequacies that become apparent soon after the filling of the reservoir. Subsequently, it is necessary to provide a drainage layer behind such a barrier in an earth dam, in order to drain out the seepage water from the embankment, which may otherwise become unstable due to increased water pressure.

5. As described Section 13.3.2, the chimney drain concept has been employed very effectively in the rehabilitation of many earth dams. In the case of embankment dams, which exhibit seepage through their downstream slope, the construction of a drainage system in the downstream zone is required. A geocomposite drain can be placed on the entire downstream slope, or a portion of it, along with a cover of a suitable backfill to mitigate the instability issues. This technique has been used successfully at the Reeves Lake Dam in the USA, which is a 13 m high dam that was repaired in 1990 by placing a geocomposite drain on the downstream slope.

REFERENCES

Artières O, Oberreiter K and Aschauer F (2009) Geosynthetic system for earth dams – 35 years of experience. *Proceedings of the 2nd International Conference on Long-Term Behaviour of Dams*, Graz, pp. 98–103.

Baldovin E (1993) Filters in geotechnical and hydraulic engineering. *New Development of Filters in Some Recent Italian Embankment Dams*, Balkema, Rotterdam, pp. 321–330.

Bao CG, Li QY and Gong LH (2006) Application of a large strain gauge to measure the deformation of composite geomembrane in the Three Gorges Project cofferdam. *Proceedings of the 8th International Geosynthetics Conference Proceedings*, Yokohama, pp. 631–634.

Bao CG, Wang ZH and Ma SD (1994) Application of geofabric in embankment engineering of Yangtze River. *Proceedings of the 5th International Conference on Geotextiles, Geomembranes and Related Products*, Singapore, pp. 612–613.

Bouazza A, Zornberg JG, McCartney JS and Nahlawi H (2006) Significance of unsaturated behaviour of geotextiles in earthen structures. *Australian Geomechanics* **41(3)**: 133–142.

Brachman RWI, Rowe RK, Arnepalli DN *et al.* (2008) Development of an apparatus to simulate the ageing of geomembranes under chemical exposure, elevated temperatures and applied stresses. *Proceedings of the First Pan American Geosynthetics Conference and Exhibition*, Cancun, pp. 444–451.

Briancon L, Girard H and Poulain D (2002) Slope stability of lining systems experimental modeling of friction at geosynthetic interfaces. *Geotextiles and Geomembranes* **20**: 147–172.

Cazzuffi D (1987) The use of geomembranes in Italian dams. *Water Power and Dam Construction* **39(3)**: 17–21.

Cazzuffi D (1988) The use of polymeric materials as mechanical reinforcement of bituminous concrete water proofing systems in earth structures. *Preprints of Euromech – Mechanical Aspects of Soil Reinforcement*, Chamrousse–Grenoble, pp. 63–64.

Cazzuffi D (1999) Behaviour of the geomembrane applied in 1959 on Contrada Sabetta dam. *Proceedings of International Symposium 'Géotextiles – Géomembranes – Produits Apparentés Rencontres 99'*, Bordeaux, pp. 305–310.

Cazzuffi D, Giroud JP, Scuero A and Vaschetti G (2010) Geosynthetic barriers systems for dams. *9th International Conference on Geosynthetics*, Guarujá, pp. 115–163.

Christopher BR and Fischer GR (1992) Geotextile filtration principles practices and problems. *Geotextiles and Geomembranes* **11**: 337–353.

Croce P (1989) Protection of earth dams from overtopping. *L' Ingegnere* **1–4**: 45–50.

Degoutte G (1987) Practical examples of geotextile used in small earth dams. *Geotextiles and Geomembranes* **5**: 239–250.

Delmas Ph, Faure YH, Farkouh B and Nancey A (1994) Long term behaviour of a geotextile as a filter in a 24-year-old earth dam: Valcros. *Proceedings of the 5th International Conference on Geotextiles, Geomembranes and Related Products*, Singapore, pp. 1199–1202.

Engemoen WO and Hensley PJ (1989) Geogrid steepened slopes at Davis Creek Dam. *Proceedings of Geosynthetics '89 Conference*, San Diego, CA, pp. 225–268.

Erlingsson S and Hauksson DR (2009) Analysis of a geomembrane face rockfill dam during earthquake loading. *Proceedings of the 17th International Conference on Soil Mechanics and Geotechnical Engineering*, Alexandria, pp. 1678–1681.

Fayoux D and Potie G (2006) Geomembranes in Tenerife and La Palma Islands – (Canary Islands). *Proceedings of the 8th International Geosynthetics Conference*, Yokohama, pp. 603–606.

Girard H, Fischer S and Alonso E (1990) Problems of friction posed by the use of geomembranes on dam slopes – examples and measurements. *Geotextiles and Geomembranes* **9**: 129–143.

Giroud JP (1992) Geosynthetics in dams: two decades of experience – Parts 1 and 2. *Geotechnical Fabrics Report* **10(5)**: 6–9; **10(6)**: 22–28.

Giroud JP (1997) Equations for calculating the rate of liquid migration through composite liners due to geomembrane defects. *Geosynthetics International* **4(3–4)**: 335–348.

Giroud JP and Bonaparte R (1989) Leakage through liners constructed with geomembranes. Part II: Composite liners. *Geotextiles and Geomembranes* **8(2)**: 71–111.

Giroud JP, Gourc JP, Bally P and Delmas P (1977) Comportement d'un textile non tisse dans un barrage en terre. *Proceedings of the First International Geosynthetics Society*, Paris, pp. 213–218.

Gourc JP and Faure Y (1990) Soil particles, water and fibres – a fruitful interaction now controlled. *Proceedings of 4th International Conference on Geotextiles, Geomembranes and Related Products*, The Hague, pp. 949–971.

Haselsteiner R, Strobl T, Heerten G and Werth K (2008) Overflow protection systems of flood embankments with geosynthetics. *Proceedings of 4th European Geosynthetics Conference, EuroGeo4*, Edinburgh, Paper No. 50, pp. 1–8.

Hollingwarth F and Druyts FH (1982) Filter cloth partially replaces and supplements filter materials for protection of poor quality materials in rockfill dam. *Proceedings of 14th International Congress on Large Dams*, Rio De Janeiro, pp. 709–725.

Hsuan YG, Schroeder HF, Rowe RK *et al.* (2008) Long-term performance and lifetime prediction of geosynthetics. *Proceedings of 4th European Geosynthetics Conference, EuroGeo4*, Edinburgh, Paper No. 327, pp. 1–40.

ICOLD (International Commission on Large Dams) (1991) Watertight geomembranes for dams. State of the art. *ICOLD Bulletin* **78**: 140.

ICOLD (1993a) Reinforced rockfill and reinforced fill for dams. State of the art. *ICOLD Bulletin* **89**: 190.

ICOLD (1993b) Embankment dams upstream slope protection. *ICOLD Bulletin* **91**: 122.

Johnson RB (1973) Spillway types in Australia and factors affecting their choice. *Proceedings of 12th International Congress on Large Dams*, Madrid, pp. 833–852.

Kaichang AS, Kun BL and Zhiyu CS (2009) The experimental research on the composite geo-membrane in the three gorges dam. *Proceedings of the 2nd International Conference on Long-Term Behaviour of Dams*, Graz, pp. 527–531.

Kern F (1977) An earth dam with a vertical downstream face constructed using fabrics. *Proceedings of International Conference on the Use of Fabrics in Geotechnics*, Paris, pp. 91–94.

Koerner RM and Hsuan YG (2003) Lifetime prediction of polymeric geomembranes used in new dam construction and dam rehabilitation. *Proceedings Association of State Dam Safety Officials Conference*, Lake Harmony, PA, p. 16.

Koffler A, Choura M, Bendriss A and Zengerink E (2008) Geosynthetics in protection against erosion for river and coastal banks and marine and hydraulic construction. *Journal of Coast Conservation* **12**: 11–17.

Legge KR (1986) Testing of geotextiles. *Proceedings of SAICE Filters Symposium*, Johannesburg.

Legge KR, Legge WCS and James GM (1994) Geotextiles as filters and transitions in fill dams. *Proceedings of the 5th International Conference on Geotextiles, Geomembranes and Related Products*, Singapore, pp. 619–624.

Lodi PC, Bueno BS, Vilar OM and Correia NS (2008) Weathering degradation of polyester and polypropylene geotextiles. *Proceedings of the 4th Asian Regional Conference on Geosynthetics*, Shanghai, pp. 35–39.

Luettich SM, Giroud JP and Bachus RC (1992) Geotextile filter design guide. *Geotextiles and Geomembranes* **11**: 355–370.

Matsushima K, Mohri Y, Yamazaki S *et al.* (2008) Design of earth dams allowing temporary overtopping based on hydraulic failure experiments and flood analysis. *Proceedings of the 4th Asian Regional Conference on Geosynthetics*, Shanghai, pp. 757–762.

McDonald LA, Stone PC and Ingles OG (1981) Practical treatments for dams in dispersive soil. *Proceedings of the 10th International Conference on Soil Mechanics and Foundation Engineering*, Stockholm, pp. 355–360.

McKenna JM (1989) Properties of core materials, the downstream filter and design. In *Clay Barriers for Embankment Dams*, Thomas Telford, London.

Navassartian G, Grource JP and Brochier P (1993) Geocomposite for dam shaft drain: La Parade dam, France. *Geosynthetic Case Histories. ISSMFE-TC9*, BiTech, Richmond, BC, pp. 16–17.

Rowe RK (2005) Long-term performance of contaminant barrier systems. 45th Rankine Lecture. *Géotechnique* **55(9)**: 631–678.

Rowe RK, Islam MZ, Brachman RWI, Arnepalli DN and Ewais AR (2010b) Antioxidant depletion from a high density polyethylene geomembrane under simulated landfill conditions. *Journal of Geotechnical and Geoenvironmental Engineering*, ASCE **136(7)**: 930–939.

Rowe RK, Quigley RM, Brachman RWI and Booker JR (2004) *Barrier Systems for Waste Disposal Facilities*. Taylor & Francis, London.

Rowe RK, Rimal S, Arnepalli DN and Bathurst RJ (2010a) Durability of fluorinated high density polyethylene geomembrane in the Arctic. *Geotextiles and Geomembranes* **28**: 100–107.

Segar CP (1935) Steel used extensively in building El Vado Dam. *Engineering News-Record* 211.

Sembenelli P and Sembenelli G (1994) Geosynthetics at New Esna earth dam. *Proceedings of 5th International Conference on Geotextiles, Geomembranes and Related Products*, Singapore, pp. 5–9.

Sherard JL, Woodward RJ, Gizienski SF and Clevenger WA (1963) *Earth and Earth–Rock Dams*. Wiley, New York.

Shu YM, Yuan QY, Chen Q, Li YL and Wu HM (2010) Technology of geomembrane on earth–rock dam for seepage prevention. *9th International Conference on Geosynthetics*, Guarujá, pp. 1265–1268.

Sowers GF and Sally HL (1962) *Earth and Rockfill Dam Engineering*. Asia Publishing House, Bombay.

Tao TK, Tang RN, Zhu HZ, Qian WC and Gu JW (1994) Engineering characteristics and application of geomembrane composite. *Proceedings of 5th International Conference on Geotextiles, Geomembranes and Related Products*, Singapore, pp. 595–600.

Thevenin J (1958) The Ghrib dam. *Travaux* 141.

Touze-Foltz N and Giroud JP (2003) Empirical equations for calculating the rate of liquid flow through composite liner due to geomembrane defects. *Geosynthetics International* **10(6)**: 215–233.

Walker FC (1958) *Development of Earth Dam Design in the Bureau of Reclamation*. US Bureau of Reclamation, Washington, DC.

Weber CT and Zornberg JG (2005) Leakage through liners under high hydraulic heads. *Proceedings of the GRI-18 at GeoFrontiers*, ASCE, Austin, TX (CD-ROM).

Wilson CB (1992) Repair to Reeves Lake Dam, Cobb County, Georgia. *Proceedings of the 1992 Annual Conference of the Association of State Dam Safety Officials*, Baltimore, OH, pp. 77–81.

Wooten RL, Powledge GR and Whiteside SL (1990) CCM overtopping protection on three Park way dams. *Proceedings of Hydraulic Engineering National Conference*, ASCE, San Diego, CA, pp. 1152–1157.

Handbook of Geosynthetic Engineering
ISBN 978-0-7277-4175-2

http://dx.doi.org/10.1680/hge.41752.279

ice
Institution of Civil Engineers
publishing

Chapter 14
Containment ponds, reservoirs and canals

D. Poulain, N. Touze-Foltz, L. Peyras and C. Duquennoi
Hydrosystems and Bioprocesses Research Unit, Cemagref, France

This chapter illustrates the use of geosynthetics, mainly geomembranes and geosynthetic clay liners, in hydraulic applications. After a brief history related to the use of geosynthetics in hydraulic applications, and the definition of the main terms related to geosynthetics used in this chapter, recommendations are given regarding the design of geosynthetic systems, taking into account the features of the underlying and overlying materials. Where necessary, a distinction is made between geomembranes and geosynthetic clay liners, which have different features, even if both perform a lining function. A summary is then given of the durability of various types of geomembranes in hydraulic applications, based on the literature review. Finally, case studies of two ponds, three reservoirs and two canals are used to illustrate the recommendations.

14.1. Introduction

The use of geosynthetics in liquid containment and conveyance applications (i.e. in containment ponds, reservoirs and canals) can be traced back to the 1940s, but their use in these roles came to prominence in the 1960s and 1970s. This chapter presents, first, the historical background necessary to better understand and estimate the importance of geosynthetics technology in this type of application. Basic concepts in the design of geosynthetic systems are given as general indications applying to generic types of containment ponds, reservoirs and canals, without taking specific features into account. Design concepts and principles are described for subgrade preparation, underliner and overliner materials, lining (barrier) systems, and such singularities as anchoring systems, access roads and connections to concrete structures. Several case studies are also included in this chapter.

14.2. Historical background

In this section, the historical background is given of geomembranes and geosynthetic clay liners (GCLs) used as lining materials, followed by a brief overview of the historical background of geotextiles and related products.

It is generally accepted that liquid containment represents the first use of geomembranes, or at least what can be considered as the historical forerunners of geomembranes. Koerner (1986), citing Staff, records the probable use of rubber liners as early as the 1930s and the use of polyvinyl chloride (PVC) liners in the 1940s. Monjoie *et al.* (1992) confirm these beginnings, citing the use of isoprene–isobutylene (better known as butyl) rubber liners in the 1940s. Almost at the same time, canal-lining techniques using the *in situ* application of spray-on bituminous coatings were developed in the USA. It is estimated that between 1947 and 1951 more than 1 million square metres of bituminous canal liners were applied in this way. Together with the application of these first-generation synthetic liners, the first problems and faults appeared, including brittleness and cracking of unprotected PVC, localised mechanical damage of unreinforced bituminous coatings, etc. These early inconveniences set the trend for technical enhancements which, from that time, took the following directions.

- *Physico-chemical enhancement of the material and its compounds.* Various types of polymer have been used over time, and specific combinations have even been designed to be used as geomembranes. Besides bituminous geomembranes, which have also been modified with bitumen–polymer combinations, the principal types of geomembrane appeared in the following chronological order: chlorinated polyethylene (CPE), chlorosulphonated polyethylene (CSPE, also known historically as Hypalon), polyvinyl chloride (PVC), polychloroprene (better known as neoprene), ethylene propylene diene monomer (EPDM), high-density polyethylene (HDPE), low-density polyethylene (LDPE), linear low-density polyethylene (LLDPE), very low density polyethylene (VLDPE) and polypropylene (PP). Various compounds, such as plasticisers, antioxidants and mineral fillers, have been developed and used to enhance the chemical stability, the durability, and the thermal, biological and mechanical properties of geomembranes. Unlike other applications where only one type of geomembrane is commonly used, many different types of commercially available products are still used in the area of liquid containment and conveyance. As will be seen later in this chapter, many geomembrane polymers are indeed suitable for use in ponds, canals and reservoirs – the final choice being influenced by such considerations as drinkability of the stored water, and the chemical compatibility, durability and mechanical characteristics of the geomembrane.

- *Inclusion of reinforcing materials*. This technique was originally used to strengthen bituminous liners. It was first used in a glass fibre fabric included in the spray-on bituminous coating. The glass fibre fabric was soon replaced by non-woven geotextiles, which appeared commercially in the 1970s (see below). Nevertheless, this *in situ* technique found its limits at the end of the 1970s, when it became obvious that the homogeneity and variability of the bituminous mixture could not be guaranteed. Bituminous liners for liquid containment and conveyance were then manufactured *ex situ*, thereby considerably enhancing the quality and its control. It must be noted that, since the beginning of the 1980s, all synthetic liners used in civil engineering, whether polymeric or bituminous, have been prefabricated, and all can be included within a broad family called 'geomembranes', a term used in analogy with 'geotextiles'. The term 'geomembranes' was originally devised to replace previous imprecise terms such as 'pond liners', 'flexible membranes', 'waterproofing sheets', etc., which can still be found in publications of the 1980s. Bituminous geomembranes are not the only ones to include reinforcing fabrics – CPE, CSPE and PVC geomembranes may also be reinforced.
- *Design of additional layers and features*. As early as the 1950s, engineers discovered that the lining of ponds, reservoirs and canals with synthetic materials does not simply consist of digging a hole and installing waterproof sheets. In fact, the early use of spray-on bitumen presaged the implementation of a soil or gravel protection layer on top of the liner. State-of-the-art design nowadays includes careful subgrade preparation, underliner drainage and overliner protection layers.

The function of lining a system can also be achieved by using GCLs. A GCL is defined by the International Geosynthetic Society (2009) as an assembled structure of geosynthetic materials and low hydraulic conductivity earth materials (clay or bentonite), in the form of a manufactured sheet, used in civil engineering applications. These materials are defined by EN ISO 10318 (ISO, 2005) as 'clay geosynthetic barriers' (GBR-Cs); however, the most widely used abbreviation of GCL is used in this chapter. There is a variety of GCLs, which differ in the type of bentonite they contain (natural sodium, natural calcium, calcium activated bentonites) and the mode of bonding of the geotextiles around the bentonite (needle-punching or stitch-bonding processes). GCLs may also have a polymeric geofilm coating on one side.

GCLs have been used in many applications, predominantly in landfill and road construction (including runways). In hydraulic applications, they have been used as a lining (initial or repair) for irrigation and retention ponds, irrigation canals and dykes (levees). This is still a very young technique (Heibaum *et al.*, 2006). The use of a GCL in a canal was first reported by Heerten and List (1990) for the Lechkanal in Germany. As for geomembranes, a number of faults appeared after the first applications of GCLs, including cation exchange, combined either with hydration–desiccation cycles or freeze–thaw cycles, internal erosion, lack of confining pressure and root penetration.

The family of geosynthetics also contains a number of porous materials, including geotextiles. Geotextiles were first used in the 1960s in coastal and river bank protection, replacing granular filters. In the 1970s, the use of geotextiles in civil engineering became widespread, the term 'geotextile' being proposed by J.-P. Giroud in 1977. The use of geotextiles as an underliner or overliner material in liquid containment and conveyance applications was first reported in the literature in 1975, and such uses have been commonplace since the end of the 1970s (see Tables 14.1–14.3 in Section 14.5).

In the 1980s and 1990s, geosynthetic products became diversified, and the term 'geosynthetic' was introduced to encompass geotextiles, geomembranes and the newly appearing related materials that could not be labelled as geotextiles or geomembranes. The various geosynthetics are now accurately defined by EN ISO 10318 (ISO, 2005) and the International Geosynthetics Society (IGS). (In this chapter, when a definition is standardised, the IGS definition is used in preference.)

- A *geostrip* is a polymeric material in the form of a strip, of width not more than 0.2 m, used in contact with soil and/or other materials in geotechnical and civil engineering applications (ISO, 2005).
- A *geonet* is a geosynthetic consisting of parallel sets of ribs overlying and integrally connected to similar sets at various angles.
- A *geospacer* is a three-dimensional polymeric structure designed to create an air space in soil, and/or other materials in geotechnical and civil engineering applications (ISO, 2005).
- The first publicised use of a *geomattress* in a canal application describes the Deschutes (USA) canal test sections, installed between 1991 and 1993 (Swihart, 1994). A geomattress is a three-dimensional, permeable geosynthetic structure, placed over the surface of a soil, and then filled with concrete mortar or soil (IGS, 2009).

A 1995 publication was the first to report the use of *geocells*, *geoarmours* and *geomats* for the protection of a canal embankment against erosion (Escaut canal, France) (Fagon *et al.*, 1997)

- A *geocell* is a three-dimensional, permeable polymeric (synthetic or natural) honeycomb, or similar cellular structure, made of linked strips of geosynthetics (ISO, 2005). After installation, they are generally filled with granular materials.
- A *geoarmour* is a permeable geosynthetic material placed over the surface of the soil, in conjunction with pattern-placed block armour units, to prevent erosion (IGS, 2009).
- A *geomat* is a three-dimensional, permeable structure, made of polymeric monofilaments and/or other elements (synthetic or natural), mechanically and/or thermally and/or chemically and/or otherwise bonded (ISO, 2005).

14.3. Design of geosynthetic systems

The following gives some general points concerning the use of geosynthetics in liquid containment and conveyance applications. They should be considered as the basics in the sense that they generally apply to any type of pond, reservoir or canal and lining material. However, they have to be adapted or supplemented when applied to a particular case, as will be seen in Section 14.5.

14.3.1 Subgrade preparation

The subgrade must be free of any vegetal or organic matter. Whenever necessary and possible, this operation can be supplemented by the use of herbicides. All elements that are potentially aggressive towards the geomembrane or the GCL (e.g. sharp stones) should be eliminated and/or avoided. The subgrade then has to be compacted in order to optimise its bearing capacity, according to state-of-the-art soil mechanics.

The bottom of the structure should form a slight slope, 2% lengthways and 3–5% sideways (SETRA/LCPC, 2000). The embankment slopes should be designed according to state-of-the-art soil mechanics; this is important, as geomembrane lining systems cannot be used to reinforce slopes. For many applications, a slope of 1V:3H (1 vertical in 3 horizontal) to 1V:2H is advised, and 2V:3H is to be considered as the maximum.

The embankment top should be wide enough to enable geosynthetic anchoring. The minimum anchoring length depends on the anchor capacity necessary to ensure the stability of the geomembrane lining system, as well as the protective layer in the case of a covered system (see Section 14.3.6).

14.3.2 Underliner drainage and protection

It is generally not recommended to lay a geomembrane directly on the subgrade, except in particular cases when the risks of the geomembrane puncturing and underliner porewater or gas pressure have been catered for. A better way to prevent the above-mentioned risks is to design specific underliner systems. Geosynthetics are particularly well suited to this application.

The purpose of underliner water drainage is to prevent the accumulation of water leaking from the pond, canal or reservoir, as well as uplift of the geomembrane lining system due to backpressure from a raised water table. Underliner water drainage can be achieved by the use of either gravel layers, gravel-filled drainage trenches or geostrips. Depending on the volume of water to be drained, perforated geopipes may supplement gravel-based drainage systems. Drain pipes are always connected to a main collecting pipe or manhole, and then to a pump or gravity outlet. Special attention must be given to filters associated with draining materials. In liquid conveyance and storage applications, filters are generally designed using geotextiles, and state-of-the-art geosynthetic filter design methods should be applied. An additional benefit of underliner water drainage, especially in containment ponds, is the possibility of using it to monitor the quantity and quality of the drained water in order to detect leaks in the lining system.

Underliner gas drainage is as essential as water drainage, especially where gas from the fermentation of organic matter or air in the pores of the compressed soil, resulting from a water-table rise, are expected. Underliner gas drainage has to be provided when fermentation may occur, especially where the total excavation of organic soils is not economically sustainable, where older storage structures may have been subject to undetected infiltration of organic liquids (e.g. at sugar factory plants or at farms), and in the case of organic liquid leaks in the geomembrane lining system. Underliner gas and water drainage have frequently been combined in the same systems, but state-of-the-art designs now tend to separate them, with water being drained in trenches or geosynthetic strips, and gas being drained in a geotextile underliner connected to gas vents passing through the geomembrane at the top of the embankments. The latter solution is particularly attractive when the same geotextile is designed to perform both underliner gas drainage and protection (see below).

Examples of underliner drainage and filtration systems are presented in the case studies given in Section 14.5, but the systems described there should not be considered as a substitute for specific studies pertaining to site-specific conditions (e.g. drainage water volume and subgrade granularity).

A protective layer may be interposed between the geomembrane and the subgrade when the latter is not smooth enough to guarantee geomembrane safety, especially below a high water head. Geotextiles are now generally preferred because of their possible combined functions of gas drainage and mechanical protection of the underliner.

When a drainage medium is used underneath a GCL, care must be taken in order to prevent internal erosion in the GCL, which would result in a loss of bentonite and, subsequently, a loss of hydraulic performance. This could occur with highly granular soils (stones larger than 25 mm and up to 150 mm) that have a low percentage of fines, or with geocomposites (Rowe and Orsini, 2003).

14.3.3 Lining systems

The core of a lining system is, of course, the impermeable material – either a geomembrane or a GCL. As seen in Section 14.2, there is a number of geomembranes or geosynthetic clay liner products available that may be used in liquid conveyance and containment structures (see the case studies in Section 14.5). It is not the purpose of this chapter to propose definitive guidelines concerning the final choice of a lining system, but rather to present the general criteria governing such a choice. As these criteria are general, they cannot be used as a substitute for site-specific conditions.

- The *economic criterion* is always critical. The cost of a geosynthetic lining system depends on many parameters, which cannot all be accurately predicted in a technical paper.
- The *hydraulic criterion* will determine only the choice between a geomembrane and a GCL, as their behaviour with regard to liquid transfer is fundamentally different. For example, mass transfer through an undamaged geomembrane can only occur as molecular diffusion, while a very small liquid flow is additionally possible through GCLs as well as through compacted clay liners (CCLs), following Darcy's law. To summarise, without using complex formulae, mass transfer will generally be smaller through geomembranes, especially for non-organic chemical species. It must be emphasised that this statement holds true for undamaged, continuous geomembrane lining systems only, which highlights the necessity to protect geomembrane liners against puncture and seam defects.
- *Mechanical criteria* are also very important, even if geomembranes and GCLs are not designed to have mechanical functions in a structure. Geomembranes and GCLs may nevertheless be subjected to mechanical stress, such as tensile stress on slopes and in the case of subgrade settlement, or puncture by gravel and an irregular subgrade. Even if mechanical stress is lessened and, in some cases, eliminated by proper design (e.g. with geotextile protection), it is generally necessary to select geomembranes with mechanical properties adapted to the expected mechanical conditions. Typical examples of geomembrane selection, governed by mechanical criteria, among others, are presented in Section 14.5. It will be seen that, on the basis of mechanical criteria, either rigid geomembranes (e.g. HDPE) or flexible geomembranes (e.g. PVC) may be selected for liquid conveyance and containment. In the case of GCLs, the overlap of panels must be increased on a slope when tensile stresses are expected, in order to prevent the possible disappearance of overlapping with time due to shrinkage of the GCL panels.
- *Chemical resistance and durability criterion.* In the case of geomembranes, this concerns both the polymer resin and the other compounds contained in the geomembrane. It is well known that HDPE is the most chemically stable polymer available; nevertheless, it may be subjected to chemical ageing phenomena, such as environmental stress cracking or oxidation, depending on the environment it is in and its physico-chemical characteristics. Other types of geomembrane may be affected by chemical degradation triggered by the non-adaptation of their compounds. For example, it is now well known that the physico-chemical durability of PVC geomembranes is highly dependent on the quality of the plasticiser used. The principal factor affecting the chemical resistance of GCLs is the ion-exchange capacity of the bentonite. In the case of GCLs, the cation exchange of the bentonite resulting from hydration–desiccation or freeze–thaw cycles will have a highly detrimental effect on the performance of the GCL. Thus, for the containment of liquid-waste, it is of upmost importance to correctly determine the expected composition of the liquid to be contained, and to compare this with the chemical resistance of the geomembrane or GCL.
- All the above-mentioned criteria are related to the intrinsic characteristics of geosynthetic liners. Nevertheless, it is important also to consider the *ease of installation and seam or overlapping performance criteria*. Indeed, the performance of the best geosynthetic lining product will always be limited by the quality of its installation and seaming. For example, such phenomena as thermally-induced wrinkling or moisture-dependent welding quality will affect HDPE geomembranes, and must be taken into account when planning the installation of geomembrane lining systems. The strict application of state-of-the-art installation procedures for each type of geomembrane is an absolute requirement, as is a state-of-the-art quality assurance and quality control programme. Despite their apparent ease of installation, GCLs are very sensitive to free swelling, and the appropriate confinement–hydration procedure must be strictly respected. The overlapping of GCLs is also a delicate process as, unlike geomembranes, GCL overlaps cannot be easily and accurately tested on site.

Besides single geomembrane or GCL lining systems, it is possible to install double lining systems, using two liners with a drainage layer in between them. This solution, which was once rare in liquid containment and conveyance applications, is now increasingly often encountered, especially in mining applications, and is only applied where the risk of leaks must be reduced considerably. An example of a triple lining system is given in Section 14.5.

14.3.4 Overliner protection and cover

One of the best ways to prevent anticipated ageing of geosynthetics in general, and geomembranes in particular, is to limit their exposure to the action of weather by covering them. The purpose of overliner layers is also to prevent liner damage caused by floating or transported solids (e.g. ice and wood), by operating vehicles or machines (e.g. mobile pumping equipment), by burrowing animals and plant roots, and by vandalism or accidental human intervention. One common design (see the case studies presented in

Section 14.5) is to protect the geomembrane with a geotextile and to cover this geosynthetic with a layer of granular material. The choice of the granular material depends on such factors as the slope and the hydraulic actions of the contained or conveyed liquid (e.g. waves). Granular layers may be composed of several sublayers of differing grain size, with the finest grained material (e.g. fine sand) being placed directly over the geosynthetic ranging up to the coarsest grained material (e.g. rip rap) placed on top of the granular layer. Other common designs may consist of concrete covers, composed of precast blocks or slabs, *in situ* poured reinforced concrete layers, or even shotcrete. Geotextile protection must also be considered and catered for in the case of concrete covers. Another purpose of overliner covers is to prevent uplift of the geomembrane lining system due to wind action. Some installation procedures may include the temporary use of ballast over the geomembrane in order to prevent uplift during installation, before the permanent overliner protection layers are installed.

The necessity of covering geomembrane lining systems differs from one application to another. The literature overview detailed in Section 14.5 shows the following trends

- The majority of geomembrane lining systems in containment pond and reservoir applications are left uncovered (24 examples of unprotected geomembrane lining systems as opposed to 16 protected systems).
- Most cover systems in containment pond and reservoir applications consist of granular layers (four granular covers as opposed to two *in situ* poured concrete covers, one shotcrete cover and one precast block cover).
- Most geomembrane lining systems in canal applications are covered (23 examples of protected geomembrane and GCL lining systems as opposed to three unprotected systems).
- A majority of cover systems in canal applications consist of *in situ* poured concrete.

The above-mentioned figures are to be considered with caution, as they may be distorted by the effect of publication – most pond and reservoir works do not lead to publication, and only reports of exceptional cases are published in the literature. Nevertheless, the tendency to leave geomembrane lining systems uncovered in this application may be considered accurate, as it is largely observed in the field. Whether or not it should be recommended to leave geomembrane lining systems uncovered is another question, as accelerated degradation of unprotected geomembranes attributed to excessive exposure has also been observed (Lambert *et al.*, 1999).

Local overliner protection is also possible wherever identified localised actions are anticipated (e.g. pumping areas, tidal range). Where mechanical action is planned, and especially where machines or vehicles will be used, precast or *in situ* poured concrete slabs are generally installed. Preventing hydraulic actions, such as fluvial erosion in the case of canals, requires the use of specifically designed systems, which are generally geosynthetic systems – geocells, geomattresses, geoarmours or geomats (see the definitions in Section 14.2). These systems may also be used alone in order to prevent bank erosion, without covering any geosynthetic lining system.

GCLs must imperatively be covered in order to perform their lining function. The thickness and the composition of the cover must be designed to ensure that no hydration–desiccation or freeze–thaw cycles will occur in the material. The thickness must thus be adapted to the local climatic conditions. The cover thickness should also be sufficient to prevent the penetration of roots to the GCL.

14.3.5 Floating covers

Many projects, especially for mining and resource processing, are being developed in areas where water is scarce and there are high rates of evaporation. There are many places that are resource rich and where the annual evaporation is considerably more than the annual rainfall. Under these circumstances there is a great incentive to provide floating geomembrane covers over water storage and process ponds in order to control valuable water loss by evaporation (Sadlier, 2009).

Conversely, temporary covers have also been used on heap leach pads in countries where high rainfall dilutes operating solutions and surplus water can trigger the need for significant water treatment, and its associated costs (Breitenbach and Smith, 2006). These covers generally include flexible geomembranes, such as PVC or PP.

The design has to take into account possible gas generation from the contained liquid, as well as water removal from the cover. The geomembrane can be specially formulated to ensure high chemical compatibility with pollutants or water for human consumption. An example of such a cover is presented in Section 14.5.

14.3.6 Singularities

Singularities always exist in liquid containment and conveyance structures. Besides localised mechanical or hydraulic actions, which have been addressed above, singularities are related to connections between geosynthetic systems and specific structures such as manholes, walls or embankment tops. Of all the possible singularities, the following are the most usual

- *Anchoring*. All geosynthetic systems must be anchored on the top of the embankment slopes or on the slope itself, depending on the overall design. The most common anchoring design is the anchor trench, which is generally a square trench in which the geomembrane is laid on one side and at the bottom. The trench is then backfilled with a non-puncturing soil. It is generally recommended that anchor trenches should be deeper and wider than 0.5 m and situated at least 0.7 m from the edge of the slope. These indications are the minimum values for pond, reservoir and canal applications, and must be confirmed for each specific structure. In particular, when the geomembrane lining system is covered, in most cases the angle of slope is too steep for the protective layer to be self-stabilising; in such conditions, stability is generally ensured by anchoring the geosynthetics at the top of the slope, and the anchor capacity this purpose must take into account. For canals in particular, the excess width of the geosynthetic related to the anchoring design may generate extra cost, and thus anchoring characteristics must be precisely derived by calculation, or alternative techniques used (e.g. tying the geomembrane to stakes). In some applications (e.g. deep reservoirs), intermediary anchoring may be required alongside the slope (see the Barlovento case study in Section 14.5.2.3).

- *Access roads and tracks* are sometimes required, especially for large containment ponds where vehicles need to access the bottom of the pond for maintenance or exploitation purposes. Special attention has to be given to the protection of geosynthetics placed under the road, and to the stability of the road over the geosynthetics. The subgrade has to be suitably shaped to take the access road into account, and extra protection of the geosynthetic lining system will need to be designed (see the Souppes-sur-Loing case study in Section 14.5.1.1). A minimum thickness of granular material must be in place before vehicles are allowed access; the French chapter of IGS recommends a minimum thickness of 0.5 m (CFG, 1995).
- *Connection to concrete structures* usually poses the problem of waterproofing continuity. Many technical solutions are available, which mainly depend on the geomembrane type. Metallic fixings are generally used in association with metallic and elastomeric plates, and/or geomembrane overlaps. In the case of GCLs, bentonite paste and an extra layer of GCL can be used to ensure continuity, by wrapping the concrete structure and ensuring sufficient overlap with the GCL lining system.

It is important to emphasise that all the above points are closely interrelated in terms of design. For example, it is impossible to select a geomembrane or a GCL without taking the characteristics of the overliner protection layer into account, and, conversely, it is impossible to design a protection layer without considering the type of lining system. A geosynthetic lining system thus has to be designed as a whole, including subgrade preparation, underliner and overliner layers, and specific features such as those described above. Moreover, different geosynthetic lining systems may be equivalent in terms of hydraulic, chemical or mechanical criteria, and the difference may be only related to installation requirements, economic criteria or availability.

As we have seen in this section, the basics of the geosynthetic systems for liquid conveyance and containment are fairly simple. However, applying them to specific works may be complex, and will require more information and experience than has been briefly presented here. Therefore, in Section 14.5, some data collected from the literature concerning what may be considered state-of-the-art geosynthetic design for containment ponds, reservoirs and canals are presented. Before dealing with those examples, the state of the art regarding the durability of geomembranes used for ponds, reservoirs and canals is described in the following section.

14.4. Evaluation of the durability of geomembranes

Heibaum *et al.* (2006) have reported that the durability of geomembranes in hydraulic applications is 20–30 years or more for exposed geomembranes, as long as their formulation is adequate, and the liner system as a whole has been well designed, built and monitored. The expected lifetime of covered geomembranes protected from the climate and mechanical action is significantly longer. However, Heibaum *et al.* found that a distinction must be made depending on the nature and the size of the facility. It seems that more faults occur in ponds and canals of modest dimensions than in dams. A possible explanation for this is that larger facilities involve significant safety issues, while for modestly sized structures the financial factor, unfortunately, often becomes dominant.

Another difference between small and large works is that, in small projects, often less experienced people are involved, even though technical challenges are similar. Indeed, good design and installation are fundamental when addressing the durability of geomembranes. The objective of this section is to discuss the durability of oxidised bituminous geomembranes, PVC geomembranes and PP geomembranes based on recent findings reported in the literature. As no data on HDPE or EPDM could be found, they are not mentioned here.

14.4.1 Durability of oxidised bituminous geomembranes

Micro-cracks can be observed at the surface of oxidised bituminous geomembranes that have been left exposed. These micro-cracks result in an increase in the flow rate through the geomembrane, as compared with a virgin material. In some cases, the performance of the exposed and unexposed geomembrane was identical, but not better than that of a 1 m-thick clay layer with a hydraulic conductivity of 10^{-9} m/s (Touze-Foltz *et al.*, 2010). When the oxidised bituminous geomembrane remains under water but without any mechanical protection, the loss of performance is less pronounced than for one exposed to air and ultraviolet radiation. In the case of covered oxidised bituminous geomembranes, no micro-cracks were observed and the initial membrane performance was maintained 30 years after installation. It is thus recommended that oxidised bituminous geomembranes do not remain exposed if their hydraulic performance is to be preserved in the long term. These results cannot be extended to elastomeric bituminous geomembranes.

14.4.2 Durability of PVC geomembranes

The long term performance of a PVC geomembrane is primarily governed by the performance of the plasticiser, which is greatly influenced by the molecular weight and molecular structure of the plasticiser (Hsuan *et al.*, 2008). A loss of plasticity induces an increase in rigidity, and brittleness at low temperatures increases. Simultaneously, an isotropic shrinkage is observed, due to the loss of volume (Carreira and Tanghe, 2008).

In a given situation, the degradation of the geomembrane will be faster if the chemical composition of the product is not optimal (Girard *et al.*, 2002). Depending on the composition of the geomembrane, the durability may be 1–2 years if poorly formulated, to over 20 years if well formulated (Carreira and Tanghe, 2008).

In the case of exposed geomembranes, weathering always results in the geomembrane becoming more rigid, shrinking and tearing (Girard *et al.*, 2002). The time taken for this phenomenon to occur will depend on the quality of the geomembrane and its exposure to ultraviolet radiation. Low temperature was also found to have a detrimental effect on the geomembrane. Tears that can then appear, and these may be star-shaped or linear.

In the case of covered geomembranes, an increase in durability was noted by Girard *et al.* (2002). Reinforced and composite (with an associated geotextile) geomembranes show reduced shrinkage associated with the manufacturing process and ageing, and may be a solution to achieving increased durability in major structures such as the Barlovento Dam (Fayoux and Potié, 2006).

14.4.3 Durability of polypropylene geomembranes

Peggs (2008) has presented a review of the use of PP geomembranes in a number of hydraulic applications: for various functions in potable water reservoirs and ponds, and as floating covers on potable water reservoirs. A number of failures occurred in the case of potable water reservoirs and covers, typically after 3–10 years of service. According to Peggs (2008), accelerated oxidation could be the result of synergy between: exposure to ultraviolet radiation, thermal exposure, the presence of chlorine, stress and the stabiliser package. Investigations of the stabiliser package of several samples showed that the distribution was far from homogeneous for a geomembrane that underwent failure, so that the protection was not homogeneous.

However, excellent performance of PP geomembranes has been observed in other weather-exposed applications and for some manufacturers in the case of potable water reservoir covers. Indeed, Wallace (2008) reported satisfactory performance of floating covers on potable water reservoirs at five years. Furthermore, different PP resins have become available, thus opening the door to potentially more efficient geomembranes than in the past.

14.5. Case studies

This section is based on the collection of 81 case studies presented in the international literature on geosynthetics (23 containment pond, 28 reservoir and 30 canal references). The cases were selected by meeting at least two of the following three criteria

- historical significance
- overall technical significance (size, volume and water head)
- representative of a specific geosynthetic.

They also had to meet the minimum criteria in relation to the quality and quantity of the available data.

The 81 cases give a good view of the design and use of geosynthetics in liquid containment and conveyance applications over the last 40 years. Of these references, seven (two containment ponds, three reservoirs and two canals) have been chosen to illustrate the basics of geosynthetic design as presented in Section 14.3. They were selected because of their technical significance, and the availability of detailed technical information and data.

14.5.1 Containment ponds

Table 14.1 gives an overview of the references on the use of geosynthetics in containment ponds.

14.5.1.1 The Souppes-sur-Loing (France) containment pond (Fayoux *et al.*, 1999)

The general features of this pond are given in Table 14.1 and shown in Figure 14.1.

The pond was designed to store waste water from sugar production processes. Waste water is purified during the ponding season, and can then be used for irrigation or recycled into the sugar beet washing process. The containment pond has been excavated in an old limestone quarry. The principal features pertaining to the geosynthetic design are as follows

- Subgrade preparation consisted of compacting and grading the soil using suitable machinery.
- Special attention was given to underliner drainage, considering the possibility of a high water table, leaks and the production of fermentation gases; 1 m-wide geospacer drain strips were laid every 25 m at the bottom of the pond and on its slopes. The strips are all connected, at the top of the slopes, to 0.1 m-high HDPE gas vents. A peripheral drainage trench (see below) was placed at the bottom of the slopes and is connected to a collection pipe by way of a manhole located outside the pond.
- A geomembrane lining system, with underliner protection (300 g/m^2 needle-punched non-woven geotextile and a 2 mm HDPE geomembrane), was placed at the bottom of the pond and on the slopes. The geomembrane was double welded using an

Table 14.1 References on the use of geosynthetics in containment ponds 1972–2010

Location	Country	Age of geomembrane: years	Use	Type and use of geosynthetic	Surface area of geosynthetic: m^2	Impoundment volume: m^3	Cover	Reference
La Glacière	France	AI	R	HDPE geomembrane (2 mm), 500 g/m^2 puncture-resistant geotextile	2 × 2500		Uncovered	Spillemaecker *et al.* (2002)
Livorno	Italy	AI	CDS	HDPE geomembrane (2 mm) non-woven needle punched geotextile 500 g/m^2	Approx. 30 000	NA	Non-woven needle punched geotextile 500 g/m^2	Michelangeli *et al.* (2006)
Malelane Mill	South Africa	AI	M	GCL and HDPE geomembrane (1.5 mm)	NA	14 800	Uncovered	Nortjé and Meyer (2006)
Atacama desert region	Chile	AI	B	PVC geomembrane on non-woven geotextile	2 × 300 000	NA	Uncovered	Berube *et al.* (2007)
Fos-sur-Mer	France	1		PP geomembrane	10 000	NA	NA	Saintot and Bilancioni (1983)
Souppes-sur-Loing	**France**	**1**	**SF**	**HDPE geomembrane (2 mm), geotextile underliner, geospacer strips for underliner drainage**	**35 000**	**160 000**	**Uncovered**	**Fayoux *et al.* (1999)**
Charente maritime	France	1	S	Needle-punched GCL with calcium-activated bentonite	4200 and 2 × 1750	NA	0.3 m calcareous soil	Touze-Foltz and Hatton (2006)
Irapuato	Mexico	2	S	PVC geomembrane (1 mm)	39 000	NA	NA	Murillo-Fernandez (1994)
Emalahleni, Witbank	**South Africa**	**2**	**B**	**HDPE geomembrane (2 mm and 1.5 mm), GCL, HDPE cuspated drains, 250 g/m^2 geotextiles**	**70 000**	**300 000**	**0.15 m of sand**	**Habte and Legg (2009)**
Mont-aux-Malades	France	5		Double lining system (polyane and elastomeric geomembrane), underliner geotextile intra-lining drainage geotextile, protection geotextile	NA	35 000	Shotcrete, precast blocks	Girollet (1983)
Lyon	France	5	R	PVC geomembrane (1.2 mm)	NA		Uncovered	Girard *et al.* (2002)
Sidi Larbi	Morocco	5	B	HDPE geomembrane (1.5 mm) on a non-woven needle-punched geotextile 500 g/m^2	20 760	130 000	NA	Ouvry *et al.* (2006)
Lyon	France	6	R	PVC geomembrane (1.4 mm)	NA		Uncovered	Girard *et al.* (2002)
Ernstein	France	7	SF	Bituminous geomembrane (4.5 mm)	60 000	133 000	Uncovered	Girollet (1983)
Saint-Martin-Pont-d'Ain	France	7		Bituminous geomembrane (4 mm), underliner geotextile on slopes	33 000	NA	NA	Breul and Herment (1995)
Lyon	France	7	R	PVC geomembrane (1 mm)	NA		Bottom granular drainage layer, uncovered on slopes	Girard *et al.* (2002)

Saint Georges-les-Baillargeaux	France	10	L		500	NA		Fayoux and Van der Sype (2000)
Sillery	France	11	SF	Butyl geomembrane	NA	NA	NA	Lescure (1983)
		11	FPP	HDPE geomembrane (1 mm) underliner non-woven PP geotextile (200 g/m^2)	14 970	NA	Uncovered	Adams and Wagner (2000)
Ontario	Canada	14	L	HDPE geomembrane (1.5 mm)	NA	2500	Uncovered	Rowe *et al.* (2003)
Lyon	France	15	R	PVC geomembrane (1.3 mm)	NA		Bottom granular drainage layer, uncovered on slopes	Girard *et al.* (2002)
Préfontaine	France	17		PVC geomembrane (1 mm) non-woven polyester needle-punched geotextile (300 g/m^2) on top of a 0.1–0.2-thick layer on slopes	40 000	NA	Uncovered	Fayoux and Van der Sype (2000)
Marcay	France	18	MPP	PVC geomembrane (1 mm)	2000	5000	Uncovered	Fayoux and Van der Sype (2000)

Bold indicates the containment ponds described in this chapter.
AI, at installation; B, brine; C, chemicals; CDS, contaminated dredging sludge; FPP, food processing plant; L, leachate; MPP, manure pit pond; NA, not available; R, runoff water; S, sewage; SE, solar evaporation; SF, sugar factory waste water.

Figure 14.1 The Souppes-sur-Loing (France) containment pond – typical cross-section and specific features (not to scale)

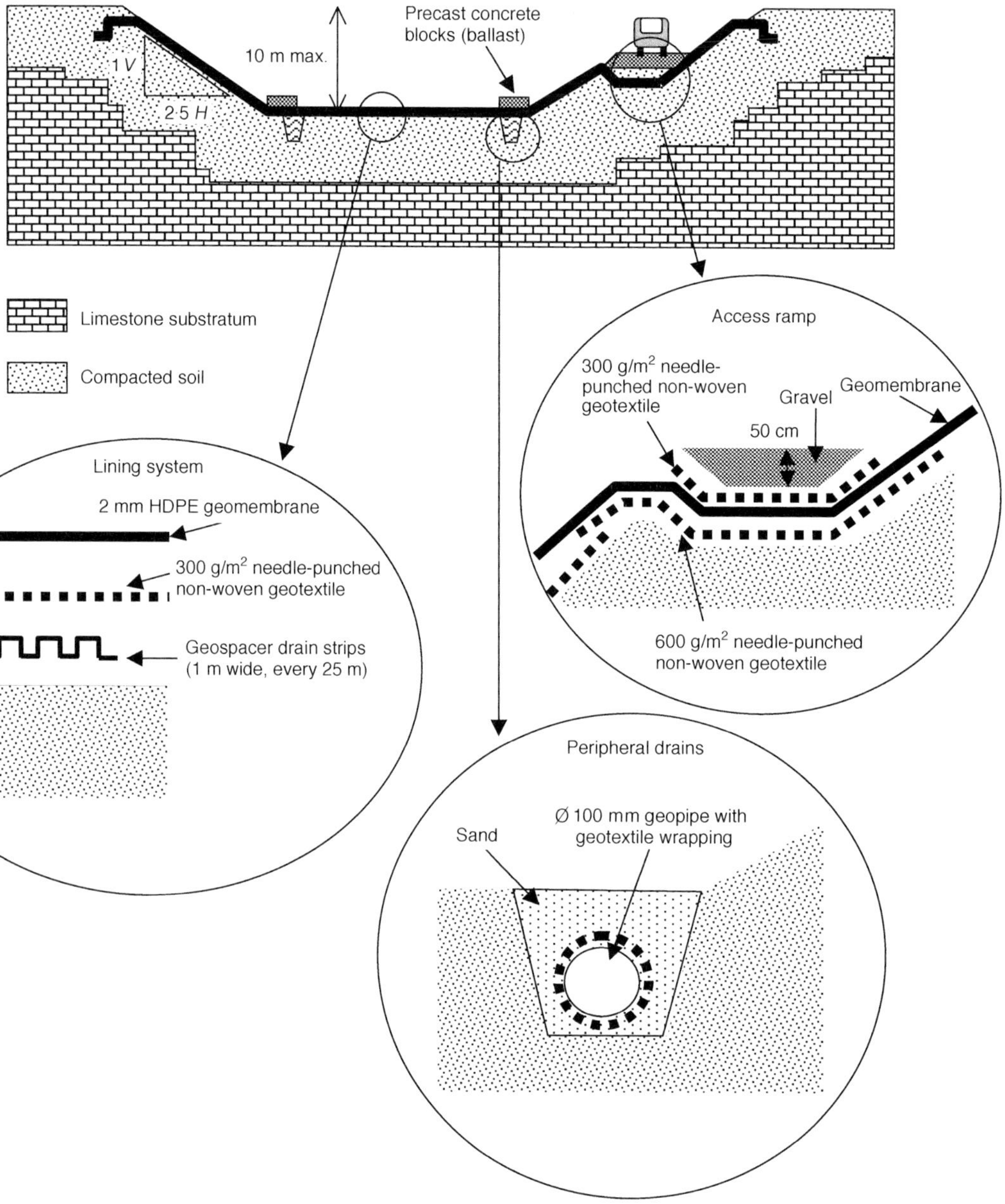

automated machine. All the welds were tested by applying air pressure (5 bar for five minutes for each weld), and weld samples were regularly submitted to laboratory peel tests. These controls were part of an overall quality assurance system for the project. As the lining system is to remain uncovered, it has been ballasted at the bottom of the slopes with precast concrete blocks, in order to avoid wind damage during periods when the pond is emptied.

- An access ramp was designed for maintenance purposes. The ramp comprises a 0.5 m-thick gravel–cement layer laid on a specially reinforced lining system, consisting of a 600 g/m^2 needle-punched non-woven underliner protection geotextile, 2 mm HDPE geomembrane and a 300 g/m^2 needle-punched non-woven overliner protection geotextile.

Most of the work was carried out within two months, the lining system being laid in 5½ weeks.

14.5.1.2 The Emalahleni water reclamation plant, brine evaporation pond, South Africa (Habte and Legg, 2009)

The general features of this pond are given in Table 14.1 and shown in Figure 14.2.

The pond was designed to store a brine liquid stream, which is classified as hazardous due to its high concentration of salts (especially sodium, S, calcium and potassium salts) and metals (such as iron, lithium, manganese and strontium). According to the minimum requirements of the South African Department of Water Affairs and Forestry, such a pond should have a liner system equivalent to that for a

Figure 14.2 The triple-liner system used for the brine evaporation pond at the Emalahleni water reclamation plant

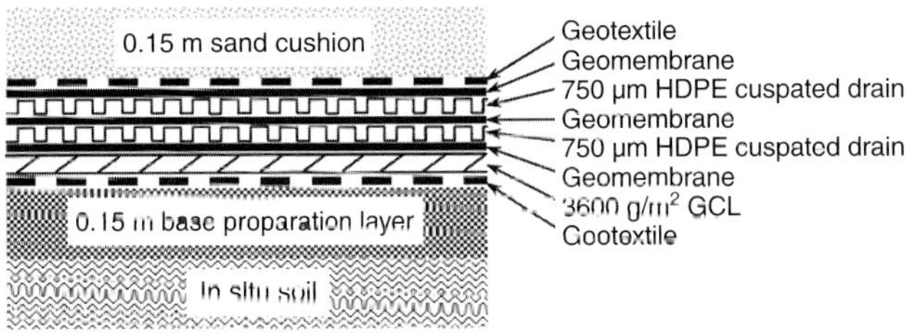

hazardous-waste lagoon – that is, a double composite liner system with a leakage detection and collection layer separating the primary and secondary composite liners. This prescriptive liner system recommends the use of either a CCL or a GCL as part of each composite liner. Neither of these clay liners could be used due to the unavailability of suitable clay in the case of a CCL and chemical incompatibility of the brine with the bentonite in the case of a GCL. Therefore, an alternative triple-liner system consisting entirely of geosynthetic components was proposed for the lining of the brine evaporation pond (Figure 14.2).

The primary and secondary liners consist of, respectively, 2 mm and 1.5 mm HDPE geomembranes with a 750 µm cuspated HDPE drainage layer between the two geomembrane liners. This primary leakage collection layer has separate collection systems for the walls and floor, draining in separate pipelines to a common brine leakage collection sump outside the walls of the pond. The primary objective of separating the wall and floor was to allow easy maintenance in the event of a leak. The secondary leakage collection and conveyance system comprises a similar 750 µm cuspated HDPE drainage layer between the secondary and tertiary HDPE liners, draining through a separate pipeline to the brine leakage collection sump. The third liner consists of a 1.5 mm HDPE geomembrane overlying a GCL. Beneath the GCL is a non-woven geotextile layer installed as part of a subsurface drainage system for perched groundwater seepage. Because the GCL on the slopes does not have an applied confining pressure, the overlap between the adjacent GCL sheets was increased to minimise the risk of panel separation that could occur (shrinkage). The purpose of the 0.15 m sand cushion layer on top of the liner is to prevent the primary HDPE geomembrane from floating in the event of a leak.

The evaporation pond was constructed partly below ground (a maximum of 3 m) and partly above ground (a maximum of 10 m) using a cut to fill operation. The walls of the facility have an outer slope of 1V:2H and an inner slope of 1V:2.5H. The facility occupies a total footprint of about 7 ha and has a storage capacity of about 330 000 m^3. At the leakage collection sump, the rate of leakage from each liner layer system can be measured, and all brine leakage collected in the sump is pumped back into the brine pond.

14.5.2 Reservoirs

Table 14.2 gives an overview of the references of the use of geosynthetics in reservoirs.

14.5.2.1 The Selvet Dam, France (Poulain *et al.*, 1997)

The general features of this reservoir are given in Table 14.2 and shown in Figure 14.3.

The Selvet Dam, designed for the storage of drinking water, is situated in the north-west of the Aveyron Department. This dam is 18 m high, with a crest length of 250 m (backfill volume 130 000 m^3, face slope 1V:2.5H), and can store approximately 600 000 m^3 of water. Geologically speaking, the subsoil consists of basalt in two or three superimposed layers, separated by levels of blocks, slag or hardened ash. The levels of permeability measured in the substratum called for a large grouting curtain. The geological and geotechnical surveys indicated that a flexible, backfill-type dam was more suitable for the site than a concrete dam. The final design choice was a homogenous backfill with coarse 0–150 mm gravel, obtained by blowing up of a basalt deposit situated on the right bank, and upstream sealing by means of a geomembrane.

A transition area (0–100 mm) was installed between the body of the backfill (0–150 mm) and the GLS shown (Figure 14.3), over a width of 3 m, following completion of the grouting curtain at the toe of the upstream face. Its surface, parallel to the slope, was reworked with an excavator, and carefully finished by hand, to eliminate all the larger pieces. This layer had to be covered with fine gravel; this solution, however, presented difficulties in implementation over a wide area (7500 m^2), due to the lack of cohesion of the gravel, which meant that the stones were likely to roll down the facing under the geomembrane when it was installed. Therefore, the support layer was made of gravel mixed with a bitumen emulsion. The gravel–bitumen emulsion mix (0–14 mm) layer, 0.15 m thick, was laid on the slope using a crawler-mounted loader, and spread with a rake. Compaction was performed according to the slope, with a tandem roller anchored to the top of the slope. The final surface was then brushed.

The piercing-proof geocomposite was unrolled over the slope, working downwards from the crest, and the 2 m-wide sections sewn together. The geomembrane, delivered to the site in 4 m rolls, was unrolled downwards from the crest, the roll being hung from a hoisting tackle mounted on a shovel loader. The smooth face of the geomembrane was laid uppermost to avoid tensile stresses from settlement of the protection layer. A detailed pattern layout enabled the sections to be adapted to the length of the slope to avoid transversal joints. The sections were joined with a blowtorch, with a 0.2 m overlap. The project manager performed a visual inspection of all the joints.

The connection of the geomembrane to the grouting curtain at the foot of the slope was an important factor for the sealing of the dam. The solution adopted (see Figure 14.3)

Table 14.2 References on the use of geosynthetics in reservoirs 1975–2010

Location	Country	Year	Use	Type and use of geosynthetic	Surface area of geosynthetic: m^2	Impoundment volume: m^3	Cover	Reference
Ile de la Réunion	France	1975	D	Butyl geomembranes (0.75, 1 and 1.5 mm), underliner geotextile	70 000	450 000	Uncovered	Loudière and Perrin (1983)
Ayron	France	1976	R	Bituminous geomembrane (4 mm), protection geotextile on upper parts	40 000 on banks	NA	Soil (0.3 m), rip rap (0.25 m) on upper parts	Alonso *et al.* (1990)
Ospedale (Corsica)	France	1979	D	Bituminous geomembrane (4.8 mm)	4000	2 860 000	Self-locking paving stones	Girard *et al.* (1996)
Mt Elbert	USA	1980	E	Reinforced CPE geomembrane (1.1 mm)	1 170 000	14 220 000	Uncovered	Frobel (1983)
Guazza	France	Prior to 1983	I	Bituminous geomembrane (3 mm)	60 000	300 000	Uncovered	Tisserand (1983), Alonso *et al.* (1990)
Various locations	Thailand	Prior to 1985	I	PE–PP geocomposite with liner coating	NA	2500 to 400 000	Various (from sand to riprap)	Wichern (1990)
Witbank	South Africa	1985	D	Reinforced CSPE floating cover	NA	60 000	Uncovered	Davies (1994)
Codole (Corsica)	France	1985	D	Geomembrane PVC-P (2 mm)	16 000	7 000 000	Concrete slab	Girard *et al.* (1996)
Genevilliers	France	1986	R	HDPE geomembrane (3 mm), geocomposite drain strips	110 000	NA	Reinforced concrete (0.1 m) on upper parts	Ialynko and Gonin (1993)
Aubrac	France	1986	D	PVC-P geomembrane (1.2 mm)	5900	1 260 000	0/25 mm crushed gravel (0.20 m) and rip rap (0.50 m)	Girard *et al.* (1996)
Eagle Rock	USA	Prior to 1990		CSPE floating cover	130 000	315 000	Uncovered	Taylor (1990)
Kuriyama	Japan	1990	E	PVC geomembrane (1.5 mm), underliner geotextile, protection geotextile	195 000	2 520 000	Sandy gravel (0.4 m) and crushed stone (0.4 m)	Yoshikoshi and Masuda (1994)
Villa Juarez	Mexico	1991	E	HDPE geomembrane (1.5 mm)	22 000	NA	NA	Murillo-Fernandez (1994)
Barlovento	**Spain**	**1991**	**I**	**On slope: reinforced PVC geomembrane (1.5 mm)** **On the bottom: unreinforced PVC geomembrane (1.5 mm)**	**250 000**	**3 120 000**	**Uncovered**	**Fayoux and Potié (2006)**
Oblatos Gorge	Mexico	1992	E	CSPE geomembranes (0.9 and 1.5 mm), underliner geotextiles, protection geotextile at bottom	140 000	1 600 000, two reservoirs	Concrete (0.15 m) at bottom	Murillo- Fernandez (1994)
Visari	Greece	1993	I	HDPE geomembrane (0.75 mm)	90 000	600 000	Sand (0.1 m) and sandy gravel (0.3 m) and rip rap (0.5 m)	Collios (1994)

Rogliano	France	1995		Reinforced PVC floating cover (1.2 mm), PVC geomembrane lining (1.5 mm), underliner geotextile	9600	45 000	Uncovered	Tisserand *et al.* (1995)
Okinawa Island	Japan	1996	E	EPDM geomembrane (2 mm), underliner geotextile	53 000	420 000	Uncovered	Shimizu and Ikeguchi (1998)
Selvet	**France**	**1996**	**D**	**Bituminous geomembrane (4 mm)**	**NA**	**600 000**	**0/31.5 mm Gravel (0.30 m) and random rockfill (0.60–1.40 m)**	**Poulain *et al.* (1997)**
Seattle	USA	2001	D	Reinforced PP geomembranes (1.13 mm)	NA	83 300	Uncovered	Wallace (2008)
Seattle	USA	2002	D	Reinforced PP geomembranes (1.13 mm)	NA	257 400	Uncovered	Wallace (2008)
Afourer	Morocco	2003	I	PVC geomembrane (1.5 mm)	NA	1 260 000 Two reservoirs	Uncovered	Fayoux and Dewalque (2006)
Berre	France	2004	D	Geomembrane PVC-P (2 mm)	>15 000	45 000	2/6 mm sand (0.10 m) 6/50 mm gravel (0.20 m), random rockfill (0.30 m)	Fons *et al.* (2006)
Tenerife Island	Spain	2005	I	On slope: reinforced PVC geomembrane (0.9 mm) On the bottom: 1.5 m-thick compacted clay	31 700	821 700	Uncovered	Fayoux and Potié (2006)
Savoie	France	2007	D	PVC geomembrane (1.5 mm)	NA	150 000	0/30 mm aggregate (0.30 m) and 0/500 mm rockfill (0.40 m)	Girard *et al.* (2010)
Adret des Tuffes, Savoie	**France**	**2008**	**S**	**Double geomembrane liner system covering the slopes and the bottom: VLDPE geomembrane (1.2 mm) and PVC geomembrane (2 mm)**	**60 000**	**400 000**	**0/30 mm aggregates (0.30 m) and 0/400 mm aggregates (0.50 m)**	**Girard *et al.* (2010)**
Corsica	France	2008/2009	D	Waterproofing: PVC geomembrane (1.5 mm) Floating cover: reinforced PVC geomembrane (1.2 mm)	34 000	80 000 Two reservoirs	Uncovered	Fayoux *et al.* (2010)

Bold indicates the reservoirs described in this chapter.
D, drinking water; E, electricity; I, irrigation; N, navigation; NA, not available; R, recreation; S, snowmaking.

Figure 14.3 The Selvet Dam (France) recreation reservoir – typical cross-section and specific features (not to scale)

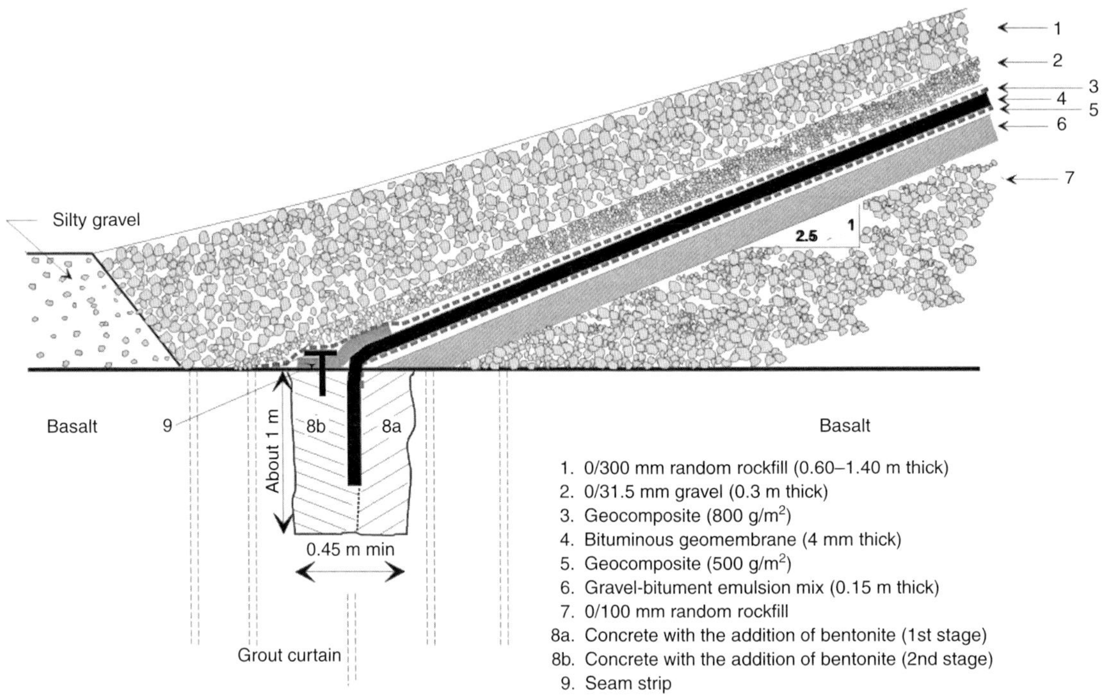

consisted of lowering the geomembrane into a trench filled with concrete with the addition of bentonite (150 kg cement, 400 litres of water with 10% bentonite, 950 kg of 0–3 mm sand and 500 kg of 8–14 mm gravel). The trench was cut in the basalt with a trencher to a depth of 1 m along the centre line of the grouting curtain; it was then filled in two phases (see Figure 14.3). The concrete was poured into the dam side of the trench (using shuttering) to obtain a smooth vertical face. The geomembrane was then installed, and the upstream part filled. A seam strip was installed to protect the geomembrane in this sensitive area and to improve the sealing of the join. The geomembrane was anchored to the crest by means of a classic trench filled with gravel.

The protection layer was installed as work progressed from the foot of the slope, using a loader and a crawler-mounted shovel loader, after the installation of the top piercing-proof geocomposite. The two machines moved over the full thickness of the protection layer (gravel layer plus random rockfill) to restrict stresses on the geomembrane and to take advantage of the stability of the random rockfill resting at the foot of the slope up against a support fill.

14.5.2.2 The Adret des Tuffes Reservoir, France (Delorme *et al.*, 2009; Girard *et al.*, 2010)

The general features of this reservoir are given in Table 14.2 and shown in Figure 14.4.

The Adret des Tuffes Reservoir in the ARCS ski resort (France) was completed in 2008. It has a capacity of 400 000 m^3 and is currently the largest reservoir of its kind in Europe. It is located at an altitude of 2200 m and, as it is upstream of residential areas, it raises public safety issues. The reservoir was created by overdeepening a natural depression and by using the moraine materials that were extracted to form two embankments, the highest of which is about 22 m above the foundation. The presence in the foundations of materials that are sensitive to water and can be eroded (dolomites) required the installation of a double geomembrane liner system covering the whole surface of the reservoir (the slopes and the bottom) (60 000 m^2).

The geosynthetic liner system is composed, from the subgrade upwards, of the following

- drainage geocomposite
- VLDPE geomembrane (1.2 mm)
- drainage geocomposite
- geocomposite barrier (2 mm PVC geomembrane laminated in the factory to a 500 g/m^2 geotextile)
- 700 g/m^2 geotextile to protect against puncturing
- cover structure (0.30 m of 0–30 mm aggregates and 0.50 m of 0–400 mm aggregates).

The advantages of this solution are that the use of the granular materials limits the risks of puncturing of the geomembranes, and that it is faster and easier to install. It meets the slope-stability criteria required of geosynthetic liner systems: (i) the stability of the cover structure is provided by the friction at each interface, the foundations taking up the shear stress due to the weight of the materials; (ii) the lowest friction value is at the interface between the upper geomembrane and the anti-puncture geotextile, with a margin of at least 3° in relation to the other interfaces to avoid the geomembranes being submitted to any tensile stresses.

Figure 14.4 The Adret des Tuffes Reservoir (France) – typical cross section and specific features (not to scale)

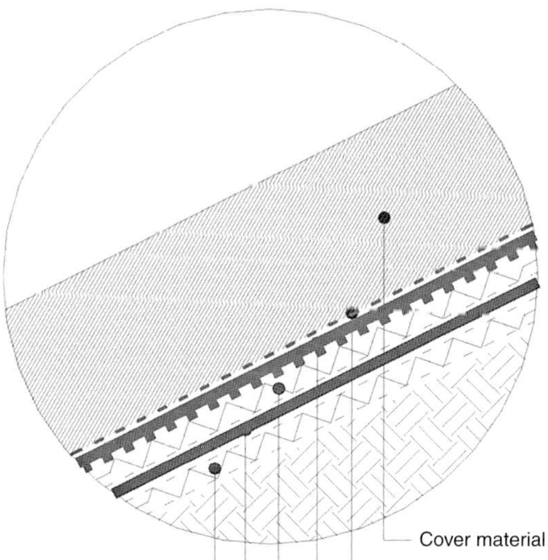

The following essential factors were taken into account in order to achieve a high-quality structure

- the power shovels were guided using satellite navigation (accurate to within centimetres) to avoid damage to the geomembrane
- great attention was paid to the subgrade and to the movements of machinery on the cover layer, with the installation of an additional thickness, to limit the puncturing strain on the geomembranes
- constant supervision of the construction work by the engineer and compliance with a demanding quality procedure
- the difficulties of undertaking installation work at high altitudes (the consequences of large day and night temperature variations on the seams of the VLDPE geomembrane, high winds requiring considerable ballasting, and the gullies furrowed by storm water in the subgrade after it has been installed) were carefully taken into account.

The total leakage flow collected perfectly by the drainage network between the two liners was less than 3 l/min when the reservoir was full.

14.5.2.3 The Barlovento Reservoir, Canary Islands, Spain (Fayoux and Potié, 2006)

The general features of this reservoir are given in Table 14.2 and shown in Figure 14.5.

The Barlovento irrigation reservoir was built between 1971 and 1975 on the island of La Palma (Canary Islands, Spain). Its storage volume is nearly 3 120 000 m^3. The substratum consists of volcanic clay with basaltic inclusions. The original liner consisted of compacted volcanic clay, but soon exhibited large leaks through major cracks. In 1991 and 1992, a PVC geomembrane lining system was installed at the bottom and up to 20 m on the 30 m-high side slopes. It was decided not to line the entire height of the reservoir in order to limit the water head on the compressible substratum, as it was suspected that settlement was one of the factors causing cracks in the volcanic clay. The total lined area of 250 000 m^2 consists of 80 000 m^2 at the bottom and 170 000 m^2 on the slopes. Slopes are thus particularly important in this case study, and the following special attention was given to their design

- No specific subgrade preparation was necessary, as the volcanic clay had already been compacted and graded for its previous use as a lining structure.
- The importance of possible runoff and groundwater on the slopes and at the bottom of the structure led to the design of an underliner drainage system. On the bottom of the reservoir, this consists of a granular drainage layer supplemented by drainage geopipes and geotextile filters. On the slopes, a porous concrete layer was added to the granular drainage layer. Underliner puncture protection is provided by a layer of sand and a spun-bonded needle-punched non-woven geotextile at the bottom, and puncture protection of the slopes is provided by the same geotextile without the sand layer. It must be emphasised that all the geotextiles were systematically stitched.
- For the reservoir bottom, a 1.5 mm unreinforced PVC geomembrane was selected, because of its puncture resistance under hydrostatic pressure, its high biaxial

Figure 14.5 The Barlovento Reservoir (Canary Islands, Spain) – typical cross-section and specific features (not to scale)

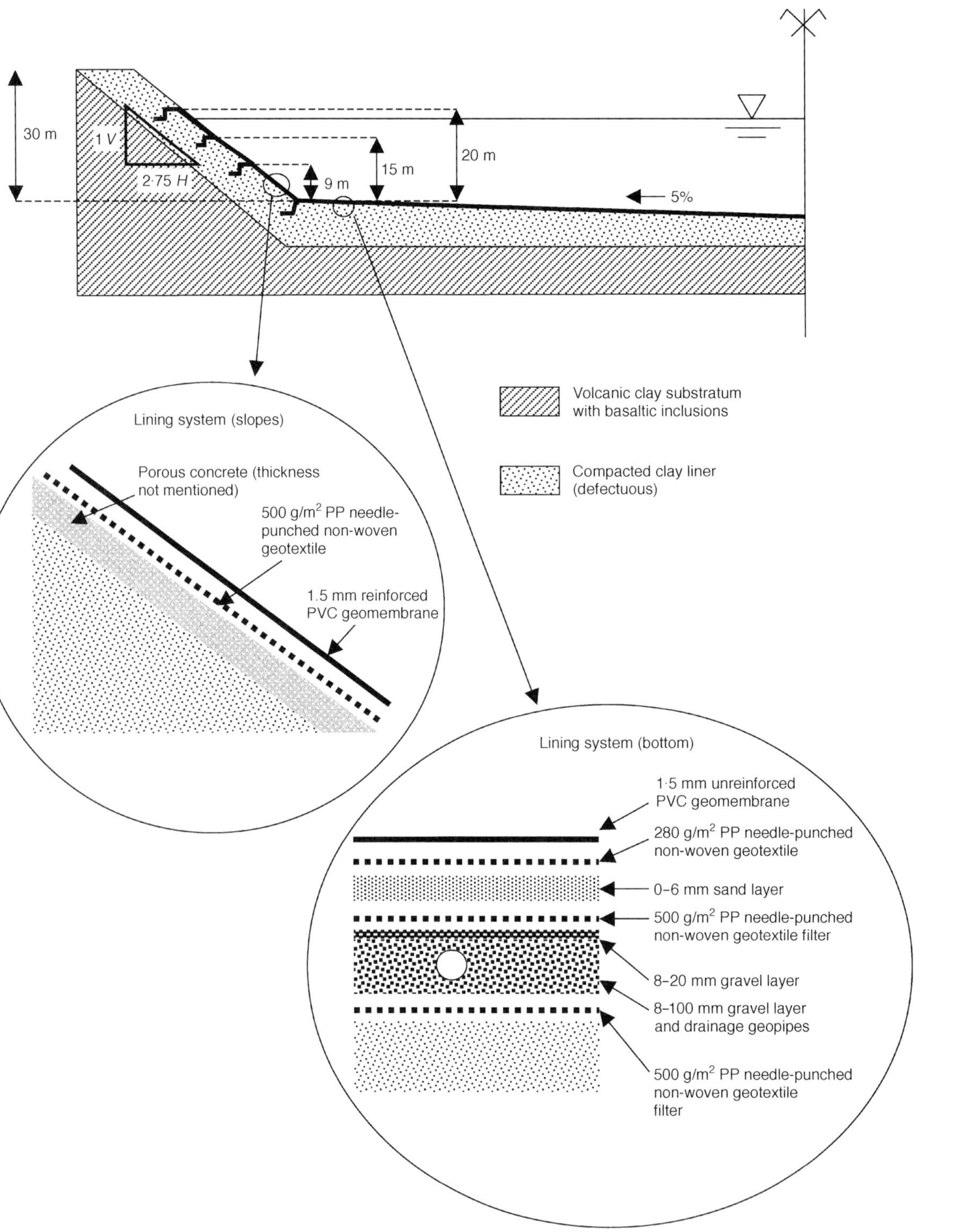

elongation in case of differential settlement, and the long-term mechanical behaviour of its joints under permanent stress. For the slopes, a 1.5 mm reinforced PVC geomembrane was selected because of its high tensile strength, which was needed for the slope-induced tensile stress. As differential settlement was less likely to occur on the slopes, high biaxial elongation was not necessary. Geomembrane welding was conducted using hot wedge automatic machines and hand-held hot air blowers for singularities. An overall quality assurance system included regular checks of the welding machines, peel tests, air lance tests and vacuum

tests of the seams. The geomembrane is anchored at the top of slopes, in three or four anchoring trenches on the slopes, and at the slope toe. Groundwater is drained at the upstream of anchoring trenches by a dedicated draining system. Liner continuity was ensured by geomembrane overlap and welding over the anchored part.
- For economic reasons, it was decided to leave the geomembrane uncovered.

14.5.3 Canals

Table 14.3 gives an overview of the references on the use of geosynthetics in canals.

14.5.3.1 The Marne–Rhin canal, France (Fagon *et al.*, 1999)

The general features of this canal are given in Table 14.3 and shown in Figure 14.6.

In 1992, the old navigation canal linking the Marne River to the Rhin River had to be repaired because of significant leaks in the dykes. Besides increasing water supply needs, the leaks endangered dyke stability and integrity. Different geosynthetic lining systems were installed to complement other techniques, such as vertical cut-off walls and concrete lining. Whereas sections were equipped with geomembrane lining systems, one canal section was lined with a geosynthetic clay liner-based system, which is described below

- Subgrade preparation consisted of restoring the profile by excavating deposited silts and filling cavities with granular material.
- A 8 mm-thick, 5 kg/m^2 geosynthetic clay liner was installed on the subgrade and mechanically confined by a 10 cm-thick overliner granular protection layer, topped with a 30 cm-thick gravel backfill layer.
- In the tidal range, a prefabricated layer of alveolar concrete elements was installed to prevent erosion and to facilitate vegetation growth.

To date, the canal section has performed correctly and without any further leaks.

14.5.3.2 The Mulhouse canal, France (Potié, 1999)

The general features of this canal are given in Table 14.3 and shown in Figures 14.7–14.9.

This channel was constructed in 1997 to convey water from the water treatment plant for the city of Mulhouse to the Hardt irrigation canal. The channel is 9 km long, with a trapezoidal section and a total output of 7 m^3/s. Some specific details are given below

- Subgrade preparation consisted of alluvial gravelly soil excavation and grading.
- A 3.5 mm bituminous geomembrane, factory-surfaced with an overliner geotextile, was installed directly on the subgrade. The geotextile was designed to drain infiltration water under the extruded concrete cover (see below). As the concrete cover was to be poured using a sliding formwork machine, sliding of the geotextile due to the action of the machine had to be prevented by bonding the geotextile and geomembrane together. The bituminous geomembrane was selected because of its puncture resistance and its ease of installation under harsh climatic conditions. Due to the surface geotextile, welding had to be carried out using 40 cm-wide bituminous geomembrane strips. Anchoring of the geomembrane on top of the embankment was not necessary because of the ballasting effect of the concrete layer.
- A protecting layer of extruded concrete was laid on the geosynthetic lining system using a sliding formwork machine, specifically designed for this work.

The rate of installation, including installation of the lining system and extruded concrete pouring, was 300 m/day.

14.6. Summary of main points

- Geomembranes have been widely used since the 1960s for the lining of ponds, reservoirs and canals; GCLs have been used in a few applications since 1990.
- The design of the lining system is crucial in every case. It must take into account the various layers and the nature of the materials in contact with the geomembrane or the GCL, and ensure that they can adequately perform their function.
- An underlying drainage layer may be required in some cases.
- Covering the geomembrane can increase its durability. Covering GCLs is necessary to ensure that they perform their lining function. The cover must be adequately designed and adapted to the lining material and the surrounding environment.
- Singularities must be studied with care, as they represent potential weak points of the lining system.
- A variety of materials can be used to line ponds, reservoirs and canals, as illustrated through the various examples presented. There is not a unique solution for all applications, and each application requires a specific design of its solution to ensure the best performance in the long term.

14.7. Some common questions and their answers

Questions

1. How do I select the best geomembrane for my application?
2. Should the geomembrane be covered?
3. What is the durability of geomembranes in hydraulic applications?
4. What is the procedure for installing a covering layer so that no damage is done to the geomembrane?
5. Why is a drainage layer required underneath the geomembrane in hydraulic applications?
6. How can leaks in a geomembrane be detected during the in-service period in hydraulic applications?
7. Can a geomembrane be installed in any climatic conditions?

Table 14.3 References on the use of geosynthetics in canals 1978–2010

Location	Country	Year of installation	Use	Type and use of geosynthetic*	Surface of geosynthetic: m^2	Section length	Cover	Reference
Canal de Bourgogne	France	1978 to 1982	N	Bituminous geomembrane	4500	NA	Gravel (0.25 m), concrete slabs	Domange (1983)
Canal du Forez	France	1979 to 1990		Bituminous and PVC geomembranes, underliner geotextile for PVC section	700	3 × 100 m	Uncovered	Duquennoi *et al.* (1995), Domange (1983)
Ishagi	Iraq	1981		Bituminous geomembrane	NA	NA	NA	Breul and Herment (1998)
Mines d'Or	Burkina Faso	1981		Bituminous geomembrane, underliner geotextile	NA	NA	Slate gravel	Breul and Herment (1998)
Esfahan	Iran	Prior to 1983		Butyl geomembrane (0.75 mm), underliner geotextile, protection geotextile	800 000 1 500 000	NA	Poured concrete (0.1 m), bituminous protection (25 mm)	Paccard (1983)
Lower main canal	Syria	1983		PVC geomembrane (0.7 mm), underliner geotextile, protection geotextile	NA	18 km	Poured concrete	Jensen *et al.* (1983)
Pedra do Cavalo	Brazil	1984	I, E	PVC geomembrane (0.8 mm), underliner geotextile, protection geotextile	180 000	13 km	Gabions	Montez and Maroni (1990)
Tanorga	India	1985		Bituminous geomembrane	64 000	NA	NA	Breul and Herment (1998)
Tungabhadia	India	1987		Bituminous geomembrane	22 000	NA	NA	Breul and Herment (1998)
Belle Fourche	USA	1987 to 1992		VLDPE (0.75 mm), LDPE (0.05 mm) and PP (0.75 mm) liner test sections with and without underliner and protection geotextiles	NA	3 × 150 m	NA	Comer (1994)
Formoso A	Brazil	1988		PVC geomembrane (1 mm), protection geotextile	82 000	10 km	Poured concrete	Montez and Maroni (1990)
Lechkanal	Germany	1989	E	Geosynthetic clay liner (4.1 kg/m^2), geotextile filter	60 000 80 000	18 km	Gravel, asphaltic layer	Heerteen and List (1990)
Coachella	USA	1989		PVC geomembrane (0.75 mm), protection geotextile; underwater	NA	300 m	Poured concrete	Morrison (1990)
Roanne–Digoin	France	1989	N	Bituminous geomembrane (3.1 mm)	NA	180 m	*In situ* poured concrete slabs	Etienne *et al.* (1995)
Deschutes	USA	1991 to 1993		VLDPE (0.75 and 1.5 mm), HDPE (1.5 and 2 mm), PVC (1 mm), CSPE (0.9 mm) geomembrane test sections with and without underliner geotextile	1500–3000 per section	18 sections: 100–300 m each	Shotcrete, geomattresses and uncovered sections	Swihart (1994)

Marne-Rhin	**France**	**1992**	**N**	**Geosynthetic clay liner (5 kg/m^2)**	**1200**	**100 m**	**Backfill, gravel, alveolar concrete elements**	**Fagon *et al.* (1999)**
Caspa District	USA	1992 to 1995		Bituminous geomembrane	150 000	NA	NA	Breul and Herment (1998)
Ponte Corvo	Italy	Prior to 1993	NA	PVC geomembrane protection geotextile	NA	NA	Shotcrete (0.07 m)	Mathieu and Fayoux (1993)
Marne Bergamo	Italy	Prior to 1993	NA	PVC geomembrane	NA	NA	Poured concrete at bottom	Mathieu and Fayoux (1993)
Marne–Rhin	France	1994	N	Bituminous geomembrane (4 mm), underliner geotextile	4400	120 m	*In situ* poured concrete slabs	Fagon *et al.* (1999), Etienne *et al.* (1995)
Jonage	France	1994	E	Concrete-filled geomattress; underwater	12 000	350 m	Alluvial material at bottom	Koffler (1995)
Niffer	France	1995	N	Bituminous geomembrane (3 mm), for upper parts	250 000	12 km	Rip rap, topsoil	Potié (1999)
Escaut	France	1995	N	Geomat, geocell, geoarmour, sand- and gravel-filled geomattresses, concrete-filled geomattresses, HDPE gabions	1250, eight test pads	260 m, eight test pads	Uncovered	Fagon *et al.* (1997)
Mulhouse	**France**	**1997**	**S**	**Bituminous geomembrane (3.5 mm), protection geotextile**	**NA**	**9**	***In situ* poured extruded**	**Potié (1999)**
Oder–Havel	Germany	1997	N	Geosynthetic clay liner (5.5 kg/m^2), sand-filled geomattress; underwater	NA	NA	Rip rap	Heibaum (1999)
Fordwah–Eastern Sadiquia	Pakistan	1997	I	VLDPE geomembrane (0.75 mm), protection geotextile	464 500	110	*In situ* poured concrete slabs	Yazdani (2001)
California Aqueduct	USA	1998	D	Bituminous protection geotextile geomembrane (3.5 mm)	NA	NA	Shotcrete	Potié (1999)
Marseille Les Aubigny	France	1999	N	Bituminous géomembrane (4 mm)	10 000	NA	Concrete (0.08–0.12 m)	Flaquet-Lacoux *et al.* (2005)
Itiquira	Brazil	About 2000	E	HDPE (1 mm)	NA	3.1	Uncovered	Abramento and Vickert (2006)
Chatillon Sur Loire	France	2003	N	Bituminous géomembrane (3 mm)	15 000	NA	Concrete (0.08 m)	Flaquet-Lacoux *et al.* (2005)

Bold indicates the canals described in this chapter.
D, drinking water; E, electricity; I, irrigation; N, navigation; NA, not available; S, sewage.
*Underwater – an underwater installation of geosynthetics was performed.

Figure 14.6 The Marne-Rhin canal (France) – typical cross-section and specific features (not to scale)

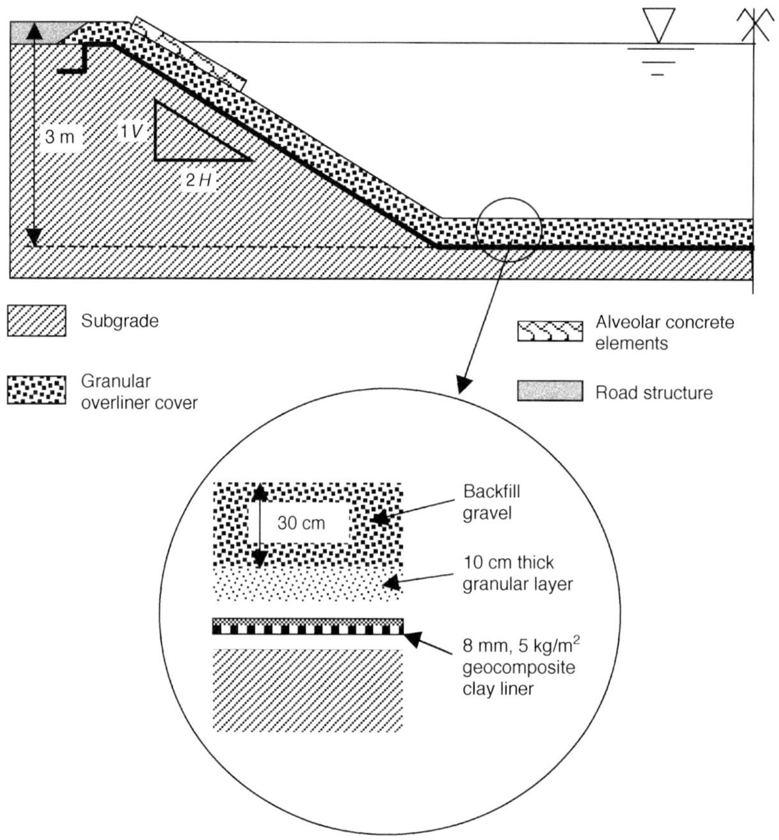

Figure 14.7 The Mulhouse canal (France) – typical cross-section and specific features (not to scale)

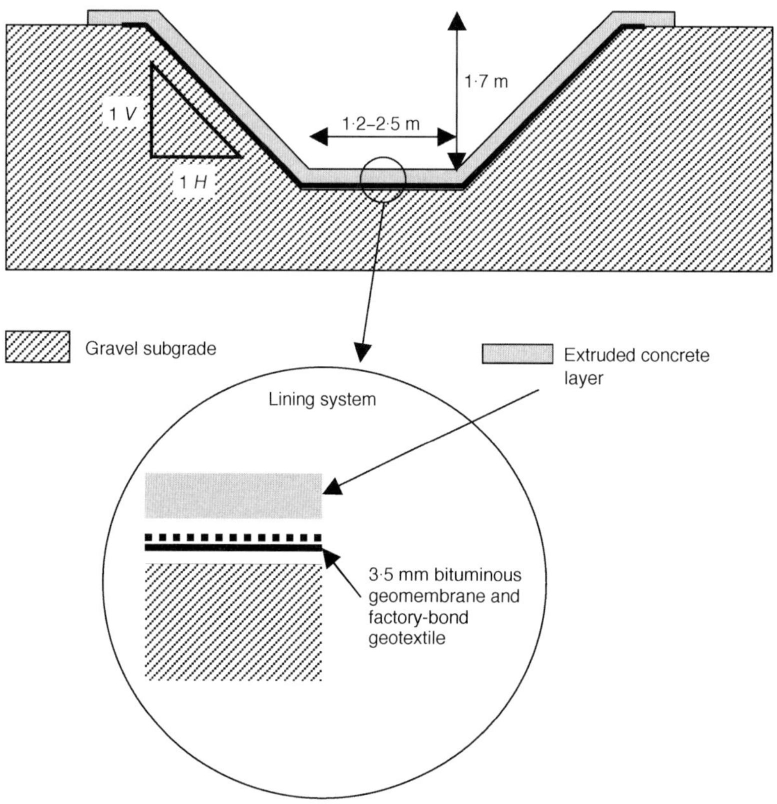

Figure 14.8 The Mulhouse canal (France) – geomembrane–geotextile composite installation
Courtesy of D. Croissant, Cemagref (1997)

Figure 14.9 The Mulhouse canal (France) – extruded concrete sliding formwork machine
Courtesy of D. Croissant, Cemagref (1997)

Answers

1. In hydraulic applications where the main limitation with regard to the choice of the geomembrane is not its chemical compatibility, what matters most are the geometric features that will impact the ease of installation. Indeed, rigid geomembranes, such as polyethylene ones, will be difficult to install in small works. In such a situation, more flexible geomembranes should be preferred.

2. The section dedicated to ageing of geomembranes has evidenced that in the case of oxidised bituminous geomembranes and PVC geomembranes, covering them ensures, in a number of cases, better durability of the geomembrane. This will also be the case in situations where objects or stones may fall on the geomembrane during the in-service period. However, one must not forget that the design and installation of the geomembrane are also possible factors affecting deterioration and that exposure is not the only source of ageing. Covering a poorly designed and installed geomembrane may not prevent its failure. The protection must be integrated in the design of the work.

3. The durability of geomembranes has been shown to be at least 30–40 years, provided that the stabiliser package used during fabrication is adequate, and that the lining system has been properly designed, installed and used. However, a number of cases were found in the literature in which this durability was not achieved, and so it cannot be accepted as the rule.

4. The implementation of a covering layer aims at avoiding damaging the geomembrane. When machines will move on the lining system, field tests are essential to determine the minimum thickness of the protection to ensure that machines can move without puncturing the geomembrane. On slopes, the protection layer must be installed as work progresses, starting from the toe of the slope, using, for example, a loader and a crawler-mounted shovel loader.

5. The section dedicated to underlining drainage systems details the two main purposes of the water drainage layer: leakage through the geomembrane has to be collected and evacuated downstream; and uplift of the geomembrane system due to backpressure from a raised water table must be prevented by the evacuation of water from the aquifer. Gas drainage is also essential. Indeed, if a gas backpressure develops due to organic fermentation or compressed soil pore air resulting from a rise in the water table, uplift of the geomembrane can be expected. Underliner gas and water drainage can usually be combined in the same drainage system.

6. In hydraulic works, the leakage is collected and evacuated downstream by the water drainage system. The system should be designed in order to allow the measurement of water flow in order to determine the level and the evolution of leakage. If this system can be divided into parts corresponding to different areas of the lining system, it is possible to determine the location of the leak. More recently, the use of optic fibres has been developed to identify leakage through liner systems. Optic fibres are installed under the lining system and are used in the measurement of the temperature underneath the geomembrane, which is related to the presence of a water flow. This type of system is relatively expensive, but allows locating of the leaks.

7. In most cases, good climatic conditions are essential for the quality of watertightness, in particular the sealing process must be done in dry conditions. When the temperature is very low, the installation of the geomembrane is more difficult, and in general it is better to avoid installation during the winter. However, conversely, very hot weather can lead to excessive dilation of the geomembrane, especially HDPE geomembranes, resulting in the formation of ripples, which may not disappear in the long term and can be a source of damage to the geomembrane.

REFERENCES

Abramento M and Vickert F (2006) Use of a geotextile as mechanical protection of a geomembrane in Itiquira dam channel, Brazil. *Proceedings of the 8th International Conference on Geosynthetics*, Yokohama, pp. 623–626.

Adams MW and Wagner N (2000) Forensic study of an HDPE liner. Evaluating physical properties of a geomembrane used at a wastewater treatment facility for 11 years. *Geotechnical Fabrics Report* **18(7)**: 1–5.

Alonso E, Degoutte G and Girard H (1990) Results of seventeen years of using geomembranes in dams and basins. *Proceedings of the 4th International Conference on Geotextiles, Geomembranes and Related Products*. The Hague, pp. 437–442.

Berube D, Diebel P, Rollin A and Stark D (2007) Massive mining evaporation ponds constructed in Chilean desert. *Geosynthetics* **25(1)**: 1–6.

Breitenbach AJ and Smith ME (2006) Overview of geomembrane history in the mining industry. *Proceedings of the 8th International Conference on Geosynthetics*, Yokohama, pp. 345–349.

Breul B and Herment R (1995) Les géomembranes bitumineuses dans la protection des sous-sols contre la pollution routière. *Revue Générale des Routes et des Aérodromes* **734** (In French).

Breul B and Herment R (1998) Bitumen geomembranes in irrigation – case histories from a range of climates. *Proceedings of the 6th International Conference on Geosynthetics*, Atlanta, GA, pp. 1133–1138.

Carreira T and Tanghe H (2008) Durabilty of PVC-P geomembranes in hydraulic structures. *Proceedings of the Eurogeo 4*, Edinburgh.

CFG (Comité Français des Géosynthétiques) (1995) Recommandations générales pour l'utilisation des géosynthétiques dans les centres de stockage de déchets. *Fascicule No. 11*.

Collios A (1994) Design and construction of an off-stream pond using geomembranes in Greece. *Proceedings of the 5th International Conference on Geotextiles, Geomembranes and Related Products*, Singapore, pp. 583–586.

Comer AI (1994) Water conservation strategies using geosynthetics. *Proceedings of the 5th International Conference on Geotextiles, Geomembranes and Related Products*, Singapore, pp. 573–578.

Davies PL (1994) Geosynthetics enable safe drinking water in developing countries. *Proceedings of the 5th International Conference on Geotextiles, Geomembranes and Related Products*, Singapore, pp. 579–582.

Delorme F, Lochu A, Kraemer P *et al.* (2009) Retenue d'altitude de l'Adret des Tuffes: une étanchéité avec double géomembrane intégralement en géosynthétiques. *Proceedings of the Rencontres Géosynthétiques 2009*, pp. 277–286.

Domange G (1983) The use of bituminous membranes in canals reclaiming. *Proceedings of the Colloque sur l'étanchéité superficielle des bassins, barrages et canaux*, Paris, pp. 263–266 (In French).

Duquennoi C, Girard H, Mathieu G and Tognetti D (1995) River and canal lining using geomembranes. *Proceedings of the Rencontres '95*, Beaune, pp. 93–99 (In French).

Etienne D, Breul B and Herment R (1995) The use of bituminous geomembranes for rehabilitating the navigation canals watertightness. *Proceedings of the Rencontres '95*, Beaune, pp. 72–78 (In French).

Fagon Y, Fouillart V, Richard F and Gourvat D (1997) Experiments of banks protection by geosynthetic processes. *Proceedings of the Rencontres '97*, Reims, pp. 84–89 (In French).

Fagon Y, Flaquet-Lacoux V, Girard H and Poulain D (1999) Record of 10 years of use of geosynthetic sealing devices in French navigable canals. *Proceedings of the Rencontres '99*, Bordeaux, pp. 187–192 (In French).

Fayoux D and Dewalque N (2006) Waterproofing by PVC geomembrane of two ponds at the pumped storage station of Afourer (Morocco). *Proceedings of the 8th International Conference on Geosynthetics*, Yokohama, pp. 611–614.

Fayoux D and Potié G (2006) Geomembranes in Tenerife and La Palma islands (Canary islands). *Proceedings of the Rencontres Géosynthétiques 2006*, pp. 473–480 (In French).

Fayoux D and Van Der Sype D (2000) Durability of PVC-P geomembranes – assessment after very long UV exposure. *Proceedings of the 2nd European Geosynthetics Conference*, pp. 651–656.

Fayoux D, Benedetti PF and Potié G (2010) Large drinking water storages with PVC geomembrane and floating cover. Experience and long term performance in Corsica. *Proceedings of the 9th International Conference on Geosynthetics*, Guaruja, pp. 1207–1212.

Fayoux D, Guérin F, Kahn JP and Ouvry JF (1999) The Souppes sur Loing sugar plant reservoir. *Proceedings of the Rencontres '99*, Bordeaux, pp. 59–64 (In French).

Flaquet-Lacoux V, Malascrabes M, Poulain D, Girard H and Lino M (2005) Rehabilitation of the watertightness of navigable canals using geomembranes. *Ingénieries Numéro Special*, pp. 159–167.

Frobel FK (1983) Quality assurance program for the Mt. Elbert Forebay flexible membrane lining installation. *Proceedings of the Colloque sur l'Etanchéité Superficielle des Bassins, Barrages et Canaux*, Paris, pp. 29–34 (In French).

Girard H, Lefranc M and Tisserand C (1996) Waterproofing and rehabilitation of dams by geomembranes in France. *Proceedings of the 1st European Geosynthetics Conference Eurogeo 1*, Maastricht, pp. 945–952.

Girard H, Poulain D, Benneton J-P and Fayoux D (2002) Durability of PVC-P geomembranes used in hydraulic structures: feedback on experience. *Proceedings of the 7th International Conference on Geosynthetics*, Nice, 2, pp. 709–712.

Girard H, Delorme F, Deroo L *et al.* (2010) Mountain reservoirs with geomembrane lining systems in France: behaviour, pathology, guidelines and outstanding projects. *Proceedings of the 9th International Conference on Geosynthetics*, Guaruja, pp. 1191–1196.

Girollet J (1983) The use of geomembranes for the construction of industrial facilities and the protection of the environment. *Proceedings of the Colloque sur l'Etanchéité Superficielle des Bassins, Barrages et Canaux*, Paris, pp. 159–163 (In French).

Habte K and Legg P (2009) Innovative liner design for brine evaporation ponds. *Proceedings of the 1st African Conference on Geosynthetics Geoafrica 2009*, Cape Town.

Heerten G and List F (1990) Rehabilitation of old liner systems in canals. *Proceedings of the 4th International Conference on Geotextiles, Geomembranes and Related Products*, The Hague, pp. 453–456.

Heibaum M (1999) Application of geosynthetic clay liners in German waterways. *Proceedings Rencontres '99*. Bordeaux, pp. 139–144 (In French).

Heibaum M, Fourie A, Girard H *et al.* (2006) Hydraulic applications of geosynthetics. *Proceedings of the 8th International Conference on Geosynthetics*, Yokohama, pp. 79–120.

Hsuan G, Schröder H, Rowe RK *et al.* (2008) Long-term performance and lifetime prediction of geosynthetics, Keynote Lecture. *Proceedings of the Eurogeo 4*, Edinburgh.

Ialynko P and Gonin H (1993) The Chanteraines artificial lake: a public park built on a landfill. *Proceedings Rencontres '93*, Joué-les-Tours, pp. 355–364 (In French).

IGS (International Geosynthetics Society) (2009) *Recommended Descriptions of Geosynthetics Functions, Geosynthetics Terminology, Mathematical and Graphical Symbols*, 5th edn. IGS, Paris.

ISO (International Organisation for Standardisation) (2005) EN ISO 10318 – *Geosynthetics – Terms and Definitions*. ISO, Geneva.

Jensen A, Roux H and Beynet JM (1983) Specifications of a PVC membrane for primary watertightness of a large canal in gypseous ground. *Proceedings of the Colloque sur l'Etanchéité Superficielle des Bassins, Barrages et Canaux*, Paris, pp. 1–6 (In French).

Koerner RM (1986) *Designing with Geosynthetics*. Prentice Hall, Englewood Cliffs, NJ.

Koffler A (1995) Renewed impermeabilization of canal de Jonage. *Proceedings Rencontres '95*, Beaune, pp. 86–91 (In French).

Lambert S, Duquennoi C and Tcharkhtchi A (1999) Use of geomembranes in high altitude applications. Examples of PVC geomembranes. *Proceedings of the Rencontres '99*, Bordeaux, pp. 285–292 (In French).

Lescure JP (1983) Sealing of sugar factory ponds: working constraints. *Proceedings of the Colloque sur l'Etanchéité Superficielle des Bassins, Barrages et Canaux*, Paris, vol. 1, pp. 137–142 (In French).

Loudière D and Perrin J (1983) Failure of facing of two reservoirs. *Proceedings of the Colloque sur l'Etanchéité Superficielle des Bassins, Barrages et Canaux*, Paris, pp. 147–152 (In French).

Mathieu G and Fayoux D (1993) Geomembranes and waterproofing of concrete structures intended for the storage and transport of liquids. *Proceedings of the Rencontres '93*. Joué-les-Tours, pp. 335–344 (In French).

Michelangeli A, Vanoni A and Vargari S (2006) Geosynthetics for disposing contaminated dredging sludge in harbour submarine basins. *Proceedings of the 8th International Conference on Geosynthetics*, Yokohama, pp. 789–792.

Monjoie A, Rigo JM and Polo-Chiapolini C (1992) *Vademecum pour la Réalisation des Systèmes d'Etanchéité – Drainage Artificiels pour les sites d'Enfouissement Technique en Wallonie*. Université de Liège, Faculté des Sciences Appliquées (In French).

Montez FT and Maroni LG (1990) Use of geotextiles and geomembranes in irrigation canals – Brazilian case histories. *Proceedings of the 4th International Conference on Geotextiles, Geomembranes and Related Products*, The Hague, pp. 449–452.

Morrison WR (1990) Use of geosynthetics for the underwater lining of operating canals. *Proceedings of the 4th International Conference on Geotextiles, Geomembranes and Related Products*, The Hague, pp. 443–447.

Murillo-Fernandez R (1994) Mexican experiences with geomembranes in hydraulic works. *Proceedings of the 5th International Conference on Geotextiles, Geomembranes and Related Products*, Singapore, pp. 565–568.

Nortjé RA and Meyer PJ (2006) The successful design and construction of a floating cover molasses storage reservoir for the TSB Malelane Mill in South Africa. *Proceedings of the 8th International Conference on Geosynthetics*, Yokohama, pp. 653–656.

Ouvry J-F, Gérard Y and Baghri K (2006) Rehabilitation of the tightness of a brine storage basin on the Sidi-Larbi site in Morocco. *Proceedings of the Rencontres Géosynthétiques 2006*, pp. 511–518 (In French).

Paccard M (1983) Utilisation of a butyl for watertightness of irrigation canals in Iran. *Proceedings of the Colloque sur l'Etanchéité Superficielle des Bassins, Barrages et Canaux*, Paris, pp. 251–255 (In French).

Peggs ID (2008) Observations and thoughts on the performance of polypropylene geomembrane liners and floating covers: towards improved specifications. *Proceedings of the Eurogeo4*, Edinburgh.

Pons C, Girard H, Jouve M, Potie G and Poulain D (2006) Geomembrane waterproofing of the Berre drinking water reservoir, *Ingénieries – E A T, Numéro Spécial Géosynthétiques Applications aux Installations de Stockage de Déchets et aux OUVRAGES HYDRAULIQUES*, pp. 77–84 (In French).

Potié G (1999) Bitumen geomembranes in canals. *Proceedings of the Rencontres '99*, Bordeaux, pp. 201–208 (In French).

Poulain D, Girard H, Mine C and Potié G (1997) Waterproofing of the Selvet dam with a geomembrane. *Proceedings of the Rencontres '97 Géotextiles et géomembranes*, Reims, France, pp. 142–148 (In French).

Rowe RK and Orsini C (2003) Effect of GCL and subgrade type on internal erosion in GCLs under high gradients. *Geotextiles and Geomembranes* **21(1)**: 1–24.

Rowe RK, Sangam HP and Lake CB (2003) Evaluation of an HDPE geomembrane after 14 years as a leachate lagoon liner. *Canadian Geotechnical Journal* **40**: 536–550.

Sadlier M (2009) Floating covers for evaporation control – an economic review. *Proceedings of the Geoafrica 2009*, Cape Town.

Saintot J and Bilancioni S (1983) Geomembrane application in the storage of aggressive waste. *Proceedings of the Colloque sur l'Etanchéité Superficielle des Bassins, Barrages et Canaux*, Paris, pp. 143–146 (In French).

SETRA/LCPC (Technical Department for Transport, Roads and Bridges Engineering and Road Safety/French Central Public Works Research Laboratory) (2000) *Etanchéité par Géomembranes des Ouvrages pour les Eaux de Ruissellement Routier* (In French).

Shimizu H and Ikeguchi Y (1998) Use of a synthetic rubber sheet for surface lining of upper pond at seawater pumped-storage power plant. *Proceedings of the 6th International Conference on Geosynthetics*, Atlanta, GA, pp. 1115–1120.

Spillemaecker P-M, Deguines J-P, Chaix C and Brousset P (2002) Waterproofing of the La Glacière controlled landfill in the French Riviera. *Proceedings of the 7th International Conference on Geosynthetics*, Nice, vol. 2, pp. 795–798.

Swihart JJ (1994) Deschutes canal lining demonstration – construction report. *Proceedings of the 5th International Conference on Geotextiles, Geomembranes and Related Products*, Singapore, pp. 553–556.

Taylor R (1990) Installing floating covers on intricately shaped reservoirs. *Proceedings of the 4th International Conference on Geotextiles, Geomembranes and Related Products*, The Hague, p. 477.

Tisserand C (1983) Reservoir of Guazza: a large reservoir of unprotected membrane. *Proceedings of the Colloque sur l'Etanchéité Superficielle des Bassins, Barrages et Canaux*, Paris, pp. 87–89 (In French).

Tisserand C, Matichard Y and Laine D (1995) Floating cover on a potable water reservoir. *Proceedings of the Rencontres '95*, Beaune, pp. 79–84 (In French).

Touze-Foltz N and Hatton R (2006) Hydraulic behaviour of a geosynthetic clay liner in three wastewater lagoons. *Proceedings of the Rencontres Géosynthétiques 2006*, pp. 539–546 (In French).

Touze-Foltz N, Croissant D, Farcas F and Royet P (2010) Quantification of oxidized bituminous geomembranes ageing through hydraulic testing. *Proceedings 9th International Conference on Geosynthetics*, Guaruja, pp. 753–756.

Wallace RB (2008) Floating covers for potable water reservoirs – two case histories. *Proceedings of the Geo-americas, The First Pan American Geosynthetics Conference and Exhibition*, Cancun, pp. 1667–1672.

Wichern HAM (1990) Use of geomembranes for community development. *Proceedings of the 4th International Conference on Geotextiles, Geomembranes and Related Products*, The Hague, p. 475.

Yazdani AM (2001) VLDPE geomembrane stops canal seepage in Pakistan. *Geotechnical Fabrics Report* **19(3)**: 36–38.

Yoshikoshi H and Masuda T (1994) Application of flexible membrane lining to a reservoir. *Proceedings of the 5th International Conference on Geotextiles, Geomembranes and Related Products*, Singapore, pp. 569–572.

Handbook of Geosynthetic Engineering
ISBN 978-0-7277-4175-2

http://dx.doi.org/10.1680/hge.41752.303

ice
Institution of Civil Engineers
publishing

Chapter 15
Hydraulic tunnels

D. Cazzuffi*, **A. Scuero†** and **G. Vaschetti†**
*IGS Past President, CESI SpA, Via Rubattino 54, Milano, Italy
†Carpi Tech BV, Via Passeggiata 1, Balerna, Switzerland

The main function of hydraulic tunnels is to convey water, under pressure or in free flow, for power and/or water supply works. Hydraulic tunnels are usually lined with steel, cast *in situ* concrete or sprayed concrete. The most common concrete linings are rarely free from defects such as voids and cracks. During operation, when the concrete linings are exposed to the dynamic action of water, they suffer from deterioration, resulting in the formation of fissures/cracks, increased permeability through body and joints, and increased roughness. The use of geomembranes allows the combination of continuous watertightness and improved friction properties with the strength of the concrete lining at a much lower cost. Geomembrane systems have reached a high degree of refinement and reliability, and, when adequately designed and installed, they are durable and reliable under high water heads and demanding environments. This chapter provides a complete description of the engineering aspects of geosynthetic-based lining systems for new and old hydraulic tunnels.

15.1. Introduction

Hydraulic tunnels are mainly used as a means to convey water, under pressure or in free flow, to be used for power and/or water supply works. With the exception of old tunnels excavated in competent self-supporting rock, hydraulic tunnels are usually lined with steel, cast *in situ* concrete or sprayed concrete. In unstable ground, the lining provides structural stability by supporting the excavation. Another important requisite of hydraulic tunnel linings is to provide a waterproofing function by preventing water from seeping out, or water and other deleterious chemicals from entering the tunnel. A further requirement is to improve the hydraulic performance of the excavation by reducing friction losses.

Traditional hydraulic tunnel linings have been steel or concrete. Steel addresses all of the above criteria; however, the lining thickness required to achieve the criteria often makes it an economically unattractive solution, being both more expensive as a material and more time-consuming in its installation. On the other hand, concrete linings are rarely free from defects such as voids and cracks that generally arise as a consequence of thermal stresses or poor construction. In addition, during operation, when the concrete linings are exposed to the dynamic action of water, they suffer from deterioration, resulting in the formation of fissures/cracks, increased permeability through the body and joints, and increased roughness leading to increased friction losses that reduce power generation and water supply (Scuero *et al.*, 1999). Movement in the slopes, due to loose or fractured soil, faults and folds, the presence of water, etc., and the quality of flowing water (pH), could also further aggravate the deterioration of the concrete lining (Figure 15.1).

The loss of water into the surrounding ground is extremely serious as it may jeopardise the stability of the ground and also that of the structure itself. This is particularly true in areas of low horizontal or vertical overburden. In steep mountain areas, and in areas of unfavourable geology, escaped water can erode the soil, and eventually start landslides or cause collapse of the subgrade on which the tunnel is founded. Exceptionally dangerous is the situation in high-pressurised concrete (reinforced and unreinforced) tunnels in rock, where leaks are usually very small; however, they can cause large failures because of the full or nearly full static head that may develop, and is transferred through cracks or joints in the rock even over great distances. The flow of water through the cracks can be close to zero, but full static pressure can be transmitted to the surrounding ground. Leakages can also cause weakening of the ground (gypsum, salt), resulting in movements affecting the tunnel. Furthermore, cracks and joints in the concrete lining allow groundwater to act as a hydraulic jack and affect the durability of the reinforcement.

The unexpected deterioration of concrete lining systems, even if stability is not at stake, entails high-maintenance costs of dewatering and periods of loss of service. In the case of leaking wastewater tunnels, a further threat is posed by potentially deleterious substances coming into contact with the groundwater. It is therefore advisable to line

Figure 15.1 Defects in pressure tunnels caused by ground movement/settlement: (a) crack development, (b) crack development and (c) fissure opening

new-built concrete tunnels with an impervious geomembrane during construction. Geomembrane waterproofing systems are a cost-effective and efficient solution to prevent design and construction delays, minimise outage loss and repair and maintenance costs, and other consequences of a failure that could be caused by a minor leak.

The development of manufacturing and installation techniques, since the late 1950s, has established the use of prefabricated synthetic geomembranes as dependable, long-term waterproofing liners for hydraulic structures. A membrane liner can be formed by one single material (geomembrane) or be a composite material (composite geomembrane), the latter consisting of a geomembrane (providing barrier properties) heat-coupled during manufacturing to a nonwoven geotextile that increases mechanical resistance and dimensional stability in respect of temperature variations, and provides some drainage capability. The first underground application of a synthetic geomembrane dates back to 1969, when it was used in a subway station in Vienna, Austria. In hydraulic tunnels, considerable experience has been acquired since the 1970s. Some Austrian applications in hydraulic tunnels and shafts with polyvinylchloride (PVC) geomembranes account for static and dynamic pressures up to about 500 and 740 m, respectively (Heigerth *et al.*, 1979; Seeber, 1982, 1985; Laufer, 1985; Scuero and Vaschetti, 1997).

This chapter describes in detail the engineering aspects of geosynthetic-based lining systems, especially geomembrane systems for the waterproofing of hydraulic tunnels in new construction, as well as in rehabilitating old ones.

15.2. Geomembrane systems

15.2.1 Applications

Geomembrane systems can be installed as a waterproofing layer for new construction, typically as a covered solution, or in rehabilitation projects as a repair measure for deteriorated leaking linings, typically as an exposed solution.

The exposed solution can be adopted in rehabilitation, as well as in new construction for water velocities of several metres per second and uplift pressures of several megapascals. In an exposed solution, the geomembrane is generally bonded to a geotextile to form a composite geomembrane (Figure 15.2) and applied to the existing lining.

The covered solution is typically adopted for new tunnels, but it can also be used in rehabilitation in the case of particularly severe service conditions. In a covered solution, the geomembrane is not bonded to the geotextile, but it is placed over an independent anti-puncturing layer of geotextile. It is typically located between the outer sprayed concrete and the inner concrete lining (Figure 15.3).

The waterproofing layer can be applied to the entire tunnel cross-section (pressure tunnels) or specifically to the problematic areas (free-flow tunnels), as shown in Figures 15.4–15.8.

The selection of the solution depends on the physical and operational characteristics of the tunnel, and on the scope of the lining. Physical and operational characteristics to consider are the water head, the static and dynamic pressure inside the tunnel, the water velocity, the negative pressure inside the tunnel due to suction during the closing and

Figure 15.2 Layering of a drained exposed system

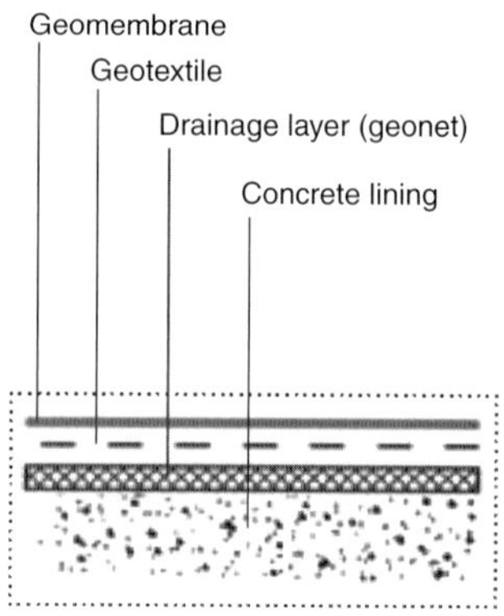

Figure 15.3 Layering of a fully covered system

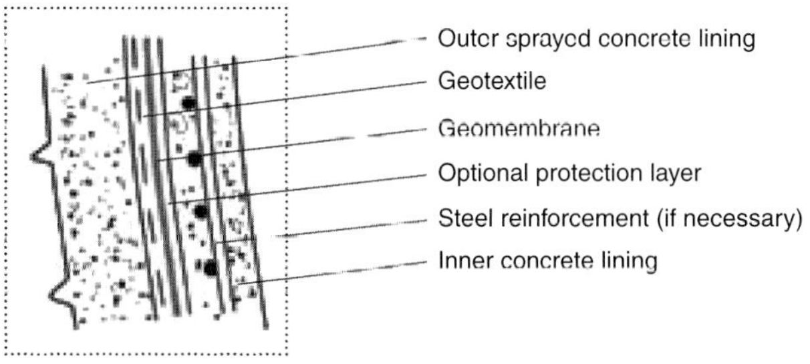

opening of turbine valves, the negative pressure outside the tunnel when the tunnel is empty (elevation of the water table in the soil surrounding the tunnel), the tunnel operation (always full, mostly empty), and the presence of chemicals aggressive to concrete.

15.2.2 Types

The experience in hydraulic tunnels consists mainly of PVC geomembranes and composite geomembranes. Some applications of polyethylene (PE) and polypropylene (PP) geomembranes have also been reported.

Extensive experience with PVC geomembranes in underground tunnels goes back for over 30 years. Since then, according to the International Tunnelling Association, PVC geomembranes have been employed in over 92% of lined tunnels all over the world. The reasons for such an impressive record are the following

- Geomembranes possess very low permeability ($k = 10^{-12}$ m/s), at least three orders of magnitude lower than new concrete.
- Geomembranes possess greater than 200% elongation, which allows the bridging of old and new cracks, and for considerable movements that can occur in slopes.

Figure 15.4 A typical pressure tunnel cross-section, where the geomembrane system is placed all around the tunnel cross-section and covered with a structural inner concrete lining

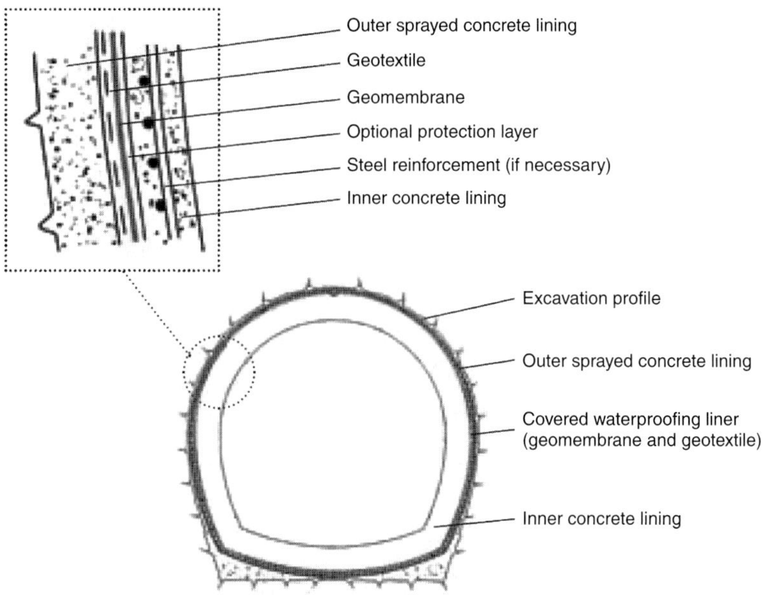

Figure 15.5 A typical pressure tunnel cross-section where the composite geomembrane is exposed

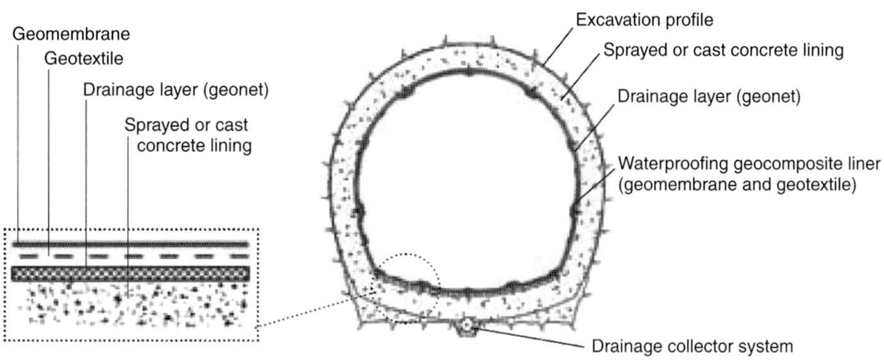

Figure 15.6 A typical free-flow tunnel cross-section where the geomembrane is placed over the walls and invert, and left exposed

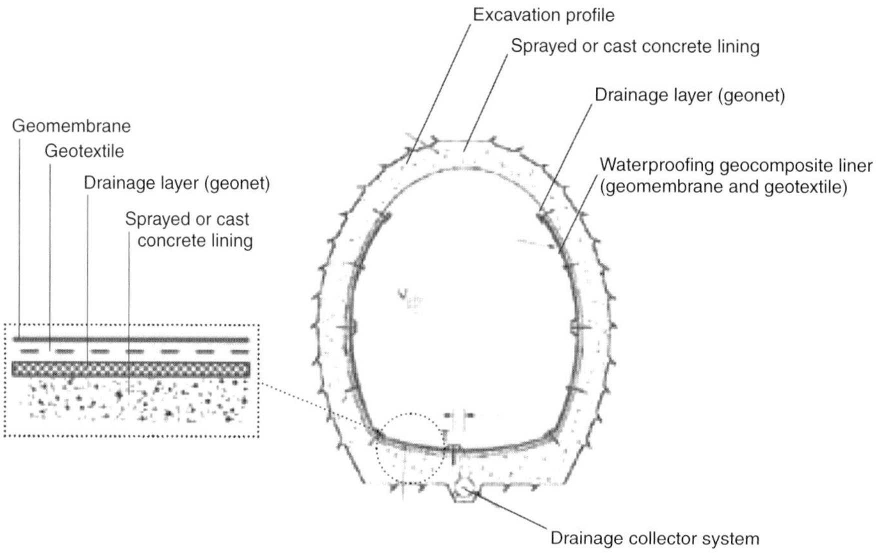

- Geomembranes are a thin (of the order of a few millimetres) lining that does not reduce the cross-section of the tunnel.
- Geomembranes possess a Manning (Gauckler–Strickler) coefficient n of less than 0.0115, which is not affected by operation. If left in contact with flowing water, they will increase the water flow compared with a concrete lining.
- Geomembranes possess excellent chemical resistance.
- Geomembranes possess excellent flexibility over a wide range of temperatures.
- Geomembranes have very good weldability even in a wet environment, and a wide weldability temperature range about three times wider than that acceptable, for example, for PE and PP membranes.
- Technologically advanced methods of welding and testing are available.
- A wide range of compatible accessories are available (waterstops, discs, grouting hoses, grouting profiles, etc.).
- Geomembranes are easy to handle to optimise installation.
- Geomembranes have performed outstandingly. Several million square metres of laid PVC geomembrane in tunnels confirm the effectiveness of the material and of the system.

Figure 15.7 Large cracks injected with acrylic resins or cementitious grout prior to the installation of the waterproofing composite geomembrane system

Figure 15.8 A geogrid being installed over active cracks, generally done in addition to or as an alternative to the injection of large cracks

15.2.3 Benefits

The role of the waterproofing liner is not necessarily only watertightness: other possible objectives are, for example, to decrease friction losses or to provide an elastic barrier to water infiltration in the event of excavations performed in expansive soil. Some of the major benefits of geomembrane systems are listed below

- *Durability*. Installations dating from the 1960s and early 1970s are still in service (Cazzuffi and Venesia, 1989). Tests have been performed on PVC geomembrane systems that indicate that the geomembrane will last more than 50 years when directly exposed to an outside environment. Geomembrane formulations have been upgraded since the early 1970s and, additionally, the thickness of the geomembrane has been increased, so that it can be safely assumed that geomembranes installed today will have an even longer service life. Considering the underground environment of hydraulic tunnels, the composite geomembrane will never be exposed to ultraviolet rays, and PVC geomembranes and composite geomembranes show no reduction in performances even over extremely long periods of immersion in water. The durability of the geomembrane is thus several decades, without the need for maintenance.
- *Waterproofing*. This refers to the elimination of inward and outward leakages. The measurement of leakage after an installation is typically so low as to be statistically insignificant. The system has no discrete joints, and forms a bridge over new cracks.
- *Hydraulic efficiency*. When left exposed, the smoothness of a PVC geomembrane can increase the velocity of water and the flow capacity of the tunnel, decreasing the maximum water level (Figure 15.9).
- *Geometry*. When left exposed, a geomembrane system does not reduce the tunnel envelope.
- *Maintenance*. In hydraulic tunnels, the reliability of the seal is of particular significance, since once in operation it is not easily accessible for subsequent repairs. Exposed geomembrane systems require no planned, routine or preventive maintenance. Removal of sediment that will deposit on the geomembrane is required for the hydraulic performance of the tunnel, not for the integrity or the performance of the geomembrane. Furthermore, the installed and exposed geomembrane allows for much easier control of future performance and cheaper maintenance during the years to come.
- *Installation costs*. The installation cost of a geomembrane system is lower compared with other lining systems. The long service life of the system further reduces the life-cycle costs. The geomembrane installation requires only simple hand tools that can operate in various weather or environmental conditions. This prevents weather from being a factor of affecting the quoted price, and assures consistent quality of the final installation. Geomembranes do not have many environmental or mechanical systems that can prevent or delay the work. Additionally, an exposed geomembrane reduces the quantity of construction material to be transported inside the tunnel and the need for preparation works when compared with steel lining. Because mobilisation does not require the movement of large equipment, the job can be undertaken over several years, with the main requirement being only the re-mobilisation of personnel.
- *Programme*. The possibility of modular installation of the geomembrane system allows installation in stages based on the customer's operating requirements, thereby saving the customer money by reducing the downtime during installation.

Figure 15.9 Increased water flow, due to the presence of a geomembrane, resulting in a decrease in the overall water level

The use of flexible PVC geomembranes furnishes the combination of continuous watertightness and the improved friction properties of a steel liner with the strength of a (reinforced) concrete lining, but at a much lower cost.

15.3. The exposed liner system

As already stated, exposed solutions, which can be adopted in rehabilitation and in new construction, for water velocities of several metres per second, and uplift pressures of several megapascals, are most frequently used for rehabilitation (Cazzuffi, 1995). In pressure tunnels, prefabricated synthetic geomembrane sheets are used to cover the entire inner surface of the tunnel, bridging existing cracks, fissures and construction joints. Underground water is usually collected in the invert or haunches by means of a drainage system (see Figure 15.5). In free-flow tunnels, the exposed composite geomembrane waterproofing system is typically applied to the lower section of the tunnel, which in existing tunnels coincides with the wet damaged area (see Figure 15.6).

15.3.1 Support system

The substrate, particularly in rehabilitation work, needs to be prepared. Where large cracks and joints are present, they should be treated and covered with a support geogrid prior to the installation of the drainage and waterproofing layers. This layer is intended to improve the efficiency of the system by spanning large openings, and preventing the geomembrane from collapsing into the gap and becoming damaged in the process (see Figure 15.8).

The waterproofing system must ensure that imperviousness is provided and maintained, notwithstanding movements at joints and opening cracks. In the case of rehabilitation work, the substrate needs to be prepared, and large cracks and joints are treated (see Figure 15.7) and covered with a geogrid (see Figure 15.8) prior to the installation of the waterproofing system. The efficiency of the system to accommodate displacements, rotations and opening of fissures without rupturing has been proven by over 30 years of field performance.

15.3.2 Drainage system

Inward external pressure caused by the hydrostatic load, applied by the groundwater entering the tunnel structure from behind the composite geomembrane, since the composite geomembrane forms an impermeable barrier, can build up. If not discharged, the built-up pressure can, in some circumstances, destroy the installation. For this reason, a method to drain behind the composite geomembrane is provided: the build-up of pressure due to groundwater is thus negated by the presence of a drainage system.

The drainage system consists of a draining layer and of a drainage discharge system. The draining layer is a highly permeable material, typically a geonet, installed over the substrate and fastened by means of impact anchors (Figure 15.10). The geonet is covered by the waterproofing composite geomembrane. The drainage system provided behind the composite geomembrane collects the water and discharges it outside the tunnel when possible; if outside discharge is not possible, patented one-way valves, made of the same composite geomembrane material, operated by pressure differentials and discharging inside the tunnel, are available. The relief valves will open only when the pressure inside the tunnel is less than the pressure from the surrounding soil, usually when the tunnel is empty. The valves discharge inside the tunnel the water conveyed by the drainage layer. Other types of geosynthetics are also used as drainage in underground applications, as described by Cazzuffi *et al.* (1986).

Figure 15.10 Geonet fastened to the substrate by means of impact nails

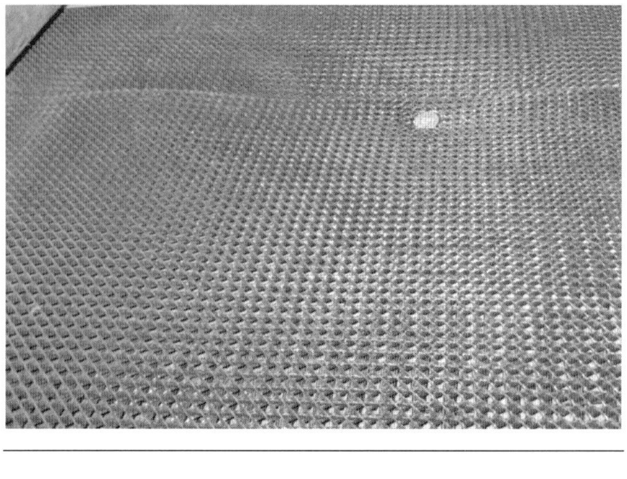

Figure 15.11 Placing of the composite geomembrane inside a pressure tunnel in the Czech Republic

15.3.3 Waterproofing liner

The waterproofing layer used in state-of-the-art projects is a very low permeability composite geomembrane, composed of a PVC geomembrane, a few millimetres thick, bonded during fabrication to a geotextile that enhances its antipuncture resistance, placed over the substrate/support/drainage layers (Figure 15.11). Particular provision, such as an additional layer of geotextile, must be made to avoid the intrusion of the waterproofing liner into possible cracks in the substrate, in the presence of high water heads. The composite geomembrane is left exposed to contact with the water flowing inside the tunnel, and is designed to withstand mechanical actions (abrasion, impact, tear) due to flowing water (Figures 15.12 and 15.13), and transported materials and sediments.

15.3.4 Fastening system

The fastening system consists of lines of stainless steel profiles, anchoring the composite geomembrane to the substrate, be it concrete or natural ground, and keeping it independent from it. The elastic geomembrane can thus freely deform over it when the tunnel is dewatered. They are typically placed along the longitudinal axis of the tunnel. There are two types of linear fastening available: tensioning profiles (originally developed and used in the rehabilitation of dams, and later adopted in water conveyance structures due to their efficiency); and flat profiles (Figures 15.14 and 15.15). If both types are used, they are generally arranged in alternate lines, with the spacing depending on the foreseen operation conditions of the tunnel. When the tunnel is in operation, the composite geomembrane will be pushed

Figure 15.12 Exposed composite geomembranes in a free-flow tunnel

against the existing surface of the concrete by the water pressure; when the tunnel is empty, the composite geomembrane will be subject to minimum deflections.

Tensioning profiles are made of two parts: the internal and the external profiles. The internal profile is anchored to the concrete structure by means of chemical anchors, whereas the external profile is installed over the PVC composite geomembrane and clamped to the internal profile. The tensioning profiles, in addition to anchoring the waterproofing liner to the existing tunnel liner, also pretension the geomembrane to eliminate wrinkles and slack areas (Figures 15.16 and 15.17), improving hydraulic efficiency. Tensioning will

Figure 15.13 Exposed composite geomembranes to resist high water velocities and turbulence

Figure 15.14 Tensioning profiles

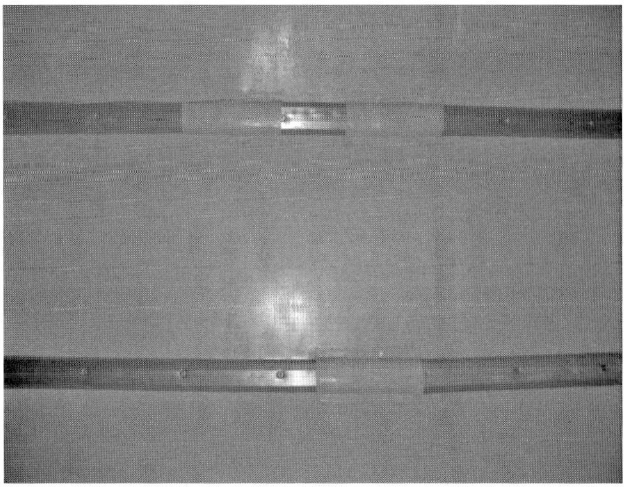

result in the most effective adherence of the composite geomembrane to the existing surface. In the event of a tear in the composite geomembrane, the tensioning profiles will keep the composite geomembrane in place, preventing it being carried away by the water flow towards, for example, turbines. Negative pressure due to suction is taken into account when designing the fastening system. The operational conditions dictate the positioning of the fastening lines, the type of anchorage method, and the need for particular provisions at sections influenced by the peak suction areas, also due to transients. The number, type and position of the fastening anchorage profiles are designed as a function of the inward pressure caused by groundwater in the case of dewatering, of suction occurring during the operation of the tunnel, and of the dynamic action of water that would result from a tear in the geomembrane.

The flat profiles are fastened by impact anchors. All profiles are covered by waterproofing PVC geomembrane strips. Additional steel plates are used in areas affected by suction due to transients, to provide support against the detaching action of the suction.

Figure 15.15 Flat and tensioning profiles arranged across a tunnel perimeter

Figure 15.16 External tensioning profiles installed over a composite geomembrane to remove any slack in a pressure tunnel in Greece

Figure 15.17 Stretching of the composite geomembrane through tightening of the external tensioning profile

If in-service conditions (high uplift, high water velocity, suction) require very closely spaced anchorages for the membrane, a covered solution may be more cost-effective.

15.3.5 Perimeter seals

The waterproofing system is sealed at the perimeter (beginning and end of the waterproofing section, and any other submersible periphery) by means of stainless steel profiles compressing the composite geomembrane against the existing surface. The distribution of compression forces is made by layers of resin bedding mortar and neoprene gaskets of adequate consistency (Figure 15.18).

15.4. The covered liner system

The covered geomembrane system described combines Austrian and Italian expertise, and construction of hydraulic tunnels with waterproofing geomembranes. Over 20 years ago, the New Austrian Tunnelling Method was developed, which is now used all over the world. The covered solution is mostly adopted for new pressure tunnels, and can be used in rehabilitation in the case of particularly severe service conditions. The covered solution is adopted when external pressure is exerted by the groundwater on the geomembrane at dewatering of the tunnel, due to water present in slopes. The covered geomembrane system provides stability of the liner under very high piezometric heads, as the inner concrete lining constitutes the permanent support of the geomembrane. This makes it necessary for the geomembrane

Figure 15.18 Perimeter seal

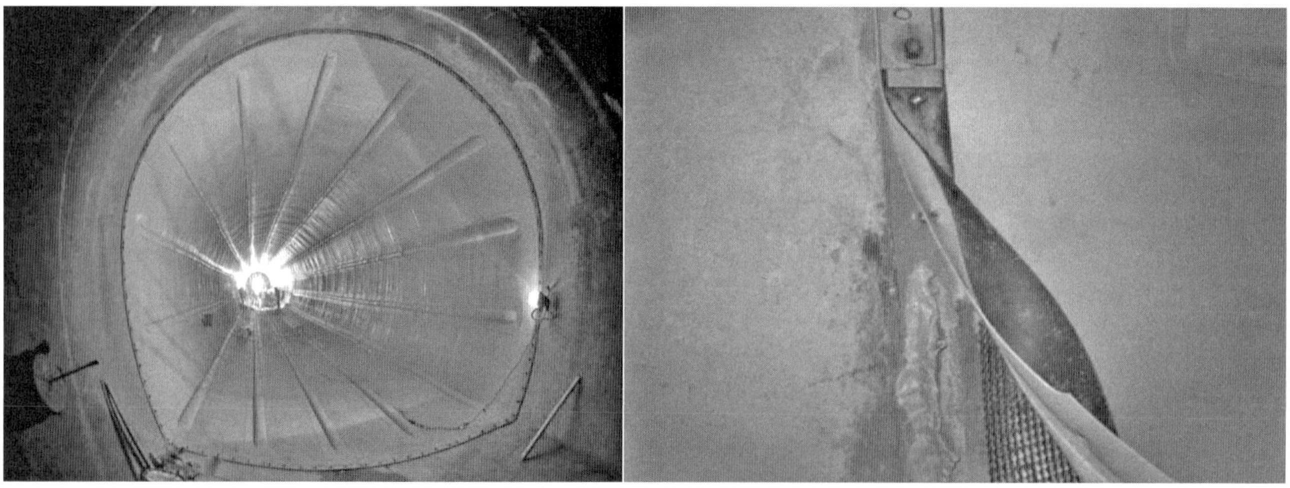

system and inner concrete lining to be designed to cope with the groundwater pressure.

In this solution, the flexible PVC waterproofing geomembrane is sandwiched between the natural excavation, smoothing shotcrete layer, outer concrete lining or the existing structure (depending if it is a new construction or rehabilitation), as described for the exposed solution, and the inner concrete (reinforced or unreinforced, according to the structural design) lining that supports the geomembrane.

The same system is adopted in shafts. From Austrian experience, the covered geomembrane system is extremely suitable for application in pressure shafts up to medium head (30–500 m). Tests have also been performed successfully up to 900 m (Heigerth, 1982). The resulting layering of the tunnel and of its waterproofing system is as follows

- natural excavation line
- structural/smoothing shotcrete (outer shotcrete lining)
- anti-puncture geotextile layer
- waterproofing geomembrane layer
- protection layer (optional)
- inner concrete lining.

15.4.1 Protective anti-puncture geotextile

The protective anti-puncture layer under the geomembrane is typically made from a material such as a geotextile, resistant to the aggression of fresh shotcrete and that can further regularise the substrate for installation of the waterproofing geomembrane. This layer is secured to the outer gunite ring by shot nails equipped with PVC roundels, specially designed to provide temporary anchorage of the PVC geomembrane, while at the same time avoiding excessive stresses being applied to it before concreting of the inner ring occurs (Figure 15.19).

Figure 15.19 Geotextile applied and secured by shotnails and PVC roundels in the foreground

Figure 15.20 Geomembrane heat welded to the roundels

15.4.2 PVC waterproofing geomembrane

The PVC waterproofing geomembrane provides a barrier to aggression of the underground environment both in the short term, to allow construction of the inner concrete ring (placement of fresh concrete), and in the long term, when it will protect the structure of the tunnel from deterioration due to infiltration of aggressive water from the ground. It will also stop any water loss through fissures that should form in the concrete. The waterproofing liner is typically a PVC geomembrane, at least 2 mm thick, placed over the geotextile and heat welded to the back surface of the roundels (Figure 15.20). The roundels are made in a colour different from that of the waterproofing liner (Figure 15.21). The geomembrane system is often composed of two layers, to improve the detection of any damage: the upper signal layer is usually a light grey colour; and the lower layer is a darker colour. Any damage affecting the integrity of the composite geomembrane will be easily detected by sight by the appearance of the darker colour of the lower layer.

15.4.3 Protective layer (optional)

The optional protection layer, which could be either a geotextile or another sacrificial sheet geomembrane of reduced properties, is installed on the waterproofing liner before the inner concrete ring is cast, to protect the geomembrane against potential mechanical damage caused by construction of the inner concrete lining, including the placement of reinforcement. It is good practice to install a sacrificial geomembrane layer over the waterproofing geomembrane at those zones that can be affected by the steel reinforcement, the invert (Figure 15.22) or by the stop ends of the formworks, or where construction of the concrete lining poses a particular risk of damaging the PVC geomembrane, such as in correspondence of the joints. A heavy duty PVC protection geomembrane 1.5 mm thick, of the same type as the waterproofing liner, is placed as a sacrificial 0.50 m-wide band over the joints. An additional sacrificial geomembrane layer will be placed on the invert to avoid possible damage during the invert concreting working phases.

Figure 15.21 Close-up of a roundel

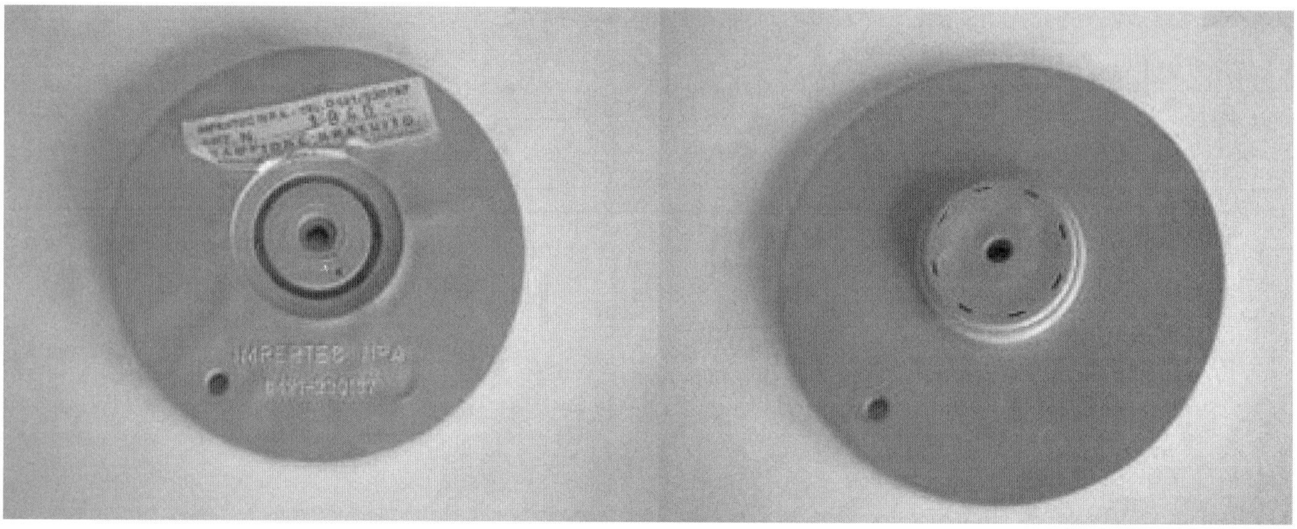

15.4.4 Protective seals and inner (reinforced) concrete lining

End fastenings are generally as described in Section 15.3.4 for the exposed geomembrane solution. The anchor bars providing attachment of the reinforcement for the inner concrete ring are secured directly on the natural rock, and they penetrate the protective anti-puncture geotextile layer, the PVC geomembrane layer and the protection layer, if applicable. The anchors are equipped with watertight fittings to avoid water seepage at crossings of the waterproofing liner. They must be designed for the maximum head to which the system is subject (water head or grouting pressure). If the reinforcement is self-supported, however, no anchoring is required. New design trends seek to avoid or minimise the penetration of the geomembrane through the use of self-supporting reinforcement cages, which do not require drilling holes in the waterproofing liner, and which should ideally be hole-free.

Figure 15.22 Geomembrane installed in the crown, walls and invert, and a protective sacrificial layer of a geomembrane (black) in the invert

Compartmentalisation is generally adopted to divide the lined area into shorter sections. Following the installation of the waterproofing geomembrane and before casting of the inner concrete lining, PVC waterstops are heat welded onto the waterproofing geomembrane at the perimeter of each compartment (Figure 15.23). The concrete is then cast, embedding the waterstops. The location of circumferential waterstops is strictly related to the length of the formwork. Longitudinal compartmentalisation will be located at the joint between the invert and the walls.

Since, as recognised by international experience, perfect placement of the concrete and filling of all voids in the

Figure 15.23 Longitudinal compartment seal made by welding the waterstop to the geomembrane. (Note: the waterstop is then embedded in the concrete lining)

Figure 15.24 Post-grouting arrangement at a waterstop

vault is impossible to achieve in practice, an allowance for grouting is made to prevent faulty concrete arising in the inner lining, particularly in the crown. Grouting is carried out once the concreting is complete, using horizontal injection hoses, previously attached to the geomembrane or placed at the top of the reinforcement (Figure 15.24). The inner concrete lining is usually constructed in two stages: one stage for the invert; and another for the crown and walls.

15.5. Summary of main points

- Geosynthetics can be successfully employed in underground construction. In particular, PVC geomembranes have been successfully used in hydraulic tunnels as barriers against water and gas infiltration from the surrounding ground, and to prevent water seeping in the ground from the tunnel. They are a key element in hydraulic tunnel design.
- Geomembrane systems have reached a high degree of refinement and reliability. Many years of successful performance of PVC geomembranes in hydraulic tunnels have shown that an adequately designed and installed geomembrane system is durable and reliable under high water heads, and in demanding environments.
- In the rehabilitation of hydraulic tunnels, when compared with steel linings, a geomembrane system provides several advantages, including financial and time savings.
- In the construction of new hydraulic tunnels, a geomembrane/composite geomembrane should be integrated in the lining system to prevent water infiltration that can cause failure of the surrounding ground and structure, and to avoid/minimise financial losses due to maintenance/repair costs and loss of service.
- In the rehabilitation of hydraulic tunnels, geomembranes are commonly selected because of their flexibility that allows accommodating potential crack growth.

15.6. Some common questions and their answers

Questions

1. Are geomembranes used in new tunnel construction, and, if so, also under high water heads?
2. How do geomembrane liners compare with more traditional steel liners?
3. Does the installation of a geomembrane allow consolidation grouting operations?
4. In the rehabilitation of tunnels, does an exposed geomembrane provide benefits in terms of hydraulic efficiency?
5. Can a covered geomembrane be repaired?

Answers

1. Geomembranes have seen extensive use in the new construction of hydraulic tunnels with high water heads, particularly in Austria. Examples are as follows.

- The Kölnbreinstollen Tunnel in the Malta scheme (Austria), subjected to maximum static and dynamic pressures of 24 and 30 bar, respectively. The geomembrane lining consisted of two single geomembranes, one being a 2 mm geomembrane made of regenerated PVC, and the other a 1.5 mm fibre-glass-reinforced PVC geomembrane.
- The Sellrain-Silz hydro-power scheme pressure shaft (Austria), under a maximum static pressure of 48 bar and a maximum dynamic pressure of 74 bar. A 3 mm PVC geomembrane with a 400 g/m^2 PP geotextile was installed as a waterproofing layer.
- The Rotenberg Tunnel on the Langenegg power scheme (Austria), subjected to a 350 m water head, employed a 5 mm-thick geomembrane for waterproofing.
- The Wallgaustollen Tunnel (Austria), with a hydrostatic head of 120 m, employed a 5 mm-thick PE geomembrane.
- The San Fiorano Tunnel (Italy), with a hydrostatic head of 103.11 m, employed a 4 mm-thick PVC geomembrane.

The most recent project is the 10.4 km third Niagara Tunnel, currently under construction and expected to be one of the largest tunnels in North America. It is waterproofed using a double waterproofing geomembrane applied to the tunnel wall in advance of the final concrete pouring. The tunnel will run under the city of Niagara Falls, parallel to the two existing 100 m-deep water tunnels but at a lower depth of 140 m.

2. The ability to eliminate leakage and the durability of steel and PVC geomembrane liners are equivalent; however, other characteristics of PVC geomembranes make them, in some cases, more attractive than their traditional counterparts.

Steel liners are generally transported inside the tunnel in sections and then assembled in place. When access to the tunnel is difficult, installation of the steel liner may require splitting of the liner into even smaller sections, which results in time-consuming assembling and welding operations inside the tunnel. In contrast, the components of a PVC lining system have dimensions and weight such that they can all be easily taken, even by hand, through very small accesses, and the installation equipment is a simple scaffolding that can be easily constructed inside the tunnel.

A steel liner is less flexible than a PVC liner: as a consequence, once the steel liner has been assembled, there may be considerable voids between the steel and the existing surface of the tunnel. It is therefore essential to take into account the time and costs required for the extensive grouting that is needed to fill the void, and provide adequate congruence between the steel liner and the supporting concrete. The time and costs associated with adequate plugging of all grouting holes should also be taken into account. A PVC geomembrane, on the other hand, being very flexible, can adapt to the existing surface of the tunnel so that fewer voids will be present and little grouting will be needed.

Steel liners, with a thickness of the order of centimetres, will cause a slight alteration in the hydraulic profile of the tunnel, while PVC geomembrane liners, with a thickness of the order of millimetres, do not alter the tunnel cross-section appreciably.

Steel liners have a lower elongation limit compared with PVC geomembrane liners, and may fail in the presence of settlements, deformations or the opening of large cracks. PVC geomembrane liners have superior performance, adapting also to active fissures and cracks, and eliminating any risk of failure from leakage.

Steel liners do not alter the natural groundwater conditions. PVC geomembrane liners with a drainage system behind can improve groundwater conditions by extracting water from the soil. The entire system must, however, be designed to minimise and resist backpressure.

Steel liners are statistically more expensive than PVC geomembrane liners. The cost of a geomembrane system is usually 20–30% of the cost of a steel liner, and the installation time is generally 15–25% of the total time required for the installation of a steel liner. This is estimated taking into account the cost of supply, but also the different design, logistics and installation requirements. Reduced outage time of a tunnel that results in, for example, higher hydropower production, and associated higher financial income, should also be added to the financial advantages offered by the geomembrane solution.

3. Consolidation grouting is typically undertaken to improve rock conditions by grouting cavities in the natural excavation, and is generally performed prior to the installation of the waterproofing system; thus, it clearly does not interfere with the placement of the geomembrane.

Additional grouting can be made behind the waterproofing geomembranes; that is, between the outer sprayed concrete lining and the geomembrane (grouting of the outer gap), and between the geomembrane and the inner concrete ring (grouting of the inner gap). Grouting is performed by installing grouting pipes that cross the geomembrane (grouting of the outer gap) or are installed on the geomembrane (grouting of the inner gap). The grouting pipes are usually placed at regular intervals, and are equipped with watertight fittings if they cross the geomembrane. This grouting system must be installed before casting of the inner concrete lining.

Grouting the outer and inner gaps aims to fill the voids between the waterproofing geomembrane and the adjacent layers. Post grouting can also be used to repair damage to the geomembrane (see Question 5).

4. The resistance of the tunnel lining to the flow of water in it is represented by Manning's coefficient, n. Manning's coefficient depends only on the surface roughness. Exposed PVC geomembrane systems in tunnels profit from the intrinsic very low hydraulic roughness of PVC geomembranes, which is maintained over time, while in concrete linings the surface roughness tends to increase over time. If the geomembrane is anchored with a system that at the same time permanently tensions the PVC liner, so as to avoid folds and the formation of slack areas, very low hydraulic roughness can be assured during construction and in the long term.

Additionally, since the PVC geomembrane liner is only a few millimetres thick, when this system is adopted there is effectively no reduction in the cross-section of the tunnel. In contrast, adopting other lining systems, such as sprayed concrete, concrete or even steel, implies a potentially significant reduction in the cross-section.

5. In hydraulic tunnels, it is generally assumed that the inner concrete lining will inevitably be fissured. The internal water pressure, applied through the fissures in the concrete lining directly to the geomembrane, pushes it against the external surface of the sprayed concrete lining. Therefore, in the case of damage to the geomembrane, the flow of water through the breached point in the geomembrane is partially reduced by the same confinement action given by the pressure.

Post-grouting of the inner gap to fill the voids between the waterproofing geomembrane and the inner reinforced concrete lining can also be used to repair any damage to the waterproofing liner, and to avoid any water leakage during future operation. The grouting system is installed before casting of the inner lining, with the grouting pipes attached to the waterproofing geomembrane on the side facing the inner concrete lining and connected to a control box placed in front of the reinforcement.

To obtain perfect filling of all voids, to stop a leak in the geomembrane, it is generally necessary to grout fine cement slurry. The setting time of the grouting material must be carefully planned. The flow, density, setting and pressure of injection must be constant in the injected area. When grouting is performed, it is important to ensure complete, uniform and fast diffusion of the grouting mix in the entire grouted area. The best control of the quantity of grout and of the efficiency of the grouting operation is achieved by reducing the area in which the grouting is made, which means creating compartments in the system. Compartments allow easier and more effective inspection, location of damage, and repair by post-grouting. Compartments are made by PVC waterstops heat welded to the waterproofing geomembrane, and then embedded in the concrete of the inner ring.

The compartmentalisation system will confine the water from accidental leakage to the area of the leak, making it possible to establish the location of the infiltration, reducing the area of investigation for leak detection, and minimising the extent of the repair. By preventing cementitious grout entering adjacent areas, the volumes of grout necessary to inject are lower, and there is a more efficient control of grout distribution. This results in a system for future repair that does not require any demolition of the inner concrete structure.

Acknowledgement

Photographic material relevant to the installation of geosynthetics in hydraulic tunnels has been provided courtesy of Carpi.

REFERENCES

Cazzuffi D (1995) Évaluation du comportement de différent types de géosynthétiques prélevés sur canaux de l'Enel. Comptes Rendus du Colloque sur les Géotextiles et les Géomembranes – Rencontres '95, vol. 2, Beaune, pp. 9–15.

Cazzuffi D and Venesia S (1989) The use of thermal analysis for the evaluation of the durability of a 24-year-old PVC geomembrane. In *Geosynthetics: Microstructure and Performance,* Orlando (Peggs ID (ed.)), American Society for Testing and Materials, West Conshohocken, PA, pp. 17–25.

Cazzuffi D, Ferrara G, Venesia S, D'Ischia A and Silvestri T (1986) Transmissivity of geotextiles and related products: laboratory tests and practical applications. *Proceedings of the 3rd International Conference on Geotextiles, vol. 3,* Vienna, pp. 879–884.

Heigerth G (1982) Determination of load bearing capacity of the rock mass and lining layout for a high stressed pressure shaft. *ISRM Symposium,* Aachen, vol. 3, pp. 1253–1260.

Heigerth G, Finger W and Schlosser K (1979) Triebwasserweg Oberstufe – Anlagenbeschreibung, Bauausführung. *ÖZE* **32(1/2)**: Jan./Feb.

Laufer H (1985) Austria's contribution Towards the Development of Water Power, Hydro Power Schemes and Large Dams in Austria. In *Die Talsperren Österreichs,* No. 29 (Simmler H (ed.)), Springer Verlag, Wien.

Scuero AM and Vaschetti GL (1997) Waterproofing a pressure tunnel with a drained polyvinylchloride membrane composite. *Proceedings Geosynthetics Asia '97,* Bangalore, pp. III.35–III.42.

Scuero AM, Wilkes J and Vaschetti G (1999) *Waterproofing Tunnels with Geosynthetics in Europe.* iGEM, University of Illinois, Urbana Champaign, IL.

Seeber G (1982) New ways for the construction of pressure tunnels. *ISRM Symposium,* vol. 2, Aachen, pp. 705–716.

Seeber G (1985) Planung und Optimierung der Triebwasserwege von Pumpspeicherwerken. *Österreichische Wasserwirtschaft* **37(1/2)**.

Handbook of Geosynthetic Engineering
ISBN 978-0-7277-4175-2

http://dx.doi.org/10.1680/hge.41752.317

Chapter 16
Geosynthetic-reinforced soil walls and slopes – seismic aspects

R. J. Bathurst*, **K. Hatami†** and **M. C. Alfaro‡**
*GeoEngineering Centre at Queen's-RMCC, Department of Civil Engineering, Royal Military College of Canada, Kingston, Ontario, Canada
†School of Civil Engineering and Environmental Science, University of Oklahoma, Norman, USA
‡Department of Civil Engineering, University of Manitoba, Winnipeg, Manitoba, Canada

The design and analysis of geosynthetic-reinforced walls, embankments and slopes are presented earlier in this book for structures under static loading conditions. This chapter focuses on the seismic aspects of these structures. It describes the properties and behaviour of cohesionless soils, geosynthetic reinforcement and facing components under dynamic and cyclic loading, and summarises the important features of current analytical and numerical methods for the seismic analysis and design of geosynthetic-reinforced soil walls and slopes. The observed performance of reinforced soil walls and slopes during earthquakes is also presented.

16.1. Introduction

The first analytical treatment of the influence of seismic-induced forces on the stability of earth-retaining structures can be traced to the work of Sabro Okabe in his landmark paper (Okabe, 1924). Since this seminal work, there has been a large body of research on the development of analytical methods that consider the potentially large forces that exert additional destabilising forces on earth-retaining walls, slopes, dams and embankments during earthquakes. The vast majority of this work has been focused on conventional earth structures. The analysis methods that have been proposed include

- pseudo-static rigid body analyses that are variants of the original Mononobe–Okabe approach
- displacement methods that originate from Newmark sliding block models
- dynamic finite-element/finite-difference methods.

However, with the growing use of geosynthetics in reinforced soil walls, slopes and embankments, the need to extend current methods of analysis for conventional structures under seismic loading to geosynthetic-reinforced systems in similar environments has developed. A concurrent need has been the requirement to select properties of the component materials that represent rapid and/or cyclic loading conditions.

The first edition of this chapter was based on an extended and updated version of a state-of-the-practice review paper by Bathurst and Alfaro (1996) that appears as a keynote paper in *Proceedings of the International Symposium on Earth Reinforcement, IS-Kyushu '96*. A more recent keynote paper with the same scope, and co-authored by the first author (Koseki *et al.*, 2006), can be found in the *Proceedings of the 8th International Conference on Geosynthetics*. This chapter presents selected published works related to the properties of cohesionless soil, geosynthetic reinforcement and facing components under cyclic loading, and summarises the important features of current analytical and numerical methods for the seismic analysis and design of geosynthetic-reinforced soil walls and slopes. The scope of the chapter is restricted to structures seated on firm foundations for which settlement and collapse of the foundation materials are not a concern. Non-surcharged structures with simple geometry are considered, and the reinforced and retained soils are assumed to be homogeneous, unsaturated and cohesionless.

An important component of recent work in the field of seismic performance of reinforced walls and slopes has been the use of carefully conducted numerical studies to gain insight into the performance of reinforced walls and slopes under simulated seismic loading. This chapter highlights numerical modelling investigations by the authors and others, and identifies the implications of the results to current design practice. Many of the examples that highlight important issues related to the seismic performance of geosynthetic-reinforced soil-retaining structures are from work by the authors and their co-workers on seismic performance of reinforced soil segmental (modular block) retaining walls. These structures have gained wide popularity in North America due to their cost-effectiveness (Bathurst and Simac, 1994). Nevertheless, these structures pose unique challenges to the designer for seismic loading conditions because of the modular dry-stacked construction of the facing column.

16.2. Material properties under dynamic loading

The properties of the components of geosynthetic-reinforced soil structures may be influenced by the rate of loading and the cyclic loading response. This section reviews data and models that have been used by the authors, their co-workers and others for the analysis, design and numerical simulation of structures under seismic loading. The complexity of the constitutive models discussed here ranges from the relatively simple for limit-equilibrium-based approaches to the relatively sophisticated for dynamic finite-element and dynamic finite-difference modelling.

16.2.1 Soil

16.2.1.1 Strength properties (Coulomb friction angle)

Pseudo-static, pseudo-dynamic and displacement (Newmark) methods, introduced later in the chapter, describe cohesionless soil strength according to the Coulomb failure criterion. The selection of an appropriate value of the soil friction angle, ϕ, becomes an issue in these methods, particularly with respect to the choice of peak, ϕ_p, or residual (constant volume), ϕ_{cv}, strength values. A review of the literature suggests that for dry cohesionless soils, the rate of loading used in direct shear or triaxial tests has negligible effect on the shear strength (Bachus *et al.*, 1993). For example, Schimming and Saxe (1964) used a direct shear device to test Ottawa sand under both static and dynamic conditions. No significant difference in strength envelopes was recorded (Figure 16.1). Conventional practice using Newmark methods is to assume that the cohesionless soil friction angle does not change during an earthquake.

Figure 16.1 The results of direct shear tests on dry Ottawa sand
Reproduced from Schwimming and Saxe (1964)

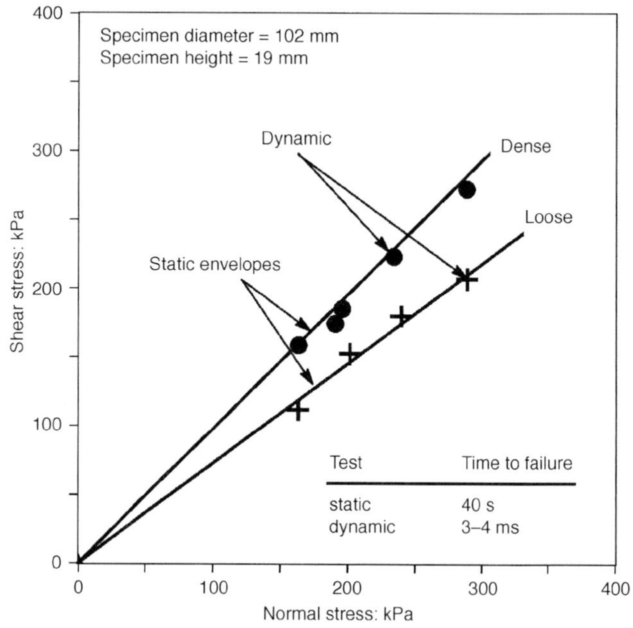

Conventional practice in pseudo-static methods of analysis for retaining walls and slopes is to relate interface friction angles, δ, to the soil friction angle, ϕ. In static stability analyses, δ is often assumed to be equal to $2\phi/3$ for internal stability analyses (facing column–reinforced soil interface) and $\delta = \phi$ for external stability analyses (reinforced soil–retained soil interface, or between wedges in two-part wedge analyses). A value of $2\phi/3$ has been shown to be applicable for wall–soil interface friction based on small-scale shaking table tests of conventional gravity wall structures (Ishibashi and Fang, 1987), and has been assumed to also be applicable for geosynthetic-reinforced retaining wall structures (Bathurst and Cai, 1995).

Peak friction angle values have been used in pseudo-static design methodology for segmental retaining walls published by the National Concrete Masonry Association (Bathurst, 1998; NCMA, 2009). The choice of peak friction angle for seismic design is consistent with the Federal Highway Administration (FHWA, 2009) and the National Concrete Masonry Association (Simac *et al.*, 1993; NCMA, 2009) guidelines for the static design of geosynthetic-reinforced soil walls. Bonaparte *et al.* (1986) have recommended that residual friction angles, ϕ_{cv}, should be used in the seismic design of slopes based on reinforcement–strain compatibility requirements. Leshchinsky *et al.* (1995) have proposed using the residual soil strength for retaining walls, but recognise that this is likely to be a conservative assumption for design.

Recommended soil friction angle values for pseudo-static stability analysis are reported by Tatsuoka *et al.* (1998). For cohesionless sands and gravels, a value of $\phi = 35°$ is recommended. They note that this value is likely to be closer to the residual friction angle of typical cohesionless soils, and, hence, the selection of this value can be argued to compensate for possible progressive failure of the backfill soil and uncertainties such as soil compaction levels. In the absence of shear test data, American Association of State Highway and Transportation Officials (AASHTO, 2010) guidelines recommend that the friction angle for select granular fills used in the reinforced soil zone should not exceed $\phi = 34°$, and, regardless of laboratory site-specific testing, should not be greater than $\phi = 40°$.

Another source of conservatism in the selection of representative soil strength values is the use of friction angles from direct shear tests versus plane strain testing. It has been demonstrated that for dilatant cohesionless soils, the true plane strain peak friction angle is greater than that determined from conventional direct shear box tests (Bolton, 1986).

It appears that, in practice, the choice of peak or residual values is either prescribed or left to engineering judgement.

16.2.1.2 Stiffness properties (stress–strain models)

Soil

Linear-elastic perfectly plastic models with the Mohr–Coulomb failure criterion have been used to simulate the dynamic hysteretic load–unload behaviour of soils in geosynthetic-reinforced soil structures, in combination with Raleigh damping (Bathurst and Hatami, 1998a,b; Zarnani and Bathurst, 2008).

For more complex modelling using dynamic finite-element/finite-difference codes, the shear behaviour of soils under seismic loading can be simulated using non-linear cyclic constitutive relationships. Ling *et al.* (2004) used a non-linear finite-element model programme with a multi-parameter plasticity model to simulate the pressure-dependent stress–strain–dilatancy behaviour of the backfill in centrifuge shaking table tests, and to carry out parametric analyses of prototype-scale geosynthetic soil-reinforced modular block walls (Ling *et al.*, 2005a, 2010). However, improved accuracy of numerical results is not guaranteed with increasing sophistication of constitutive models adopted in numerical simulations (e.g. Huang *et al.*, 2009; Zarnani and Bathurst, 2009a).

A model of intermediate complexity that has been used by some researchers employs Masing behaviour for the hysteretic unloading and reloading of cohesionless soils, and is illustrated in Figure 16.2 (Finn *et al.*, 1986; Yogendrakumar *et al.*, 1991, 1992; Cai and Bathurst, 1995; Kramer, 1996a).

The relationship between the soil shear stress, τ, and the shear strain, γ_ε, for the initial loading phase (backbone curve) is assumed to be hyperbolic, and is given by

$$\tau = f(\gamma_\varepsilon) = \frac{G_{max}\gamma_\varepsilon}{1 + (G_{max}/\tau_{max})|\gamma_\varepsilon|} \quad (16.1)$$

Figure 16.2 Non-linear hysteretic loading paths

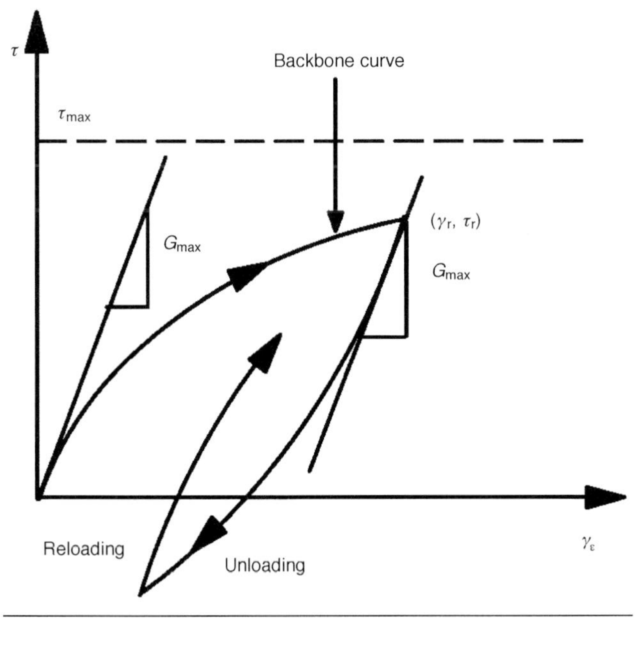

where G_{max} is the maximum shear modulus and τ_{max} is the maximum shear strength. The equation for the unloading curve from the point (γ_r, τ_r) at which the loading reverses direction is given by

$$\frac{\tau - \tau_r}{2} = f\left(\frac{\gamma_\varepsilon - \gamma_r}{2}\right) \quad (16.2)$$

or

$$\frac{\tau - \tau_r}{2} = \frac{G_{max}\left(\frac{\gamma_\varepsilon - \gamma_r}{2}\right)}{1 + (G_{max}/\tau_{max})|\gamma_c|} \quad (16.3)$$

The shape of the unloading–reloading curve is shown in Figure 16.2. The tangent shear modulus, G_t, for a point on the backbone curve is given by

$$G_t = \frac{G_{max}}{[1 + (G_{max}/\tau_{max})|\gamma_\varepsilon|]^2} \quad (16.4)$$

and, at a stress point on an unloading or reloading curve,

$$G_t = \frac{G_{max}}{[1 + (G_{max}/2\tau_{max})|\gamma_\varepsilon - \gamma_r|]^2} \quad (16.5)$$

The response of the soil to a uniform confining pressure is assumed to be non-linear elastic and dependent on the mean normal stress. Hysteretic behaviour, if any, is neglected in this mode. The tangent bulk modulus, B_t, is expressed in the form

$$B_t = K_b P_a \left(\frac{\sigma_m}{P_a}\right)^n \quad (16.6)$$

where K_b is the bulk modulus constant, P_a is the atmospheric pressure in units consistent with the mean normal effective stress σ_m, and n is the bulk modulus exponent.

The equivalent-linear method (ELM), first introduced by Seed and Idriss (1969), is similar to the approach described above. A description of the ELM is provided by Zarnani and Bathurst (2009a), who used the model to simulate the dynamic behaviour of granular soil located behind geofoam seismic buffers. The ELM requires strain-dependent shear stiffness degradation curves and damping curves, typically obtained from resonant column testing for small shear strain amplitudes and cyclic triaxial tests for larger strain amplitudes.

Zarnani and Bathurst (2008, 2009a) carried out a series of numerical simulations of physical seismic buffer tests. They showed that up to about 0.7g peak base excitation there was no practical advantage of using the ELM for the soil, compared with a simple linear-elastic plastic Mohr–Coulomb model. The explanation for this apparent contradiction is that hysteretic unload–reload behaviour is a natural outcome at post-yield using the linear-elastic plastic Mohr–Coulomb model.

The textbook by Kramer (1996a) is a useful reference for further reading on this topic.

Soil–geosynthetic composites

et al. (1996) carried out dynamic triaxial tests on reinforced sand samples. They found that the shear modulus of reinforced specimens increases with effective confining pressure. The effect of reinforcing material on the improved stiffness of the soil was found to be more beneficial under low confining pressure, but the equivalent modulus of the composite specimens did not increase proportionally with the stiffness of the reinforcing material.

Chen and Chen (1998) studied the response of reinforced soils under cyclic loading using an equivalent homogeneous approach. They simulated the results of dynamic triaxial tests on reinforced soil specimens using the finite-difference-based programme FLAC (Itasca, 1998). The soil was Ottawa sand (type C109) consisting of rounded quartz particles, and the reinforcement was a polyethylene sheet. The composite specimen was assumed to be an equivalent homogenous and transversely isotropic material. The soil was modelled using the Duncan and Chang (1970) hyperbolic model, and the reinforcement was assumed to be a linear-elastic material. They found that their numerical simulation results using the equivalent composite approach and hyperbolic soil model predicted hysteretic behaviour for reinforced soil samples. However, the results of actual tests showed little hysteretic response. Otherwise, the predicted behaviour from numerical simulations and the experimental tests were in general agreement.

The results of the work reported by Chen and Chen may not apply to prototype-scale reinforced soil structures since the reinforcement spacing in the field is typically much larger than that in laboratory-scale triaxial tests. Conventional practice in the numerical modelling of actual reinforced soil structures is to treat the soil and reinforcing layers as discrete components, rather than to use the homogeneous approach (see Section 16.3.4).

16.2.2 Geosynthetic reinforcement

16.2.2.1 In-isolation monotonic load–strain behaviour

In-isolation monotonic load testing of high-density polyethylene (HDPE) and woven polyester (PET), geogrid reinforcement materials have been reported by Bathurst and Cai (1994), and Shinoda and Bathurst (2004a,b). The results for a constant rate of loading (monotonic loading tests) showed that HDPE geogrids were sensitive to the rate of loading, while PET geogrids were less sensitive (Figure 16.3).

Bernardi and Paulson (1997) and Greenwood (1997) summarised observations from the results of index tensile tests carried out on geosynthetic-reinforcement materials after long-term creep loading. They concluded that the rupture strength reduction in PET and polyolefin reinforcement products does not vary linearly with logarithm of time. Rather, the residual index strength of polymeric reinforcement products is always greater than what is assumed based on conventional log-linear creep–rupture curves. Residual strength curves for materials with an index tensile

Figure 16.3 Influence of the strain rate on the monotonic load extension behaviour of typical geogrid reinforcement products
Reproduced from Bathurst and Cai (1994), Industrial Fabrics Association International

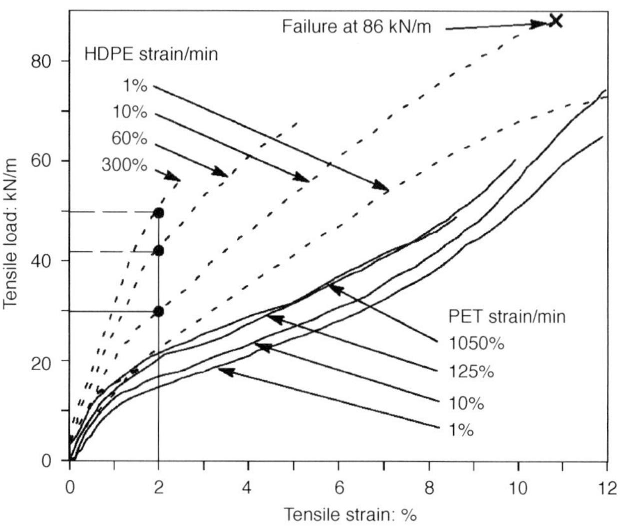

strength, T_0, are illustrated in Figure 16.4. The residual strength curves are assumed to intersect the conventional creep-reduced strength curve at static and dynamic design strength values, T_{DS} and T_{DD}, respectively. In North American practice, the design load under seismic loading can be increased by 33%. Hence, $T_{DD} > T_{DS}$ in this figure. Importantly, a reinforcement layer at a value of T_{DD} can be expected to have an available residual strength $T_{RDS} \gg T_{DD}$. This additional strength is not considered in current limit-equilibrium methods of design and is a potential source of conservatism.

An implication of observations reported in this section to seismic design is that the available strength and stiffness of geosynthetic reinforcement products under earthquake loading is not less than conventional estimates of available reinforcement strength in static load environments, and may indeed be very much greater.

16.2.2.2 In-isolation cyclic load testing

In order to determine cyclic load parameters for reinforcement models used in dynamic finite-element modelling, in-isolation cyclic load tests were carried out on typical polymeric geogrid reinforcement materials (Bathurst and Cai, 1994; Cai and Bathurst, 1995). Example results are presented in Figure 16.5 for an HDPE geogrid. The cyclic load–strain behaviour of typical HDPE and woven PET geogrid reinforcement materials exhibited two distinct features

- non-linear hysteresis unload–reload loops
- a load–strain cap that is tangent to all initial unload–reload hysteresis curves.

At low strains or to simplify numerical computations, the hysteretic behaviour of the geosynthetic may be ignored. In this case, the relationship between the axial load and the

Figure 16.4 The concept of residual strength available to the reinforcement layer under dynamic loading

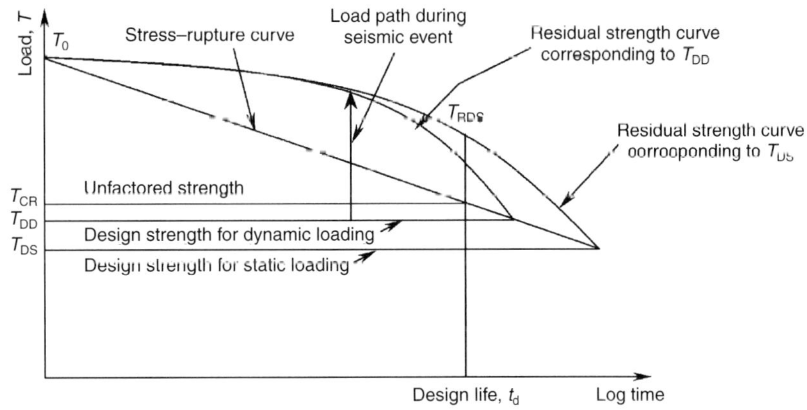

axial strain for the initial loading can be assumed to take a non-linear quadratic form (Chalaturnyk *et al.*, 1988; Yogendrakumar *et al.*, 1991), expressed as

$$T_a = J_i \varepsilon_a \left(1 - \frac{\varepsilon_a}{2\varepsilon_{af}}\right) \tag{16.7}$$

where T_a is the axial load per unit width (e.g. kN/m), J_i is the initial load modulus, ε_a is the axial strain, and ε_{af} is the axial strain at failure. The details of the model parameters are shown in Figure 16.6(a). The tangent load modulus, J, on the initial loading curve is calculated as

$$J = \frac{dT_a}{d\varepsilon_a} = J_i\left(1 - \frac{\varepsilon_a}{\varepsilon_{af}}\right) \tag{16.8}$$

During the analysis, compression is not allowed in the polymeric geosynthetic reinforcement, and the hysteresis during unloading and reloading is not modelled. The unloading and reloading portions are approximated as straight lines, and the unload–reload modulus is defined as

$$J_{ur} = KJ_i \tag{16.9}$$

where J_{ur} is the unload–reload modulus and K is a constant.

The load–strain cap/hysteretic unload–reload model, described earlier for soil materials, has been modified for polymeric reinforcement materials by Yogendrakumar and Bathurst (1992), and is illustrated in Figure 16.6(b). The relationship between the axial load and the tensile strain

Figure 16.5 An in-isolation cyclic load test on an HDPE geogrid
Reproduced from Cai and Bathurst (1995), Industrial Fabrics Association International

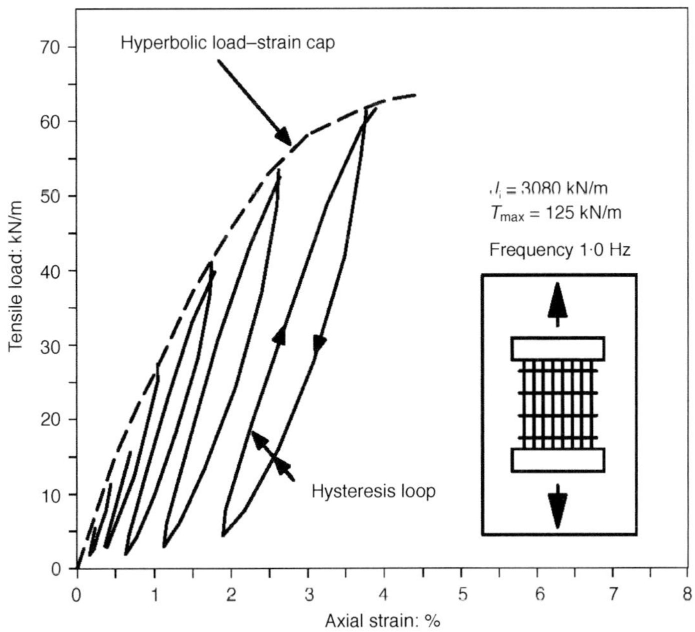

Figure 16.6 Cyclic unload–reload models for polymeric reinforcement: (a) the non-linear model with non-hysteretic unload–reload behaviour (after Yogendrakumar *et al.*, 1991) and (b) the non-linear model with hysteretic unload–reload behaviour
Reproduced from Yogendrakumar and Bathurst (1992), National Academy of Science

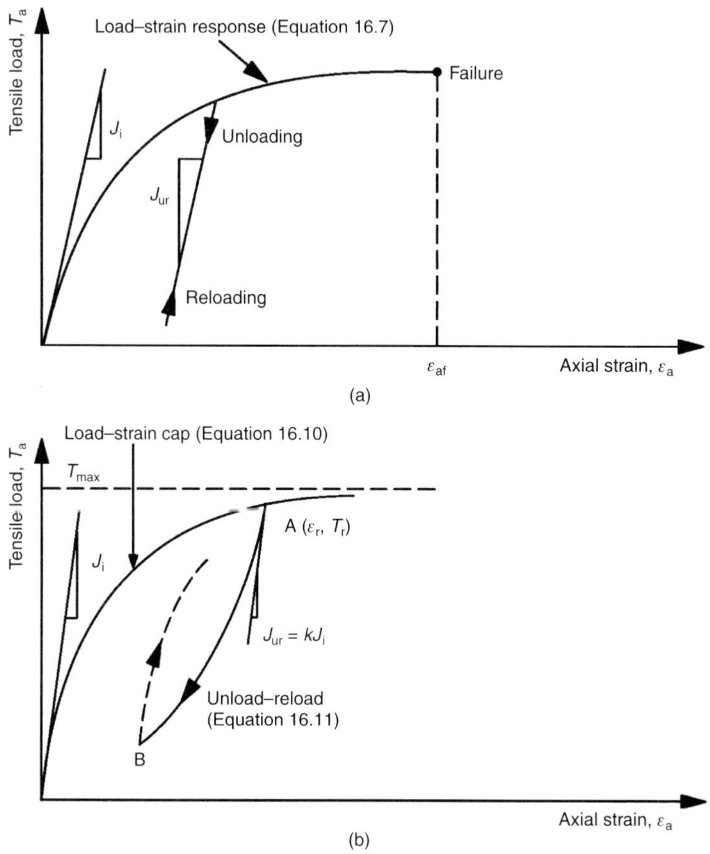

for the load–strain cap (backbone curve) is expressed by

$$T_a = \frac{J_i \varepsilon_a}{1 + (J_i/T_{max})|\varepsilon_a|} \tag{16.10}$$

where T_a is the axial tensile load per unit width of a specimen, ε_a is the axial strain, J_i is the initial modulus, and T_{max} is the extrapolated asymptotic ultimate strength of the reinforcement material. The data in Figure 16.7 show that the initial stiffness, J_i, and the shape of the load–strain cap, are sensitive to the loading frequency for HDPE geogrids and essentially frequency-independent for woven PET geogrids. During an unload–reload cycle, the reinforcement model is assumed to follow the Masing rule. The equation for the unloading curve from point A (ε_r, T_r), or for the reloading curve from point B at which the load reverses direction, is given by

$$\frac{T_a - T_r}{2} = \frac{J_{ur}(\varepsilon_a - \varepsilon_r)/2}{1 + (J_{ur}/2T_{max})|\varepsilon_a - \varepsilon_r|} \tag{16.11}$$

where J_{ur} is the unload stiffness defined in terms of the initial load stiffness according to $J_{ur} = kJ_i$, and k is a constant.

Bonaparte *et al.* (1986) cautioned that the strain at rupture for HDPE geogrids will decrease with an increasing rate of loading, and, hence, influence the choice of rupture load in limit state design. Only one of the tests shown in Figure 16.3 was taken to rupture due to equipment limitations; so, possible rate effects cannot be quantified here. The reduction in the rupture load capacity for HDPE geogrids under high rates of loading also has implications for Newmark sliding block methods of analysis where large cumulative displacements may be computed (see Section 16.3.3).

Moraci and Montanelli (1997) also carried out cyclic load tests on HDPE specimens at frequencies in the range 0.1–1 Hz and at different load amplitudes. They found that for the HDPE material, the unload–reload stiffness value decreased with increasing load amplitude and increased with greater loading frequency. They observed that the unload–reload stiffness of the HDPE reinforcement materials for load amplitudes less than 60% of the reference tensile strength was approximately 1.5–2 times the secant stiffness from monotonic tensile strength tests. This observation can be used to estimate the unload–reload stiffness of

Figure 16.7 Load–strain cap secant stiffness versus frequency of loading for HDPE and PET geogrid specimens
Reproduced from Bathurst and Cai (1994), Industrial Fabrics Association International

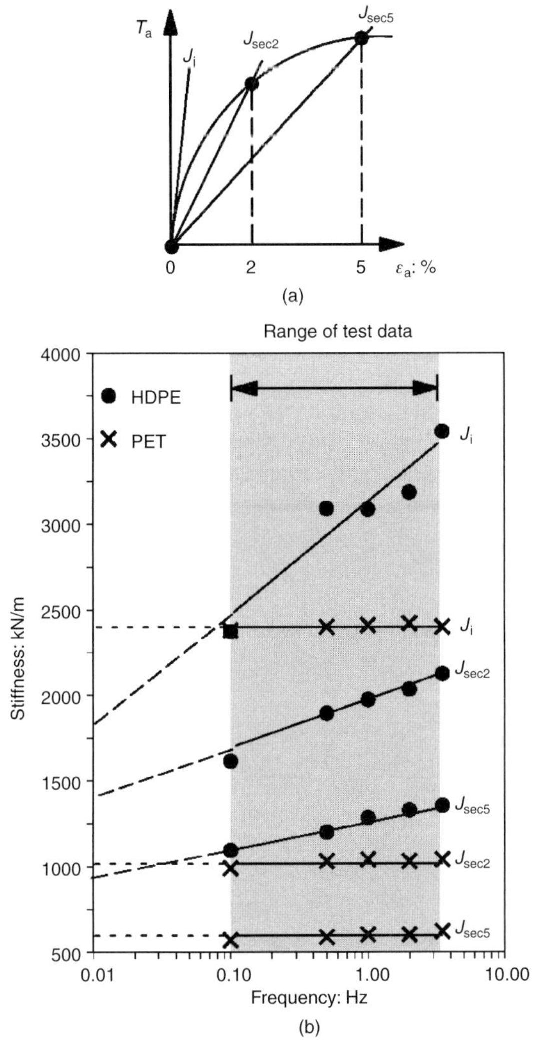

the specific HDPE geogrid investigated. They also found that the stiffness value of HDPE specimens was greater for specimens that were cycle loaded from a minimum load equal to 20 or 40% of the maximum applied load (i.e. prestressed specimens) than for specimens that were fully unloaded during each load cycle.

Ling *et al.* (1998) carried out strain-controlled cyclic loading tests on virgin and prestressed specimens of three commonly used geogrids manufactured from HDPE, polypropylene and woven PET. The cyclic strain rate was kept at the 10%/min rate in conformance with the ASTM D4595 method of testing (ASTM, 1996). They found that the reload stiffness of all the polymeric materials examined at any given load level increased with the number of loading cycles. The magnitude of the stiffness increase was greater at higher load levels. They also concluded that the index strength load–strain curve from static loading tests was in reasonable agreement with the backbone curve for each material under low-frequency cyclic loading. The results of cyclic loading tests on the three different reinforcement materials showed that the index strength of polypropylene geogrid specimens was not changed significantly as a result of cyclic loading. In contrast, the post-cyclic tensile strength values of HDPE and woven PET-type geogrids increased with the number of cyclic loads and load amplitude. Finally, Ling *et al.* (1998) proposed the following hyperbolic formula to estimate the accumulated reinforcement strain, $\varepsilon - \varepsilon_0$, from the number of load cycles, N_1, for a given load intensity level:

$$\varepsilon - \varepsilon_0 = \frac{N_1}{\kappa + \zeta N_1} \tag{16.12}$$

where ε_0 is the strain developed during primary loading, and κ and ζ are constants. In subsequent modelling, Ling *et al.* (2004) simulated the uniaxial cyclic behaviour of geogrid reinforcement layers using a nine-parameter bounding surface model.

An implication to seismic design of the results of standard monotonic loading wide-width tensile tests, and the cyclic load data reviewed here, is that initial and secant stiffness values for uniaxial HDPE and woven PET geogrids used for static loading may greatly underestimate reinforcement stiffness and tensile working strength of these materials under dynamic loading. Furthermore, the *in situ* unload–reload stiffness values of uniaxial HDPE and woven PET geogrids may be higher than those assumed from cyclic test results of the type reported by Bathurst and Cai (1994). This is a result of the prestressing effect that occurs during static loading prior to an earthquake event.

Data on the cyclic load response of geotextile reinforcement products are sparse. Guler and Biro (1999) reported results of uniaxial cyclic testing on two nonwoven geotextile reinforcement products. They found that the accumulation of strain in the reinforcement during the cyclic loading of a needle-punched geotextile was larger than that recorded for a spun-bonded geotextile.

As a general rule, the strain at rupture for nonwoven geotextiles can be expected to be much greater than that for typical geogrid products used today in reinforced wall and slope applications. Consequently, the applicability of limit-equilibrium models that assume that collapse of the structure occurs as a result of rupture of the reinforcement can be challenged when highly extensible geotextile products are used.

16.2.2.3 In-soil reinforcement cyclic load testing

Some guidance on the effect of soil confinement on geotextile load–strain deformation may be inferred from the results of in-soil tensile tests using monotonic constant rates of displacement. The work of McGown *et al.* (1982), Ling *et al.* (1992) and Wilson-Fahmy *et al.* (1993) indicates that increasing confining pressure increases the modulus of needle-punched nonwoven geotextiles, and may increase the ultimate strength as well. The in-air modulus and ultimate strength of woven geotextiles was shown to be unchanged due to soil confinement (Wilson-Fahmy *et al.*, 1993). A useful summary of the reinforcement strain and stiffness behaviour of geosynthetics, and the influence of soil confinement during monotonic loading, can be found in the paper by Walters *et al.* (2002).

Results on the in-soil cyclic load testing of geosynthetic reinforcement materials are sparse. McGown *et al.* (1995) performed a series of low-frequency, in-soil cyclic load tests on a stiff uniaxial HDPE geogrid similar to that reported by Bathurst and Cai (1994). McGown *et al.* (1995) illustrated that stresses and permanent strains may be 'locked in' the reinforcement due to repeated tensile loading, resulting in a stiffer reinforcement response than that for in-air tests.

Taken together, the implications for seismic design is that cyclic in-air tests of the type reported by Bathurst and Cai (1994) may represent a lower bound on reinforcement stiffness values (J_i and J_{ur} values) at working stress levels for stiff HDPE geogrids, but confinement is likely to have a negligible influence on stiffness for woven geotextiles and geogrids.

16.2.3 Interface properties

Geosynthetic–soil interface sliding and pull-out of reinforcement within anchorage zones are potential failure mechanisms in reinforced walls, slopes and embankments. A conventional approach is to quantify the shearing resistance at these interface locations by an interaction coefficient, C_i, that is defined as the ratio of the interface friction coefficient to soil friction coefficient ($C_i = \tan\phi_{ds}/\tan\phi$). The interaction coefficient is usually evaluated using a direct shear test (e.g. ASTM, 2008) and/or a pull-out test (e.g. ASTM, 2001a). These two tests differ significantly in loading path and boundary conditions, and interaction coefficients for nominally identical specimen conditions may vary between tests (Juran *et al.*, 1988).

16.2.3.1 Soil–geosynthetic interface

Shear strength tests

A large body of work has been reported on the interface shear characteristics of soil–geosynthetic interfaces using a variety of direct shear methodologies and apparatus (Takasumi *et al.*, 1991). The work is restricted almost exclusively to monotonic loading. Myles (1982) reported values of sand–geotextile interface coefficients in the range 0.81–0.97 for three different types of geotextiles. Miyamori *et al.* (1986) reported interaction coefficients in the range 0.72–0.87 for dry sand/nonwoven geotextile interfaces. Myles argues that loading rate effects are not a concern for cohesionless sands, but recommends that the residual interface shear strength should be used for design with geotextiles, to be consistent with the notion that full mobilisation of the shear strength in reinforcement applications occurs at large geotextile strains. Cancelli *et al.* (1992) reported interaction coefficients in the range 1.04–1.12 for a number of different stiff HDPE geogrids in combination with sand and gravel. Cancelli *et al.* argue that the interface shear for geogrids is controlled by the soil–soil interface shear strength. There are no published reports of cyclic interface direct shear tests on geotextiles and geogrids. However, a limited number of repeated direct shear tests on a single specimen of HDPE sheet, in combination with Ottawa sand at low confining pressure, showed that there was no reduction in the interface shear strength with the number of shear applications (O'Rourke *et al.*, 1990). Based on the data presented above and the expectation that the soil–soil interface shear capacity for dry cohesionless soils is independent of the rate of loading, it is reasonable to use the results of monotonic loading direct shear tests for limit-equilibrium-based seismic design.

Fakharian and Evgin (1995) describe the results of cyclic shear tests between sand and fine steel mesh surfaces, which showed that monotonic and cyclic direct shear tests gave the same values of peak and residual interface shear strength. However, under cyclic shear conditions, there was evidence that peak and post-peak behaviour may occur at displacement amplitudes that are less than the displacement required to fail the interface under monotonic loading conditions. The applicability of this result to geotextile–soil interfaces has not been investigated.

Pull-out tests

The simplest pull-out model in limit-equilibrium-based methods of analysis takes the form (e.g. Public Works Research Institute (PWRI), 1992)

$$T_{pull} = 2L_a C_i \sigma_v \tan\phi \qquad (16.13)$$

where T_{pull} is the pull-out capacity, L_a is the anchorage length, σ_v is the vertical stress acting over the anchorage length, ϕ is the friction angle of the soil, and C_i is the interaction coefficient that is interpreted from the results of pull-out tests. In the USA, a combination of terms are used to calculate default values of C_i based on the type of geosynthetic, aperture size, and d_{50} of the confining soil. The reader is referred to the FHWA (2009) guidelines for details.

A large amount of data can be found on the pull-out behaviour of geotextiles and geogrids in combination with cohesionless soils (e.g. Farrag, 1990; Huang and Bathurst, 2009). Bachus *et al.* (1993) reported the results of constant rate of displacement (static) pull-out test results on four different geogrids in sand. Most tests gave interaction coefficient values equal to, or slightly in excess of, 1.0. Increasing the rate of loading from 1 to 150 mm/min did not result in significant changes in interaction coefficient values.

Relatively few investigations have been performed that examine the effect of a repeated tensile load application using conventional pull-out box devices. Bathurst and McLay (1996) carried out large-scale, repeated load pull-out tests on 1.6 m-long specimens of a stiff uniaxial HDPE geogrid in combination with standard #40 laboratory silica sand. Tensile loads were applied to the specimens subjected to constant surcharge pressures ranging from 25 to 73 kPa. The cyclic load amplitude ranged from about 25 to 50% of the index strength of the material. Pull-out of the specimens was not achieved with the cyclic load amplitudes and overburden pressures used in this test series – even after 90 000 load applications in some tests. The mobilised length of the reinforcement was observed to increase linearly with the log number of load applications. Full shear mobilisation along the 1.6 m lengths of reinforcement was not observed in tests with overburden pressures greater than 25 kPa. The rate of displacement of the front (in-soil) end of the reinforcement was observed to diminish linearly with the number of load applications (log–log scale), indicating that the anchorage system was intrinsically stable under repeated loading. The qualitative features of the test programme by Bathurst and McLay (1996) are in agreement with similar work by Hanna and Touahmia (1991). Min *et al.* (1995) and Yasuda *et al.* (1992) carried out repeated load pull-out tests on stiff uniaxial and biaxial polyolefin geogrids subject to constant surcharge pressure. Raju (1995) carried out cyclic load tests on both uniaxial HDPE and woven PET geogrids at small strain amplitudes representative of working load levels in the field. Raju (1995) and Yasuda *et al.* (1992) report that the magnitude of the peak cyclic load to cause pull-out failure is greater than the load required for static pull-out failure. Min *et al.* (1995) carried out repeated load pull-out tests using a biaxial polypropylene geogrid, and concluded that the interaction coefficient, C_i, was reduced by about 20% due to repeated loading compared with the interaction coefficient back-calculated from single-load pull-out tests.

The conflicting results with respect to C_i for limit-equilibrium calculations may be attributed to the interpretation of anchorage length used to back-calculate interaction coefficient values. In addition, the interpretation of pull-out results is sensitive to test size, set-up and execution (Juran *et al.*, 1988). Finally, it can be noted that the AASHTO (2010) and FHWA (2009) guidelines recommend a reduction in the pull-out interaction coefficient to 80% of the values used for static design. This recommendation appears to be based on the results of pull-out tests on steel strip reinforcement. However, this reduction is more than compensated for in current AASHTO and FHWA load and resistance factor design codes that recommend higher resistance factors against pull-out failure in limit-equilibrium-based design compared with static design values.

Cyclic and shaking table tests

Shaking table tests, used to measure the dynamic interface coefficient for geotextile–geomembrane interfaces, have been reported by Zimmie *et al.* (1994). This technique offers possibilities for the characterisation of interface shear properties for soil–geotextiles under light surcharges, since the frequency of horizontal shear loading can be chosen to match the frequency and duration of typical seismic events (Figure 16.8). The critical acceleration, a_c, required to initiate slip can be used to calculate interface

Figure 16.8 The dynamic interface shear test using a shaking table: (a) schematic of the test set-up and (b) block acceleration versus table acceleration

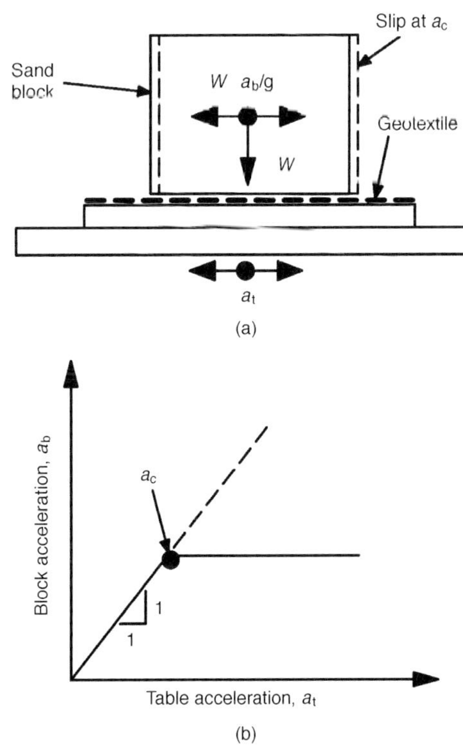

friction coefficients according to $C_i = \tan \phi_{ds} = a_c/g$. To examine interface shear resistance at greater surcharges, Zimmie *et al.* placed a shaking table apparatus in a centrifuge. Additional investigations to quantify the interface shear behaviour of interfaces formed from different combinations of geotextiles and geomembranes have been reported by De and Zimmie (1997, 1998a,b, 1999), and Yegian and Kadakal (1998). However, to the best of the authors' knowledge, similar lines of investigation focused on soil–geotextile and soil–geogrid dynamic interaction using shaking tables are yet to be carried out.

16.2.3.2 Facing connection tests

Geosynthetic-reinforced segmental retaining walls comprise dry-stacked columns of modular concrete blocks that may be solid or infilled with granular soil. The connection between the facing column and the reinforced soil mass is typically formed by extending the reinforcement layers between facing units to the front face of the wall. This connection must carry greater loads during seismic shaking, and additional shear forces may be transmitted between modular block units. The performance of the connection between dry-stacked modular concrete units, and the interface shear between facing units with and without the inclusion of a geosynthetic reinforcement layer, can be evaluated by adapting test protocols originally proposed by the senior author and his co-workers (Bathurst and Simac, 1993; Simac *et al.*, 1993; ASTM, 2001b; Bathurst and Huang, 2010) for static load environments. Example

Figure 16.9 A cyclic load test on a woven PET geogrid–solid block connection (T_{ult} is the index strength of the geogrid using ASTM D4595)

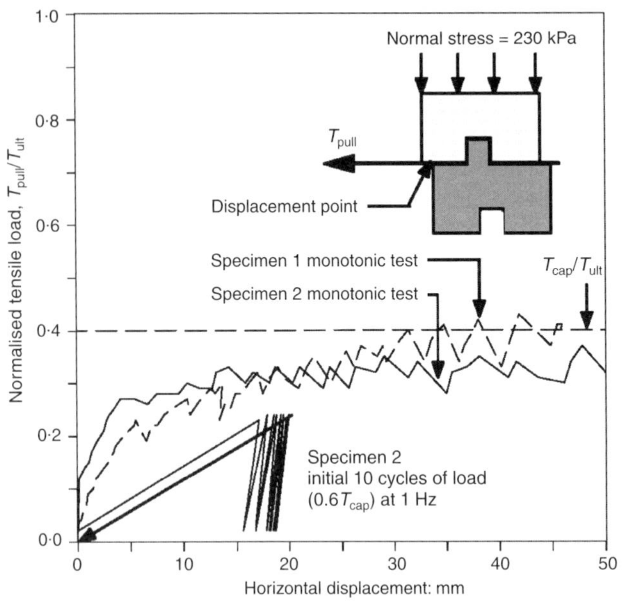

Figure 16.10 The interface slip model

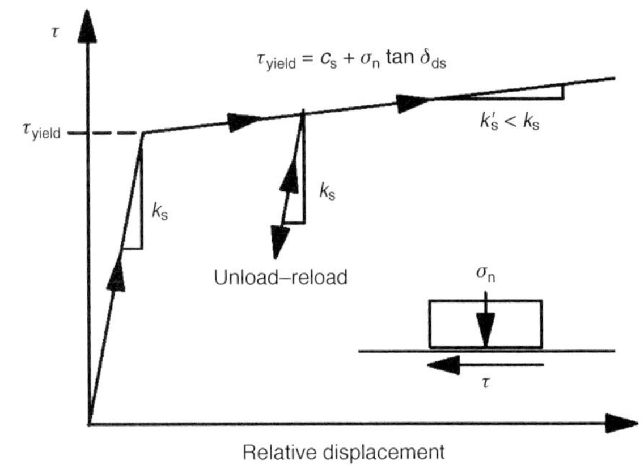

test results for a particular connection system prior to and after (load-controlled) cyclic loading are illustrated in Figure 16.9. The reinforcement in this particular test was a woven PET geogrid, and the block was a solid concrete unit with a continuous concrete shear key. A constant rate of displacement load test was carried out on a virgin specimen of geogrid. A second identical test was carried out after the connection had been subjected to ten load cycles at 60% of the ultimate connection capacity, T_{cap}. For this particular system there was no degradation of the connection based on ultimate strength or capacity after 18 mm displacement measured at the back of the block. This result cannot be assumed for all block–geosynthetic systems on the market today, or at lower normal stresses. Further research on repeated load connection testing is required.

Repeated load interface shear tests can also be carried out using the NCMA (Simac *et al.*, 1993; ASTM, 2003) methodology for block–block shear response. Static testing shows that interface shear behaviour can be influenced by the presence of a geosynthetic layer and test procedure (Bathurst *et al.*, 2008). Cai and Bathurst (1996a) assumed that static interface shear values were reasonable in sliding block analyses for systems that provide positive interlock in the form of shear keys, pins and other forms of connectors (see Section 16.3.4.1).

16.2.3.3 Interface shear-displacement modelling

The shear transfer at reinforcement–soil and soil–facing unit interfaces (or in the case of segmental retaining walls, reinforcement–concrete block and block–block interfaces) can be modelled in dynamic finite-element/finite-difference codes using a slip element model proposed by Goodman *et al.* (1968) (Figure 16.10). The failure (yield) state of the slip element is assumed to obey the Mohr–Coulomb failure criterion, where τ_{yield} is the shear strength at which slip occurs for the first time, c_s is the apparent cohesion, σ_n is the normal stress, and δ_{ds} is the interface friction angle at the yield state. When the applied shear stress exceeds the yield strength, the shear stiffness of the slip element is reduced to a fraction of the original value, and slip is initiated. In the normal direction, the stress, σ_n, is assumed to vary linearly with the average relative nodal displacement. Separation of the contact interface is assumed to occur when the normal stress, σ_n, is tensile (which may occur at facing column–soil interfaces). The interpretation of physical test results to obtain example property values can be found in the paper by Cai and Bathurst (1995).

16.3. Seismic analysis and the design of walls and slopes

Analytical and numerical approaches for the seismic analysis of reinforced walls, slopes and embankments can be divided into the following categories

- pseudo-static methods
- displacement methods
- dynamic finite-element/finite-difference methods.

In this chapter, global stability modes of failure for walls are not addressed.

16.3.1 Pseudo-static methods

Pseudo-static methods extend conventional limit-equilibrium methods of analysis for earth structures to include destabilising body forces that are related to assumed horizontal and vertical components of ground acceleration.

16.3.1.1 Mononobe–Okabe approach

Pseudo-static rigid body approaches that use the Mononobe–Okabe (M–O) method to calculate dynamic earth forces (Okabe, 1924; Mononobe and Matsuo, 1929), acting on earth-retaining structures (typically walls), are well established in geotechnical engineering practice (e.g. Seed and Whitman, 1970; Richards and Elms, 1979; Koseki *et al.*, 1998a). The M–O method can be recognised as an extension

Figure 16.11 Forces and geometry used in the pseudo-static seismic analysis of segmental retaining walls

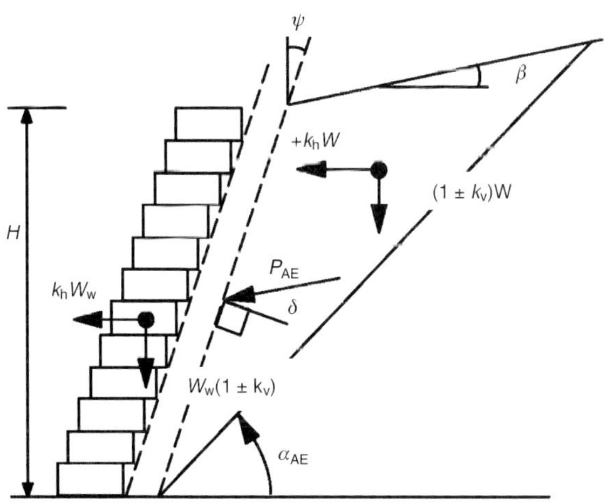

of classical Coulomb wedge analysis. The total active earth force, P_{AE}, imparted by the backfill soil is calculated as (Seed and Whitman, 1970)

$$P_{AE} = \tfrac{1}{2}(1 \pm k_v)K_{AE}\gamma H^2 \qquad (16.14)$$

where γ is the unit weight of the soil, H is the height of the wall and K_{AE} is the seismic active earth pressure coefficient. The application of force, P_{AE}, against the facing column of a segmental retaining wall structure is illustrated in Figure 16.11. The earth pressure coefficient, K_{AE}, can be calculated as follows:

$$K_{AE} = \frac{\cos^2(\phi + \psi - \theta)/\cos\theta\cos^2\psi\cos(\delta - \psi + \theta)}{1 + \sqrt{\dfrac{\sin(\phi + \delta)\sin(\phi - \beta - \theta)}{\cos(\delta - \phi + \theta)\cos(\psi + \beta)}}} \qquad (16.15)$$

where ϕ is the peak soil friction angle (ϕ_{peak}), ψ is the wall–slope face inclination (positive in a clockwise direction from the vertical), δ is the mobilised interface friction angle at the back of the wall (or back of the reinforced soil zone), β is the backslope angle (from the horizontal), and θ is the seismic inertia angle given by

$$\theta = \tan^{-1}\frac{k_h}{1 \pm k_v} \qquad (16.16)$$

The quantities k_h and k_v are horizontal and vertical seismic coefficients, respectively, expressed as fractions of the gravitational constant; for example, Seed and Whitman (1970) decomposed the total (active) earth force, P_{AE}, calculated according to Equations 16.14 and 16.15, into two components representing the static earth force component, P_A, and the incremental dynamic earth force due to seismic effects, ΔP_{dyn}. Hence

$$P_{AE} = P_A + \Delta P_{dyn} \qquad (16.17)$$

or

$$(1 \pm k_v)K_{AE} = K_A + \Delta K_{dyn} \qquad (16.18)$$

where K_A is the static active earth pressure coefficient and ΔK_{dyn} is the incremental dynamic active earth pressure coefficient. Closed-form approximate solutions for the orientation of the critical planar surface from the horizontal, α_{AE}, have been reported by Okabe (1924) and Zarrabi (1979). These solutions can be expressed as follows:

$$\alpha_{AE} = \phi - \theta + \tan^{-1}\left(\frac{-A_\alpha + D_\alpha}{E_\alpha}\right) \qquad (16.19)$$

where

$$\begin{aligned} A_\alpha &= \tan(\phi - \theta - \beta) \\ D_\alpha &= \sqrt{A_\alpha(A_\alpha + B_\alpha)(B_\alpha C_\alpha + 1)} \\ E_\alpha &= 1 + [C_\alpha(A_\alpha + B_\alpha)] \\ B_\alpha &= 1/\tan(\phi - \theta + \psi) \\ C_\alpha &= \tan(\delta + \theta - \psi) \end{aligned} \qquad (16.20)$$

Equation 16.19 can be used to calculate the orientation of the assumed active failure plane within the reinforced soil mass and in the retained soil. However, the result of pseudo-static analyses of the type described here have been shown to lead to excessively long reinforcement lengths if reinforcement layers are required to extend beyond the internal failure plane. Current practice in North America is to assume that the orientation of the internal failure plane for reinforcement design is described by static load conditions (i.e. α_{AE} ($k_h = k_v = 0$)) (Bathurst, 1998; FHWA, 2009; NCMA, 2009; AASHTO, 2010). Koseki *et al.* (1998a) and Tatsuoka *et al.* (1998) have proposed a pseudo-static design method that results in internal failure planes that are steeper than those calculated using a rigorous interpretation of the extended Coulomb wedge approach.

Bathurst and Cai (1995) have proposed the total active earth pressure distribution illustrated in Figure 16.12 for external, internal and facing stability analyses of reinforced segmental retaining walls. The normalised elevation of the resultant total earth force varies over the range $1/3 < m_d < 0.6$, depending on the magnitude of ΔK_{dyn}. The assumed pressure distribution is based on a review of the literature for conventional gravity retaining wall structures in North America, where the dynamic increment is typically taken as acting at $0.6H$ above the base of the wall. The total pressure distribution is identical to that recommended for the design of flexible anchored sheet pile walls under seismic loads (Ebeling and Morrison, 1993), and is used in the AASHTO (2010) and FHWA (2009) design guidelines for reinforced soil wall structures. In the latest edition of the NCMA (2009) design guidelines, the trapezoidal dynamic earth pressure contribution (Figure 12(b)) has been simplified to a uniform (rectangular) distribution. In the absence of ground acceleration, the distribution reduces to the triangular active

Figure 16.12 Calculation of the total earth pressure distribution due to soil self-weight: (a) the static component, (b) the dynamic increment and (c) the total pressure distribution
Reproduced from Cai and Bathurst (1995), Industrial Fabrics Association International

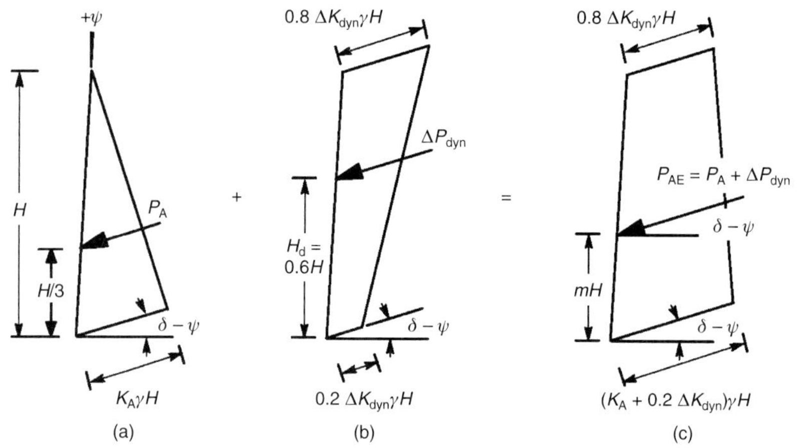

earth pressure distribution due to soil self-weight. The influence of reinforcement stiffness and ground motion, on the distribution and line of action of active earth forces under static and dynamic loading, has been investigated through numerical modelling by Bathurst and Hatami (1999a), and is discussed in Section 16.3.4.2.

16.3.1.2 Selection of seismic coefficients

In conventional pseudo-static methods of analysis, the choice of the horizontal seismic coefficient, k_h, for design is related to a specified horizontal peak ground acceleration for the site, a_h. The relationship between a_h and a representative value of k_h is nevertheless complex, and there does not appear to be a general consensus in the literature on how to relate these parameters. For example, Whitman (1990) reports that values of k_h from 0.05 to 0.15 are typical for the design of conventional gravity wall structures, and that these values correspond to one-third to one-half of the peak acceleration of the design earthquake. Bonaparte *et al.* (1986) used $k_h = 0.85a_h/g$ to generate design charts for geosynthetic-reinforced slopes under seismic loading using the two-part wedge method of analysis.

However, the results of finite-element modelling of reinforced soil walls (Segrestin and Bastick, 1988; Cai and Bathurst, 1995), limited half-scale experimental work (Chida *et al.*, 1982) and FLAC modelling (Bathurst and Hatami, 1998a) have shown that the average acceleration of the composite soil mass may be equal to or greater than a_h, depending on a number of factors, including the wall height, the wall toe boundary (i.e. the degree of toe restraint), the base acceleration intensity, the ratio of the ground motion predominant frequency to the wall fundamental frequency, f_g/f_1, soil properties and, to a lesser extent, the reinforcement stiffness.

The FHWA (2001) guidelines use an equation proposed by Segrestin and Bastick (1988) that relates k_h to a_h according to

$$k_h = \frac{a_h}{g}\left(1.45 - \frac{a_h}{g}\right) \qquad (16.21)$$

This formula results in $k_h > a_{h/g}$ for $a_h < 0.45g$. However, as clearly stated by Segrestin and Bastick, Equation 16.21 should be used with caution because it is based on the results of finite-element modelling of steel-reinforced soil walls up to 10.5 m high that were subjected to ground motions with a very high predominant frequency of 8 Hz. Recommendations to reduce k_h can be found in the FHWA (2001) guidelines to mitigate excessive design conservatism that results when using k_h computed from Equation 16.21 in the M–O formulations applied to relatively flexible reinforced soil walls. The results of finite-element modelling reported by Cai and Bathurst (1995), for a 3.2 m-high geosynthetic-reinforced segmental retaining wall with $a_h = 0.25g$ and a predominant frequency range of 0.5–2 Hz, gave a distribution of peak horizontal acceleration through the height of the composite mass and retained soil that was, for practical purposes, uniform and equal to the base peak input acceleration. These observations are consistent with the results of Chida *et al.* (1982), who constructed 4.4 m-high steel-reinforced soil wall models and showed that the average peak horizontal acceleration in the soil behind the walls was equal to the peak ground acceleration for ground motion frequencies less than 3 Hz.

The general solutions to pseudo-static methods of analysis admit both vertical and horizontal components of seismic-induced inertial forces. The choice of positive or negative k_v values influences the magnitude of dynamic earth forces calculated using Equations 16.14 and 16.15. In addition, the resistance terms in factor of safety expressions for internal and external stability of walls and slopes, that include the vertical component of seismic force, are influenced by the choice of sign for k_v. An implicit assumption in many of the papers on pseudo-static design of conventional gravity wall structures cited in the literature is that the vertical component of seismic body forces acts upward. However, the designer must evaluate both positive and negative values of k_v to ensure that the most critical condition is considered in dynamic stability analyses if non-zero values of k_v are assumed to apply. For example, Fang and Chen (1995) have demonstrated in a series of example calculations that the magnitude of P_{AE} may be 12% higher for the case when the vertical seismic force acts downward ($+k_v$) compared with the case when it acts upward ($-k_v$). Nevertheless, selection of a non-zero value of k_v implies that peak horizontal and vertical accelerations are time coincident, which is an unlikely occurrence in practice. For example, Madabhushi (1996) investigated the arrival time of horizontal and vertical stress waves to selected recording sites. He concluded that since the horizontal and vertical waves arrive at different times, the design ground acceleration coefficients for retaining walls do not need to be combined at their maximum values. The assumption that peak vertical accelerations do not occur simultaneously with peak horizontal accelerations is made in the current FHWA and AASHTO guidelines for the seismic design of mechanically stabilised soil retaining walls, as well as in the Japanese design and construction manual for reinforced soil structures (PWRI, 1992).

Seed and Whitman (1970) have suggested that $k_v = 0$ is a reasonable assumption for the practical design of conventional gravity structures using pseudo-static methods. Wolfe *et al.* (1978) studied the effect of combined horizontal and vertical ground acceleration on the seismic stability of reduced-scale model reinforced earth walls using shaking table tests. They concluded that the vertical component of seismic motion may be disregarded in terms of practical seismic stability design. Their conclusion can also be argued to apply to geosynthetic-reinforced walls. Nevertheless, significant vertical accelerations may occur at sites located at short epicentral distances, and engineering judgement must be exercised in the selection of vertical and horizontal seismic coefficients to be used in pseudo-static seismic analyses.

In order to address specific concerns raised by Allen (1993) related to the facing stability of geosynthetic-reinforced segmental retaining walls during a seismic event that includes vertical ground accelerations, parametric analyses were carried out by Bathurst and Cai (1995) to investigate the combined effect of horizontal and vertical acceleration using the range $k_v = -2k_h/3$ to $+2k_h/3$. The upper limit on the ratio k_v to k_h is equal to the calculated ratio of the peak vertical ground acceleration to the peak horizontal ground acceleration, from seismic data recorded in the Los Angeles area (Stewart *et al.*, 1994). The results are shown in Figure 16.13, and illustrate that for $k_h < 0.35$ the effect on the total dynamic earth pressure is not significant.

Based on experience with the performance of conventional and reinforced soil retaining walls during the Kobe earthquake, Tatsuoka *et al.* (1998) reviewed the choice of the

Figure 16.13 Influence of the seismic coefficients, k_h and k_v, and the wall inclination angle, ψ, on dynamic earth force, P_{AE}
Reproduced from Cai and Bathurst (1995), Industrial Fabrics Association International

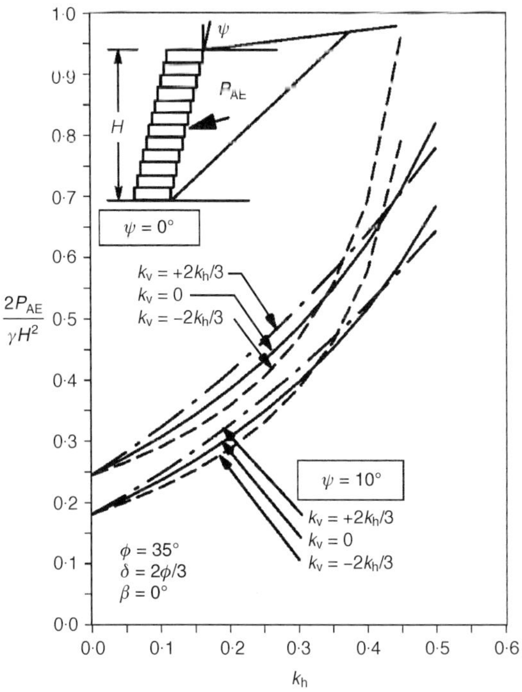

horizontal seismic coefficient value used in pseudo-static design methods in Japan. They suggested that the design k_h value for geosynthetic-reinforced soil walls with full-height rigid facings should be taken as 0.3. This value is less than their recommended value of 0.35 for unreinforced cantilever walls, and considerably less than their recommended value of 0.4 for conventional gravity-type retaining walls. They attributed the selection of the design value of $k_h = 0.3$ for reinforced soil wall structures to

- typically conservative assumptions for soil strength
- positive structural dynamic effects (e.g. wall ductility and flexibility)
- a global factor of safety value that is normally taken to be larger than unity.

In practice, the final choice of k_h may be based on local experience, or prescribed by local building codes or other regulations. The magnitude of a_h for a particular location in the USA can be found in the United States Geological Survey (2000), AASHTO (2010) and Building Seismic Safety Council (BSSC) (1994) guidelines. Similar data are published by the Canadian Geotechnical Society (2006) for Canada. Readers may refer to the book by Paz (1994) for information on seismic codes for most other countries. The textbooks by Kramer (1996a) and Okamoto (1984), and agency documents by the AASHTO and the BSSC, provide valuable information on the effect of foundation conditions on attenuation or amplification of bedrock source ground motion.

Finally, the FHWA (2001) guidelines for reinforced soil wall structures caution that pseudo-static design methods should be restricted to sites where the peak horizontal ground acceleration is not expected to exceed 0.29g. For more intense earthquakes, large structure displacements may occur, and the services of a specialist are recommended. As a minimum requirement, retaining wall structures should be analysed using a Newmark-type sliding block approach (see Section 16.3.3.1). For reinforced soil slopes as flexible structures, the FHWA (2009) guidelines allow peak horizontal ground acceleration values to be reduced by 50%.

16.3.1.3 External stability calculations for walls

External stability calculations for factors of safety against base sliding and overturning of geosynthetic-reinforced retaining walls are similar to those carried out for conventional gravity structures. For reinforced structures, the gravity mass is taken as the composite mass formed by the reinforced soil zone. For segmental retaining walls, the gravity mass includes the facing column, since it may comprise a significant part of the gravity mass, particularly for low-height structures (and, hence, they generate additional inertial forces during a seismic event). The earth pressure distribution shown in Figure 16.12 is used to calculate the destabilising forces in otherwise conventional expressions for the factor of safety against sliding along the foundation surface and overturning about the toe of the structure. The simplified geometry and body forces assumed in these calculations for the case of segmental retaining walls are illustrated in Figure 16.14. The term W_R in the figure is the weight of the reinforced zone, plus the weight of the facing column used to calculate resisting terms in factor of safety expressions for base sliding and overturning. The quantity P_{IR} denotes the horizontal inertial force due to the gravity mass used in the external stability factor of safety calculations. Different strategies have been proposed in North America to compute $P_{IR} < k_h W_R$ to ensure reasonable designs. The justification is based on the expectation that horizontal inertial forces induced in the gravity mass and the retained soil zone will not reach peak values at the same time during a seismic event. Christopher *et al.* (1989) proposed the following expression for horizontal backfills:

$$P_{IR} = 0.5\eta k_h \gamma H^2 \tag{16.22}$$

where $\eta = 0.6$ based on recommendations for reinforced walls that use steel reinforcement strips (Segrestin and Bastick, 1988). Cai and Bathurst (1995) proposed an expression that gives similar results for typical L/H ratios for segmental walls:

$$P_{IR} = \eta k_h W_R \tag{16.23}$$

where $\eta = 0.6$. The AASHTO (2010) proposes that P_{IR} be calculated using Equation 16.22 with $\eta = 1$ and that the external dynamic active earth force component, ΔP_{dyn}, be reduced by 50%. North American practice is to reduce dynamic factors of safety against sliding and overturning to 75% of the static factor of safety values, in recognition of the transient nature of seismic loading. The calculation method for P_{IR} and the reduction in the static factors of safety described above for AASHTO has been adopted for pseudo-static seismic design of reinforced segmental retaining walls by the National Concrete Masonry Association (NCMA, 2009).

Dynamic factors of safety are also reduced in Japan (Geogrid Research Board (GRB), 1990; Koga and Washida, 1992; PWRI, 1992). However, factor of safety calculations for wall-base sliding in Japan do not consider any reduction in the inertial force, P_{IR} (i.e. Equation 16.23 is used with $\eta = 1$). In order to further reduce conservatism in the Japanese approach for base sliding, Fukuda *et al.* (1994) have proposed ignoring the dynamic force increment, ΔP_{dyn}, and restricting seismic loading contributions to the gravity mass term, P_{IR}, only. Overturning criteria for walls are restricted to ensuring that the resultant force acting at the base of the reinforced mass, W_R, falls within $L/3$ of the base midpoint for walls subject to earthquake. The FHWA (2001) guidelines for geosynthetic-reinforced walls also omit overturning as a potential failure mode for geosynthetic-reinforced soil walls. However, to be consistent with the current static design of reinforced segmental retaining walls (Simac *et al.*, 1993), overturning is considered for seismic design of this class of structure (Bathurst, 1998). Bathurst *et al.* (1997) used the NCMA pseudo-static method to produce design charts for the preliminary evaluation of seismic resistance of segmental reinforced soil-retaining walls on firm foundations. The charts are presented as the ratio of dynamic to static safety factor values for peak horizontal ground accelerations up to 0.5g and soil friction angle values in the range $25° < \phi_{peak} < 45°$.

Figure 16.14 Forces and geometry for external stability calculations for base sliding and overturning

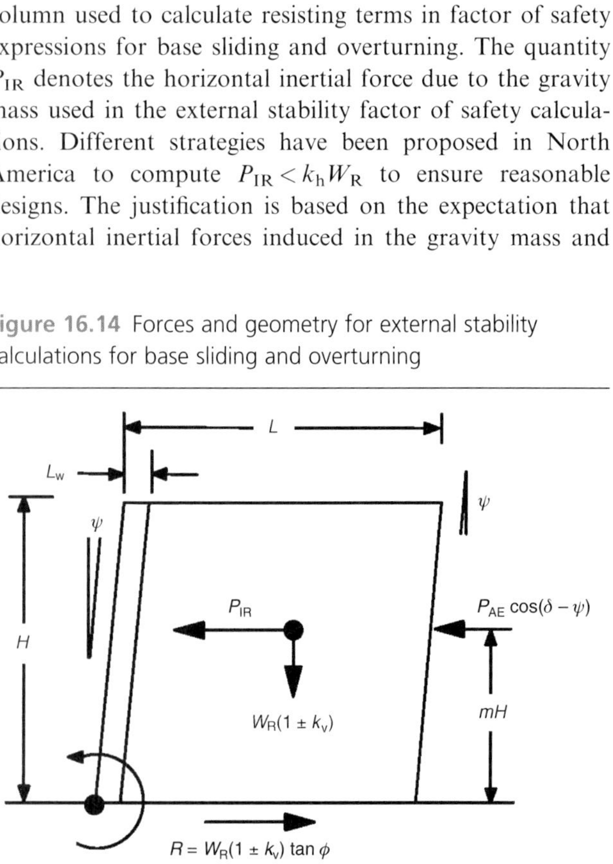

16.3.1.4 Internal stability calculations for walls and slopes

Pseudo-static/dynamic methods for walls, which involve an assumed distribution of internal earth pressure (e.g. Figure 16.12), require that each reinforcement layer carry a

Figure 16.15 Calculation of the tensile load, T_i, in a reinforcement layer due to dynamic earth pressure and wall inertia for segmental retaining walls
Reproduced from Cai and Bathurst (1995), Industrial Fabrics Association International

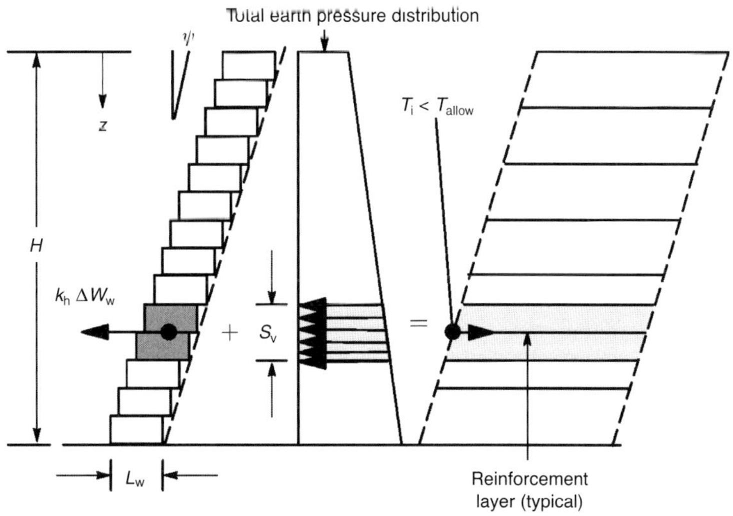

portion of the integrated earth pressure over a tributary area, S_v, as illustrated in Figure 16.15. The magnitude of the tensile force must not exceed the allowable design load in the reinforcement based on the tensile over-stressing, facing connection strength and pull-out capacity of the layer. In North American practice, factors of safety against these modes of failure are reduced to values that are typically 75% of static values. Figure 16.15 also demonstrates that the inertial force due to the tributary portion of the facing column should be added to the reinforcement forces under seismic loading in the case of segmental walls. An important implication of the assumed earth pressure distribution using the pseudo-static M–O method described earlier is that the relative proportion of load to be carried by the reinforcement layers closest to the crest of a wall with uniform reinforcement spacing increases with increasing horizontal acceleration. This may require a greater number of layers towards the top of the wall than is required for static load environments. A similar conclusion was reached by Vrymoed (1989) using a tributary area approach that assumes that the inertial force carried by each reinforcement layer increases linearly with the height above the toe of the wall for equally spaced reinforcement layers. Bonaparte *et al.* (1986) applied the tributary area method to walls and slopes, but recommended a uniform distribution for the dynamic earth pressure increment (i.e. $H_d = 0.5H$ in Figure 16.12). Nevertheless, they concluded that the combination of higher available reinforcement strength and reduced factors of safety used for seismic loading cases will often result in no requirement to increase the number of reinforcement layers required for static loading cases. The FHWA (2001) guidelines use the procedure shown in Figure 16.16 to assign reinforcement forces for over-stressing and pull-out calculations. In this method, the static earth force, P_A, is calculated using the Rankine earth pressure theory with a Rankine failure plane ($\alpha = \pi/4 + \phi/2$) for vertical walls, and the Coulomb theory with a Coulomb angle according to Equation 16.19 (using $k_h = k_v = 0$ in Equation 16.16) for walls with a facing batter greater than 10°. The dynamic earth force is calculated as $\Delta P_{dyn} = k_h W_A$, , where W_A is the weight of the static internal failure wedge. The distribution of the dynamic tensile reinforcement load increment, ΔT_{dyn}, is weighted based on the total anchorage

Figure 16.16 Calculation of the tensile load, T_i, in a reinforcement layer for reinforced soil walls with extensible reinforcement using the FHWA (2001) method

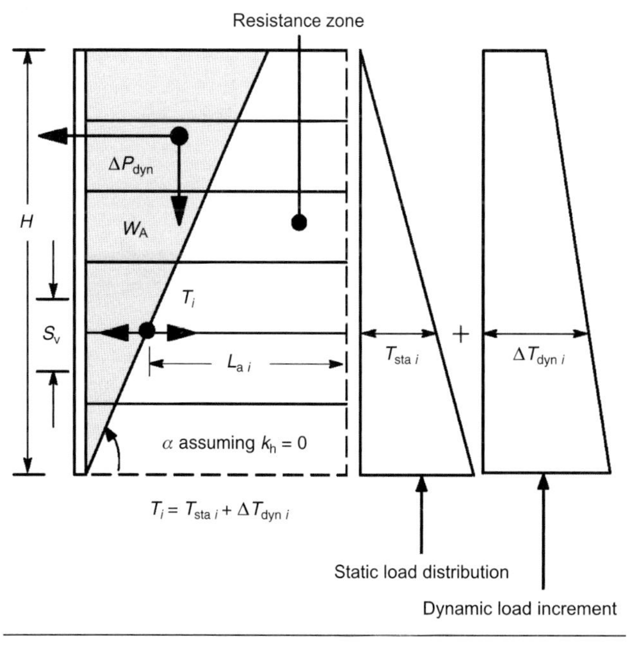

length in the resistance zone according to

$$\Delta T_{\mathrm{dyn}\,i} = \frac{\Delta P_{\mathrm{dyn}} L_{\mathrm{a}\,i}}{\sum_{j=1}^{N} L_{\mathrm{a}\,j}} \tag{16.24}$$

where N is the number of reinforcement layers and L_a is the anchorage length. This approach leads to the redistribution of the dynamic force to the lower reinforcement layers for internal stability calculations in structures with a uniform reinforcement length. This strategy is based on the results of the finite-element modelling of reinforced walls that used (inextensible) steel strips (Segrestin and Bastick, 1988). However, the dynamic increment force distribution shows the opposite trend to that used for external stability calculations in the same FHWA (2001) guidelines (see Figure 16.12). Although not demonstrated, it is clear that the FHWA method is the least conservative for the design of reinforcement forces, T_i, of all the methods reviewed. Furthermore, the FHWA approach is less likely to result in an increased number of reinforcement layers at the top of reinforced wall structures and increased reinforcement lengths to accommodate shallower internal failure surfaces with increasing horizontal acceleration, which is often the case using a rigorous interpretation of the M–O theory.

16.3.1.5 Two-part wedge failure mechanism

The general solution for a trial, two-part wedge failure mechanism in a slope subjected to horizontal and vertical acceleration components is illustrated in Figure 16.17. The horizontal and vertical forces P_1 and V_1 acting on wedge 2 from wedge 1 are, respectively,

$$P_1 = \frac{(1 \pm k_v)W_1 + B_1 A_1 k_h W_1}{\lambda \tan\phi_f + B_1 A_1} \tag{16.25}$$

$$V_1 = \lambda P_1 \tan\phi_f \tag{16.26}$$

Figure 16.17 Two-part wedge analysis: (a) free-body diagram and (b) with reinforcement forces

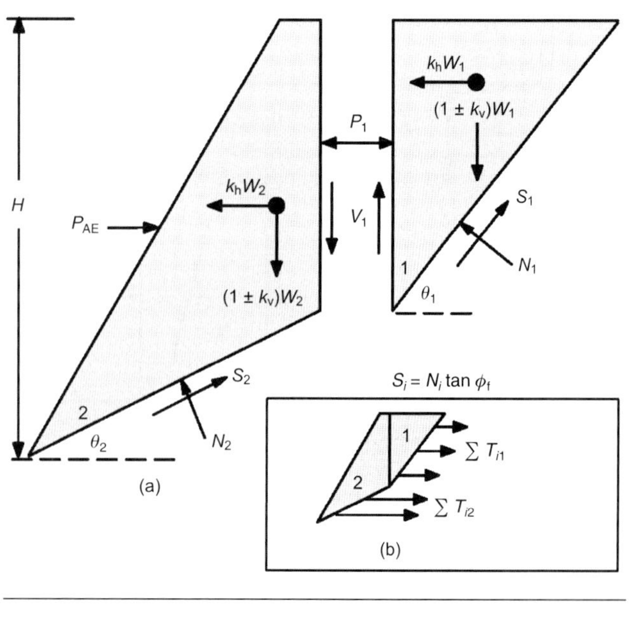

where

$$A_1 = \frac{1}{\sin\theta_1 - \tan\phi_f \cos\theta_1} \tag{16.27}$$

$$B_1 = \tan\phi_f \sin\theta_1 + \cos\theta_1 \tag{16.28}$$

The quantity λ is the inter-wedge shear mobilisation ratio, and varies over the range $0 \le \lambda \le 1$. The parameter ϕ_f is the factored soil friction angle, expressed as

$$\phi_f = \tan^{-1}(\tan\phi/\mathrm{FS}) \tag{16.29}$$

where FS is the factor of safety.

The horizontal out-of-balance force, P_{AE}, is calculated as

$$P_{AE} = P_1 + k_h W_2 - B_2 A_2[(1 \pm k_v)W_2 + V_1] \tag{16.30}$$

where

$$A_2 = \frac{1}{\tan\phi_f \sin\theta_2 + \cos\theta_2} \tag{16.31}$$

$$B_2 = \tan\phi_f \cos\theta_2 - \sin\theta_2 \tag{16.32}$$

By setting FS = 1 (i.e. $\phi = \phi_f$), an equivalent total active earth pressure coefficient for the most critical trial geometry (i.e. a trial search that yields a maximum value for P_{AE} in the slope) can be calculated as

$$K_{AE} = 2P_{AE}/\gamma H^2 \tag{16.33}$$

This approach has been used by Bonaparte *et al.* (1986) to produce seismic design charts for geosynthetic-reinforced soil slopes. The total required design strength of the horizontal layers of reinforcement is taken as $\sum T_i = P_{AE}$. The two-part wedge approach with $\lambda = 0$ is used by the GRB (1990) to calculate K_{AE} according to Equations 16.30 and 16.33 for internal stability calculations.

The two-part wedge analysis degenerates to a single-wedge analysis by restricting trial searches to $\theta_1 = \theta_2$ and setting $\lambda = 0$. All three solutions (M–O, single and two-part wedge) give the same solution for the horizontal component of the total earth force when $\lambda = \psi = 0$. In addition, direct sliding mechanisms, including those generated at the base of the reinforced soil mass or along reinforcement layers, can be analysed using the two-part wedge approach.

An alternative strategy that extends the general approach used by Woods and Jewell (1990) for statically loaded slopes to the seismic case (Bathurst, 1994) is to rewrite Equation 16.30 as

$$P_{AE} = P_1 - \frac{B_1 A_1 \sum T_{i1}}{\lambda \tan\phi_f + B_1 A_1} + k_h W_2 - \sum T_{i2} - B_2 A_2[(1 \pm k_v)W_2 + V_1] \tag{16.34}$$

The factor of safety for a given two-part wedge geometry corresponds to the value that yields $P_{AE}=0$. The factor of safety for a slope corresponds to its minimum value from a search of all potential failure geometries. It should also be noted that in this approach, the same global factor of safety is applied to the reference design tensile strength of the reinforcement and pull-out capacity defined by Equation 16.13. Equation 16.34 illustrates that the value of the factor of safety against collapse is independent of the location of the reinforcement layers for $\lambda=0$.

Ling *et al.* (1996) presented design charts for calculating geosynthetic-reinforcement strength and length against direct sliding using a two-part wedge mechanism. Ling *et al.* (1997) maintained that the direct sliding failure mode governs reinforcement length design for lower layers as the seismic acceleration increases.

Tatsuoka *et al.* (1998) concluded that the two-part wedge geometry is a valid failure geometry for geosynthetic-reinforced soil walls with a full-height rigid facing and short reinforcement lengths based on shaking table tests. The pattern and location of the failure shape is controlled by the reinforcement length. They proposed a modified two-part wedge method, and concluded that the size of the failure wedge from the modified two-part wedge method was typically smaller than what would be predicted from conventional two-part wedge analysis, and more realistic according to experimental observations.

Ismeik and Guler (1998) considered the contribution of vertical, full-height concrete panel facing rigidity on wall stability using a two-part wedge analysis. Their method allows the contribution of the facing rigidity to be included explicitly to reduce the reinforcement loads and reinforcement lengths that would otherwise be larger without the contribution of the structural facing.

16.3.1.6 Log spiral failure mechanism

Log spiral failure mechanisms (Figure 16.18) have been used to calculate the out-of-balance force to be carried by horizontal reinforcement layers in slopes and walls under seismic loading (Leshchinsky *et al.*, 1995). An advantage of this method is that moment equilibrium is also satisfied (i.e. the problem is statically determinate). The trace of a log spiral surface is given by

$$R = Ae^{\xi(-\tan\phi_f)} \quad (16.35)$$

For an assumed surface (i.e. for any three independent parameters defining a log spiral, x_p, y_p and A), the moment equilibrium equation about the pole, P, can be explicitly written as

$$\sum M_p = (1 \pm k_v)W(x_c - x_p) + k_h W(y_p - y_c) - P_{AE}(y_p - y_{AE}) = 0 \quad (16.36)$$

Figure 16.18 Log spiral analysis: (a) free-body diagram and (b) with reinforcement forces

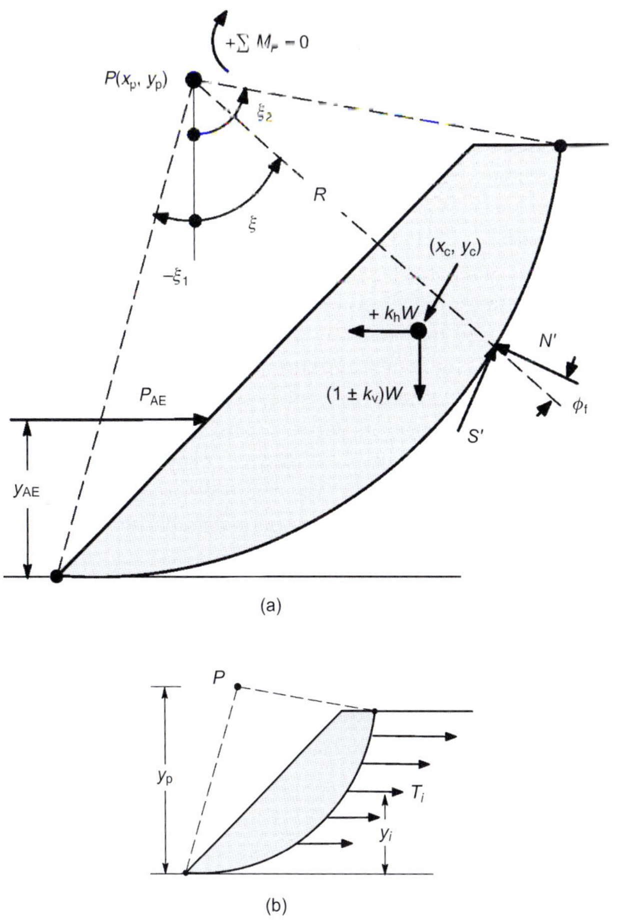

Note that the moment about the log spiral pole is independent of the distribution of normal and shear stresses over the log spiral because their resultant must pass through the pole. The point of application of the components of seismic inertial forces is taken at the centre of the failure mass. The critical mechanism corresponds to the trace that yields the maximum value of P_{AE} required to satisfy Equation 16.36. Clearly, the elevation, y_{AE}, of the equivalent out-of-balance horizontal force P_{AE} influences the magnitude of P_{AE}. Here, it is assumed, *a priori*, that $y_{AE}=H/3$. The equivalent dynamic active earth pressure coefficient, P_{AE}, can be calculated using Equation 16.33 with FS = 1 (i.e. $\phi=\phi_f$).

In practice, the factor of safety against collapse of a reinforced slope can be determined by replacing $P_{AE}(y_p - y_{AE})$ with $\sum T_i(y_p - y_i)$ in Equation 16.36 and finding the minimum value for the factor of safety from a search of all potential failure geometries that yields $\sum M_P=0$. This value corresponds to the minimum factor of safety for the reinforced soil slope (Leshchinsky *et al.*, 1995). The formulation of Equation 16.36 illustrates that the factor of safety against collapse is a function of the location of the reinforcement layers.

Ling *et al.* (1997) used a log spiral failure pattern in tie-back internal stability calculations. Ling and Leshchinsky (1998) extended the method to calculate the stability and permanent displacement of geosynthetic-reinforced soil walls under the combined effect of horizontal and vertical ground acceleration. They considered three different modes of failure in their analysis

- tie-back/compound failure
- direct sliding
- pull-out.

They assumed a log spiral failure shape in their pseudo-static analyses of the tie-back–compound failure mechanism. As in all pseudo-static methods of analyses, the method can be expected to result in conservative design because a momentary acceleration-induced force is assumed to act permanently on the wall. However, they argue that the inherent conservatism in the method is required since possible acceleration amplification is disregarded.

Figure 16.19 Circular slip analysis: (a) circular slip geometry and (b) method of slices

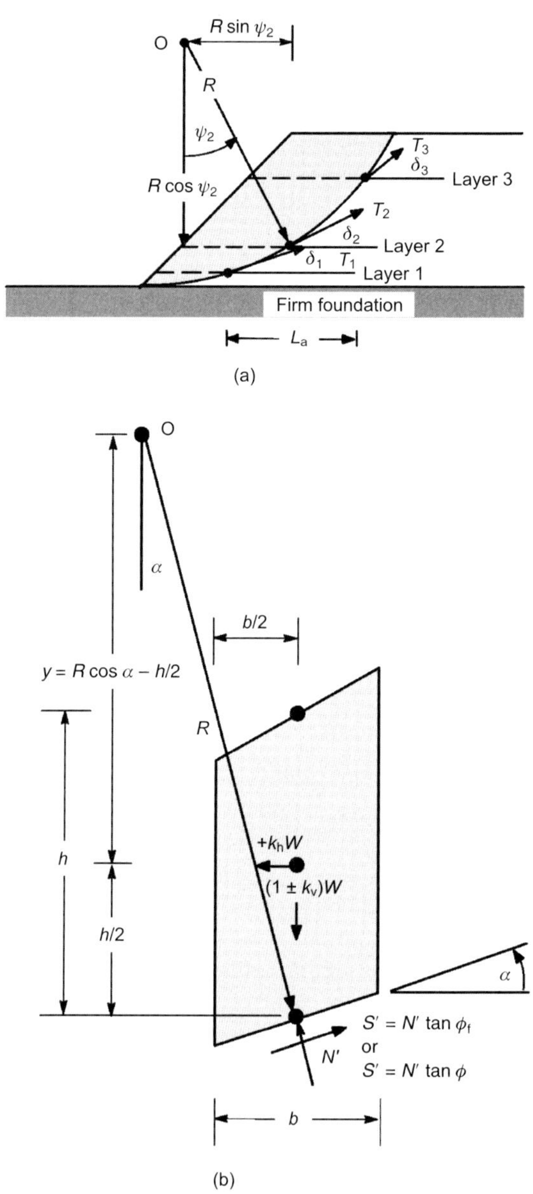

16.3.1.7 Circular slip failure mechanism

Conventional methods of slices can be modified to account for the additional restoring moment due to reinforcement layers. The general case can be referred to in Figure 16.19. Moment equilibrium leads to the following equation to calculate the factor of safety against collapse:

$$FS = \frac{M_R + \Delta M_R}{M_D} \qquad (16.37)$$

where M_R is the moment resistance due to the soil shear strength, ΔM_R is the increase in the moment resistance due to the reinforcement and M_D is the driving moment. Introducing k_v into the derivations for Bishop's simplified method (e.g. Fredlund and Krahn, 1976) results in the driving moment, calculated as

$$M_D = \sum W[(1 \pm k_v) R \sin\alpha + k_h y] \qquad (16.38)$$

The moment resistance due to cohesionless soil shear strength is

$$M_R = (1 \pm k_v) R \sum \frac{W \tan\phi \sec\alpha}{1 + \tan\alpha \tan\phi f} \qquad (16.39)$$

The additional resisting moment due to the tensile capacity of the reinforcement is calculated as

$$\Delta M_R = R \sum T_i \cos(\psi_i - \delta_i) \qquad (16.40)$$

The summation term in Equation 16.40 considers the available reinforcement tensile force in each layer (lesser of the tensile reinforcement strength based on over-stressing or pull-out) and the orientation, δ_i, of the force with respect to the horizontal. For flexible geosynthetic reinforcement products, the restoring force, T_i, can be argued to act tangent to the slip surface at the incipient collapse of the slope. This assumption leads to the summation term in Equation 16.40 becoming $\sum T_i R$. This approach is used in the FHWA (2001) guidelines, together with $k_v = 0$. It is important to note that, in the above formulation, the influence of the reinforcement capacity, T_i, and the horizontal acceleration term, k_h, on base sliding resistance is not considered.

An alternative strategy is to modify the 'ordinary method' (e.g. Fredlund and Krahn, 1976). In this approach, equations for the vertical and horizontal equilibrium of slices include forces due to acceleration components and reinforcement forces. Hence, these parameters directly affect base sliding resistance. The resisting moment term in Equation 16.37 becomes

$$M_R = R \sum [(1 \pm k_v) W \cos\alpha - k_h W \sin\alpha] \tan\phi \qquad (16.41)$$

and the incremental resisting moment due to reinforcement layers is

$$\Delta M_R = R \sum T_i[\cos(\psi_i - \delta_i) + \sin(\psi_i - \delta_i)\tan\phi] \quad (16.42)$$

where the summation term in Equation 16.42 is with respect to reinforcement layers. An advantage of the modified 'ordinary method' is that the right-hand side of Equation 16.37 is a linear function of the factor of safety. This approach is used by PWRI (1992) in Japan, with $\delta_i = 0$ for retaining walls and $\delta_i = \psi_i$ for slopes. In the Japanese approach, the distribution of the total reinforcement load is assumed to be uniform with depth for slopes less than 45° from the horizontal. For steeper slopes, including walls, the static portion of the required reinforcement load is assumed to increase linearly with the depth below the crest, while the additional seismic portion is assumed to be distributed uniformly. The FHWA (2001) guidelines allow the global factor of safety to be as low as 1.1 for the seismic design of slopes using pseudo-static methods.

16.3.1.8 Comparisons between selected pseudo-static methods

A comparison of the total active earth forces, calculated using wedge and log spiral pseudo-static methods, is illustrated in Figure 16.20(a) for frictionless soil–facing interfaces ($\delta = 0$). In these calculations, fully mobilised inter-wedge friction was assumed ($\lambda = 1$), and the point of equivalent total earth force application was taken as $H/3$. Figure 16.20(a) shows that for vertical faced slopes and walls ($\psi = 0$), the magnitude of P_{AE} from different pseudo-static methods is the same. However, for shallow slopes, there can be a significant difference between the methods. In particular, the M–O method may be non-conservative at high horizontal ground accelerations. For walls, the choice of earth pressure theory is not a concern, but, for slopes, the choice of theory must be considered carefully. In conventional tie-back methods of design, it is necessary that reinforcement lengths extend beyond the assumed active failure volume in order that pull-out resistance is available for each layer. This is of particular concern towards the top of reinforced wall and slope structures. All rigorous pseudo-static methods consistently predict that the minimum

Figure 16.20 Comparison of wedge and log spiral pseudo-static methods (L_{min} is the minimum length of reinforcement to contain failure volume; note that L_{min} may not be at the top of the reinforced mass): (a) the normalised active earth force, (b) the maximum width of the failure volume (vertical face) and (c) the maximum width of the failure volume (sloped face)

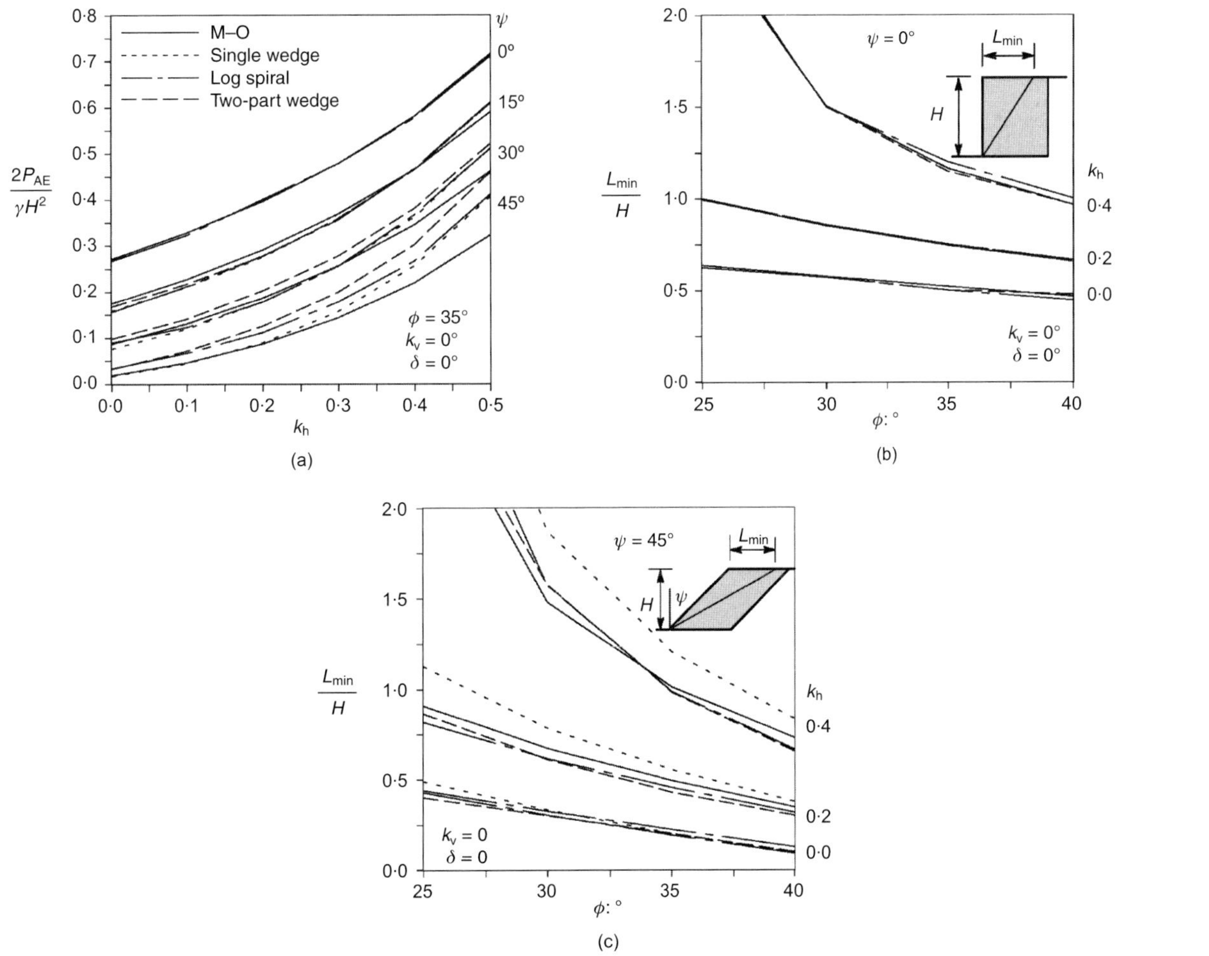

required reinforcement length will increase with increasing horizontal ground acceleration (Figures 16.20(b) and 16.20(c)) and, hence, reinforcement lengths may have to be increased for reinforced soil structures, particularly towards the crest. The observed cracking at the back of the reinforced soil mass in some wall structures has been attributed to this deficiency in post-earthquake surveys reported in the literature (see Section 16.6).

16.3.2 Pseudo-dynamic methods

A pseudo-dynamic earth pressure theory has been proposed by Steedman and Zeng (1990) to account for the influence of phase difference over the height of a vertical retaining wall. The approach recognises that a base acceleration input will propagate up through the retained soils at a speed that corresponds to the shear velocity of the soil. The general approach has been extended to the case of cohesionless slopes by Sabhahit *et al.* (1996). Introducing an interface friction angle, δ, and setting $k_v = 0$, leads to a further refinement (Figure 16.21). The horizontal acceleration is assumed to vary as

$$a(z,\ t) = a_0 \sin\left[\omega\left(t - \frac{H - z}{V_s}\right)\right] \tag{16.43}$$

where ω is the angular frequency, V_s is the shear wave velocity of the cohesionless soil, a_0 is the peak base acceleration and t is time. Horizontal slices of the assumed failure wedge with a linear failure surface, α, have incremental mass, calculated as

$$m(z) = \frac{\gamma}{g}(H - z)(\cot\alpha - \tan\psi)\,\mathrm{d}z \tag{16.44}$$

The total active earth force is computed as

$$P_{AE}(t) = \frac{Q_h(t)\cos(\alpha - \phi)}{\cos(\delta - \alpha + \phi)} + \frac{W\sin(\alpha - \phi)}{\cos(\delta - \alpha + \phi)} \tag{16.45}$$

Figure 16.21 The pseudo-dynamic method

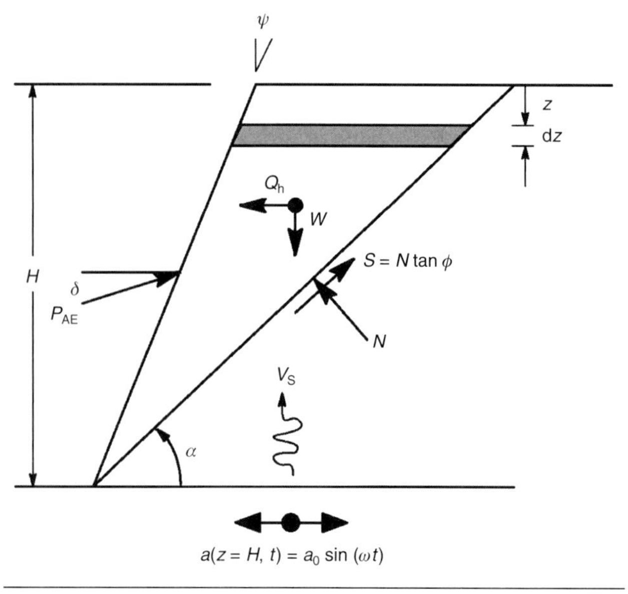

Figure 16.22 The point of application of the dynamic force increment

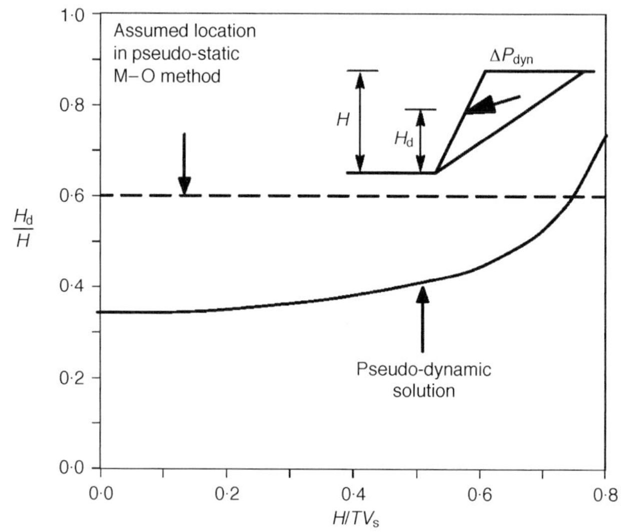

where

$$Q_h(t) = \int_0^H m(z)a(z,t)\,\mathrm{d}z \tag{16.46}$$

The calculation of an equivalent dynamic coefficient of earth pressure, K_{AE}, follows from Equation 16.33. The pseudo-dynamic approach leads to values of $P_{AE}(t)$ that in the limit $V_s \to \infty$ give the pseudo-static value according to the M–O theory. The pseudo-dynamic approach allows the location, H_d, of the dynamic force increment ΔP_{dyn} (the first term in Equation 16.45) to be determined numerically for a range of base motion frequencies. The solution is independent of the soil friction angle, ϕ, and the slope angle, ψ, but is dependent on the shear velocity (soil density and shear modulus) and the period, T_p, of the assumed sinusoidal horizontal acceleration function. The results of the calculations are illustrated in Figure 16.22, and show that, for low-frequency excitation, the point of application is at $H_d = H/3$ above the toe of the soil mass, but will increase at higher frequencies. It appears that the pseudo-static M–O method is reasonable for overturning/base eccentricity design calculations for a wide range of base motion frequencies.

16.3.3 Displacement calculations

As with all limit-equilibrium methods of analysis, pseudo-static approaches cannot explicitly include wall or slope deformations. This is an important shortcoming, since the failure of geosynthetic-reinforced soil walls, in particular, may be manifested as unacceptable movement without structural collapse. The permanent displacement of a geosynthetic-reinforced soil structure due to horizontal sliding/shear mechanisms can be estimated using one of the two general approaches, as described below.

16.3.3.1 Newmark's method and variations

For a given input acceleration time history, Newmark's double-integration method for a sliding mass can be used

Figure 16.23 Calculation of permanent displacements (unidirectional displacement) using Newmark's method

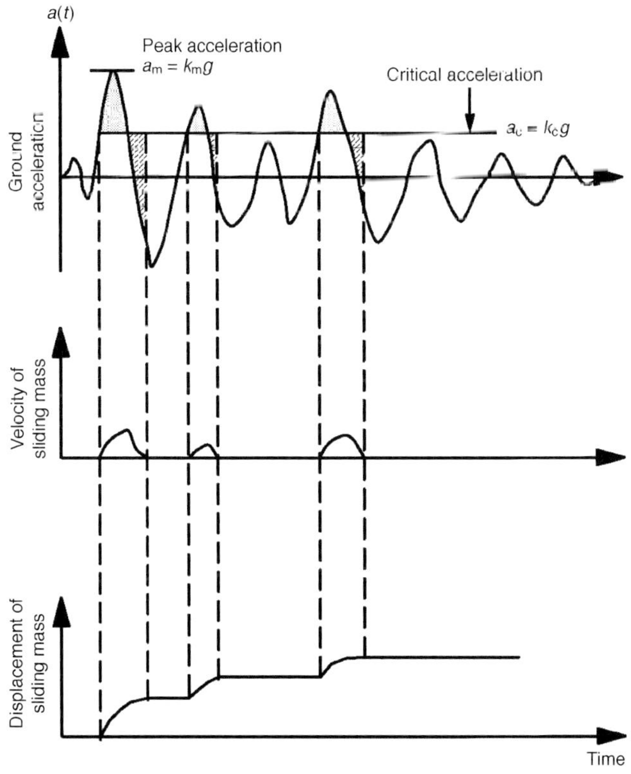

to calculate permanent displacement (Newmark, 1965). According to Newmark's theory, a potential sliding body is treated as a rigid-plastic monolithic mass under the action of seismic forces. Permanent displacement of the mass takes place whenever the seismic force induced on the body (plus the existing static force) overcomes the available resistance along a potential sliding/shear surface. Newmark's method requires that the critical acceleration, k_c, to initiate sliding or shear failure be determined for each translation failure mechanism. The value of k_c can be determined by searching for values of k_h that give a factor of safety of unity in the pseudo-static factor of safety expressions. The critical acceleration is then applied to the horizontal ground acceleration record at the site, and double integration is performed to calculate cumulative displacements, as illustrated in Figure 16.23, where g is the gravitational constant, $a(t)$ is the horizontal ground acceleration function with time t, $a_m = k_m g$ is the peak value of $a(t)$, and $a_c = k_c g$ is the critical horizontal acceleration of the sliding block. For a given ground acceleration time history and a known critical acceleration of the sliding mass, the earthquake-induced displacement is calculated by integrating those portions of the acceleration history that are above the critical acceleration and those portions that are below, until the relative velocity between the sliding mass and the sliding base reduces to zero.

A number of researchers have postulated that the critical acceleration value to initiate slip should be based on the peak shearing resistance of the soil (e.g. ϕ_{peak}), but thereafter residual strength values should be used (e.g. Elms and Richards, 1990; Chugh, 1995). Alternatively, conservative (for design) estimates of seismic-induced displacements should be based on residual strength values if a single value of ϕ is adopted to simplify analyses.

Bathurst *et al.* (2007b) demonstrated a simple displacement-type block model to compute the compression–load–time response of an idealised seismic buffer placed against a rigid wall and used to attenuate earthquake-induced dynamic loads. The seismic buffer is modelled as a linear-elastic material, and the soil wedge shear surface by a stress-dependent linear spring. The model was shown to capture the trends observed in four physical reduced-scale model shaking table tests carried out with similar boundary conditions up to a base excitation level of about $0.7g$ (Bathurst *et al.*, 2007a). In most cases, quantitative predictions were in reasonable agreement with physical test results.

16.3.3.2 Empirical approaches

If the input acceleration data at a site are specified by characteristic parameters such as the peak ground acceleration and the peak ground velocity, then empirical methods that correlate the expected permanent displacement to the characteristic parameters of the earthquake, and a critical acceleration ratio for the structure, are required. Alternatively, if the tolerable permanent displacement of the structure is specified, based on serviceability criteria, the wall can then be designed using an empirical method so that expected permanent displacements do not exceed specified values. Newmark's sliding block theory has been widely used to establish empirical relationships between the expected permanent displacement and characteristic seismic parameters of the input earthquake by integrating existing acceleration records. The critical acceleration ratio, which is the ratio of the critical acceleration, $k_c g$, of the sliding block to the peak horizontal acceleration, $k_m g$, of the earthquake, has been shown to be an important parameter that affects the magnitude of the permanent displacement. Thus, the seismic displacement of a potential sliding soil mass computed using Newmark's theory has been traditionally correlated with the critical acceleration ratio, k_c/k_m, and other representative characteristic seismic parameters, such as the peak ground acceleration, $k_m g$, the peak ground velocity, v_m, and the predominant period, T, of the acceleration spectrum (e.g. Newmark, 1965; Sarma, 1975; Franklin and Chang, 1977).

Cai and Bathurst (1996b) have reformulated a number of existing displacement methods based on non-dimensionalised displacement terms that are common to the methods, and divided them into two separate categories based on the characteristic seismic parameters referenced in each method. Example relationships between the dimensionless displacement term, $d/(v_m^2/k_m g)$, where d is the actual expected permanent displacement, and the critical acceleration ratio, are shown in Figure 16.24. Other curves are available in the literature, but it should be noted that any empirical curve will be influenced by the earthquake data that are

Figure 16.24 Summary of proposed relationships between the non-dimensionalised displacement term and the critical acceleration ratio
Reproduced from Cai and Bathurst (1996b), with permission from Elsevier

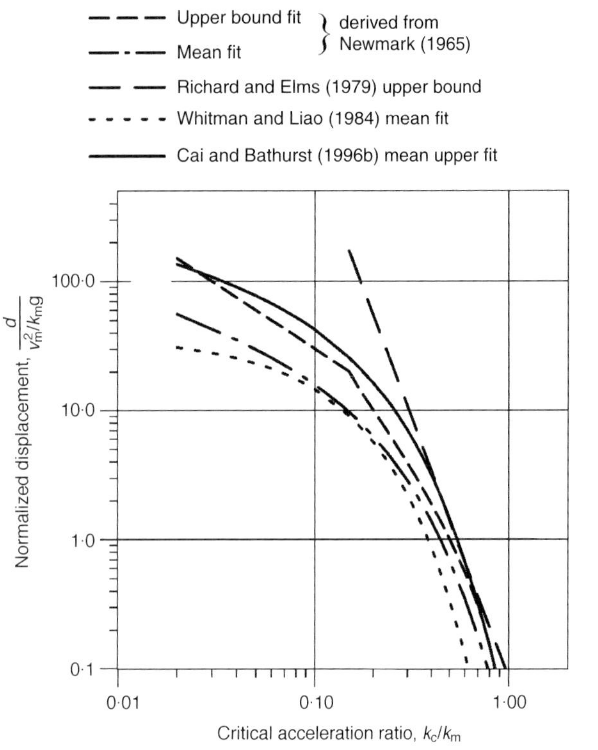

used to establish the curve and the interpretation of the original data.

16.3.3.3 Example applications

Newmark's methods have been applied to unreinforced slopes (Chang *et al.*, 1984). Vrymoed (1989) used Newmark's method to estimate the cumulative base sliding displacement of a rectangular reinforced soil mass for a single cycle of the base acceleration record. Ling *et al.* (1996, 1997) have proposed a method to calculate reinforcement loads and anchorage lengths under horizontal seismic loads using a two-part wedge-sliding block model. Cai and Bathurst (1996a) demonstrated the application of Newmark's method and empirical approaches to geosynthetic-reinforced soil segmental retaining walls. Analyses are restricted to horizontal sliding or shear mechanisms

- external sliding along the base of the total structure, which includes the reinforced soil mass and the facing column (Figure 16.14)
- internal sliding along a reinforcement layer and through the facing (Figure 16.25(a))
- interface shear between facing units with or without the presence of a geosynthetic inclusion (Figure 16.25(b)).

A summary of calculation results for the geosynthetic-reinforced soil wall structure shown in Figure 16.26 is given in Table 16.1, assuming $\phi = 35°$. The material properties for the facing units have been taken from large-scale

Figure 16.25 Newmark's sliding block method applied to geosynthetic-reinforced soil segmental retaining wall structures: (a) internal sliding and (b) facing column shear

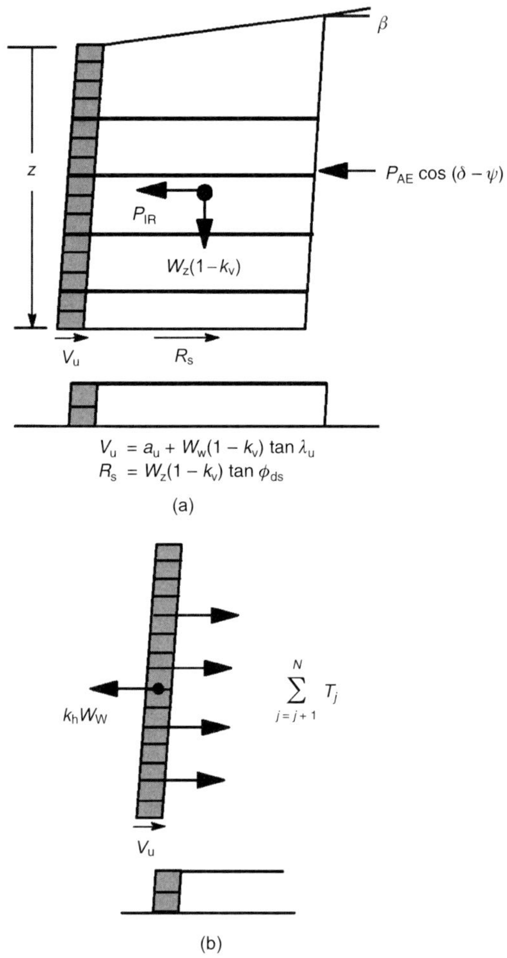

Figure 16.26 Geogrid-reinforced soil segmental retaining wall used in the displacement method example
Reproduced from Cai and Bathurst (1996a), © Canadian Science Publishing or its licensors

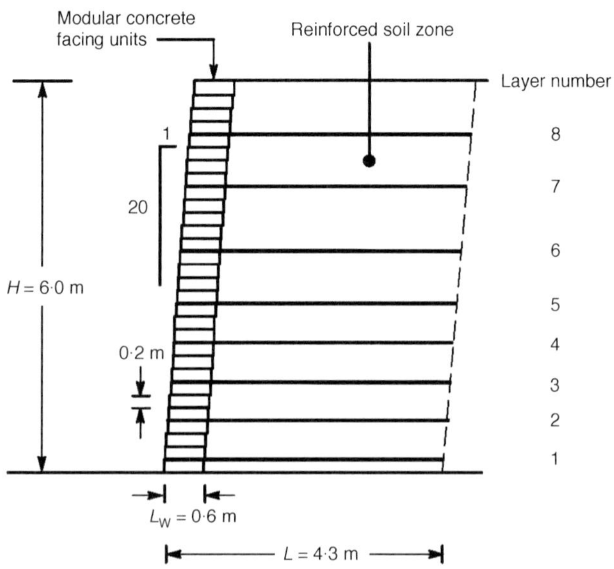

Table 16.1 Total permanent displacement considering all displacement mechanisms

Layer	Displacement: mm	
	Newmark	Empirical
8	154[a]	206[a]
7	47[a]	70[a]
6	29[a]	49[a]
5	25	41
4	25	41
3	25	41
2	24	36
1	21	29
Base sliding	11	15

[a] Controlling mechanism is facing shear; otherwise, internal sliding controls.

laboratory tests carried out at the Royal Military College of Canada (RMCC). The block–geosynthetic interface shear properties (a_u, λ_u) were selected to represent a system with relatively low interface shear capacity in order to generate a worst case set of displacement predictions. The east–west (90°) horizontal ground acceleration component, recorded at Newhall Station (California Strong Motion Instrumentation Program) during the 17 January 1994 Northridge earthquake ($M = 6.7$), was used as the input earthquake data. The record shows a peak horizontal ground acceleration of $k_m = 0.60$. The total permanent displacement at the wall face at each elevation from the initial static position was estimated by adding the layer displacement to the cumulative displacement below that layer. The layer displacement was taken as the larger of the column shear displacement or internal sliding at that layer. The data in Table 16.1 shows that large displacements are possible at the top of the wall using $k_h = k_m = 0.6$ in the pseudo-static seismic stability analysis. This is an extreme loading condition that was used to illustrate the general approach. Similar calculations with a higher-quality fill (i.e. $\phi = 40°$) resulted in displacements that were restricted to the top two facing courses. Furthermore, analyses with better block–geosynthetic properties resulted in insignificant or no displacements at all elevations. This last result is consistent with observations made at the site of two segmental retaining walls after the Northridge earthquake that showed no detectable shear movement of the facing column units despite significant horizontal ground accelerations that were estimated to be as high as $0.5g$ (Bathurst and Cai, 1995). The table illustrates that the order of magnitude accuracy of the empirical method (compared with Newmark's method) is satisfied for all large displacement results. Predicted displacements must be viewed as order-of-magnitude estimates rather than accurate predictions. Engineering judgement plays an important role in the interpretation of results using any empirical approach.

You and Michalowski (1999) maintained that the rotational failure mechanism is more crucial for slopes than walls. They used Newmark's sliding block method to calculate the displacement of a rotating block of soil mass with a log spiral failure mechanism subjected to horizontal excitation. Their proposed analysis approach included the following steps

(*a*) determine the critical acceleration for a given slope
(*b*) calculate a coefficient that is related to the slope failure mechanism
(*c*) integrate the earthquake acceleration record.

The product of the calculated coefficient and double integral value above the critical threshold yields the horizontal displacement of the slope at the toe. They presented the calculated total displacement results for their model slopes subjected to a number of selected ground motion records in the form of design charts. The charts included failure mechanisms at the toe and through the foundation. You and Michalowski concluded that the through-foundation failure mode is more critical for slopes with shallow face angles and low-friction angles.

Matasovic *et al.* (1997) modified Newmark's sliding block method to account for the cyclic degradation of the soil shear strength. The work was focused on sliding systems comprising geosynthetic interfaces; however, the implications of the work to reinforced wall systems are clear. The constant yield acceleration assumption in the conventional Newmark method was replaced by a variable-yield acceleration approach in the modified method. They concluded that classical Newmark analysis with the yield acceleration that is based on soil residual shear strength is conservative. They argued that a degrading strength model from peak to residual strength is more reasonable.

Ling *et al.* (1997) proposed a displacement calculation analysis based on the direct sliding of a two-part wedge mass. They assumed that ground acceleration in the half cycle away from the backfill does not affect the magnitude of permanent displacement. The stiffness and damping terms in the equation of motion of the mass block were not included in their analysis. They concluded that a tolerable permanent displacement can be selected using their proposed charts and a given seismic coefficient for cases with limited construction space, or where the geosynthetic reinforcement length required to resist direct sliding becomes excessively long.

16.3.4 Dynamic analysis using numerical techniques

In this section, numerical modelling techniques based on finite-element and finite-difference methods are reviewed. In addition, selected results from numerical studies, as they relate to the analysis, design and performance of reinforced soil walls and slopes, are highlighted. A survey of numerical methods adopted for stability analysis and design of reinforced soil walls, with a brief review of modelling approaches, can also be found in the paper by Otani *et al.* (1997).

16.3.4.1 Finite-element method

The attraction of properly formulated finite-element methods is that they can implement complex models for the component materials, such as the non-linear cyclic behaviour of the soil, and reinforcement materials using models such as those described in Section 16.2.

Reinforced slopes

The dynamic response of reinforced and unreinforced soil slope models with $c - \phi$ properties resting on a firm foundation was determined by the senior author and his co-workers using a modified version of the TARA-3 program (Finn *et al.*, 1986). The slopes were 12 m high with a side slope of 1:1 (Yogendrakumar *et al.*, 1991). One slope was lightly reinforced with 12 m-long polymeric reinforcement layers with a vertical spacing of 2 m. The finite-element representation of the reinforced soil slope is shown in Figure 16.27(a). The reinforcement was modelled using the non-linear quadratic equation with the linear (non-hysteretic) unload–reloading behaviour described in Section 16.2.2.2. The slope was subjected to the first 9.6 seconds of the north–south component of the 1940 El Centro earthquake scaled to 0.2*g*. The base was assumed to be rigid, and the nodes on the left and right vertical boundary were supported on horizontal rollers for the dynamic analysis. A static analysis was first conducted to establish the stress–strain field prior to the earthquake excitation. The program simulated the incremental construction process of the slope. The results of the analyses showed that the time history of the slope displacement was not significantly influenced by the presence of the reinforcement (Figure 16.27(b)) for the duration of shaking applied. However, it must be noted that a perfect bond was assumed between the reinforcement and the soil (which is consistent with $C_i = 1$ for many geogrid reinforcement products in cohesionless soils). These results suggest that reinforcement may not reduce seismic-induced displacements of slopes that do not require reinforcement to prevent collapse. This particular result deserves further investigation.

Figure 16.27 Dynamic finite-element analysis of a reinforced soil slope: (a) the finite-element representation of the reinforced soil slope and (b) the time history of horizontal displacement at the slope crest
Reproduced from Yogendrakumar *et al.* (1991)

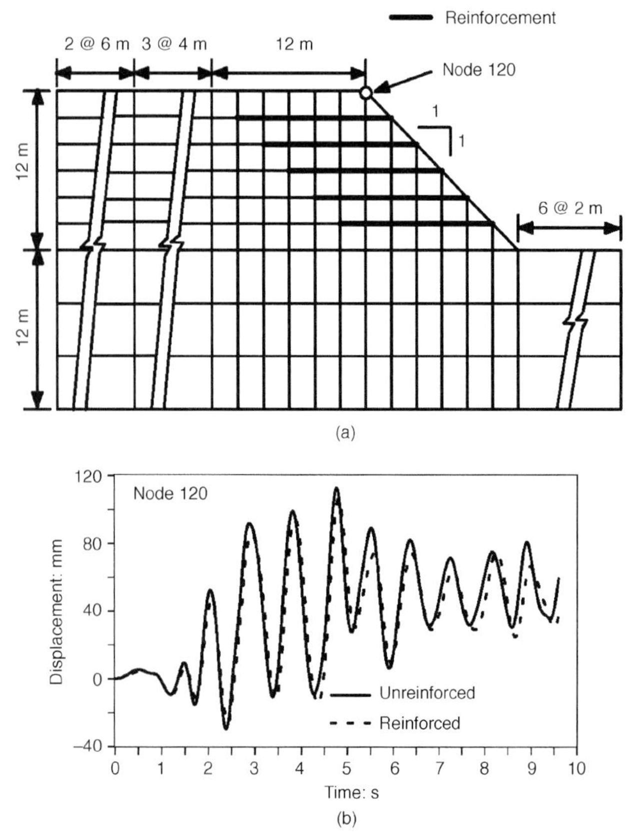

Lin *et al.* (1996) analysed the seismic response of a 15 m-high slope using the finite-element program FLUSH that implements the equivalent linear method. They examined the influence of fixed versus transmitting truncated backfill boundaries on the magnitude of slope displacement and the acceleration response. They concluded that the fixed-boundary condition would result in a larger response in the slope than the transmitting truncated boundary. However, they found that the fundamental frequency of the slope model remained unchanged between the models with different truncated boundary conditions. The influence of the treatment of model boundaries on the results of dynamic numerical modelling is discussed further in Section 16.3.4.2.

Reinforced soil walls

Finite-element modelling has been used to gain insight into the behaviour of geosynthetic-reinforced soil walls under static loading conditions (Wu, 1992; Rowe and Ho, 1993; Karpurapu and Bathurst, 1995).

The use of dynamic finite-element modelling for reinforced earth structures is much more limited. Segrestin and Bastick (1988) and Yogendrakumar *et al.* (1991) used the programs SUPERFLUSH and TARA-3, respectively, to study the seismic response of reinforced soil walls that used inextensible reinforcement (steel strips). Bachus *et al.* (1993) used the program DYN3D to simulate the blast response of geosynthetic-reinforced soil walls constructed with incremental concrete facing panels.

The results of the finite-element parametric analyses of RECO systems (Segrestin and Bastick, 1988) have been the principal source of analysis and design guidelines for reinforced soil walls with inextensible (steel strip) reinforcement. Because of the lack of similar parametric data for extensible reinforced structures, the data for simulated RECO walls have been adopted by the FHWA and AASHTO in the USA for the design of geosynthetic-reinforced soil walls.

Cai and Bathurst (1995) carried out dynamic finite-element modelling of geosynthetic-reinforced segmental retaining walls in order to investigate the load–deformation response of an example system under simulated earthquake loads. The modified TARA-3 code mentioned in the previous section was used together with the hysteretic soil and reinforcement models described in Sections 16.2.1.2 and 16.2.2.2. The results of large-scale interface shear and connection tests

Figure 16.28 Facing column lateral displacement at the end of excitation history
Reproduced from Cai and Bathurst (1995), Industrial Fabrics Association International

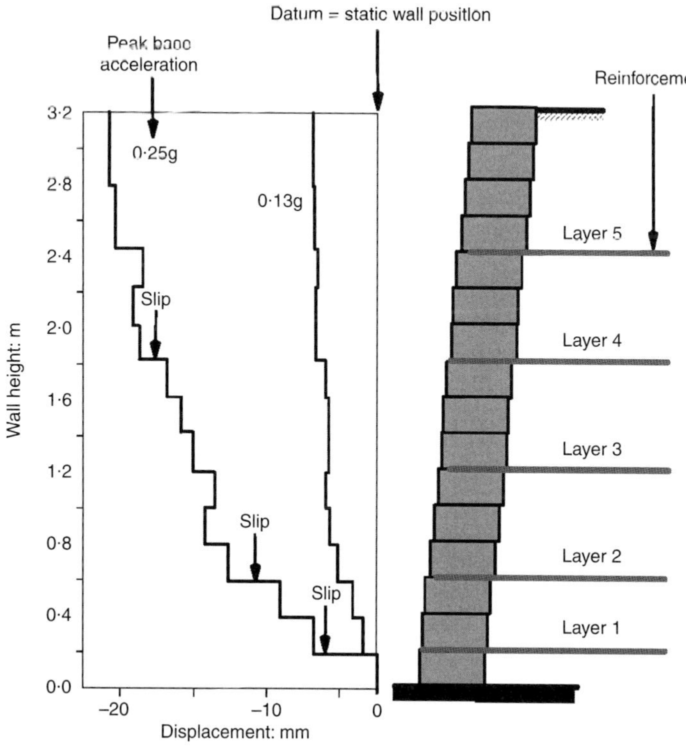

were used to provide parameters for the modelling of the facing column. The interface shear capacities that were used are considered to be relatively poor for segmental retaining wall systems, based on a large amount of test data gathered by the first author. The scaled 1940 El Centro earthquake record was used as the reference base acceleration time history. Spectrum analysis of the input acceleration record gives a dominant frequency range of 0.5–2 Hz. Predicted cumulative lateral deformations through the height of the facing column at the end of two scaled base input records are illustrated in Figure 16.28. The relative displacements are largest at the reinforcement elevations where the locally greatest interface shear loading occurred. While the potential for interface shear leading to collapse of these structures is clear, it is worth noting that the vertical out-of-alignment displacement is less than 1% of the height of the wall. In practice, this amount of relative displacement is within the limits usually achieved during construction (Bathurst *et al.*, 1995, 2010) and, hence, from practical considerations it may be judged to be insignificant. The results suggest that for the range of peak accelerations and the duration of excitation applied to this low wall height, the structure performed well, despite relatively poor interface shear characteristics. An important observation made by the authors was that reinforcement forces predicted by the finite-element model were consistently lower than those computed using the pseudo-static M–O approach, as illustrated in Figure 16.29. This result is consistent with the opinion of many practitioners that the M–O theory is conservative for typical soil-retaining wall structures. In addition, for this low wall height, the maximum incremental dynamic reinforcement forces were observed towards the top of the wall, which is consistent with the pseudo-static model

Figure 16.29 Distribution of peak reinforcement forces
Reproduced from Cai and Bathurst (1995), Industrial Fabrics Association International

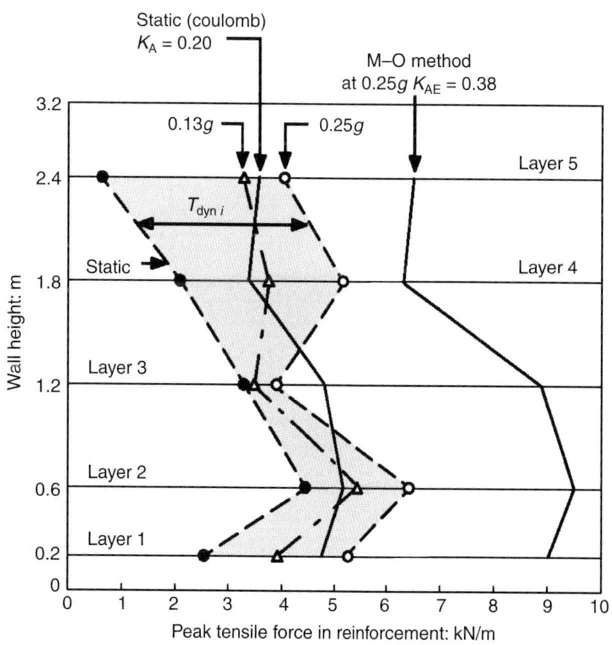

proposed by Bathurst and Cai (1995) for segmental retaining walls with extensible reinforcements. Finally, the data in Figure 16.29 show that reinforcement loads were low even under seismic shaking, and likely to be well within the expected reinforcement load limits.

16.3.4.2 Finite-difference method

The seismic response of reinforced walls and slopes can be analysed using explicit dynamic finite-difference methods implemented in computer codes, such as FLAC (Itasca Consulting Group, 2005). FLAC (Fast Lagrangian Analysis of Continua) is based on the Lagrangian calculation scheme that is well suited for modelling large distortions and material collapse. Complete descriptions of the numerical formulation are reported by Cundall and Board (1988). Several built-in constitutive models are available in the FLAC package (e.g. the equivalent linear method – ELM), and can be easily modified by the user. For example, geosynthetic reinforcement layers can be represented as either cable, beam or pile elements. One advantage of using FLAC in seismic analysis is the simplicity of applying seismic loading anywhere within the problem domain.

Reinforced soil slopes

Examples of preliminary FLAC analyses for reinforced soil slopes, in which the reinforcement layers were modelled using cable elements, are shown in Figures 16.30 and 16.31. The duration of base shaking for the slope in Figure 16.31 was 2.5 seconds, with a horizontal sinusoidal base acceleration having a peak amplitude of 0.6*g* and a frequency of 2 Hz. The results of preliminary analyses by the authors using the slope in Figure 16.30 show that the behaviour of the slope is very dependent on the stiffness of the foundation materials (soil or rock). For example, the effectiveness of reinforced soil masses to minimise cumulative lateral displacements during horizontal ground shaking increases with decreasing depth to the bedrock.

Reinforced soil walls

Bathurst and Hatami (1998a,b) used the FLAC program to investigate the influence of reinforcement design (i.e. length and stiffness) and the toe restraint condition on the dynamic response of geosynthetic-reinforced soil retaining walls with a propped-panel facing to a variable-amplitude harmonic base acceleration (Figure 16.32). In order to reduce the computation time and to simplify the interpretation of numerical results in all dynamic finite-difference modelling by the writers, the reinforcement was modelled using linear-elastic (FLAC) cable elements (i.e. J = constant). Furthermore,

Figure 16.30 Example of the FLAC analysis of a reinforced soil slope: (a) base input velocity and crest response and (b) the deformed slope

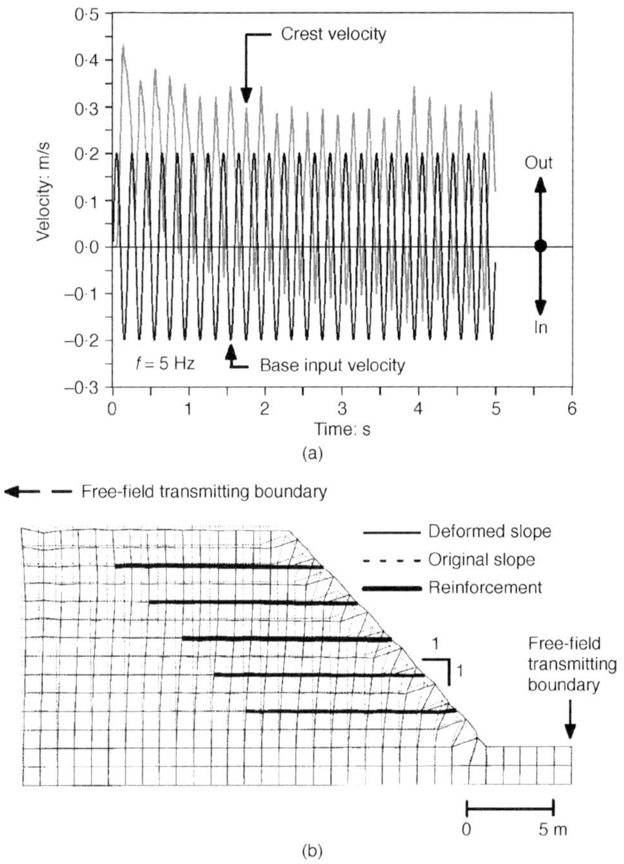

Figure 16.31 Example of the FLAC analysis of a wrapped-face reinforced soil slope
From Kramer (1996b)

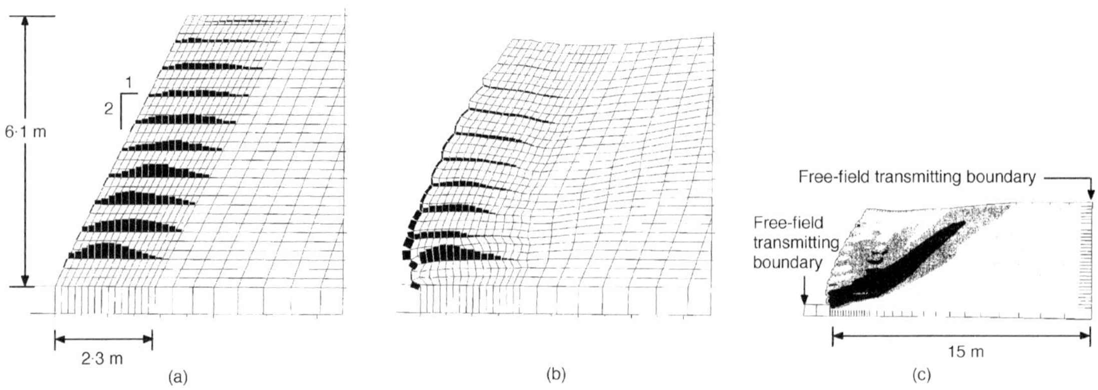

Figure 16.32 FLAC simulation of a propped panel wall under base excitation: (a) the numerical grid and (b) variable-amplitude harmonic base acceleration
Reproduced from Bathurst and Hatami (1998b), Industrial Fabrics Association International

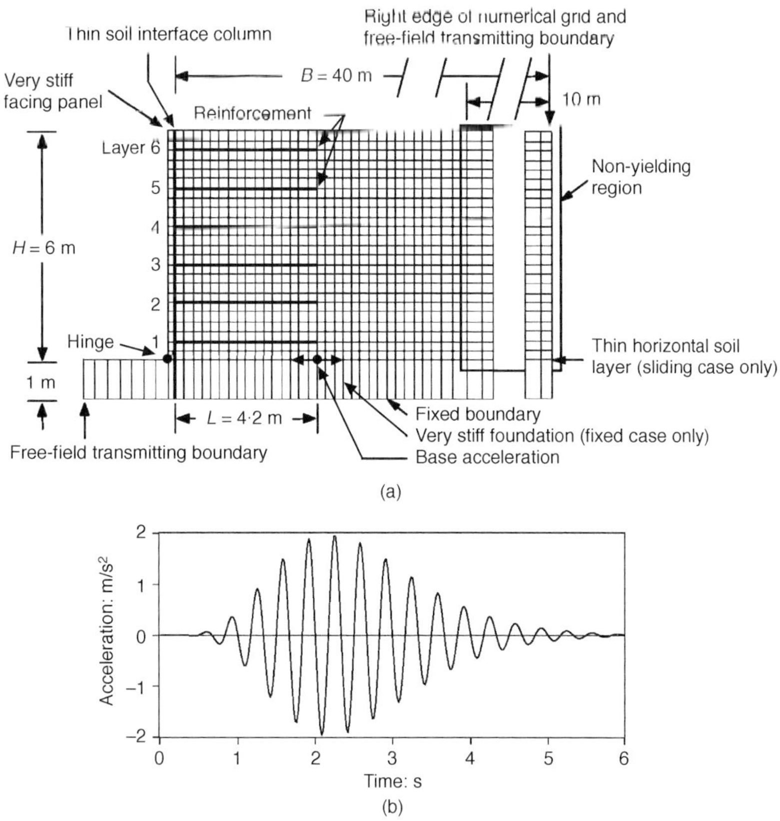

the reinforcement was assigned a high tensile strength value and was fully bonded to the backfill so that the possibility of reinforcement rupture or pull-out was avoided.

The frequency of the input acceleration ($f_g = 3$ Hz) was chosen close to the fundamental frequency of the reference model wall ($f_1 = 3.4$ Hz), to induce significant response magnitude. Bathurst and Hatami found that the facing panel dynamic displacement amplitudes were relatively small compared with the permanent outward displacement at the end of shaking. Wall models with stiffer or longer reinforcement layers developed smaller facing lateral displacement. However, the wall toe restraint condition showed the largest influence on wall lateral displacement. Bathurst and Hatami found that the wall displacement and the reinforcement load gradually accumulated with time. These results are in qualitative agreement with similar results reported by Cai and Bathurst (1995), who used a dynamic finite-element code (see Section 16.3.4.1). The reinforcement load was typically greatest at the reinforcement-facing connections. The wall models with a sliding toe (i.e. a pinned toe with a horizontal degree of freedom) developed higher reinforcement loads. The maximum reinforcement incremental load distribution along the wall height was almost linear for the sliding toe case. However, it was practically uniform for the fixed-toe condition, which is different from the distribution suggested by the AASHTO (2010) (Figure 16.33). The reinforcement load distribution with height was significantly more sensitive to reinforcement stiffness than reinforcement length. They found that reinforcement load and dynamic reinforcement load amplitude increased with reinforcement stiffness, especially in lower reinforcement layers.

Bathurst and Hatami (1998c) suggested that a bilinear load distribution over the wall height would better represent the variation of reinforcement incremental load with height compared with the linear trend recommended by the AASHTO, which is more appropriate for stiffer metallic reinforcement types.

The geometry of the failed mass was a combined two-part wedge (Figures 16.34(a) and 16.34(b)), which is similar to the observed failure geometry for walls with a similar reinforcement-to-height ratio in shaking table studies reported by Tatsuoka *et al.* (1998) (Figure 16.34(c)). The top wedge in the numerical modelling cited here extended beyond the reinforced zone at an angle that was consistent with the predicted value from the M–O theory considering acceleration amplification over the depth of backfill.

Bathurst and Hatami (1998b) carried out two groups of parametric analyses on physical and numerical model

Figure 16.33 Influence of the reinforcement stiffness, J, reinforcement length, L, and base condition on the reinforcement dynamic load increment, ΔT: (a) $L/H = 0.7$, fixed-base condition, (b) $L/H = 1$, fixed-base condition, (c) $L/H = 0.7$, sliding-base condition and (d) $L/H = 1$, sliding-base condition

Reproduced from Bathurst and Hatami (1998b), Industrial Fabrics Association International

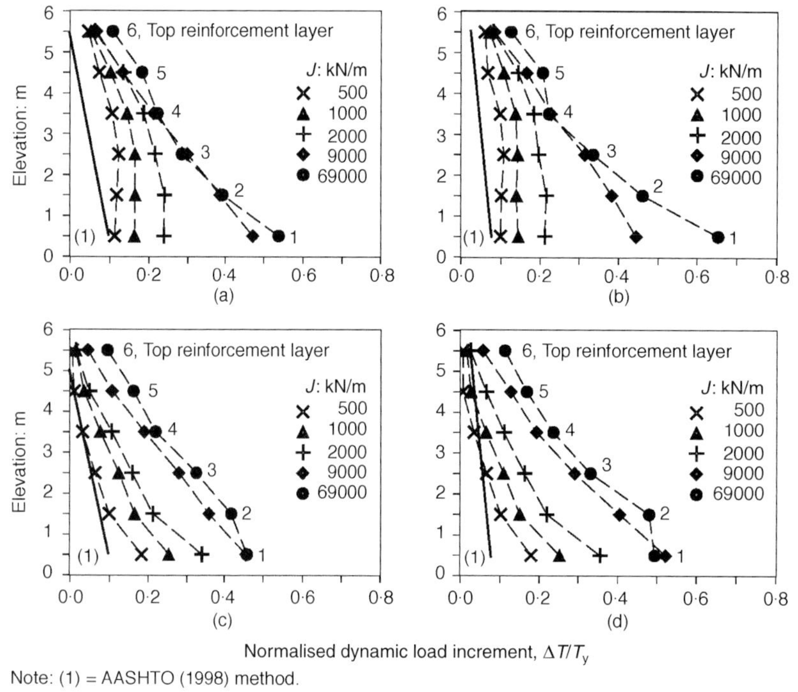

parameters. In the parametric analyses on physical parameters, the reinforcement stiffness values ranged from very stiff geogrids to metallic reinforcement. Bathurst and Hatami compared the reinforcement load distribution behind the facing with the distribution predicted from the Coulomb and Rankine earth pressure theories. They found that the load distributions for geosynthetic reinforcement materials in the lower stiffness range (e.g. $J < 2000$ kN/m) were essentially uniform over the height of the wall with the fixed-toe condition, and deviated from the linear distribution predicted from the two earth pressure theories (Figure 16.35). The effect of reinforcement stiffness on the load distribution behind the wall has an important implication for the pseudo-static seismic design of geosynthetic-reinforced soil walls. The dynamic load distribution behind metallic-reinforced soil walls is triangular, whereas it is essentially uniform for the case of less stiff, polymeric reinforcement. The load distribution determines the local failure mode of the facing in segmental retaining walls. Bathurst and Hatami found that the influence of reinforcement stiffness and length on the wall response was larger for the fixed-toe case compared with a toe that was pinned but free to slide laterally. They also found that acceleration amplification in the backfill was slightly greater for the fixed-toe condition compared with the case where the toe was free to slide.

The numerical model parameters investigated included the backfill far-end boundary condition, backfill width and viscous damping ratio. Bathurst and Hatami found that the influence of backfill width on the calculated response of the wall was significant for both toe restraint conditions when $B/H < 5$, where B is the width of the numerical grid and H is the height of the wall facing.

Bathurst and Hatami (1999a) examined the change in elevation of the reinforcement load resultant with reinforcement stiffness under static (end of construction) and seismic (end of input ground motion) loading conditions using numerical simulation. Figure 16.36 shows that the resultant reinforcement load elevation under both static and dynamic loading conditions is generally less for higher reinforcement stiffness values. Furthermore, for static load conditions and a given reinforcement stiffness, the normalised elevation of the load resultant, m_s, is lower in taller wall models. An important implication of the trend in the data in Figure 16.36(a) to limit-equilibrium-based design of walls under static loading is that the assumption of a triangular load distribution may be most applicable for very stiff reinforcement systems (i.e. steel-strip-reinforced walls), and may not be applicable for extensible reinforcement systems (i.e. geosynthetic-reinforced soil walls). The curves in Figure 16.36(b) show that the normalised elevation of the load resultant during

Figure 16.34 Failure mechanisms observed in physical and numerical models illustrating the development of a two-part wedge: (a) the FLAC model, fixed toe, (b) the FLAC model, sliding toe (dark shading indicates relatively large shear strains) and (c) the reduced-scale shaking table test
Reproduced from (b) Bathurst and Hatami (1998b) and (c) Tatsuoka *et al.* (1998), Industrial Fabrics Association International

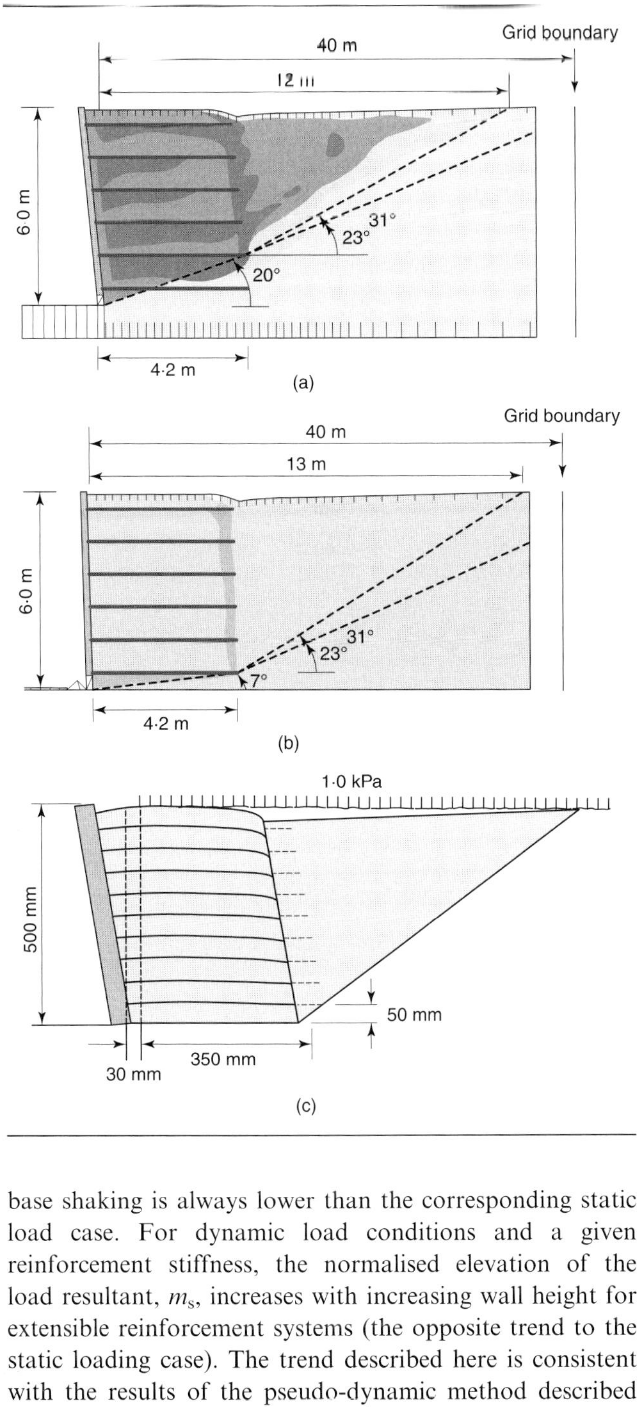

base shaking is always lower than the corresponding static load case. For dynamic load conditions and a given reinforcement stiffness, the normalised elevation of the load resultant, m_s, increases with increasing wall height for extensible reinforcement systems (the opposite trend to the static loading case). The trend described here is consistent with the results of the pseudo-dynamic method described in Section 16.3.2.

It should be noted that the value of m_d for the dynamic case in Figure 16.36(b) is based on total reinforcement loads recorded along each reinforcement layer during base shaking. The peak loads in different reinforcement layers are not necessarily time coincident. The numerical results discussed here have potential implications for conventional, limit-equilibrium seismic design of reinforced soil walls with propped panel wall facings. The AASHTO (2010) and FHWA (2009) guidelines recommend that a trapezoidal distribution is assumed for the total dynamic pressure distribution in reinforced retaining wall design. This distribution is used to partition apparent dynamic earth pressures to individual reinforcement layers using a tributary area approach. The method limits m_d to $0.33 \leq m_d \leq 0.6$. The data in Figure 16.36(b) fall within this range, but illustrate how the selection of m_d may be refined to consider the effect of wall height and reinforcement stiffness.

Influence of model boundaries and damping ratio

Bathurst and Hatami (1999b), and Hatami and Bathurst (1999b), discussed the numerical modelling aspects of the response to harmonic ground motion of reinforced soil walls on rigid foundations. They concluded that the predicted seismic response of reinforced soil walls using the FLAC program is strongly influenced by the width and damping ratio of the backfill model, as well as the type of far-end boundary condition adopted for the backfill. The use of excessively narrow retained soil models to reduce the numerical grid size and the computation time could result in an underestimation of the wall displacement and the reinforcement loads. The simulation cases with the free-field boundary condition option in the FLAC program resulted in larger displacement values and reinforcement forces than cases with a fixed far-end boundary and a forced (prescribed uniform acceleration) far-end boundary when a loading frequency below the fundamental frequency of the reinforced soil wall models was applied. Hatami and Bathurst (1999b) concluded that adopting a larger damping ratio in the model did not change the qualitative response of the retaining walls. However, lower values for the wall lateral displacement and reinforcement load were obtained when larger viscous damping ratios were used in the model. It appears that greater viscous-type damping ratios will reduce wall response in addition to response reduction effects due to soil plasticity.

Influence of the fundamental frequency on geosynthetic-reinforced soil walls

Madabhushi (1996) pointed out that the fundamental frequency of a retaining wall structure could be an important parameter determining the magnitude of the wall response. He proposed that the fundamental frequency of reinforced wall systems would increase with reinforcement stiffness.

Bathurst and Hatami (1998c) used the FLAC program to investigate the influence of the reinforcement stiffness on the reinforced soil wall response. They found that the reinforced soil wall frequency response was practically independent of reinforcement stiffness, but that the wall response magnitude was measurably influenced by the reinforcement stiffness.

The investigation of fundamental frequency effects was extended by Hatami and Bathurst (1999a, 2000), and Hatami *et al.* (2005), to include the influence of the wall

Figure 16.35 Influence of the reinforcement stiffness, J, reinforcement length, L, and base condition on the maximum load in each reinforcement layer: (a) $L/H = 0.7$, fixed-base condition, (b) $L/H = 1$, fixed-base condition, (c) $L/H = 0.7$, sliding-base condition and (d) $L/H = 1$, sliding-base condition (note: T_{max} is the maximum tensile load recorded along the entire length of the reinforcement layer, and $T_y = 200$ kN/m is the yield strength of the reinforcement)

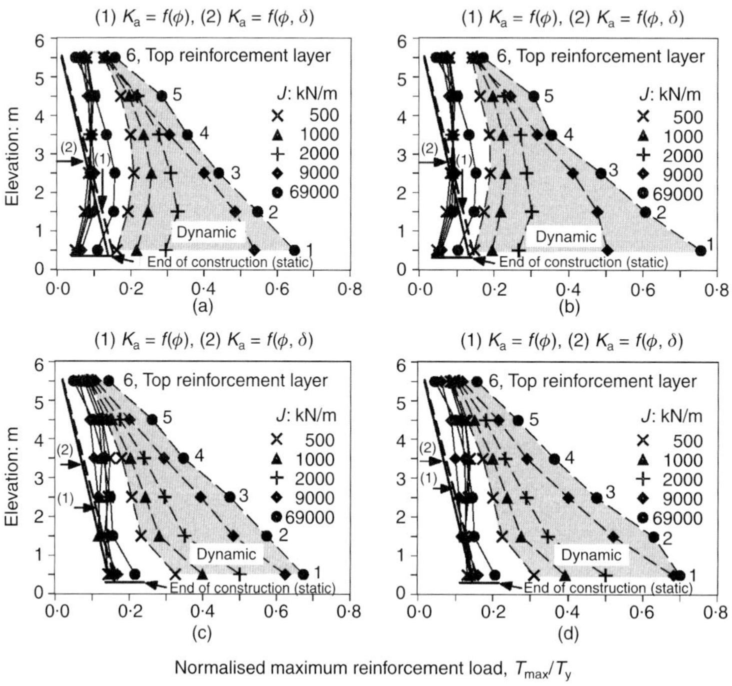

Figure 16.36 Influence of reinforcement stiffness, J, and spacing, S_v, on the location of the reinforcement resultant: (a) end of construction (static) and (b) during shaking (dynamic)
Reproduced from Bathurst and Hatami (1999a), Industrial Fabrics Association International

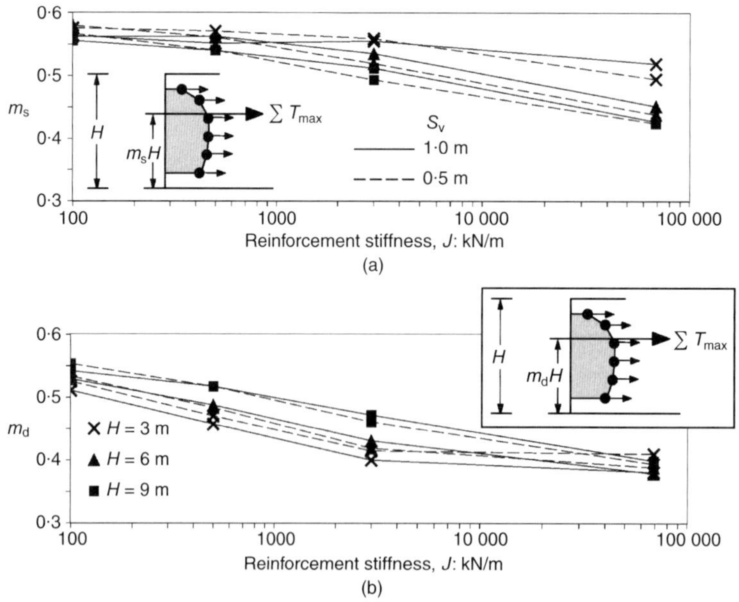

height, reinforcement stiffness and length, and the toe restraint condition, on the wall fundamental frequency. They found that the fundamental frequency of a reinforced soil wall of a given height and backfill width can be estimated with reasonable accuracy from linear-elastic theory in spite of evidence of plasticity and the non-linear response of backfill material. However, the wall resonance frequency with a strong ground motion (e.g. peak ground acceleration (PGA) $\geq 0.2g$) can be noticeably lower than its (small-amplitude) fundamental frequency. Possible reasons for this phenomenon are the reduction in the soil modulus and the increase in damping with an increase in the soil shear strain. Hatami and Bathurst concluded that reinforcement design (i.e. length and stiffness) has little influence on the predicted wall fundamental frequency.

Hatami and Bathurst (2000) proposed that the seismic response of retaining walls of typical heights (e.g. $H < 10$ m) is dominated by their fundamental frequency. They summarised the available theoretical solutions for predicting the fundamental frequency of retaining walls in the form

$$f_{11} = \mathrm{GF} \times f_1 \tag{16.47}$$

where f_{11} denotes the fundamental frequency predicted from two-dimensional solutions and f_1 is the fundamental frequency of an infinitely long soil layer of uniform height, H, and shear wave speed, V_s:

$$f_1 = \frac{V_s}{4H} \tag{16.48}$$

$\mathrm{GF} = f(v, B/H)$ is a geometric factor that represents the modification of f_1 to obtain the fundamental frequency of a two-dimensional retaining wall model from the one-dimensional frequency formula for an infinitely long uniform soil layer. Hatami and Bathurst compared the fundamental frequency values from theoretical solutions to the values inferred from the numerical simulation of reinforced soil retaining wall models subjected to harmonic base excitation. They carried out parametric seismic analyses to investigate the influence of different structural components on the retaining wall fundamental frequency. The structural components included the reinforcement stiffness and length, the restraining condition at the toe (footing) of the facing panel, and the friction angle of the granular backfill soil. The geometry parameters included the wall height and the backfill width. The intensity of the ground motion, characterised by peak ground acceleration, was also varied.

The results of the analyses showed that Equation 16.47 provided a reasonable estimate for the fundamental frequency of reinforced soil-retaining wall systems with wide uniform backfill subjected to moderately strong ground motion (e.g. $\alpha_g = 0.2g$ in their study). Among the two-dimensional approaches examined, the frequency formula proposed by Wu and Finn (1996, 1999) gave the closest agreement to the fundamental frequency value inferred from numerical

Figure 16.37 Influence of the frequency of the harmonic base input record on the wall response: (a) comparison of the fundamental frequency value from numerical simulation and the predicted value from the theoretical solution (after Bathurst and Hatami, 1998c) and (b) the influence of the backfill width-to-height ratio on the wall fundamental frequency
Reproduced from Hatami and Bathurst (2000), with permission from Elsevier

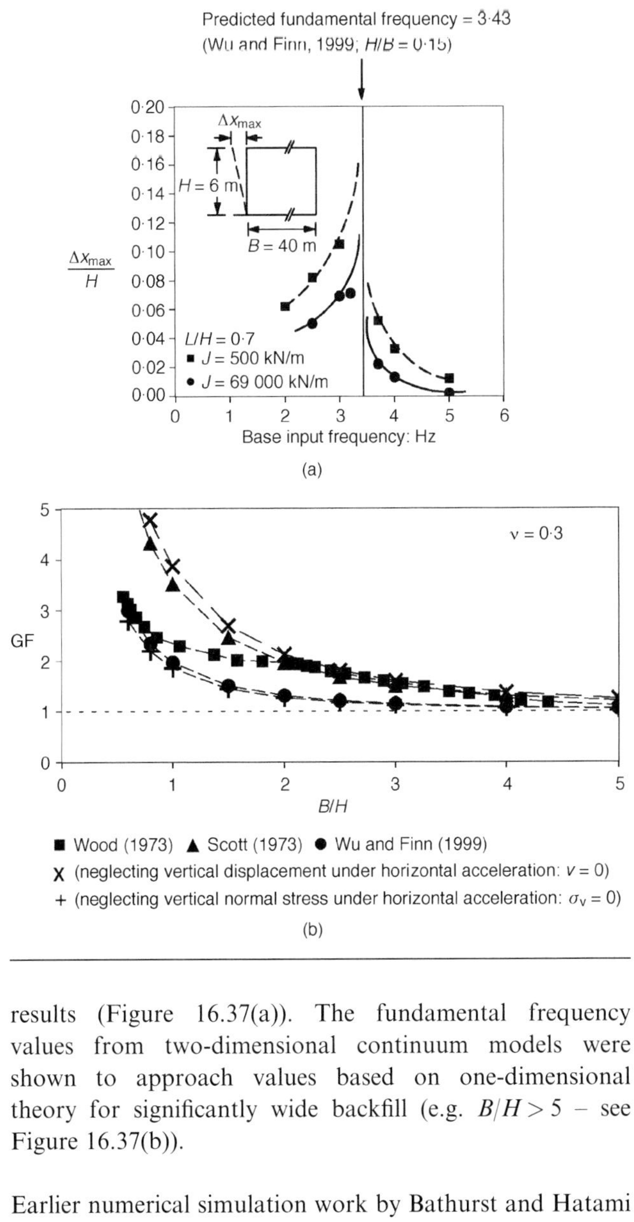

results (Figure 16.37(a)). The fundamental frequency values from two-dimensional continuum models were shown to approach values based on one-dimensional theory for significantly wide backfill (e.g. $B/H > 5$ – see Figure 16.37(b)).

Earlier numerical simulation work by Bathurst and Hatami (1998b) had demonstrated that the reinforcement stiffness, reinforcement length and toe restraint condition could have a significant influence on the magnitude of the reinforcement forces and wall displacements of reinforced soil wall models during a simulated seismic event. However, the results of the study by Hatami and Bathurst (2000) using the same numerical models demonstrated that these variables did not significantly affect the fundamental frequency of reinforced soil wall models with a wide range of structural component values. Hatami and Bathurst found the numerical results of model wall fundamental frequency to be less sensitive to the backfill width compared with theoretical closed-form predictions. They attributed the reason for the reduced effect of the backfill width partly to the soil plasticity in the near field behind the facing panel, which would reduce the geometrical effect of a purely elastic backfill on wall response.

Hatami *et al.* (2005) examined the acceleration amplification behind reinforced soil-retaining walls with a hinged propped panel face with three different heights ($H = 3$, 6 and 9 m) using a FLAC numerical model and two different reference ground motions. It was found that the magnitude of the acceleration amplification generally increased with the elevation above the foundation in the backfill of each wall model and increased with

- increasing wall height for a given input-to-structure frequency ratio (f_ω/f_{11})
- a decreasing value of $|f_\omega - f_{11}|$ (i.e. closer proximity of the input acceleration predominant frequency (f_ω) and the fundamental frequency of the structure (f_{11}))
- decreasing ground motion intensity (for walls taller than 3 m)
- a decreasing backfill damping ratio value.

16.4. Physical testing of model walls and slopes

Model tests for seismic studies typically fall into two categories

- reduced-scale shaking table tests
- centrifuge tests subjected to base shaking.

Both shaking table and centrifuge model tests share certain drawbacks, among the most recognised of which are similitude and boundary effects.

16.4.1 Gravity (1*g*) shaking and tilt table tests

The advantage of reduced-scale shaking table tests is that they are relatively easy to perform. The principal disadvantages are related to problems of similitude between reduced-scale models and equivalent prototype-scale systems (Fairless, 1989). Similitude rules have been proposed by Sugimoto *et al.* (1994) and Telekes *et al.* (1994). Of particular concern is the difficulty of 1*g* models to scale non-linear soil strength and stress–strain properties that vary with confining pressure. An important consequence of these difficulties is that failure mechanisms observed in reduced-scale models may be different from those observed at the prototype scale.

Large-scale 1*g* physical models (2.8 m-high modular block faced) have been reported by Ling *et al.* (2005b), and these are attractive candidates for numerical modelling. Nevertheless, despite complex constitutive models for the component materials, the accuracy of predictions was not always satisfactory (Lee *et al.*, 2010; Ling *et al.*, 2010). Another important factor not discussed for these physical tests is the truncated soil volume extending behind the facing units (i.e. $B/H = 1.3$, where B is the toe to far-field boundary and H is the height of the structure). Inertial effects during shaking can be expected to increase with the size of the retained soil volume, as demonstrated in parametric analyses reported by Bathurst and Hatami (1998b).

Nevertheless, the summary of investigations given in Table 16.2 identifies important performance features of reinforced soil structures under dynamic loading. Most investigators have noted amplification of the base input acceleration over the height of structures, particularly at the top of the structures. These observations give support to design methodologies that either incorporate empirical acceleration profiles directly (Steedman and Zeng, 1990) or indirectly (Bathurst and Cai, 1995), and lead to the requirement, in some cases, to increase the number and length of reinforcement layers close to the top of reinforced wall structures based on limit-equilibrium design.

Bathurst *et al.* (1996) and Pelletier (1996) have reported the results of a series of shaking table tests that examined the seismic resistance of model-reinforced segmental retaining walls. The tests were focused on the influence of interface shear properties on facing column stability, which was identified as an important design consideration based on pseudo-static methods of analysis (Bathurst and Cai, 1995). A set of 1/6-scale model walls were constructed inside a Plexiglas box, and were 2400 mm long by 1400 mm wide by 1020 mm high. Similitude rules proposed by Iai (1989) were used to scale the model components and geometry. A typical test configuration is illustrated in Figure 16.38. The models were constructed with concrete blocks 100 mm wide (toe to heel) by 160 mm wide by 34 mm high. Five layers of a weak geogrid (HDPE bird fencing) were used to model the reinforcement. The backfill was standard laboratory silica #40 sand prepared at a relative density of 67%. The four test configurations used are summarised in Table 16.3. The differences between the tests are related to the interface shear capacity and the wall batter. Interfaces identified as frictional in Table 16.3 derive their shear capacity solely from sliding resistance at the interface. These interfaces represent a very poor facing column detail with respect to shear capacity. In two of the tests, the interfaces were fixed at some locations in order to simulate systems with high shear capacity at all or selected facing column interfaces (i.e. positive interlock due to effective shear keys, pins or other types of connectors). Each test was subjected to a staged increase in the base input motion, resulting in the acceleration–time record shown in Figure 16.39. The base input frequency was kept constant at 5 Hz. At the prototype scale, this frequency corresponds to 2 Hz.

The influence of the interface shear capacity and the facing batter can be seen in Figure 16.40. The vertical wall with fixed interface construction (high shear capacity at each interface) required the greatest input acceleration to generate large wall displacements during staged shaking (test 4). The vertical wall with poor interface shear at all facing unit elevations performed worst (test 1). However, the resistance to wall displacement was improved greatly for the weakest interface condition by simply increasing the wall batter (test 3). The vertical wall with poor interface properties

Table 16.2 Shaking table studies on geosynthetic-reinforced soil walls

Reference	Model details	Observed behaviour and implications for design and analysis
Koga *et al.* (1988), Koga and Washida (1992)	1.0–1.8 m-high models with vertical and inclined slopes at 1/7 scale. Sandbags with wrapped-face facing. Nonwoven geotextile, plastic nets and steel bars with sandy silt backfill	Deformations decreased with increasing reinforcement stiffness and density, and decreasing face slope angle. Failure volumes were shallower for reinforced structures. Relative reduction in the deformation of reinforced structures compared with unreinforced structures increased with the steepness of the face. The circular slip method agrees well with experimental results except for steep face models
Murata *et al.* (1994)	2.5 m-high 1/2-scale model walls with gabion/rigid concrete panel walls. Geogrid with dry sand backfill. Horizontal shaking using sinusoidal and scaled earthquake record. Base accelerations up to 0.5*g* at 3.4 Hz	Increase in reinforcement forces due to shaking was very small. Reinforcement loads increased towards the front of the wall. Acceleration amplification was negligible up to the mid height of the wall, but increased to about 1.5 at the top. Amplification behaviour was similar for reinforced and unreinforced zones. The reinforced zone behaved as a monolithic body. The sinusoidal base input resulted in greater deformations than the scaled earthquake record. Rigid facing adds to the wall seismic resistance
Sugimoto *et al.* (1994), Telekes *et al.* (1994)	1.5 m-high model embankment with sand bags and wrapped-face slope surface. Geogrid reinforcement with sand backfill. Model scales 1/6 and 1/9. Sinusoidal and scaled earthquake record. Base acceleration up to 0.5*g* at 40 Hz	Reinforced models more stable than unreinforced models. Proposed similitude rules for small- and large-strain deformation modelling. The largest amplification was recorded at the crest of models. Failure of structures was progressive from the top of the structure downward. Reinforcement forces increased linearly with acceleration up to the start of failure. The failure mechanism was difficult to predict using the proposed scaling rules. Under seismic loading conditions, there was a tendency for shallow slopes to fail compared with steeper ones. Scale effects due to vertical stress and the apparent cohesion of backfill soil influenced the relative performance of steep-faced and shallow-faced models
Budhu and Halloum (1994)	0.72 m-high model wall with wrapped-face facing. Geotextile with dry sand backfill. Base acceleration in increments of 0.05*g* at 3 Hz	Sliding progressed with increasing acceleration from the top geotextile/sand interface to the bottom layer. No consistent decreasing trend of critical acceleration was observed with an increasing spacing-to-length ratio. The critical acceleration was proportional to the soil/geotextile interface friction value
Sakaguchi *et al.* (1992), Sakaguchi (1996)	1.5 m-high model walls. One wrapped-face and four unreinforced rigid concrete panel walls. Geogrid with dry sand backfill. Sinusoidal loading with base acceleration up to 0.72*g* at 4 Hz	The wrapped-face wall behaved as a rigid body, and failed at a higher acceleration than unreinforced structures. However, at smaller accelerations (due to stiff facing panels), the displacements of the unreinforced structures were less. A base input acceleration of 0.32*g* delineated stable wall performance from yielding wall performance for the reinforced structure. Residual strains were greatest closest to the face. It was concluded that more-rigid lightweight modular block facings may be effective in reducing reinforcement loads
Koseki *et al.* (1998b)	0.5–0.53 m-high propped-panel models, phosphor-bronze reinforcement strips (with $L/H = 0.4$) connected together in a grid form. One uniform-length model and one model with extended reinforcement length at the top. 5 Hz sinusoidal base acceleration with stepwise increase in amplitude	Overturning was observed to be the main failure mode. Simple shear deformation of the reinforced zone was observed. The ratio of observed and predicted critical seismic coefficients (corresponding to 5% lateral displacement) was about 1.05 for the uniform reinforcement model and 1.15 for the model with an extended reinforcement layer length at the top. These ratio values were larger than the values for conventional retaining wall models (values <1) tested in the same study. Walls on shaking tables were more stable than on equivalent tilting tables. The observed failure plane angle was steeper than the predicted value

Table 16.2 Continued

Reference	Model details	Observed behaviour and implications for design and analysis
Matsuo *et al.* (1998)	1–1.4 m-high models with hard facing panel. Reinforcement length, $L/H = 0.4$ and 0.7. One model with inclined facing. 5 Hz sinusoidal base acceleration with stepwise increase in amplitude. In addition, recorded ground motion was applied	Walls showed a larger margin of safety when subjected to recorded ground motion compared with sinusoidal base acceleration. Did not observe failure of the model walls in spite of predicted factors of safety <1
El-Emam and Bathurst (2004, 2005, 2007)	Fourteen 1 m-high models subjected to 5 Hz sinusoidal stepped-amplitude excitation. Parameters varied: (a) facing panel toe boundary condition (i.e. hinged or sliding); (b) reinforcement layer length, spacing and axial stiffness; (c) facing panel thickness (mass) and inclination angle; and (d) input motion predominant frequency	Experimental results showed that the facing toe boundary condition had a significant effect on the seismic response of reinforced soil walls. The magnitude of the cumulative facing lateral displacement under base excitation was less for nominal identical walls with: (a) less facing panel mass; (b) an inclined facing; (c) longer reinforcement layers; (d) stiffer reinforcement; and (e) smaller reinforcement vertical spacing
Ling *et al.* (2005b)	Three 2.8 m-high modular-block geosynthetic-reinforced soil walls with horizontal component of Kobe earthquake motions and vertical record for one of the walls	Negligible deformation under moderate excitation (peak acceleration 0.4*g*). Wall behaviour improved with a longer reinforcement length at the top of a wall. Existing design codes in the USA underestimate the seismic capacity of flexible wall systems

only at the geosynthetic layer elevations (test 2) gave a displacement response that fell between the results of walls constructed with uniformly poor interface shear properties (test 1), and the nominally identical structure with uniformly good interface shear properties (test 4). The resistance of the facing column to horizontal base shaking improved with increasing shear capacity between dry-stacked modular blocks or by increasing the wall batter.

The results of this study confirmed that measured accelerations were not uniform throughout the soil–wall system. Large acceleration amplifications as high as 2.2 were recorded, particularly at the top of the unreinforced portion of the facing column. Observed critical accelerations to cause failure of the wall models were compared with predictions based on the analysis method proposed by Bathurst and Cai (1995). The measured peak acceleration at the middle wall height or at the top of the backfill surface was shown to give more accurate estimates of critical acceleration to be used in pseudo-static analysis. The total load in the reinforcement layers was estimated to be only a very small percentage of the tensile capacity of the reinforcement layers. The test results showed that, while critical accelerations to cause incipient collapse of the wall models could be predicted reasonably well, the actual failure mechanism was difficult to predict. For example, pull-out of the top reinforcement layer was identified as a critical mechanism when, in fact, the observed failure mechanism was toppling at the top of the unreinforced facing column.

Figure 16.38 Example of a shaking table model of a reinforced soil segmental retaining wall

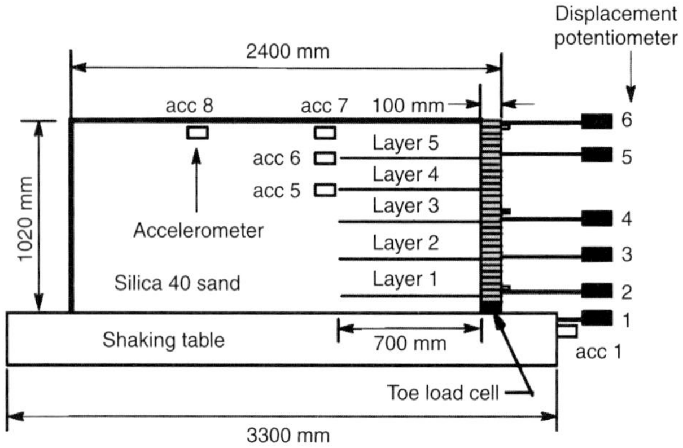

Tatsuoka *et al.* (1998) observed an important difference in the seismic response of reduced-scale retaining walls using different testing methods in the laboratory. They found

Table 16.3 Model test configurations
Data from Bathurst *et al.* (1996)

Test No.	Facing batter interface	Block–block interface	Block–geosynthetic
1	Vertical	Frictional	Frictional
2	Vertical	Fixed	Frictional
3	8°[a]	Frictional	Frictional
4	Vertical	Fixed	Fixed

[a] From the vertical.

Figure 16.39 Base input acceleration record for shaking table tests

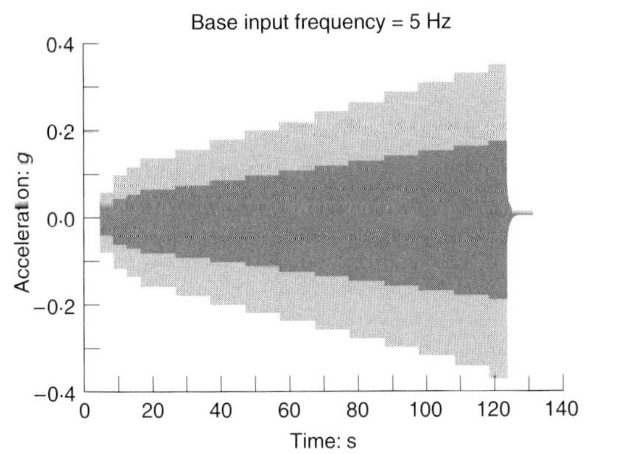

that, in static tilting tests, an active failure wedge developed, and the wall failed after the tilting angle was increased slightly. On the other hand, in shaking table tests, collapse of the wall occurred at a later stage (compared with the tilting case) after considerable deformation and development of an active failure plane. They attributed this difference to 'positive effects' inherent in dynamic loading, and concluded that, as in the pseudo-static methods of analysis, static tilting tests are not appropriate to evaluate the seismic stability of retaining walls in the field. Tatsuoka *et al.* (1998) observed different failure patterns for unreinforced (straight failure plane) versus reinforced backfills (two-part wedge) in model walls on a shaking table. They concluded that unreinforced and reinforced retaining wall systems should be designed using different pseudo-static approaches. They did not find any clear correlation between the failure plane angle and the critical k_h value in their shaking table tests. They also concluded that the failure plane angle was not totally controlled by the critical k_h value, and attributed the difference between the observed and predicted failure plane angles to the dynamic characteristics of ground motion loading.

Koseki *et al.* (1998b) carried out tilting and shaking table tests on 0.5 m-high conventional and geosynthetic-reinforced soil retaining wall models. The reinforced soil model walls were of the propped-panel type, and phosphor-bronze reinforcement strips (with $L/H = 0.4$) were used to model the geosynthetic reinforcement. The reinforcement strips were connected together in a grid form. The reinforced soil test walls included one uniform-length model and one model with extended reinforcement length at the top. The input ground motion was modelled with a 5 Hz sinusoidal base acceleration, which was increased in amplitude in a stepwise manner. Koseki *et al.* (1998b) found that overturning was the main failure mode in both conventional and reinforced soil. However, the model walls in their tests were placed on a soft foundation (200 mm thick, pluviated Toyoura sand), which increased the likelihood of overturning as the major failure mode of the wall models. An implication of this observation to current practice recommended in the FHWA (2009) guidelines is that overturning modes of failure for reinforced soil walls should not be ignored if structures are seated on weak and/or compressible foundation soils (see Section 16.3.1.3). In reinforced soil walls, Koseki *et al.* observed that the reinforced zone underwent simple shear deformation, and suggested that this shear deformation should be included in the displacement calculation of reinforced soil walls. The ratio of observed and predicted critical seismic coefficients (corresponding to 5% lateral displacement) was about 1.05 for the uniform-reinforcement model and 1.15 for the model with an extended reinforcement layer length at the top. These ratio values were larger than the values obtained for conventional retaining wall models (where the ratio values were less than 1) tested in the same study. They found that the walls were more stable when tested on the shaking table, compared with the equivalent tilting table tests. In addition, they also found that the observed failure plane angle in all model walls, except in the reinforced soil wall with the extended reinforcement layer at the top, was steeper than the predicted value from the pseudo-static theory.

Figure 16.40 Displacement close to the top of the wall versus the peak base input acceleration

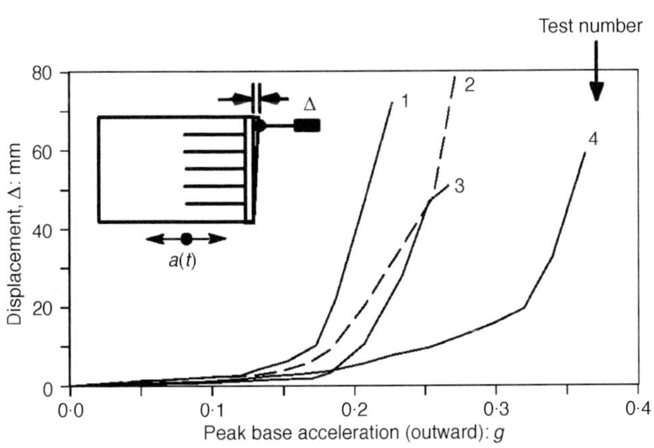

Matsuo *et al.* (1998) carried out shaking table tests on six reduced-scale, reinforced soil wall models. They used a polypropylene material for the geogrid reinforcement. The variables in the test walls were reinforcement length ($L/H = 0.4$ and 0.7), wall height ($H = 1$ and 1.4 m), wall facing type (incremental and propped) and facing slope (vertical and battered). The input base acceleration was sinusoidal, with a frequency of 5 Hz and with a stepwise increase in amplitude. In addition, the north–south component of the recorded ground motion at Kobe Maritime Observatory was applied to one test model. The model walls showed larger margins of safety when subjected to recorded ground motion, compared with sinusoidal base acceleration. Matsuo *et al.* (1998) suggested that the predominant frequency of the base accelerations also contributed to the difference in the wall response magnitude to sinusoidal versus recorded base accelerations. They predicted the magnitude of the wall horizontal displacement subjected to recorded input acceleration using sliding block and cumulative damage concepts. The predicted permanent displacement magnitude from the sliding block approach was found to be only about one-quarter of the measured displacement. The predicted displacement magnitude from the cumulative damage approach was about one-fifth of the measured value. They pointed to the effect of ground motion predominant frequency (not included in the above approaches) and shear deformation in the reinforced zone among the possible reasons for the difference between the predicted and measured values for the wall displacement in their tests. They also observed that the model walls subjected to base acceleration remained stable in spite of predicted factors of safety that were less than unity. They attributed the stability of the test walls in spite of low factors of safety to the ductile behaviour of the walls. They found that increasing the reinforcement length ratio L/H from 0.4 to 0.7 was the most effective method to reduce the wall deformation. In addition, the horizontal displacement at the top of the walls with a continuous facing panel was greater than the corresponding displacement in discrete facing walls. They found this result unexpected. However, a large lateral displacement at the top of propped-panel walls, with a fixed-toe condition subjected to base acceleration, has also been observed in numerical simulation studies (e.g. Bathurst and Hatami, 1998b).

El-Emam and Bathurst (2004, 2005, 2007) reported the results of 14 reduced-scale shaking table model walls with height $H = 1$ m. Similitude rules proposed by Iai (1989) were used in the experimental design so that the geosynthetic reinforcement arrangement and stiffness satisfied a model-to-prototype scale ratio of 1/6. The experimental design was selected to focus on the influence of the following parameters on the wall response during base shaking: (a) facing panel toe boundary condition (i.e. hinged or sliding); (b) reinforcement layer length, spacing and axial stiffness; (c) facing panel thickness (mass) and inclination angle; and (d) input motion predominant frequency.

The toe restraint condition of the model facing panels was either hinged-type (i.e. restrained in the vertical and horizontal directions, while free to rotate) or sliding-type (i.e. restrained in the vertical direction only and free to rotate and slide horizontally). This performance feature was investigated because the numerical study by Bathurst and Hatami (1998b) showed that the toe boundary condition had a major influence on wall displacements, and the distribution and magnitude of reinforcement loads during simulated seismic excitation. It can be argued that field walls fall between these two idealised conditions because field walls are typically constructed with toe embedment and a footing that provide some horizontal compliance (Huang *et al.*, 2010). The reinforcement length, L, was varied to give $L/H = 0.6$ or 1.0. The reinforcement vertical spacing was varied to isolate the effect of reinforcement density on the dynamic response of the model walls. Two facing panel systems with different thicknesses were used to investigate the influence of the facing panel thickness (mass) on the wall response. The influence of facing inclination was investigated in two of the model walls by constructing the facing column at 10° from the vertical. Inclined facings are a typical feature of segmental retaining walls (Bathurst and Simac, 1994). Two different geogrid reinforcement materials were used to examine the influence of the reinforcement axial stiffness on the dynamic behaviour of reinforced soil walls. Measurements from full-scale structures (e.g. Allen *et al.*, 2003; Bathurst *et al.*, 2009) and numerical parametric analyses, reported by Bathurst and Hatami (1998b), demonstrate that the response of reinforced soil walls with a very stiff reinforcement (e.g. metallic materials) is fundamentally different from nominal identical structures constructed with a relatively extensible reinforcement (i.e. geosynthetics). The backfill was poorly graded sand. The sand was placed and gently vibro-compacted (using the shaking table) in 100 mm lifts, while the wall facing was externally braced. At the end of soil placement, the external braces were removed, and an initial static load condition established.

A unique feature of this experimental programme is that an instrumented footing was used to decouple the vertical and horizontal toe-load components of the facing panel. For the sliding toe model, the horizontal load cells at the wall toe were removed after construction. The horizontal movement of the wall facing was measured using displacement potentiometers mounted against the facing panel on the shaking table platform. Reinforcement strains were measured using strain gauges that were bonded directly to the PET geogrid longitudinal members and extensometers that were attached to selected geogrid junctions. The acceleration response during shaking was measured using two accelerometers attached to the facing panel and four accelerometers buried at different locations in the backfill.

The target horizontal base acceleration was a stepped-amplitude sinusoidal function with a frequency of 5 Hz (e.g. see Figure 16.39). The excitation record was applied in 0.05g increments of five seconds duration. The excitation stages were applied until excessive model deformation occurred. A stepped-acceleration amplitude base excitation record has been shown to be more aggressive than actual earthquake records scaled to the same peak acceleration in physical shaking table tests (Murata *et al.*, 1994) and numerical modelling (Hatami and Bathurst, 2001, 2005). However, a stepped record simplifies interpretation of the dynamic wall response.

Experimental results showed that the facing toe boundary condition had a significant effect on the seismic response of reinforced soil walls. The combination of a restrained toe and a stiff facing panel was demonstrated to attract a significant portion of the seismic-induced lateral horizontal earth force under both static and dynamic loading conditions. The vertical load developed at the footing was found to be larger than the facing self-weight, indicating that down drag-forces were significant. While a horizontal restrained toe condition reduced wall lateral deformations, greater base acceleration amplification was observed compared with otherwise identical models constructed with a sliding toe. The experimental results showed that the magnitude of cumulative facing lateral displacement under base excitation was less for nominally identical walls, with: (a) less facing panel mass; (b) an inclined facing; (c) longer reinforcement layers; (d) stiffer reinforcement; and (e) smaller reinforcement vertical spacing.

16.4.2 Centrifuge shaking table tests

The scaling difficulties identified for 1*g* shaking table tests can be overcome theoretically using centrifuge testing. Sakaguchi *et al.* (1992, 1994) and Sakaguchi (1996) mounted a shaking table on a centrifuge apparatus. Sakaguchi and his co-workers examined the response of a segmental retaining wall constructed with lightweight facing units. Three 150 mm-high models were accelerated in the centrifuge to simulate walls 4.5 m high. Three different geotextiles were used, having a range of tensile strengths, and walls were built with three different reinforcement lengths. The qualitative performance of the centrifuge tests was similar to that recorded for the 1*g* shaking table tests (see Table 16.2). The results showed that, up to a limiting value of the reinforcement length ($L/H = 2/3$), there was a corresponding reduction in wall displacements with increasing reinforcement length. The geotextile strength for the range of materials used did not influence the wall deformation.

Nova-Roessig and Sitar (1998) carried out dynamic centrifuge tests on 150 mm-high reinforced soil, wrapped-face slope models (1/48 scale). The reinforcement layers were made of a nonwoven geotextile material, with the reinforcement length-to-wall height ratio equal to 0.7. Both sinusoidal and recorded ground motions were used as the input base acceleration. Nova-Roessig and Sitar found that the amplification of acceleration in the backfill depended on the amplitude of the input base acceleration. They measured an acceleration amplification as great as 2.5 for 0.1*g* base acceleration. The model slopes showed de-amplification when they were subjected to stronger (e.g. PGA > 0.35*g*) input accelerations. These results were consistent with the observations of Matsuo *et al.* (1998) on 1*g* shaking table tests on walls with a hard facing. Nova-Roessig and Sitar found that the model slopes under base acceleration deformed in a ductile manner, with a considerable amount of shear deformation near the crest and with no distinct failure surface. This observation is also consistent with the observations by Matsuo *et al.* (1998) for reinforced soil walls in 1*g* shaking table tests (see Section 16.4.1). Nova-Roessig and Sitar suggested that the lack of a well-defined shear failure surface in reinforced soil slopes subjected to base acceleration contradicts the routine assumption of a distinct failed mass behind the reinforced zone in limit-equilibrium-based design methods. They proposed that deformation-based approaches should be adopted for the seismic design of reinforced soil walls and slopes.

16.5. Seismic buffers

The generic term 'geofoam' has entered geosynthetic terminology to describe expanded foams used in geotechnical applications (Horvath, 1995). Horvath proposed that geofoam panels could be used against rigid wall structures (e.g. basement walls) to reduce seismic-induced stresses that would otherwise overstress rigid wall structures. Hatami and Witthoeft (2008) proposed that the technique can also be applied at the boundary between the reinforced soil zone and the retained soil zone in geosynthetic-reinforced soil walls.

To the best of the authors' knowledge, the first application of this technology in North America for seismic applications was reported by Inglis *et al.* (1996). Panels of low-density expanded polystyrene (EPS) from 450 to 610 mm thick were placed against rigid basement walls up to 10 m in height at a site in Vancouver, British Columbia. Analyses using the FLAC program showed that a 50% reduction in lateral loads could be expected (Figure 16.41) during a seismic event compared with a rigid wall solution. The design challenge using this technique is to optimise the thickness of the buffer panels for a candidate geofoam material so that the horizontal compliance under peak loading is just sufficient to minimise lateral earth pressures without excessive lateral deformation of the backfill or crushing of the geofoam buffer.

Bathurst *et al.* (2007a), and Zarnani and Bathurst (2007), reported the results of a series of physical reduced-scale

Figure 16.41 Results of FLAC analyses on seismic load reduction using a geofoam buffer
Reproduced from Inglis *et al.* (1996)

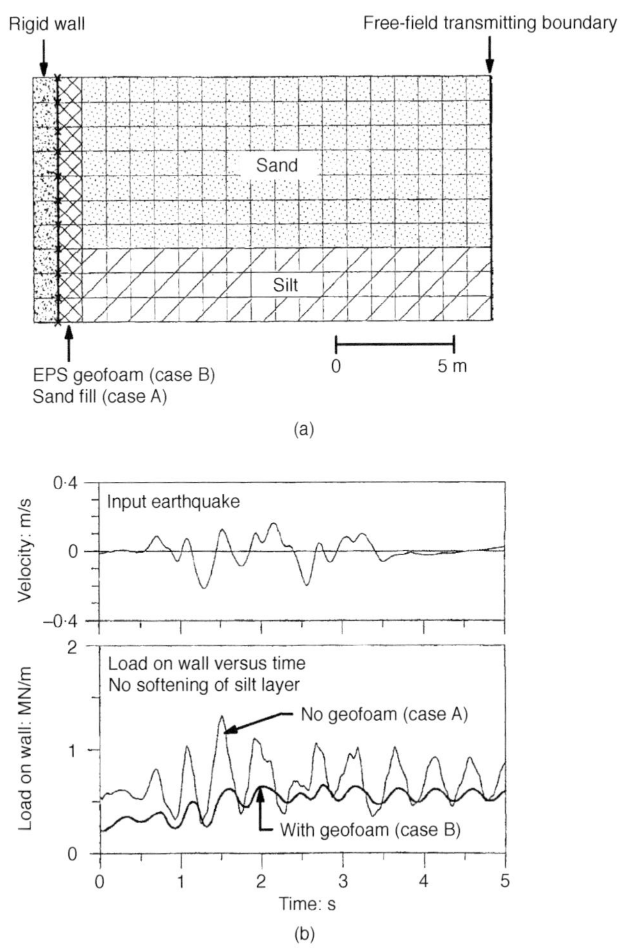

Figure 16.42 Example design charts for the selection of geofoam seismic buffers
Reproduced from Zarnani and Bathurst (2009b), © Canadian Science Publishing or its licensors

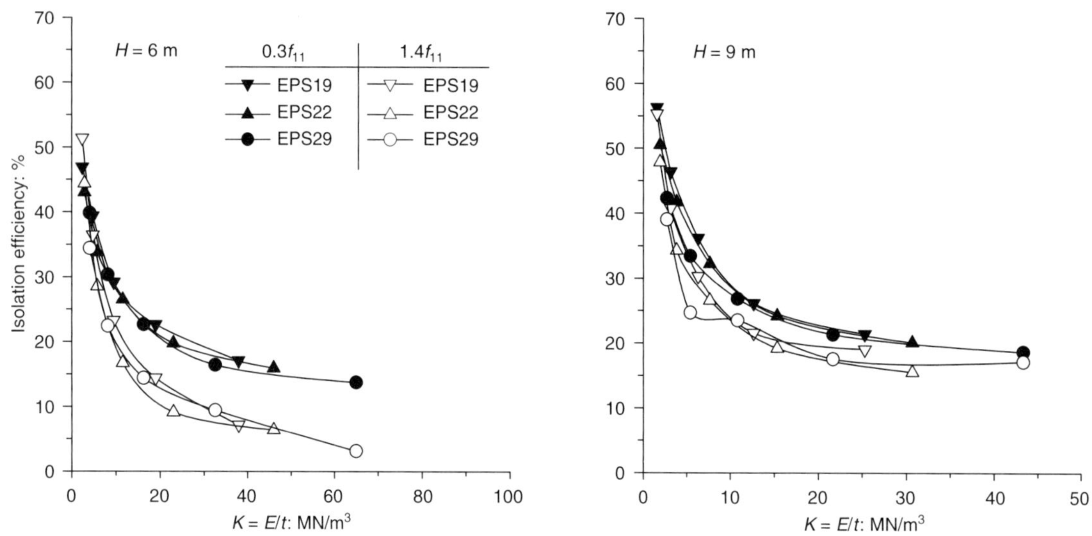

shaking table tests that demonstrated a seismic buffer 'proof of concept'. The experimental results showed that by placing a geofoam compressible inclusion at the back of a rigid wall, the seismic earth forces at a peak base excitation amplitude of about 0.7*g* could be reduced by 15–40% compared with the rigid wall case. The greatest force reduction occurred for the geofoam buffer with lowest density, corresponding to the material with the lowest initial elastic tangent modulus. Zarnani and Bathurst (2008, 2009a) used the results of the physical shaking table tests to verify numerical models that were then used to generate a set of design charts for EPS seismic buffers in combination with rigid walls and granular backfill soils (Zarnani and Bathurst, 2009b). Example design charts are reproduced in Figure 16.42 for two wall heights (H). The independent parameter in these charts is the seismic buffer stiffness, defined as $K = E/t$, where E is the 'index' elastic modulus determined from the initial slope of a conventional monotonic load compression versus strain plot and t is the buffer thickness. The index elastic modulus value does not imply that this value is representative of the actual dynamic modulus of the geofoam material; rather, it is used as a convenient reference input parameter to use the design charts. Isolation efficiency is defined as the ratio of the change in wall force between rigid and seismic buffer cases, divided by the peak wall force without the buffer. The response curves are reasonably independent of EPS type (density) and type of excitation record (i.e. an actual earthquake accelerogram or synthetic record). However, the influence of the proximity of the fundamental frequency (f_{11}) (Equation 16.47) of the seismic buffer system above or below the predominant frequency of the input accelerogram increases as the height of the wall decreases. For the range of parameters investigated, a buffer stiffness value less than $K = 50\ \text{MN/m}^3$ was judged to be the practical range for the design of these systems.

16.6. Observed performance of reinforced soil walls and slopes during earthquakes

16.6.1 North American experience (Northridge, 1994; Loma Prieta, 1989)

Sandri (1994a,b) conducted a survey of reinforced soil segmental retaining walls greater than 4.5 m in height in the Los Angeles area immediately after the Northridge earthquake of 17 January 1994 (moment magnitude = 6.7). The results of the survey showed no evidence of visual damage to nine of 11 structures located within 23–113 km of the earthquake epicentre. Two structures (Valencia and Gould Walls) showed tension cracks within and behind the reinforced soil mass that were clearly attributable to the results of seismic loading. Bathurst and Cai (1995) analysed both structures, and noted that minor cracking at the back of the reinforced soil zone could be attributed to the flattening of the internal failure plane predicted using the M–O theory. The facing columns for all walls were intact, even though peak horizontal ground accelerations as great as 0.5*g* were estimated at one site.

A similar survey of three geosynthetic-reinforced walls, and four geosynthetic-reinforced slopes by White and Holtz (1996) after the same earthquake, revealed no visual indications of distress. Stewart *et al.* (1994) report that slope indicator measurements at the toe of a 24 m-high geogrid-reinforced slope, which was estimated to have sustained peak horizontal ground accelerations of 0.2*g*, showed no movement. Some unreinforced crib walls and unreinforced segmental walls were observed to have developed cracks in the backfill during the same survey by Stewart *et al.* They concluded that concrete crib walls may not perform as well as more flexible retaining wall systems under seismic loading. Similar good performance of several geosynthetic-reinforced soil walls and slopes during the 1989 Loma Prieta earthquake (Richter magnitude = 7.1) was reported by Eliahu and Watt (1991), and Collin *et al.* (1992).

16.6.2 Japanese experience (Hanshin, 1995)

Tateyama *et al.* (1995) reported on the seismic performance of traditional unreinforced wall structures after the Great Hanshin earthquake of 17 January 1995 (moment magnitude = 6.8). Concrete and masonry walls suffered serious failures, including collapse. Conventional reinforced concrete cantilever structures suffered some cracking and limited displacement.

Tatsuoka *et al.* (1995, 1997) reported on the performance of a 6.2 m-high geosynthetic-reinforced soil retaining wall with a full-height rigid-facing construction. The peak ground acceleration at the site was estimated to have been as great as 0.7*g*. The structure was observed to have moved 260 mm at the top and 100 mm at ground level, but was otherwise undamaged. Tatsuoka *et al.* concluded that shortening of the reinforcement lengths due to site constraints was a likely cause of the observed tilting of the wall.

Nishimura *et al.* (1996) surveyed ten geogrid-reinforced soil walls and steepened slopes after the same event. All structures survived the earthquake even though peak ground accelerations were estimated in the range 0.3–0.7*g*. They determined critical accelerations for these structures using the GRB (1990) and PWRI (1992) methods of analysis, and found that predicted critical acceleration coefficient (k_h) values were as low as 0.1. They concluded that both methods are very conservative. Where minor damage was observed, it was related in one instance to minor separation between an unattached concrete facing column, and in another case there was cracking at the back of the reinforced soil mass, although this last observation may be the result of poor base foundation conditions. The results of stability calculations using the GRB and PWRI methods led Nishimura *et al.* to conclude that the length of the reinforcement layers at the top of the reinforced soil structures should be increased in order to capture critical failure volumes generated under even modest horizontal seismic accelerations. Nishimura *et al.* argued that the phase lag between retained backfill and the reinforced zone adds to the seismic stability of reinforced soil retaining walls by enabling the reinforcement to resist active earth pressure behind the facing.

Tatsuoka *et al.* (1998) reported that, according to the results of field observation and laboratory tests, the seismic stability of geosynthetic-reinforced soil walls with propped panel facing is marginally higher than the stability of conventional reinforced concrete retaining walls and considerably higher than that of gravity-type walls. In addition, the observed size of the failure zone is not predicted satisfactorily from pseudo-static analysis methods.

16.7. Summary of main points

Largely qualitative observations of the performance of geosynthetic-reinforced slopes and walls in both the USA and Japan suggest that these structures perform well during seismic events when located on competent foundation soils and above the water table. The relatively flexible nature of reinforced soil walls, constructed with extensible and inextensible reinforcement, is routinely cited as the reason for the good performance of these structures during a seismic event. However, it is becoming more apparent that the combination of a structural facing (i.e. a concrete facing) and a reinforced backfill is a viable strategy for earthquake-resistant design that combines the advantages of ductility of the reinforced soil mass with the benefit of soil containment and uniform wall deformation by a structural facing.

Nevertheless, the geotechnical engineer requires seismic design tools and representative component properties for geosynthetic-reinforced soil walls and slopes, in order to optimise the design of these structures in seismic environments. The review of the literature and the work by the authors and their co-workers leads to the following conclusions and research needs.

- The depth, strength and stiffness of the foundation soil may have a greater influence on the internal and external stability of reinforced soil slopes and walls than the design of the reinforced mass in isolation. Parametric analyses are required to investigate the influence of the foundation condition on seismic performance.
- The design methodologies that are currently used in the USA for geosynthetic-reinforced soil walls have been based largely on the results of the numerical modelling of reinforced structures constructed with inextensible reinforcement (steel strips). Numerical studies by the authors confirm that the general approach is not valid for reinforced soil wall structures constructed with relatively less-stiff geosynthetic products. Further numerical and experimental work is required to investigate the validity of pseudo-static analysis methods that predict increased reinforcement lengths at the top of reinforced walls and slopes.
- Ground motion amplification (or attenuation) through retained soils plays a major role in generating additional dynamic loads on geosynthetic reinforcement and wall-facing components. More work is required to offer guidance on the appropriate distribution of incremental seismic forces to be applied to extensible reinforcing elements, and to establish the influence of system stiffness (i.e. the combined effect of reinforcement stiffness, number of reinforcement layers, facing stiffness and height of structure) on this distribution. Numerical models calibrated against the results of carefully conducted large shaking table tests or small-scale centrifuge tests are possible research strategies to meet this goal.
- The single most important characteristic determining the seismic response of reinforced soil walls is the fundamental frequency of the structure, namely the

predominant frequency of the design seismic event. The calculation of the fundamental frequency of a reinforced wall structure in a seismic area should be part of the analysis and design process. Simple expressions are available to carry out this evaluation.

- A number of design methodologies have been proposed in the USA and Japan for the seismic design of walls and slopes that can lead to important differences in the required number/strength, location and length of reinforcement layers. Comparative analyses should be carried out to examine the relative conservatism (or non-conservatism) of the proposed methodologies.
- Geosynthetic-reinforced segmental retaining walls in seismic areas offer unique challenges to the designer because of their modular facing column construction. These structures involve analyses not required for other retaining wall systems. The experience of the authors is that the economic potential of these systems in seismic areas will not be fully realised until confidence is developed through proven design methodologies for these structures.
- The design engineer will continue to be attracted to relatively simple seismic design tools based on pseudo-static and displacement methods for the design and analysis of routine walls and slopes under modest seismic loads. Nevertheless, the results of sophisticated numerical models carried out by experienced modellers offer the possibility of refining simple models to minimise unwarranted conservatism.

16.8. Some common questions and their answers

Questions

1. Is peak friction angle of a granular backfill soil under seismic loading conditions different from its value under static loading conditions?

2. How is the horizontal seismic coefficient (k_h) selected for geosynthetic-reinforced soil wall structures?

3. When should the conventional Mononobe–Okabe (M–O) method for seismic analysis and design be used, and when should displacement methods be used?

4. Why have geosynthetic-reinforced soil walls behaved better than conventional gravity structures?

5. What are some advantages and disadvantages of centrifuge modelling and 1g reduced-scale modelling of reinforced soil walls using shaking tables?

6. How do I select a representative ground acceleration record for design?

7. Are there other emerging technologies that hold promise to improve the performance of earth-retaining wall structures during earthquakes?

Answers

1. There is no evidence that the intrinsic peak friction angle of granular soils under cyclic loading at frequencies typical of earthquakes is less than the value under static loading conditions. Some designers prefer to use the residual (large-strain) friction angle of the soil in limit-equilibrium (e.g. Mononobe-Okabe) analyses to be conservative (i.e. render safer designs). However, this can lead to excessively conservative design outcomes.

2. In practice, the choice of k_h is based on local experience, or prescribed by local building codes or other regulations that are applicable to the project soil and foundation conditions.

3. There is a growing consensus among experienced designers that for peak horizontal ground accelerations of $0.3g$ or more, the M–O method does not apply; indeed, analytical solutions may not be possible. Design outcomes using solutions from M–O analyses have been shown to be excessively safe in many cases. The Newmark-type (displacement-based) analyses are becoming more common, and are recommended for seismic design, particularly for structures subjected to peak ground accelerations in excess of $0.3g$.

4. Geosynthetic-reinforced soil walls have proven to be more ductile and flexible during an earthquake event. In addition, current limit-equilibrium-based methods for the internal stability design of these structures under static loading conditions are often very conservative. This introduces additional margins of safety for these structures when subjected to seismic loading.

5. Both laboratory methods can give important qualitative insight into the behaviour of reinforced soil walls under simulated earthquake loading. A major challenge in both methods is proper scaling of the properties of the soil and geosynthetic. Many 1*g* reduced-scale models reported in the literature have been carried out with models that are over-reinforced when compared with the prototype-scale systems that are being simulated. A limitation of 1*g* shaking table tests is the influence of stress-level-dependent properties of the frictional soils used in the model. A limitation of centrifuge testing is that the reinforcement geometry and thickness when scaled up to prototype scale do not represent field materials.

6. Conventional practice is to carry out stability analysis using actual near-site earthquake accelerograms with a scaled peak acceleration value matching local code recommendations for the site location and ground conditions. In the USA and Canada, there are site-specific synthetic accelerograms for different locations. A minimum of (say) three of these records should be applied to the structure. The designer can then use the most critical outcome for design, based on actual scaled earthquake records or synthetic records for the site.

7. In Japan, structures have been built recently with cement-stabilised soil and cement–fibre–soil mixtures placed in the reinforced soil zone, together with horizontal geogrid reinforcement layers (Koseki *et al.*, 2006; Izawa *et al.*, 2009). These hybrid structures offer the combined benefit of improved soil properties and geosynthetic reinforcement. Another relatively recent innovation is the use of integral bridge construction (reinforced concrete bridge deck and abutments are constructed without joints) with geosynthetic-reinforced soil abutment backfills (Tatsuoka *et al.*, 2009).

REFERENCES

AASHTO (American Association of State Highway and Transportation Officials) (2010) *LRFD Bridge Design Specifications*, 5th edn. AASHTO, Washington, DC.

Allen TM (1993) Issues regarding design and specification of segmental block-faced geosynthetic walls. *Transportation Research Record* **1414**: 6–11.

Allen TM, Bathurst RJ, Walters DL, Holtz RD and Lee WF (2003) A new working stress method for prediction of reinforcement loads in geosynthetic walls. *Canadian Geotechnical Journal* **40(5)**: 976–994.

ASTM (American Society for Testing and Materials) (1996) D4595. Standard test method for tensile properties of geotextiles by the wide-width strip method. ASTM, West Conshohocken, PA.

ASTM (2001a) D6706-01. Standard test method for measuring geosynthetic pullout resistance in soil. ASTM, West Conshohocken, PA.

ASTM (2001b) D6683. Standard test method for determining the connection strength between segmental concrete units and geosynthetic reinforcement. ASTM, West Conshohocken, PA.

ASTM (2003) D6916-03. Standard test method for determining the shear strength between segmental concrete units (modular concrete blocks). ASTM, West Conshohocken, PA.

ASTM (2008) D5321-08. Standard test method for determining the coefficient of soil and geosynthetic or geosynthetic and geosynthetic friction by the direct shear method. ASTM, West Conshohocken, PA.

Bachus RC, Fragaszy RJ, Jaber M, Olen KL, Yuan Z and Jewell R (1993) *Dynamic response of reinforced soil systems*. Engineering Research Division, US Department of the Air Force Civil Engineering Support Agency, March 1993, 1 and 2. Report ESL-TR-92-47.

Bathurst RJ (1994) Reinforced soil slopes and embankments. *Technical Notes for Computer Programs GEO-SLOPE and GEOPLOT*.

Bathurst RJ (1998) *NCMA Segmental Retaining Wall Seismic Design Procedure – Supplement to Design Manual for Segmental Retaining Walls*. National Concrete Masonry Association, Herdon, VA.

Bathurst RJ and Alfaro MC (1996) Review of seismic design, analysis and performance of geosynthetic reinforced walls, slopes and embankments. *Proceedings of the International Symposium on Earth Reinforcement*, Fukuoka, pp. 887–918.

Bathurst RJ and Cai Z (1994) In-isolation cyclic load-extension behavior of two geogrids. *Geosynthetics International* **1(1)**: 3–17.

Bathurst RJ and Cai Z (1995) Pseudo-static seismic analysis of geosynthetic-reinforced segmental retaining walls. *Geosynthetics International* **2(5)**: 787–830.

Bathurst RJ and Hatami K (1998a) Influence of reinforcement stiffness, length and base condition on seismic response of geosynthetic reinforced retaining walls. *Proceedings of the 6th International Conference on Geosynthetics*, Atlanta, GA, pp. 613–616.

Bathurst RJ and Hatami K (1998b) Seismic response analysis of a reinforced soil retaining wall. *Geosynthetics International* **5(1–2)**: 127–166.

Bathurst RJ and Hatami K (1998c) Influence of reinforcement properties on seismic response and design of reinforced-soil retaining walls. *Proceedings of the 51st Canadian Geotechnical Conference*, Edmonton, pp. 479–486.

Bathurst RJ and Hatami K (1999a) Numerical study of the influence of base shaking on reinforced-soil retaining walls. *Proceedings of Geosynthetics '99*, Boston, MS, pp. 963–976.

Bathurst RJ and Hatami K (1999b) Earthquake response analysis of reinforced-soil retaining walls using FLAC. *Proceedings of the International FLAC Symposium on Numerical Modelling in Geomechanics*, Minneapolis, MN, pp. 407–415.

Bathurst RJ and Huang B (2010) A geosynthetic modular block connection creep test apparatus, methodology and interpretation. *ASTM Geotechnical Testing Journal* **33(2)**: 103–111.

Bathurst RJ and McLay MJ (1996) *Repeated Load Pullout Testing of a HDPE Geogrid*. Department of Civil Engineering, Geotechnical Research Group Internal Report, Royal Military College of Canada, Kingston.

Bathurst RJ and Simac MR (1993) Laboratory testing of modular unit-geogrid facing connections. *STP 1190 Geosynthetic Soil Reinforcement Testing Procedures* (Cheng SCJ (ed.)), ASTM, West Conshohocken, PA, pp. 32–48.

Bathurst RJ and Simac MR (1994) Geosynthetic reinforced segmental retaining wall structures in North America. Keynote paper. *Proceedings of the 5th International Conference on Geotextiles, Geomembranes and Related Products, Singapore*, pp. 1275–1298.

Bathurst RJ, Althoff S and Linnenbaum P (2008) Influence of test method on direct shear behavior of segmental retaining wall units. *ASTM Geotechnical Testing Journal* **31(2)**: 157–165.

Bathurst RJ, Simac MR and Sandri D (1995) Lessons learned from the construction performance of a 14 m high segmental retaining wall. In *Geosynthetics: Lessons Learned from Failures* (Giroud JP (ed.)). Industrial Fabrics Association International, St Paul, MN, pp. 21–34.

Bathurst RJ, Cai Z and Pelletier MJ (1996) Seismic design and performance of geosynthetic reinforced segmental retaining walls. *Proceedings of the 10th Annual Symposium of the Vancouver Geotechnical Society*, Vancouver.

Bathurst RJ, Cai Z and Simac MR (1997) Seismic performance charts for geosynthetic reinforced segmental retaining walls. *Proceedings of Geosynthetics '97, Long Beach*, CA, pp. 1001–1014.

Bathurst RJ, Miyata Y and Allen TM (2010) Facing displacements in geosynthetic reinforced soil walls. Keynote paper. *Proceedings of Earth Retention Conference 3 (ER2010)*, ASCE Geo-Institute, Bellevue, WA, pp. 442–459.

Bathurst RJ, Nernheim A, Walters DL, Allen TM, Burgess P and Saunders DD (2009) Influence of reinforcement stiffness and compaction on the performance of four geosynthetic-reinforced soil walls. *Geosynthetics International* **16(1)**: 43–59.

Bathurst RJ, Zarnani S and Gaskin A (2007a) Shaking table testing of geofoam seismic buffers. *Soil Dynamics and Earthquake Engineering* **27**: 324–332.

Bathurst RJ, Keshavarz A, Zarnani S and Take A (2007b) A simple displacement model for response analysis of EPS geofoam seismic buffers. *Soil Dynamics and Earthquake Engineering* **27**: 344–353.

Bernardi M and Paulson J (1997) Is creep a degradation phenomenon? *Proceedings of the International Symposium on Mechanically Stabilised Backfill*, Denver, CO, pp. 289–294.

Bolton MD (1986) The strength and dilatancy of sands. *Géotechnique* **36(1)**: 65–87.

Bonaparte R, Schmertmann GR and Williams ND (1986) Seismic design of slopes reinforced with geogrids and geotextiles. *Proceedings of the 3rd International Conference on Geotextiles*, Vienna, pp. 273–278.

BSSC (Building Seismic Safety Council) (1994) *NEHRP Recommended Provisions for Seismic Regulations for New Buildings and Other Structures*, parts 1 and 2. BSSC, Washington, DC.

Budhu M and Halloum M (1994) Seismic external stability of geotextile reinforced walls. *Proceedings of the 5th International Conference on Geotextiles, Geomembranes and Related Products*, Singapore, pp. 529–532.

Cai Z and Bathurst RJ (1995) Seismic response analysis of geosynthetic reinforced soil segmental retaining walls by finite element method. *Computers and Geotechnics* **17(4)**: 523–546.

Cai Z and Bathurst RJ (1996a) Seismic-induced permanent displacement of geosynthetic reinforced segmental retaining walls. *Canadian Geotechnical Journal* **33**: 937–955.

Cai Z and Bathurst RJ (1996b) Deterministic sliding block methods for estimating seismic displacements of earth structures. *Soil Dynamics and Earthquake Engineering* **15**: 255–268.

Canadian Geotechnical Society (2006) *Canadian Foundation Engineering Manual*, 4th edn. BiTech, Richmond.

Cancelli A, Rimoldi P and Togni S (1992) Frictional characteristics of geogrids by means of direct shear and pull-out tests. *Proceedings of the International Symposium on Earth Reinforcement Practice, IS-Kyushu '92*, Fukuoka, pp. 51–56.

Chalaturnyk RJ, Scott JD, Chan DH and Richards EA (1988) Stresses and deformations in a reinforced slope. *Proceedings of the 3rd Canadian Symposium on Geosynthetics*, Kitchener, pp. 79–89.

Chang CJ, Chen WF and Yao JT (1984) Seismic displacement in slopes by limit analysis. *Journal of Geotechnical Engineering, ASCE* **110(7)**: 860–875.

Chen RH and Chen TC (1998) Numerical simulation of dynamic behaviour of soil with reinforcement. *Proceedings of the 6th International Conference on Geosynthetics*, Atlanta, GA, pp. 1083–1086.

Chen TC, Chen RH, Lee YS and Pan JC (1996) Dynamic reinforcement effect of reinforced sand. *Proceedings of the International Symposium on Earth Reinforcement, IS-Kyushu '96*, Fukuoka, pp. 25–28.

Chida S, Minami K and Adachi K (1982) Test de stabilité de remblais en Terre Armée. (Unpublished report translated from Japanese.)

Christopher BR, Gill SA, Giroud JP, Juran I, Schlosser F, Mitchell JK and Dunnicliff J (1989) *Reinforced Soil Structures*, vol. I. *Design and Construction Guidelines*, Federal Highway Administration, Washington, DC. Report No. FHWA-RD-89-043.

Chugh AK (1995) Dynamic displacement analysis of embankment dams. *Géotechnique* **45(2)**: 295–299.

Collin JG, Chouery-Curtis VE and Berg RR (1992) Field observations of reinforced soil structures under seismic loading. *Proceedings of the International Symposium on Earth Reinforcement Practice, IS-Kyushu '92*, Fukuoka, pp. 223–228.

Cundall P and Board M (1988) A microcomputer program for modelling large-strain plasticity problems. *Proceedings of the 6th International Conference on Numerical Methods in Geomechanics*, Innsbruck, pp. 2101–2108.

De A and Zimmie TF (1997) Factors influencing dynamic frictional behaviour of geosynthetic interfaces. *Proceedings of Geosynthetics '97*, Long Beach, CA, pp. 837–849.

De A and Zimmie TF (1998a) A study of slip displacements caused by dynamic loading at geosynthetic interfaces. *Geotechnical Earthquake Engineering and Soil Dynamics III*, ASCE, Seattle, WA, pp. 997–1007.

De A and Zimmie TF (1998b) Estimation of dynamic interfacial properties of geosynthetics. *Geosynthetics International* **5(1–2)**: 17–39.

De A and Zimmie TF (1999) Estimation of dynamic frictional properties of geonet interfaces. *Proceedings of Geosynthetics '99*, Boston, MS, pp. 545–558.

Duncan JM and Chang CY (1970) Nonlinear analysis of stress and strain in soils. *Journal of Soil Mechanics and Foundation Engineering, ASCE* **96**: 1629–1653.

Ebeling RM and Morrison EE (1993) *The Seismic Design of WATERFRONT Retaining Structures*. Naval Civil Engineering Laboratory, Port Huenene, CA. Technical Report ITL-92-11 NCEL TR-939.

El-Emam M and Bathurst RJ (2004) Experimental design, instrumentation and interpretation of reinforced soil wall response using a shaking table. *International Journal of Physical Modelling in Geotechnics* **4(4)**: 13–32.

El-Emam M and Bathurst RJ (2005) Facing contribution to seismic response of reduced-scale reinforced soil walls. *Geosynthetics International* **12(5)**: 215–238.

El-Emam M and Bathurst RJ (2007) Influence of reinforcement parameters on the seismic response of reduced-scale reinforced soil retaining walls. *Geotextiles and Geomembranes* **25(1)**: 33–49.

Eliahu U and Watt S (1991) Geogrid-reinforced wall withstands earthquake. *Geotechnical Fabrics Report* **9(2)**: 8–13.

Elms DG and Richards R (1990) Seismic design of retaining walls. *ASCE Specialty Conference: Design and Performance of Earth Retaining Structures*, Ithaca, NY, pp. 854–871.

Fairless GJ (1989) *Seismic Performance of Reinforced Earth Walls*. Department of Civil Engineering, University of Canterbury, New Zealand.

Fakharian K and Evgin E (1995) Simple shear versus direct shear tests on interfaces during cyclic loading. *Proceedings of the 3rd International Conference on Recent Advances in Geotechnical Engineering and Soil Dynamics*, St Louis, MT, pp. 13–16.

Fang YS and Chen TJ (1995) Modification of Mononobe–Okabe theory. *Géotechnique* **45(1)**: 165–167.

Farrag K (1990) Interaction properties of geogrids in reinforced soil walls – testing and analysis. PhD thesis, Louisiana State University, Baton Rouge, LA.

FHWA (Federal Highway Administration) (2001) *Mechanically Stabilised Earth Walls and Reinforced Soil Slopes Design and Construction Guidelines* (Elias V, Christopher BR and Berg RR (eds)), FHWA, Washington, DC. FHWA-NHI-00-043.

FHWA (2009) *Design and Construction of Mechanically Stabilized Earth Walls and Reinforced Soil Slopes*, vols 1 and 2 (Berg RR, Christopher BR and Samtani NC (eds)), FHWA, Washington, DC. FHWA-NHI-10-024.

Finn WDL, Yogendrakumar M and Yoshida N (1986) *TARA-3: A Program to Compute the Response of 2-D Embankment and Soil–Structure Interaction Systems to Seismic Loading*. Department of Civil Engineering, University of British Columbia, Vancouver.

Franklin AG and Chang FK (1977) *Permanent Displacement of Earth Embankments by Newmark Sliding Block Analysis*. Soil and Pavements Laboratory, US Army Engineering Waterways Experimental Station, Vicksburg, MS. Miscellaneous Paper S-71-17.

Fredlund DG and Krahn J (1976) Comparison of slope stability methods of analysis. *Canadian Geotechnical Journal* **14**: 429–439.

Fukuda N, Tajiri N, Yamanouchi T, Sakai N and Shintani H (1994) Applicability of seismic design methods to geogrid reinforced embankment. *Proceedings of the 5th International Conference on Geotextiles, Geomembranes and Related Products*, Singapore, pp. 533–536.

Goodman RE, Taylor RL and Brekke TL (1968) A model for the mechanics of jointed rock. *Journal of the Soil Mechanics and Foundation Engineering Division, ASCE* **94**: 637–659.

GRB (Geogrid Research Board) (1990) *Geogrid Construction Method Guidelines*, 1 and 2. GRB, Fukuoka (in Japanese).

Greenwood JH (1997) Designing to residual strength of geosynthetics instead of stress-rupture. *Geosynthetics International* **4(1)**: 1–10.

Guler E and Biro MST (1999) A dynamic uniaxial wide strip tensile testing of tow geotextiles in isolation. *Geotextiles and Geomembranes* **17(2)**: 67–79.

Hanna TH and Touahmia M (1991) Comparative behavior of metal and Tensar geogrid strips under static and repeated loading. *Proceedings of Geosynthetics '91*, Atlanta, GA, pp. 575–585.

Hatami K and Bathurst RJ (1999a) Frequency response analysis of reinforced-soil retaining walls. *Proceedings of the 8th Canadian Conference on Earthquake Engineering (8CCEE)*, Vancouver, pp. 341–346.

Hatami K and Bathurst RJ (1999b) Dynamic response of reinforced-soil retaining walls to ground motion, part II: parametric analysis. *Proceedings of the 17th Canadian Congress of Applied Mechanics, CANCAM 99*, McMaster University, Hamilton, pp. 89–90.

Hatami K and Bathurst RJ (2000) Effect of structural design on fundamental frequency of reinforced-soil retaining walls. *Soil Dynamics and Earthquake Engineering* **19(3)**: 137–157.

Hatami K and Witthoeft A (2008) A numerical study on the use of geofoam to increase the external stability of reinforced soil walls. *Geosynthetics International* **15(6)**: 452–470.

Hatami K, Bathurst RJ and El-Emam M (2005) Acceleration amplification in the backfill of reinforced soil walls of different heights. *Proceedings of the 3rd Biot Conference on Poromechanics, Norman, OK* (Abousleiman YN, Cheng AHD and Ulm FJ (eds)), Balkema, London, pp. 725–731.

Horvath JS (1995) *Geofoam Geosynthetic*. Horvath Engineering, Scarsdale, NY.

Huang B and Bathurst RJ (2009) Evaluation of soil-geogrid pullout models using a statistical approach. *ASTM Geotechnical Testing Journal* **32(6)**: 489–504.

Huang B, Bathurst RJ and Hatami K (2009) Numerical study of reinforced soil segmental walls using three different constitutive soil models. *Journal of Geotechnical and Geoenvironmental Engineering* **135(10)**: 1486–1498.

Huang B, Bathurst RJ, Hatami K and Allen TM (2010) Influence of toe restraint on reinforced soil segmental walls. *Canadian Geotechnical Journal* **47(8)**: 885–904.

Iai S (1989) Similitude for shaking tests on soil–structure–fluid models in 1g gravitational fields. *Soils and Foundations* **29(1)**: 105–118.

Inglis D, Macleod G, Naesgaard E and Zergoun M (1996) Basement wall with seismic earth pressures and novel expanded polystyrene foam buffer layer. *Proceedings of the 10th Annual Symposium of the Vancouver Geotechnical Society*, Vancouver.

Ishibashi I and Fang YS (1987) Dynamic earth pressures with different wall movement modes. *Soils and Foundations*, JSSMFE **27(4)**: 11–22.

Ismeik M and Guler E (1998) Effect of wall facing on the seismic stability of geosynthetic-reinforced walls. *Geosynthetics International* **5(1–2)**: 41–53.

Itasca Consulting Group (1998) *FLAC: Fast Lagrangian Analysis of Continua*, version 3.4. Itasca Consulting Group, Minneapolis, MN.

Itasca Consulting Group (2005) *FLAC: Fast Lagrangian Analysis of Continua*, version 5. Itasca Consulting Group, Minneapolis, MN.

Izawa J, Ito H, Saito T, Ueno M and Kuwano J (2009) Development of rational seismic design method for geogrid-reinforced soil wall combined with fibre-mixed soil–cement and its applications. *Geosynthetics International* **16(4)**: 286–300.

Juran I, Knochenmus G, Acar YB and Arman A (1988) Pullout response of geotextiles and geogrids (synthesis of available experimental data). *Proceedings of the Symposium on Geosynthetics for Soil Improvement*, Nashville, TN, pp. 92–111.

Karpurapu R and Bathurst RJ (1995) Behavior of geosynthetic reinforced soil retaining walls using the finite element method. *Computers and Geotechnics* **17(3)**: 279–299.

Koga Y and Washida S (1992) Earthquake resistant design method of geotextile reinforced embankments. *Proceedings of the International Symposium on Earth Reinforcement Practice, IS-Kyushu '92*, Fukuoka, pp. 255–259.

Koga Y, Itoh Y, Washida S and Shimazu T (1988) Seismic resistance of reinforced embankment by model shaking tests. *Proceedings of the International Geotechnical Symposium on Theory and Practice of Earth Reinforcement, IS-Kyushu '88, Fukuoka*, Balkema, Rotterdam, pp. 413–418.

Koseki J, Bathurst RJ, Güler E, Kuwano J and Maugeri M (2006) Seismic stability of reinforced soil walls. Invited keynote paper. *Proceedings of 8th International Conference on Geosynthetics, Yokohama* (Kuwano J and Koseki J (eds)), Millpress, Rotterdam.

Koseki J, Tatsuoka F, Munaf Y, Tateyama M and Kojima K (1998a) A modified procedure to evaluate active earth pressure at high seismic loads. *Soils and Foundations (Special Issue)*, 209–216.

Koseki J, Munaf Y, Tatsuoka F *et al.* (1998b) Shaking and tilt table tests of geosynthetic-reinforced soil and conventional-type retaining walls. *Geosynthetics International* **5(1–2)**: 73–96.

Kramer SL (1996a) *Geotechnical Earthquake Engineering*. Prentice-Hall, Upper Saddle River, NJ.

Kramer SL (1996b) Personal communication.

Lee KZZ, Chang NY and Ko HY (2010) Numerical simulation of geosynthetic-reinforced soil walls under seismic shaking. *Geotextiles and Geomembranes* **28(4)**: 317–334.

Leshchinsky D (1995) *Design Procedure for Geosynthetic Reinforced Steep Slopes*. Waterways Experiment Station, US Army Corps of Engineers, Vicksburg, MS. Technical Report REMR-GT-120 (temporary number).

Leshchinsky D, Ling HI and Hanks GA (1995) Unified design approach to geosynthetic reinforced slopes and segmental walls. *Geosynthetics International* **2(5)**: 845–881.

Lin DY, Lin SS and Kuo SH (1996) Predicting seismic performance of geogrid-reinforced slopes. *Proceedings of the International Symposium on Earth Reinforcement, IS-Kyushu '96*, Fukuoka, pp. 791–795.

Ling HI and Leshchinsky D (1998) Effects of vertical acceleration on seismic design of geosynthetic-reinforced soil structures. *Géotechnique* **48(3)**: 347–373.

Ling HI, Leshchinsky D and Perry EB (1996) A new concept on seismic design of geosynthetic-reinforced soil structures: permanent-displacement limit. *Proceedings of the International Symposium on Earth Reinforcement, IS-Kyushu '96*, Fukuoka, pp. 117–122.

Ling HI, Leshchinsky D and Perry EB (1997) Seismic design and performance of geosynthetic-reinforced soil structures. *Géotechnique* **47(5)**: 933–952.

Ling HI, Mohri Y and Kawabata T (1998) Tensile properties of geogrids under cyclic loadings. *Journal of Geotechnical and Geoenvironmental Engineering, ASCE* **124(8)**: 782–787.

Ling HI, Wu JTH and Tatsuoka F (1992) Short-term strength and deformation characteristics of geotextiles under typical operational conditions. *Geotextiles and Geomembranes* **11(2)**: 185–219.

Ling HI, Liu H, Kaliakin VN and Leshchinsky D (2004) Analyzing dynamic behaviour of geosynthetic-reinforced soil retaining walls. *Journal of Engineering Mechanics* **130(8)**: 911–920.

Ling HI, Liu H, Kaliakin VN and Mohri Y (2005a) Parametric studies on the behavior of reinforced soil retaining walls under earthquake loading. *Journal of Engineering Mechanics* **131(10)**: 1056–1065.

Ling HI, Mohri Y, Leshchinsky D, Burke C, Matsushima K and Lui H (2005b) Large-scale shaking table tests on modular-block reinforced soil retaining walls. *Journal of Geotechnical and Geoenvironmental Engineering* **131(4)**: 465–476.

Ling HI, Yang S, Leshchinsky D, Liu H and Burke C (2010) Finite-element simulations of full-scale modular-block reinforced soil retaining walls under earthquake loading. *Journal of Engineering Mechanics* **136(5)**: 653–661.

Madabhushi SPG (1996) Importance of strong motion in the design of earth reinforcement. *Proceedings of the International Symposium on Earth Reinforcement, IS-Kyushu '96*, Fukuoka, pp. 239–248.

Matasovic N, Kavazanjian E and Yan L (1997) Newmark deformation analysis with degrading yield acceleration. *Proceedings of Geosynthetics '97*, Long Beach, CA, pp. 989–1000.

Matsuo O, Tsutsumi T, Yokoyama K and Saito Y (1998) Shaking table tests and analyses of geosynthetic-reinforced soil retaining walls. *Geosynthetics International* **5(1–2)**: 97–126.

McGown A, Andrawes KZ and Kabir MH (1982) Load-extension testing of geotextiles confined in soil. *Proceedings of the 2nd International Conference on Geotextiles*, Las Vegas, NV, pp. 793–798.

McGown A, Yogarajah I, Andrawes KZ and Saad MA (1995) Strain behavior of polymeric geogrids subjected to sustained and repeated loading in air and in soil. *Geosynthetics International* **2(1)**: 341–355.

Min Y, Leshchinsky D, Ling HI and Kaliakin VN (1995) Effects of sustained and repeated tensile loads on geogrid embedded in sand. *Geotechnical Testing Journal, ASTM* **18(2)**: 204–235.

Miyamori T, Iwai S and Makiuchi K (1986) Frictional characteristics of nonwoven fabrics. *Proceedings of the 3rd International Conference on Geotextiles*, Vienna, pp. 701–705.

Mononobe N and Matsuo H (1929) On the determination of earth pressure during earthquake. *Proceedings of the World Engineering Congress*, Tokyo, pp. 177–185.

Moraci N and Montanelli F (1997) Behaviour of geogrids under cyclic loads. *Proceedings of Geosynthetics '97*, Long Beach, CA, pp. 961–976.

Murata O, Tateyama M and Tatsuoka F (1994) Shaking table tests on a large geosynthetic-reinforced soil retaining wall model. *Recent Case Histories of Permanent Geosynthetic-reinforced Soil Walls, Seiken Symposium* (Tatsuoka F and Leshchinsky D (eds)), Tokyo, pp. 259–264.

Myles B (1982) Assessment of soil fabric friction by means of shear evaluation. *Proceedings of the 2nd International Conference on Geotextiles*, Las Vegas, NV, pp. 787–791.

NCMA (National Concrete Masonry Association) (2009) *Design Manual for Segmental Retaining Walls*, 3rd edn (Bernardi M (ed.)), NCMA, Herndon, VA.

Newmark NM (1965) Effect of earthquakes on dams and embankments. *Géotechnique* **15(2)**: 139–159.

Nishimura J, Hirai T, Iwasaki K, Saito Y and Morishima M (1996) Earthquake resistance of geogrid-reinforced soil walls based on a study conducted following the southern Hyogo earthquake. *Proceedings of the International Symposium on Earth Reinforcement, IS-Kyushu '96*, Fukuoka, pp. 439–444.

Nova-Roessig L and Sitar N (1998) Centrifuge studies of the seismic response of reinforced slopes. *Proceedings of the 3rd Geotechnical Engineering and Soil Dynamics Conference*, ASCE, Seattle, WA, pp. 458–468.

Okabe S (1924) General theory on earth pressure and seismic stability of retaining wall and dam. Doboku Gakkai. *Journal of the Japan Society of Civil Engineers* **10(6)**: 1277–1323.

Okamoto S (1984) *Introduction to Earthquake Engineering*. University of Tokyo Press, Tokyo.

O'Rourke TD, Druschel SJ and Netravali AN (1990) Shear strength characteristics of sand-polymer interfaces. *Journal of Geotechnical Engineering, ASCE* **116(3)**: 451–469.

Otani J, Yamamoto A, Kodoka T, Yasufuku N and Yashima A (1997) Current state on numerical analysis of reinforced soil structures. *Proceedings of the International Symposium on Earth Reinforcement, IS-Kyushu '96*, Fukuoka, pp. 1159–1170.

Paz M (1994) *International Handbook of Earthquake Engineering*. Chapman and Hall, New York.

Pelletier MJ (1996) Investigation of the seismic resistance of reinforced segmental walls using small-scale shaking table testing. MEng thesis, Department of Civil Engineering, Royal Military College of Canada, Kingston.

PWRI (Public Works Research Institute) (1992) *Design and Construction Manual for Reinforced Soil Structures Using Geotextiles*. PWRI, Ministry of Construction, Tsukuba. Internal Report No. 3117 (in Japanese).

Raju M (1995) Monotonic and cyclic pullout resistance of geosynthetics. PhD thesis, University of British Columbia, Vancouver.

Richards R and Elms DG (1979) Seismic behavior of gravity retaining walls. *Journal of the Geotechnical Engineering Division, ASCE* **105(GT4)**: 449–464.

Rowe RK and Ho SK (1993) A review of the behavior of reinforced soil walls. *Proceedings of the International Symposium on Earth Reinforcement Practice, IS-Kyushu '92*, Fukuoka, pp. 801–830.

Sabhahit N, Madhav MR and Basudhar PK (1996) Seismic analysis of nailed soil slopes – a pseudo-dynamic approach. *Proceedings of the International Symposium on Earth Reinforcement, IS-Kyushu '96*, Fukuoka, pp. 821–824.

Sakaguchi M (1996) A study of the seismic behavior of geosynthetic reinforced walls in Japan. *Geosynthetics International* **3(1)**: 13–30.

Sakaguchi M, Muramatsu M and Nagura K (1992) A discussion on reinforced embankment structures having high earthquake resistance. *Proceedings of the International Symposium on Earth Reinforcement Practice, IS-Kyushu '92*, Fukuoka, pp. 287–292.

Sakaguchi M, Yamada K and Tanaka M (1994) Prediction of deformation of geotextile reinforced walls subjected to earthquakes. *Proceedings of the 5th International Conference on Geotextiles, Geomembranes and Related Products*, Singapore, pp. 521–524.

Sandri D (1994a) Retaining walls stand up to the Northridge earthquake. *Geotechnical Fabrics Report* **12(4)**: 30–31.

Sandri D (1994b) Personal communication.

Sarma SK (1975) Seismic stability of earth dams and embankments. *Géotechnique* **25(4)**: 743–761.

Schimming BB and Saxe HC (1964) Inertial effects of soil strength criteria. *Proceedings of the Symposium on Soil–Structure Interaction*, Tucson, AZ, pp. 118–128.

Scott RF (1973) Earthquake-induced earth pressures on retaining walls. *Proceedings of the 5th World Conference on Earthquake Engineering*, Rome, pp. 1611–1620.

Seed HB and Idriss IM (1969) Influence of soil conditions on ground motion during earthquakes. *Journal of the Soil Mechanics and Foundations Division, ASCE* **95**: 99–137.

Seed HB and Whitman RV (1970) Design of earth retaining structures for dynamic loads. *Proceedings of the ASCE Specialty Conference: Lateral Stresses in the Ground and Design of Earth Retaining Structures*, Ithaca, NY, pp. 103–147.

Segrestin P and Bastick MJ (1988) Seismic design of reinforced earth retaining walls – the contribution of finite element analysis. *Proceedings of the International Geotechnical Symposium on Theory and Practice of Earth Reinforcement, IS-Kyushu '88, Fukuoka*, Balkema, Rotterdam, pp. 577–582.

Shinoda M and Bathurst RJ (2004a) Strain measurement of geogrids using a video-extensometer technique. *Geotechnical Testing Journal, ASTM* **27(5)**: 456–463.

Shinoda M and Bathurst RJ (2004b) Lateral and axial deformation of PP, HDPE and PET geogrids under tensile load. *Geotextiles and Geomembranes* **22**: 205–222.

Simac MR, Bathurst RJ, Berg RR and Lothspeich SE (1993) *National Concrete Masonry Association Segmental Retaining Wall Design Manual*. Earth Improvement Technologies, Fort Mill, SC.

Steedman RS and Zeng X (1990) The influence of phase on the calculation of pseudo-static earth pressure on a retaining wall. *Géotechnique* **40(1)**: 101–112.

Stewart JP, Bray JD, Seed RB and Sitar N (1994) *Preliminary Report on the Principal Geotechnical Aspects of the January 17, 1994 Northridge Earthquake*. Earthquake Engineering Research Centre, University of California at Berkeley, CA. Report UCB/EERC-94/08.

Sugimoto M, Ogawa S and Moriyama M (1994) Dynamic characteristics of reinforced embankments with steep slope by shaking model tests. *Recent Case Histories of Permanent Geosynthetic-Reinforced Soil Walls, Seiken Symposium*, Tokyo, pp. 271–275.

Takasumi DL, Green KR and Holtz RD (1991) Soil–geosynthetics interface strength characteristics: a review of state-of-the-art testing procedures. *Proceedings of Geosynthetics '91*, Atlanta, GA, pp. 87–100.

Tateyama M, Tatsuoka F, Koseki J and Horii K (1995) Damage to soil retaining walls for railway embankments during the Great Hanshin-Awaji Earthquake, January 17, 1995. *Proceedings of the 1st International Conference on Earthquake Geotechnical Engineering, IS-Tokyo '95, Tokyo*, Balkema, Rotterdam, pp. 49–54.

Tatsuoka F, Koseki J and Tateyama M (1995) Performance of geogrid-reinforced soil retaining walls during the Great Hanshin-Awaji Earthquake, January 17, 1995. *Proceedings of the 1st International Conference on Earthquake Geotechnical Engineering, IS-Tokyo '95, Tokyo*, Balkema, Rotterdam, pp. 55–62.

Tatsuoka F, Tateyama M, Uchimura T and Koseki J (1997) Geosynthetic-reinforced soil retaining walls as important permanent structures. Mercer Lecture. *Geosynthetics International* **4(2)**: 81–136.

Tatsuoka F, Koseki J, Tateyama M, Munaf Y and Horii K (1998) Seismic stability against high seismic loads on geosynthetic-reinforced soil retaining structures. Keynote lecture. *Proceedings of the 6th International Conference on Geosynthetics*, Atlanta, GA, pp. 103–142.

Telekes G, Sugimoto M and Agawa S (1994) Shaking table tests on reinforced embankment models. *Proceedings of the 13th International Conference on Soil Mechanics and Foundation Engineering*, New Delhi, pp. 649–654.

United States Geological Survey (2000) *Earthquake Hazards Program – National seismic hazard mapping project*. USGS, Golden, CO (http://geohazards.cr.usgs.gov/eq/).

Vrymoed J (1989) Dynamic stability of soil-reinforced walls. *Transportation Research Record* **1242**: 29–38.

Walters DL, Allen TM and Bathurst RJ (2002) Conversion of geosynthetic strain to load using reinforcement stiffness. *Geosynthetics International* **9(5–6)**: 483–523.

White DM and Holtz RD (1996) Performance of geosynthetic-reinforced slopes and walls during the Northridge, CA earthquake of January 17, 1994. *Proceedings of the International Symposium on Earth Reinforcement, IS-Kyushu '96*, Fukuoka, pp. 965–972.

Whitman RV (1990) Seismic design and behavior of gravity retaining walls. *ASCE Specialty Conference: Design and Performance of Earth Retaining Structures*, Ithaca, NY, pp. 817–842.

Whitman RV and Liao S (1984) Seismic design of gravity retaining walls. *Proceedings of the 8th World Conference on Earthquake Engineering*, San Francisco, CA, pp. 533–540.

Wilson-Fahmy RF, Koerner RM and Fleck JA (1993) Unconfined and confined wide width tension testing of geosynthetics. In *Geosynthetic Soil Reinforcement Testing Procedures* (Cheng SCJ (ed.)), ASTM, Philadelphia, PA, pp. 49–63.

Wolfe WE, Lee KL, Rea D and Yourman AM (1978) The effect of vertical motion on the seismic stability of reinforced earth walls. *Proceedings of the ASCE Symposium on Earth Reinforcement*, Pittsburgh, PA, pp. 856–879.

Wood JH (1973) *Earthquake-induced earth Pressures on Structures*. California Institute of Technology, Pasadena, CA. Report EERL 73-05.

Woods RI and Jewell RA (1990) A computer design method for reinforced soil structures. *Geotextiles and Geomembranes* **9(3)**: 233–259.

Wu G and Finn WDL (1996) Seismic pressures against rigid walls. *ASCE Specialty Conference on Analysis and Design of Retaining Structures against Earthquakes*, Washington, DC, pp. 1–18.

Wu G and Finn WDL (1999) Seismic lateral pressures for design of rigid walls. *Canadian Geotechnical Journal* **36**: 509–522.

Wu JTH (ed.) (1992) Geosynthetic-reinforced soil retaining walls. *Proceedings of the International Symposium on Geosynthetic-Reinforced Soil Retaining Walls*, Denver, CO.

Yasuda S, Nagase H and Marui H (1992) Cyclic pullout tests of geogrids in soils. *Proceedings of the International Symposium on Earth Reinforcement Practice, IS-Kyushu '92*, Fukuoka, pp. 185–190.

Yegian MK and Kadakal U (1998) Geosynthetic interface behaviour under dynamic loading. *Geosynthetics International* **5(1–2)**: 1–16.

Yogendrakumar M and Bathurst RJ (1992) Numerical simulation of reinforced soil structures during blast loads. *Transportation Research Record* **1336**: 1–8.

Yogendrakumar M, Bathurst RJ and Finn WDL (1991) Response of reinforced soil slopes to earthquake loadings. *Proceedings of the 6th Canadian Conference on Earthquake Engineering*, Toronto, pp. 445–452.

Yogendrakumar M, Bathurst RJ and Finn WDL (1992) Dynamic response analysis of a reinforced soil retaining wall. *Journal of Geotechnical Engineering, ASCE* **118(8)**: 1158–1167.

You L and Michalowski L (1999) Displacement charts for slopes subjected to seismic loads. *Computers and Geotechnics* **25(1)**: 45–55.

Zarnani S and Bathurst RJ (2007) Experimental investigation of EPS geofoam seismic buffers using shaking table tests. *Geosynthetics International* **14(3)**: 165–177.

Zarnani S and Bathurst RJ (2008) Numerical modeling of EPS seismic buffer shaking table tests. *Geotextiles and Geomembranes* **26(5)**: 371–383.

Zarnani S and Bathurst RJ (2009a) Influence of constitutive model on numerical simulation of EPS seismic buffer shaking table tests. *Geotextiles and Geomembranes* **27(4)**: 308–312.

Zarnani S and Bathurst RJ (2009b) Numerical parametric study of EPS geofoam seismic buffers. *Canadian Geotechnical Journal* **46(3)**: 318–338.

Zarrabi K (1979) Sliding of gravity retaining wall during earthquakes considering vertical acceleration and changing inclination of failure surface. MSc thesis, Department of Civil Engineering, Massachusetts Institute of Technology, Cambridge, MS.

Zimmie TF, De A and Mahmud MB (1994) Centrifuge modelling to study dynamic friction at geosynthetic interfaces. *Proceedings of the 5th International Conference on Geotextiles, Geomembranes and Related Products*, Singapore, pp. 415–418.

Handbook of Geosynthetic Engineering
ISBN 978-0-7277-4175-2

http://dx.doi.org/10.1680/hge.41752.365

ice
Institution of Civil Engineers
publishing

Chapter 17
Geosynthetic applications – general aspects and selected case studies

S. K. Shukla
Discipline of Civil Engineering, School of Engineering, Edith Cowan University, Perth, Australia

The successful application of geosynthetics in engineering projects depends on many general application aspects in addition to methods of analysis, design and construction of geosynthetic-reinforced structures, which have been described in detail in the previous chapters. This chapter discusses general aspects of geosynthetic applications, namely general application guidelines, quality control and *in situ* monitoring, cost analysis, and some general application problems. Many case studies have already been described in previous chapters, but some selected additional case studies are presented in this chapter to help readers and geosynthetic users develop confidence in geosynthetic applications.

17.1. Introduction

Geosynthetics have pervaded many areas of civil engineering, including geotechnical engineering, environmental engineering, hydraulic engineering and transportation engineering. It is now no longer possible to work without geosynthetics in these areas. Geosynthetics perform several functions in a variety of field applications, as explained in Chapter 1. Their major field applications have been described, in detail, in previous chapters, and many case studies have been presented there. There are several application-related general aspects, namely general guidelines on the application of geosynthetics, quality control and *in situ* monitoring, cost analysis, and general problems in applications, that are essential knowledge for users of geosynthetics if applications are to be successful. This chapter provides information on all these aspects, which have not been dealt with in detail in the previous chapters. Several additional case studies are also described here so that readers can easily develop confidence in the applications of geosynthetics.

17.2. General guidelines

In all the field applications of geosynthetics, the common objective is to install the correct geosynthetic in the correct location without impairing its properties during the construction process. Several general and specific guidelines have been suggested concerning this common objective (John, 1987; Ingold and Miller, 1988; Koerner, 2005; Shukla and Yin, 2006). Within the scope of this book, some general guidelines are presented below.

Care and consideration

In many projects, environmental factors during on-site storage, and mechanical stresses during construction and initial operation, place the most severe conditions on a geosynthetic during its projected lifetime. The successful installation of a geosynthetic is, therefore, largely dependent on the construction technique and the management of construction activities. Thus, the installation of geosynthetics in practice requires a degree of care and consideration.

In the past, most of the geosynthetic-related failures were reported to be construction related, and a few design related. The construction-related failures were caused mainly by the following problems

- poor construction
- loss of strength due to ultraviolet (UV) exposure
- lack of proper overlap of geosynthetic layers
- inferior-quality materials
- high installation stresses.

Although the general nature of installation-induced damage to geosynthetics (e.g. cuts, tears, splits and perforations) can be assessed by site trials, no test methods have yet been derived by which the same type and degree of damage can be reproduced consistently in the laboratory. However, the strength reduction due to damage during installation might be partially, or completely, avoided where the damage is found to be most severe by considering carefully the following elements

- firm, rock or frozen subgrades
- thin lift thickness using heavy equipment
- large particle size, poorly graded cover soil
- lightweight, low strength, geosynthetics.

If the type of subgrade cannot be changed, the options remain to change the construction practice or to modify the geosynthetic being used for a specific application. However, one may attempt to do both by recommending less severe construction practices and adopting a set of criteria on geosynthetic strength, such as reductions in the values of strength and strain to be taken into account when assessing the design tensile capacity of the geosynthetics.

Due care has also to be taken during spreading and compaction of the fill materials on geosynthetic layers, particularly for very soft subgrades and/or very coarse fill materials (stones, rockfill, etc.), in order to avoid or minimise mechanical damage to geosynthetics.

Geosynthetic selection

The selection of the geosynthetic type to function as a reinforcement should be done by keeping in mind the general objectives of its use, which are to increase the stability of the soil (bearing capacity, slope stability and resistance to erosion) and to reduce its deformation (settlement and lateral deformation). In order to provide stability, the geosynthetic has to have adequate strength; and to control deformation, it has to have suitable force–elongation characteristics, measured in terms of modulus (the slope of the force elongation curve), as explained in Section 1.6.2.

The selected geosynthetic should have a certain minimum strength and stiffness so that it is fit to withstand the effects of placement on the ground, and the loads imposed by equipment and personnel during installation. In other words, construction engineers should consider the field survivability and workability requirements of geosynthetics during their selection. These requirements can be expressed in terms of grab strength, puncture strength, burst strength, impact strength and tear strength. The actual values of these survivability properties of geosynthetics should be decided on the basis of the expected degree of damage (low, moderate, high or very high) on their installation in a specific field application.

Many times, the cost and availability of geosynthetics may govern their selection.

Test methods

If the test methods for determining the geosynthetic properties are not completely field simulated, the test values must be adjusted, as discussed in Section 1.6.

Protection before installation

When delivered, all the rolls of a geosynthetic should be wrapped in a protective layer of plastic to avoid any damage/deterioration during transportation.

Storage areas should be located as close as possible to the point of end use, in order to minimise subsequent handling and transportation. It is usually adequate to stack the rolls with a protective outer wrapper directly on the ground, provided that this is even, well-drained and free from sharp projections, such as rock pieces, stumps of trees or bushes. If the rolls are to be stored for a long period, some form of shading is required, unless the wrapper is made of opaque material, to give protection against UV light attack. If the wrapper is damaged and beyond repair, the roll should be stored by making a suitable arrangement to prevent ingress of water. Without this, the geotextile, particularly the nonwoven types, may collect water in voids, thereby causing the weight of the roll to increase, possibly to an unmanageable level. It must be noted that geosynthetics are generally hydrophobic (i.e. they repel water), so they have no wick-like action. Where geosynthetics are to be used as filters, it is important to keep the wrapper intact in order to provide protection against the ingress of dust and mud.

Site preparation

The original ground level may need to be graded to some predetermined formation level. During site preparation, sharp objects, such as boulders, stumps of trees or bushes, that might puncture or tear the geosynthetic should be removed. Disturbance of the subgrade should be minimised where the soil structure, roots in the ground and light vegetation may provide additional bearing strength.

Geosynthetic installation/placement

When handling rolls of geosynthetic manually or by some mechanical means at any stage of installation, the load, if any, should not be taken directly by the geosynthetic. It should be rolled/unrolled into place rather than dragged. An overlap between adjacent sheets must be provided when unrolling the geosynthetic into position after site preparation. The overlap used is generally a minimum of 30 cm, but, in applications where the geosynthetic is subject to tensile stresses, the overlap must be increased or the sheets of geosynthetic sewn/bonded.

Joints

Where necessary, geosynthetics, particularly geotextiles, are jointed by sewing or some other mechanical means before placing. High-strength geosynthetics, employed for their reinforcing potential, should normally be sewn. Overlapping, by 0.3–1 m, may be employed if relatively small tensile forces are developed in geosynthetic layers. Overlaps should not be employed in the primary tensile direction in geosynthetic applications. Stapling may be used with geotextiles to make temporary joints, but should never be used for structural jointing.

It should be noted that seams and overlaps are most probably the weakest link in geosynthetic-reinforced soil structures. Hence, joints should be formed to have the highest mechanical and durability efficiency possible, compared with the performance characteristics of the parent materials.

The most important aspect of construction with geomembranes is the seam. Without proper seams, the concept of using a geomembrane liner as a liquid or vapour barrier is meaningless. A geomembrane seam, in service, must maintain its leak-free condition. Metal hog rings should never be used when geonets are used in conjunction with geomembranes.

Cutting of geosynthetics on site

The cutting of geosynthetics is a labour-intensive, time-consuming operation, which, in most cases, can be avoided by forward planning, although the total width of an area to be covered will rarely be an exact multiple of available widths. The maximum geosynthetic width is generally 5.3 m. There is less wastage of time and money, if slightly larger overlaps or wraparound are allowed to take up the excess width, than if the geosynthetic is cut on site. In the case of walls and steep-sided embankments, the wraparound may enhance compaction at the edges and also helps to reduce erosion, and may assist in the establishment of vegetation.

Protection during construction and service life

The damage due to UV light exposure can usually be avoided by not laying more geosynthetics in a day than can be covered by fill in that same day. Unused portions of rolls must be rerolled and protected immediately. It is to be noted that when the geosynthetic is UV-stabilised, degradation is largely reduced but not entirely eliminated.

Protection of the wall face against degradation due to UV light exposure and, to some extent, against vandalism, can be provided by covering the geosynthetic with gunite (shotcrete), asphalt emulsion, asphalt products or other coatings. A wire mesh anchored to the geosynthetic may be necessary to keep the coating on the face of the wall. In the case of walls constructed from geogrids, vegetation can easily grow through (or be placed behind) the large openings, and UV degradation of the relatively thick ribs is significantly lower. Thus, the need to cover the wall face is not as compelling as with geotextiles, and the front of geogrid walls is sometimes left exposed.

The chemical resistance of the geosynthetic liner to the contained liquid must be considered for the entire service life of its installation. The minimum thickness of the geomembrane liner is usually recommended to be 20 mils (0.50 mm), irrespective of design calculations; however, this lower limit may be 80 mils (2.0 mm) in the case of the containment of hazardous materials. When the secondary liner is also a geomembrane, it must be of the same thickness as the primary liner.

Geosynthetic clay liners are extremely sensitive to damage during and after construction, owing to their thinness and small mass of bentonite. So, great care is required in applications with geosynthetic clay liners.

Once the geosynthetic is laid, it should not be trafficked until an adequate layer of fill is placed over it, thus affording some protection. One exception to this rule is where a heavy-weight geosynthetic is used, which is specifically designed to be directly trafficked by vehicles. For the initial stages of construction, low ground pressure and small dump-trucks should be used. For very soft formations, it is necessary to use special low-bearing-pressure tracked vehicles for spreading the fill over the geosynthetic layer. During filling operations, the blades or buckets of the construction plant must not be allowed to come into contact with the geosynthetic. In the case of road and embankment construction, the minimum fill cover should be 200–300 mm, depending on aggregate size and weight of trucks. A maximum lift thickness may be imposed in order to control the size of the mud wave (bearing failure) ahead of the dumping due to excessive fill weight. A further lift may be placed after consolidation of the subgrade has increased its strength. Compaction of the first aggregate layer is usually achieved by the construction equipment alone. A continued build-up of cover material will allow vibratory rollers to be used. Proof rolling by a heavy rubber-tyred vehicle may provide pre-tensioning of the geosynthetic by creating initial ruts, which are subsequently refilled and levelled.

Figure 17.1 Effects of heavy compaction
Reproduced from Voskamp *et al.* (1990), Industrial Fabrics Association International

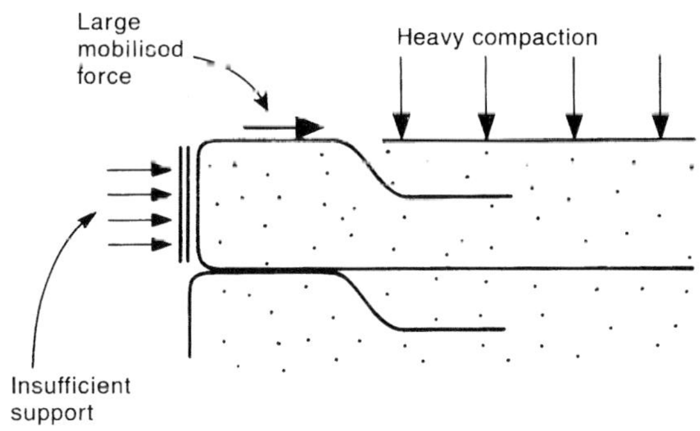

Proper care must be given during compaction of the top layers of wraparound reinforced steep slopes. This is required because very high compaction results in very high stresses in the geosynthetic reinforcement due to movement of the fills in the wraparound sections, as shown in Figure 17.1, and such situations are not desirable.

All vehicles, and all construction equipment weighing more than 1500 kg, should be kept at least 2 m away from the face of the wall or steep slope. The fill within 2 m of the face of the wall or steep slope should be compacted using a vibro-tamper, a vibrating plate compactor with a mass not exceeding 1000 kg, or by a vibrating roller with a total mass not exceeding 1500 kg.

If it is necessary to use poorly graded aggregate fill, and heavy construction equipment for placement and compaction, it may be prudent to place a cushioning layer of sand above the geosynthetic.

If a geosynthetic is used in conjunction with bituminous material, care must be taken to ensure that the temperature of the bituminous material is well below that of the melting point of the geosynthetic.

Figure 17.2 (a) Simple run-out and anchor trenches, (b) a rectangular trench, (c) a V trench and (d) a narrow trench
Reproduced from Hullings and Sansone (1997), with permission from Elsevier

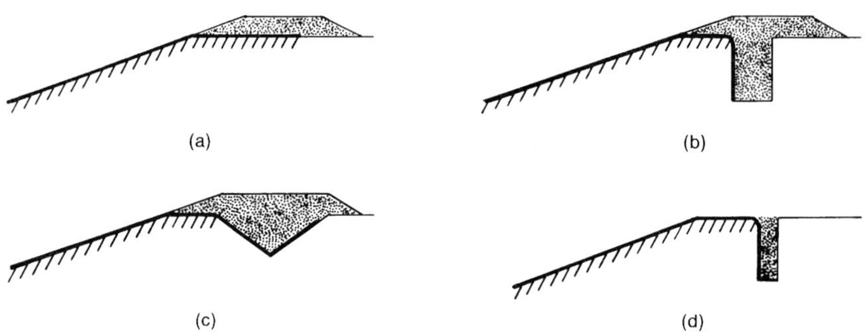

In the case of a liquid containment pond, to shield the geomembrane liner from ozone, UV light, temperature extremes, ice damage, wind stresses, accidental damage and vandalism, a soil cover of at least 30 cm thickness may be provided. For proper design of the containment, a geotextile should be used beneath the geomembrane, placed directly on the prepared soil subgrade before liner placement. The cover soils over geomembranes, installed on sloping ground, can unravel and slump very easily, even under static conditions. To alleviate this situation somewhat, it is common practice to taper the cover soil, laying it thicker at the bottom and gradually thinner going towards the top.

Damage correction

For more critical structures, such as reinforced soil walls and embankments, it is safest to remove the damaged section, if any, of the geosynthetic entirely and replace it with undamaged geosynthetic. In these applications, a certain degree of damage may be acceptable, provided that this has been allowed for at the design stage. In low-risk applications, where the geosynthetic is not subject to significant tensile stress or dynamic water loading, it is permissible to patch the damaged area.

Anchorage

To maintain the position of the geosynthetic sheet before covering with soil/fill, the edges of the sheet must be weighted or anchored in trenches, thereby providing significant pull-out resistance (Figure 17.2). Anchorage selection depends on site conditions. In the case of unpaved roads, the geosynthetic should be anchored on each side of the road. The bond length (typically around 1.0–1.5 m) can be achieved by extending the geosynthetic beyond the required running width of the road (Figure 17.3(a)), or by providing an equivalent bond length by burying the geosynthetic sheet in shallow trenches (Figure 17.3(b)) or by wraparound (Figure 17.3(c)). Similar approaches can also be adopted in some other applications.

Prestressing

Simple procedures, such as prestressing the geosynthetic, may enhance the reinforcement function in some applications. For example, to specifically add reinforcement for paved roads on firm subsoils, a geosynthetic prestressing system may be required. By prestressing the geosynthetic, the aggregate base will be placed in compression, thereby providing lateral confinement, and will effectively increase its modulus over the unreinforced case. Prestressed geosynthetic layers with their edges anchored have been found to be effective in foundations and embankments (Shukla and Chandra, 1994; Shukla and Kumar, 2008; Lovisa *et al.*, 2010). Therefore, the geosynthetic layers should be prestressed in such field applications, provided the prestressing technique can be adopted economically.

Maintenance

All geosynthetic-reinforced soil structures should be subjected to a regular programme of inspection and maintenance. A habit should be developed to keep records of the inspections and any maintenance carried out.

It should be noted that geosynthetic applications in the field may require many specific construction guidelines. Many such guidelines have already been discussed for the applications described in previous chapters. Readers can obtain further details from the relevant available codes of

Figure 17.3 Use of geosynthetics in unpaved road construction
Reproduced from Ingold and Miller (1988)

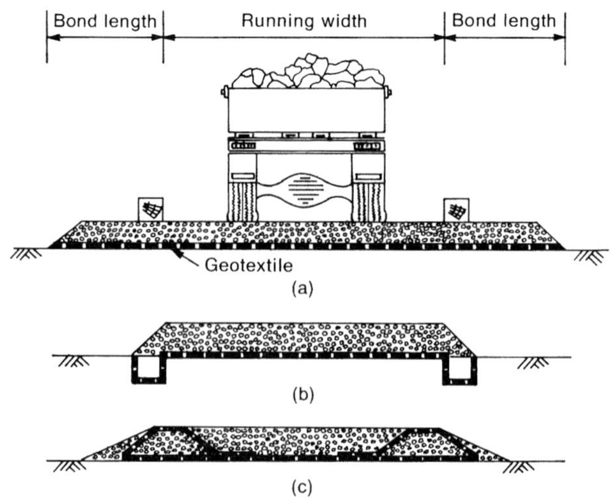

Table 17.1 Codes of practice and standards for some applications of geosynthetics

Designations of codes of practice/standards	Topics
BS 8006:1995	Strengthened/reinforced soils and other fills
BS EN 13249:2001	Characteristics required for use in the construction of roads and other trafficked areas (excluding railways and asphalt inclusion)
BS EN 13250:2001	Characteristics required for use in the construction of railways
BS EN 13251:2001	Characteristics required for use in earthworks, foundations and retaining structures
BS EN 13252:2001	Characteristics required for use in drainage systems
BS EN 13253:2001	Characteristics required for use in erosion control works (coastal protection and bank revetments)
BS EN 13254:2001	Characteristics required for use in the construction of reservoirs and dams
BS EN 13255:2001	Characteristics required for use in the construction of canals
BS EN 13256:2001	Characteristics required for use in the construction of tunnels and underground structures
BS EN 13257:2001	Characteristics required for use in solid waste disposals
BS EN 13265:2001	Characteristics required for use in liquid waste containment projects
ASTM D6088-06	Installation of geocomposite pavement drains
ASTM D5889-97(2008)	Quality control of geosynthetic clay liners
ASTM D7239-06	Specification for hybrid geosynthetic paving mat for highway applications
IS 14715:2000	Specification of jute geotextiles
IS 15872:2009	Application of coir geotextiles for rain water erosion control in roads, railway embankments and hill slopes
IS 15871:2009	Use of coir geotextiles in unpaved roads

practice/standards, some of which are mentioned in Table 17.1. Many geosynthetic manufacturers have developed design charts and graphs, as well as construction guidelines, for geosynthetic-reinforced structures. If a specific geosynthetic product is to be used, these guidelines can be considered. However, it should be noted that they are product-specific in their assumptions regarding allowable strength, the factor of safety, etc.

17.3. Quality control and *in situ* monitoring

The 'quality' of a geosynthetic is the confidence that can be placed in it consistently meeting the numerically claimed variation limits in properties taken into account by the design engineer, and extrapolated into the *in situ* installation and functioning of the product (Donckers, 1994). Quality control is strictly the statistical control of the product during the manufacturing process. To achieve this, the manufacturer needs a quality assurance system, of which quality control is only a part.

Quality control on construction sites is done by index testing, which has been discussed in Section 1.6.2. Index testing involves the use of very simple techniques, which do not provide definitive design parameters for a geosynthetic but do give reproducible results, suitable for quality control and comparison of geosynthetics. Users should always check for the type and quality of the geosynthetics being delivered. In order to identify each roll or package of geosynthetic, the following basic information might be provided (Ingold and Miller, 1988)

- manufacturer's name
- commercial name of the geosynthetic
- method of manufacture and the constituent materials
- mass per unit area
- nominal thickness
- dimensions
- weight of the geosynthetic in a roll.

A simple check on the mass per unit area may be made using basic equipment, such as a balance and a scale. In the case of high-risk applications, such as the use of geosynthetic filters in dams and geosynthetics as a soil reinforcement, testing of every roll, or at least every other roll, should be performed. In such demanding applications, the most important property (see Chapter 1, Table 1.8) should be determined in addition to the basic index properties mentioned above. In the case of low-risk applications, such as the use of the geosynthetic as a separator in unpaved roads, only the basic index tests need to be carried out for every one in 10 or 20 rolls. It is thus noted that the frequency and degree of quality control testing are generally functions of application and the risk involved in that application.

The *in situ* monitoring of geosynthetics and the geosynthetic-related system usually has two goals. One addresses the integrity and safety of the system, whereas the other provides guidance and insight into the design process. It is important to conceive and execute a monitoring plan with clear objectives in mind. Dunnicliff (1988) provides a methodology for organising a monitoring programme in geotechnical instrumentation. The checklist of specific steps that are recommended is as follows

(*a*) Define the project conditions.
(*b*) Predict the mechanism(s) that control behaviour.

(*c*) Define the question(s) that need answering.
(*d*) Define the purpose of the instrumentation.
(*e*) Select the parameter(s) to be monitored.
(*f*) Predict the magnitude(s) of change.
(*g*) Devise remedial action.
(*h*) Assign relevant tasks.
(*i*) Select the instruments.
(*j*) Select the instrument locations.
(*k*) Plan for factors influencing the measured data.
(*l*) Establish procedures for ensuring corrections.
(*m*) List the purposes of each instrument.
(*n*) Prepare a budget.
(*o*) Write an instrument procurement specification.
(*p*) Plan the installation.
(*q*) Plan for regular calibration and maintenance.
(*r*) Plan for data collection, processing, presentation, interpretation, reporting and implementation.
(*s*) Write the contractual arrangements for field services.
(*t*) Update the budget as the project progresses.

The above checklist should be considered in planning for the *in situ* monitoring of geosynthetics whenever permanent and/or critical installations are under consideration, or are being otherwise challenged (Koerner, 1996).

Presently, there are a wide range of *in situ* monitoring methods/devices that have generally resulted in reliable data. Table 17.2 provides a summary of the monitoring methods/devices as presented by Koerner (1996). In this table, monitoring methods or devices are somewhat arbitrarily divided into recommended and optional categories. Table 17.3 gives a further description of the various methods/devices listed in Table 17.2. Since the monitoring is site-specific, its cost must be assessed on a case-by-case basis.

17.4. Cost analysis

The design engineer is usually confronted with an important task: whether a conventional solution or geosynthetic-related solution should be preferred in a particular civil engineering project at a specific site. In order to make a rational decision, data related to the following aspects should be compared carefully (Durukan and Tezcan, 1992)

- relative economy
- cost–performance efficiency
- factors of safety
- feasibility
- availability of materials
- relative speed of construction.

In the case of reinforced soil walls, it is generally accepted that, under normal circumstances, and especially after a wall height of about 6 m, they become more economical, and also they are relatively easier and faster to build than their conventional counterparts (Ingold, 1982). Reinforced soil retaining walls are almost indispensable when normal slopes cannot be constructed due to property line constraints, high expropriation costs, the existence of important structures, or land being reserved for future structures. In order to arrive at a scientific conclusion, however, a comparative cost analysis should be performed.

The *rate of relative economy* (E_r) is defined as

$$E_r = \left(\frac{C_c - C_r}{C_r} \times 100\right)\% \qquad (17.1)$$

where C_c is the cost of the conventional soil structure and C_r is the cost of the geosynthetic-reinforced soil structure.

A general idea of the cost–performance efficiency of a geosynthetic, or any other element of reinforced soil structure, can be represented by the *normalised cost* (C_m). In the case of geosynthetic-reinforced soil retaining walls, C_m can be defined as

$$C_m = \frac{C}{T} \qquad (17.2)$$

where C_m is the normalised cost of geosynthetic reinforcement carrying a safe tensile load of 1 kN on a 1 m run wall, C is the cost of 1 m^2 geosynthetic within a 1 m run wall, and T is the safe tensile resistance of one layer geosynthetic for a 1 m run wall. For any other reinforcing element of the structure, C_m can be defined similarly in view of the function served by that element.

It has been determined by a group of researchers in the UK that the rate of relative economy of reinforced-soil walls increases steadily with the height of the wall, as shown in Figure 17.4. A similar cost-effectiveness study of reinforced soil embankments by Christie (1982) in the UK showed that when space restrictions or high land-acquisition costs necessitated steep walls, it was almost unavoidable to use soil reinforcement. Murray (1982) also reported that a repair project for a cutting using reinforced *in situ* soil saved about 40% when compared with conventional replacement techniques. It was reported by Bell *et al.* (1984) that the total cost of a series of geotextile-reinforced retaining walls varied between US $118 and US $134 per square metre of the wall surface. The average cost breakdown is shown in Table 17.4. In a blast protection embankment in London in the UK, it was established by Paul (1984) that the geogrid-reinforced design was the most economical choice when compared with either a conventional reinforced concrete wall or an unreinforced soil wall. The relative costs for a 1 m run of the wall are shown in Table 17.5.

To evaluate the direct cost effect of geotextile applications in India, four typical geotextile uses were examined by Ghoshal and Som (1993) for four different regions of the country. The costs of the material, labour and land were collected for the metropolitan cities of Mumbai, Bangalore, Delhi and Kolkata. For identical soil data and design parameters, the variation in the cost with and without geotextiles for the selected functions was determined. An examination of the economic analysis reveals that the use of geotextiles depends significantly on the unit cost of different inputs.

Table 17.2 Summary of monitoring methods/devices
Reproduced from Koerner (1996)

Geosynthetic type	Function or application	Recommended	Optional
Geotextiles	Separation	Water content measurements Porewater transducers	Level surveying Earth pressure cells Inductance gauges
	Reinforcement	Strain gauges Movement surveying Inclinometers Extensometers	Earth pressure cells Inductance gauges Pore water transducers Water content measurements Settlement plates Temperature
	Filtration	Water observation wells Pore water transducers	Flow meters Turbidity meters Probes for pH, conductivity and/or dissolved oxygen
	Drainage	(Same as geotextile filtration)	
	Barrier (e.g. reflective cracking)	Surface defections Level surveying Surface roughness measurements Profilometery (for rut depths)	Water content measurements
Geogrids	Walls	Strain gauges Inclinometers Extensometers Monument surveying	Earth pressure cells Piezometers Settlement plates Probes for pH Temperature
	Slopes	Strain gauges Inclinometers Extensometers	Earth pressure cells Piezometers Monument surveying
	Foundations	Strain gauges Level surveying Extensometers	Earth pressure cells Piezometers Settlement plates
Geonets	Drainage	Flow meters Turbidity meters	Probes for pH, conductivity and/or dissolved oxygen Piezometers
Geomembranes	Tensile stress	Strain gauges	
	Temperature	Temperature measurement	
	Global leak monitoring	Flow meters Downgradient wells	Turbidity meters Probes for pH, conductivity and/or dissolved oxygen
Geosynthetic clay liners	Global leak monitoring	Flow meters Downgradient wells	Turbidity meters Probes for pH, conductivity and/or dissolved oxygen
	Shear strength	Extensometers Deformation telltales	Gypsum cylinders Fibre-glass wafers Strain gauges (inductance coils)
Geocomposites	Separation (e.g. erosion control)	Flow meters Turbidity meters	Level surveying
	Reinforcement	(Same as geotextiles and geogrids)	
	Drainage (e.g. edge drains)	Flow meters Turbidity meters	Probes for pH, conductivity and/or dissolved oxygen
	Barrier	(Same as geotextiles, geomembranes and geosynthetic clay liners)	

Table 17.3 Selected description and commentary on the methods and devices listed in Table 17.2
Reproduced from Koerner (1996)

Category	Methods/device	Resulting value/information
Surveying	Monument surveying	Lateral movement of vertical face
	Level surveying	Vertical movement of surface
	Settlement plates	Vertical movement at depth
Deformation	Telltales	Measures movement of fixed rods or wires Can accommodate any orientation
	Inclinometers	Measures vertical movement in a casing Inclined movements up to 45°
	Extensometers	Measures changes between two points in a borehole
Strain measurement	Electrical resistance gauges: ■ bonded foil ■ weldable	Measures strain of a material over gauge length, typically, 0.25–150 mm
	Inductance gauges (coils): ■ static measurements ■ dynamic measurements	Measures movement between two embedded coils up to 1000 mm distance apart
	LVDT gauges	Measures movement between two fixed points 100–200 mm apart
Stress measurement	Earth pressure cells: ■ diaphragm type ■ hydraulic type	Measures total stress acting on the cell, can be placed at any orientation, can also measure stress (pressure) against walls and structures
Soil moisture	Water observation wells	Measures stationary groundwater level
	Gypsum cylinders	Measures soil moisture content up to saturation
	Fibre-glass wafers	Measures soil moisture content up to saturation
Groundwater pressure	Piezometers: ■ hydraulic type ■ pneumatic type ■ vibrating wire type ■ electrical resistance type	Measures pore water pressures at any depth Can be installed as single point or in multiple point array Can be placed in any orientation
Temperature measurement	Bimetal thermometer	Measures temperature in adjacent area to ±1.0°C
	Thermocouple	Measures temperature at a point to ±0.5°C
	Thermistor	Measures temperature at a point to ±0.1°C
Liquid quantity	Tipping buckets	Measures flow rates (relatively low values)
	Automated weirs	Measures flow rates (relatively high values)
	Flow meters	Measures flow rates (very high values)
Liquid quality	Turbidity meters	Measures suspended solids
	pH probes	Measures pH of liquid
	Conductivity probes	Measures conductivity of liquid

The apparent cost–benefit derived by using geotextiles is not uniquely determined on the basis of the cost of the geotextiles alone. For example, where the land cost is high, as in Mumbai, the economy of using geotextiles becomes more predominant in the slope stability function than in the separation function (Tables 17.6 and 17.7). On the other hand, the separation function appears to give greater economy in Kolkata than in Mumbai because of the higher cost of stone aggregate in Kolkata.

For erosion control projects, Li *et al.* (2009) investigated the relationship between the cost and performance of slope erosion control products in the Texas Department of Transportation's Approved Product List. The study shows that the product price is highly associated with predicted product longevity, but that the predicted longevity does not correlate with the initial soil loss performance of the products. The lifetime product performance correlates well with product price, which suggests that marketed product prices accurately represent the lifetime worth of the products.

It should be noted that the total cost of a geosynthetic-reinforced soil structure depends not only on the relative costs of individual elements, but is also influenced by the geometry of the reinforced soil structure. For the purpose of determining the relative economy as well as the cost efficiency of reinforced soil structures, a comprehensive cost analysis should be performed by taking into account

Figure 17.4 The rate of economy in reinforced soil walls
Reproduced from Anon (1979), Transport Research Laboratory

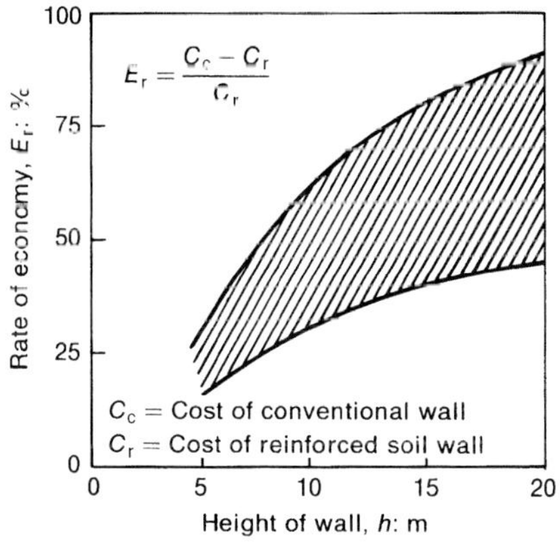

the costs (both direct and indirect) of various elements of the specific application.

17.5. General problems

Geosynthetics are being used in varying quantities in different parts of the world. In the author's opinion, there are some factors that inhibit the use of geosynthetics in certain projects, especially on a large scale

- *High cost.* This is mainly due to the high price of local raw materials, the high rate of duties levied by a government, the uneconomical scale of production due to lack of demand, the lack of modernisation of the production units and high overhead costs. In fact, the high cost of geosynthetics in relation to costs of labour and conventional materials has limited the application and widespread use of geosynthetics in some developing countries.
- *Lack of awareness.* Many people/project owners are not aware of the benefits of using geosynthetics. A survey, conducted by the author, has indicated that a large

Table 17.4 Geotextile-reinforced soil walls
Reproduced from Bell et al. (1984), National Academy of Sciences

Item	Cost: US $/m^2	Share: %
Geotextile	23	19
Labour	7	6
Equipment	7	6
Fill	53	44
Facing	30	25
Total	120	100

Table 17.5 Cost comparisons of an embankment
Reproduced from Paul (1984)

Wall type	Land width: m	Cost: US $/m
Reinforced concrete wall	18.9	2625
Geogrid-reinforced embankment	13.5	1775
Unreinforced embankment	32.5	1911

Table 17.6 Separation function: comparison of the cost of an unpaved road with and without a geotextile (base course thickness of 800 mm (for 1 m^2 surface))

Place	Without geotextile		With geotextile		Ultimate saving: rupees
	Amount of loss per year: rupees	Amount of ultimate loss: rupees	Quantity of geotextile required: m^2/m	Cost of geotextile per m^2/m: rupees	
Bangalore	8.80	26.40	1	50	−23.40
Mumbai	12	36	1	50	−14.00
Kolkata	32	96	1	50	46.00
Delhi	18.40	55.20	1	50	5.20

Table 17.7 Slope stability: comparison of the cost for an embankment with and without a geotextile for one side of a slope (height of the embankment is 8 m)

Place	Slope without geotextile		Slope with geotextile		Saving in land area per metre run on one side: m^2	Cost of land saving: Rupees	Saving per metre run of embankment: rupees
	Angle of slope: °	Cost: rupees	Angle of slope: °	Cost: rupees			
Bangalore	27	1701	45	3364	7.7	23 100	24 763
Mumbai	27	3465	45	4260	7.7	38 500	39 295
Kolkata	27	1575	45	3300	7.7	15 400	17 125
Delhi	27	3024	45	4036	7.7	30 800	31 812

number of field engineers are not very familiar with the applications of geosynthetics. The main reason is that geosynthetics have not been included in undergraduate programmes in engineering and technical colleges, unlike other construction materials such as brick, stone, timber, steel and concrete – at least until recently.

- *Lack of confidence*. Confidence has not developed among people/project owners. The reason for this is that large-scale research and development programmes on geosynthetic applications, as well as field demonstrations, are insufficient. Also, field monitoring and performance studies of geosynthetic-reinforced structures are not being carried out adequately, resulting in a lack of reporting on geosynthetics at a regional, national and international level.
- *Vandalism*. This is particularly troublesome in areas of uncontrolled site access. Some people also have an irrational fear of vandalism of geosynthetics used in some near-surface applications, such as erosion control.
- *Unavailability of geosynthetics*. Geosynthetic products are not always available in local markets, and many products are imported from other countries. Hence, they are not easily procured. Getting such products is both time-consuming and costly. Even for research work, orders may need to be placed in advance, and several weeks required for delivery.
- *Unavailable of national standards*. Standards on various aspects of geosynthetics and their applications are not available in all countries. In many countries, specific standards are still under preparation.
- *Habitual use of conventional methods*. It is a human trait to avoid adopting new solutions to problems. It always takes time for new methods to become popular among a large group of people.

Owing to the above factors inhibiting the use of geosynthetics, geosynthetic-like products, manufactured from natural raw materials such as jute, bamboo, wood, etc., are still being used in several areas of civil engineering, especially in low-income countries. In erosion control applications, where vegetation is considered to be the long-term solution from an environmental point of view, short-term erosion control is efficiently performed from a technical viewpoint in a diverse set of environments and soil conditions by jute products (called geojutes), as described in Chapter 10. The low cost (despite the significant costs of transportation) and the inherent variability in soil applications suits a natural fibre product. In the past few decades, this fact has been well recognised by jute producers, product suppliers, researchers, consultants and traders. A complete description of all the aspects related to the development and potential of natural products, in detail, is beyond the scope of this chapter. However, readers can find useful information in the work of Datye and Gore (1994), and Ranganathan (1994).

17.6. Selected case studies

Many case studies have been presented in the previous chapters. In this section, selected case studies in some application areas are described.

17.6.1 Retaining walls and steep slopes

Geosynthetic-reinforced soil walls have gained considerable attention as retaining structures, and provide a valuable alternative to traditional concrete walls. No footing of any kind is required in the case of retaining walls, and the lowest geosynthetic layer can be placed directly on the foundation soil. With respect to the concrete walls, geosynthetic reinforced soil walls present a good ratio between cost and effectiveness and low environmental impact.

A geosynthetic-reinforced soil-retaining wall with segmental facing panels was constructed on the Mumbai–Pune Expressway (Panvel bypass – package I) in 2001 by the Maharashtra State Road Development Corporation Limited, Mumbai, India. The height of the retaining wall is up to 13 m. The extensive use of Tensar connectors gives the perfect connection between the wall facing panels and the Tensar geogrids. A non-woven geotextile was used to wrap over the perforated pipe, to allow free drainage. The design was carried out using the tie-back wedge method, which has been described in Chapter 4. The construction sequences adopted were based on extensive model experiments, experience and technical justifications. Figure 17.5 shows the details of the wall at one of its cross-sections, along with soil and reinforcement properties. A portion of the wall during the construction stage is shown in Figure 17.6. The retaining wall has performed well without any noticeable problem since its completion.

Limited space for construction means there is a growing need to construct steep slopes. The most important consideration may be to achieve a natural look even during construction. de Niet (1996) reported a case study of a steep slope, reinforced with a geosynthetic and having natural ground cover. The construction work started with the positioning of temporary formwork to the angle of slope. The first layer of the geogrid was laid and pinned to the slope with steel pins. The grids were tensioned manually, and fill was deposited and compacted. Turfblocks, which were 0.30 m wide and about 0.05 m thick, were stacked to the angle set by the formwork. When a fill layer of 0.60 m had been compacted, the next layer of the grid was pinned to the blocks using steel pins. While doing this, it was important that the grids did not protrude from the slope. Once again, the grids were tensioned and the fill material was placed. In this way, the reinforced soil construction was built up, layer by layer. The result was a noise barrier, approximately 200 m long and 7 m high, on the building side, the appearance of which catches the eye due to the natural ground cover. Figure 17.7 shows the cross-section of the reinforced soil wall.

Lee *et al.* (1996) carried out a full-scale field experimentation of a new technique, called 'green coating', for protecting steep 'mudstone' slopes in south-western Taiwan. Mudstone is a weak sedimentary rock, formed during the Miocene to Pliocene and Pleistocene. Many forms of geological damage, such as erosion, mud flow and slope failure, were often seen in the mudstone area during the rainy season. The new technique consisted of three main elements

Figure 17.5 Cross-section of the geosynthetic-reinforced retaining wall on the Mumbai–Pune Expressway (Panvel Bypass – Package I)
Courtesy of Netlon India (2001)

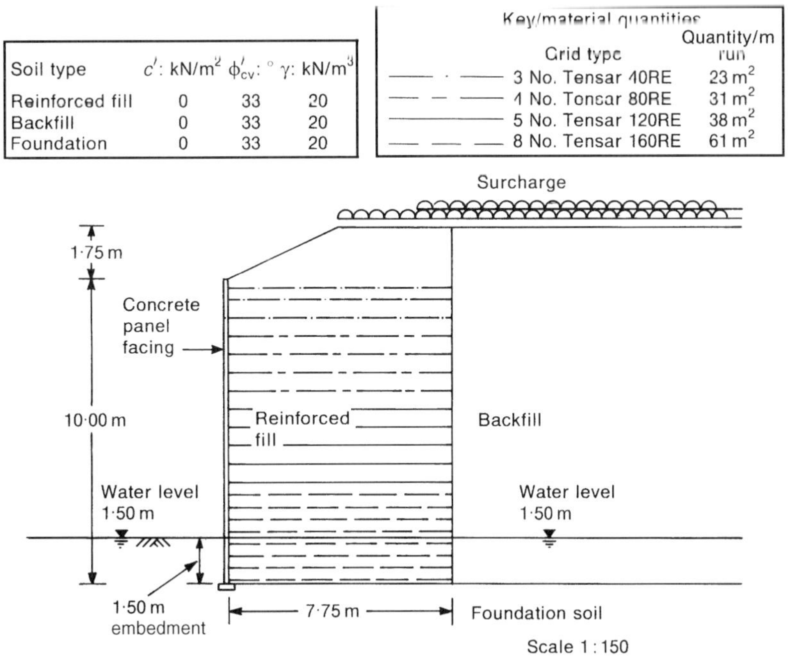

- cutting the natural mudstone slope into a multistage slope with a steep angle and a short height in each stage
- spraying RC-70 liquid asphalt on the slope and covering it with green geotextile sheets
- placing concrete platforms on the top of each stage of the slope for drainage and vegetation.

The total surface area of the cut slopes treated with the 'green coating' technique was about 630 m^2. The construction

Figure 17.6 A portion of retaining wall on the Mumbai–Pune Expressway (Panvel Bypass – Package I) during its construction stage
Courtesy of Netlon India (2001)

Figure 17.7 Cross-section of a geosynthetic-reinforced noise barrier in the Netherlands
Reproduced from de Niet (1996)

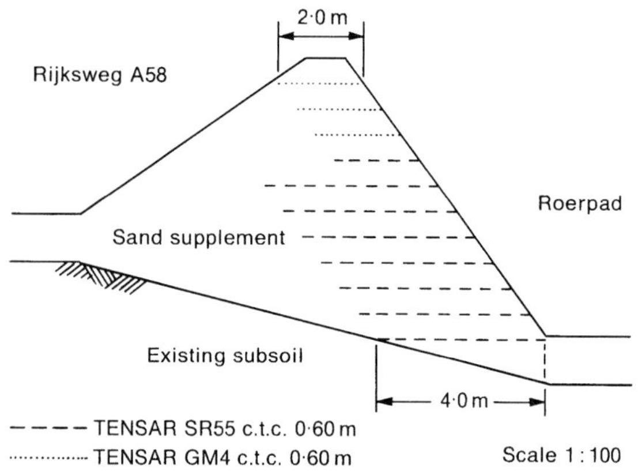

began at the top of the hill (test site) and gradually worked towards the bottom. Immediately after each slope stage was completed, the waterproofing and drainage work was carried out. The first step in this work was to clean up the slope surface, and to remove loose and broken rock. The clean surface was then sprayed with RC-70 asphalt. This asphalt coating serves two purposes

- preventing water from entering the mudstone
- providing adhesion between the mudstone and the geotextile sheet.

It was observed that the sprayed asphalt firmly stuck to the surface of the newly excavated mudstone and, thus, was effective in preventing erosion. The drainage strips were installed next on the slope surface. Finally, the slope was covered by geotextile sheets, which had two layers – the inner layer was an asphalt coating and the outer layer was a geotextile. The width of each geotextile sheet was 1 m, with a 10 cm overlap with the next sheet. Steel nails were used to fasten the sheets to the mudstone surface, and waterproofing treatment of these nails by asphalt coating was carried out immediately. Meanwhile, excavation continued for the next stage of the slope. During the excavation, it was evident that many joints were present in the test hill. To prevent erosion of the newly excavated slope surface by water (which could reduce the strength of mudstone and cause the failure of the slope), no excavation was allowed on rainy days. The construction of the designed slopes began in January 1992 and ended in March 1992. Two types of measurement were made to record the movement and the inclination of the treated slopes. Based on the field observations and measurements, the treated slopes did not show any signs of significant erosion or movement.

In North America, segmental retaining walls (SRWs) have gained wide acceptance as an economical alternative to both conventional cast-in-place concrete retaining walls and mechanically stabilised earth (MSE) walls, using metallic reinforcements. Collin (2001) reported the failure of a hybrid segmental retaining wall system that was constructed using both steel and geosynthetic reinforcement. The failure took place in 1998, and the failure was analysed by Collin with respect to design and construction to determine the causes(s) of the failure. The analysis identified the connection between the soil reinforcement and the segmental concrete units as the primary cause of the failure. The design of the wall used a proprietary software program that did not consider the connection strength of the hybrid system used. The remediation included the removal of the wall face and reconstruction increasing the connection capacity between the segmental concrete units and the reinforcement by a factor of three.

Yoo (2004) reported signs of distress and unexpected large lateral movements of a geosynthetic-reinforced segmental retaining wall, constructed during late 1994 through early 1995 at a chemical processing plant located in the southern part of Korea. The entire wall was approximately 300 m long, with a height ranging from 3.6 to 9.8 m. The wall was constructed using modular blocks (320 mm × 460 mm in plan and 200 mm in height), high-density polyethylene geogrids and weathered granite soil backfills. It was reported that the significant wall movements occurred due to poor construction quality control.

Kazimierowicz-Frankowska (2005) presented the results of 33 months of monitoring a 1.0 m-high reinforced experimental model wall with geosynthetic (polypropylene needle-punched geotextile reinforced with polyethylene geonet, mass per unit area = 608 g/m^2, average ultimate tensile strength = 19.5 kN/m at 8.04% strain) wrap-around facing with three reinforcement layers and gravelly sand backfill (unit weight = 19.3 kN/m^3, cohesion = 0 kN/m^2, angle of internal friction = 34°, Young modulus = 32 MPa, Poisson's ratio = 0.2). The wall was exposed to natural weather conditions (temperature, humidity, UV light). The experimental study shows that the greatest displacements occurred in the top layer of the wall. The rate of deformation was the greatest immediately after the end of the construction, and reduced progressively with time. The greatest (4%) elongation of the reinforcement was indicated in the bottom layer, and the smallest (0.5%) in the top one, 33 months after completion of the wall. The specimens of geosynthetic taken from the inside the wall after 33 months of working as reinforcement were in very good condition. Their basic mechanical parameters were nearly the same as the initial results: there was no reduction in the tensile strength for specimens taken from the top and middle layers. The results of additional tensile tests of the geosynthetic specimens exposed to the UV rays indicated a loss in the tensile strength ranging from 83 to 97% over a period of 33 months. This fact shows that exposure of geosynthetics to UV rays has a substantial negative impact on the geosynthetic strength.

17.6.2 Landfills

High-water-content and fine-grained sludges have posed formidable disposal problems for engineers throughout history. Usually, their low shear strength, combined with the proposed overburden loads, require the sludge to be stabilised before it can be covered. Numerous techniques, generally categorised as ground modification (e.g. soil mixing and grouting), are available. However, these techniques are site-specific, costly and time consuming. Geotextiles can successfully be used as a reinforcement and separation layer to facilitate the construction of a landfill closure over 'zero-strength' sludge with an accelerated schedule.

Guglielmetti *et al.* (1996) reported a case study of an instrumented geotextile-reinforced landfill cap for a process sludge landfill located in Wilmington, Delaware, USA, near the confluence of the Christina and Delaware rivers. The landfill cell served as a sludge disposal site for the DuPont Edge Moor facility for about ten years. The cell was approximately rectangular, with sides 293 m × 119 m for a total area of 34 803 m^2. The average depth of the sludge was 7.6 m. The contained process sludge was ferric chloride, and had an average pH of 2. Prior to placement of the sludge into the cell, the sludge was neutralised in pits using granular dolomite. The geotextile used to reinforce and separate the sludge at this facility was a woven polypropylene (ultimate tensile strength (cross-machine direction = 75 kN/m, optimum seam efficiency = 62%)). The geotextile was instrumented with foil strain gauges, and the sewn seams were instrumented with extensometers. It was probably the first attempt to measure seam deformation. After placement of the geotextile, low-ground-pressure bulldozers began placing a 0.6 m layer of stabilised sludge material over the geotextile, beginning at the south end. The sludge fill was placed in a finger–palm configuration, to allow tensioning of the geotextile perpendicular to the seams. The fill pattern is shown in Figure 17.8. The sludge was a fine-grained material with a high moisture content and a permeability in the range 1×10^{-7} to 1×10^{-8} m/s. The sludge was used in an effort to save costs by eliminating select granular fill material. The sludge proved to be stable, and did not allow for adequate drainage of expelled water from the underlying sludge. Cracking in the backfill material and tears in the geotextile seams appeared behind the leading edge of the backfilling zone as fill placement proceeded north. Backfilling was immediately halted. The backfill operation moved to the north end of the cell, and a structural granular fill (average unit weight = 19.7 kN/m^3) was used in place of the stabilised sludge. Large settlements were observed under the weight of the granular fill. Because of concerns about seam stressing and the cost of the additional volume of fill required, a lightweight fill was then substituted for the granular fill. The lightweight fill was power-plant bottom ash, having an average unit weight of 11.8 kN/m^3 and a permeability of approximately 1×10^{-6} m/s. In addition, the ash layer served as a drainage layer above the geotextile. The final grade was made with a pug-milled residual material (average unit weight = 11.8 kN/m^3). The attempt to monitor the field seam performance was unsuccessful, but did present an innovative technique that may prove effective in the future.

Designing a constructible composite liner system, for the side slopes of a landfill, is a challenging task. To meet this challenge at the Lopez Canyon Sanitary Landfill, Los Angeles, USA, an entirely geosynthetic composite liner, and a leachate collection and removal system (LCRS), was developed in 1991 (Snow *et al.*, 1994). A schematic cross-section of the geosynthetic alternative, developed for the side slopes of the disposal area, is shown in Figure 17.9. The veneer of concrete was specified to have a compressive strength of 170–205 kPa, and was sprayed on to the graded canyon side slopes to provide support and a smooth surface for the composite liner. A polyethylene geonet was used in lieu of granular soil to provide an LCRS on the side slopes. The primary advantages of the geonet are simple installation and a high drainage capacity, resulting in a low liquid head on the composite liner. Construction of the geosynthetic side slope liner system was subjected to large temperature variations, high winds and the steep slopes at the site. The familiarity of the person installing the geosynthetics with these conditions from past experience on other landfills in the area was a significant benefit to the project. A total of about 15 500 and 77 000 m^2 of geosynthetic composite side slope liner system was placed during phases I and II of the liner system construction, respectively. Phase I geosynthetic clay liner joints were simply overlapped with no additional preparation, while the phase II geosynthetic clay liner joints were prepared by the addition of powdered bentonite at a rate of 1.5 kg/m^2 in the overlap areas. Performance of the liner system under dynamic loading was observed

Figure 17.8 Schematic aerial diagrams showing how fill was placed over the deployed geotextile: (a) anchoring geotextile after positioning it, (b) filling in a finger–palm manner from south to north, (c) filling in a longitudinal road configuration and (d) distributing the fill and making the final grade
Reproduced from Guglielmetti *et al.* (1996), with permission from Elsevier

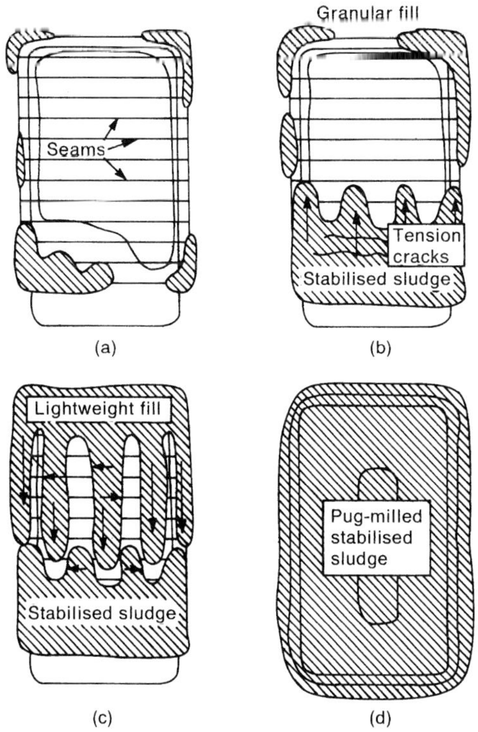

Figure 17.9 A geocomposite liner system for steep canyon landfill side slopes in Los Angeles, USA
Reproduced from Snow *et al.* (1994), Industrial Fabrics Association International

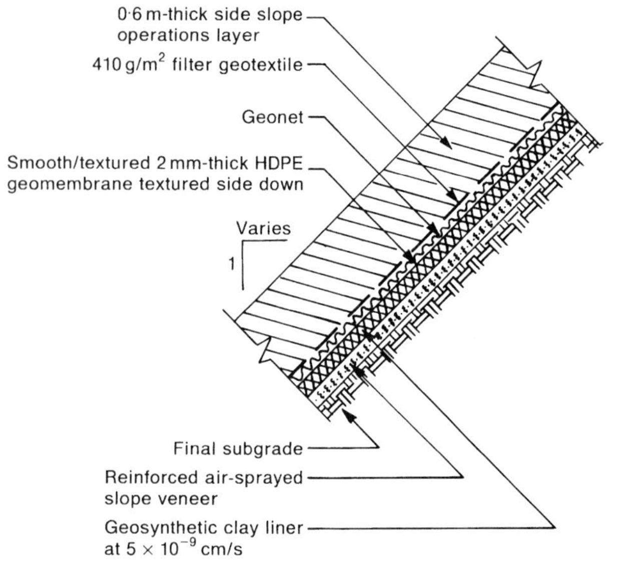

during the Northridge Earthquake, Richter magnitude 6.6, which struck Los Angeles on 17 January 1994. The Lopez Canyon site is located less than 15 km from the earthquake epicentre. Nearby recording stations measured horizontal peak ground accelerations of up to 0.44*g*. Observations, made that same day, indicated that the geosynthetic side slope liner system performed very well.

17.6.3 Pipeline and drainage systems

In pipeline and drainage systems, as well as in erosion control systems, geosynthetics, especially geotextiles and geocomposites, are being widely used as substitutes for the traditional aggregate layer. When geosynthetics are used in these applications, one of their major functions is filtration. It is a misconception that the geosynthetic can replace the function of a granular filter completely. A granular filter also serves other functions, which relate to its thickness and weight. The performance of geosynthetics as a filter is significantly affected by the interaction between the geosynthetic opening size and the soil particle size. Geosynthetics can be selected for filter applications in accordance with the filter criteria presented in Section 1.6.3.

Vertical strip drains (also called prefabricated vertical band drains or wick drains) are geocomposites, used for land reclamation and stabilisation of soft grounds. They accelerate the consolidation process by reducing the time required for the dissipation of excess porewater pressure, as described in Section 5.3. The efficiency of the drains is partly controlled by the transmissivity (i.e. discharge capacity), which can be measured, using the drain tester, to check their short-term and long-term performance. The discharge capacity of drains is affected by several factors, such as the confining pressure, the hydraulic gradient, the length of the specimen, the stiffness of the filter and the duration of loading. The effect of confining pressure has been discussed in Section 1.6.3. The experimental study, conducted in the laboratory by Broms *et al.* (1994), suggests that the effect of the length of the drains and the duration of loading on the discharge capacity of the drain is small, whereas the stiffness of the filter of the drain can have a considerable effect. The discharge capacity of the drain decreases with decreasing stiffness of the filter.

For a stretch of more than 2 km on the A64 Malton Bypass, England, where it cuts through limestone, North Yorkshire County Council has installed soakaway drains, which have been designed to accumulate surface storm water and allow it to percolate into the permeable limestone. In order to protect the soakaway stone from contamination by fines washed through from the surface or from the sub-base (the result of which would be to progressively reduce its storage capacity), the drainage medium was wrapped in Terram 700 (a thermally bonded non-woven geotextile). First, Terram 700 was used to line the excavated trench, then single-sized stone was filled to the level required to meet the design capacity, and then – before completing the filling of the trench – the Terram filter was folded over the top of the backfill to enclose it completely and, thus, provide the necessary protection (Figure 17.10).

Prefabricated drainage geocomposites are currently designed for structures requiring vertical drainage, such as bridge abutments, building walls and retaining walls. The composite normally consists of a spacer/core sandwiched between two geotextile layers. This construction combines in a single flexible sheet. The core is usually either a geonet or a semi-rigid extruded plastic sheet. In the past 30 years, prefabricated drainage geocomposites have become a common method of groundwater drainage for a wide variety of purposes. In many cases, drainage geocomposites are cheaper, easier to install, and require less space than conventional drains

Figure 17.10 Cross-section of the soakaway drain on the A64 Malton Bypass, England
Courtesy of Terram® (2000)

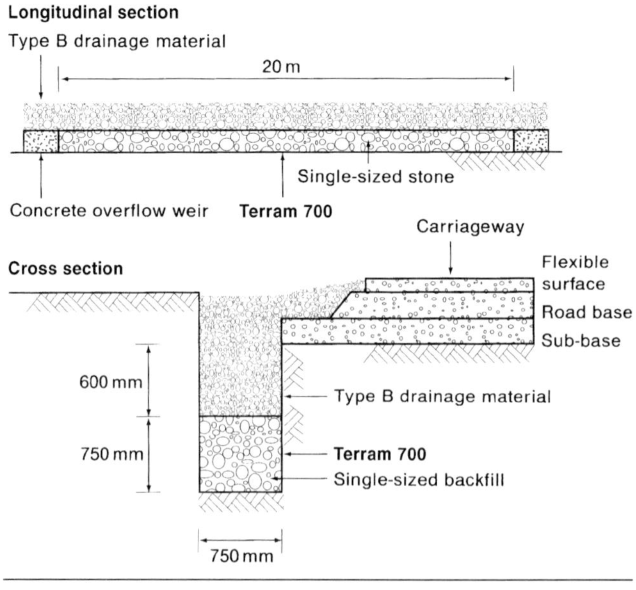

constructed using granular materials wrapped in a geotextile. McKean and Inouye (2001) presented the field evaluation of the long-term performance of three geocomposite sheet drains located in Oregon and Northern California, USA, at sites on roads administered by the US Department of Agriculture, Forest Service. The design objective of the drains was to intercept all of the shallow groundwater flow and to keep porewater pressures low in the immediate area downslope of the drain, even during extreme hydrological events. One drain was placed 2.4 m deep in a vertical trench in soil that varied from clay and clayey gravel to silty sand. At a second site, a drain was installed 8 m deep in a silty sandy soil behind a retaining wall. At the third site, a drain was installed to a maximum depth of 2.4 m in a vertical trench in a silty sandy soil. It was reported that over a three-year period, the first drain did not allow a rise in the water table even during extreme groundwater events. The performance of the second drain was consistently excellent over a 14-year period. The performance of the third drain was marginal because a trench wall collapsed during drain installation, which caused an uneven drain slope towards the outlet. Despite the construction problems, this drain was partially dewatering the road prism just down-gradient from geocomposites.

17.6.4 Slopes – erosion control

Erosion consists of the loosening and transportation of soil particles. The land and vegetation disturbances due to deforestation, mining and construction create conditions for accelerated erosion. Erosion control methods that are of particular relevance to civil engineers have been discussed in Chapter 10.

Riverbank and coastline erosion is counteracted by the protection of the surface to resist the forces generated by flow and waves. The method widely used is to install a layer of stone pitching on the bank to stop the loss of soil. The rise and fall of the water level, as well as the wave action, causes water to flow into the pitched bank and then drain away. This two-way flow, known as dynamic flow, is capable of dislodging and carrying away soil lying below/behind the stone protection and, ultimately, causing the revetment to fail. The soil erosion may be reduced by installing a filter between the stone layer and the soil. Traditionally, a granular layer is used as a filter, which allows the water to pass through freely, but not the soil particles. The design and choice of a suitable granular material for this filter can be made, but it is not an easy task to accurately realise the function of the filter. The use of geosynthetic filters in such cases has proven to be an attractive alternative.

Kolkata Port Trust Authorities used a jute geotextile as a revetment filter for river bank protection at Nayachor Island, Haldia, India, during 1992 (Sivaramakrishnan, 1994). The eroded site was first prepared to form a uniform one vertical to one horizontal slope. The bare jute geotextile was unrolled over the slope of the embankment, starting from the top of the bank. The geotextile was anchored at the top as well as at the sides in a trench of 1 m × 1 m. The overlappings were nailed with 254 mm-long iron pegs at intervals of 1 m. The bottom portion of the jute geotextile was fabricated in such a manner that it had multiple pockets that could be filled with sand. This was done to anchor the geotextile in place, and to protect against erosion by reverse current and eddies. After the entire area was covered with the jute geotextile, small laterite boulders were placed over the jute geotextile. The small laterite boulders were laid to provide a cushion effect on the geotextile. On the top of the small laterite boulders, large laterite boulders, weighing approximately 15–20 kg, were pitched to a height of 457 mm, as shown in Figure 17.11. The entire operation was carried out during low tide. A good amount of siltation, up to a height of 600 mm, was observed after a period of eight months, which indicates that the jute geotextile was effective in protecting the slope.

Figure 17.11 River bank protection at Nayachor Island, Haldia, India
Reproduced from Sivaramakrishnan (1994), Industrial Fabrics Association International

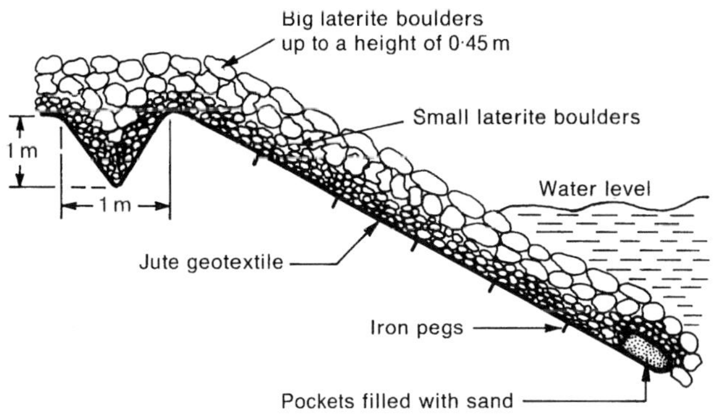

A permanent and self-supporting vegetative cover has been found to be an ideal solution to most erosion problems. Natural and biodegradable fibres, such as coir and jute in the form of nettings, aid in vegetative turfing along slopes. These nettings provide a physical barrier between the soil particles and rainwater, and thus protect the seeds and the soil until the sites are permanently stabilised with vegetative cover. Lekha (2004) presented the slope erosion control method for hill slopes using the coir geotextile. The method was implemented in a typical degraded hilly area of about 0.4 ha in Kerala, India. After installation of the coir geotextile, growth of lemon grass and a thick vegetative cover was established in the area within a period of one year. Consequently, soil erosion at the site was reduced substantially. Figure 17.12 shows a view of the coir-geotextile-laid vegetated slope one month and one year after laying the coir geotextile. The annual soil loss from one protected plot was found to be 94.9% less compared with a nearby non-protected plot. Analysis of the ultimate tensile strength of coir geotextile showed that after seven months of laying, the field-laid coir net has retained 22% of the strength of the fresh sample. This fast rate of biodegradation of the coir geotextile was not of concern since the lemon grass developed a well-defined root system and was able to adequately hold the soil in place within a period of

Figure 17.12 View of the coir-geotextile-laid vegetated slope: (a) after one month and (b) after one year
Reproduced from Lekha (2004), with permission from Elsevier

(a) (b)

six months. Moreover, the thickly growing canopy can reduce the impact of rainwater, reduce surface run-off and increase infiltration. It was also reported that there was a substantial improvement in parameters such as organic carbon, soil water content and vegetative growth.

17.6.5 Irrigation channels and reservoirs

Irrigation systems generally demand the construction of channels and reservoirs that require watertight linings. Among the various types of linings available, geosynthetics, especially geomembranes, in combination with geotextiles and geosynthetic clay liners, provide technically valid options and permit rapid installation. Many case studies on irrigation channels and reservoirs, as well as containment ponds, are presented in Chapter 14.

17.6.6 Earth dams

A lake, covering 30 000 m^2, was created in France for tourist purposes by means of an earth fill dam, with a maximum height of 15 m above the natural ground level and a crest length of approximately 100 m (Alonso *et al.*, 1990). The work was carried out during the summer of 1988. A Teranap 431 bituminous geomembrane was used, as shown in Figure 17.13. Continuity of watertightness at the toe of the initial fill was ensured by means of a shallow clayey trench, and then by a 0.50 m-wide trench dug down to the very hard schists a few metres below. After installation of the geomembrane down to the bottom of the trench, the trench was filled with concrete at 250 kg of cement per cubic metre. The waterproofing system was completed by a 1 m-wide covering membrane, bonded to the concrete and to the geomembrane. Near the geomembrane, the loose schist was compacted by the wheels of the trucks bringing the material (the compacting equipment was a heavy sheepsfoot roller). During the filling operations for the upstream one vertical to three horizontal slope, no tensile stress was observed on the geomembrane. The lake had been full since the spring of 1989. Measurements recorded evidence of the good watertightness of the structure. There has been no flow in the pipes linked with the vertical drain. The pipes installed in the drainage blanket (consisting of 5 m-wide strips, 5 m apart) had a total flow of 0.15 l/s.

Many case studies on geosynthetic applications in dams are presented in Chapter 13.

17.6.7 Roads

The use of geosynthetics as reinforcements in unpaved roads on soft soils makes it possible to increase their load-bearing capacity. A geosynthetic layer in an unpaved road allows the passage of heavily loaded vehicles over granular fill of reduced thickness, placed on the soft subgrade. This, in turn, results in the reduced use of materials, transport expenses and duration of construction. Geosynthetics have

Figure 17.13 Cross-section of the Valence d'Albi dam, France
Reproduced from Alonso *et al.* (1990), Industrial Fabrics Association International

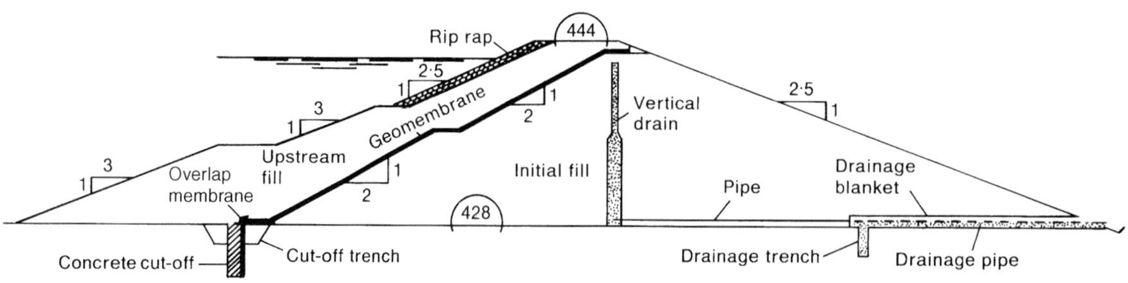

also been used at the soil subgrade level to function as a separator/filter/drainage layer (Shukla, 2006). Applications of geosynthetic in paved roads include their installation as paving fabrics at the base level of the bituminous overlays (Shukla, 2008).

One of the first roads in the former USSR, where a geotextile was first used, was a temporary road in the Smolenskaya region (Kazarnovsky and Brantman, 1993). Construction of the road had to be accelerated to evacuate populated localities from areas that were to be flooded, when the reservoir of the Vazuzskaya hydro system was being filled with water, and also to allow for the movement of construction vehicles. A temporary road, about 20 km long, was to be constructed within the shortest period of time and with a minimum thickness of fill. The site was characterised by soft plastic loam soils, by a high groundwater level, and by a prolonged stagnation of water above the ground. Construction procedures for the road sections on which the geotextile ('Dornit' $\phi -1$) was used included

- a rough grading of the soft subgrade by a bulldozer going back and forth with a lowered blade
- unrolling the geotextile across the fill axis with 300 mm overlaps between adjacent rolls
- filling and grading of a 450–500 mm-thick medium-grained sand layer (containing gravel and 2% of silty and clayey particles), followed by compaction with a lightweight roller.

The difference in driving conditions on the sections with the geotextile and without it could be observed immediately after installation of the geotextile. It was actually impossible to perform work after eight to ten passes of the dump trucks along the same ruts in the section where the geotextile was not used, and the sand fill thickness was limited to 400 mm for comparison. On the road section where the geotextile was placed under the sand fill, the rut depth did not exceed 100–120 mm, and intermixing of the fill sand with the subgrade did not occur. Every working day, 300 vehicles, mostly dump-trucks, travelled in both directions. After the road was used for several months, the ruts on the road section without the geotextile had to be graded continually. Every morning, before the main traffic drove on the road, both the sections with and without the geotextile were graded with a bulldozer. Measurement of the ruts showed that, on the section without a geotextile, ruts from 200–250 mm deep were formed, and the traffic speed slowed to 5 km/h. On the road section with the geotextile, the rut depths were only 100–120 mm, in spite of the fact that the dump trucks travelled along the same ruts (Figure 17.14). The traffic was able to maintain a speed of 25–30 km/h, and the vehicles could pass each other using the whole width of the fill.

If the pavement of a road is only surface dressed or resurfaced, the deep-seated cracks, if any, will emerge to degrade the resurfacing. The crack resistance of the pavement can be improved by installing a geosynthetic between

Figure 17.14 Change in the rut depths depending on the number of vehicles passing (sand-fill thickness = 400 mm)
Reproduced from Kazarnovsky and Brantman (1993)

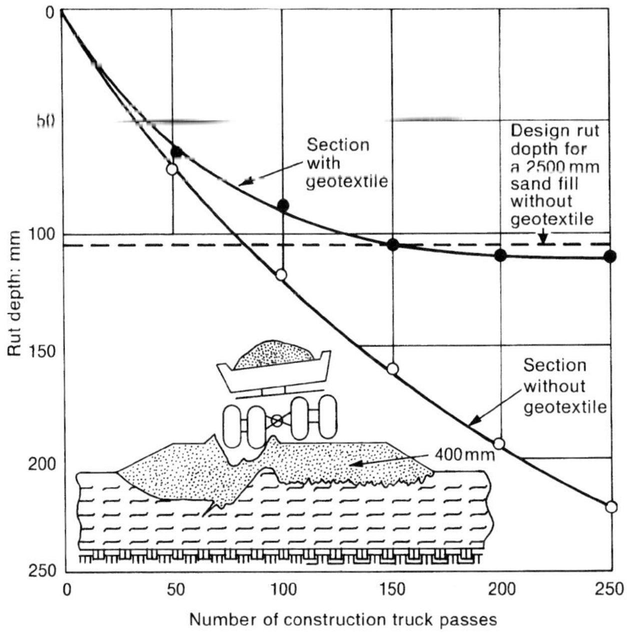

the old and new surfacing. The geosynthetic may also prevent excessive moisture reaching and softening the sub-base and subgrade, provided that the geosynthetic used is of low permeability. In installing the geosynthetic, it is vital that a good bond be achieved between the old and new works. This involves first coating the old surfacing with a hot tar spray, or an emulsion, before rolling out the geosynthetic. Once this is in place, the new surfacing is placed in the normal manner (Ingold and Miller, 1988). It is to be noted that the geosynthetic, used in the prevention of reflection cracks in distressed pavement, should have high strength, high modulus and low creep.

A section of Federal Highway B 180, more than 20 m long, at Neckendorf, near Eisleben, Germany, was destroyed across its entire width in 1987 by a sink hole approximately 8 m in diameter, located almost on the road axis below 30 m depth. Although the hole was filled with fill material, the danger of a new cave-in due to caverns deep underground still existed. To allow the roadway to return to operation, the opening had to be bridged over sufficiently to allow subsidence of no more than 10 cm over 30 m of roadway, even under heavy truck traffic. The 20 m-long weak section was bridged over with a geogrid-reinforced gravel/sand layer (Figure 17.15). The layer was about 60 cm thick by 60 m long and approximately 11 m wide. This layer supported the entire road surface. The geogrid reinforcement was installed in three layers. The bottom layer consisted of two 5 m geogrid strips laid longitudinally side by side. The second layer consisted of a transverse geogrid strip, completely encapsulated and overlapped, resulting in a third layer. The design provided effective reinforcement against longitudinal and transverse deflection as well as

Figure 17.15 Geogrid-reinforced gravel layer bridges a sink-hole on Federal Highway B 180 near Eisleben, Germany
Courtesy of Huesker Synthetic GmbH & Co., Germany

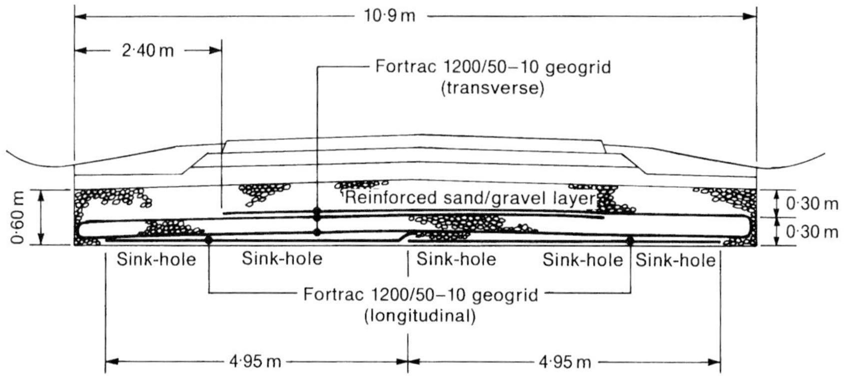

torsion. The flexible geogrid was composed of very low-elongation, low-creep Aramid fibres with Fortrac 1200/50-10, having total tensile strength of 1200 kN/m and only 3% elongation. The mesh size was 10 mm × 10 mm. The reinforced layer was prepared within a few days in October 1993.

Shukla *et al.* (2004) reported the use of a geotextile layer between the soil subgrade and the granular sub-base course in the national roadway project undertaken in the Varanasi zone, India during early 2003 (Figure 17.16). This project formed a part of the Golden Quadrilateral, connecting four metropolitan cities of India, namely Delhi, Mumbai, Chennai and Kolkata, with a total length of about 6000 km. The needle-punched non-woven geotextile (grab strength = 900 N, tear strength = 350 N, puncture strength = 350 N, burst strength = 1700 kPa, AOS = 0.60 mm, permittivity = 0.02 s^{-1}, UV stability = 50% after 500 hours of exposure) layer was installed between the soil subgrade and the granular sub-base course, forming a part of the bituminous concrete pavement, in many sections of the roadway. By using the geotextile layer as a separation and drainage

Figure 17.16 A non-woven geotextile layer installed at the soil subgrade and underlying the granular sub-base material
Reproduced from Shukla *et al.* (2004)

layer, pavement layers were constructed conveniently and economically even over poor soil subgrade. The geotextile layer contributed to the construction speed of the granular lifts. Local shear failures during construction could also be reduced significantly with a geotextile separator. The performance of the roadway during 2003–2007 was found to be satisfactory in most of the roadway sections.

17.6.8 Tunnels

The waterproofing of tunnels can be successfully carried out using geosynthetics. Even a completely submerged tunnel can be waterproofed. In most of the reported case studies (Benneton *et al.*, 1993; Davies, 1993), the following points comprise the major construction steps

- excavation of rock and/or soil
- grouting to stop/minimise inflowing water, if present
- supporting the exposed surface by shotcrete (gunite)
- fastening the thick needle-punched non-woven geotextile, as a protective screen as well as a drainage medium, to the shotcrete by means of polyvinylchloride (PVC) plastic discs (plates) and fasteners (nails)
- fixing the geotextile to underdrains on each side of tunnel base
- placement of a geomembrane (usually PVC) to PVC plastic discs by means of hot air welding
- spot-bonding of a protective shield (3 mm-thick PVC) to the geomembrane
- placement of the concrete liner against the geomembrane
- providing additional seals (consisting of an expansion product, e.g. butyl bentonite) at concrete restart points.

Figure 17.17 shows the cross-section of a tunnel vault with the general arrangement of the lining system. Chapter 15 describes many practical aspects of geosynthetic applications in hydraulic tunnels.

17.6.9 Containers

A range of river bank and costal protection projects uses geotextiles as soil/sand containers. These projects cover several areas such as scour protection, groynes, berms,

Figure 17.17 Cross-section of a tunnel vault showing the general arrangement of the lining system

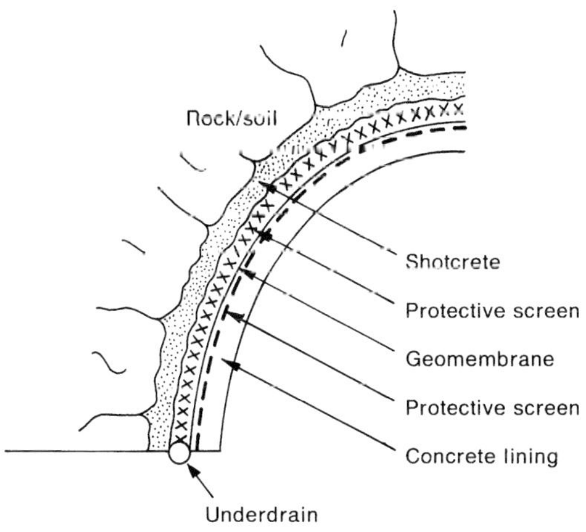

artificial reefs and containment of hazardous materials. Restall *et al.* (2002) presented case studies showing the growth and development of geotextile sand containers from an Australian perspective. Figure 17.18 shows a sand-filled geotextile groyne, constructed of 100 m-long tubes encapsulated in geotextile envelopes. Similar geotextile structures can be successfully used to solve conventional and non-conventional challenges. Non-woven sand-filled geotextiles structures adapt and conform readily to changing site conditions. The approach to design and construction requires consideration of site-specific conditions and geosynthetic techniques. Factors such as vandalism potential, abrasion potential of the sand, interface friction, etc. must be addressed in the designs. Several companies/contractors are involved in the manufacture and installation of geotextile containers in a variety of forms, and they have long experience in the field.

17.7. Summary of main points

- Most aspects discussed in this chapter are general, and so they should be considered on a site-specific basis. Manufacturers have been taking the lead in the area of geosynthetics by producing new products regularly. It is up to readers to update their understanding of geosynthetics in order to solve their field problems most effectively.

Figure 17.18 A sand-filled geotextile groyne
Reproduced from Restall *et al.* (2002), with permission from Elsevier

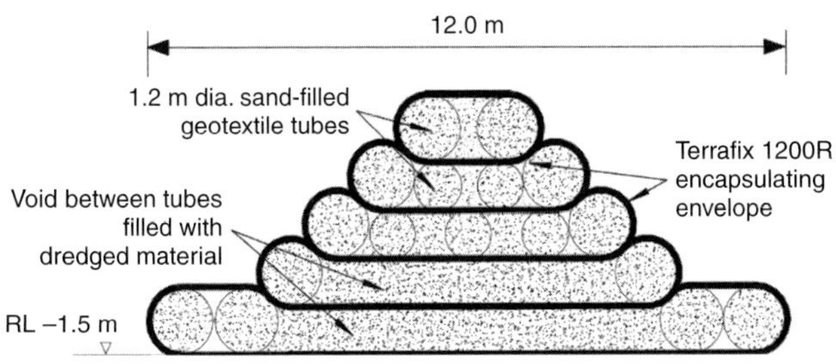

- Many geosynthetics manufacturers have design methods for use with their particular products. These methods use their own background theory based on laboratory tests and field observations. Since a wide variety of geosynthetics are available, a method that considers them on the basis of a specific, well-defined property should be considered.
- Damage to geosynthetics during construction is a major concern in the design and application of geosynthetics. Therefore, field trials should be performed to investigate possible damage to geosynthetics before their application, especially in critical field projects.
- There is a vast application potential for a wide range of geosynthetics throughout the world. To tap into this demand, it will be necessary to develop products that will meet the specific needs of users. Geosynthetics must demonstrate resistance to environmental and chemical factors, while still retaining the physical and mechanical properties inherent to specific design applications. Manufacturers should develop geosynthetic products that will be easy to use, safe and cheap, so that developing countries can use them in large quantities to solve varied civil engineering problems.
- A careful cost comparison must be made to finalise the alternative solutions (with or without a geosynthetic) for any civil engineering problem. However, from an economic viewpoint, the alternative solutions for any problem should be advantageous to both the client and the construction partnership.
- Case studies, reported in this chapter as well as in other chapters, indicate a positive trend in the applications for geosynthetics, as predicted by several researchers and practising engineers during the initial developments of geosynthetic engineering.

17.8. Some common questions and their answers

Questions

1. Why should geosynthetics be handled and stored properly?
2. What is the most important aspect of construction with geomembranes?
3. If a geosynthetic is damaged during its installation, what steps should be taken?
4. What is the purpose of obtaining a quality certification for geosynthetics?
5. When is the rate of wall deformation a maximum during its life?
6. What is the main cause of several reported reinforced segmental wall failures in the past?
7. What is the main application of geonaturals, and how do they perform this role?
8. Can sand-filled geotextile containers be used for recreational facilities such as beaches?

Answers

1. To ensure that their specified properties are retained, and therefore the geosynthetics can perform their intended functions as per the project needs.

2. Ensuring proper seams, without which the whole concept of using geomembranes as a fluid barrier becomes meaningless.

3. It is always safest to remove the damaged section of the geosynthetic. In some applications, a certain level of installation damage may be acceptable, provided that this has been allowed for at the design stage.

4. A quality certificate ensures that the geosynthetics delivered meet the design requirements.

5. Laboratory and field measurements have shown that the rate of wall deformation is the greatest immediately after the end of construction, and that it decreases progressively with time.

6. Poor construction quality control has been the main cause of these failures.

7. Geonaturals such as geojutes/geocoirs are very effective in erosion control applications. The case studies show that geonatural netting spread over seeded slopes/grounds shields the soil and seeds from the impact of raindrops, minimises run-off and slows down its velocity, maintains the capacity of soil to absorb water, holds the soil particles and seeds in place, and retains water content.

8. Yes, they are ideal for these applications.

REFERENCES

Alonso E, Degoutte G and Girard H (1990) Results of seventeen years of using geomembranes in dams and basins. *Proceedings of the 4th International Conference on Geotextiles, Geomembranes and Related Products*, The Hague, pp. 437–442.

Anon (1979) *Reinforced Earth and Other Composite Techniques*. Transportation and Roads Research Laboratory, London. Supplementary Report 457.

Bell JR, Barrett RK and Ruckmann AC (1984) Geotextile earth-reinforced retaining wall tests. *Transportation Research Record* **916**: 59–69.

Benneton JP, Mahuet JL and Gourc JP (1993) Geomembrane water-proofing of dry tunnel under two rivers, Lyon metro tunnel, France. In *Geosynthetics Case Histories* (Raymond GP and Giroud JP (eds)), Technical Committee TC9, International Society for Soil Mechanics and Foundation Engineering, London, pp. 118–119.

Broms BB, Chu J and Chora V (1994) Measuring the discharge capacity of band drains by a new drain tester. *Proceedings of the 5th International Conference on Geotextiles, Geomembranes and Related Products*, Singapore, pp. 803–806.

Christie IF (1982) Economic and technical aspects of embankments reinforced with fabric. *Proceedings of the 2nd International Conference on Geotextiles*, Las Vegas, NV, pp. 659–664.

Collin JG (2001) Lessons learned from a segmental retaining wall failure. *Geotextiles and Geomembranes* **19(7)**: 445–454.

Datye KR and Gore VN (1994) Application of natural geotextiles and related products. *Geotextiles and Geomembranes* **13**: 371–388.

Davies PL (1993) Geomembrane waterproofing of wet hydroelectric tunnel, Drakensberg hydroelectric tunnel, South Africa. In *Geosynthetics Case Histories* (Raymond GP and Giroud JP (eds)), Technical Committee TC9, International Society for Soil Mechanics and Foundation Engineering, London, pp. 114–115.

de Niet R (1996) A steep slope reinforced with geosynthetic reinforcement and with natural ground cover. *Proceedings of the 1st European Geosynthetics Conference – Eurogeo 1*, Maastricht, pp. 501–502.

Donckers F (1994) Quality control and quality assurance procedure for geosynthetics. *Proceedings of the 5th International Conference on Geotextiles, Geomembranes and Related Products*, Singapore, pp. 1101–1104.

Dunnicliff J (1988) *Geotechnical Instrumentation for Monitoring Field Performance*. Wiley, New York.

Durukan Z and Tezcan SS (1992) Cost analysis of reinforced soil walls. *Geotextiles and Geomembranes* **11**: 29–43.

Ghoshal A and Som N (1993) Geotextiles and geomembranes in India – state of usage and economic evaluation. *Geotextiles and Geomembranes* **12**: 193–213.

Guglielmetti JL, Koerner GR and Battino FS (1996) Geotextile reinforcement of soft landfill process sludge to facilitate final closure: an instrumented case study. *Geotextiles and Geomembranes* **14**: 377–391.

Hullings DE and Sansone LJ (1997) Design concerns and performance of geomembrane anchor trenches. *Geotextiles and Geomembranes* **15**: 403–417.

Ingold TS (1982) *Reinforced Earth*. Thomas Telford, London.

Ingold TS and Miller KS (1988) *Geotextiles Handbook*. Thomas Telford, London.

John NWM (1987) *Geotextiles*. Blackie, London.

Kazarnovsky VD and Brantman BP (1993) Geotextile reinforcement of a temporary road, Smolenskaya region, USSR. In *Geosynthetics Case Histories* (Raymond GP and Giroud JP (eds)), Technical Committee TC9, International Society for Soil Mechanics and Foundation Engineering, London, pp. 194–195.

Kazimierowicz-Frankowska K (2005) A case study of a geosynthetic reinforced wall with wrap-around facing. *Geotextiles and Geomembranes* **23(1)**: 107–115.

Koerner RM (1996) The state-of-the-practice regarding *in-situ* monitoring of geosynthetics. *Proceedings of the 1st European geosynthetics Conference – Eurogeo 1*, Maastricht, pp. 71–86.

Koerner RM (2005) *Designing with Geosynthetics*, 5th edn. Prentice Hall, Englewood Cliffs, NJ.

Lee DH, Tien KG and Juang CH (1996) Full-scale field experimentation of a new technique for protecting mudstone slopes, Taiwan. *Engineering Geology* **42**: 51–63.

Lekha KR (2004) Field instrumentation and monitoring of soil erosion in coir geotextile stabilised slopes – a case study. *Geotextiles and Geomembranes* **22(5)**: 399–413.

Li MH, Yi YJ and McFalls J (2009) Lifetime cost and performance of slope erosion control products for construction. *Geosynthetics International* **16(3)**: pp. 139–146.

Lovisa J, Shukla SK and Sivakugan N (2010) Behaviour of prestressed geotextile-reinforced sand bed supporting a loaded circular footing. *Geotextiles and Geomembranes* **28(1)**: 23–32.

McKean J and Inouye K (2001) Field evaluation of the long-term performance of geocomposites sheet drains. *Geotextiles and Geomembranes* **19(4)**: 213–234.

Murray RT (1982) Fabric reinforcement of embankments and cuttings. *Proceedings of the 2nd International Conference on Geotextiles*, Las Vegas, NV, pp. 707–713.

Paul J (1984) Economics and construction of blast embankments using Tensar geogrids. *Proceedings of the Conference on Polymer Grid Reinforcement*, London, pp. 191–197.

Ranganathan SR (1994) Development and potential of jute geotextiles. *Geotextiles and Geomembranes* **13**: 371–388.

Restall SJ, Jackson LA, Heerten G and Hornsey WP (2002) Case studies showing the growth and development of geotextile sand containers: an Australian perspective. *Geotextiles and Geomembranes* **20(5)**: 321–342.

Shukla SK (2006) Geosynthetic at subgrade level in paved roadways – design practice and installation. *Civil Engineering and Construction Review, New Delhi, India* **19(3)**: 78–84.

Shukla SK (2008) Engineering concepts of the paving geosynthetic application in pavement construction. *Civil Engineering and Construction Review, New Delhi, India* **21(2)**: 58–68.

Shukla SK and Chandra S (1994) The effect of prestressing on the settlement characteristics of geosynthetic-reinforced soil. *Geotextiles and Geomembranes* **13(8)**: 531–543.

Shukla SK and Kumar R (2008) Overall slope stability of the prestressed geosynthetic-reinforced embankment on soft ground. *Geosynthetics International* **15(2)**: 165–171.

Shukla SK and Yin JH (2006) *Fundamentals of Geosynthetic Engineering*. Taylor and Francis, London.

Shukla SK, Chauhan HK and Sharma AK (2004) Engineering aspects of geotextile-reinforced roadway of the National Highway, NH-2, Varanasi zone. *Civil Engineering and Construction Review, New Delhi, India* **17(12)**: 56–61.

Sivaramakrishnan R (1994) Jute geotextiles as revetment filter for river bank protection. *Proceedings of the 5th International Conference on Geotextiles, Geomembranes and Related Products*, Singapore, pp. 899–902.

Snow MS, Kavazanjian Jr E and Sanglerat TR (1994) Geosynthetic composite liner system for steep canyon landfill side slopes. *Proceedings of the 5th International Conference on Geotextiles, Geomembranes and Related Products*, Singapore, 1994.

Voskamp W, Wichern HAM and van Wijk W (1990) Installation problems with geotextiles, an overview of producer's experience with designers and contractors. *Proceedings of the 4th International Conference on Geotextiles, Geomembranes and Related Products*, The Hague, pp. 627–630.

Yoo C (2004) Performance of a 6-year-old geosynthetic-reinforced segmental retaining wall. *Geotextiles and Geomembranes* **22(5)**: 377–397.

Handbook of Geosynthetic Engineering
ISBN 978-0-7277-4175-2

http://dx.doi.org/10.1680/hgc.41752.387

ice
Institution of Civil Engineers
publishing

Chapter 18
Geosynthetic applications – sustainability aspects

A. Bouazza* **and G. Heerten†**
*Monash University, Melbourne, Australia
†NAUE GmbH & Co KG, Espelkamp, Germany

The global reduction of energy consumption and reduced emission of greenhouse gases are big challenges for the construction industry. To encourage consideration of sustainable development in the construction industry, sustainability components, such as energy consumption and the emission of greenhouse gases, should be determined, compared and calculated for different designs and construction choices to identify the most sustainable construction solution. Tools such as life-cycle assessment or embodied-energy calculations can help perform such comparative evaluations. This chapter presents a comparison between classical construction techniques and geosynthetic construction alternatives. The cumulated energy demand and carbon dioxide emission of primary, prefinished and finished products, their transport to the manufacturer and to the construction site, and their installation, are discussed. The cases presented in this chapter indicate that geosynthetics can provide environment-friendly and energy-efficient solutions in the construction industry.

18.1. Introduction

In recent years, there has been renewed public debate about the sustainability and development of our society in general, due to strong public sentiment about environmental and social issues surrounding engineering activities among others. In the original context, 'sustainable development' is not only restricted to the requirements of saving energy and raw materials, but also comprises a model of sustainable development based on the protection of natural resources, and the achievement of social and economical balance.

The will and ability to achieve these sustainability aims will decide our future, completely independently of any hypothetical climate-change scenario, and regardless of the degree of anthropogenic influence and the degree of predictability regarding regional impacts. The worldwide climate-protection approach to save energy and to reduce radically the emissions of greenhouse gases is an important component of successful sustainable development, and is therefore to be supported without restriction.

Greenhouse gas emissions attributable to the construction sector, in particular the built environment, in Australia are expected to almost double by early 2010 onward, increasing by 95% (from 32 Mt of CO_2-e to 63 Mt of CO_2-e per annum between 1990 and 2010) according to the Australian Greenhouse Office (AGO, 1999). The greatest increase in emissions is projected to be from the building industry. This sector relies extensively on electricity for its total energy requirements. In 1990, electricity accounted for 65% of energy requirements and 89% of total greenhouse gas emissions from commercial buildings, a scenario that AGO has forecasted to remain relatively constant up to 2010 and beyond. The situation is not different in other parts of the world, where the construction sector was found to be a major contributor to greenhouse gas emissions. A report from the Intergovernmental Panel on Climate Change (IPCC, 2007) has indicated that, in 2004, the building sector worldwide represented almost a quarter of the total global carbon dioxide (CO_2) emissions, including those produced through electricity use. This represents a tremendous potential for an improvement in the state of practice for sustainability in the construction industry, including geotechnical, hydraulic and coastal engineering industries. Innovations in the construction industry have been essentially influenced and driven by cost minimisation. To more effectively encourage sustainable development in the construction industry, sustainability components such as energy consumption or emission of greenhouse gases should be determined and quantified accurately.

In this respect, civil engineers can play an important role in embedding sustainability into the development process. However, it is interesting to refer to the classification proposed by the International Federation of Consulting Engineers (FIDIC, 2004) for sustainable development

Table 18.1 Outline sustainable development classifications
Condensed from FIDIC (2004); taken from Jefferis (2008), ASCE

Scale	Description	Example
Global/regional	Overall assessment of the current state of the world mapped to Agenda 21 and response to local Agenda 21	Millennium assessment
Industry/NGOs	Sustainability of the operations of an organisation. Indicators of how an organisation is performing in terms of a set of indicators for sustainable development	Global reporting initiative
Project risk assessment	Investor-based indicators. Principles, processes and indicators for assessing project risks	The equator principles
Financial performance	Any published index that tracks the financial performance of companies that have committed to sustainability principles	Dow Jones sustainability index, FTSE4Good
Project screening	Project-based indicators. Indicators for project as to their likelihood of achieving sustainability outcomes	World Bank, the equator principles
Project performance	Contribution a project makes towards sustainable development. Includes efforts made in the construction phase	SPeAR, CRISP, BEQUEST Network

BEQUEST Network, a forum for pan-European research, training and practical action in the quality assessment of the urban environment; CRISP, Construction Research and Innovation Strategy Panel; NGO, non-governmental organisation; SPeAR, Sustainable Project Appraisal Routine.

indicators according to the scale of the problem addressed. Table 18.1 gives an outline of this classification. It starts, as indicated by Jefferis (2008), with indicators at the global/regional scale, which are appropriate when considering the current state of the world. Smaller scale indicators are then considered, but it is only at the foot of the table, at the 'project performance' scale, that the construction phase is considered. Jefferis (2008) has also indicated that few geotechnical engineers are regularly involved at the stages above 'project performance' listed in Table 18.1; most will be involved at or below this level (i.e. at the design or construction phase), where decision-making on sustainability issues is, in most cases, too late.

However, there are steps that an engineer can take to improve the sustainability of a given project. In particular, the engineer needs to be aware that material selection and site processes have direct impacts at the local site level, as well as on a wider scale. Tools such as life-cycle assessment (LCA) and/or embodied-energy assessment are available to help engineers assess the impacts of the materials and energy used in projects works. For example, an LCA of alternative construction methods can be made. Furthermore, the energy demand and the associated emission of greenhouse gases can predict any potential impacts on the environment. The cumulated energy demand (CED) and CO_2 emission projections can be used as a 'short LCA' for the ecological evaluation. The LCA for the extraction and production of construction materials used, their loading, transport and installation, the CED, and the projected CO_2 emissions, can be determined for each of the construction alternatives, enabling comparison for sustainability decision-making.

For a sustainable evaluation and comparison of the selection of alternative processes or materials, the money equivalent of the projected CO_2 emissions would have to be determined and the costs of CO_2 emission certificates then added to the quoted price for the construction work; thus the most sustainable offer could, in principal, be accepted.

18.2. Economical and environmental advantages of using geosynthetics

The use of geosynthetics in civil engineering applications is widespread and growing. There are numerous types of geosynthetics, and each has a specific function, as discussed in earlier chapters. The economical and ecological advantages of using geosynthetics in geotechnical and hydraulic engineering applications are well established (Koerner *et al.*, 1998). When traditional construction methods are compared with those using geosynthetics, it is found that use of the latter results in a considerable reduction in construction costs and/or construction time, as well as in considerably lower volumes and masses that need to be excavated, transported and installed. More recently, Chau and Soga (2007) (from O'Riordan *et al.*, 2011) reported that various ground-improvement methods, including geotextile-reinforced embankments, were considered for highway embankment construction in Tokyo that had to accommodate high seismic forces. Their study indicates that geotextile-reinforced embankment provides the most energy-efficient solution compared with methods such as deep soil mixing and sand column piles. In other words, the embodied energy was the lowest for the geotextile-based embankment. Embodied energy is defined as the amount of energy used to make a material or product, to transport it to suppliers and onto site, and in the on-site construction process. The results will depend on where in the world the product is made, what processes are employed, plant efficiencies, the scale of manufacture, the sources of materials, etc.

In addition, there are environmental advantages of using geosynthetics. For example, they allow for the possibility of erosion-controlling, carbon-sink plantings on geogrid-reinforced slopes, or the natural settlement of marine flora and fauna in coastal protection measures, as has been shown at the artificial reef consisting of 'geotextile containers' near the Gold Coast in Australia (Heerten *et al.*, 2000) and the riverbank revetments constructed of sand-filled geotextile bags in Bangladesh (Oberhagemann and Hossain, 2011). In addition, using geosynthetic reinforcement (geotextile or geogrid) allows a reduced thickness of unbound road construction, and therefore reduces the mass of aggregate required for extraction (Belton *et al.*, 2009). Significant reduction in the carbon footprint associated with the extraction process and up to transportation to site is also achieved. From an economical viewpoint, there are also several examples related to waste-containment facilities which indicate the gain in savings that could be achieved by changing material selection. For example, the cover sealing of an old and closed landfill (Neu Wulmstorf landfill) located near Hamburg was one of the first big projects (320 000 m^2/32 ha) in Germany where a classic/traditional cover system composed of a compacted clay liner and a gravel drainage layer was replaced by a geosynthetic alternative (Heerten, 1998). The cover system consisting of a geosynthetic clay liner (GCL), a high-density polyethylene (HDPE) geomembrane and a geosynthetic drainage layer was installed on a gas-permeable mineral levelling layer, which directly covered the waste. For this project, it was established that

- instead of 21 000 truckloads of clay and gravel, only 165 truckloads were needed for the transport of the GCL and geosynthetic drainage layer
- instead of the projected construction costs of approximately 36 million Euros, the actual costs were only approximately 25 million Euros (i.e. a 30% reduction in the cost of the project)
- instead of a construction period of approximately three years, the project was completed within two years.

Another project, completed in 1999, the Lichte landfill near Stuttgart, Germany, showed that a cost reduction of 47% was achieved by using alternative geosynthetic construction methods (Haubrich, 2002, 2010). In this case, a geocomposite drain was used as a landfill gas-collecting layer, a geomembrane–GCL composite liner as a sealing layer and a geocomposite drain to drain water from the cover soil component.

18.3. Life-cycle assessment

A life-cycle assessment (LCA) addresses the economic sustainability of potential environmental impacts, and human health and resource concerns associated with a product, including raw material acquisition, through manufacture, use, end-of-life treatment, recycling and final disposal. It denotes the systematic analysis of the environmental impact of products during their entire life cycle. This comprises any environmental impact during the production, utilisation and disposal of the product, as well as any upstream and downstream processes connected with these phases (e.g. the production of raw and process materials). Environmental impact may include any ecologically relevant extraction from the environment (e.g. raw oil, soil, ore), as well as emission into the environment (e.g. waste, CO_2 emissions). The principal steps involved are: (i) an inventory of inputs and outputs; (ii) evaluation of the impacts of the inputs and outputs; and (iii) interpretation of the results. LCA does not provide data on where the impacts occur. LCA is completed through international standards, namely ISO 14040 and ISO 14044 or equivalent standards.

Figure 18.1 Constituents of a life-cycle assessment (pursuant to DIN EN ISO 14040 2006-10/14044, 2006–2010)

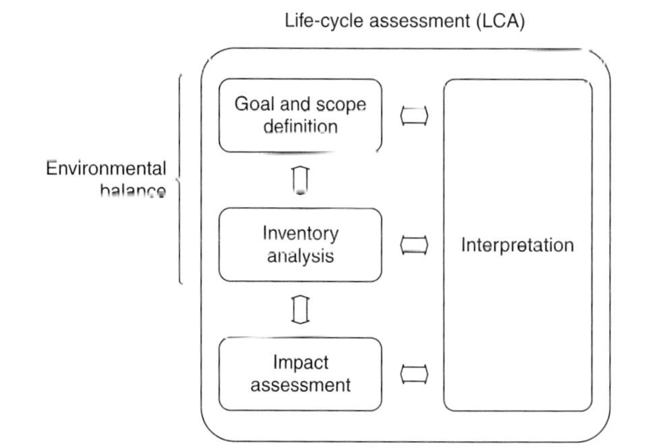

Figure 18.1 shows the phases of an LCA, and the correlation between the terms 'life-cycle inventory analysis', 'impact assessment' and 'evaluation'. Direct applications of LCAs comprise, for instance, the development and improvement of products, strategic planning, political decision-making processes, marketing, etc.

In the interests of a subsequent, and possibly comprehensive, evaluation, it is reasonable to conduct an impact-related accounting exercise between the sole life-cycle inventory analysis and an evaluation of the environmental balance. The flow and inventory of parameters collected during the life-cycle inventory analysis are described or assessed with regard to their potential effects. Their impact on selected global and regional or local environmental factors is considered. The effects as assessed in the life-cycle inventory analysis may be gauged with regard to the following environmental categories (SETAC, 1993)

- resource depletion
- land use
- global warming
- ozone depletion
- photochemical ozone creation
- acidification
- eutrophication, nutrification
- toxicological effects
- ecotoxicological effects
- waste
- modification of ecosystems and landscapes.

To establish comparative life-cycle balances for different products, the following preconditions are necessary

- same scope of use
- same state-of-the-art technology
- same range of functions.

The comparison is conducted by means of the following environmental balance factors

- excavation of raw materials (e.g. soil, sand, gravel, limestone, marl, clay, iron ore, crude oil)
- transport of the raw materials to the site or the manufacturer
- production of the primary products (e.g. cement, lime, structural steel, polypropylene granulates)
- transport of the primary products to the manufacturer or the construction site
- manufacturing of the products (e.g. concrete, geogrid, geotextiles)
- transport of the products to the construction site
- integration of the products (e.g. distribution, milling, consolidation, laying).

Considering these factors, it is then possible to calculate the CED, which can be stated in the following units

- MJ/kg in relation to the product; or
- MJ/m^3 in relation to the compacted/stabilised soil; or
- MJ/m^2 in relation to the sealed surface.

As an indication for the environmental impact, the CO_2 emissions are indicated in kilogram per kilogram of the applied product, or in kilogram per cubic metre of stabilised soil, or in kilogram per cubic metre of sealed area with regard to the global warming potential.

If a large part of the environmental impact results from the supply and consumption of energy, in many cases the CED may be used as a first rough check (i.e. a short LCA). It provides at least some initial information that can assist in the environmental evaluation of a given project (i.e. evaluation of energy, transport and material services). Even though the CED also requires data, the energy data may be collected and standardised easily. For example, the RIEE (Research Institute for Energy Economy) in Germany offers most of these energy data.

18.4. Comparison of different construction methods

When comparing traditional construction methods with the use of geosynthetics on the basis of CED or CO_2 emissions, it is necessary to evaluate the excavation and production of all the different construction materials needed, in addition to the transportation to the construction site, and the installation of the construction material or product. Case studies related to slope protection, road construction, a mechanically stabilised earth wall and a landfill-capping system are provided in the following sections (for further details see: Egloffstein, 2009a,b; Heerten, 2010, 2011; Rafalko *et al.*, 2010). Detailed information of specialty materials, such as geogrids, geotextiles, different soils, fine ground lime, cement, ready-mixed concrete and structural steel, have been collected and considered for comparison of the CED and CO_2 emissions for the cases presented in Sections 18.4.1, 18.4.2 and 18.4.4 (from Germany). A comparison of the LCA and embodied-energy has been used for the case presented in Section 18.4.3 (from the USA).

18.4.1 Slope protection

In the vicinity of Frankfurt/Main, Germany, a new connecting road (Gänsbergspange Idstein) was proposed to improve local traffic conditions. The design put out to tender asked for a vertical gravity wall to support a section of road. An alternative design employing a geogrid-reinforced slope was presented (Figure 18.2) and finally built, providing the background to compare these alternatives with regard to their environmental impact.

The comparison, between a slope-protection system with geogrids, including a vegetated steep slope protection system, with an anti-erosion mat and a vertical gravity-retaining wall made of reinforced concrete, showed considerably

Figure 18.2 The geogrid-reinforced steep slope road construction alternative at Gänsbergspange Idstein, Germany

Table 18.2 Cumulated energy demand (CED) for all components involved in the geogrid-reinforced steep slope alternative (Gänsbergspange Idstein, Germany)

Geogrid-reinforced steep slope	Data	Mass	Data	CED: MJ
Production of grid 1 60/60 Q6				
Built-in area	3750 m²			182 978
Mass per unit area	0.62 kg/m²	2325 kg	78.70 MJ/kg	
Production of grid 1 60/60 Q6				
Built-in area	1600 m²			
Mass per unit area	0.58 kg/m²	928 kg	78.70 MJ/kg	73 034
Transport to construction site, distance to manufacturer	400 km	3253 t	1.5 MJ/t km	1952
Delta Green wire grating reinforcement system for outer slope	630 m²			
Production of PP geotextiles per m² of slope reinforcement	0.16 kg/m²	100.8 kg		
Production of PP granules	0.16 kg/m²		65.50 MJ/kg	6602
Production of PP fibres	0.16 kg/m²		1.908 MJ/kg	192
Production of non-woven PP	0.16 kg/m²		0.324 MJ/kg	33
Production of wire grating reinforcement; front/bottom grating	6.95 kg/m²	4380	27.69 MJ/kg	121 282
Transport of non-woven PP to construction site, distance to manufacturer	270 km	0.1008 t	2.5 MJ/t km	68
Transport of wire grating to construction site, distance to manufacturer	220 km	4.38 t	1.5 MJ/t km	1445
Installation by excavator wheel loader (Geogrid + Delta Green)	5980 m²		1.04 MJ/m²	6219
Soil exploitation by backhoe, 8500 m³	8500 m³		7.6 MJ/m³	64 600
Transport of 17 000 t soil, transportation distance 15 km	15 km	17 000 t	2.5 MJ/t km	637 500
Soil spreading by bulldozer, layer thickness 0.30 m	8500 m³		8.98 MJ/m³	76 330
Soil compaction by roller compactor, layer thickness 0.30 m	8500 m³		4.80 MJ/m³	40 796
Production of anchored reinforced concrete ridgepole (0.5 m × 0.5 m × 140 m)	35 m³	2.4 t/m³	658 MJ/t	55 272
Production of reinforcing steel ST 500/500	77 kg/m³	2.7 t	27.960 MJ/t	75 352
Transport of concrete from concrete plant to construction site	20 km	84 t	2.5 MJ/t km	4200
Transport of structural steel to construction site	50 km	2.7 t	2.5 MJ/t km	337
Production of concrete ridgepoles (excavator, wheel loader, concrete pump, etc.)	35 m³	2.4 t/m³	15 MJ/m³	525
Faced brickwork of natural stones $A = 41$ m², $t = 0.2$ m; density 2.7 t/m³	41 m²	8.2 m³		
Exploitation of faced brickwork of natural stones in quarry		22.14 t	40 MJ/t	886
Transport of faced brickwork from quarry to construction site	35 km	22.14 t	2 MJ/t km	44
Production of faced brickwork (excavator, wheel loader, etc.)	35 m³		15 MJ/m³	525
Total CED = 1350 GJ				

PP, polypropylene.

better results for the geosynthetics solution with regard to both the CED and the CO_2 emissions.

The geosynthetics solution required not only 5350 m² of geogrid layers made of polypropylene (PP), but also a 0.5 m × 0.5 m × 140 m-long reinforced concrete top beam, forming the upper cover of the slope construction. In addition, 41 m² of faced brickwork made of natural stone and 630 m² of a slope-protection system, referred to as Delta-green, with almost 7 kg of steel wire per square metre, as well as a 200 g/m² non-woven geotextile made of PP, had to be integrated into the slope-protection system. Within the context of an energy and CO_2 balance, the system components had to be considered individually for comparison with the vertical gravity-reinforced concrete retaining-wall construction. Furthermore, the geogrid-reinforced embankment construction required approximately 40% more soil to be excavated, transported and installed compared with the vertical gravity-retaining wall. The details are given in Table 18.2.

The geosynthetics solution has proven to be not only considerably less energy consuming, but also more sustainable with regard to CO_2 emissions than the production, transport and installation of 1107 m³ of concrete with approximately 77.5 t of steel reinforcement. In this process, the steel reinforcement contributed considerably to the higher CED and CO_2 emissions, as can be seen in Table 18.3. The CED

Table 18.3 Cumulated energy demand (CED) for all components of the vertical retaining wall alternative (Gänsbergspange Idstein, Germany)

Angular retaining wall	Data	Mass	Data	CED: MJ
Total length	150 m			
Average height	5.5 m			
Angular retaining wall, average cross-sectional area (min. = 1.8 m^2; max. = 9.5 m^2)	7.38 m^2			
Production of standard concrete C20/25	1107 m^3	2.4 t/m^3	657 MJ/t	1745.518
Transport of concrete from concrete plant to construction site	20 km	2.657 t	2.5 MJ/t km	132 840
Production of reinforcing steel ST 500/500, 70 kg/m^3 for angular retaining wall	77.49 t		27 690 MJ/t	2145.698
Transport of structural steel to construction site	50 km	77.49 t	2.5 MJ/t km	9686
Production of angular retaining wall (excavator, wheel loader, concrete pump, etc.)	1107 m^3		15 MJ/m^3	16 605
Soil exploitation by backhoe	5146 m^3		7.6 MJ/m^3	39 110
Transport of 10 292 t soil, transportation distance 15 km	15 km	10 292 t	2.5 MJ/t km	385 950
Soil spreading by bulldozer, layer thickness 0.30 m	5146 m^3		8.5 MJ/m^3	46 213
Soil compaction by roller compactor, layer thickness 0.30 m	5146 m^3		5.40 MJ/m^3	27 790
Total CED = 4552 GJ				

for the geosynthetics solution amounted to only 30% and the CO_2 emissions to approximately 18% of the traditional alternative vertical gravity retaining wall. Figure 18.3 gives more details, showing the CED and CO_2 emissions of the alternative construction solutions. The CED and the CO_2 emissions related to the geosynthetic solution are, respectively, about 3.5 times and 5.4 times less than the traditional construction approach with a vertical gravity retaining wall construction.

18.4.2 Road construction

This case compares traditional construction work with a modern geosynthetic solution for a sub-base or base course improvement in district road construction close to the city of Aix-la-Chapelle, Germany. Figure 18.4 shows the base-soil improvement achieved by mixing a small percentage of binder (lime/cement) into the soil. Figure 18.5 shows the geogrid alternative for base-soil reinforcement.

In the construction of the district road, the geogrid solution provided a considerable advantage over the use of lime, cement and concrete in the form of a considerably lower CED and CO_2 emissions. One of the main reasons for this is the very small amount of PP geosynthetics required, and the high specific CED per kilogram of PP. Only 0.67 kg of PP per cubic metre of soil was required for the equivalent of 1 m^3 of stabilised soil, on the basis of a geogrid mass per unit area of 200 g/m^2 PP and a geogrid layer distance of 0.3 m.

In order to allow for a comparative assessment, an average value of 3.5% by mass for the addition of fine-ground lime and of 4% by mass for the addition of lime hydrate was selected according to Recommendation 551 of the German Road and Transportation Research Association (FGSV-Merkblatt 551, 2004). This corresponds to 70–80 kg of fine-ground lime or lime hydrate per cubic metre of soil at a moist density of approximately 2000 kg/m^3. The production of the much larger amount of fine-ground lime required has obvious negative effects on the CED and the CO_2 emissions (Figure 18.6), and the geogrid alternative has a CED about 5.4 times lower and CO_2 emissions about 27 times lower than the corresponding values for the lime/cement option. Furthermore, there is no projected impact on groundwater or air pollution when geogrids are used for base-soil reinforcement, unlike with the use of lime/cement dust.

18.4.3 A mechanically stabilised earth wall versus a gravity wall

Rafalko *et al.* (2010) reported an LCA comparison of a mechanically stabilised earth (MSE) wall system and an equivalent concrete gravity wall. The actual MSE wall, located at the Washington Dulles International Airport, USA, is 131 m long and has an average height of 4.6 m. The evaluation included the imported backfill, the facing elements and soil reinforcements. The results of the LCA study for the four stages of the MSE wall life cycle is presented in Table 18.4, which indicates that the bulk of the environmental impact is caused by the manufacturing process, with the greatest part being due to the imported backfill material.

For all four stages, the analysis showed that the backfill material had the greatest environmental impact, as MSE walls consist mostly of backfill. In addition to the backfill material, the concrete panels and soil-reinforcing strips significantly contributed to the environmental impact of the MSE wall in the manufacturing stage.

Rafalko *et al.* (2010) performed a comparison of the environmental impact of the manufacturing material for a gravity wall (of the same dimensions and soil properties as the MSE wall) and the MSE wall (backfill, facing elements and

Figure 18.3 Cumulated energy demand (CED) and CO_2 emissions at Gänsbergspange Idstein, Germany

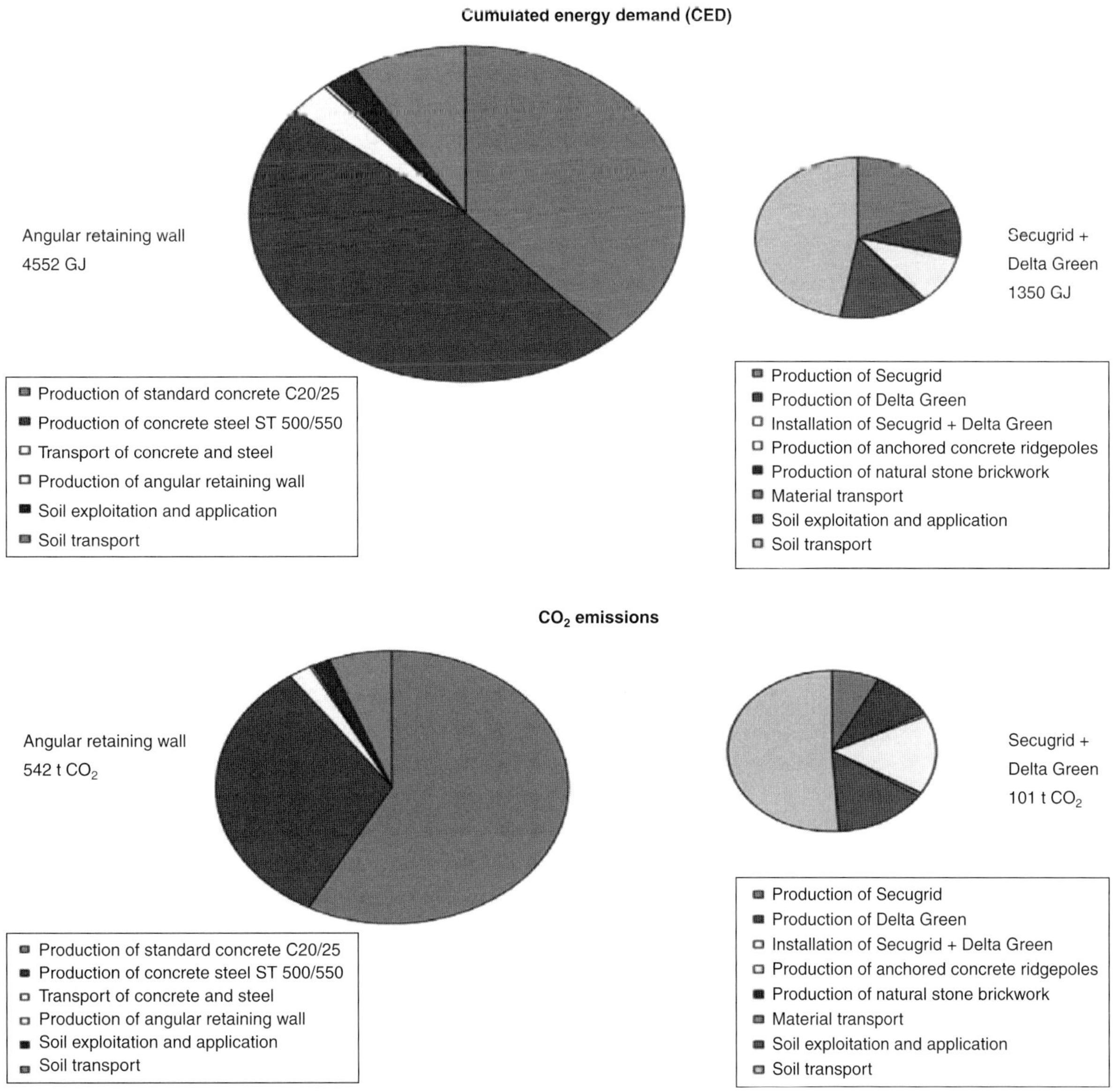

Figure 18.4 Base-soil improvements achieved by mixing lime/cement into the soil

Figure 18.5 Base-soil reinforcement achieved using a biaxial polypropylene geogrid

Figure 18.6 Cumulated energy demand (CED) and CO_2 emissions at district road K34, district Aix-la-Chapelle, Germany (2008)

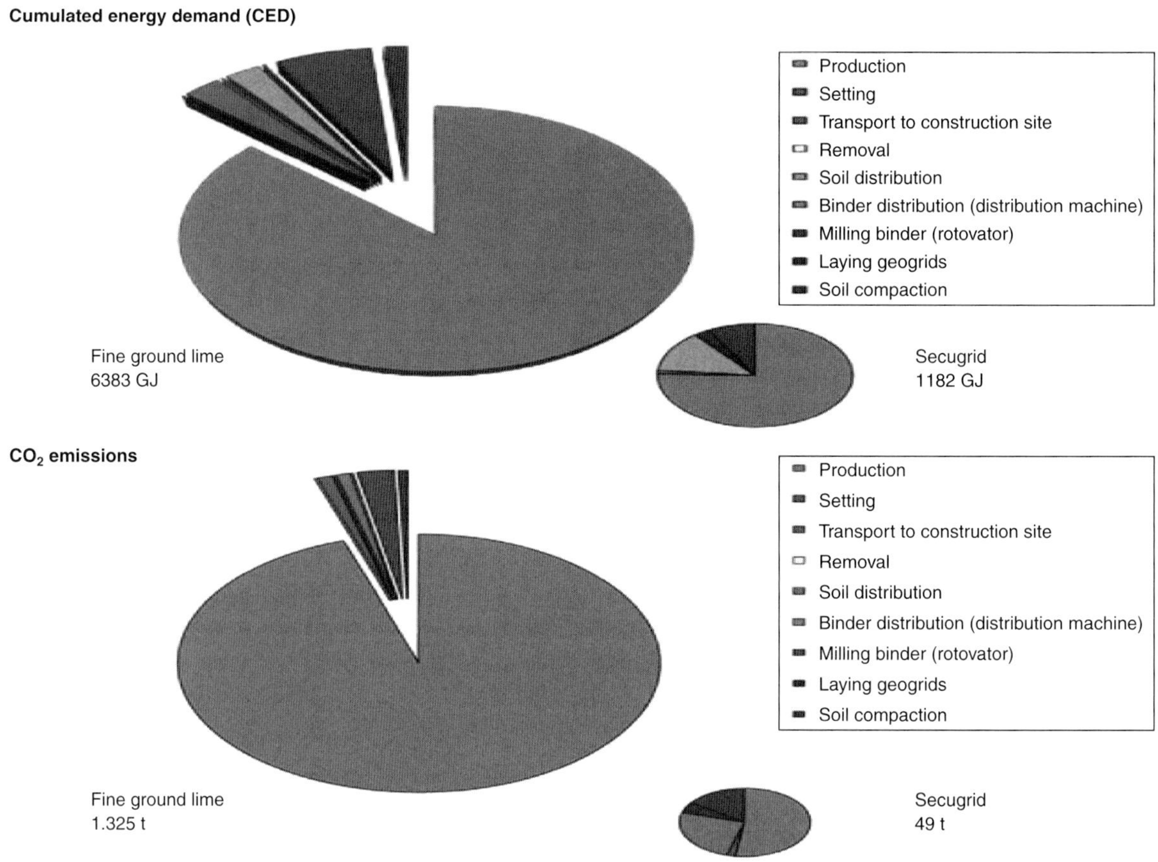

soil reinforcements). Not surprisingly, the energy associated with the manufacturing of the gravity wall was found to be much higher than the energy associated with the MSE wall (Table 18.5). The gravity wall is a concrete wall requiring the use of large amounts of high embodied energy materials such as cement and aggregates for its manufacturing process, thus reducing its energy efficiency.

18.4.4 Capping system for a landfill site

This assessment comprises a short LCA or CED for the capping system of a landfill site (Hillern landfill) located in Northern Germany. The assessment compares a geosynthetics-based solution (a geosynthetic clay liner (GCL) and a geocomposite drain) with the traditional soil-based solution (a compacted clay liner (CCL) and a mineral gravel drain). It should be noted that the comparison is site specific.

When comparing traditional construction methods with the use of geosynthetics-based methods on the basis of CED or CO_2 emissions, it is necessary to evaluate the excavation and production of all different construction materials needed. Transportation to the construction site and installation of the material or product must also be considered.

Table 18.4 Environmental impact of the four main stages of an MSE wall, and embodied energy
Data from Rafalko *et al.* (2010)

Stage	Inputs		Output air emissions		
	Energy consumption: GJ	Abiotic depletion: kg Sb eq.	Photo-oxidant formation: kg C_2H_4 eq.	Acidification: kg SO_2 eq.	CO_2 emissions: kg CO_2 eq.
Manufacturing	860	700	69	540	99 000
Transportation	129	34	20	100	11 100
Construction	18.8	9.7	1.6	9.4	1600
Dismantling	4.6	2.3	0.6	2.8	450

Table 18.5 Environmental impact of manufacturing material and embodied energy for an MSE wall and a concrete gravity wall
Data from Rafalko *et al.* (2010)

Wall system	Inputs		Output air emissions		
	Energy consumption: GJ	Abiotic depletion: kg Sb eq.	Photo-oxidant formation: kg C_2H_4 eq.	Acidification: kg SO_2 eq.	CO_2 emissions: kg CO_2 eq.
MSE wall	860	700	69	540	99 000
Gravity wall	2200	910	147	1310	420 000

Detailed information on the geosynthetics, the bentonite used in the GCL, the clay used in the base soil and the gravel have been collected and considered for CED comparison. The quantities of different materials have been considered as being put out to tender, defined by a special offer/construction proposal, or as recommended by the German landfill directive. The comparison in this case is limited to the sealing and drainage components of the surface sealing system, because all other components (e.g. levelling layer gas drain on top of the waste and cover and vegetation layer) are the same for both options.

A 0.5 m thick CCL is compared with a needle-punched GCL, which consists of needle-punched non-woven cover layer (300 g/m^2), sodium bentonite (4700 g/m^2) and needle-punched non-woven/woven composite (350 g/m^2). A thick gravel drain (16/32 mm grain size) is compared with a geocomposite drain composed of a needle-punched nonwoven filter layer (300 g/m^2), a coarse monofilament drainage core (600 g/m^2) and a needle-punched non-woven carrier layer (300 g/m^2). A comparison of the energy consumption of these components is given in Figure 18.7.

The total energy requirement for the geosynthetics solution is 114 MJ, compared with 192 MJ for the traditional mineral solution (a 69% increase in energy demand compared with the geosynthetics solution).

Figure 18.7 (a) Geosynthetic system (geosynthetic clay liner and geocomposite drain) and (b) traditional soil-based system (compacted clay liner and mineral gravel drain)

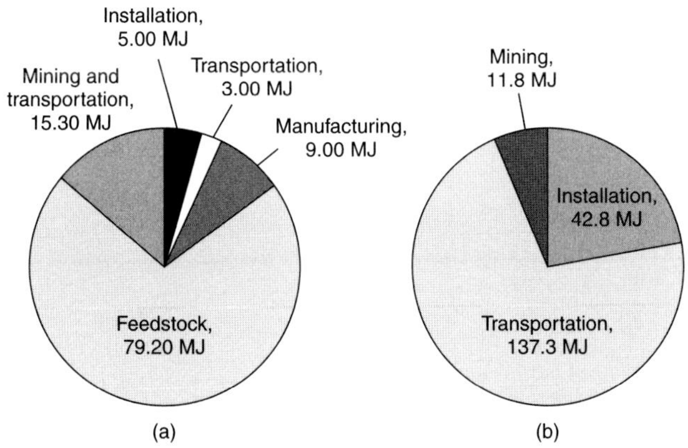

18.5. Summary of main points

- The global demand for reduced energy consumption and reduced emissions of greenhouse gases is a big challenge for the construction industry. Today, innovations and developments are essentially influenced and driven by reducing overall costs — the economic approach.
- To include ecological improvements and to encourage consideration of sustainable development in the construction industry, sustainability components such as energy consumption and the emission of greenhouse gases should be determined, compared and calculated for different designs and construction choices, in order to identify the most sustainable construction solution. Tools such as life-cycle assessment (LCA) or embodied-energy calculations can help when undertaking such comparative evaluations.
- The field application cases presented in this chapter indicate that geosynthetics can provide environment-friendly and energy-efficient solutions in the construction industry.

18.6. Some common questions and their answers

Questions

1. What is a carbon footprint of an organisation or activity?
2. Why do we need to evaluate energy efficiency?
3. What is life-cycle assessment (LCA)?
4. What is embodied energy?
5. How can I compare the energy efficiency of construction processes or design?

Answers

1. The carbon footprint is the amount of greenhouse gas emissions that an organisation or activity is responsible for.

2. An evaluation of energy is needed to identify the most sustainable construction solution and/or design.

3. LCA is a technique used to assess the environmental impact of products over their entire life cycle. This comprises any environmental impact during the production, utilisation and disposal of the product, as well as any upstream and downstream processes connected to it (e.g. the production of raw and process materials).

4. Embodied energy is the amount of energy used to make a material or product, to transport it to suppliers and onto site, and involved in its use in the on-site construction process.

5. Energy efficiency is compared and calculated for different designs and construction processes by determining components such as energy consumption and the associated emission of greenhouse gases. This process needs to be done for each step of the construction process or design.

REFERENCES

Australian Greenhouse Office (1999) *Australian Commercial Building Sector Greenhouse Gas Emissions 1990–2010*. AGO, Canberra.

Belton J, Hall C, Humphreys M and Gallagher E (2009) Potential savings in carbon dioxide emissions by using geogrid reinforcement in unbound roads. *Proceedings of the 1st Regional African Geosynthetics Conference*, Cape Town.

Egloffstein T (2009a) Ökologischer Vergleich zwischen Bauweisen mit mineralischen Baustoffen und Bindemitteln sowie Bauweisen mit Geokunststoffen. *Geotechnik* **32(2)**.

Egloffstein T (2009b) Soil stabilisation with lime and cement admixes or retaining walls – ecological comparisons with geosynthetic solutions. *Proceedings of the Geotechnical Seminar on Geosynthetics Applications*, University of Gdansk.

FGSV-Merkblatt 551 (2004) *Merkblatt für Bodenverfestigungen und Bodenverbesserungen mit Bindemitteln*. Veröffentlichung der Forschungsgesellschaft für Strassen- und Verkehrswesen, Cologne.

FIDIC (2004) *Project Sustainability Management Guidelines*. International Federation of Consulting Engineers, Geneva.

Haubrich E (2002) Oberflächenabdichtungssysteme, Stand der Technik – Systemauswahl – technische und wirtschaftliche Bewertungskriterien. Fachtagung Deponietechnik. Abschlussplanung und Sicherung von Deponien, LAU Halle.

Haubrich E (2010) *Oberflächendichtung vollständig aus Geokunststoff – Praxisbeispiel Deponie Lichte (BW)/Vergleich zweier Bauabschnitte*. Heute und vor 10 Jahren, 6. Leipziger Deponiefachtagung.

Heerten G (1998) Environmental geotechnics with geosynthetics. *Proceedings of the 11th Danube–European Conference on Soil Mechanics and Geotechnical Engineering*, Porec.

Heerten G (2010) The new German landfill directive and environmental advantage of using geosynthetics in landfill sealing systems. *Proceedings of the 6th International Congress on Environmental Geotechnics*, New Delhi, pp. 282–295.

Heerten G (2011) Reduction of climate damaging gases in geotechnical engineering practice using geosynthetics. *Geotextiles and Geomembranes*, in press, http://dx.doi.org/10.1680/hge.41752.39610.1016/J.Geotexmem.2011.01.006.

Heerten G, Jackson A, Restall S and Saathoff F (2000) New developments with megasand containers of nonwoven needle-punched geotextiles for the construction of coastal structures. *Proceedings of the 27th International Conference on Coastal Engineering*, Sydney.

IPCC (2007) Climate change 2007: mitigation of climate change. In *4th Assessment Report of the Intergovernmental Panel on Climate Change* (Metz B, Davidson OR, Bosch PR, Dave R and Meyer LA (eds)), Cambridge University Press, Cambridge.

Jefferis SA (2008) Moving towards sustainability in geotechnical engineering. Geosustainability and geohazards mitigation. *Proceedings of the Geocongress 2008*, New Orleans, LA, pp. 844–851.

Koerner J, Soong T-Y and Koerner RM (1998) *Retaining Wall Costs in the USA*. GRI report No. 20, Geosynthetic Institute, Folsom, PA.

Oberhagemann K and Hossain MM (2011) Geotextile bag revetments for large rivers in Bangladesh. *Geotextiles and Geomembranes* **29(4)**: 402–414, http://dx.doi.org/10.1680/hge.41752.396 10.1016/J.Geotexmem. 2010.12.003.

O'Riordan N, Nicholson D, Hughes L and Phear A (2011) Examining the carbon footprint and reducing the environmental impact of slope engineering options. *Ground Engineering* **44(22)**: 28–30.

Rafalko SD, Sankey JE and Freitag N (2010) Sustainability measures for MSE walls and baseline environmental impact evaluations. *Proceedings of the 2010 Earth Retention Conference*, Bellevue, WA.

SETAC (1993) *Guidelines for Life Cycle Assessment – A Code of Practice*. Society for Environmental Toxicology and Chemistry, SETAC Workshop held at Sesimbra, Portugal, Edition 1, Brussels and Pensacola, FL.

Handbook of Geosynthetic Engineering
ISBN 978 0 7277 4175 2

http://dx.doi.org/10.1680/hge.41752.397

Index

Page locators in *italics* refer to figures separate from the corresponding text.

AASHTO *see* American Association of State Highway and Transportation Officials
abrasion, 30, 199
absorption functions, 9
accelerated weathering tests, 32
access roads, 284
access shafts, 256
acrylic resin, 304, *306*
Adret des Tuffes Reservoir, 292–3
ageing of erosion control geosynthetics, 217–18
ageing mechanisms
  earth dams, 269
aggregate base course, 165
agricultural erosion control, 211
agronomic erosion control, 210–11
AGS *see* anchored geosynthetic system methods
Aix-la-Chapelle, 392, *394*
alternative liners, *247*, 248
American Association of State Highway and Transportation Officials (AASHTO), 318, 325, 329–30, 343, *344*, 345
  paved roads, 183–4
American Railway Engineering Association (AREA), 198
American Society for Testing and Materials (ASTM) methods, 73, 74, 76, 323
analytical failure solutions
  embankments, 108–9
anchorage
  applications of geosynthetics, 368
  embankments, 110, *112*, 113
  shallow foundations, 131, 143
  singularities, 283–4
  unpaved roads, 165
  see also pull-out
anchored geosynthetic system (AGS) methods, 234–5
anti-puncture geotextiles, 311
AOS *see* apparent opening size
apertures, 1, 4
apparent opening size (AOS), 22 3, 71, 198
application areas, 33–5, 36–7, *38–9*
applications of geosynthetics, 365–85
  case studies, 365, 374–83
    containers, 382–3
    drainage systems, 378–9
    earth dams, 380
    irrigation channels, 380
    landfills, 376–8
    pipeline systems, 378–9
    reservoirs, 380
    retaining walls, 374–6
    roads, 380–2
    slope erosion control, 379–80
    steep slopes, 374–6
    tunnels, 382
  cost analyses, 370, 372–3
  general guidelines, 365–9
    anchorage, 368
    care and consideration, 365–6
    construction protection, 367–8
    cutting, 367
    damage correction, 368
    geosynthetic selection, 366
    installation, 366
    joints, 366–7
    maintenance, 368–9
    placement, 366
    pre-installation protection, 366
    prestressing, 368
    service life, 367–8
    site preparation, 366
    test methods, 366
  general problems, 373–4
  in situ monitoring, 369–70, 371, 372
  quality control, 369–70, 371, 372
  sustainability aspects, 387–96
    construction method comparisons, 390–5
    economics, 388–9
    environment, 388–9, 394, 395
    life-cycle assessment, 388, 389–90, 392, 394–5
Approved Product List
  erosion control, 212–13, 218–19
arching in fill material, 119, 120
AREA *see* American Railway Engineering Association
articulate concrete blocks, 273, *274*
artificial radiation, 32
asphaltic concrete slabs, 265–6
ASTM *see* American Society for Testing and Materials methods
Auchenhowie Road, 156
Australia
  railway track case studies, 201–2
automatic seaming machines, 255
axisymmetric unpaved road reinforcement, 170

backfilling, 377
bagged stone walls, 239–40
ballast
  earth dams, 268
  railway structures, 193, 195–6
BAM *see* Bundesanstalt für Materialforschung und-prüfung
bamboo fascine mattress foundations, 157
bank protection, 274
Barlovento Reservoir, 293–5

basal lining systems, 247–50, 249, 251, 253–4
basal seals, 256
base course lateral restraint, 181
base-course reduction ratios (BRR), 182, 187–9
base reinforcement, 185–9, 392, *393*
base-soil reinforcement, 392, *393*
bathtub drainage conditions, 203
BCR *see* bearing capacity ratios
beam theory, 122–3
bearing capacities
  retaining walls, 93–4
  shallow foundations, 132
    reinforced clay, 136, 142
    reinforced granular fill, 137–8, 140–2
    reinforced granular fill-soft foundation soil systems, 142–5
  slope stabilisation, 230–1
  soil–geosynthetic interaction interface resistance, 49
  unpaved roads, 170–1, *172*
  see also load bearing capacities; ultimate bearing capacities
bearing capacity ratios (BCR), 132, *133*, 134
Ben Boyd Dam, 269, *270*
bench erosion control scales, 215–16
bending moment, 116–17
berms, 113
BFMs *see* bonded fibre matrices
biaxial geogrids, *3*
  definition, 1
  piled embankments, 119
  reinforced clay slopes, 230, *231*
  sustainability aspects, 392, *393*
biaxial geonets, 252
biaxial tensile tests, 15, *16*
Bilancino Dam, 268
biochemical oxygen demand (BOD) values, 73, 74
biodegradable geotextiles, 11
biological clogging, 73, 74
biological degradation, 32
biological erosion control, 210–11
bi-oriented geogrids, 202
biotechnical mats, 273
biotechnical soil erosion solutions, 211, 212
Bishop methods, 110
bituminous concrete layers, 272–3
bituminous geomembranes, 203, 284
bituminous materials
  applications of geosynthetics, 367
blanket drains, 270
  embankments, 113–14
  landfills, 247–8, 251, 252, 255–6
blanket layers
  railway tracks, 195
blinding phenomena, 31, 69, *70*
blocking phenomena, 31, 69
bodkin joints, 18, *19*
bonded fibre matrices (BFMs), 212
bonded geogrids, 1, *3*
bottom liner systems, 254
bottom-sea dredged sediments, 79
boxing railway tracks, 195
BPT *see* bubble point test methods
Brazil
  railway tracks, 202
breaking of geosynthetic layers, 132
bridge approach embankments, 155
bridge piers, 35, *38*
bridging layers, 26
bridging networks, 69
brine evaporation ponds, 288–9
British Standards Institute (BSI), 118–20
BRR *see* base-course reduction ratios
Brugnens earth dam, 269
BSI (British Standards Institute), 118–20
BSSC *see* Building Seismic Safety Council
bubble point test (BPT) methods, 22
Building Seismic Safety Council (BSSC), 329
Bundesanstalt für Materialforschung und-prüfung (BAM), 250
burst strength, 20

calendering, 13
California bearing ratios (CBR), 186
Canada
  railway tracks, 200–1
canals, 279–302
  case studies, 285, 295, 296–7, *298*, *299*
  design, 281–4
  geomembrane durability, 284–5
  historical background, 279–81
Canary Islands, 268, 293–5
capping systems, 247, 256, *257*, 258, 394–5
carbon dioxide (CO_2) emissions, 387, 388, 390–2, *393*, 394–5
case studies, 285, 289–95, 365, 374–83
  canals, 285, 295, 296–7, *298*, *299*
  containers, 382–3
  containment ponds, 285–9
  drainage systems, 378–9
  earth dams, 380
  irrigation channels, 380
  landfills, 376–8
  pipeline systems, 378–9
  reservoirs, 380
  retaining walls, 374–6
  roads, 380–2
  slope erosion control, 379–80
  steep slopes, 374–6
  tunnels, 382
cast-in-place concrete cover layers, 271, *272*
catenary layers
  embankments, 123–4
CBIP *see* Central Board of Irrigation and Power
CBR *see* California bearing ratios
CCLs *see* compacted clay liners
CED *see* cumulated energy demand
cellulose fibre mulches, 212
cement, 392, *393*
cementitious grout, 304, *306*
Central Board of Irrigation and Power (CBIP), 7
centrifuge modelling, 225
centrifuge shaking table tests, 349–50, 353
chemical bonding, 12
chemical clogging, 73
chemical erosion control geosynthetic degradation, 217
chemical resistance
  endurance and degradation, 32
  liner systems, 282
  railway tracks, 199
chemical stabilisation methods, 223
chimney drains, 269, 270

chlorinated polyethylene (CPE), 10
chlorosulphonated polyethylene (CSPE), 10
circular failure surfaces, 103
circular holes, 204
circular slip analyses, 225–6
circular slip failure mechanisms, 334–5
clamping of specimens, 57
classifications, 1–5
clay core stone ballast, 268
clay liners, 4, *6*, 14
  applications of geosynthetics, 367, 377–8
  landfills, 247, 254, 257–8
  see also compacted clay liners; geosynthetic clay liners
clay loam, 215
clay silt subgrades, 183
clay soils, 144, 156–7
  reinforced, 136, 142
Climate Change Panel (IPCC), 387
clogging, 26, 68, 69–70, 72, 73
clogging ratios *see* gradient ratios
coastline erosion, 379
Codole rockfill dam, 271, *272*
cofferdams, 268
cohesion influence, 103, 110–11
cohesive soils, 25–6
coir-geotextile-laid vegetated slopes, 379–80
collection pipes, 256
combined failure surface mechanisms, 104–5
combined geosynthetic functions, 185
commercially available geosynthetics, 32–3
compacted clay liners (CCLs), 248, 249, 254, 256, 258
  applications of geosynthetics, 394, 395
  liner systems, 282
compaction of aggregate base course, 165
compaction of drains, 233–4
composite liner systems, 247, 250–53, 254, 377–8
compressibility, 14–15
compressive stresses, 25
concrete
  blocks, 273, *274*
  cover layers, 271
  footings, 134–5
  layers, 271, 272–3
  linings, 304, *305*, 312–13
  panels, 238
  slabs, 264, 265–6, 267
  sliding formwork machines, 295, *299*
  structure connections, 284
concrete test tanks, 229, *230*
confinement effect, 130–1
confinement functions, 197
confinement stresses, 52, *53*
conservation practice factors, 212
consolidation ratios, 114–15
constant-head permeability test methods, 25
constriction sizes, 71
construction expedients, 270
construction of liners, 253–5
construction quality assurance (CQA), 248, 255, 258
construction quality control (CQC), 255, 258, 259
construction survivability, 31
containers, 382–3
containment functions, 9
containment ponds, 279–302
  case studies, 285–9
  design, 281–4
  geomembrane durability, 284–5
  historical background, 279–81
contamination of ballast, 196
continuous-filament non-woven geotextiles, 270
Contrada Sabetta Dam, 267
corrugated PVC pipes, 236–7
cost analyses, 370, 372–3
Coulomb friction angles, 318
coverage ratios, 96
cover layers, 271, *272*
cover liner systems, 310–13
cover systems
  landfills, 256–9
  overliners, 282–3
CPE *see* chlorinated polyethylene
CQA *see* construction quality assurance
CQC *see* construction quality control
crack resistance, 381
creep, 28–30, 110, 132
critical circle radii, 106
critical slip circle, 105
crop factors, 212
cross-drains, 205, *206*
crushed rock drains, 234
CSPE *see* chlorosulphonated polyethylene
cumulated energy demand (CED), 388, 390–2, *393*, 394–5
cushioning functions, 9
cyclic loading
  unpaved roads, 165, 167, 175
cyclic load testing
  geosynthetic-reinforced soil slopes and walls, 320–4
cyclic tests
  soil–geosynthetic interface, 325

damping ratios, 345
dams, 35, *39*, 274
  see also earth dams
Darcy's coefficient, 23
Davis Creek Dam, 272
deep footings, 140, 141
definitions, 1–5
deflection profiles, 151–3, *154*
deformation compatibility conditions, 147
deformation of drains, 233
deformation fields, 165, *166*
degradation properties, 28–33
depth of drains, 233
design by function, 9
desktop analyses, 274–5
dewatering, 78–9, 258–9
digital image analyses, 225
dikes, 274
direct shear tests, 20–1, 55, 56–60, 318
direct sliding, 47–9, 50
displacement, 336–9
  Empirical approach, 337–8
  example applications, 338–9
  geosynthetic-reinforced soil slopes and walls, 318
  Newmark's method, 336–7, *338*, 339
  slope physical tests, 348, 350, *351*
  soil–geosynthetic interaction, 54, 62, 326
  wall physical tests, 348, 350, *351*

distress features
paved roads, 179–81
ditch drains, 203, 204
Don Martin Dam, 265
double composite liners, 248
double-stitch line seams, 18, *19*
drainage, *39*, 67, 80–3
applications of geosynthetics, 378–9
composites, 4, *6*
earth dams, 264, 265–6, 267–8, 269
embankments, 113–16
functions, 8, 10, 181, 184, 185, 197
geocomposites, 82–3
geonets, 82–3
geotextiles, 80–2
hydraulic tunnels, 304, 308
landfills, 247–8, 251, 252–3, 255–6, 258–9
paved roads, 180, 181, 184, 185
railway tracks, 202–5, *206*
separators, 4, *6*
shallow foundations, 155–6
slope stabilisation, 231–4, 236–7
spacing of drains, *232*, 233
strip drains, 13
subsoil pipes, 236–7
trench drains, 35, *39*, 205
underliners, 281
dry sieving test methods, 22
dual hot wedge fusion, 255
DuPont Edge Moor facility, 377
durability, 31
liner systems, 282, 307
DYN3D program, 340
dynamic creep, 29–30
dynamic filtration tests, 76, 79
dynamic interlock, 165
dynamic loading, 318–26
dynamic vertical stresses, 186

earth dams, 263–78
bank protection, 274
case studies, 380
conventional material use, 264–6
design considerations, 274–5
geosynthetic use, 266–74
river beds, 274
earth pillows, 274
earthquakes, 354–5
see also seismic analyses
earth-retaining walls, 89, *91*
earth walls, 392, 394, 395
ease of installation criteria, 282
economic criteria, 282
EF *see* erodibility factors
EKG *see* electrokinetic geosynthetics
elastically embedded membrane theory, 120–1
elastic rebound, 167
electrokinetic geosynthetics (EKG), 5, 10
ELM *see* equivalent-linear methods
El Vado Dam, 265
Emalahleni water reclamation plant, 288–9
embankments, 101–27
anchorage lengths, 110, *112*, 113
basal reinforcement, 101–13
cost analyses, 372, 373
design approach, 103–9
drainage, 113–16
earth dams, 265, 268, 274, 275
generalised failure, 104–9
piled embankments, 117–24
railway tracks, 203
reinforcement choice, 109–10
shallow foundations, 155–7
on soft foundation soil, 35, *38*
vertical drains, 116–17
embodied energy, 388, 392, 394
Empirical displacement calculation approach, 337–8
end-bearing piles, 119
endurance properties, 28–33
engineered erosion control, 211
enhanced confinement, 164–5
EOS (equivalent opening size), 22–3
EPS *see* expanded polystyrene
equivalent contact pressure, 169
equivalent-linear methods (ELM), 319
equivalent opening size (EOS), 22–3
erodibility factors (EF), 213–15
erosion control, 35, *38*, 209–22
case studies, 379–80
classification, 210–11
cost analyses, 372
design approach, 211–13
earth dams, 273–4
lifetime cost analyses, 218–19, *220*
performance testing, 215–18
product performance, 218–19, *220*
rain erosion, 209–10
river erosion, 209–10
short-term yield factors, 213–15
surface erosion mechanics, 210
Erosion Control Laboratory of the Texas Transportation Institute, 212, 215
Europe
landfills, 250, 251, 259
railway track, 201
European Landfill Directive, 246–7
European standards
soil–geosynthetic interface resistance quantification, 57, 60
expanded polystyrene (EPS), 10, 13, 353, 354
external stability calculations, *328*, 330
external stability of retaining walls, 89, *90*, 93–6
extra-heavy non-woven geotextiles, 198
extruded concrete sliding formwork machines, 295, *299*
extruded geogrids, 1, *3*, 12
extruded polystyrene (XPS), 10, 13
extrusion processes, 13

fabric-wrapped crushed rock drains, 234
face-to-face seams, 18, *19*
facing connection tests, 325–6
factors of safety, 28, 29
embankments, 103–10
erosion control, 210
slope stabilisation, 225–6, 227, 228
failure approach
Lowe, 105–6
failure mechanisms
pseudo-static methods, 333–5

retaining walls, 93–4
shallow foundations, 132
failure tests
hydraulic overflow-induced, 274
fastening systems, 308–10
Fast Lagrangian Analysis of Continua (FLAC) systems, 342–3, 348, 353
fatigue limit pressures, 167
fatigue strength, 20
Federal Highway Administration (FHWA), 318, 325, 328, 330–2, 335, 345
Federal Institution for Material Research and Testing, 250
FI *see* fouling indices
fibrillated yarns, 11
FIDIC *see* International Federation of Consulting Engineers
field tests
erosion control, 216–17
filaments, 11
fill active thrust, 119
fill material arching, 119, 120
fill subgrade railway track structures, 193–4
filtration, 8, 26, 67–80
design, 76–8
earth dams, 266–7, 269–71, 275
geotextiles, 67–74, 78–80
high water content geomaterials, 78–80
laboratory filtration performance tests, 74–6
landfills, 247–8, 255–6, 259
paved roads, 181, 184
railway tracks, 196–7
filtration opening size (FOS), 71
fin drains, 13, 204–5
fine-fraction filtration tests, 76, 79
finite-difference methods, 342
finite-element analyses
embankments, 116, *117*
landfills, 256
numerical dynamic analyses, 339–42
shallow foundations, 139, 141, 144–6
slope stabilisation, 227
fixation of specimen influence, 59
flat profiles, 308–10
flexibility of geosynthetics, 14
flexible mattresses, 274
floating covers, 283
fluid barriers, 8, 264, 267–9
fluid transmission functions, 8
fly ash slopes, 230–1
fly ash slurry, 79
footings, 134, 135, 141–2, 145, 146, 230–1
forestation erosion control, 211
formation subgrade railway track structures, 193–4
FOS *see* filtration opening size
fouling of ballast, 196
fouling indices (FI), 196
foundation soil expulsion, 103–4
foundation soil systems, 35, *38*, 136–9, 142
free-flow tunnels, 304, *306*, 307, *309*
free soil dilatance, 47
Fresh Kill Landfill leachate, 73–4
frictional resistance-dependent reinforcement, 9–10
friction piles, 119
front block tying, *237*, 238
frost-susceptible clay silt subgrades, 183
full erosion control scales, 215–16

gabions, 273–4
gas venting regulation, 256–7
GCLs *see* geosynthetic clay liners
geoarmours, 280, 281
Geoblock type geogrid-reinforced concrete wall systems, *238*, 239
geocells, 5, 138–9, *138*, 156–7, 280–1
geocomposites, *6*
definition, 4
drains, 82–3, 204–5, 378–9
earth dams, 266, 267–8, 269
edge drains, 204–5
function, 9, 10
geosynthetic-reinforced soil slopes and walls, 320
hydraulic tunnels, 304, *305–6*, 308–9, *309–10*
landfills, 247, 257, 259, 377, *378*
manufacturing processes, 13
pipeline systems case studies, 378–9
railway tracks, 201–2, 204–5
raw materials, 13
selection, 9, 10
shallow foundations, 139
geofabrics, 4, *156*, 157, 231–4
geofabric-wrapped drain (GWD) methods, 231–4
geofoams
definition, 4
function, 10
manufacturing processes, 13
raw materials, 13
seismic buffers, 354–5
selection, 10
geogrid-reinforced modular concrete block wall systems (GRMCBWSs), 238–9
geogrid-reinforced soil
interlocking mechanism, 1, 4
geogrids, *3*
applications of geosynthetics, 381–2
definition, 1
earth dams, 272
function and selection, 9–10
geosynthetic-reinforced soil slopes and walls, *321*, 352
hydraulic tunnels, 304, *306*
mechanical properties, 17
paved roads, 183, 186
piled embankments, 119
railway tracks, 202
retaining walls, 89, *90*, 96–7
shallow foundations, 135, 144, 156
slope stabilisation, 226, 229, 230–1, *231*, 237, 238–9
soil–geosynthetic interaction, 47–9, 52, *53*, 54, 56, 62, 63
sustainability aspects, 392, *393*
total strain versus log *N* plots, 29–30
unpaved roads, 167
geomaterials
high water content filtration, 78–80
geomats, 4, *6*, 211–12, 251, 273, 280–1
geomattresses, 273, 274, 280
geomembranes
canals, 295, *299*
definition, 4
durability, 284–5
earth dams, 264, 265, 266, 267–9, 271, *272*

geomembranes (*continued*)
function and selection, 10
hydraulic tunnels, 304–7
applications, 304–5
benefits, 307
cover liner systems, 310–11
inner concrete linings, 313
protective seals, 313
PVC waterproofing, 311, *312*
types, 305–6
landfills, 247–52, 254–6, 258, 259
manufacturing processes, 13
piled embankments, *117*, 118–22
railway tracks, 203
raw materials, 13
seams, 18, *19*, 20
shallow foundations, 143
unpaved roads, 164
geomeshes, 4–5
geonaturals, 5
geonets, 280
definition, 4
drains, 82–3
function, 10
hydraulic tunnels, 308
landfills, 252, 259, 377–8
manufacturing processes, 12–13
mechanical properties, 17
raw materials, 12–13
selection, 10
geopipes, 5
geoproducts, 5
geospacers, 5, 280
geostrips, 5, 280
geosynthetic clay liners (GCLs), 248, 249, 250, 256
applications of geosynthetics, 389, 394, 395
historical background, 280
landfills, 257–8
liner systems, 282
geosynthetic-reinforced cohesive shallow foundations, 145
geosynthetic-reinforced granular fill-soft soil systems, 146–7, 148, *149*
geosynthetic-reinforced soil slopes and walls, 317–63
design, 326–48
dynamic loading, 318–26
earthquake performance, 354–5
material properties, 318–26
physical testing, 348–53
seismic analyses, 326–48
seismic buffers, 353–4
geotextiles, *2*
cost analyses, 370, 372, 373
definition, 1
drains, 80–2
earth dams, 266–73, *272*, *274*
filters, 67, 78–80
affecting factors, 70–4
long-term performance, 72–4
mechanisms, 67–70, 73
function, 9–10
history, 5
hydraulic tunnels, 311
joints, 366–7
landfills, 255–6
manufacturing processes, 10–11
mechanical properties, 25, 26
paved roads, 184
per cent open area, 71–2
pore size distributions, 71
railway tracks, 196–202, 204
raw materials, 10–11
retaining walls, 89, *90*, 91–6
external stability, 93–6
internal stability, 92–3
road case studies, 381, 382
selection, 9–10
shallow foundations, 135–6, 149–51, 155–7
slope erosion control, 379
slope stabilisation, 225, 226, 237
soil–geosynthetic interface resistance quantification, 56
total strain versus log *N* plots, 29–30
tubes, 27
unpaved roads, 168, 169
geotubes, 5
geowebs, 139
German piled embankment methods, 120–2
Germany, 251, 252, 258–9
Ghrib Dam, 265–6
Giroud and Han unpaved road reinforcement design, 174–5
Giroud and Noiray unpaved road reinforcement design, 168–70, 172, 173–4
global stability analyses, 227, 228, *228*
GR *see* gradient ratios
grab tensile tests, 15, *16*
grade-crossing rehabilitation, 200–1
graded filter zones, 69
gradient ratios (GR), 30–1, 74, 79
granite bed-rock formation, 236
granular fill drains, 233
granular filters, 270–1
granular soil behaviour, 45–6
grassed topsoil cover, 273, *274*
gravity drainage, 80, 81–2
gravity shaking tests, 348–53
gravity walls, 392, 394, 395
green-faced structures, 237–8
Grenada Dam, 273–4
GRMCBWSs (geogrid-reinforced modular concrete block wall systems), 238–9
grouting, 304, *306*, 312–13
GWD *see* geofabric-wrapped drain methods

hanging bag tests (HBT), 27–8
Hans Strijdom Dam, 270–1
hard facing precast concrete panels, 238
hazardous-waste landfills, 247, 248
HBT *see* hanging bag tests
HCR *see* hydraulic conductivity ratios
HDPE *see* high-density polyethylene
heat-bonded non-woven geotextiles, 135–6
heavy compaction effects, 367
heavy non-woven geotextiles, 198
high-density polyethylene (HDPE), 5, 10
applications of geosynthetics, 389
earth dams, 272
geosynthetic-reinforced soil, 320, *321*, 322–5
landfills, 250, 251, 258
soil–geosynthetic interactions, 51–2

high water content geomaterials, 78–80
honey comb geocell structures, 138–9, *138*
horizontal equilibria, 170, *171*
horizontal force, 59–60
Houlsby and Jewell unpaved road reinforcement design, 173–4
Houzhou embankment, 274
hydraulic conductivity, 68, 70, 73, 76, 259
hydraulic conductivity ratios (HCR), 76
hydraulic criteria, 282
hydraulic overflow-induced failure tests, 274
hydraulic properties, 21–8
Hydraulics, Sedimentation, and Erosion Control Laboratory of the Texas Transportation Institute, 212, 215
hydraulic tunnels, 303–15
  cover liner systems, 310–13
    inner concrete linings, 312–13
    protective anti-puncture geotextiles, 311
    protective layers, 311, *312*
    protective seals, 312–13
    PVC waterproofing geomembranes, 311, *312*
  exposed liner systems, 307–10
    drainage systems, 308
    fastening systems, 308–10
    perimeter seals, 310
    support, 308
    waterproofing liners, 308
  geomembrane systems, 304–7
    applications, 304–5
    benefits, 307
    types, 305–6
hydrodynamic test methods, 22
hydroelectric power stations, 268
hydrological stabilisation methods, 223
hydro-seeding, 211

image analyses, 22, 225
impact strength, 20
inclined plane tests (IPT), 21
index tests, 9, 18, 32
India
  railway track case studies, 201
induced deformations concepts, 129
induced-stress concepts, 129
infinite slope analyses, 210
in-isolation cyclic load testing, 320–3
in-isolation monotonic load–strain behaviour, 320
inner concrete linings, 312–13
in-plane permeability, 25
  see also transmissivity
in-plane water flow capacity, 259
in situ biotechnical erosion control, 211
in situ monitoring, 369–70, 371, 372
in situ slope stabilisation, 234–5
in-soil reinforcement cyclic load testing, 323–4
instability mechanisms, 101, *103*
insulation functions, 9
interface friction angles, 269
interfacial shear stress membrane support, 143
Intergovernmental Panel on Climate Change (IPCC), 387
interlayer functions, 9
internal erosion, 69
internal stability analyses
  pseudo-static methods, 330–2
  slope stabilisation, 227, *228*
internal stability of retaining walls, 89, 92–3
International Commission on Large Dams (ICOLD), 274
International Federation of Consulting Engineers (FIDIC), 387–8
IPCC (Intergovernmental Panel on Climate Change), 387
IPT (inclined plane tests), 21
irrigation channels, 380

Japan
  earthquakes, 355
joints
  applications, 366–7
jute geotextiles, 379

Kaniraj failure approach, 106, 107–8, *107*
kinematic approach, 168, 227
knitted geotextiles, *2*
  definition, 1
  manufacturing processes, 12
  raw materials, 12

laboratory filtration performance tests
  dynamic filtration tests, 76
  filters, 74–6
  fine-fraction filtration tests, 76
  gradient ratio tests, 74
  long-term flow tests, 74–5
  permittivity tests, 76
laboratory simulation tests, 32
laboratory tests
  erosion control, 216–17, 218
  shallow foundations, 134, 135, 137–8, 139, 147–8
  slope stabilisation, 229–31
  soil–geosynthetic interface resistance, 55–64
landfills, 35, *39*, 245–61
  applications of geosynthetics, 376–8, 394–5
  basal lining systems, 247–8
  categories, 246–7
  cover systems, 256–9
  dewatering, 258–9
  geotextile filters, 73–4
  leachate collection and removal, 255–6
  multi-barrier concepts, 246
  solid-waste, 246–7, 248, 253–4, 256
  South Africa, 252–3
  sustainability aspects, 394–5
La Parade homogeneous earthfill dam, 269
lapped seams, 18, *19*
large box shear tests, 269
large erosion control scales, 215–16
large-scale tests
  landfills, 256
  shallow foundations, 139, 147–8
  unpaved roads, 167
lateral earth pressures, 89, 91
lateral restraint, 164–5
layer number of drains, 233
LCA *see* life-cycle assessment
leachate
  geotextile filters, 73–4
  landfills, 247, 249, 252, 255–6
life-cycle assessment (LCA), 388, 389–90, 392, 394–5
life-cycle cost analyses, 189
lifetime cost analyses, 218–19, *220*

lime mixing, 392, *393*
limit equilibrium methods, 110, 225–6
limit stability analyses methods, 226–7
liner systems
  canals, 282
  containment ponds, 282
  earth dams, 269
  hydraulic tunnels, 304, *305*, 307–13
  landfills, 247–9, 251, 253–5, 377–8
  reservoirs, 282
  tunnel case studies, 382, *383*
liquid containment ponds, 368
liquid reservoirs, 35, *39*
load bearing capacities
  shallow foundations, 139–45, 149–51
    reinforced granular fill, 139–42
load cycles
  paved roads, 186
load diffusion models, 168–9, 170, *171*
load spreading angles
  shallow foundations, 141, 143
load–strain behaviour, 320
load-strain curves, 16–17
load-transfer platforms, 122–4
load transfers
  piled embankments, 120–1
log *N* versus total strain plots, 29–30
log spiral failure mechanisms, 333–4
longevity
  geosynthetic properties, 31
long-term flow tests, 74–5
lower-bound plasticity theory, 143
Low *et al.* failure approach, 105–6
low-height earth dams, 266–7

McKay Dam, 264
maintenance benefits, 307
manufacturing processes, 10–14, 248
manufacturing quality assurance (MQA), 248
Marne–Rhin canal, 295, 296–7, *298*
mass per unit area, 14, *15*
mats
  landfills, 4, *6*, 211–12, 251, 273, 280–1
mattresses, 273, 274, 280
MDPE *see* medium-density polyethylene
mechanical bonding, 12
  see also needle-punched geosynthetics
mechanical criteria
  liner systems, 282
mechanical erosion control, 210–11
mechanical foundation models, 146–7, *147*
mechanically stabilised earth (MSE), 229, 376, 392, 394, 395
mechanical properties of geosynthetics, 14–21
mechanical stabilisation methods, 223
medium-density polyethylene (MDPE), 10
membrane effect, *130*, 131, 144
membrane tension, 164
mercury intrusion methods, 22
mesh-based biotechnical erosion control, 211
microbial activity, 217
micro-organisms, 255–6
Miel earth fill dam, 267
mineral sealing layers, 257
mixed scheme reinforcement, *237*, 238
mobilised interface friction angles, 144
moderately graded recycled ballast, 201–2
modular concrete blocks (MCBs), 238
Mohr–Coulomb strength criteria, 225
monofilament-on-monofilament geotextiles, 24–5
monofilament yarns, 11
Mononobe–Okabe pseudo-static approach, 326–8
monotonic loading interactions, 164–5, *166*
monotonic load–strain behaviour, 320
Moochalabra Dam, 273
MQA *see* manufacturing quality assurance
MSE *see* mechanically stabilised earth
Mulhouse canal, 295, 296–7, *298*, *299*
multi-barrier landfill concepts, 246
multifilament woven polyester geotextiles, 155
multifilament yarns, 11
multilayer sealing systems, 248
municipal waste landfills, 247, 248, 255, 256

Nanyang Technological Institute (NTI), 233–4
National Concrete Masonry Association (NMCA), 318, 326, 330
natural fibre nets, 214, *215*
natural ground subgrade railway track components, 193–4
natural UV radiation, 32
needle-punched geosynthetics, 12, 25, 199–201, 269
net-based biotechnical erosion control, 211
New Austrian Tunnelling Method, 310–11
New Esna Barrage Dam, 271
Newmark's displacement calculation method, 318, 336–7, *338*, 339
New South Wales (NSW), 201–2
NMCA *see* National Concrete Masonry Association
noise barriers, 374, 376
non-biotechnical soil erosion solutions, 211, *211*, 212
non-elastic deformations, 167
non-linear hysteretic loading paths, 319
non-woven filter layers, 259
non-woven geotextiles, *2*
  compressibility, 14–15
  definition, 1
  earth dams, 266–7, 268, 269, 270
  filter clogging criteria, 69
  landfills, 255–6
  manufacturing processes, 12
  mechanical properties, 25
  paved roads, 184
  railway tracks, 198, 199–200, 201, 202
  raw materials, 12
  road case studies, 382
  shallow foundations, 135–6
normal stress membrane support, 143
North America
  earthquakes, 354–5
  geosynthetic growth, 34
  see also USA
NSW *see* New South Wales
NTI *see* Nanyang Technological Institute
numerical dynamic analyses, 339–48
  damping ratios, 345
  finite-difference methods, 342
  finite-element methods, 339–42
  fundamental frequencies, 345, 347–8
  model boundaries, 345

offset moduli, 17
offtake drains, 203
optimal filter performance requirements, 26
orientation of drains, *232*, 233
outdoor weathering tests, 32
overflow dikes, 274
overlapping performance criteria, 282
overliner protection and cover, 282–3
over-steep geogrid-reinforced slopes, 238
overtopping phenomena, 273, *274*
overturning checks, 93
Oxford unpaved road reinforcement design, 170–4
oxidised bituminous geomembrane durability, 284

PA *see* polyamid
Paduli Dam, 274
Panci Toll Road Project, 156
parabolas, 151, 152–3
particle filters, 67
passive anchorage, 165
passive interface resistance, 49
passive resistance-dependent reinforcement, 9–10
paved roads, 35, *38*, 179–92
  base reinforcement, 185–9
  distress features, 179–81
  geosynthetic functions, 181–5
pavements
  surface dressed, 381
PE *see* polyethylene
per cent open area (POA), 22, 71–2
performance tests, 9, 18
perimeter seals, 310
permanent deformations, 167
permanent radial strain, 186
permanent surface deformations, 186
permanent use groupings, 270
permanent vegetative cover, 379–80
permeability, 23, 24–5, 26, 27
permittivity, 21, 23–4, 25, 76, 198
PET *see* polyester (PET)
physical model tests
  shallow foundations, 149, *150*
physical properties of geosynthetics, 14
piled embankments, 117–24
pipeline systems, 378–9
placed soil subgrade railway track structures, 193–4
plain strain tensile tests, 15, *16*
plain weaves, 11–12
plastic friction models, 146–7
plasticiser-to-resin bonding, 269
plasticity theory, 143, 227
plastic-reinforced rock-fill rafts, 156–7
ply-soil, 129
POA *see* per cent open area
Poisson's ratio, 17
polyamide, 14
polyamid (PA), 10, 28
polyester (PET), 10, 11, 14
  creep tests, 28
  earth dams, 269, 272–3
  geosynthetic-reinforced soil, 320, 322–3, 325, 352
  shallow foundations, 155
polyethylene (PE), 10, 14, 28, 258, 377–8
polyisobutylene geomembrane, 267
polypropylene (PP), 10–11, 14, 392, *393*
  creep tests, 28
  durability, 285
  earth dams, 268
  railway tracks, 202
  sustainability, 391, 392, *393*
polystyrene (PS), 13
polyvinyl chloride (PVC), 10, 14
  durability, 285
  earth dams, 267–8, 269, 271, *272*, 273–4
  history, 5
  hydraulic tunnels, 304, 305–6, 307, 311, 312, *312*
  slope stabilisation, 236–7
pore-size distributions, 71
porosity, 21, 22, 71–2
porous concrete, 265–6
PP *see* polypropylene
prayer seams, 18, *19*
pre-cast steel-reinforced concrete footings, 134–5
prefabricated concrete slabs, 267
prefabricated drainage geocomposites, 378–9
prefabricated highway edge drains, 83
prefabricated vertical band drains, 378
prefabricated vertical drains (PVDs), 82, 204–5, 378
pressure drainage, 80, 81
pressure tunnels, 304, *305*, 307
prestressing, 147–51, 368
product evaluation results
  erosion control, 212–13
properties of geosynthetics, 14–33
protection functions, 9
protection systems, 251–2, 271, *272*, 281–3, 311, *312*
protective anti-puncture geotextiles, 311
protective geomembrane layers, 251–2
protective layers
  earth dams, 271, *272*
  hydraulic tunnels, 311, *312*
PS *see* polystyrene
pseudo-dynamic methods, 318, 336
pseudo-static methods, 318, 326–36
  circular slip failure mechanisms, 334–5
  external stability calculations, *328*, 330
  internal stability calculations, 330–2
  log spiral failure mechanisms, 333–4
  method comparisons, 335–6
  Mononobe–Okabe approach, 326–8
  seismic coefficient selection, 328–30
  stability analyses, 227
  two-part wedge failure mechanisms, 332–3
pull-out
  mechanical properties, 20–1
  shallow foundations, 132
  soil–geosynthetic interactions, 48–9, 50–6, *53*, *54*, 60–4, 324–5
puncture strength, 20
PVC *see* polyvinyl chloride
PVDs *see* prefabricated vertical drains

quality assurance, 248, 255, 258
quality control, 369–70, 371, 372
quantification of soil–geosynthetic interface resistance, 55–64

radial drainage, 115
radial strain, 186

Rail Infrastructure Corporation (RIC) of New South Wales, 202
railway ballast reinforcement, 167
railway-switch rehabilitation, 200–1
railway tracks, 35, *38*, 193–208
components, 193–6
design procedures, 199–200
drainage, 202–5, *206*
requirements, 203
side drains, 203–5
geosynthetic functions, 196–7
geosynthetic installations, 200
geosynthetic properties, 197–9
stabilisation case histories, 200–2
substructure, 193–6
rain erosion, 209–10
rainfall erosivity indices, 212
rain splash, 216–17
Rankine active earth pressure theory, 89
raw materials, 10–14
RECO systems, 340
rectangular patterns of pile distribution, 121
recycled ballast, 201–2
reduced-area shear-type tests, 57–8
reduction factors, 109, 113
regular non-woven geotextiles, 198
reinforced soil structures (RSSs), 235–40
reinforcement
drainage separators, 4, *6*
earth dams, 264, 271–3
economy rates, 370, *373*
embankments, 101–13, 116–17
functions, 7–8, 129–31, 181–2, 197
hydraulic tunnels, 312–13
interface resistance, 47–9
mechanisms, 46, 47, 129–31
paved roads, 181–2, 185–9
railway tracks, 197
shallow foundations, 129–42, 147–51
slope stabilisation, 224–9
soil–geosynthetic interactions, 48
sustainability aspects, 392, *393*
unpaved roads, 164–75
see also geosynthetic-reinforced soil slopes and walls
repeated loading effects, 165, 167
reservoirs, 279–302
case studies, 285, 289–95, 380
design, 281–4
earth dams, 268
geomembrane durability, 284–5
historical background, 279–81
resin-treated geotextiles, 200–1
Reslope software, 229
restricted soil dilatance, 47
resurfaced pavement, 381
retaining walls, 35, *38*, 89–99
applications of geosynthetics, 374–6
design, 89–91
geogrids, 96–7
geotextiles, 91–6
lateral earth pressures, 89, 91
stability, 89, *90*
tie force, 91
geogrids, 96–7
geotextiles, 91–6
self-evaluation questions, 97–8
retention criteria, 26, 27, 68–9, 70
retention requirements, 68
RF *see* runoff factors
riverbank erosion, 379
river beds, 274
river erosion, 209–10
RMCC *see* Royal Military College of Canada
roads
case studies, 380–2
reinforcement design, 170–4
singularities, 284
sustainability, 392, *393*
unpaved, 35, *38*, 163–77, *166*, 368
rock drains, 234
rockfill dams, 267, 268–9, 271, *272*
rolled erosion control products (RECPs), 212, 217–18, 219
root-reinforcing mats, 212
rope sheet methods, 148
rough elastic membranes, 148, *149*
roundels, 311, *312*
Royal Military College of Canada (RMCC), 200, 338–9, 341
RSSs *see* reinforced soil structures
rubber tyre walls, 237
runoff, 213–15, *215*, 216–18
runoff factors (RF), 213–15, *215*
rutting factors, 143

safety factors *see* factors of safety
sand-filled mats, 251
satin plain weaves, 11
screening functions, 9
scupper drains, 233–4
sealing systems, 248
seams, 18, 19, *19*, 255, 282
secant moduli, 17
Sedimentation, and Erosion Control Laboratory of the Texas Transportation Institute, 212, 215
segmental retaining walls (SRWs), 376
seismically active regions, 272–3
seismic analyses, 227, 268–9, 326–48
seismic buffers, 353–4
self-supporting vegetative cover, 379–80
Selvet Dam, 289–92
separation effect, 131
separation functions, 185
cost analyses, 372, 373
paved roads, 181, 182–4
railway tracks, 196
serge plain weaves, 11
settlement analyses, 145–7
settlement behaviour, 149–51
settlement profiles, 147, *148*
shaking table tests, 325
shallow foundations, 129–61
deflection profiles, 151–3, *154*
field applications, 153–7
load-bearing capacity analyses, 139–45
reinforced clay, 142
reinforced granular fill, 139–42
reinforced granular fill-soft foundation soil systems, 142–5
model tests, 132–9

modes of failure, 131–2
prestressed geosynthetic-reinforced foundations, 147–51
reinforcement mechanisms and functions, 129–31
reinforcing patterns, 131
settlement analyses, 145–7
strain analyses, 151–3, *154*
shear-displacement modelling, 326
shear stresses, 170–1, *172*, 181
shear stress reduction effects, 129–30
shear tests, 47, 48, 57–8, 269, 324
sheet drain materials, 82–3
short-term erosion yield factors, 213–15
Shuikou hydroelectric power station, 268
side drains, 203–5
single-sized soils, 25–6
single-stitch line seams, 18, *19*
singularities, 283–4
sink-holes, 381, *382*
skin friction, 46–7, 56, 60
slab effect, 130–1
sleeve influence, 64
sliding block methods *see* Newmark's displacement calculation method
sliding checks, 93
slip-line stability analyses, 227
slit films, 11
slopes, 317–63
applications of geosynthetics, 374–6, 379–80, 390–2
design, 326–48
dynamic loading, 318–26
earthquake performance, 354–5
erosion control, 35, *38*, 209–22, 379–80
geosynthetic-reinforced soil, 317–63
length factors, 212
material properties, 318–26
numerical dynamic analyses, 340, 342
physical testing, 348–53
seismic analyses, 326–48
seismic buffers, 353–4
stabilisation, 35, *39*, 223–43
stability, 258, 372, 373
steepness factors, 212
sustainability, 390–2
slotted pipes, 204
sludge-drying beds, 79
sludge specimens, 76
slurry specimens, 76
small erosion control scales, 215–16
small-scale model tests
shallow foundations, 134, 135, 144
soft clays, 156–7
soft foundation soil systems, 35, *38*, 136–9
soft soil vertical stress transfers, *117*, 118
soft subgrade condition stabilisation, 185
soil
cake, 31
density, 53–4
earth dams, 274–5
erodibility indices, 212
erosion control, 210–11
filters, 26, 77–8
geosynthetic interactions, 20, 45–66, 320, 324–5
granular soil behaviour, 45–6
height influence, 63
influencing factors, 49–55
interface resistance, 55–64
landfills, 256–7
loss equation, 211–12
loss indices, 219
mechanisms, 46–7
numerical dynamic analyses, 342
particle sizes, 49–52, *53*
piled embankments, 118
railway tracks, 196–7
shallow foundations, 156
stiffness properties, 319–20
strength properties, 318
tightness criteria, 26
vertical stress transfers, *117*, 118
soil–iron sheet shear tests, 58
soil–soil friction, 56, 60
soil–soil shear tests, 58
soil walls
economy rates, 370, *373*
numerical dynamic analyses, 340–5, *346*, 347–8
solid-waste landfills, 246–7, 248, 253–4, 256
Souppes-sur-Loing containment pond, 285–6
South Africa
containment ponds, 288–9
earth dams, 270–1
landfills, 252–3
spacing of drains, *232*, 233
specialist geosynthetics, 11
specimen clamping, 57
specimen fixation influence, 59
specimen size influence, 62–3
spider netting methods, 234–5
splash cup tests, 215
spray-on products, 212
spread-coating processes, 13
spun yarns, 11
square footings, 134, 141
SRWs *see* segmental retaining walls
stabilisation, 35, *39*
earth dams, 274
earth walls, 392, 394, 395
geosynthetic functions, 185
railway tracks, 200–2
slopes, 223–43
geosynthetic orientations, 224
geosynthetic types, 224
model tests, 229–31
in practice, 231–40
reinforcement, 224–9
stability analyses, 225–7
stability
applications of geosynthetics, 372, 373
retaining walls, 89, *90*
stability analyses
embankments, 111–12
pseudo-static methods, *328*, 330–2
slope stabilisation, 225–7
surficial, 227, 228
unreinforced embankments, 106
standards, 35, 40, 118–20
applications of geosynthetics, 374
landfills, 252–3
soil–geosynthetic interface resistance, 57, 60

stapled seams, 18, *19*
staple fibres, 11
static tests, 139
steel plates, 265
steel-reinforced concrete footings, 134–5
steep-sided embankments, 35, *38*
steep slopes, 226, 374–6
stiffness of geosynthetics, 14
stitch-bonded geotextiles, 1
stormwater runoff, 217–18
strain analyses, 29–30, 151–3, *154*
strands, 11
strength profiles, 106
stress relaxation, 29
stress–strain models, 319–20
stretched rough elastic membranes, 148, *149*
strip drains, 13
strip footings, 134, 135, 141–2, 145, 146, 230–1
sub-ballast railway structures, 193, 194–5
subgrade deformation kinematics, 168
subgrade preparations, 281
subgrade railway track components, 193–4, 203, 204, 205
submerged mattresses, 274
subsoil-drainage pipes, 236–7
suitability tests, 250–1
SUPERFLUSH systems, 340
support systems, 308
surface deformations, 186
surface dressed pavement, 381
surface erosion control mats, 4, *6*
surface erosion mechanics, 210
surficial stability analyses, 227, 228
sustainability
  geosynthetic applications, 387–96

TARA-3 systems, 340–1
TASi *see* Technische Anleitung Siedlungsabfall
TBR *see* traffic benefit ratios
tear strength, 20
Technische Anleitung Siedlungsabfall (TASi), 246–7
Tenax TT301 SAMP geogrids, 239
Tensar geogrids, 229
tensile force, 102–3
tensile membrane action, 164
tensile moduli, 16–17
tensile strength, 15
  slope stabilisation, *232*, 233
tensile tests, 15, *16*
tensioning profiles, 308–9, *310*
tension membrane theory, 123
Terramesh Systems, 239–40
test methods, 14–33, 139, 250–1
  triaxial, 165, 167
Texas Department of Transportation (TxDOT), 212–13, 216, 218–19
thermal bonding, 12
thickness of geosynthetics, 14, 15
Three Gorges Dam project, 268
tie force, 91
tilt table tests, 348–53
total strain versus log *N* plots, 29–30
tracks
  singularities, 284
traffic benefit ratios (TBR), 182, 187–9
transition placement, 247–8
transmissivity, 21, 23, 24, 25, 81–2, 198
trench drains, 35, *39*, 205
triangular patterns of pile distribution, 121–2
triaxial geonets, 252
triaxial tests, 165, 167
tunnels
  case studies, 382
  see also hydraulic tunnels
twill plain weaves, 11
two-dimensional pipeline theory, 120
two-part wedge failure mechanisms, 332–3
TxDOT *see* Texas Department of Transportation

ultimate bearing capacities
  shallow foundations, 134, 135, 140–1, 144
ultraviolet (UV) light, 32, 217, 367
underliner drainage and protection, 281
undisturbed soil zones, 26
uniaxial geosynthetics, 1, *3*, 119
uniform strip footings, 145
United States of America (USA), 200–1, 251
  see also North America
United States Army Corp of Engineers (USACE) methods, 110
universal soil loss equation (USLE), 211–12
unpaved roads, 35, *38*, 163–77, *166*, 368
unsaturated earth dams, 266–7
USACE *see* United States Army Corp of Engineers methods
USLE *see* universal soil loss equation
UV *see* ultraviolet light

vandalism, 374
variational extremisation stability analyses methods, 226
vegetation establishment, 216
vegetative cover, 379–80
vertical displacement measurements, 60
vertical drains, 114–17, 155–6
vertically driven relief piles, 155
vertical shaft structures, 256
vertical stresses
  embankments, *117*, 118, 120, *121*, 123
  paved roads, 181–2, 186
  shallow foundations, 144
  soil–geosynthetic interface resistance quantification, 59–60
  unpaved roads, 171, *172*
vertical strip drains, 378
very low-density polyethylene (VLDPE), 10, 258
vibro-tampers, 367
viscoelastic behaviour, 17
VLDPE (very low-density polyethylene), 10, 258
void filters, 67
volume flow rates, 23

walls, 237, 317–63
  applications of geosynthetics, 374–6
  design, 326–48
  dynamic loading, 318–26
  earthquake performance, 354–5
  economy rates, 370, *373*
  geosynthetic-reinforced soil, 317–63
  material properties, 318–26
  numerical dynamic analyses, 340–5, *346*, 347–8
  physical testing, 348–53

  seismic analyses, 326–48
  seismic buffers, 353–4
warp yarn directions, 11
water, 202–3
water channels, 35, *39*
waterproofing, 304, *306*, 307, 308, 311, *312*
waterproofing subgrades, 203
water reclamation plant, 288–9
waterstops, 312, 313
water-to-plasticiser bonding, 269
weaves, 11–12
weft yarn directions, 11
welded steel plates, 265
well-graded soils, 25–6
Western Europe estimated consumptions, 34, *35*
wet sieving test methods, 22
wick drains, 82, 204–5, 378
wide-slab mechanisms, 140, 141
wide-width strip tensile tests, 15
wool mat exercises, 214–15
working moduli, 17
woven geogrids, 1, *3*
woven geotextiles, *2*
  definition, 1
  earth dams, 272–3
  history, 5
  manufacturing processes, 11–12, *12*
  monofilament-on-monofilament, 24–5
  paved roads, 184
  raw materials, 11–12, *12*
  shallow foundations, 155
  slope stabilisation, 237
  total strain versus log *N* plots, 29–30
woven products
  railway tracks, 198–9
wraparound geotextiles, 204
wraparound techniques, 237–8

Xixiayuan reservoir, 268
XPS *see* extruded polystyrene
X ray observations
  unpaved roads, 165, *166*

yarns, 11
yield factors (YF), 212, 213–15, 216, *217*

Zhusou reservoir, 268